Chimpanzee

The chimpanzee is one of our planet's best-loved and most instantly recognizable animals. Splitting from the human lineage between four and six million years ago, it is (along with its cousin, the bonobo) our closest living relative, sharing around 99 percent of our genes. First encountered by Westerners in the seventeenth century, virtually nothing was known about chimpanzees in their natural environment until 1960, when Jane Goodall traveled to Gombe to live and work with them.

Accessibly written, yet fully referenced and uncompromising in its accuracy and comprehensiveness, this book encapsulates everything we currently know about chimpanzees: from their discovery and why we study them to their anatomy, physiology, genetics, and culture. The text is beautifully illustrated and infused with examples and anecdotes drawn from the author's 30 years of primate observation, making this a perfect resource for students of biological anthropology and primatology as well as non-specialists interested in chimpanzees.

Kevin D. Hunt is Professor of Anthropology and an affiliate of the Stone Age Institute at Indiana University, Bloomington. He is also Founder and Director of the Semliki Chimpanzee Project, which was established in 1996 to study and preserve the chimpanzees in the Toro-Semliki Wildlife Reserve. Broadly trained in various anthropological disciplines, much of Professor Hunt's published work has centered on functional morphology and what chimpanzee locomotion, posture, and ecology can tell us about what led humans to diverge from apes and what advantage bipedalism gave our chimpanzee-like ancestors roughly five million years ago.

"*Chimpanzee: Lessons From Our Sister Species* condenses over 60 years of chimpanzee research into an informative and entertaining book. Drawing on his own first-hand experience, the research of other scientists and historic accounts, Kevin Hunt describes the fascinating lives of chimpanzees in the wild, as well as the research methods used by leading experts in the field. If you want to know just how alike we truly are to our closest living relatives then you will get a very good idea from reading this book."

Jane Goodall, PhD, DBE, Founder of the Jane Goodall Institute & UN Messenger of Peace

"Hunt skillfully weaves anecdotes and history into this scientific compendium of the behavioral ecology, biology, and evolution of chimpanzees. The book is generously illustrated, and each chapter includes extensive references. It is written in an accessible, conversational style that could only be achieved by someone with Hunt's first-hand experiences in the field and encyclopedic perspective. It will make a valuable reference for anyone interested in what is known and not yet known about one of our closest living relatives."

Karen B. Strier, Vilas Research Professor & Irven DeVore Professor, Department of Anthropology, University of Wisconsin-Madison, USA

"An exceptional book that delivers on every promise in its table of contents. Grounded in Hunt's 30+ years of chimpanzee field work and his commanding knowledge of others' research, he gives us a state-of-the-art research volume that will become an essential reference for primatologists, and anyone who wants to understand the true nature of our sister species.

Hunt's writing is lucid, scholarly and wide-ranging as he carefully explains chimpanzee evolution, biology, social behavior, and so much more. Hunt skillfully embeds his own field observations to help readers grasp concepts like chimpanzee positional behavior, personality, maternal behavior, cognition and communication, hunting and aggression. He balances this perspective with a wealth of laboratory and captive findings.

The extensive references for each chapter provide an outstanding resource for students, teachers and readers who choose to delve further. The volume is generously illustrated with photos, line drawings and abundant figures that enrich the text."

Linda F. Marchant, Professor Emerita, Miami University, Oxford, Ohio, USA

Chimpanzee
Lessons from our Sister Species

KEVIN D. HUNT
Indiana University, Bloomington

CAMBRIDGE
UNIVERSITY PRESS

University Printing House, Cambridge CB2 8BS, United Kingdom

One Liberty Plaza, 20th Floor, New York, NY 10006, USA

477 Williamstown Road, Port Melbourne, VIC 3207, Australia

314–321, 3rd Floor, Plot 3, Splendor Forum, Jasola District Centre, New Delhi – 110025, India

103 Penang Road, #05-06/07, Visioncrest Commercial, Singapore 238467

Cambridge University Press is part of the University of Cambridge.

It furthers the University's mission by disseminating knowledge in the pursuit of education, learning, and research at the highest international levels of excellence.

www.cambridge.org
Information on this title: www.cambridge.org/9781107118591
DOI: 10.1017/9781316339916

© Kevin D. Hunt 2020

This publication is in copyright. Subject to statutory exception and to the provisions of relevant collective licensing agreements, no reproduction of any part may take place without the written permission of Cambridge University Press.

First published 2020
3rd printing 2022

Printed in the United Kingdom by Print on Demand, World Wide

A catalogue record for this publication is available from the British Library.

Library of Congress Cataloging-in-Publication Data
Names: Hunt, Kevin D., author.
Title: Chimpanzee : lessons from our sister species / Kevin D. Hunt, Indiana University, Bloomington.
Description: Cambridge, United Kingdom ; New York, NY : Cambridge University Press, 2020. | Includes bibliographical references and index.
Identifiers: LCCN 2019055992 (print) | LCCN 2019055993 (ebook) | ISBN 9781107118591 (hardback) | ISBN 9781107544413 (paperback) | ISBN 9781316339916 (epub)
Subjects: LCSH: Chimpanzees.
Classification: LCC QL737.P94 H86 2020 (print) | LCC QL737.P94 (ebook) | DDC 599.885–dc23
LC record available at https://lccn.loc.gov/2019055992
LC ebook record available at https://lccn.loc.gov/2019055993

ISBN 978-1-107-11859-1 Hardback
ISBN 978-1-107-54441-3 Paperback

Cambridge University Press has no responsibility for the persistence or accuracy of URLs for external or third-party internet websites referred to in this publication and does not guarantee that any content on such websites is, or will remain, accurate or appropriate.

To Marion
the sine qua non

*Half the author's profits from this book
will be donated to chimpanzee research
www.indiana.edu/~semliki*

CONTENTS

Foreword   page ix
Richard Wrangham
Preface   xi
Acknowledgments   xv

1. Sister's Keeper: Humans and Chimpanzees   1
2. Wild Lesson: Why Study Animals in Nature?   9
3. A Most Surprising Creature: The Discovery of the Chimpanzee   20
4. Kin: The Chimpanzee's Place in Nature   40
5. Scratching Out a Living in an Unforgiving World: Habitat and Diet   61
6. Guts, Glorious Guts, Large Stomach, and Colon: Plant Chemistry, Fruit Ripening, Digestive Physiology, and Gut Anatomy   80
7. Thews, Sinews, and Bone: Chimpanzee Anatomy and Osteology   97
8. Arboreal Gathering, Terrestrial Traveling: Locomotion and Posture   119
9. Forged in Nature's Cauldron: Engineering the Chimpanzee   131
10. Up from the Protoape: The Evolution of the Chimpanzee   158
11. Building a Natural Wonder: Growth, Development, and Life History   188
12. The Source of Similarity: Chimpanzee Genetics   217
13. Making Your Way in the Great Wild World: Chimpanzee Senses   242
14. The Grim Reaper in the Forest Primeval: Wild Chimpanzee Diseases and Lessons for Healthy Living   252
15. Powering Life: Endocrinology and Physiology   274
16. Shelter from the Storm: Chimpanzee Mothering   298
17. Meat-Seeking Missiles: Chimpanzees as Hunters   312
18. The Mind of the Chimpanzee: Reasoning, Memory, and Emotion   323
19. The Brain of the Chimpanzee: The Mind's Motor   357
20. Tired Nature's Sweet Restorer: Chimpanzee Sleep   372
21. Chimpanzee Thought Transfer: Communication and Language   383
22. Ape Implements: Making and Using Tools   404
23. Wisdom of the Ages: Chimpanzee Culture   419
24. The Daily Grind: Within-Group Aggression   427
25. A Nation at War with Itself: Defending a Community of the Mind   437
26. The Sporting Chimpanzee: Dominance without Destruction   449

27 **The Passion of *Pan*: Sex and Reproduction**   458

28 **Into the Light: Semliki Chimpanzees**   473

29 **The Other Sister, Bonobos: The Monkey Convergence Hypothesis**   499

30 **Sister Species: Lessons from the Chimpanzee**   517

Appendix 1: Taxonomy of the Primates   536
Appendix 2: Professional Grade Chimpanzee Testable Hypotheses   553
Index   559

Color plates can be found between pages 268 and 269.

FOREWORD

In every society, for as long as we have had records, people have grappled with the great existential question. Why are we here?

In 1859, Charles Darwin published *On the Origin of Species*, and the theory of evolution by natural selection was born. Within a century, biologists were increasingly confident that a single evolutionary tree linked all life on Earth. The exciting questions then became how to explain the evolution of particular species and their characteristics – even humans.

Humans proved to have evolved from a forest-living African ape living some seven million years ago. That ancestor was closely related to gorillas, and it spawned two lines. One led to the chimpanzees (including bonobos, also known as pygmy chimpanzees); the other led to us.

Chimpanzees had long been of interest, but in the second half of the twentieth century two scientific leaps made these apes especially fascinating. The advances happened independently of each other. They were both thoroughly unexpected.

First was a series of reports from Jane Goodall and Toshisada Nishida that chimpanzees in Tanzania practice various behaviors that had been thought of as markers of humanity. The apes modify objects into tools that they use to help get their food. They hunt for fresh meat and share the kill with each other. Males conduct raids into neighboring territories, where they stalk and kill lone victims. And gestures and facial expressions used by chimpanzees can be so similar to those used by humans – such as an outstretched hand and an intense stare when begging – that many are intuitively understandable. In short, chimpanzees are unnervingly human-like in the wild. They prove to share more behaviors with us than does any other wild species.

Second was the discovery that chimpanzees are more closely related to humans than they are to gorillas. The idea was at first so startling that many people did not believe it. The skepticism is easily understood: Chimpanzees look so like gorillas that they are easily confused with them. So, naturally enough, the traditional assumption was that chimpanzees' closest relatives were gorillas. But eventually the genetic data became overwhelming. Humans really are genetically closer to chimpanzees than are any other species, even gorillas.

What does it mean that chimpanzees behave more like us, and are also more closely related to us, than to any other species? *Chimpanzee* answers that question by giving a rich account of the evolution of both chimpanzees and humans.

To do so, Hunt examines the relationship between ape and human from a broader perspective than any book has done. His eclectic approach takes courage. To succeed as a scientist, one needs to be an expert on a narrow slice of the natural world. So it is easy (and safe) to forget the big picture. But Kevin Hunt has bucked the trend by going beyond his background as a functional morphologist. He treats the natural history of these apes like a detective story, following leads wherever they happen to take him.

His quest begins with problems arising from his own initial fieldwork. He observed individual chimpanzees at close quarters for a year. He saw their social lives giving way to the search for food, especially the daily need to find scarce patches of freshly ripe tree-fruits as places for morning meals. That observation provoked questions that Hunt follows up. What aspects of digestive physiology commit the apes to a diet of tree-fruits? What does that commitment mean for their anatomy? Why do they differ from humans in these ways? The answers to such questions took him into many corners of evolutionary biology. Eventually he combined them into a big, satisfying picture of the species' adaptations to its environments.

Chimpanzee's far-reaching perspective embraces other species, too. Hunt's book is mostly about the more widespread species of chimpanzee, which is known simply as "the chimpanzee." It lives in some 20 countries on the northern, right-hand bank of the

Congo, including Tanzania and Uganda, where Hunt has done his major fieldwork. But Hunt also writes about the other species of chimpanzee, the bonobo. Bonobos live on the southern, left-hand bank in a much smaller geographical range, entirely within the Democratic Republic of the Congo. Their differences from their sister species provide many new evolutionary problems to consider.

So, too, do the apes of the past. *Chimpanzee* ventures back in time as well, exploring the biology and adaptations of the australopithecine apes that dominated the evolutionary history of the pre-human lineage from around seven million to two million years ago.

Ever since the seventeenth century, when chimpanzees were first brought to Europe, writers have suspected that apes have a story to tell about human life and our prehistoric origins. Year by year the details of that story are being worked out better and better. *Chimpanzee* is a terrific account from the leading edge.

Richard Wrangham
Ruth Moore Research Professor of Biological Anthropology, Department of Human Evolutionary Biology Harvard University

PREFACE

The history of chimpanzee scholarship has been one of generally gradual but occasionally sudden shifts in the perception of how closely related humans and chimpanzees are. Here I will try to bend that perspective a little bit to view chimpanzees more as a counterpoint to human uniqueness, taking care to point out how they differ from humans and how explaining those differences helps us to understand human evolution, including our frailties and failings. To draw out those insights, I will attempt an unhurried grand tour through the natural sciences, always with chimpanzees in the forefront, always contrasting their biology and behavior to that of humans. I intend this volume to be current and thoroughly grounded in the latest research, only one step from formal scholarship, but at the same time stripped of the jargon-heavy phraseology that weighs down many similar efforts. Where I use obscure terms, I will attempt to define them on the fly.

I see my audience as scholars from other fields, laypersons with a sincere interest in the behavior, biology, and history of primates, doctoral students in human evolution, human biology, and primatology, and colleagues in scholarly specialties neighboring my own. If successful, it will be a volume worth dipping into for scholars in any natural science field, even the chimpanzee expert. I hope a primatologist might turn to this volume if he or she wants to dabble in functional anatomy or endocrinology. While I realize this is probably too much to hope for, I sought to make the book engaging enough that a colleague or student might be drawn into exploring an area they had not anticipated investigating. Ideally, this volume would be an accessible version of the six-volume G.H. Bourne volume *The Chimpanzee*, published from 1969 to 1973.

As I worked, I found that sometimes the most thorough investigation of a subject – for example, balance, hearing, ape cognition, aspects of gross anatomy – is half a century old. In the course of discovering the chimpanzee, researchers sometimes got it right the very first time, in which case I cited the classic scholarship that shaped a particular issue, occasionally skipping a recent publication that added a minor flourish to the subject. I would like this volume to be something of the corrective to the tendency I have seen recently among younger scholars to cite the most recent publication on some subject, no matter how tangential, ignoring truly foundational work with which they ought to be familiar. Other times experts long labored under misapprehensions that have been corrected only recently, in which case I focused on the recent scholarship with only a nod to ancient errors. My shorthand way of summarizing my emphasis on studying micro- and macrobiology, as well as behavior, is that while the bricks are indeed interesting, they are most interesting when they help us understand the castle.

I began this project in a small way in 1996, when I had just begun to teach a class at Indiana that – in a modest way – surveyed the breadth of chimpanzee research. I called it *Sister Species: Lessons from the Chimpanzee*. The text for the class was a coursepack made up of review articles and a few specialist research articles. The readings that were most successful with students, I learned from student evaluations, were often those from *Evolutionary Anthropology*. These were professional-grade offerings, current and cite-able, but still accessible to committed neophytes. I have tried to work on that model in this book.

The first few times I taught *Sister Species*, the first reading in my coursepack was a brilliantly written news article by Claire Martin (1994), a short work that focused on chimpanzee and human cognitive similarities. Students loved it. As evocative as this work was, I was left feeling that the reading for my first lecture should have a slightly broader perspective and more robust scholarly citations. That same year, 1996, I started research on chimpanzees at the Toro-Semliki Wildlife Reserve, Uganda. One day, problems with my vehicle left me stuck in a chimpanzee-less part of the reserve for two hot, humid days with nothing to do but sweat. During this time, I wrote a short exposition that quoted extensively from Martin's article but went on to cover a broader constellation of ideas that I hoped would best set the stage for my first lecture. It would have done its job well except that almost as soon as I finished it the intense heat fried my hard drive. I lost everything but a couple of paragraphs I had copied onto a floppy disk.

Anyone who has lost a piece of work like this knows how disheartening it can be, and how the virtuosity of the lost work swells in the imagination, discouraging a rewrite. Although I still had the framework in my head, it was almost 10 years before I rewrote that chapter. When I brought the finished work to students they welcomed it enthusiastically, encouraging me to write a few more chapters to fill gaps in my reading list. As I wrote those short chapters I began to piece together plans for this volume. I aimed to review chimpanzee functional anatomy research (which it will be apparent I revel in) in such a way that my colleagues who specialize in social behavior might be tempted to look into it for links to their own work; for instance, I hoped my review might allow primate paleontologists to think more deeply about the role social behavior plays in anatomical evolution. And *vice versa*. I blocked out the table of contents in 2005 during a delightful sabbatical at the Max Planck Institute in Leipzig, during which time I emerged with 4 of the 30 chapters, enough to get a book contract. I only got serious about writing after signing a contract in 2013.

As I completed parts of the manuscript, I occasionally passed on a chapter to a "civilian," a friend or colleague who had no background in primatology. Their responses were encouraging enough that it was brought home to me in a way it had not been before that chimpanzees are charismatic enough to serve as a sort of hook or ice-breaker that might rather painlessly introduce the breadth of natural science to new audiences. As I wrote subsequent chapters I found myself adding a flourish here and an elaboration there that I felt gave a narrative boost to the "story," leavening the stultified scientific language that always threatens to smother scholarly writing. I hoped that a personal anecdote or amusing fact here and there might make wading through the more painstaking scholarly concepts easier and more rewarding.

Buried in each chapter are a few original insights that I hope someone will latch onto and expand upon, but which, regretfully, are not very likely to reach specialists in that area, who are naturally unlikely to pick up a survey book like this. I hope advanced students who happen upon this volume will recognize those original and informed speculations, which I have summarized in Appendix 2, and carry them to their doctoral supervisors as possible research or dissertation topics.

As we embark on this journey, let me offer, at the suggestion of one of my editors, a few personal details. I am a third-generation Gombe researcher, though not a typical one. My doctoral and postdoctoral supervisor was Richard W. Wrangham, a name that will appear many times in the pages to follow. He came to Gombe in 1974 and became the first researcher to make the all-day, 13-hour, nest-to-nest follow the foundation of his research. It was through his encouragement that I backed into chimpanzee research, but not before trying to avoid it. I may hold the distinction of being the most reluctant beneficiary of the research world Jane Goodall inspired at Gombe and Toshisada Nishida at Mahale. My research is focused on the evolution of bipedalism, and I began my doctoral work just as a reassessment of the Lucy *Australopithecus afarensis* fossils was underway. Early on, australopith publications by paleontologists Donald Johanson and colleagues emphasized the human-like morphology of Lucy, particularly her very human – though not completely human – pelvis and lower limb (Johanson & Taieb, 1976; White et al., 1981). Further evidence that australopiths were human-like came from tooth development research that suggested early hominins had a long, human-like juvenile period (we call this a "slow life history"), perhaps an indication of advanced cognitive abilities.

This "miniature human" or homunculus hypothesis, as some derisively referred to it, was soon challenged. With more fossils to work with, B. Holly Smith found that the rate of maturation of australopiths was chimpanzee- rather than human-like, and not just a little like chimpanzees, but much more ape-like than human-like (Smith, 1987). Two landmark articles reassessed Lucy's anatomy, one by anatomists and paleontologists Jack Stern and Randy Susman (Stern & Susman, 1983) and a second by French paleontologists Brigitte Senut and Christine Tardieu (Senut & Tardieu, 1985). These showed that while Lucy and other members of her species had human-like hips and legs, they also exhibited arboreal anatomy wherever it might not interfere with upright walking – australopiths have long toes, short legs, and flexible joints, for instance. Above the waist, australopiths were far more chimpanzee- than human-like. As a shorthand, their studies suggested that australopiths are better described as heavy-faced chimpanzees from the waist up and human-like from the waist down.

But even if the anatomy of Lucy is ape-like, what does that mean? It turned out that we had not yet gathered quantitative information on locomotion and posture in chimpanzees that could link a particular anatomical detail, a cone-shaped thorax, for instance, with a behavior. Yes, Lucy's ribcage, **robust** (this term will come up again and again: "heavily built") fingers, elbows, and toes were chimpanzee-like, but what did that mean for her behavior? Did australopiths live in the trees? Were these traits linked to climbing, hanging by an arm, walking on the ground, or what? We did not know, because we had little data on chimpanzee locomotion. Knowing what chimpanzees used their unusual anatomy for was critical – presumably Lucy used her body the same way. But chimpanzees hold another key. Did chimpanzees ever stand up like humans, and if so what motivated them to stand? Both bits of information would help us understand first how Lucy lived and, extrapolating, from that information, why humans evolved bipedalism.

I approached Richard Wrangham to see if I could mine these data from his records. If not, could we talk someone at Gombe into making some notes on chimpanzee locomotion? Richard realized from the beginning this make-do plan was half-baked. With no idea how much effort it required to accomplish it, or how much it would affect my scholarly life, I mulled over his exhortations that I should gather this information firsthand. When I finally agreed, after an idiotically long time considering it, he arranged with the late Toshisada Nishida and Jane Goodall for me to work at these two world-famous study sites. I will be forever grateful to this trio.

When I began watching chimpanzees I quickly realized, as I should have beforehand, that their dietary needs were behind much of their minute-to-minute movements. I had to pay attention to their food to understand their locomotion and posture. Access to the best feeding sites seemed to be related to social rank and sex. High-ranking males moved differently than low-ranking males, and females differently as well, so that I had to understand their social relationships to fully understand their locomotion. Their social relationships were grounded in larger societal issues like territorial defense. I found myself being sucked ever deeper into the study of the social world of chimpanzees and I came to understand the force of John Muir's famous quotation: "when we try to pick out anything by itself, we find it hitched to everything else in the Universe." Following Wrangham's lead, I began to interest myself in things like food chemistry and male violence. I found Goodall's work on the challenges of mothering and the evolution of advanced cognition to be more interesting. While I started my chimpanzee research only interested in what it could tell me about fossils and the evolution of bipedalism, after working at Gombe and Mahale I found these other issues every bit as exciting. I moved on to a postdoctoral fellowship with Wrangham and helped his team gather data for his pioneering work on food chemistry and ape and monkey adaptations (Wrangham et al., 1998), an area of interest I have continued to pursue (Hunt, 2016).

As I worked in a two-year postdoc position with Wrangham, I applied for a job at Indiana University, an unusual job that seemed well-suited to me. In what I have come to know as typical Indiana frugality, the university balked at hiring both a primatologist and a human paleontologist and sought to cram two or three jobs into one – at Harvard, where I had been a postdoctoral fellow, there had been three professors and two other staff in these two fields. I accepted the position at Indiana and thus began a very atypical teaching career that spanned primatology and human paleontology. With this somewhat forced diversification, I found myself free to pursue my interest in virtually every aspect of chimpanzee research. As I taught human paleontology and primate social behavior year after year, and as I continued research on wild chimpanzees, I thought more than I might have otherwise about what lessons chimpanzees can teach us about human evolution, what our place is in this world, and what it means to be human; and to be chimpanzee.

While I may have come to chimpanzee fieldwork reluctantly, it has become my life's work. As you enter the fascinating world of chimpanzees, you may see why. *Pan in sempiternum!*

References

Hunt KD (2016) Why are there apes? Evidence for the co-evolution of ape and monkey ecomorphology. *J Anatomy* 228, 630–685.

Johanson DC, Taieb M (1976) Plio-Pleistocene hominid discoveries in Hadar, Ethiopia. *Nature* 260, 293–297.

Martin C (1994) A question of humanity. *Denver Post Magazine*, 18 December, p. 12.

Senut B, Tardieu C (1985) Functional aspects of Plio-Pleistocene hominid limb bones: implications for taxonomy and phylogeny. In *Ancestors: The Hard Evidence* (ed. Delson E), pp. 193–201. New York: Alan R. Liss.

Smith BH (1987) Maturational patterns in early hominids. *Nature* 328, 673–675.

Stern JT, Jr, Susman RL (1983) The locomotor anatomy of *Australopithecus afarensis*. *Am J Phys Anthropol* 60, 279–317.

White TD, Johanson DC, Kimbel WH (1981) *Australopithecus africanus*: its phyletic position reconsidered. *S Afr J Sci* 77, 445–470.

Wrangham RW, Conklin-Brittain NL, Hunt KD (1998) Dietary response of chimpanzees and cercopithecines to seasonal variation in fruit abundance: I. Antifeedants. *Int J Primatol* 19, 949–970.

ACKNOWLEDGMENTS

I am particularly grateful to the three scholars who made my work with chimpanzees possible, all stratospherically famous now, but well-known in a much less spectacular way early in my career: Jane Goodall, the late Toshisada Nishida, and my mentor, Richard Wrangham. Richard lobbied Goodall and Nishida to let me work at Gombe and Mahale and it is only through his persistence and encouragement that I was enabled to conduct the research that made this volume possible. Jane and Toshi graciously welcomed me at Gombe and Mahale, even though I took up valuable space while doing work that had little to do with their own interests. I am very grateful. Observations at Mahale, Gombe, and Kibale provide many of the examples I offer to exemplify the scientific principles that anchor this work, observations made possible by the generosity and foundational work of these much-appreciated mentors. Habituating chimpanzees and creating the infrastructure that makes research possible can seem inevitable in retrospect, but such tasks are monumental.

In addition to welcoming me to Kibale-Kanyawara and hiring me into my first professional academic job, Richard introduced me to primatology and shared with me many anecdotes, ruminations, and speculations on primate evolution that I treasure. Our lively discussions taught me as much about chimpanzees and primates as his brilliant lecturing. The influence of his scholarship will be evident in nearly every chapter of this volume.

After Richard, Loring Brace had the most profound influence on my scholarly worldview, and his perspective – one that values history, logical, consistency and attention to detail – strongly influenced this work, particularly the discussion of fossils and the evolution of bipedalism. With Richard, he co-supervised my doctoral education. More important than the details of fossils, his dynamic perspective on the forces of evolution and his brilliant historical vision inspired me.

Three scholars at my home institution, Indiana University, cooked up the idea of hiring some young scholar into a position where his or her teaching responsibility would combine primatology and paleontology. It was this unconventional pairing that allowed me to keep a foot in each field over my career. I am grateful to Della C. Cook, Paul L. Jamison, and Robert J. Meier. Once at IU, I was quickly drawn into the orbit of paleoanthropologists Kathy D. Schick and Nicholas P. Toth, both of whom have influenced and encouraged my scholarship since my first days in Bloomington.

Of course, none of this would have mattered without the early mentorship of my undergraduate advisor, Fred Smith. Somehow, he put up with my numerous sophomoric (in this case, quite literally) shenanigans and pushed me along to a fulfilling (and published) Honors thesis and then into a graduate career. I flatter myself that he and I think alike, and his scholarship on Neanderthals and adaptation, in particular, have strongly influenced me.

Every scholar finds a few colleagues with whom he can have the sorts of conversations that flow easily and push ideas forward in such a contagious way that insights pile up far out of proportion to the time spent in discussions. For me, Lucia Allen, Bill McGrew, Blaine Morgan, Esteban Sarmiento, Tom Schoenemann, and Susannah K.S. Thorpe have been such collaborators. I thank them.

Many colleagues, friends, acquaintances, and even strangers have shared their narratives with me over the years. Sometimes their anecdotes crystalized into a scientific finding, but even when not, they stimulated my curiosity and sparked my imagination. In addition to those named above, I am particularly grateful to Kelly Baute, Don Byrd, Ron Clarke, Richard Connor, Debby Cox, the late Irv DeVore, Todd Disotell, Susan Ford, Jason Heaton, Kim Hill, Mike Huffman, Teresa Hunt, Andy Kramer, Tetsuro Matsuzawa, Michael McCourt, Henry McHenry, John Mitani, Jim Moore, Blaine Morgan, David Pilbeam,

Sue Savage-Rumbaugh, Anne Russon, Craig Stanford, Karen Strier, the late Alan Walker, David Waterman, and David Watts.

Several scholars whom I have spoken with only sporadically over the years (or in one case not at all) nevertheless loom large in my intellectual life in that they have published works of such importance that I have returned to them again and again for insights and inspiration: John Fleagle, Sarah Blaffer Hrdy, Cliff Jolly, Rich Kay, Jay Kelley, Mike Rose, the late Adolf Schultz, and Russ Tuttle are among these influential scholars. They have helped establish my perception of the roots of primatology and the core of functional anatomy research.

This volume is the lineal descendent of five similar works. It will be apparent to any astute reader that Goodall's *Chimpanzee of Gombe* was a crucial resource and inspiration, in both organization and content. Her chapters on foraging, diet, cognition, tool use, and society were particularly foundational. Yerkes and Yerkes' (1929) *The Great Apes: A Study of Anthropoid Life* was the source that first introduced me to a historical perspective on chimpanzee scholarship, and my history chapter relies on it and Vernon Reynolds' *The Apes*, as well. Adolf Schulz's scholarship, particularly *The Life of Primates*, was a powerful influence on my understanding of primate biology. His brilliant illustrations are sometimes reproduced here, but influenced my approach to many of my original images as well. G.H. Bourne's *The Chimpanzee* filled in important gaps and its comprehensiveness influenced early planning.

The work of my excellent editors at Cambridge University Press was made much less onerous by the unexpected and therefore all the more appreciated editing of Debra Pekin. She made time to read the entire manuscript and helped not only with grammar, spelling, and style, but also big-picture tasks like consistency, tone, transitions, and accessibility. She helped me find the right places to remind readers of the purpose of the volume. Thank you, Deb. My daughter, M. Alison Hunt, is an accomplished editor who is somehow able to find that perfect word to express my intent better than I did; several chapters benefited from her expertise.

Given the breadth I have aspired to here, some chapters relied heavily on one or two pieces of scholarship, particularly large, comprehensive reviews, or on comments and editing by only one or two colleagues. I must mention some of them here.

Chapter 1 was inspired by a wonderful article on human-reared chimpanzees by Claire Martin, a work I still reread over 20 years later. Alyce Miller helped to focus the chapter more cleanly. Harry Raven's 1932 and 1933 articles on Meshie provided the foundation for the chapter (see individual chapters for the references).

The works of C.R. Carpenter (1942), Irv DeVore (1965), and the massive *Primate Societies* (Smuts et al., 1986) were important resources for Chapter 2, updated of course with more recent offerings. That said, when McGrew's 2016 review of chimpanzee field studies appeared I incorporated important ideas from it.

Chapter 3, on the history of chimpanzee science, has at its foundation Yerkes & Yerkes (1929), Hill (1969), Reynolds (1967), and Goodall (1986). Dale Peterson read this chapter early on and saved me from several mistakes and errors of emphasis. Decades of interaction with Loring Brace sharpened my appreciation of the importance of historical depth when attempting to understand any scholarly area, and his views on the Great Chain of Being and the influence of European colonialism on the discovery of the chimpanzee are apparent.

I am grateful to the late Colin Groves for patiently answering questions about taxonomy that vastly improved Chapter 4 and the taxonomy appendix. Betty Rose Nagle, a Professor Emerita of Classical Studies here at IU, helped me with some of the Latin inherent in taxonomy.

R.W. Wrangham's enormous body of scholarly work on primate diets, nutrition, foraging, and plant chemistry inform Chapters 5 and 6 – those that cover diet, habitat, and digestion. Alain Houle and I had several conversations about these chapters and his advice improved them. John Mitani and especially Nick Newton-Fisher helped compile information on diet for Chapter 5. Almost all of the information on tropical forests in this chapter was based on T.C. Whitmore's stylish and informative *Introduction to*

Tropical Forests. Chapter 6 owes much to David Southgate's wonderful 1995 digestion review, but also to work by Chivers, Givannoni, Gautier-Hion, Hladik, Lambert, Milton, and Wrangham. Jim Giovannoni read and improved the section on fruit ripening.

Chapters 7, 8, and 9 on anatomy, functional morphology, locomotion, and posture are the focus of my own scholarship, but my understanding of these topics grew out of study of the works of Charles Oxnard and publications by and conversations with Mary Marzke, John Fleagle, John Cant, Rich Kay, Bill Jungers, Esteban Sarmiento, and Mike Rose. Kris Carlson and I engaged in many thought-provoking discussions of function and form both during his days as my graduate student and since. David Pettifer commented helpfully on all three of these chapters while swatting tsetse flies in the sweltering heat of Semliki.

Some of Chapter 10 started out as a grant proposal I wrote with Kelly Baute. Other parts began as an article I published in 2016. Susannah K.S. Thorpe commented on early versions the chapter. As it took a more final shape, Jay Kelley meticulously ferreted out errors of fact, emphasis, grammar, and terminology, and shared with me his different perspective on ape and monkey competitive relationships, or the lack thereof. He and I exchanged enough word-volume via email to make a short book. I am very grateful. I treasure colleagues who can engage in spirited debates without rancor and with collegiality even when there are fundamental differences in outlook, and Jay is one of those colleagues. However, he failed to convince me of the primate paleontological community's certainty that *Proconsul/Ekembo* is merely a stem hominoid, only indirectly related to living apes; I take the more liberal view that if not *Proconsul*, something very like it and closely related to it was at the root of the ape lineage. He takes a more conservative perspective on chimpanzee and OWM competitive relationships and on any number of small details where I went out on a limb, if you can excuse the pun. He should not be blamed for my divergent viewpoint.

Chapter 11, on growth and development, drew heavily on Adolf Schultz's work and Paul Jamison's as well; Paul edited the chapter. Shawn Hurst edited and commented as well, and helped with the brain figure. My summary of evo-devo and homeobox genes depended mostly on Sean Carroll's excellent book on the topic; the late Rudy Raff commented helpfully and his wonderful *The Shape of Life* helped immensely.

Robert J. Meier very kindly edited Chapter 12 and John F. McDonald commented on apoptosis. Randy Brutkiewicz and Yansheng Du helped me understand the function of the *IL* gene family and interleukins.

Diana Kewley-Port patiently explained the physics of sound to me over drinks one afternoon at Yogi's Bar and Grill, which helped tremendously for Chapter 13 on the senses.

As with Chapter 10, some of Chapter 14 started out as a grant proposal I wrote with Kelly Baute. Sean Prall very kindly provided extensive and very helpful comments and edits on the entire chapter. Thank you, Sean.

The scholarly work of my valued former IU colleague, Michael Muehlenbein, and Michael's and Richard Bribiescas' review chapter (see Muehlenbein, 2015) served as the basis for Chapter 15 on hormones. Michael provided comments on Chapter 14 as well.

Jane Goodall and Anne Pusey founded chimpanzee mothering research, and Chapter 16 is packed with their insights. Katherine A. Cronin generously shared her wonderful photographs of a mourning mother with me.

Craig Stanford very graciously set aside time to comment on Chapter 17, my hunting review, even while he was simultaneous putting the finishing touches on a competing (drats!) volume.

Tom Schoenemann coached me through some of the intricacies of brain anatomy and helped with Chapter 19, on the mind, as well.

Shawn Hurst reviewed Chapters 18 and 19 and provided some of the figures. More importantly, his interest in the brain as he blocked out his dissertation research inspired me to learn more about it. His lively imagination and thoughtful perspective on the mind and brain strongly influenced the chapter. Linda Marchant and I have had many discussions about handedness/laterality, tool use, and primate brains and I benefited enormously from her perspective. Colin Allen worked through the entire chapter on the mind, added important insights, and pointed out some glaring omissions. Discussions with Shawn, Tom Schoenemann, Peter Todd, Kathy Schick, Nick Toth, Sally Boysen, and Ralph Holloway were also extremely helpful and clarified many of the critical issues in these chapters.

Chapter 20 on sleep draws heavily on my former doctoral student David Samson's pioneering research and historical perspective, not to mention our many lively discussions about sleep and its evolution. He edited the chapter from top to bottom and improved it enormously.

Numerous discussions with Steve Franks and Tom Schoenemann sharpened my perspective on chimpanzee language, presented in Chapter 21. Steve made several editing passes at the chapter and helped me to understand the linguistic perspective on the evolution of language, one a little different from my own. Phil LeSourd's comments and advice added some nuance to the chapter. Tom Van Cantfort saved me from an embarrassing error and helped to form my writing on Washoe and research surrounding her. Mary Lee Jensvold shared her most recent research findings with me and helped to refine the chapter.

Chapter 22 is mostly a condensed and updated version of Bill McGrew's *Chimpanzee Material Culture*. His influence and research can be seen throughout this chapter and indeed throughout the entire volume. I wish I could write as clearly as he does. He generously edited the chapter adding important facts and perspectives, saving me from a number of embarrassing mistakes.

Chapters 23 and 24 are based on numerous conversations with Richard Wrangham and on Goodall's foundational work. Bill McGrew has been interested in chimpanzee culture for decades and his work is much cited as well.

Chapter 24 owes a great debt to Martin Muller's 2002 chapter on aggression as well as the work of Mike Wilson and Richard Wrangham.

Chapters 24, 25, and 26 grew out of the research and discussions with Richard Wrangham and fewer, though still valuable, back-and-forths with Bill McGrew.

Conversations with Tobias Deschner informed Chapter 27, and he generously supplied me with images that help make sense of sex and reproduction in chimpanzees. My first exposure to reproductive endocrinology came when I worked with the late Pat Whitten as a teaching assistant in the 1980s. Thanks, Pat.

Chapter 28 is quite short considering that I have spent over 20 years working at Semliki and nearly three years actually on-site. I am grateful to the ownership and staff of the Semliki Safari Lodge. Jonathan Wright and his ecotourist company, WildPlaces, have provided support and collaboration for my research at Semliki from its beginning. Tim White graciously shared some of the images in the chapter. Many of my staff, employees of the Semliki Chimpanzee Project, have worked to sustain the project, but I am particularly grateful to the late Karamajong Kule, to Eriik Kasutama, and to former camp manager Moses Comeboy. I have benefited from the help of nearly three dozen assistant project managers – local supervisors – over the years. I cannot name them all, but some who were particularly indomitable, industrious, resilient, *and* faithful to their contracts (a six-month stint in a tent is no picnic) are Rachel Weiss, Jim Latham, Sacha Cleminson, Esther Bertram, James Fuller, Teague O'Mara, Chris Wade, Jim Reside, David Inglis, Jess Tombs, Carmen Vidal, Alissa Jordan, Tim Webster, David Samson, Maggie Hirschauer, Will Symes, Jeremy Borniger, Corey Mitchell, Caro Deimel, Luke Louden, Katie Gerstner, Ben Lake, Wendy Craft, Jaycee Chapman, and Steven Wade. Clint Schipper held things together one six-month period when ADF terrorists plagued us the most. For many years Melanie Ebdon has supported my research with monthly donations to SCP; thank you, Melanie. Michael McCourt stepped in with financial support when I needed it most. Daniel and Nancy Farlow have helped out and were welcome visitors in 2018. The NSF funded the project early on. I am grateful to the Ugandan Government, the Uganda Wildlife Authority, the staff of the Toro-Semliki Wildlife Reserve, and the rangers and staff of UWA, too numerous to name, who made this research possible.

In addition to the research cited in Chapter 29, much of my perspective on bonobo society, social behavior, and distinctiveness was sharpened in conversations with Caro Deimel. Important conversations with Frans de Waal, Barb Fruth, Gottfried Hohmann, and Gil Ramos sharpened my understanding of bonobo society. Richard Wrangham made extremely helpful suggestions – though I failed to convert him to the Monkey Convergence Hypothesis. Bill Jungers pointed me to some essential data on chimpanzee body masses and very generously performed a statistical test on his data, which he allowed me to cite here.

In the very last stages of writing Bill McGrew helped enormously with Chapter 30. Thanks, Bill.

As I related above, my two doctoral co-supervisors were Loring Brace and Richard Wrangham, which means my academic pedigree is C.L. Brace–E.A. Hooton/W.W. Howells–A. Keith and R.W. Wrangham–R.A. Hinde–D. Lack/N. Tinbergen. I am proud.

Thanks to Jodi Pope Johnson and Olivia Pope Pfingston who helped with early cover design and artwork.

I am grateful to the editors and staff at Cambridge University Press for their unfailing encouragement and guidance as I put this volume together. Commissioning Editors Megan Keirnan and Ilaria Tassistro served as valued sounding boards as the book took shape. Editorial Assistant Olivia Boult provided helpful motivation as I reached the light-at-the-end-of-the-tunnel stage. I am an author who gives a copy-editor things to do and Gary Smith did it; thank you for your diligent work on this difficult task. Content Managers Annie Toynbee and Jenny van der Meijden helped me keep my eye on the big picture and took care of the little picture at the same time.

Lastly, my wife Marion Gewartowski Hunt read, corrected, and commented on nearly the entire manuscript, some chapters several times, even though her scholarly interests have little to do with primatology. I wrote her a letter every day when I was at Gombe and Mahale, and she kept them all, preserving a written record that was an essential aid to my memory of my time at Gombe and Mahale. Marion took up the slack in our family life when I was in Africa and also when holed up writing on this volume. I am more than grateful – I am forever indebted.

1 Sister's Keeper
Humans and Chimpanzees

From *Natural History*, March-April 1932 and November-December 1933, copyright © Natural History Magazine, Inc., 1932, 1933, with permission

1.1 Close Encounters with Chimpanzees

There may be only two groups of humans who truly understand how chimpanzees look at the world. The first is those who have studied chimpanzees in the wild, trailing along behind them day after day, watching as males manipulate both friends and enemies in their struggle to ascend to alpha status; or looking on in sympathy as a heavily pregnant female, tired after a long day of gathering, calculates which feeding site might pay-off best for the day's last meal. After perhaps a year of such snooping, an observer might begin to understand what it is like to inhabit the mind of the chimpanzee, to think about the world and other chimpanzees as a chimpanzee does.

The second group is made up of people who have reared a chimpanzee from infancy, treating her as they might have treated their own infant, learning her abilities, desires, and fears as she experiences the joy, loneliness, anxiety, frustration, and confusion of childhood.

The approach followed by the first group has led to incredible scientific discoveries and vivid biographies of individual chimpanzees. That followed by the second group is heavily discouraged nowadays, and thankfully so. It has always (at least, in every case I know of) played out as a heart-wrenching tragedy entailing shattered relationships, forlorn isolation, and premature death. To quote Jane Goodall, "Every primate belongs in an environment that is as close to a wild setting as possible. They are beautiful and intelligent animals, but highly complex with very specific needs."

Most people belong to neither of these groups, but see chimpanzees in a third context: captivity. Ceding control of your life to a jailer, no matter how benign the jailer, shrinks the spirit and shrivels the intellect. Captive chimpanzees are diminished beings, not the striving, calculating, problem-solving personalities I know from the wild.

Those who have inhabited the chimpanzee world come to know that while chimpanzees truly are intelligent, their intelligence is as unlike ours – but also as like ours – as their anatomy. For some cognitive tasks their intellect and psychology mirror ours almost perfectly; for others it is a fun-house mirror version of the human condition, similar but strangely warped. For still other intellectual tasks their abilities are surprisingly meager. These differences mean that it takes time and effort to inhabit their minds. This volume means to introduce a diverse audience to aspects of chimpanzee behavior and biology they may have neglected. While the rest of this book will focus on chimpanzee research, both in the wild and in captivity, I want to introduce chimpanzee nature to you by profiling a few chimpanzees whose lives were among humans, and I will wrap it up back in the wild, where we will spend most of our time in this volume.

1.2 Nkuumwa

The Uganda Wildlife Education Center (the Entebbe Zoo[1] to some) of July 1997 was half green hope and half dusty depression. The startling transformation that has rendered it the green and shady animal sanctuary it is today had only just begun. Yet, even then you could see signs of what was to come. There were new, spacious nature-mimicking exhibits under construction everywhere you looked. Yet still amid these emerging oases of wildness you could find (in a kiddie pool-sized mud puddle) a lone, tattered, and bloated-looking crocodile whiling away eternity, nothing about the creature reminiscent of the sharp danger and vivid action of the wild animal.

Strolling the zoo's dusty paths with Debby Cox, one of the driving forces behind the zoo's current zenith, I wended my way past steel-meshed, concrete-floored monkey cages and topped a hill to catch a glimpse of what I had come there to see, the green, spacious, and spanking-new chimpanzee habitat. This modern enclosure had been built as a home for some of nature's most appealing castaways. Or – not castaways – but *unfortunates*, wild beings kidnapped from nature. Here lived chimpanzees who had narrowly escaped the cooking pot – or if not that, the short, miserable life of a shackled pet. Debby proudly showed off the enclosure. It was hilly, clothed in green, alive with healthy vegetation, and situated so that the restraining fences were as little noticed as possible. I circled the compound, listening to the clean-toned, high-decibel hoots of the chimpanzees, watching their faces and mannerisms for some hint of their harrowing past. Sadly, some clearly showed it.

Yet, others seemed happy, virtually indistinguishable from wild chimpanzees.

Near the enclosure was a modest structure that held the offices and workspace for the ape caretakers. There I met the zoo's newest castaway, an infant chimpanzee named Nkuumwa (Figure 1.1). Possibly a year old, perhaps two, the orphan Nkuumwa was too young to live on solid food alone. Chimpanzees mature only slightly faster than humans. If she were in the wild she would still be getting all her nutrition from nursing and would continue to do so for another couple of years. She was here because she was still too small and weak to cope with the nearby raucous chimpanzees who would one day become her companions. Instead, she was being hand-reared by Debby and the zoo staff.

Nkuumwa had been confiscated five months earlier from a cook at the US Marine House in Kampala, where she had been kept as a sort of mascot, regularly passed from person to person during social events, a circumstance that could have been no more comfortable for her than it would be for a human baby.

If Nkuumwa had a typical experience during her orphaning she would have seen her mother shot and killed, and as horrific as it is to contemplate, it is not at all unlikely that she watched as her mother was butchered by poachers. Then she would have been carried to a roadway, chained, offered for sale, and eventually bought by a passing stranger. Since wild chimpanzees of Nkuumwa's age are rarely out of their mother arms, her capture would have been the first

Figure 1.1 Nkuumwa at four years old. Courtesy of Ngamba Wildlife Reserve.

---

[1] Entebbe is a city in Uganda, East Africa, best known as the site of Uganda's international airport; the capital, Kampala, is some 45 miles distant.

time she had ever been more than an arm's length from her mother.

Someone notified the authorities that Nkuumwa was at the Marine House, and it fell to the Entebbe Zoo staff to confiscate her. When zoo personnel first gathered her up she was so covered with lice that her normally black fur appeared white. She was too weak to stand, and when she was examined by medical staff she was found to have pneumonia. At first Nkuumwa refused to eat or drink. Pneumonia is extremely dangerous for chimpanzees, and for infants it is fatal more often than not. Nkuumwa was lucky. After only a brief time in the care of the knowledgeable zoo staff, she rallied. She gained strength, then gained weight, and soon it was apparent she would survive. By the time I met her she looked healthy and well-fed. At a cost. Debby was woken up several times a night to feed Nkuumwa on demand, just as her mother would have done.

Nkuumwa's luck has continued up to now. She is 21 years old as I write this, living in the Ngamba Island chimpanzee sanctuary in Uganda, a well-adjusted, lively young female, seemingly bearing no scars from her harrowing life-journey.

While Nkuumwa's story is a disturbing one, I can hardly blame the Marine who bought her and took her home. He was young and infant chimpanzees are heart-breakingly cute; perhaps he did what many of us would have done in his place: He saw an appealing and sad animal he wanted to comfort. This is exactly what so many sympathetic chimp owners in history did.

I recounted Nkuumwa's biography because it says something about people and chimpanzees that I find telling. For, as I was meeting Nkuumwa, her caretaker asked if I would like to hold her. As a fieldworker I had been fully indoctrinated in the "avoid-contact" code all wild primate researchers follow. I was inculcated to avoid close proximity to a chimpanzee, and touching them on purpose was out of the question. The no-contact rule is an important one. Not only might contact pass on diseases, if chimpanzees interact too intimately with humans the overexposed individuals tend to treat humans just as they might treat another group member. Despite what you may have heard, that does not mean sensitively grooming them and patting them on the head – it often means inclusion in the violent world of chimpanzee physical dominance, a rigid social hierarchy we will discuss in detail in Chapters 24, 25, and 26. So this was a first for me. In all my years of living among chimpanzees I could count on one hand the incidental physical contacts I had had with them.

While it would never have occurred to me to ask to hold Nkuumwa, I made a quick readjustment as her keeper looked at me expectantly. I reached for her and as I did, she held her arms out to be gathered up, exactly as a human infant would. There was something endearing about the way she gripped my shirt and arm so tightly, even though there was no indication she was fearful. I pulled one of her hands away from my shirt and held it between my thumb and finger, examining her surprisingly narrow light brown hand with its long, thick fingers, wrinkled, warm, soft on the outside, but with hard strong sinews and bone underneath. I looked down at her face and she gazed back at me steadily, her eyes darting back and forth between each of mine, something that somehow made her seem unsettlingly like a human baby.

I had received some potent doses of infant chimpanzee cuteness in the past, but holding this tiny, clinging baby in my arms evoked a much more powerful emotion than I had experienced before. It was not just that Nkuumwa was cute; I felt embarrassed that I immediately felt a strong protectiveness and possessiveness toward her, as if I might have trouble handing her back to her keeper. At that instant I felt a brief stab of guilt, a guilt I have felt before when cuddling someone else's baby, the hollow feeling that somehow my parental feeling toward Nkuumwa was betraying my own children. This feeling was quickly followed by the more objective realization that some important distinction I had always assumed existed between humans and animals had been short-circuited.

1.3 Meshie

While this episode was unusual, I am just one of many to have experienced it. Soon after this incident, I read an article by Claire Martin that told a tragic tale with which I was vaguely familiar, but she told it in a more personal way than anything I had read before (Martin, 1994).

It was the story of a scientist and a chimpanzee – a frail orphan purchased on impulse. It was a story that featured the same species-confusion I had experienced holding Nkuumwa. It was the story of the eminent scientist Harry Raven and his adopted chimpanzee daughter, Meshie (Figures 1.2–1.4).

Figure 1.2 Meshie eating ice cream. Raven reports that she learned to use a spoon very quickly. With permission of Natural History Magazine.

Figure 1.3 Meshie drinking grape juice, using a straw. With permission of Natural History Magazine.

Figure 1.4 Harry Raven's daughter Mary, held by Meshie. "When they took the baby back the first time she held her, Meshie's hands were all sweaty. She was terrified of anything happening to the baby" (Martin, 1994). With permission of Natural History Magazine.

Raven's is a well-known story (Raven, 1932, 1933), told in two articles that were sensations at the time among those who were interested in primates, retold many times even before Martin's retelling. Writer Preston Douglas reported on Raven's story and was so intrigued by it he turned it into a captivating novel that was widely reviewed and read, *Jennie* (Preston, 1994).

Raven, Douglas, and Martin tell Meshie's story. Her name was given to her by children in an African village Raven was visiting. "Meshie Mungkut," Martin wrote, was said to mean "little chimpanzee who fluffs her hair up to look big" (chimpanzee hair stands on end when they are excited or aggressive, a phenomenon known as **piloerection**).

In 1930 and 1932, while he was curator at the American Museum of Natural History, Harry Raven traveled from his tidy home on Long Island to Cameroon, West Africa on an expedition "collecting" primate specimens for anatomical study; you probably suspect what collecting means – "specimens" are not normally found lying on the forest floor.

Two men walked into Raven's camp carrying the tiny Meshie. One of the men had killed her mother with a poison dart, and the two had eaten her. Meshie, seeing Raven, stretched her arms toward him and he could not resist taking her; she held on tight "as if she feared she might fall," and stroked the hair on his arm. He must have felt the same impulse I felt when holding Nkuumwa, the powerful impulse to protect. Raven bought Meshie and traveled with her around Africa, Meshie sitting next to him in the front seat of his vehicle as he drove. He managed to get her back to the USA. While she slept for a time in a treehouse Raven made for her, making a nest of blankets each night, for the most part she was reared in his home and treated as a sibling to Raven's other three children (Figure 1.5); afterward this rearing-as-if-human came to be known as **cross-fostering**. In Raven's home, Meshie soon learned to eat at the table, pedal a kiddie-car, and eventually to ride her tricycle around with neighborhood children (Figure 1.6).

In short, Meshie was allowed about as much freedom as a chimpanzee can be allowed in human society. Perhaps because Raven was a prominent scientist, nobody seemed to question his custom of allowing her the run of his home and of the wider neighborhood beyond.

You have read my own species-confusion story, but in the case of Meshie we have more than mere written descriptions with which to comprehend Meshie's

Figure 1.5 Meshie comforts a crying Mary. With permission of Natural History Magazine.

Figure 1.6 An older Meshie leads a wheeled parade of local children. With permission of Natural History Magazine.

intimate relationship with her human family. We have home movies. Martin describes one of these films with these powerful words:

Meshie is about five years old. She is perching on a stool next to the Ravens' six-month-old baby, Mary, who is in a high chair. With one thick, wrinkled hand, Meshie holds a bowl of what looks like rice cereal. With the other hand, Meshie patiently and deftly spoon-feeds the baby, who is blasé about being fed by a chimpanzee but fascinated by the camera, constantly twisting to stare curiously at the bright lights. The scene cuts to the baby grabbing a sloppy handful of cereal and smearing it on the high chair's tray. Meshie, who was fastidiously neat during her own meals, is visibly appalled. She hops down, fetches a rag and wipes the baby's messy face. Then she cleans up the cereal, scrubbing until she's satisfied that the tray is spotless.

This visual image brought home to Douglas and Martin the similarity of chimpanzee and human minds in a way that nothing else had. In an interview with Martin, Douglas remembered his response to the film: "I thought 'My God! This is an animal feeding a human! With a spoon! Feeds her with a spoon, very carefully, and then wipes up the mess afterward. This isn't at all like a cat or a dog or a horse; we're talking about an animal so close to being human that a real confusion is going on here.'"

For some time Meshie seemed to slip almost seamlessly into the family, but as she matured she became increasingly willful and uncontrollable. Eventually she became too dangerous to be allowed to roam freely. Raven estimated that as a 46-pound four-year-old she was already as strong as an adult human.

At his wit's end, Raven sadly realized she had to be restrained somehow. At first, he tied her up with a rope, but she untied the knot. He tried a chain, but before long she learned to break the links or unbuckle the collar. Raven's frustration at Meshie's repeated escapes is palpable in his retelling, but so also is a suppressed delight in her skill at untying knots and slipping off collars. Finally, he had a cage custom-built in his basement, a comfortable cage, but still too much like a prison to be a happy thing.

Reading Raven's 1933 article, his love for Meshie (Figure 1.7) bubbles to the surface as he describes the loving persistence and remarkable steadfastness with

Figure 1.7 Harry Raven and Meshie. With permission of Natural History Magazine.

which he resignedly repaired one after another of Meshie's devastations. She ripped out electrical wires; she bent gas lines; she toppled paint cans – all this from *inside* her cage. She escaped to play in the coal bin, tracking black footprints everywhere. Not only did she have incredible strength, she was as nimble as a circus acrobat; once out of her cage, she climbed everything climbable, making recapturing her an exhausting trial. Raven unflinchingly bore it all – the financial costs, the phone calls summoning him home from work early, and the stress of an increasingly chaotic home life that his put-upon wife bore even more heavily than he. His tale begins to sound much like that of any exhausted parent of a wayward son or daughter who is constantly in trouble with the authorities.

1.4 Meshie's Fate

Ultimately, even with restraints, Meshie proved too dangerous to both herself and to children in the neighborhood to remain in the Raven home, and in the end she suffered the fate of the vast majority of pet chimpanzees: She was sent to a zoo, confined to a sturdy steel cage at the Brookfield Zoo in Chicago. Having sat watching her mother butchered and eaten, she was now torn from her loving adoptive family and doomed to confinement, a circumstance that undoubtedly left her lonely and frightened. Zoo staff saw nothing cute or appealing in Meshie. In her confusion and anger she struck out at her caretakers with shockingly aggressive displays. To zoo staff she was a vicious, violent, dangerous beast, nothing like the loving "child" who had cuddled her baby sister and fed her with a spoon.

Douglas and Martin report that Raven visited her once, some years later. He ignored concerned keepers' warnings that she would rip him apart and entered Meshie's cage where the "vicious beast" and her adoptive father had a tender reunion. Sadly, love does not conquer all. Despite Raven's affection for Meshie, the reunion was short-lived; it was simply impossible for him to take her home.

I wish there were a happy ending to this story, but there is not. Meshie remained in the zoo as an animal on exhibit. In her anger and confusion she never fully integrated with the other chimpanzees and interacted with them little. She did assimilate well enough to mate with a male and ultimately became pregnant. She died giving birth. Motherhood can calm captive primates; having a helpless, appealing baby to fawn over gives their drab lives some meaning; it is the ultimate tragedy of Meshie that she never had this last chance at adult happiness. Raven is said to have grieved Meshie's death for years.

There is a lesson to Raven's and Meshie's tragedy. Their heart-breaking separation is one that ultimately divides all chimpanzees and their civilian caretakers. Chimpanzees cannot be pets. As appealing as they are as infants, they are best reared among other chimpanzees from birth, if they are born in captivity.

Raven's home movies and narratives from the 1930s are important when placed in context. They are part of the history of discovery of chimpanzee cognitive abilities. They provide a less scientific and more personal addition to a whole procession of astonishing scientific discoveries about – if I can phrase it this way – the humanity of the chimpanzee, much of it played out in those few celluloid seconds of Raven family life. We see tool use. An animal that clearly understands the somewhat fuzzy concept of "neat," and the seamless (for the moment) splicing of a member of different species into a human family. Rule following. Mental mapping (a **mental map** is a cognitive representation of a geographic area) of unobserved and distant objects. Empathetic concern for loved ones (Figure 1.5). Is it any wonder that even decades later Mary Raven Hockersmith, the child being fed in the movie, displayed in her home a photograph of herself as an infant, being held by Meshie, her doting and ultimately stolen sibling?

For scientists, Meshie's story adds texture to publications that appeared around the time of Meshie's adoption but that largely spoke to a scientific audience (Köhler, 1925; Kellogg & Kellogg, 1933; Ladygina-Kohts, 1935).

1.5 Vicki

At the risk of lingering overlong on cross-species adoptions, let me relate two more anecdotes that show the adoption of human cultural traditions by Meshie was not unusual. Martin also reported on another human-reared chimpanzee named Vicki; she was a further model for Douglas's novel *Jennie*. One of Vicki's favorite playthings was a pull-toy she towed around her house with great enjoyment. The toy occasionally got hung up on furniture or fixtures, particularly as she raced around the toilet. Surprisingly, when the pull-toy went missing Vicki continued to "play" with it, trailing an imaginary pull-toy behind her as she ran through the house. The nonexistent pull-toy got hung up on the toilet, just as did the real toy, at which point Vicki patiently stopped to untangle it.

Now and then someone wonders whether human-reared chimpanzees might be more intelligent – more human-like – than wild chimpanzees: Distinguished lab primatologist Duane Rumbaugh asked this of me once. This view arises from the idea that life in the wild is less intellectually challenging than life in the lab. I hold a diametrically opposed view. Other than superficial behaviors like using a spoon, the basic "humanity" of chimpanzees is just as startlingly apparent in the wild. In Chapter 18 I will discuss the close correspondence between the chimpanzee mind and the cognitive demands they face in the wild. There, in their natural home, chimpanzees exhibit all their intellectual strengths, and many are similar to those that distinguish humans from other primates: In the forest their excellent memories, superbly tuned sense of geographical space, social subtlety, and intense mother–infant bonds closely match human intelligence. They are like us not because they mimic us, but because they are a close relative, a sister species.

1.6 Kakama

As an example of Vicki- or Meshie-like behavior among *wild* chimpanzees, Richard Wrangham (Wrangham & Peterson, 1997) tells the story of a boisterous eight-year-old male, Kakama and his sluggish, pregnant mother. Kakama plucked a log, a largish one half his size, from the forest floor and carried it off and on for hours, snuggling with it, juggling it while he lay on his back in a day-nest, placing it carefully beside him when he fed, frolicking with it as if it were a baby. Fate had placed Kakama, an outgoing, playful young male, with a somewhat stodgy, antisocial mother. Could he have been anticipating the birth of a playmate? Wrangham was left wondering. Several months later, just weeks before Kakama's mother gave birth, two of Wrangham's assistants saw behavior just like that he had seen. When he abandoned the log the assistant brought it back to camp and stapled to it a label that interpreted what they had seen in a delightfully straightforward manner: "Kakama's toy baby." There was no question that Kakama invented his doll without human interference.

1.7 Understand Chimpanzees, Understand Human Evolution

Because we are **Sister Species**, chimpanzees have important lessons to teach us about nature, and about ourselves. Of all species on the planet, surely this one, and the other human sister species, **bonobos** (*Pan paniscus*, also known as pygmy chimpanzees) are the species from which we learn the most and the species for which we should move heaven and earth to save in the wild – for purely selfish reasons if nothing else. We have much more to learn about chimpanzees and bonobos (Chapter 29), especially if you consider this: As I wrote this book, I consulted a large number of experts (see Acknowledgments) and there were a multitude of disagreements, questions about the reliability of this fact or that detail; there were gentle accusations of "speculation" or "overinterpretation." Maybe so, but the important thing is that only wild chimpanzees can answer the scholarly questions my colleagues raised. My particular interest in chimpanzees is what they can tell us about human evolution. You may be surprised at how human-like chimpanzees are as you read this volume, but you may be even more surprised as you gradually become aware of how many mysteries about human origins we can solve by understanding chimpanzees.

References

Kellogg WN, Kellogg, LA (1933) *The Ape and the Child: A Study of Environmental Influence upon Early Behavior.* New York: McGraw-Hill.

Köhler W (1925/1959) *The Mentality of Apes,* 2nd ed. New York: Viking.

Ladygina-Kohts, NN (1935) *Infant Ape and Human Child.* Moscow: Museum Darwinianum.

Martin C (1994) A question of humanity. *Denver Post Magazine,* December 18, 12.

Preston D (1994) *Jennie.* New York: St Martin's Press.

Raven HC (1932) Meshie, the child of a chimpanzee. *Nat Hist* 32, 158–166.

Raven HC (1933) Further adventures of Meshie. *Nat Hist* 33 607–617.

Wrangham RW, Peterson D (1997) *Demonic Males: Apes and the Origins of Violence.* New York: Houghton Mifflin.

2 Wild Lesson
Why Study Animals in Nature?

Photo by author

PART I: THE IMPORTANCE OF WILD ANIMALS

The first time I saw chimpanzees in the wild, I was filled with wonder. I had studied them in the zoo (Chapter 18) before arriving at Mahale,[1] but seeing them in the wild was a different experience. As a male walked past me I looked in awe at the muscles in his lower leg and how the pad of his foot conformed to the uneven ground; it struck me how improbable it was that this seemingly simple environment could produce something as complicated, intelligent, intense, and powerful as a chimpanzee. I imagined a sort of chimpanzee-mist rising out of the soil and coalescing into the individual in front of me. A fanciful thought, but as incredible as it seems, an environment very much like that of Mahale *did* produce chimpanzees.

The story of Meshie and her adoptive father, Harry Raven, resonates with us all the more because in these personal narratives we see chimpanzees struggling to negotiate a human world that – while comfortable and familiar to us – is a challenge to them. We pity them. A chimpanzee in the wild is an altogether different thing. This is where we see the true chimpanzee. While a chimpanzee cleaning up after a messy baby is fascinating, I find the ingenuity of chimpanzees in the wild even more compelling. Wild behaviors are home-grown; they could only have been invented by the chimpanzees themselves, a fact that makes their complexity and human-ness more impressive.

2.1 The Romance of Fieldwork

Studying primates in the wild is worth it, but such research is not easy. I spent most of 1986 and 1987 gathering data for my dissertation, shuttling back and forth between Mahale and Gombe on whatever water transport I could find on Lake Tanganyika. Back then, transport for a poor doctoral student was the ubiquitous water taxi, dozens of which plied the waters of Lake Tanganyika at any one moment. These 50-foot wooden boats could hold 40 people comfortably. I say "comfortably," but typically they were loaded far beyond comfort. It was more common to see them weighed down to the Plimsoll line and beyond with as many as 100 people. Every once in a while, the newspaper would report that an overloaded boat had capsized, inevitably resulting in several drownings. I was happy to be a strong swimmer.

---

[1] The Mahale Mountains National Park is the site of the second oldest continuously operating chimpanzee research project, founded by Toshisada Nishida of Japan in 1965. The Gombe National Park hosts the oldest continuous chimpanzee study site, founded by Jane Goodall in 1960

Water taxis are powered by ridiculously inadequate outboard motors that putt-putt along so slowly you hardly seem to be moving. The trip from Kigoma to Gombe is only 15 miles, but with stops it often took us two hours and more. On one trip my local assistant, Hamisi Katinkila, and I piled onto one of these precarious vessels, manhandling my two large stuffed-to-bursting backpacks as we did, struggling to keep them out of the couple of inches of filthy water sloshing in the bottom. There is always a wait as the owner crowds more and more passengers onto the already oversubscribed boat, but at last we shoved off for Gombe, me fearing as always that at any moment all my Tanzanian possessions, including my camera and tape recorder, would topple into the lake.

We puttered along until halfway into the trip, at which point the motor made a disturbing whining noise, sputtered, and stopped. The crew implacably paddled us to shore where everyone debarked. Hamisi and I waded to shore with my two backpacks, each bearing 30 kg of gear. Then, miraculously, I saw that from among the other passengers' luggage rice, fish, charcoal stoves, and sufuria cooking pots materialized, and they began cooking their lunches. Meanwhile the boatmen had heaved the motor onto a plastic tarp covering the lakeside foliage and had begun disassembling it. Eventually, Hamisi, milling around, overheard the news that a repairman was on his way.

In due course a teenager on a bicycle hauled up with an older, graying fellow on the back, somewhat overdressed in suit jacket and wool trousers. He unstrapped a canvas packet of clanking tools from the bike and joined the boat owners disassembling the greasy boat motor. Hamisi and I stood by, swatting mosquitoes. Eventually I pushed closer to the bowl-shaped depression in which the toiling repairmen labored. Arranged on the tarp like a museum display, I saw pieces of engine housing and a wide variety of rods, oily cogs, bolts, and washers. In the middle was the still-suited "fundi," examining a worn-looking cog the size of a hockey puck. He drew an outline of it on a piece of wood and began carving. "You cannot be serious," I thought. "He thinks he can copy this hardened steel, delicately balanced cog? By hand? Out of wood? It will take hours and then it will last ten seconds at which point the whole contraption will explode, slaying six."

Trailing my first-world arrogance like a storm cloud, I informed the astonished Hamisi that we were going to walk the rest of the way. It was still early in the afternoon. We could hardly get lost since all we had to do was follow the shoreline and we would eventually hit Gombe. Surely we could walk almost as fast as the water taxi could putter along, anyway, and I doubted it would be puttering at all anytime soon. I figured we were over halfway there, which left us, what, seven, eight miles? No way was I going to wait on this ill-fated repair.

We staggered off with the two huge backpacks, wading through the fine pea gravel along the shoreline. It was tough going. With each step we sunk several inches into the gravel, making each stride essentially a half-step. So the long the afternoon wore on. The sun drew low on the horizon, and still we saw nothing familiar. It grew dark, still with no sign of the park boundary. I pulled out a flashlight and we continued for another hour, then two, and soon my cheap Tiger Head Chinese batteries (if you were ever in East Africa around this time, you know them well!) were exhausted. *We* were exhausted, and in any case, it was too dark to make our way any longer without a flashlight. We would sleep on the beach and continue the next day.

We extracted from my luggage two thin blankets I hoped would provide some minimal warmth and perhaps protection from the voracious mosquitoes. At last we slept. A few hours later, after the whirring insects had quieted, we were awoken by a sound, the putt-putt-putt of a water taxi. Out on the lake we saw, illuminated by the light of a pressure lantern swinging from a spar in front, an unexpected but heartening sight. Our water taxi! How could the motor be running with that wooden cog? No matter, it was! Belatedly we began to shout to stop "Hoo, hoo! Here, here!" hoping to see the bow turn toward us. No luck. We resignedly returned to the body-shaped impressions we had made in the pea-gravel, destined to swat mosquitoes for a few more hours.

At first light we emerged from our bedding, me itching fiercely from my mosquito bites. Hamisi seemed unaffected "mosquitoes like the soft skin of mzungus more than we Africans," he told me. Out of curiosity, I counted the itchy lumps, starting on the back of one hand. I had not yet reached my shoulder when my count reached 100. It turns out we were frustratingly close to Gombe. Some fisherman happened by in a canoe and we shouted and waved them down. We quickly agreed on a very reasonable fee to paddle us to chimp-camp. In the crisp morning stillness, the lake was like glass. We watched the sun come up over the hills above Gombe as the friendly fisherman paddled, chatting amiably. We were there in 20 minutes.

I have another 10 or 20 stories like this with which I could bore you, and many of my colleagues have much harsher tales to tell involving bandits, rogue buffaloes, vehicle breakdowns, dehydration, disease, gaping wounds, hunger, and – worst of all – missed flights home. In short, fieldwork is not much like lab work. There is no commute home at day's end to greet your spouse, kids, and dog. It is a 24-hour-a-day commitment – and for months or even years at a time. So you may be wondering, "Is it truly worth it, as you claim?" Laboratories and museums are safe, climate-controlled, bug-free, commutable environments. Is there really so much to be learned from studying wild primates that it justifies the pain? Can we not learn just as much – *more* – in a lab?

2.2 Wild Studies Show *Why*

Wild studies have something much more important to offer us than simply showing that chimpanzees are clever, as captive studies already had done. Ingenious experiments in captivity demonstrated that chimpanzees had complex mental maps in their heads, that they could collate information from numerous senses and sources to solve problems and that they could understand human language. While this is remarkable, lab work has its limits; it can show us *what* chimpanzees can do, but only study of wild chimpanzees can tell us *why* they evolved these abilities. It goes beyond that; wild chimpanzees do not simply explain lab abilities, wild chimpanzees solve problems we never imagined and demonstrate complex cultural traditions of which we had no inkling when we knew them only from their behavior in captivity (Carpenter, 1942; DeVore, 1965; Smuts et al., 1986).

We can freeze a chimpanzee. We can photograph it. We can put its bones in a museum to be studied 100 or even 1000 years from now. We could get DNA from a chimpanzee who died in a Dutch zoo 200 years ago. We have already sequenced their genome, and we have enough blood and tissue samples lying around that we probably have most of their genetic disabilities and diseases "on file." We can even find out something about their diet from trace elements preserved in hair and bones.

All true, but it is behavior we will have lost when chimpanzees go extinct. True, some behaviors are genetically programmed – but chimpanzees also have traditions, cultural practices learned at mother's knee that no lab study, no study of primates in captivity, can tell us (McGrew & Tutin, 1978; Whiten et al., 1999; McGrew, 2004). No one suspected wild chimpanzees could make tools until Jane Goodall (Figure 2.1).

Figure 2.1 Jane Goodall founded what has become the longest continuously active animal research study when she began work at Gombe in 1960. Photo: Ted Hersman.

discovered it (Goodall, 1964), much less that tool manufacture and use would come to be seen as a universal among chimpanzees (McGrew, 1992, 2010), and it was only the discovery of tool-making in the wild that convinced scientists that it was not merely due to interaction with humans (Chapter 22). Likewise for many other surprising dispatches from the field. It was from the study of wild chimpanzees that we learned of the heart-breaking period of mourning a chimpanzee mother goes through when her baby dies, carrying and grooming the dead body for days (Chapter 16). Jane Goodall discovered *that*, too (Goodall, 1965, 1968, 1971). We have already heard Richard Wrangham's story of a juvenile who imagined a piece of wood as a doll, a toy he kept with him for hours, cuddling and playing with it as if it were a baby (Chapter 1). Rosalind Alp saw chimps crush leafy branches, grip them between their big and lateral toes, and walk on thorny branches using these rough-and-ready sandals to avoid injury (Alp, 1997). Nobody suspected that chimpanzees dug drinking holes centimeters from seemingly drinkable water until I saw it in 1997; we still little understand why they do it.

2.3 Emergent Phenomena

Behaviors that are seen in the wild but not captivity are often **emergent properties**. Emergent properties are *complex phenomena that arise from the accumulation or interaction of numerous simpler phenomena, often as they are repeated over time; they cannot be predicted from the simpler phenomena*. The complex behavior of territorial patrolling emerges from the less complex inclination for males to form close bonds with childhood friends, combined with their tendency to be unfriendly toward strange males. We would never predict organized conflicts that some have compared to war, however, from the simple preference males have for familiar males, and their antipathy toward strange males.

While behavior in captivity makes us expect some pretty spectacular things from wild chimpanzees, the exact form of that breathtaking behavior is not predictable. In captivity we see that chimpanzees clearly value one another's company. They groom one another. They turn to one another for a hug when something unpleasant happens. They defend friends during social conflicts. In captivity both males and females form close social bonds, but they also have violent and sometimes even deadly conflicts. Yet, nothing in these behaviors would lead anyone to expect the bizarre and murderous world of wild chimpanzee intercommunity combat, nor the dramatic differences between males and females in the patterns of aggression and violence. Looking at two males sitting close to one another in a cage, how could we possibly know that in the wild groups of males occasionally lock eyes as they sit grooming quietly after a **feeding bout** (an uninterrupted stretch of food gathering and eating), then stand up, march silently and quickly to the edge of their community territory, and there seek out neighboring males to attack and kill. Wild study is valuable because wild chimpanzees display complicated behaviors that only appear in the correct circumstances. These are the challenging, complicated, multilayered social circumstances that are only found among large groups dispersed over huge areas – as in the wild.

Wild chimpanzees have traditions – culture – that have emerged over generations as daughter has watched mother, nuances that accumulated slowly over generations, incrementally advancing solutions to complicated problems. Safari ants are a tasty treat for a chimpanzee, but they defend themselves with a stinging bite. How do you get them? Pick them out one at a time, crushing them before they bite? Too slow. Stick your hairy arm in among them and then eat the ones that grab on one at a time? Maybe, but still too slow. Instead, chimpanzees at Gombe have a complicated strategy they have developed over generations. Pick a long stick, say two or three feet long, and pluck off any protruding branches to make it a smooth wand. Find a convenient vine or small tree next to the safari ant colony. Hang from a vine nearby, dip the wand into the ants, and wait until hundreds swarm onto the stick. Then, gripping one end with your left hand, wrap your index finger tightly around the base of the wand. Finally, in one quick swoop run your hand up the wand, pulling the end toward your mouth, piling up the ants into a seething angry formican ball that is pulled off the end of the ant-wand just as your hand reaches your mouth. Pop the ball into your mouth and chew quickly (Chapter 22). Not something a chimp figured out the first time he or she came upon an ant colony!

Once chimps are extinct in the wild, we will make no such new discoveries. There is still much, much to be learned about chimpanzee behavior, and learned behavior – culture – is the most fragile of things. Once the species is extinct, such cultural behaviors are gone forever. And it may well be that every single community of chimpanzees has its own secrets, its own culture; and there are likely several thousand communities left.

You might guess that it was the study of wild chimpanzees that inspired us to bring chimpanzees into captive situations where we could find out exactly what was going on by designing controlled studies. In fact, it was the opposite. It was chimpanzee behavior in labs and zoos that inspired wild studies. Chimpanzees are skillful mimics. Remember Meshie feeding baby Mary with a spoon? They are adept acrobats who can ride bikes, walk tightropes, and do backflips with only the sketchiest of training. Their cognitive skills are even more surprising than their physical skills. When they are given tasks similar to the game of "Pairs" or "Concentration" we see their short-term memory is at least as good as like-aged humans, and probably better (Tinklepaugh, 1932). They can solve complicated brain-teaser puzzles. Later, as we began to study chimpanzee language capacity, it became obvious that it was difficult to define language so as to leave chimpanzees out.

2.4 Captive Studies Demonstrated Astonishing Abilities

In 1925, Wolfgang Köhler's book, *The Mentality of Apes*, shocked a psychological establishment that viewed even the brainiest of primates as simple automatons. Köhler found chimpanzees could stack boxes to reach treats suspended high above their heads. They used sticks to scoop in desirable items otherwise out of their reach, and then extended their raking tools by attaching one to the other when bait was move farther out of reach. In Nadia Ladygina-Kohts' 1935 book, *Infant Chimpanzee and Human Child: Comparative Study of Ape Emotions*, she documented the very human-like emotional life of "Johnny," a chimpanzee she reared in her home from 1913 to 1916: She showed that a chimpanzee can reach into a bag and feel an object and pick out an identical object they could see – cross-modal transfer (Chapter 18). Kohts' and Köhler's anecdotes made it clear that chimpanzees interpreted human facial expressions and gestures in very nearly the same manner as people. They formed attachments to their caretakers that were spookily human-like – remember Meshie and Nkuumwa? – and they seemed to feel very human-like empathy and affection.

You might well have known all these things already, but I remind you of them to point out how they drove research on wild primates. As early as the mid-1930s, at the time of Hitler's Beer Hall Putsch, Köhler and Kohts had made the scientific establishment aware that many primates had sophisticated emotional and intellectual lives. But here lies the problem: Why? Chimpanzees live in the forest primeval. Their home is in the tropics, where there is no need to find shelter from the cold. Nurturing rain falls from the sky daily, bestowing primates, it might seem, with luscious tropical fruits, leafy salads, and healthful nuts. Only leopards and lions mar this idyllic world, and chimpanzees are rarely pursued by even these threats. Why might apes in particular and primates in general be endowed with such marvelous intelligence when their lives are about as challenging that of a pet gerbil? Incredible abilities, Darwin tells us, are the product of intense evolutionary pressure. Where is the need for incredible intelligence when the tropical forest provides abundant food, a leafy, comfortable bed at night, and a congenial climate?

You probably already know that the life of a chimpanzee is a lot more challenging than the fanciful paradise depicted here. If no other menace threatened, there would still be fierce competition with other members of their own species who also need to eat, sleep, and provide for their offspring.

Table 2.1 **Active chimpanzee research sites**

Site	Dates	Principal investigator(s)
Boussou, Guinea	~1976 to present	Tetsuro Matsuzawa
Budongo, Uganda	~1990 to present	Vernon Reynolds
Fongoli, Senegal	2001 to present	Jill Pruetz
Gombe, Tanzania	1960 to present	Jane Goodall
Goulago, Congo (Br.)	1999 to present	Crickette Sanz and David Morgan
Kalinzu, Uganda	1994 to present	Takeshi Furuichi
Kibale Ngogo, Uganda	1983 to present	John Mitani and David Watts
Kibale Kanyawara, Uganda	1989 to present	Richard Wrangham
Mahale, Tanzania	1965 to present	Toshisada Nishida
Semliki, Uganda	1996 to present	Kevin Hunt
Taï, Ivory Coast	1976 to present	Christophe and Hedwige Boesch
Ugalla, Tanzania	2001 to present	Jim Moore

The study of wild primates came about as scientists realized that unique and unknown challenges must face chimpanzees and other primates in the wild for them to have acquired the cognitive powers demonstrated in captivity.

A psychologist once asked me whether it might be that the extraordinary faculties of captive chimpanzees are a product of their interaction with humans, rather than some innate ability. Could it be that interaction with humans stimulated brain growth and multiplied connections among cognitive centers in the brain? Might it be that the seeming disconnect between chimpanzees' intellectual abilities and intellectual needs in the wild might be simpler than we think? Might it be that wild chimps are, well, stupid? Maybe there really *was no disconnect* between abilities and needs: Wild chimps were dumb and captive chimps were smart. I found this astonishing. In fact, I often find it painful to look at chimpanzees in some captive situations, especially when housed in small groups or singly. They can look depressed and desperate, like death row inmates. This is not true of all primates, by the way. To me, orangutans and gorillas consistently look much happier in captivity.

Figure 2.2 Toshisada Nishida founded and directed the Mahale Mountains Wildlife Research Center for over 40 years, beginning in 1965. Photo credit: Agumi Inaba; used with permission.

That dullness in captive chimpanzees might well be because captives are not forced to strain their brain the same way wild chimpanzees are. In the wild, chimpanzees range over huge areas with dramatic differences in topography and vegetation, rocky bluffs, marshy swamps, startling overlooks, grassy areas where grazing animals might be found, and scenes with majestic trees towering over darkened forest floors. And they do not wander randomly; chimpanzees appear to have an intimate knowledge of these landscapes and the resources in them, and they appear always to be aware of where they are in their territory. In captivity, in contrast, their mapping challenges are pathetically tiny. In the wild, chimpanzees eat hundreds of different food types, and merely planning out the foraging route is an intellectual exercise. They have to remember which foods are edible and to judge which of the fruits they see in front of them provides the most calories for the least effort. Their memories must be sharp; some fruits appear as much as two years apart. They must remember processing techniques because different foods are harvested differently, from grubs dug up in the soil, to ants extracted with a stick, to gooey tree vascular tissue, scraped from the inside of bark. In captivity a big bucket of monkey chow is deposited in their cage, possibly supplemented by some orange slices.

In the wild there are as many as 150-plus individuals in a social group, some of them allies, some enemies, some with predictable natures, others explosively violent without warning or provocation. Negotiating life in this complex social environment is much more complicated than figuring out what is

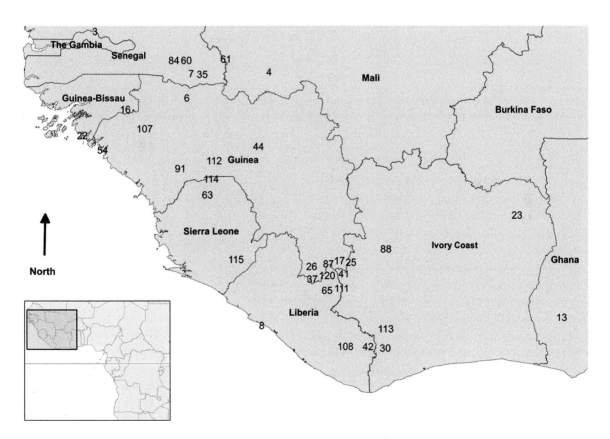

Figure 2.3 Sites of chimpanzee field study in West Africa (*Pan troglodytes verus*); well-known sites include 17 Bossou, 35 Fongoli and 84 Mt. Assirik, 113 is Taï. Maps by Alex Piel from McGrew, 2016; refer to the same for keys to site numbers.

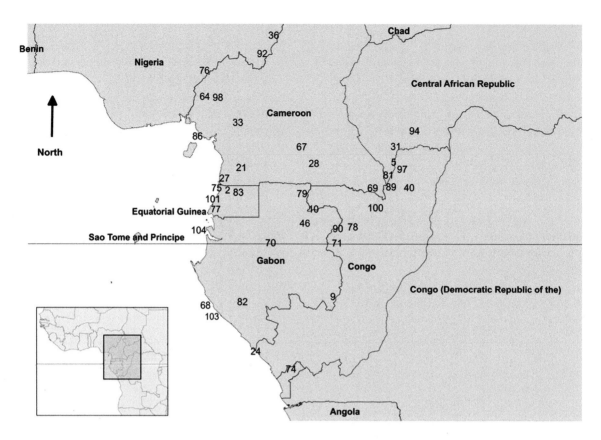

Figure 2.4 Sites of chimpanzee field study in Central Africa (*Pan troglodytes troglodytes*); well-known sites include 5 Bai Houkou, 31 Dzanga Ndoki, 40 Goualougo and 70 Lope. Maps by Alex Piel and from McGrew, 2016; refer to same for keys to site numbers; used with permission.

going on with the two other individuals with whom you share a cage. In the wild, unpleasant though it may be, balancing the rewards and risks of travel in areas that may have wonderful foods but also dangerous enemies is a cognitive puzzle that chimpanzees confront on a daily basis.

Then there is the emotional component. Wild chimpanzees are highly motivated to help their loved ones, to avoid murderous males from neighboring communities, to stave off hunger while finding time to socialize. And what is it like when they succeed? There must be a rush of relief and satisfaction when they solve the complicated calculus of finding the right foods while avoiding dangers. Wild chimpanzees desperately search for creative ways to solve complicated problems – their lives literally depend on it.

PART II: WHERE ARE WILD CHIMPANZEES STUDIED?

2.5 Wild Chimpanzees Research Sites

In 1960 Jane Goodall famously began what has since become the longest continuous field study of chimpanzees at what is now the Gombe Stream National Park, Tanzania. Chimpanzees are so ubiquitous on television that it may be surprising that there have only been 120 attempts to study chimpanzees in the wild (McGrew, 2016), and at the moment there are only a dozen chimpanzee study sites that have operated continuously for more than a decade (Table 2.1). Soon after Goodall began her work at Gombe, Toshisada Nishida (Figure 2.2) initiated

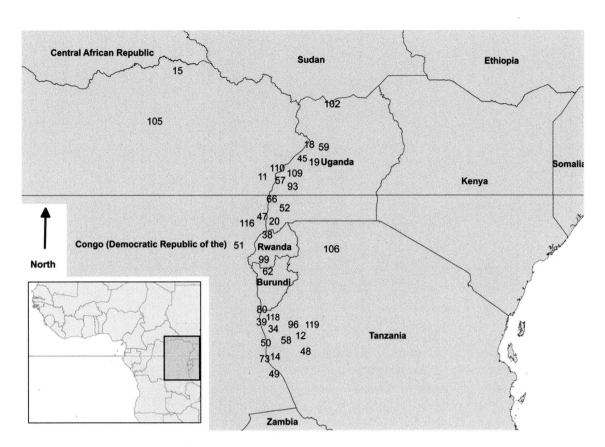

Figure 2.5 Sites of chimpanzee field study in East Africa (*Pan troglodytes schweinfurthii*); well-known sites include 18 Budongo, 39 Gombe, 57 Kibale-Kanyawara, 73 Mahale, 93 Kibale-Ngogo, 110 Semliki and 119 Ugalla; used with permission.

research at Mahale, also in Tanzania, in 1965. Frances and Vernon Reynolds began their work at Budongo in Uganda in the same year, and it continues to this day, though research has been interrupted several times. In West Africa, Christophe and Hedwige Boesch began studying chimpanzees at the Taï National Park in 1974 (Figures 2.3 and 2.4).

Many wonder why there are so few studies of this romantic, charismatic species. Part of the answer is the challenge of working in remote and sometimes ungoverned areas. Perhaps the most compelling story along these lines happened at Gombe itself; it is certainly the most poignant when it comes to the effect of politics on primatology. Or perhaps "political upheaval" is a better characterization.

In 1975, four young researchers at Gombe were kidnapped at gunpoint by Congolese terrorists (Aronstam, 1998). Beaten and bound, they were held captive for weeks, and one was held for two and a half months, even after a ransom was paid. The kidnappers were members of the *Marxist Parti de la Révolution Populaire*, a rebel group under the leadership of one Laurent Kabila. If you are a well-informed Africa hand you might recognize that name. He parlayed the ransom he chiseled out of those poor families, a million dollars back in the day when that was big money, into his successful political career. He ruled Congo from 1997 until 2001, at which time he was killed by his own bodyguard. His son, Joseph Kabila, succeeded him, serving as president until 2019.

Before the kidnapping, Gombe had grown to be more like a miniature university than a typical field site, with dozens of scholars pursuing a diverse catalog of research initiatives ranging from parenting to social dominance, from diet to intercommunity

Figure 2.6 Christophe and Hedwige Boesch (here with son Lucas) began work at Taï, Ivory Coast in 1976; research continues today. Photo courtesy of Christophe Boesch; used with permission.

interactions. Afterwards, numbers were drastically reduced. Goodall herself soldiered on despite the danger, training her local assistants to take on all of the data-collection duties. A few foreigners – mostly Europeans – also worked there: Tony Collins soon returned to help out with the lesser-known but extremely important baboon project. I am proud that the first new initiative after the kidnapping involving a foreigner was my doctoral research. This means the egregious impact of this horrific event is apparent in the date my research began: 1986, over a decade after the kidnapping.

My own field site at the Toro-Semliki Wildlife Reserve (site 110 in Figure 2.5) has not been immune from these sorts of depredations. Among the many glimpses of evil I have seen is an assault on the Kichwamba Technical School, contiguous with the reserve, by a terrorist organization I refuse to even validate with a name. On June 9, 1998, these monsters locked 85 children and adolescents in their dormitory and burned it to the ground, killing all.

2.6 The Challenge of Persisting

The risk of physical harm is one barrier to the lifespan of long-term field sites. Then there is money. Nearly

Figure 2.7 Richard Wrangham was one of the early researchers to join Goodall at Gombe in the early 1970s (depicted here after an all-day follow). He went on to found the Kibale Chimpanzee Project in the 1980s, an initiative that has proven to be one of the most successful study sites. Photo: David Bygott; used with permission.

every time I tell a large audience how few active studies there are, someone comes up to me afterward to express their deep astonishment that the scientific world – or the wider world for that matter – chooses not to invest more in such studies. Our sister species? How can the resources *not* be there? The truth is, projects are sustained by a single researcher who simply refuses to give up, despite lack of funds, malaria, and long hours in the field.

Possibly some site founders are deterred by – believe it or not – the boredom. That *National Geographic* special you saw on chimpanzees was exciting, I know. However, the breath-taking leaps, the hunting parties, the dramatic battles among aggressive males, those are things that make riveting images, but they happen for a few minutes a day. The other 12 hours and 50 minutes of a 13-hour day is taken up with your research target moving slowly and quietly through the forest, chewing, grunting now and then, passing gas, and staring into space.

Despite it all, Jane Goodall, the late Toshisada Nishida, Vernon Reynolds, Christophe Boesch (Figure 2.6) and Richard Wrangham (Figure 2.7) all returned to the field again and again, each for over 40 years. I first stepped onto the shore of Gombe 34 years ago. It is a lifetime commitment.

References

Alp R (1997) "Stepping-sticks" and "seat-sticks": new types of tools used by wild chimpanzees (*Pan troglodytes*) in Sierra Leone. *Am J Primatol* 41, 45–52.

Aronstam BC (1998) Out of Africa. *Stanford Magazine* July/August.

Carpenter CR (1942) Sexual behavior of free ranging rhesus monkeys (*Macaca mulatta*): I. Specimens, procedures and behavioral characteristics of estrus. *J Comp Psychol* 33, 113–142.

DeVore I (1965) *Primate Behavior: Field Studies of Monkeys and Apes*. New York: Holt, Rinehart & Winston.

Goodall J (1964) Tool-using and aimed throwing in a community of free-living chimpanzees. *Nature* 201, 1264–1266.

Goodall J (1965) Chimpanzees of the Gombe Stream Reserve. In *Primate Behavior* (ed. DeVore I), pp. 425–473. New York: Holt, Rinehart & Winston.

Goodall J van Lawick (1968) The behaviour of free-living chimpanzees in the Gombe Stream Reserve. *Anim Behav Monogr* 1, 161–311.

Goodall J van Lawick (1971) *In the Shadow of Man*. London: Collins.

Köhler W (1925/1959) *The Mentality of Apes*, 2nd ed. New York: Viking.

Ladygina-Kohts, NN (1935) *Infant Ape and Human Child*. Moscow: Museum Darwinianum [taDP].

McGrew WC (1992) *Chimpanzee Material Culture: Implications for Human Evolution*. Cambridge: Cambridge University Press.

McGrew WC (2004) *The Cultured Chimpanzee: Reflections on Cultural Primatology*. Cambridge: Cambridge University Press.

McGrew WC (2010) Chimpanzee technology. *Science* 328, 579–580.

McGrew WC (2016) Field studies of *Pan troglodytes* reviewed and comprehensively mapped, focusing on Japan's contribution to cultural primatology. *Primates*. DOI: 10.1007/s10329-016-0554-y.

McGrew WC, Tutin CEG (1978) Evidence for a social custom in wild chimpanzees? *Man* 13, 234–251.

Smuts BB, Cheney DL, Seyfarth RM, Wrangham, RW, Struhsaker TT (1986) *Primate Societies*. Chicago, IL: University of Chicago Press.

Tinklepaugh OL (1932) Multiple delayed reaction with chimpanzees and monkeys. *J Comp Psychol* 13, 207–243.

Whiten A, Goodall J, McGrew WC, et al. (1999) Cultures in chimpanzees. *Nature*, 399, 682–685.

3 A Most Surprising Creature
The Discovery of the Chimpanzee

Credit: Grafissimo / DigitalVision Vectors / Getty Images

We have known that chimpanzees exist for centuries (see Yerkes & Yerkes, 1929 for the definitive review of great ape historical scholarship; also, Reynolds, 1967; Hill, 1969; Goodall, 1986). In the course of that time they have been discovered, rediscovered, confused with humans, confused with orangutans, portrayed as mythical monsters, and transported across seas merely to amuse royalty. Many of them ultimately achieved a sort of immortality as museum specimens.

In 1699 **Edward Tyson** described their anatomy thoroughly enough as to leave no doubt about their place in nature: They were among our closest relatives, startlingly like us in some details of anatomy. Just as important, Tyson's work left it beyond a doubt that they were not humans, a tantalizing possibility at the time (Tyson, 1699). The story of their discovery is interwoven with the strands of European history and the history of Western philosophy in the Age of Reason. It was tied up in the ruthless imperialism of the time and its grand military strategies, with the wealth of nations, with blatant racism, and perhaps most surprisingly to a contemporary reader, it was driven, at least for some actors in the drama of discovery, by unbridled religious zeal.

Over centuries following the medieval period, Europe grew wealthy, explored the world, founded colonies and freed colonies, and all the while the chimpanzee lived in the imagination of Europeans as a pseudo-human anatomical curiosity. As Europe struggled to understand the place of the chimpanzee in nature, their most remarkable attribute – their human-like behavior – remained a secret sealed away from human understanding by the inaccessibility of their habitats, abetted by human preconceptions. When their wild behavior was at last described, it was so very human in form that many experts either ignored the reports or dismissed them as nonsense. It was only filmed images of chimpanzees hunting and using tools at Gombe that finally broke down the preconceptions of primatologists and began the final phase of the discovery of the chimpanzee – the documentation of their sophisticated behavior.

The discovery of chimpanzees was no accident, yet when we sort through the intellectual and psychological aspects of European culture that went into discovering them, we might as well be looking at a box full of random keepsakes, none linked in any recognizable way to the other. But if you plucked up the Idea-of-Ape from that box of meme-trinkets, you would find them as joined as pearls on the same string, linked to historical trends such as colonialism, nationalism, religious worship, creation myths, the nature of God, and incredibly even the cultural tradition of the Sunday Drive in the Country.

PART I: EXPLORATION

3.1 Chimpanzees: First Signs

European and Mediterranean culture may have been aware of chimpanzees as early as 700 BC. A silver bowl discovered in Italy in 1876 depicts what seems likely to be a chimpanzee (Figure 3.1). A tailless hairy but human-like figure holds foliage in one hand and some solid object, perhaps a stone, in the other (Reynolds, 1967; Hill, 1969). The creature's face is ape-like.

Two things suggest that it is a chimpanzee, and together the evidence is compelling. First, it is more or less a hairy human, and such a depiction seems unlikely if it were a monkey. Onto the human form the artisan has grafted a very unhuman pattern of body hair, a projecting snout, a distinctive beard and long head hair. Yet, there is no divergent big toe. One can imagine the returning explorer seeking out an artist to depict the amazing creature he had encountered. "It resembles a human, but with hair over the entire body and an almost dog-like snout." We are left with the image in Figure 3.1.

Second, the individual is standing erect, gripping branches, and throwing some large football-sized object – classic chimpanzee social or charging display behavior. Chimpanzees throw stones and rip up foliage during social displays, and social display is often bipedal. No other primate exhibits this precise set of behaviors. Note that next to the displaying figure we find a column bearing diamond-shaped cross-hatching much like that on palm trees. It is not a certainty, but this seems quite likely to be the first historical record of chimpanzees.

3.2 Hanno

The first written record of chimpanzees is probably the 470 BC diary of the Carthaginian general **Hanno**, who was assigned the formidable task of circumnavigating Africa with a small armada, establishing colonial outposts as he went. Hanno's is a surprising story, and one that illustrates the cultural forces that can drive exploration (from the writings of the Greek historian Periplus; Yerkes & Yerkes, 1929; Reynolds, 1967).

Figure 3.1 A silver bowl discovered in Italy and dated to 700 BC appears to depict a chimpanzee. A hairy human-like animal grasps what appears to be foliage in one hand, much as do chimpanzees during social display. With permission of Museo Nazionale Etrusco di Villa Giulia.

From its founding in 850 BC (Falconer, 1797; Hill, 1969), citizens of the city of Carthage, near modern-day Tunis, lived on the rim of the immense African continent with only the vaguest awareness of the extraordinary animals, plants, and exotic landscapes to their south (Figure 3.2). Despite its African geography, culturally Carthage was Mediterranean, a peer first of Greek city-states early on, later of Rome. Carthage grew into a flourishing trading center endowed with a powerful military, eventually becoming the richest city in the Mediterranean (Durant, 1944). As their strength and reach increased, Carthaginians became – more than the scholarly city of Athens or militaristic Sparta – African explorers.

Historical records document Hanno's impressive leave-taking as he began his circumnavigation attempt, sailing at the head of a fleet of 60 ships bulging with an astonishing 30,000 sailors and future colonists, 500 people per sailing ship. In addition to establishing colonies, he was tasked with recording new geographical features, documenting new animal species, and acquiring samples of valuable trade items. He was ordered to map the coastline and the shape of the continent (Falconer, 1797; Warmington, 1960; Hill, 1969; Harden, 1963).

Hanno sailed toward and then through the Straits of Gibraltar, dropping many of his horde of 30,000 at early ports of call in present-day Morocco, possibly because he had no idea how big the continent was. If he had known, he might have had second thoughts about the circumnavigation.

He sailed down the west coast of Africa, recording geographical landmarks and collecting plants and animals as he went. Deep into the journey he sailed past an immense mountain, or rather an erupting volcano, spewing lava, lighting up the sky at night, and igniting intensely hot brush fires, a detail that gives us some idea of the extent of his travels, since the volcano is likely to have been Mount Cameroon (Falconer, 1797; Harden, 1963). Mt. Cameroon still erupts periodically (seven times in just the last 100

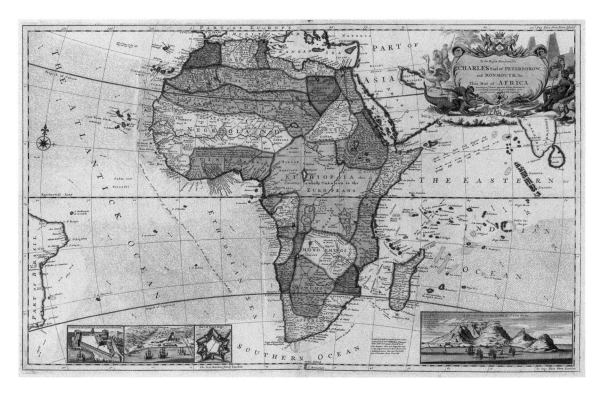

Figure 3.2 A 1710 map of Africa. The immense country taking up most of the continent is "Aethiopia." (A black and white version of this figure will appear in some formats. For the color version, please refer to the plate section.)

years; Suh et al., 2003) and might easily have been active in 470 BC. Hanno eventually arrived at an island on which he found a lake, and in that lake was still another island.

Hanno and crew explored the island, collecting samples of whatever exotic plants, animals, and minerals they found of interest. This was no small task, particularly the collection of animals, which tend to disappear whenever some galumphing rabble of foreigners spills over the horizon. His landing crew was really a hunting party. They discovered, in Hanno's words, a tribe of savages:

far the greater part of whom were women, whose bodies were hairy, and whom our interpreters called "gorillæ." Though we pursued the men, we could not catch any of them; but all escaped us, climbing over the precipices, and defending themselves with stones. Three women were, however, taken; but they attacked their conductors with their teeth and nails, and could not be prevailed upon to accompany us. (Translation by Falconer, 1797)

We cannot be certain what these creatures were, but we need not be too put off by the word "gorilla"; there were numerous false starts before names finally got attached to the animals they are attached to now. In the first published comprehensive anatomical description of the chimpanzee, the specimen was called an orangutan (Tyson, 1699). We can dismiss right off the idea that this "tribe" was made up of people. The ancients expected, unlike later European scholars, that there was some sort of continuity between humans and animals, and so it would have been quite natural to refer to human-like animals as savages. The creatures certainly were not monkeys; Mediterranean people were quite familiar with baboons, which makes it highly unlikely Hanno could have failed to recognize their similarity to baboons if they were monkeys. Cameroon is in the range of both chimpanzees and gorillas, so that is no real help; and his local guides might easily have mistaken one for the other.

Throwing stones is something chimpanzees do, not gorillas or baboons. More critically, descriptions of gorillas invariably mention their immense size. Hanno noted the sex of the various "savages" he was chasing, so it seems unlikely that he would have failed to notice and record that the males were the size of a grand piano.

Furthermore, gorillas, even female gorillas, could never have been manhandled as Hanno describes his hunting party manhandling these creatures. A gorilla female who attacked a man "with teeth and nails" would not have left a pretty corpse; instead of writing that the creatures escaped, we might expect a complaint about the difficulty of gathering up the stray body parts of the dead.

When Hanno found he was unable to capture the females alive, he did not settle for just jotting down a few notes on their hair color and ear shape. He "killed them, and flayed them, and brought their skins ... to Carthage" so that they could be examined by scholars.

Can we trust Hanno's records? When we read of 60 ships, 30,000 colonials, volcanoes spitting fire, and hairy savages, we might be inclined to think Hanno was padding his resume just a little bit. While this is certainly possible, Hanno's narrative, including the number of ships, people, the colonies founded, and so on, were engraved on tablets that were publicly displayed in the Temple of Chronos, in Carthage. These events had a very public vetting.

I cannot resist adding this last historical note. Although Carthage flourished for centuries, it was not to persist as did its competitors Athens and Rome. Carthage was such a threat to mighty Rome that for years the Roman statesman Cato the Elder (234–149 BC) ended each of his speeches in the Roman Senate with the words "Carthago delenda est" (Little, 1934): *Carthage must be destroyed.* Cato's tireless campaign succeeded at last in 146 BC, when Roman soldiers were instructed to destroy the city so completely that no two stones should be left one on top of the other (Durant, 1944; Warmington, 1960; Harden, 1963). While they could not have been that thorough, they were said to have sown salt in the fields to ensure crops would never again grow to nourish Rome's enemy. Hanno's chimpanzee skins, after surviving over 300 years, were lost in this event, obliterated with the rest of Carthage. The modern city Tunis, Tunisia is nearby, but not on the site of the ancient Carthage.

After the time of Hanno, we find an occasional mention in Greek and Latin writings of primates that might well be chimpanzees, but nothing very informative. Hundreds of years passed before a Greek scholar, Agatharchides, wrote in a treatise on "Ethiopia" in 100 BC – Ethiopia being a place name used for pretty much all of East Africa at the time – describing a primate with limbs that were human-like but elongated in their entirety, including fingers, hands, and feet (Reynolds, 1967). Agatharchides called these beings wood-eaters (hylophagoi), after their habit of eating bark. They "made war with each other," and lived to the age of 50, he wrote, at which point they developed cataracts and stumbled to a quick death. How Agatharchides' informants went about estimating the age of wild animals is anybody's guess, but the imagery neatly describes chimpanzees, which do eat bark (other African primates do so only rarely), form large groups to attack one another, and do develop cataracts as they age. Later, Aristotle describes what might well have been an ape in his *Historia Animalium* (Thompson, 1910; Yerkes & Yerkes, 1929), an animal with human-like teeth, ears, nostrils, and eyelashes.

These brief descriptions are all we see of chimpanzees in ancient writings. Soon the Dark Ages intercede and people who could write spent their time and ink puzzling over knotty religious problems. Scholarship in Greek and Roman societies waned, leaving deep-thinking to people far from the tropics, and therefore far from chimpanzees.

3.3 The Renaissance: First Definitive Evidence

Eventually the Renaissance washed over Europe, and to the regret of the many folks worldwide who were, to coin a phrase, "explored to death," a new age of exploration emerged, and seagoing nations such as Spain and Portugal ranged far enough afield to log reports of sub-Saharan Africa and its wildlife. Renaissance Europeans, it turns out, shared with powerbrokers in Carthage an interest in the fauna and flora of Africa. In 1598 a Portuguese sailor, Duarte Lopez, visited the Congo and described an animal that was almost certainly a chimpanzee, and he provided artwork that clinches it (Figure 3.3). Lopez described a man-like creature that was prone to lurking around his camp and snatching stray items that were then used to mimic humans (Reynolds, 1967). He even provided a picture of a tailless primate with a very chimpanzee-like face trying on boots and marching around (Figure 3.3).

In 1625 Englishman Andrew Battell's description of two creatures in Angola, a larger and smaller version of an "Ape-Monster," certainly seem to be gorillas and chimpanzees (Purchas, 1905; Reynolds, 1967). His description of the larger ape, called "Mpondwe" locally, was extensive if a little fanciful (for example, they were said to come into villages and sit by fires until they went out, and to attack elephants with clubs). If the larger ape was a gorilla, the smaller, "Nshiego," was a chimpanzee. Battell somehow heard "Mponwe" as "Pongo," not too much of a surprise since in Angolan geography we find lands called Bongo and Longo; he might have had -ongo on the brain. "Nshiego" was heard by Battell as "Engeco." They were said to have a face that "differeth not from a man," whereas their legs differed "for they have no calfe." I take this to mean their calf muscle is not as

Figure 3.3 Depiction of chimpanzees stealing boots from Duarte Lopez's camp. From Yerkes & Yerkes, 1929; used with permission.

globular or full as that of a human, which is accurate. Both terms, Pongo and Engeco, were used for years, and *Pongo* eventually became the Latin name for the orangutan.

3.4 Chimpanzees in Europe

Explorers often return from their exploits with stirring tales of dangerous encounters with exotic wildlife. Such tales must have conferred some prestige on the returning voyager, but of course there are always doubters who might wonder out loud how large a grain of salt should be taken with the traveler's tale. The doubters could be sent packing if you were able to produce that live animal, and in 1641 that is exactly what happened; the first live chimpanzee was transported to Europe, probably from Congo or Gabon. "This will show them," the intrepid explorer might think, "'Probably a large rat,' you said! 'No primate is that big,' you claimed. Well, take a gander at THIS!" You will remember Hanno's first attempt to capture one of his mysterious animals alive; he only killed them when that failed.

Capturing and bringing a living ape to Europe is a more impressive undertaking than you might think. Pursuing chimpanzees over hill and dale, dogs yapping at their heels, might take some of the starch out of a chimpanzee's ability to resist capture, but not much. In fact, capturing a live chimpanzee makes spearing or shooting them look like child's play, as Hanno discovered. Still, they accomplished it somehow, possibly by capturing an infant. Unfortunately for the first chimpanzee to reach Europe, she was hardly better off than if she *had* been killed. She was presented as a gift to the Prince of the Netherlands, probably a groveling ploy on the part of the explorer to motivate the prince to sponsor a later expedition.

If the prince's interest was keen, he would have been well-advised to make a quick examination of the chimpanzee, because these first apes transported to Europe lived only the briefest of times, often dying of respiratory illnesses – something even wild chimps suffer from in the wet weather of the rainy season.

3.5 Confusion with Orangutans

The carcass of this first-transported chimpanzee was examined by a Dutch anatomist, Nicolaas Tulp (Figure 3.4). Tulp knew the South Asian ape, the orang-outang, which from reports seemed to have been a red-haired version of the specimen lying on his dissecting table. Perhaps it is no accident that his illustration looks as much like a chimpanzee-orangutan hybrid as a chimp (Figure 3.4). He marveled over how human-like its ears, arms, legs, thumbs, and feet were, as like those of humans as "one egg [is to] another," he wrote (Yerkes & Yerkes, 1929; Reynolds, 1967).

Tulp's illustrator, while apparently attempting to depict the thickening of the waist characteristic of

Figure 3.4 Tulp's illustration of the 1641 juvenile chimpanzee, looking very much as if his knowledge of orangutans had influenced him. From Yerkes & Yerkes, 1929, with permission.

great apes, instead offers us a hairy human who appears to have swallowed a beach ball.

Scientists confused chimpanzees and orangutans for centuries, a confusion less surprising than it appears at first glance. An anatomist who has just received an expired orangutan for dissection and has only his copy of Tulp's description of an orangutan-like "orang-outang" with "long arms, short legs and black hair," the eighteenth-century anatomist who concludes that the red-haired, long armed, short legged specimen on his dissecting table is the same thing is not the fool he first appears. Perhaps red and black "orang-outangs" should no more be placed in different species than should a redheaded Swedish woman and her black-haired Norwegian date.

3.6 Edward Tyson

In 1698 yet another juvenile chimpanzee was brought to the northern hemisphere, to England this time. He quickly responded to the wonders of the Western world as had other transportees, and died. This chimpanzee, however little he might have appreciated it, achieved a kind of immortality, becoming the subject of the first thorough, truly scientific study of chimpanzee anatomy, published by Edward Tyson in 1699 (Figure 3.5). While he called the specimen an orangutan, he knew it was not an Asian ape. Tyson dissected the juvenile and produced accurate, detailed drawings of chimpanzee anatomy. As many would after him, Tyson compared chimpanzees to monkeys and humans and found – surely to his surprise – that of the 81 traits he examined, 47 were most similar to humans, while only 34 were more similar to monkeys (Tyson, 1699).

His careful records and excellent illustrations became the first objective description of the chimpanzee, a critical standard since other scientists could now use this detailed description to determine whether specimens they collected were the same thing or something new and different (Yerkes & Yerkes, 1929; Reynolds, 1967). From this time forward it was clear exactly what a chimpanzee was, and precluding confusing them with orangutans.

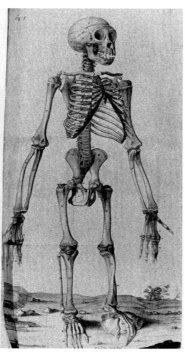

Figure 3.5 Plate from Tyson's *Orang-outang, sive* Homo sylvestris: *or the anatomy of a pygmie compared with that of a monkey, an ape and a man.*

3.7 Chimpanzee

The work of Tyson and others made European scientists familiar enough with chimpanzees that Darwin was comfortable pronouncing that human ancestors must have been similar to them. As the nineteenth century drew to a close it had not escaped the notice of naturalists that not only did the bodies of chimpanzees resemble humans, their **neuroanatomy** did as well. This similarity of brains suggested that chimpanzees might serve as stand-ins for ancient humans, that they might be a living fossil, as it were, and that they might tell us what proto-human psychology and behavior were like before the bulging mass of learned behavior we all carry around with us overwhelmed – or at least disguised – our "true nature."

3.8 Taxonomic Naming

We knew what chimpanzees were long before they had a name. Nearly 40 years after the publication of Tyson's monograph, the term chimpanzee was used for the first time, in 1738 in *The London Magazine*, which published this description: "a most curious creature is brought over ... taken in a wood in Guinea. She is a female of the creature which Angolans call chimpanzee, or mockman." The name caught on.

While the object of our study now had an informal name, two centuries would pass before the chimpanzee got its proper scientific name, *Pan troglodytes*. In 1758, the founder of taxonomy, Linnaeus (of whom we will hear more later), called chimpanzees *Homo troglodytes*, considering them a primitive, cave-dwelling human – troglodyte is Latin for cave-dweller – primitive, but a kind of human nonetheless. While his broadmindedness in including such a primitive being in our own genus was quite in keeping with tradition at the time, other scholars scratched their heads over this scheme and wondered whether we really wanted to have such a smelly, sex-crazed, hairy being right next door, rather than somewhere just vaguely in our taxonomic neighborhood.

The next player in our taxonomic drama is Johann Friedrich Blumenbach. He parsed species so finely that a good day of tanning might change your taxonomic classification. With humans falling into distinct subdivisions in Blumenbach's fussy taxonomy, apes were naturally booted right out of town. He put them in their own genus, *Simia*, in 1775, making them formally *Simia troglodytes*. But the *Simia* genus name failed to catch on. *Simia* was, however, quickly coopted to use at a broader taxonomic level to mean both apes and monkeys.

The taxonomy of monkeys and apes became so confused and so contentious that the official taxonomic naming organization, the International Commission on Zoological Nomenclature (ICZN), thought it would be better to simply start over. Thus, it came to be that in 1929 the ICZN, officially suppressed *Simia* – they declared it an invalid name. This was much to the good, since many scientists had already begun to use a different genus name, *Pan*, for the chimpanzee, a name created by Lorenz Oken in 1816. Things would have been resolved there, except that Oken goofed by not presenting both a genus and species name when he introduced *Pan*, in violation of ICZN protocol. It was 1985 before the ICZN at last ruled in favor of the genus name everybody had been using for decades, *Pan*, and chimpanzees received their final, officially approved name, *Pan troglodytes*. Nevertheless, the rules of nomenclature treat Blumenbach's publication as the official naming of chimpanzees, which means the most precise taxonomic name for chimpanzees is *Pan troglodytes* (Blumenbach 1775).

3.9 European Scholars Get to Know the Chimpanzee in Captivity

While academics dithered over taxonomy, the 1800s wore on, the 1900s arrived, and all the while chimpanzees were being captured and transported to Europe in increasing numbers; by 1900 as many as 200 chimpanzees had been exhibited to an eager public. Typically, those exhibited were juveniles, in part because the best way to capture a chimpanzee was to shoot a mother and grab her kid.

When a chimpanzee died, it was not unusual for a local scholar to be allowed to examine the corpse, after which the skeleton might end up in a museum. Skeletons of wild chimpanzees are a treasure to anatomists and osteologists (**osteology** is the study of bones), since details of **morphology** (shape, anatomical detail) reflect the impact that life in the wild has on the skeleton, so at least a bit of scientific value was gained from murdering all those chimpanzees.

While the anatomy of chimpanzees was quite well known to European science by 1900, the fleeting lives of captive chimpanzees left observers little time to document their behavior. By 1915, zoos had finally gotten the knack of keeping chimpanzees alive and a milestone was reached: The first chimpanzee was born in captivity. This is no small feat. Zookeepers had managed to get both a male *and* a female, most of whom came to Europe as easier-to-subdue-and-capture juveniles or infants, to survive to reproductive age while keeping them sane and healthy enough to procreate. Some captive species are so disturbed and depressed by captivity that they have no interest in copulating; others are so stressed out that something goes wrong during gestation and they abort. Zookeepers had accomplished something quite difficult.

The task of figuring out the cognitive capacities and natural behavior of chimpanzees might seem a snap at this point: Just watch them. Yet it was scarcely that easy; biases and unnatural conditions in zoos and labs skewed the behavior of captives. Biases came in part from accounts of ape behavior in the wild, in retrospect much of it focused on the gorilla, but at that time it was still a little unclear that chimpanzees and gorillas were different animals (Reynolds, 1967: 57). Early documentation of wild behavior naturally relied on firsthand accounts of folks who had seen them, which meant hunters. Hunters' tales are a lot like fish tales. The fish, or the ape in this case, gets larger and more vicious as the tale gets retold. We might read in the nineteenth-century equivalent of *Guns and Ammo* something like this: "For days I had trekked ever deeper into the gloomy, sweltering forest, when one afternoon with a deafening roar the bloodthirsty creature charged at me from the foliage.

I scarcely aimed, but my bullet found its mark, and the stinking beast expired at my very feet. Otherwise, I would have surely been devoured. I had conquered science's first example of the feared vervet monkey." Fifteen pounds of fury, quelled!

I made all that up, but a "collecting expedition" could not have looked very scientific: Five sweating white guys weighed down by heavy elephant guns, slapping at bugs, cursing as they pushed themselves along a trail behind 10 gun porters, 15 supply porters, two skinners, and far out in front two local trackers and a half-dozen dogs, all pursuing prey to exhaustion over the course of hours or even days. Once cornered in an isolated tree, with the dogs barking and lunging to keep the exhausted ape at bay, the five hunters pull up at last, dash the sweat from their brow, catch their breath, and blast away at the Bloodthirsty Creature. Hunter's tales were more or less useless.

Robert Garner's study in 1890 was the first real attempt to study chimpanzees and gorillas in their wild home. We infer that he might have been just a wee bit skittish, because as a safety precaution he installed a room-sized steel cage in the forest from which to observe the apes (Figure 3.6). For 112 days Garner sat in his cage, peering into the forest, taking notes on wildlife. In that time he was fortunate enough to see 22 gorillas and five chimpanzees. His longest observation was of a female and infant, and lasted four minutes.

3.10 Early Captive Studies

While information on wild chimpanzees was uninformative, by the 1920s psychologists had begun creative experiments with captive chimpanzees that remain informative to this day. **Nadia Ladygina Kohts'** work was notable not only for her amazing discoveries, but for the fact that they were made in Stalinist Russia, and by a woman. She and **Wolfgang Köhler,** a German, conducted experiments testing chimpanzee memory and problem-solving ability, and along the way they documented startlingly human aspects of social demeanor, observations you are almost certainly familiar with, even if the names

Figure 3.6 Garner and his ape observation cage.

Kohts or Köhler are unfamiliar to you. We will learn more about their work later. By the time **Robert and Ada Yerkes** (1929) published their now-famous compendium of great ape research, *The Great Apes*, the English-speaking world had at its fingertips sensational evidence that chimpanzees and humans shared more than anatomy – we shared many intellectual attributes as well. Between the World Wars, Yerkes organized an expedition to West Africa to capture study subjects to be brought to the USA. Leading this expedition was Henry Nissen, who had just completed the first field research on wild chimpanzees in French Guinea, a two-and-a-half-month study conducted in the late dry season. The work of the Yerkeses hammered home to the scientific world that the extraordinary cognitive abilities of chimpanzees – and other primates – could only be explained by investigating the intellectual challenges they faced in the wild (Reynolds, 1967: 115). Nissen noted that chimpanzees were only partly arboreal and that they slept at night in tree nests (1931). The scientific world had no idea what the chimpanzee social structure was. It had become received wisdom among naturalists that they lived in what they had assumed was the human norm, mother–father–offspring family groups – with no evidence whatsoever to support this assumption. Science had no idea whether chimpanzees were bark-eaters or fruit-eaters, whether they fought for dominance status, or interacted peacefully.

3.11 Wild Chimpanzees

Jane Goodall answered those questions and more beginning in 1960, and it was her pioneering study of wild chimpanzees that began a revolution in the way chimpanzees were perceived by the scientific community and the public at large. As Goodall's observations were published in the 1960s and other workers followed, the era of discovery of the chimpanzee – a 300+ year span of time – ended. We might mark the end of this early chimpanzee behavior history as 1986, when Goodall's *Chimpanzees of Gombe* appeared, the scientific documentation of a quarter-century of study at Gombe. Certainly it was a personal milestone for me. I bought it in a bookstore in New York in August 1986 when I was on my way to Gombe and Mahale to begin my doctoral research.

3.12 Why Discover Chimpanzees?

This discovery narrative is part of a larger fabric of the history of the Age of Discovery, and of the development of science, and as such it is something of a reflection of its time. Does anything about this narrative strike you as odd? Imagine a contemporary wealthy industrial nation dispatching a navy warship to a tropical port. A landing craft is deployed and roars away on a mission to … capture local wildlife and bring the animals back to the ship alive? Then, at great effort and expense, the living creatures are shipped home to be presented to the head of government? The scenario seems preposterous in current circumstances. It is worth considering how it might have come about in the past.

As you have seen already, it is no easy task finding wild animals in their habitat – in early December 1986 at Mahale a female chimpanzee I was studying gave me the slip. It was three weeks before I saw another chimpanzee, despite searching 6–8 hours each day. Wild animals tend to hide when a hunting party comes for them, but killing an animal is relatively easy compared to capturing them, as Hanno found in ancient times. Once captured, keeping a wild animal alive on a long ocean journey is no easy task either; depressed, stressed, and fed a strange diet, they are prone to illness. What could have motivated ships' captains from Hanno to Captain FitzRoy of Darwin's *Beagle*, to put so much effort into collecting animals and plants? Why did military expeditions spend so much energy documenting the world's landscape and its topography?

We should not discount the curiosity of powerful people. Despots, kings, and emperors were powerful enough to act on whims if they were interested in a specific creature. Still, that was likely only a small part of the equation (Gunn & Codd, 1981). Operations like those of Hanno had an economic motivation; they were fact-finding and mapping missions aiming to make getting where you were going safer, easier, and more profitable. Warmington (1960) suggests that Hanno's mission was meant to securely establish a route for gold trading. Many such colonizing and exploring efforts were meant to provide an edge to military commanders, an edge that might make warring less deadly – for the ones holding the maps, at least.

Conquest and profit you probably expected. You might be surprised to find that after the Renaissance there was a new spur to the exploration and species-naming impulse: religion. Educated gentlemen, not excepting naval officers, were trained not just in literature, art, and the sciences, but in religion as well. Darwin's work was part of an intellectual upheaval that cleft a schism between some religious teaching and science, but prior to that schism religion was the foundation on which study of the natural world was based.

If this is your first exposure to this surprising bit of history, you might well be thinking, "surely that is an exaggeration; surely he wrote that just for effect." Far from it. In the time of Darwin it was not just that there was no antagonism between science and religion, it was that religion was often the *motivation* for scholars to do science. Not so long ago, study of the natural world was commonly referred to as "natural history" (as it is still today at times) or "natural philosophy" when it tended toward theory. While the two labels for the study of nature are appropriate, before the time of Darwin another appellation, **natural theology**, was more appropriate. Natural theology sought to explain how feathers, sharp teeth, horns, predator–prey relationships, and camouflage functioned, not by appealing to evolution, but by imagining what God's plan was for the world (e.g., Paley, 1802).

Educated Europeans believed nature was made more interesting because God had created it. Natural theology can be traced back to medieval Catholic philosopher-clerics who built on the knowledge of ancient Greek philosophy. In some sense, nearly *every* scholarly discipline starts with the ancient Greeks. In some fields, this is a bit of a stretch, but in this case, it is not at all. We cannot understand Darwin until we understand Plato.

PART II: THE GREAT CHAIN OF BEING

3.13 Plato, His Cave, and His Ideals

Plato sought to understand how people know things, and among his interests was the mental process involved in identifying and naming things (Figure 3.7). He wanted to understand why it is that we categorize so quickly and unambiguously. He believed that categories were distinct in a much deeper way than we do today. His concept of the world was one where the world we can see is strongly influenced by an unseen, unseeable parallel world that is not simply very real, but even *more* real than the world we do see (Lovejoy, 1936; Brace, 1988, 2005). Plato's teachings are so ancient that one might expect that – outside of Greek and Roman philosophy classes – there are no traces of them to be found in contemporary society. Not so; you are probably familiar with the literary device Plato used to illustrate his ideas, his famous "Plato's Cave."

Figure 3.7 Essentialist Greek philosopher Plato (~428–348 BC). (A black and white version of this figure will appear in some formats. For the color version, please refer to the plate section.)

In Plato's philosophy, nothing we see around us is truly real. Things in our world are imperfect representations of fixed and unchanging "*ideals*" or "types" found in another, unseen, ideal world. Imagine a cave that is being used as a prison of sorts, and chained to chairs, facing the back wall of the cave, are prisoners (Figure 3.8). They are so tightly bound that they cannot turn around to look out the mouth of the cave. They are doomed to live their lives staring at the back of the cave. In Plato's analogy a bright fire burns at the mouth of the cave, and between the prisoners and the fire is a band of puppeteers performing a little play. They are holding aloft characters from their performance – a cat, a table, a chair. The fire casts shadows on the rear wall of the cave. The prisoners can see shadows of the performance, yet they are unable to see the puppets themselves. The shadows in Plato's cave are analogous to the things we can see and touch in our world while the puppets are analogous to the "real things," the things that exist in the world of ideals (Figure 3.8). Plato called the real things, the things that (somewhat counter-intuitively) exist only in an unseen world, "**eidos.**" In the original Greek script eidos is spelled ειδος. We translate eidos as "ideals" or more literally, **forms.**

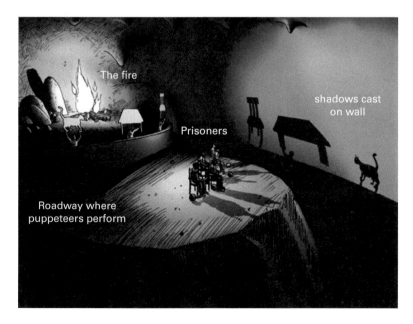

Figure 3.8 Plato's cave. Illustration by John D'Alembert.

This bizarre constellation of postulates had a profound impact on Western philosophy. "Idea" is derived from the Greek word eidos. Forms are sometimes called "types," and the terms "typical" and "stereotype" come from this root. Medieval philosophers were to call forms, types, or ideals **essences.** Ideals are perfect and unchanging, and they exist permanently in a perfect, fixed world invisible to us. Things in our world are imperfect copies of things in the perfect world. Ideals influence things in our own world (Lovejoy, 1936).

3.14 Plato's Essentialism

Why was this satisfying to Plato? He felt that understanding how humans label and categorize things was a fundamental philosophical problem. How is it that when we see a cat it is instantly recognizable as a cat, without any doubt? We do the same things with chairs, tables, vases, olives, and flowers. Plato believed that we do that by identifying some essence of the ideal objects in the everyday things we can see, and we make some connection with that ideal through the object we see or feel. We identify and categorize things by a sort of sympathetic vibration between our essence – medieval philosophers would see that as our soul – and the essence in the vase or flower. The concept of an ideal (nowadays we can call it a **meme**; Dawkins, 1976), of an unseen and unchangeable world that affects our own world, influencing the shape of objects we can see and feel, is called "**essentialism**" or "**typology.**"

3.15 Making Room for God in Platonism

Fast-forward 500 years. In the waning years of the ancient Western Roman Empire, scholarship came to be more and more bound up in religious teaching and study (Russell, 1945), a focus that continued through the Fall of Rome in AD 476 and throughout the medieval period (~476–1500). European philosophy sought to explain God, God's relationship to humans, and God's influence on the physical world. Roman Catholic clerics – cleric being a general term for priests, bishops, monks, archbishops, and so on – did the heavy lifting in this philosophical world.

A Roman follower of Plato, Plotinus (c. 205–270), sought a deeper understanding of Plato's dichotomy between the world of forms and our world. He marks the beginning of the injection of God into Platonic philosophy. He maintained that there was a force in the universe, a supreme, transcendent, immutable "One," the source of everything that defies description and yet is perfect. While he danced around this conclusion, you and I would call this thing "God." In Plotinus' cosmology there was a gradation of perfection starting with the transcendent One, moving down to the less perfect angels, the still less-perfect humans, then to the "prime" or first-most of the animals, the primates, to simpler animals, to minerals, and at the base of this glorious ladder of worth and perfection were the base metals (Deck, 1967). Plotinus and his adherents further added the idea of *worth* or value to Plato's "real things." With the inclusion of this concept of a hierarchy of perfection and worth, and its further development by medieval philosophers, Platonic philosophy morphs into **Neoplatonism** – "neo" being Latin for new, combined with Plato's name (Dillon & Gerson, 2004).

Saint Augustine (AD 354–430) began his scholarly training outside the Catholic sphere, but found himself irresistibly drawn into religious debates. He converted to Catholicism and his writings are now considered foundational in Catholic philosophy. For our purposes, he was important as a vessel that carried Neoplatonism through the transition from classical (in the sense of Roman and Greek) times to medieval times.

As Plotinus, Saint Augustine, and the Neoplatonists injected God into Platonism, they increasingly sought to reconcile Plato's view of the world with the text of the Bible. Their job, in essence, was to figure out the exact relationship between God and the world. Perhaps the most profound "discovery" they made was the location of Plato's ideals (or forms, or types, or eidos, whichever you prefer). They were, of course, in the mind of God and with that revelation it all fell into place for Neoplatonism. God has a constant influence on this world; his thoughts are a force that helps to determine the "form" of things in this world. However, while God's thoughts exert a divine pressure on this world, it is still imperfect Earth; perfect things exist only in heaven. In our world, there are only imperfect representations of the forms or types or ideals in God's mind. This is why, in the view of Neoplatonists, two different goats or two different chimpanzees are not identical. While no single thing in this imperfect world looks exactly like God's "ideal," some come closer than others. The world, in other words, is an imperfect reproduction of God's thoughts.

3.16 Plenitude and Perfecting the Great Chain

God planned this Earth to the last detail, and it is an intricately interdependent creation, fitted together like cogs in a machine. Individual things in this world are dependent on one another, they are *linked* in a chain, a great chain stretching from God at the top, holding everything up, down to the base metals at the base (Figure 3.9). This is the **Great Chain of Being**. The chain is also known as the **Scala Naturae**.

A further important subtlety: Not only are organisms linked, they are arrayed in such a way that every gap is filled. This is the concept of **plenitude**, the concept that every type of organism that can exist, does exist.

3.17 Saint Thomas Aquinas and Adaptation

To this wondrous Neoplatonic edifice we must add a final but important Corinthian column, that of design. **Saint Thomas Aquinas** (1225–1274) reasoned that objects in the Great Chain are not suited to their role in the workings of the world by accident, they were endowed with their form and role by God, on purpose (Figure 3.10). Animals that burrowed to escape predators were manifesting God's wisdom; God endowed moles with the instinct and ability to burrow to reach safety. The keen eyesight of predators was given to them by God to allow them to hunt effectively. The fact that grass so precisely meets the nutritional needs of buffaloes and that buffaloes naturally feed on grass was seen as part of the miracle of creation. On the other hand, grass is of no use to bats, but bats show no inclination to eat foliage anyway – another example of just how perfect God's world was. Every animal has a place and every animal has a role in God's great scheme. In modern idiom we would call this **adaptation**.

3.18 Neoplatonic Essentialism Begets Natural Theology

Here is the link between **Neoplatonic essentialism** and science: Because the world is a physical expression of God's thoughts, it is an expression of God's love and wisdom. Achieving a better understanding of the world and the relationships among species can give us mere mortals insight into God's mind that we can get in no other way. Can you see where this is going? In Aquinas' view, the practice of what we would call "science" today was an act of religious devotion. Studying nature and gaining a better insight into how bees pollinate plants and how homing pigeons navigate across huge distances is a way of praising God. Explaining God's plan for nature by investigating how species interact with one another, how organisms grow, how they cope with natural challenges, and just about any other scholarly investigation of nature, was a scientific endeavor called **natural theology**. Natural theology sought to explain God's plan for all living things.

This prayer, written by St. Thomas Aquinas, expresses sentiments little different from those that might successfully guide students even today:

Student's Prayer by St. Thomas Aquinas
Creator of all things, true source of light and wisdom, origin of all being, graciously let a ray of your light penetrate the darkness of my understanding.
Take from me the double darkness in which I have
  been born,

Figure 3.9 The Great Chain of Being. God reigns at the top of the chain, followed by angels, humans, animals, plants, and nonliving things. Artwork by Diego de Valadés 1579.

an obscurity of sin and ignorance.
Give me a keen understanding, a retentive memory, and the ability to grasp things correctly and fundamentally.
Grant me the talent of being exact in my explanations and the ability to express myself with thoroughness and charm.
Point out the beginning, direct the progress, and help in the completion.
I ask this through Christ our Lord.

All the concepts are in place now. Species are **distinct** from one another because they match God's categories. God knew exactly what he meant when he conceived a zebra. Species are linked to one another in an irrevocable way; we are all interdependent, bound together in a complicated, Godly scheme; grass grows to feed the Uganda kob, which in turn feeds the lion, and so on. And God would never allow his perfect world to change. If one species were to go extinct the whole world would collapse, since all

Figure 3.10 St. Thomas Aquinas surrounded by the doctors of the Catholic Church. *The Apotheosis of St. Thomas Aquinas* 1631, Franccisco De Zurbaran. With permission from The Museum of Fine Arts (Museo Provincial de Bellas Artes), Seville. (A black and white version of this figure will appear in some formats. For the color version, please refer to the plate section.)

species are interdependent. Thus, we say that a principle of Neoplatonic essentialism is that the chain is **immutable**, or unchangeable. Species are **hierarchical**, with the more perfect beings resembling God; the highest species on the chain are both more worthwhile and more perfect than those below. The world is a less-than-perfect physical expression of the perfect ideals that exist in God's mind.

In short, the world is an intricately planned and immutable physical expression of God's thoughts (Lovejoy, 1936). The first complete expression of the principles of natural theology was **William Paley**'s (1743–1805) *Natural Theology: or, Evidences of the Existence and Attributes of the Deity*, published in 1802. It was the apotheosis of Christian principles concerning the organization of the natural world. It is still good reading today.

It is quite likely that you have already heard one of Paley's pithy logical proofs of God's existence. Let us say you come upon a stone. If you examine it closely, Paley writes, nothing about it would preclude it from having been there since the beginning of time. However, he goes on, instead of a stone,

"suppose I had found a watch upon the ground, and it should be inquired how the watch happened to be in that place, I should hardly think of the answer which I had before given – that, for anything I know, the watch might have always been there. Yet why should not this answer serve for the watch as well as for the stone?... For this reason, and for no other ... that, when we come to inspect the watch we perceive (what we could not discover in the stone) that its several parts are [put together] for a purpose [description of watch omitted] ... This mechanism being observed ... the inference, we think, is inevitable, that the watch must have had a maker... who comprehended its construction, and designed its use."

Well, that, as they say, is that.

God's wondrous world was considered to be so complicated that no one human could truly understand it, but by studying nature we mere humans might at least here and there have the same thoughts God did. It is a wonderful world, and exploring it can only bring one closer to God by bringing one's thoughts into closer accord with those of God.

PART III: NATURAL THEOLOGY AND TAXONOMY

3.19 Linnaeus: Science Serves God

Perhaps it was only a matter of time before someone conceived the ambition of trying to find out precisely what is was that God had in mind when he created this great living world. That man was Carl Linné or Carolus Linnaeus in the Latinized version of his name, the name by which we know him best (Groves, 2001). Linnaeus was a paragon of religious intellectualism, a man who saw studying plants and animals as a transcendentally religious experience. Linnaeus imagined himself as a modern-day Adam, assigned the task of naming every living thing in God's creation, or at least giving it a good try. One imagines he was not much of a cut-up at parties. Linnaeus realized that one of the barriers to identifying a species is the possibility of mistaking it for another animal or plant. He felt the need for a foolproof, systematic method that would make such mistakes impossible, so he invented one. Today we call the scientific naming of living things **taxonomy**, and we call the scheme Linnaeus invented the **binomial system**.

Linnaeus published a rough draft of this scheme, his *Systema Naturae*, in 1735. It was really just a pamphlet. I remember my disappointment when I first saw it, a thin volume with little detail, just a barebones outline. By his tenth (1758) edition, though, you can see his all-encompassing ambition realized. He produced a thick, detail-rich, confident tome that he may have felt was beginning to come close to his goal of naming all living things. We know now he was not even close, but he is said to have believed that there were perhaps 10,000 species in the world, and he could identify 6000 on sight. Today nearly two million species have been classified, and more are being added each year. We doubt the end is anywhere in sight, but Linnaeus could not have known that.

Variation was a fly in the ointment for Linnaeus, just as it was for Neoplatonists; he realized that some lions have longer than average tails, some have wider snouts, some have more extensive manes, and so on. This variation was a problem because it made it difficult to list features precisely. Should the six-banded armadillo (*Euphractus sexcinctus* LINNAEUS 1758) be defined as having six bands, or seven? Occasionally, they were even found with eight (Smith, 2012). Linnaeus dealt with this ambiguity by selecting a specimen that his broad experience and sound instincts led him to believe best represented the ideal or type that God had intended for the species; he called it the **type specimen**, the one specimen that was utterly typical of the species. You have probably noticed the similarity of this concept and term **type** that Neoplatonists used for the ideals in God's mind. He had a growing collection of type specimens in his home. One can imagine that he felt his own personal type specimens were somehow holy. These types, most of them, are kept secure in three places: the Zoology Department of Uppsala University, the Stockholm Natural History Museum, and in Linnaeus' own house, now essentially a museum (Groves, 2001). Now all the pieces were in place, not just for identifying species – he even had a method for naming new species. If a specimen came to his attention that he deemed too different from his type specimen for the species, he named a new species.

His life work had come to full fruition, and one wonders whether he consciously thought he had secured for himself a place of honor in heaven. He wrote with no small pride of this masterwork: "I followed in the footsteps of God, and was thrilled." Not bad work if you can get it.

Linnaeus was not just organizing nature for his own heavenly reward and satisfaction. He wanted everyone to be able to identify any animal or plant they came upon, and he saw his ambition realized. In its day, *Systema Naturae* was on the short list of books with which any educated person was expected to be familiar. Its principles were a core part of the Victorian culture, of which Darwin was a part. Identifying animals and plants was simply something in which every educated Englishman was expected to be interested. Collecting animals, naming animals, studying their habits, identifying beautiful plants, and pressing flowers were not merely respectable hobbies in the Victorian age, they were activities that placed people closer to God and his great works. Keying out a species using *Systema Naturae* was not unlike prayer.

If understanding nature was a way of paying tribute to God's wisdom, what could be more appropriate that carrying your copy of *Systema Naturae* with you on a nature walk? And all the more appropriate that it be on a Sunday. God said keep the Sabbath holy, so what else could you do on a Sunday to entertain your friends? Keying out beetles, birds, and flowers was fun and pious. For Darwin and thousands of others, a jaunt in the countryside on Sunday after church was just the thing to round out the Sabbath religious experience.

If the sympathetic vibration with God's mind you got from keying out beetles was not enough of a glimpse of eternity for you, you could grasp it in another way. If you found something and failed to key it out with your copy of *Systema Naturae*, you could ship the thing off to Linnaeus for a diagnosis. If your specimen turned out to be truly unique, if it differed from Linnaeus' vast collection of type specimens, he would have to give it a new name, and he might well name it after you. Thusly, your name would be blazoned across the heavens for eternity as part of an undying natural world. In Latin, a word ending with one or two of the letter "i" means "of" whatever it's attached to. The "ii" at the end of *Pan troglodytes schweinfurthii* (a subspecies of chimpanzee) means "chimpanzee of Schweinfurth." As you might expect, there are many, many species named *darwini*.

3.20 Darwin Takes a Walk

Of all the Victorians on God's green Earth, perhaps none rivaled Charles Darwin in the sheer joy he wrung out of keying out beetles. He loved tromping across meadows in search of new beetles and splashing in the water to collect a specimen. If you had discovered that Darwin's family mapped out a religious life for him by encouraging him to become a country parson, that may have seemed an odd career choice for him without the context of this chapter. I hope it seems quite natural now. Darwin's enthusiasm for understanding nature was in perfect harmony with a career in the Church – in the first analysis he is a natural natural theologist. It turned out differently. Little did his family know, his work would sound the death knell for natural theology, eventually replacing it with the field of natural history.

Whatever your religious inclinations, it may seem a pity the tradition of the Sunday nature walk has lapsed, but it is not completely gone. In Darwin's day, if you wanted to cover a lot of ground, and if you were well-to-do, your Sunday nature walk might have included a carriage ride in the country. If you were infirm or lazy, maybe *only* the carriage ride. With the passage of time and the replacement of the carriage by the automobile, the tradition morphed into the Sunday drive in the country. In my childhood at least (a time period rapidly becoming history, I realize), the Sunday drive was still something families did now and then. Thus, do we find one of those delightful circumstances where the culture of Darwin's time reaches right out of the past and touches us today, even if only a few of us realize it. It is a withered tradition, of course. The evolution of the automobile from an open, horseless carriage to a sleek glass-and-steel capsule means we are glassed off from nature, so that the species-identifying part of the exercise is pretty much dead.

PART IV: LESSONS

There were four reasons explorers were so eager to document plants and animals. First, it was good for business. Mapping new geographic areas by mapping the plants and animals that were found there was essential to finding your way around as you sought access to new trade goods and new markets. Second, what we often call nationalism these days was a strong motivation. Being the first to discover some important animal, plant, or geographic area boosted national pride. Good maps could help you know where to site that new colony, and locating important resources was good for the economic health of the country. Third, good maps are quite handy during war. When you are attacking another city-state, you want to know where it is, where best to attack it, and what the best route is for the assault. If the siege fails, you will want to know the best path of retreat. Lastly, in the Renaissance and thereafter, Europeans who explored the natural world were also driven by religious zeal. Possibly, these four driving forces behind the discovery of the chimpanzee were not the four you might have guessed when you started this chapter.

The scientific community learned a number of lessons about how to understand nature in the process of following in God's footsteps. Early on, scientists neither incorporated earlier research into their science very well, nor did they communicate their results very effectively. While we university faculty often decry the publish-or-perish culture of academia, there is much to be said for writing stuff down and sharing it with your colleagues. The chimpanzee was, in essence, discovered a number of times. Scientists, beginning with our Dutch friend Tulp, may have studied their subjects closely, but their published records were so ambiguous, sketchy, and poorly illustrated that later scholars could not be certain whether they were looking at the same animal or not.

Perhaps another interesting lesson is how we humans view ourselves in relation to nature. It has not been constant. Early scientists had no idea whether the chimpanzees and orangutans they examined were humans or animals. Most of us draw a crisp, clear line between humans and animals – so that when most folks are told that chimpanzees are not monkeys, and they are no more closely related to monkeys than are humans, it fails to register. Yet if I say "even though your cousin is hairy and small-brained, he is no more closely related to monkeys than you are," we have no problem.

Renaissance anatomists went the opposite direction of most casual observers of primatology; rather than seeing apes as human-like monkeys, they more often saw them as hairy people. The sharp distinction we make between humans and animals came later, and observers who were not too enmeshed in Neoplatonism were happy to envision advanced people (i.e., whatever group the writer belonged to), less advanced people (others), primitive people (mostly folks poorly observed and vaguely described by travelers), advanced apes, and so on in a smooth unbroken line right down to the sponges.

Yet another lesson is that religion is a powerful motivator. For a deeply religious person, the prospect of taking a peek inside the mind of God must have been an intoxicating proposition. It certainly thrilled Linnaeus. Nowadays we try to understand the function of a wing or a kidney by thinking about the evolutionary advantage these organs confer on the animal, and scientists do their research for the inherent satisfaction of solving a vexing problem. Also, admittedly, for the public recognition, to get grants to support your lab or to get tenure.

Imagine how much more satisfying it must have been to feel that your research was also praising God and placing you in a state of grace. And it might even gain you some eternal reward. In short, until *The Origin of Species* appeared, and to this day in many quarters (Ruse, 2004), science and religion were not at odds, but were in a very real sense one and the same thing. The science–religion dichotomy is largely a twentieth-century, Christian evangelical, American phenomenon. Read this discussion from a Yale-educated neuroscientist – and Catholic priest – if you have a deeper interest (Pacholczyk, 2008).

History matters. We still find many references to and remnants of the Great Chain of Being in contemporary culture. When we talk about baser instincts, we are referring to the primitive behavior of animals at the base of the Great Chain. "Base behavior" has the same origin, as does the concept of "debased." Higher thought is thought that is higher on the Great Chain, more God-like. We derive the term "missing link" from Great Chain of Being thought: It is the idea that humans are so special that there seems to be a gap in the chain between the next-to-the-angels perfection of humans and the hairy, debased apes. There must be a missing link in the chain between us. The term "bone of contention" was coined to describe the **premaxilla**, the absence of which some contended defined humans (Holdrege, 2014). Who would guess – without knowing the history of the Great Chain of Being – that we owe the Sunday drive in the country to Plato?

Our modern notion of chimpanzee society is an extremely recent phenomenon. When Jane Goodall stepped off her water taxi and onto the beach at Gombe, we still believed they might be found to live in monogamous pairs, even though we had a quite modern understanding of their osteology and anatomy. Now you and I can hardly imagine that we might ever view chimpanzees very differently than we do now. We are just adding a detail here and a flourish there – we know chimpanzees. History tells us that this is not necessarily true. A glance at a primate journal or the occasional headline tells us we are still discovering the chimpanzee.

References

Brace CL (1988) Punctuationism, cladistics and the legacy of Medieval Neoplatonism. *Human Evolution* 3, 121–138.

Brace CL (2005) *"Race" Is a Four-Letter Word: The Genesis of the Concept*. Oxford: Oxford University Press.

Dawkins R (1976) *The Selfish Gene*. Oxford: Oxford University Press.

Deck JN (1967) *Nature, Contemplation and the One: A Study in the Philosophy of Plotinus*. Toronto: University of Toronto Press.

Dillon JM, Gerson LP (2004) *Neoplatonic Philosophy: Introductory Readings*. New York: Hackett.

Durant W (1944) *Caesar and Christ*. New York: Simon & Schuster.

Falconer T (1797) *The Voyage of Hanno: Translated, and Accompanied with the Greek Text, Explained from the Accounts of Modern Travellers, Defended Against the Objections of Mr. Dodwell and Other Writers, and Illustrated by Maps from Ptolemy, D'Anville, and Bougainville*. T Cadell, Jun. & Davies.

Goodall J (1986) *The Chimpanzees of Gombe: Patterns of Behavior*. Cambridge, MA: Harvard University Press.

Groves CP (2001) *Primate Taxonomy*. Washington, DC: Smithsonian Institution Press.

Gunn M, Codd, LEW (1981) *Botanical Exploration of Southern Africa: An Illustrated History of Early Botanical Literature on the Cape Flora: Biographical Accounts of the Leading Plant Collectors and Their Activities in Southern Africa from the Days of the East India Company Until Modern Times*. Boca Raton, FL: CRC Press.

Harden D (1963) *The Phoenicians*. Harmondsworth: Penguin Books.

Hill WCO (1969) The discovery of the chimpanzee. In *The Chimpanzee*, Vol. I (ed. Bourne GH), pp. 1–21. Basel: Karger.

Holdrege C (2014) Goethe and the evolution of science. *Context* 31: 10–23.

Little CE (1934) The authenticity and form of Cato's saying "Carthago Delenda Est." *Classical J* 29, 429–435.

Lovejoy AO (1936) *The Great Chain of Being: A Study of the History of an Idea*. Cambridge, MA: Harvard University Press.

Nissen HW (1931) A field study of the chimpanzee: observations of chimpanzee behavior and environment in western French Guinea. *Comp Psychol Monogr* 8, 122.

Paley W (1802) *Natural Theology*. New York: American Tract Society.

Pacholczyk T (2008) Are science and religion really enemies? Catholic Education Resource Center, www.catholiceducation.org/en/religion-and-philosophy/apologetics/are-science-and-religion-really-enemies.html.

Purchas S (1905) *Hakluytus Posthumus or Purchas his Pilgrimes, contayning a History of the World in Sea Voyages and Lande Travells, by Englishmen and others (1625)*, Vol. 6. Glasgow: Robert Maclehose & Co.

Reynolds V (1967) *The Apes*. London: Cassell.

Ruse M (2004) *Can a Darwinian be a Christian? The Relationship between Science and Religion*. Cambridge: Cambridge University Press.

Russell B (1945) *A History of Western Philosophy*. New York: Simon & Schuster.

Smith P (2012) *FAUNA Paraguay Handbook of the Mammals of Paraguay*, Vol. 2, www.faunaparaguay.com/mammhb2.html.

Suh CE, Sparks RSJ, Fitton JG, et al. (2003) The 1999 and 2000 eruptions of Mount Cameroon: eruption behaviour and petrochemistry of lava. *Bull Volcanol* 65, 267–281.

Thompson DW (1910) *Historia Animalium. Book XI*.

Tyson E (1699) *Orang-outang, Sive Homo sylvestris: or the Anatomy of a Pygmie Compared with that of a Monkey, an Ape and a Man*. London: Osborne.

Warmington BH (1960) *Carthage*. London: Pelican Press.

Yerkes RM, Yerkes AW (1929) *The Great Apes: A Study of Anthropoid Life*. New Haven, CT: Yale University Press.

4 Kin
The Chimpanzee's Place in Nature

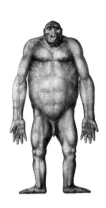

From *The Life of Primates*, Adolf H. Schultz, 1969, copyright © Orion Publishing Group, London, with permission

4.1 The Linnaean Binomial System: Taxonomy

We had an introduction to the bold ideas of naturalist Carolus Linnaeus in the previous chapter (Figure 4.1). You will remember that he wanted first to identify unambiguously every living thing on the planet and second to organize the living world into as simple a system as possible. He coined the term **taxonomy** as a slightly irregular mash-up of the Greek words **taxis**, meaning arrangement, and **nomos**, meaning method. Thus, the modern science of describing and classifying living things began with Linnaeus and – with fewer modifications than you might guess – his system is still used today. No wonder. The structure he conceived was positively brilliant. His revolutionary innovation was the invention of a rule-based, hierarchical system that unequivocally differentiated species from one another, gave every organism a unique name, and specified the relationships among species.

Prior to Linnaeus, when a new animal was described in a scientific publication it was given a Latin name that was much more of a description of the organism that just happened to be in Latin than it was the formal binomial we use today. A red squirrel might be called *sciurius rubicunda*, "red squirrel" in Latin, but there was no rule about how many names you could give it. One might give a cave-living common red squirrel the name *sciurius troglodytes rubicunda vulgaris*, Latin for "common cave-dwelling red squirrel," and that was fine, the four Latin terms notwithstanding. Linnaeus required that all animals be referred to by two and only two Latin words (nowadays the Latin is no longer required), which is why his scheme is called the binomial system; **bi** is Latin for two, and **nomial** is the Latin word for name, **nomen**.

In his book *Systema Naturae* (first edition, 1735; Figure 4.2), Linnaeus specified strict rules for assigning these two names. The first was to be a general name, and any number of closely related but distinct organisms could be placed under this general name, so long as the combination of the two names was unique. Linnaeus called this name the organism's **genus**, the root for the word "general." Each organism also had a more specific name, its **species** name. Either name, genus or species, could be used more than once, but the two together, genus plus species, had to be unquestionably unique for every living thing. It will be clearer with an example. The species name *troglodytes* (derived from the Greek for *one who crawls in holes*) is common as a species name; it is used for *Pan troglodytes* (the chimpanzee), *Hadogenes troglodytes* (the flat rock scorpion), and even *Troglodytes troglodytes* (the winter wren). *Pan*, the chimpanzee's genus name, is also used for the bonobo (*Pan paniscus*). In this way, the name *Pan* has been used for more than organism (bonobo and chimpanzee) and the name *troglodytes* has as well, but the two together, the *Pan troglodytes* binomial, is unique.

Linnaeus wanted as streamlined a system as possible. He wanted to make it user-friendly, and that

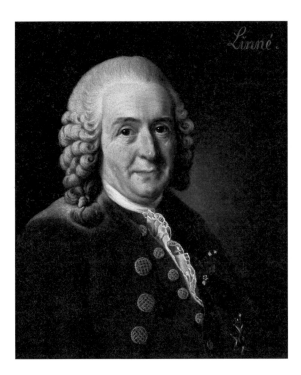

Figure 4.1 Carolus Linnaeus (Swedish botanist, 1707–1778), the "Father of Taxonomy." By Alexander Roslin; Nationalmuseum, Sweden.

CAROLI LINNÆI
EQUITIS DE STELLA POLARI,
ARCHIATRI REGII, MED. & BOTAN. PROFESS. UPSAL.;
ACAD. UPSAL. HOLMENS. PETROPOL. BEROL. IMPER.
LOND. MONSPEL. TOLOS. FLORENT. SOC.

SYSTEMA
NATURÆ
PER
REGNA TRIA NATURÆ,
SECUNDUM
CLASSES, ORDINES,
GENERA, SPECIES,
CUM
CHARACTERIBUS, DIFFERENTIIS.
SYNONYMIS, LOCIS.

TOMUS I.

EDITIO DECIMA, REFORMATA.

Cum Privilegio S:æ R:æ M:tis Sveciæ.

HOLMIÆ,
IMPENSIS DIRECT. LAURENTII SALVII,
1758.

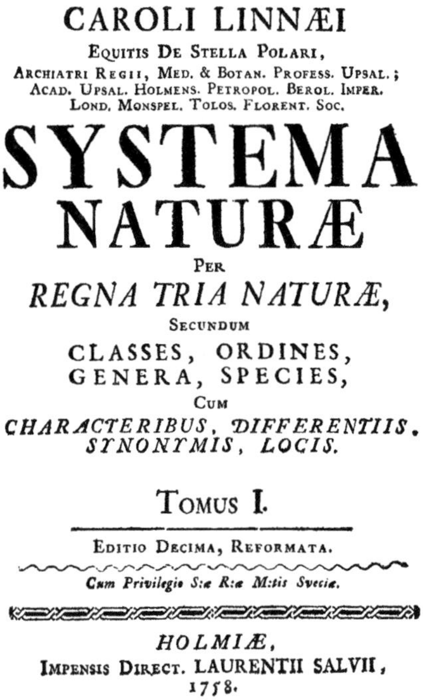

Figure 4.2 Title page of the definitive 1758 tenth edition of Linnaeus' *Systema Naturae*.

meant squeezing out ambiguity. He hoped to achieve this simplicity by finding the absolute minimum number of features necessary to identify each of his species. He called these **diagnostic features** (sometimes called "killing" features). As an example, I will use his system to unambiguously diagnose a well-known animal here. We can do it with only two features. Does it have a "trunk," and just to be unambiguous, by that I mean a well-muscled extension of its nose approximately the length of one of its legs? If yes, then species is either *Loxodonta africana* (the African elephant) or *Elephas maximus* (the Asian elephant). Look at the tip of the trunk. If there are two finger-like projections it is *Loxodonta africana*, if only one, then *Elephas maximus*. These two readily observed features distinguish these two species from every other of the millions of species on the planet.

Linnaeus had unshakeable confidence in his taxonomic system, so much so that he viewed it, quite literally, as God-given – eternal and unambiguous. As a foundational concept in the binomial system, it behooves us to define the term "species" well. Here is one definition, one that reflects the principles Linnaeus held dear, but really there is no agreement on what the definition should be: *A species consists of those individuals who make up an actually or potentially interbreeding natural population, reproductively isolated from other such groups.* We know now, as Linnaeus did not, that there are many gradations when it comes to the distinctness of species. Some pairs of species closely resemble one another morphologically, but cannot interbreed – the gorilla and the chimpanzee, for example. These are good species. Other species pairs are easily distinguished, yet *can* interbreed, and some even interbreed and produce fertile offspring. The redtail/blue monkey species pair is one example. With its two-toned blondish-gray body, white nose spot, and orange tail, the redtail is easily distinguished from the blue monkey, which looks like a giant redtail that has been dipped in a bucket of battleship gray paint; the

Table 4.1 **Taxonomy**

1. **Kingdom** – Animalia; other kingdoms are Plantae (plants), Monera (bacteria), Protista (e.g., algae), Fungi (e.g. molds, yeasts, and mushrooms)
2. **Phylum** – Chordata (other phyla include Molusca, Arthropoda)
3. **Class** – Mammalia (vs. Pisces, Amphibia, Reptilia, Aves; there are ~5400 species of mammals)
4. **Order** – 13 orders among mammal – Primates (vs. Carnivora, Perissodactyla, Artiodactyla, Proboscoidea, Rodentia, Cetacea, Chiroptera, Dermoptera, Insectivora, Sirenia, Edentata, Lagomorpha)
Suborder – Strepsirhini, Haplorhini
Superfamily – (Lemuroidea, Lorisoidea, Tarsiioidea, Ceboidea, Cercopithecoidea, Hominoidea)
5. **Family** – Hominidae for us
6. **Genus** – (66 primate genera)
7. **Species** – (about 340)

two are easily distinguished, yet they interbreed and produce fertile offspring.

In Linnaeus' original first-edition system he recognized only four taxonomic levels: class, order, genus, and species. We call a taxonomic name, the term we use to identify a group of related living things, a **taxon**; the plural is **taxa**. There are conventions about how we refer to taxonomic names, and it is worth taking time here to mention them. You might be tempted to try to figure them out on your own by deducing the rules from reading about different taxa; you would probably fail miserably. The system has order, but it is complex.

Taxonomic level names such as class or genus are not capitalized. However, when you name an *actual* taxon, e.g., Mammalia or Colobinae, that must be capitalized. The taxonomic levels to which we humans belong are as follows: our kingdom is Animalia (notice how I capitalized Animalia but not kingdom); other kingdoms are Monera (bacteria), Protista (e.g., algae), and Fungi (molds, yeasts, and mushrooms).

Our phylum is Chordata; members of the Chordata have a **notochord**, a rigid backbone-like cartilaginous structure that develops into a spinal column in animals that have vertebrae. Other phyla in the kingdom Animalia (the plural of phylum is phyla) are Molusca (squids, snails, and clams), and Arthropoda (insects, crustaceans, and spiders).

Our class is Mammalia; we think mammals are important, but there are only 5400 or so of them still living (Wilson & Reeder, 2005), and nearly half of those are rodents. Other classes include Pisces (fish), Amphibia (frogs, toads, newts, and salamanders), Reptilia (turtles, geckos, lizards, snakes, and crocodiles) and Aves (birds).

Thus the seven taxonomic levels are kingdom, phylum, class, order, family, genus, species. Memorize them. Years ago I came across an article entitled something like "100 things every college graduate should know." The author conceded that his list was arbitrary, but charged ahead anyway, asserting that if you were ignorant of any of his 100 things, your education was second-rate. One of the items was the seven Linnaean taxonomic levels (Table 4.1).

4.2 Primate Taxonomy

Our order, Primates, is one of 13 in the class Mammalia. Nearly all of these orders had their origin either just before or early in the Cenozoic, which means it is far from clear to which other order primates are most closely related. There are three likely candidates. Tree shrews (order: Scandentia) have basicranial and foot anatomy that resembles early primates (Bloch et al., 2007), and they fall closest to primates in some genetic studies (reviewed in Murphy et al., 2001). Flying lemurs, also known as colugos (order: Dermoptera; Figure 4.3) are linked to primates by an even greater number of genetic studies (Murphy et al., 2001; Janečka et al., 2007; Perelman et al., 2011) and a resemblance to early fossil primates. The fruit bats (suborder Megachiroptera – not the smaller, echo-locating bats) share brain, eye, forelimb, and some external genital morphology with primates. Some few molecular studies also link them to primates (Pettigrew, 1991); however, genetics suggests these connections are red herrings (Murphy et al., 2001). Flying monkeys – intriguing, but perhaps a bit too bizarre.

Returning to taxonomic levels, the seven basic levels failed to parse species in finely enough, so systematists came up with a convention for creating more taxonomic levels. Primatologists have determined that a number of taxonomic families (in the Linnaean sense) are more closely related than others, so we group a number of families into superfamilies. Superfamilies almost always end in -oidea. Table 4.1 shows two "extra" taxonomic levels, a suborder level (splitting the order Primates into two groups) and a superfamily level.

4.3 Primate Superfamilies

Among the primates there are six superfamilies, Lemuroidea, Lorisoidea, Tarsiioidea, Ceboidea, Cercopithecoidea, and Hominoidea. In the taxonomy in Figure 4.4 the "family" taxonomic level is left out; if you feel yourself becoming a taxonomy geek the more complete taxonomy in Appendix 1 will appeal to you (adapted from Groves, 2001).

We often refer to taxonomic groups informally, using part of the taxonomic name. When we use a taxonomic name as an informal adjective, it is not capitalized. Members of the Cercopithecoidea, for instance, can be called "cercopithecoids." The Ceboidea, monkeys confined to South and Central America, can be referred to informally as "ceboids." The Ceboidea are also often referred to as **New World Monkeys**, while the cercopithecoids are **Old World Monkeys**.

Linnaeus felt that taxonomy should reflect God's plan; contemporary biologists attempt to make their taxonomy reflect the evolutionary history of the groups under consideration. Looking at Figure 4.4, my taxonomic diagram reflects the fact that tarsiers are more closely related to monkeys than they are to lemurs and lorises. This requires them to be in the taxon Haplorhini, with the monkeys and apes.

4.4 Primate Diversity

Primates are an unusual order because they are extraordinarily diverse, and it is not just overweening pride in my own taxonomic order that leads me to write that. They range in size from 30 g (a single ounce) for the world's smallest primate, the pygmy mouse lemur, to 180 kg (400 lb) for male gorillas. Their brains range in size from 1.7 cm^3 (Stephan et al., 1981), the size of a gumball and the same as a hamster (Nieuwenhuys et al., 1998), to 470 cm^3 for the gorilla, the size of a grapefruit. Human brains measure 1300 cm^3.

Body shapes range from squirrel-like for the marmosets and tamarins, with their long bodies, short legs, and "claws," to long-armed, short-legged species like the orangutan, to stout, short-limbed, and heavily

Figure 4.3 Colugo, or "flying lemur." Colugos are not lemurs, but their face is very lemur-like. Photo: Tim Laman; with permission.

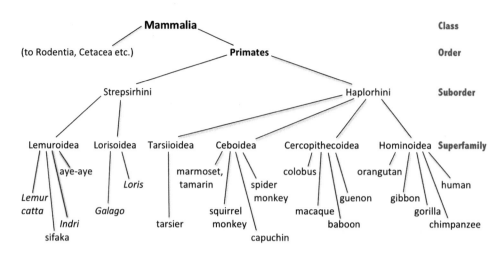

Figure 4.4 Simplified primate taxonomy.

muscled primates like the capuchin. Copulation frequency ranges from something like once a year for the female gorilla to six times a day for male chimpanzees. Tails range in size from the longest, most muscular, and most dexterous in the animal kingdom (spider monkeys) to no tail at all (apes, some macaques, the indri). Diets range from practically all insects among many strepsirhines, to hardly anything but leaves for the black and white colobus, to as much fruit as they can get for the chimpanzee.

Looking across the other orders, bats, elephants, ungulates, carnivores, whales, and rabbits all have much more consistent body plans than primates; and their body sizes are less variable as well. We are a very special order, we primates.

4.5 Primate Trends and Traits

When we want to distinguish one order from another, we inevitably turn to things the members share in common: Cetaceans are aquatic; Chiroptera have wings, and so on. The primates are more difficult, but we do find traits they share; these traits tend to be related to their arboreal (tree-adapted) lifeway (Napier & Napier, 1967). While there are other mammals that are competent in the trees (squirrels and raccoons come to mind), their movements seem clumsy compared to the graceful, inventive movements of the supremely arboreally adapted members of our order.

Primates (Figure 4.5) appear to have made the move to life in the trees almost as soon as fruit- and flower-bearing trees came into existence. That means for 65 million years they have been evolving into the acrobatic arborealists we know today. Primates had a good reason for climbing into the trees – they went where the food was. Their diets are dominated by arboreal foods: fruits, leaves, flowers, gums, bark, and for many species, the insects that feed on these foods. It is the evolutionary accommodation to an arboreal life that gave these large-ish mammals their distinctive physical traits. However, the diversity of primates means not all primates express traits most characteristic of the order in equal measure. The exceptions are numerous enough that primatologists have come to call primate traits "trends" to reflect the fact that they are not universal (Table 4.2).

Our first trend is the retention of a generalized skeleton. Unlike horses, which have lost several digits on their feet (they have only one toe) and other skeletal elements in their legs (e.g., the fibula), primates have retained nearly the full complement of ancestral skeletal elements. When rigid structures are reduced in number, the number of joints is reduced as well, reducing flexibility but improving efficiency and/or power. Primates have retained flexibility.

While many primates, like lemurs (Figure 4.6) and vervets, have body proportions hardly different from the body plan of dogs or cats, primates have hands rather than dog- or cat-like paws, even those primates

Figure 4.5 Diversity among the primates. Top, left to right, bald uakari, white-fronted spider monkey, Guatamalan black howler, lar gibbon, pygmy marmoset (on gibbon's head), red-shanked douc langur (under gibbon), DeBrazza's monkey, common squirrel monkey, mantled guereza (or eastern black and white colobus), Coquerel's sifaka and (hanging from tail) Philippine tarsier.
Bottom, left to right: mandrill, proboscis monkey, ring-tailed lemur, black-headed night monkey, Eastern gorilla, golden lion tamarin, chimpanzee, Bornean orangutan, Japanese macaque and slow loris. Illustration by Michael Boardman, Coyote Graphics. Used with permission.

with long, dog-like faces, such as the Lemuroidea (note the hands in Figure 4.6). Long, flexible fingers that can move independently of one another endow primates with a very versatile tool that can interact with their environments in a variety of ways: primate hands can assume multiple **grips**. That is, they can position their fingers in different ways to hold onto odd shaped and odd-angled supports, or to grip and process oddly shaped foods. John Napier (1980) brought terms like **key grip, power grip** (how you grip a baseball bat), and **precision grip** (baseball grip) into widespread usage – your pet dog is incapable of these grips.

Primates also tend to have fingernails rather than claws. Claws can be pressed into the surface of large-diameter supports too large for the hand to encircle, like tree trunks. Claws are ineffective on tiny twigs; imagine a cat trying to sink its claws into a pencil. Primates, then, tend to have gripping hands (and feet) and broad, stiff nails that give some rigidity to their fleshy, sensitive fingertips. These organs are terrific for maneuvering among small branches, combing through a companion's fur, or pulling open thick-husked fruits.

One of the most noticeable things about primates – a trait that can give some of the less human-like of the primates a creepily human affect – is their tendency to hold their torso upright (Figure 4.6). This posture frees the hands to manipulate fruits, pick the wings off insects, or groom a fellow group member.

Table 4.2 **Primate evolutionary trends**

1.	Generalized skeleton
2.	Manual adaptations for dexterity and acuity: long, mobile digits; grasping ability; a mobile thumb and ability to oppose thumb to other digits; nails instead of claws; sensitive, fleshy fingertips
3.	Tendency to erectness in upper body
4.	Eye specializations; eyes oriented forward (**frontated**); overlapping fields of vision = **stereoscopy**; color vision; *postorbital bar*, **postorbital closure**
5.	High-quality diet: fruit, omnivory; simple stomachs and intestines; low proportions of antifeedants
6.	Reduced **olfactory** acuity and reduced snout; **kyphosis** (snout not between the eyes)
7.	Simple teeth; no elaborate infolding of enamel as in elephants and bovids; all four tooth types; incisor, canine, premolar, molar; loss of 3-1-4-3 **dental formula**
8.	Large brain, especially neocortex; elaborate social organization; dependence on learned behavior to find food; recognition of individuals
9.	High parental investment; efficient placenta; long gestation; single births; late maturity; long lifespan
10.	**Diurnal** (active during daylight hours) rather than nocturnal (active at night) activity

Figure 4.6 *Lemur catta*; notice the upright torso and human-like hand; the black, hairless area on the forearm is the antebrachial scent gland. Image courtesy of OpenCage.

Primates tend to have keen vision. Not only do most species have good color perception, their eyes tend to face forward, giving them great depth perception (we call that **stereoscopy**), a faculty that comes in very handy when leaping among branches. Because the color of many food items is correlated with its chemical composition – ripe bananas are yellow – color vision is useful for identifying how nutritious fruits, leaves, and other items are. The importance of vision for primates has resulted in the evolution of a bony feature in the skull, called a **postorbital bar**, a bar of bone found just behind the eye (Figure 4.7) that helps to stabilize the visual field; primates are among the very few mammals that have this skeletal element. An elaboration of this adaptation is a wall behind the eyeball that isolates the eye from the chewing muscle, the temporalis, helping to keep the field of vision stable during chewing (Ross, 1995; Figure 4.7). All primates have a postorbital bar but only haplorhines have the bony wall also known as **postorbital closure**. Monkeys spend much of their day chewing, and postorbital closure allows them to monitor the social landscape or to walk among unstable branches while chewing.

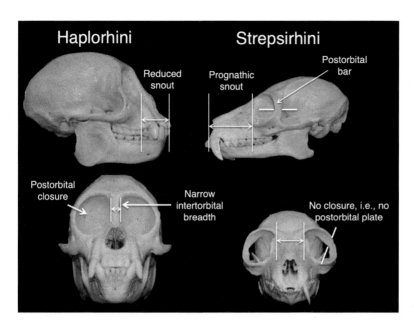

Figure 4.7 Looking at the lemur skull on the right, you can pass a pencil into the eye orbit and past the braincase, because lemurs have a bony bar enclosing the eye in a ring, but they lack *postorbital closure,* a bony plate behind the eye. Illustration by the author.

A further primate trend is a tendency to eat nutritionally dense foods – in stark contrast to animals like the panda that specialize on a single or several low-quality dietary items. Primates have taken the opposite path. They eat a wide variety of foods, but focus on nutritionally dense food items, foods that are high in fat, sugar, and protein and low in fiber. Everything is relative, of course, even dietary quality – you and I would quite literally die on a lemur's diet, but in turn they would die on a goat's diet.

Anatomically, primate stomachs, small intestines, and colons are short and simple compared to those of animals with lower-quality diets.

Unlike your pet dog, primates tend to rely on their keen vision rather than their noses to make their way around the world and consequently tissue inside the nose and its supporting bony structures are reduced compared to many mammals. Acute vision is a requirement for moving quickly among small branches in trees, but monkeys and apes also use their keen eyesight for communication, often relying on facial expressions and eye contact to communicate, rather than odor. It is no coincidence that primates that rely on olfaction rather than vision have long snouts, extensive olfactory tissue filling up the nasal cavity, and no postorbital closure. Many primates (lemuroids are the exception) have not only reduced structures related to olfaction, but their snouts have been shifted to start below rather than between the eyes, a facial configuration that allows their eyes to be close together; the convergence of visual fields creates stereoscopic, or 3D, vision.

Because primate food items vary in size, shape, and texture, primates have retained all four tooth types – incisors for biting off chunks of food, premolars (bicuspids, according to your dentist) for breaking up the food into a few large chunks, and relatively broad, flat molars for grinding food into small, easily digestible particles. Canines are mostly used for gripping, ripping, and puncturing other primates during fighting, though they are also important food-processing teeth in species such as capuchins.

Primates have large brains compared to other mammals, almost certainly to keep track of the complex social relationships that characterize the order (Dunbar, 1998), but possibly also to deal with their sophisticated foraging strategies (Milton, 1981). Primates must remember the physical characteristics, processing requirements, and locations of many different foods, and they switch from one food to another, choosing the most nutritious available at any one moment. They rely on learning when it comes to finding food, whereas other animals have pre-programmed food-finding senses and instincts. Animals with single-item diets like pandas and koala bears that are culinary simpletons also fail to impress in the realm of intelligence.

Large brains require not only a high-quality diet, but a long period of growth, which in turn requires a heavy investment by the parent(s), beginning before birth. Primates have a particularly complicated and efficient placenta that allows intense energy, waste, and gas exchange between mother and offspring. Gestation is extended among the primates, and they tend to have single, heavy infants. A long gestation is followed by a long juvenile period and late maturity. The payoff is a long life. This whole phenomenon is known as a **slow life history**.

Primates tend to be active during the daylight hours. This **diurnality** means they can rely on vision. It is probably no surprise that the primates with the most sensitive noses are active at night (**nocturnal**) or at dawn and dusk (**crepuscular**).

4.6 Suborder Strepsirhini

The taxonomic group, the lorises and lemurs, is least related to humans and chimpanzees, and the least likely to win a beauty contest, is the suborder Strepsirhini (Figure 4.4). Strepsirhines seem hardly to deserve their place in the lofty company of primates more familiar to you, such as orangutans and baboons, and their taxonomic name, Strepsirhini, reflects their primitive morphology.

The Greek roots of the taxonomic term Strepsirhini refer to their upper lip, the area underneath the nose where you and I have a broad, relatively flat expanse of skin. Strepsirhines lips are much like those of a dog or cat, with an empty space under the nose and a left and right lip to the outside, somewhat like an inverted V (see Figure 4.6). "Strep" comes from the Greek root for "curved" and "rhine" is Greek for nose (like rhinoplasty). It is actually their lip (Greek: chelios) that has curves, and while "Strepsichelia" might be more accurate, Strepsirhini it is. Their snout is tipped by a **rhinarium**, an area of highly sensitive, roughened, moist tissue – a wet nose. The internal nose of strepsirhines, lemuroids more than lorisoids, is packed with heavily vascularized, highly enervated tissue, much of it attached to large, bony nasal conchae, the shape of which maximizes contact between tissue and inspired air, giving them an extremely sensitive sense of smell (Mogicato et al., 2012).

Strepsirhines use their snout for more than smelling. Whereas a monkey might reach out to gather fruit with a hand and would comfort a friend by grooming their fur with her hands, strepsirhines eat more often with their mouth and they use their incisors and canines, especially the lower ones, as a grooming tool. The forward tilted, comb-like constellation of teeth is quite appropriately called a **toothcomb**. Rather than using their hands to groom, they draw their toothcomb through the fur of their grooming partners. When lemurs and lorises self-groom, what they cannot handle with their teeth is dealt with using a specialized nail or claw on their second toe, called a **toilet claw**. Your pet dog probably wishes she had one, because it is used to scratch in a rather dog-like manner.

Loris and lemur brains are smaller than those of the average primate (Stephan et al., 1981). Strepsirhines typically give birth to two (or more) infants. Compared to monkeys, they have better night-vision at the cost of poorer day vision. They have behind their retina a **tapetum lucidem** that reflects light, essentially giving their photoreceptor cells two passes at perceiving light entering the eye. The tapetum will be familiar to you because it gives animals eye shine at night. Their color vision is variable – they have a greater profusion of rods than cones, endowing them with excellent night vision. Many of them are nocturnal.

Strepsirhines are small compared to monkeys, more the size of cats or mice than chimpanzees. As all primates do, strepsirhines have a postorbital bar (Figure 4.7), but they lack the bony plate behind the eye, the postorbital closure, that monkeys and apes have.

4.7 Superfamily Lemuroidea

The superfamily **Lemuroidea** has 51 living species. They are perhaps most similar of all living primates to the last common ancestor of the entire order, which makes it even more baffling that they are now, despite an extraordinary diversity, losers in the global competition for a place in the primate world. Lemuroids are confined to Madagascar, an island off the southeastern coast of Africa. It is a big island, the size of California, but for an animal whose ancestors once roamed the world it seems a humble legacy. If they are plotting to retake the world, it remains their secret, and given their modest brain size and preoccupation with smearing their stinky secretions everywhere (as we will discuss in more detail), chimpanzees are unlikely to lose sleep worrying about it.

They have legs slightly longer than their arms and a generalized anatomy that leads us to call them **generalized quadrupeds.** They get around on all four limbs and have a very cat- or dog-like locomotor repertoire, with running, walking, and leaping predominating. Of course, their gripping hands and a foot with a divergent and powerfully muscled big toe (Figure 4.6) are very handy for climbing. Among the lemuroids with this basic body plan is the ring-tailed lemur, *Lemur catta*, and a variety of lesser-known lemurs such as the sportive lemur, ruffed lemurs, brown lemurs, dwarf lemurs, and mouse lemurs (Figure 4.8). We must not forget the small-in-size but long-in-name pygmy mouse lemur, the world's smallest primate (*Microcebus berthae*; Dammhahn & Kappeler, 2005). Already diminished by having "mouse" preceding your name, it seems a little over the top to be further belittled by having "pygmy" wedged into your moniker but the diminutives are fitting for the tiny one-ounce pygmy mouse-lemur.

You may be wondering what it is that lemurs and other strepsirhines are smelling with all that machinery for acute olfaction; it is mostly one another. Lemurs (and lorisoids as well) engage in **scent marking.** They excrete aromatic oils from various body parts, some more disgusting than others. It varies by species, but scent glands can be located in the armpit, forearm, anogenital region, or all three, augmented by broadly dispersed glands more like sweat glands (Hill, 1953–1970; Jolly, 1966). They rub these secretions all over themselves, over others, and over practically anything unable to run away when a

Figure 4.8 Goodman's mouse lemur, *Microcebus lehilahytsara*. Photo courtesy of Robert Zingg/Zoo Zürich.

lemur rubs its bottom on it. If you have a strong stomach, you may not be nauseated yet, but I can remedy that; they also urinate on their hands and feet, allowing them to leave stinky footprints during their daily perambulations. They spew out all these scents to communicate fear, excitement, and other emotions. If lemurs could talk, a good conversation-starter might be "you smell pretty happy today."

The heavyweight of the lemuroid world is the indri (Figure 4.9), weighing in at 9 kg (20 lb), the weight of a hefty cat or a terrier. Indris, conveniently known by the same name as their genus name, have long hindlimbs and get around by a type of locomotion known as **vertical clinging and leaping** (Napier & Walker, 1967). They grasp relatively vertical branches with their hands and feet, push off with their long, powerful hindlimbs, and touch down – many meters away from the starting point – by contorting their body in a pirouette so that their hindlimbs contact the landing tree first. They make their point of view known with haunting, high-pitched, siren-like wails that sound almost like a very loud cat. *Indri indri*

Figure 4.9 *Indri indri*. Note the long legs and lack of a tail. Photo: Rhett Butler; with permission.

Figure 4.10 The aye-aye. Painting by Joseph Wolf, 1863.

stands alone as the single species in its genus, but the similar but smaller sifaka (genus *Propithecus*) is more speciose, with seven species in total. The sifakas are very like indris in having long hindlimbs and getting around by vertical clinging and leaping, but they retain a tail.

For the sake of completeness, we must not leave the Lemuroidea without mentioning the **aye-aye**, *Daubentonia madagascarensis* (Figure 4.10). You will hardly believe they are real. Their goggle-eyed, buck-toothed, disheveled appearance is so unsettling that locals believe that merely setting eyes on one is bad luck. Like rodents, they have ever-growing incisors – and they need them. Their powerful jaws and teeth make them more like a power tool than a primate. They can gnaw into the middle of a two by four in three or four bites. Their diet is rich in insect larvae, which they locate by tapping on tree trunks, listening for the subtle sound of an empty space underneath

(the crypt for insect larvae; Erickson, 1991; Sterling, 1994), and then biting through the wood surrounding it. Once they have managed to open a larval crypt they make use of another bizarre anatomical curiosity, their wire-like third digit. They insert their middle finger into the just-exposed hole, tickling up the worm-like larva inside, which obligingly does exactly what the aye-aye wants it to do; in an attempt to escape, it arches away from the aye-aye's finger, but unfortunately for it, twisting itself into a convenient donut shape around the aye-aye's finger only allows it to be pulled out. You may be disgusted, but this is fine dining for the aye-aye – larvae are packed with calories in the form of fat.

4.8 Lorisoidea

The second superfamily in the Strepsirhini is the **Lorisoidea**. There are 28 species of lorisoid, but here we will focus on two genera as representative of the taxon, the bushbaby (genus *Galago*), and the slow loris. Lorisoids are found across sub-Saharan Africa, in India, and in Southeast Asia. They are less variable than lemuroids when it comes to body size, ranging in weight from 70 g (3 oz) to 1.5 kg (3 lb). All are nocturnal. Insects make up most of the diet for most of the species. Their face is not so snouty as that of lemuroids.

Bushbabies (Figure 4.11) include fruit and gums in their diet, but their quick, darting movements and dexterous hands are well-adapted to snatching up rapidly moving insects, which make up a large part of their diet. They have bat-like pointy ears that they can rotate independently as they listen for insect prey on the wing. Their eyes are large, as is typical among nocturnal primates. They tend to be hoppers and leapers, and accordingly their legs are longer than their arms, though the larger species engage in a more generalized quadrupedal locomotion (Bearder, 1987; Nash et al., 1989).

Lorises (see Nekaris & Bearder, 2011) are larger than bushbabies and more stoutly built, with arms and legs approximately equal in length (Figure 4.12). They are immediately distinguishable from bushbabies by their lack of liveliness as much as by their stouter body form. They avoid fast movements such as leaping, running, and even normal-speed walking; instead, they move slowly and stealthily in the trees in a manner of progression called **slow climbing**. When in posture they prefer to have all four prehensile hands/feet firmly gripping their perch and seem to regret having to move at all, so much so that when moving they are said to resemble four moveable clamps. They eat more poisonous, slow-moving insects than do bushbabies.

They can orient their hands and feet precisely to whatever angle supporting branches happen to offer and as a consequence they have rotatable wrists, much like yours and mine (Chapter 7). The halting, swaying movement of the slender loris resembles that of chameleons, to my eye. Other lorisoids are the potto and the aptly named slow loris, which lives in video fame principally for its endearing "stick-em-up" response to being tickled. The red slender loris goes by what might be the most delightful binomial of any animal, *Loris tardigradus*, which, if you recall your Latin, you will translate as the "slow-moving loris."

Figure 4.11 The bushbaby *Galago moholi*. Photo: Gerald Doyle; with permission.

Figure 4.12 The slow loris, *Nyctecebus coucang*. Courtesy of WikiMedia Creative Commons.

4.9 Suborder Haplorhini

The Haplorhini, from the Greek for "simple-nosed," are the monkeys, apes, and, inconveniently, the enigmatic tarsiers. The Haplorhini is divided into four superfamilies. You may have noticed that the two superfamilies you met in the Strepsirhini got their names by tacking "-oidea" onto the colloquial name for a common species in the taxon, giving us the taxa Lemuroidea and Lorisoidea. Superfamilies in the Haplorhini are slightly more complicated, but not imposingly so. They consist of the Tarsiioidea, the Ceboidea (derived from the type genus *Cebus*), the slightly more challenging-to-pronounce Cercopithecoidea (type genus *Cercopithecus*; pronounced SARE-coe-PITH-uh-cuss), and at last our own superfamily, the Hominoidea, which takes its name from our own taxon, the hominins.

Historically, taxonomists have recognized three species of tarsiers (Figure 4.13), distributed among various islands in Malaysia, Indonesia, and the Philippines, but there is considerable diversity in the taxon, and the three species are believed by some experts to represent as many as three genera and eight or even nine total species (Groves & Shekelle, 2010). We will focus on one genus, *Tarsius*, which Groves (2001) breaks into six species.

Figure 4.13 The Philippine tarsier, *Tarsius syrichta*. Photo: WikiMedia Creative Commons.

4.10 Tarsiioidea

Like the lemurs, tarsiers were once much more common and much more widely distributed than they are now. Their name is derived from the two quite distinctive ankle bones, **tarsal** bones, that are bizarrely elongated, giving them what amounts to an extra joint in their legs. Tarsiers are vertical clingers and leapers *par excellence*. The two bones just below the knee, the fibula and tibia, are fused together to better withstand the great forces generated during their spectacular leaps. They are insectivores. Their legs are much longer than their arms, and like galagos they have dexterous hands they use to snatch flying insects from the air, sometimes in mid-leap.

Tarsiers are tiny, 140 g (5 oz) or less. They are nocturnal and have enormous eyes, giving them excellent night vision, despite lacking the tapetum lucidem of the strepsirhines. They hunt for insects among small-diameter, vertical supports, mostly within 3 m of the ground (MacKinnon & MacKinnon, 1980), where they first listen for prey with their large, mobile ears and then home in on them with their keen vision. In accord with the importance of their vision, they have near-complete postorbital closure, which as we have already learned helps to isolate the eye from the disrupting movement of chewing muscles (Ross, 1995). They also possess very unmonkey-like toilet claws, but they redeem themselves with a very large brain, a feature typical of the haplorhines. To service that large brain during fetal development they have a complicated placenta that allows a more intimate association between the blood supply of the mother and fetus. This more intricate placentation allows for a more efficient transfer of gases and nutrients between mother and fetus. Their small size, insectivorous diet, nocturnality, rapid maturation, and toilet claw make them resemble strepsirhines, but their genetics, large brain, nearly complete postorbital closure, single births, lack of rhinarium, and intricate placentation leave no doubt that they belong in the Haplorhini (Shoshani et al., 1996; Hartig et al., 2013).

4.11 Ceboidea

Monkeys, apes, and humans tend to have short snouts (Figure 4.7), stereoscopic vision (with frontated eyes and a narrow interorbital breadth), postorbital closure, a tendency to hold the torso erect during posture, larger brains, and (with a few exceptions) single births. Most but not all of the New World monkeys, the **Ceboidea**, the ceboids are in accord with this generalization. The ceboids are found in South and Central America; they are a diverse group of 16 genera and 107 species (see Appendix 1), in many ways much more diverse than the Old World monkeys. No ceboid species is well adapted for ground living – though capuchins do spend a lot of time on the ground – and many are so thoroughly arboreal that they are born, live, and die without stepping foot on soil. Only one genus, *Aotus*, appropriately named the night monkey, is nocturnal. Their nostrils tend to be oriented laterally, and in many species they are so widely separated they leave a distinct flat-ish area near the center of the face. Some ceboids, the members of the family Atelidae, have prehensile tails that can grasp almost like hands; atelids are the only primates that can hang by their tails.

For the sake of simplicity, the Ceboidea will be divided into three groups (though New World Primate specialists may feel this is just a bit too simple): the tiny **callitrichids** (marmosets and tamarins), the generalized quadrupeds, a group that retains standard-issue primate morphology. The generalized quadrupeds you may be familiar with are the capuchin and the squirrel monkey, primates that were once common in the pet trade.

The third group, the atelids, are so distinctly convergent on apes that the 14 species of spider monkeys and their relatives (subfamily Atelinae) share with chimpanzees not only anatomy (Larson, 1998), but also a social system. Like the apes they are suspensory, but have not only kept their tails, they can use them almost like hands; they are the only primates that can hang by their tails.

Marmosets and tamarins, members of the family Callitrichidae, are tiny, almost squirrel-like primates weighing from 60 g (2 oz) to 500 g (1 lb). In contrast to tarsiers, who cling and leap from vertical stems not much bigger than your thumb, the equally tiny callitrichids use sharp, claw-shaped nails to cling to tree trunks that are more the diameter of your desk. Marmosets (genera *Callitrhrix* and *Callimico*) are adapted to high-impact bark-gouging using their stout, procumbent incisors; they are wood-pecker primates. Tamarins are a little less peck-ish.

Skeletally, the callitrichids are not terribly diverse – it takes a specialist to differentiate among the various species when working from the skeleton alone – but in life they are delightfully diverse in their pelage (fur), and easy to tell apart. Individual species sport an extravagant puff of fur on the top of the head, or on the ears, or on the upper lip, or in a fringe around the head (Figure 4.14).

Squirrel monkeys and **capuchin monkeys** are exemplars of the generalized quadrupeds of the New World. The squirrel monkey (Figure 4.15) may be the cutest monkey on the planet. They are tiny, but to my eye much more human-like than any of the callitrichids, with their completely forward-facing eyes and tall forehead. All that cuteness is just a ruse. The males are extremely aggressive during mating season, during which time they put on fat around their shoulders, get very short-tempered, and attempt to tear one another to bits. During mating season, males often sport multiple wounds in various stages of healing, and males who are veterans of multiple mating seasons can be a patchwork of scars.

Capuchins – not, as you might expect, the chimpanzee-like spider monkeys – have the biggest brains of the ceboids (Figure 4.15), though one wishes they did something a little more socially acceptable with all that gray matter. They have dreamed up any number of creative social traditions that I would have preferred they kept to themselves. They sniff one another's hands, pick one another's noses, poke their

Figure 4.14 (a) Cotton-top tamarin (*Saguinus oedipus*; photo by Terry Waters), (b) common marmoset (*Callithrix jacchus*; photo by Raimond Spekking) and (c) the emperor tamarin (*Saguinus imperator;* photo by TheBrockenInaGlory). (A black and white version of this figure will appear in some formats. For the color version, please refer to the plate section.)

Figure 4.15 (a) Squirrel monkey (*Saimiri sciurius*) and (b) brown-capped capuchin monkeys (*Cebus apella*). Photos: Frans de Waal; with permission.

fingers deep into one another's eyes, and playfully rip out tufts of hair from one another's shoulders (Perry et al., 2003). What could be cuter! They also recognize their images in mirrors, open nuts with stone tools, understand the symbolism of "money," and can be trained to perform a long list of human activities as service animals for quadriplegics.

Spider monkeys have long, thin arms and legs, long fingers and toes (Figure 4.16), remarkably flexible shoulders, powerful, dexterous hand-like tails, and a chimpanzee-like male-bonded, fission–fusion social system – the only primate other than chimpanzees with this system (Robinson & Janson, 1987). Their tails are long, muscular, and dexterous; they have a hairless area on the inside of the tip of the tail that has **dermal ridges**, what would be called fingerprints if they were found on the digits. Spider monkeys can pick up an individual peanut with the tip of their tail. The suspensory Atelidae are divided into two subfamilies, the Atelinae – spider monkeys and their kin – and the Alouattinae – the raucous howler monkeys.

Howlers have diets rich in leaves; when feeding they often hang from their tails, rarely using their arms. Spider monkeys, in contrast, are at least as much arm-hangers as tail-hangers.

4.12 Cercopithecoidea

While I have felt guilty giving short shrift to the amazing Ceboidea, I feel even guiltier trying to limit the Cercopithecoidea to just a few pages. There are 21 genera and 129 species in the Cercopithecoidea, spread across all of Africa, into the Middle East on the Arabian Peninsula, and on into Asia, India, China, and many other Asian countries, including temperate-zone Japan. They range in size from the 2.5 kg (5 lb) for the Chihuahua-sized talapoin monkey to the nearly chimpanzee-sized Chacma baboon, where males can exceed 30 kg (65 lb; Smith & Jungers, 1997). They are all diurnal. They all share the basic monkey solution to the physical world, possessing narrow shoulders, arms and legs of approximately equal length, a reduced snout, eyes completely frontated, dexterous hands, and gripping feet, though their big toe is less divergent than that of apes.

Whenever you read a scientific article and the study subject is described as "the monkey," the authors are referring to the rhesus macaque (*Macaca mulatta*) (Figure 4.17). While cercopithecoids tend to have the reduced, almost human-like face that we all think of when we hear the word monkey, there is tremendous variation; the olive baboon (*Papio anubis*) has an

Figure 4.16 The spider monkey (*Ateles geoffroyi*). Photo: Roy Fontaine.

Figure 4.17 Japanese macaques, close relatives of "the monkey," *Macaca mulatta*. Courtesy of WikiMedia.

enormous snout fitted with flanges on either side, almost like fins on a 1950s cruiser, but baboons are snouty in a different way than the lemur. Their eyes are completely frontated, leaving the snout to protrude below them. The baboon-like mandrills are probably a familiar species, with their bright red, white, and blue faces.

Cercopithecoids are long-lived primates – lifespans into the forties are not unusual – that give birth to single offspring that mature slowly, meaning monkey babies sustain relationships with their mothers over many years. The Old World monkeys are remarkably consistent in their social structure, which typically has as its stable center a group of closely bonded, related females, with only rare exceptions. Females have **visible estrus**. **Estrus** is a period of physical and behavioral change timed to coincide with ovulation. In the cercopithecoids, estrus is accompanied by a **sexual swelling**, the tumescence of sexual skin surrounding the vagina. The swelling advertises the period of maximum fertility.

The Cercopithecoidea is separated into two groups at the taxonomic level of the subfamily; the two taxa are the Cercopithecinae (including baboons, macaques, mandrills, redtails, vervets, and all members of its type genus *Cercopithecus*, which are also known as **guenons**; see Appendix 1) and the Colobinae, which includes langurs, members of the genus *Colobus*, and other leaf-eating monkeys. The colobines have evolved specialized stomachs, intestines, and body chemistry for digesting leaves. In fact, the langurs are often called leaf monkeys.

Savanna baboons (Figure 4.18(a)) make a striking contrast to the Tonkin snub-nose monkey (Figure 4.18(c)), which has no snout at all and instead looks like an Ewok-faced plush toy so preternaturally cute that they risk veering off into creepy. The proboscis monkey, a relative of the snub-nose monkey, has gone in the completely opposite direction; it has a bulbous floppy, cucumber shaped nose (Figure 4.18(d)). Yet another relative, the golden snub-nosed monkey (Figure 4.18(b)), has a blue face and has given up on a nose altogether, just making do with two holes in its face. Black and white colobus are the howler monkeys of the Old World, sharing a croaky, froggy vocalization with their New World counterparts.

Figure 4.18 (a) A Gombe olive baboon female with a kidnapped infant that later died of dehydration (photo by author), (b) the golden snub-nosed monkey (*Rhinopithecus roxellana*), (c) the Tonkin snub-nose monkey (*Rhinopithecus avunculus*; photo by Tito Nadler), (d) the proboscis monkey (*Nasalis larvatus*; photo by Charlesjsharp), and (e) the black and white colobus (*Colobus guereza;* photo by author). (A black and white version of this figure will appear in some formats. For the color version, please refer to the plate section.)

Baboons are a group characterized by their propensity to spend time on the ground. The gelada baboon (*Theropithecus gelada*) is virtually 100 percent terrestrial, a stark contrast to the redtail monkey, which rarely comes to the ground. The typical cercopithecoid inclines toward redtails, spending considerably more time in the trees than on the ground. The arboreal cercopithecoids get around walking quadrupedally in the trees unless confronted by a large gap in the tree canopy, in which case they exhibit extraordinary leaping ability. They sit so much of their time, including when they sleep, that they have evolved a built-in seat cushion, **ischial pads (or ischial callosities)**, a hairless area, underlain by fibrous tissue much like the palm of your hand, the bony attachment of which are the ischial tuberosities of the pelvis.

4.13 Hominoidea

This brings us at last to our own superfamily, the **Hominoidea**, of which we are the type genus. You may know the Hominoidea by its colloquial term, ape. There are 21 species among the apes, most of them (14) in the family Hylobatidae, the so-called lesser apes. Apes share anatomy to be discussed in later chapters, the most obvious of which is that all apes have shoulders – their torsos are broad and their shoulders are mobile. All apes except humans have arms longer than their legs (Figure 4.19). Apes have the biggest brains of the primates, the longest lives, and the nearest-to-human intellectual abilities.

Gibbons and **siamangs**, the hylobatids, are our most distant relatives, weighing in at 6–12 kg (15–30 lb). They have bizarrely elongated arms, long enough quite literally to drag along the ground. While they are clumsy on the ground, their movement in trees is every bit as graceful as any Olympic gymnast. As they swing by their arms in a locomotor mode we call **brachiation**, they can seem to float in the air, and when they move rapidly their rhythmic propulsion with first one arm then the other looks as much like a balletic dance as anything in nature.

The gibbon face is reduced and pulled under the skull, human-like enough that when *Homo erectus* was discovered some paleontologists, perhaps preferring not to have a species as unsavory looking as *Homo erectus* in the family, protested that it was not a fossil human but only a "giant gibbon." Gibbons are often black with white highlights here and there, though in some species the females are buff-colored, almost white. As human-like as their large braincase and reduced faces make them, when it comes to intelligence they resemble monkeys as much as chimpanzees.

Orangutans (*Pongo pygmaeus*) are found only in the forests of Borneo and Sumatra, and may be teetering on the edge of extinction (Figure 4.20). Females weigh 35 kg (75 lb) and males over twice that 78 kg (175 lb) or more; some males continue to put on fat and muscle well into adulthood. Despite their immense size, they spend the vast majority of their time in the trees. Fully grown adult males develop **cheek flanges** (Figure 4.20), the function of which is the object of lively debate, though whether they evolved to direct calls, fend off biting attacks from other males, or some other purpose we are not certain. The females rarely come to the ground, and males typically come down only for their daily "long jaunt" – long if you call 500 m long

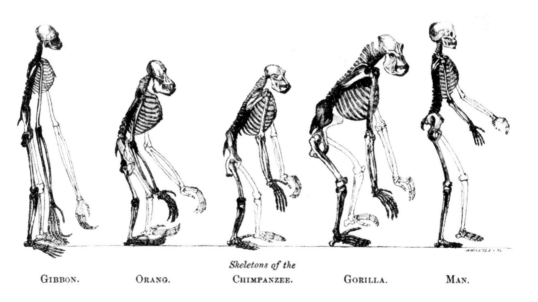

Figure 4.19 T.H. Huxley's famous illustration supporting Darwin's argument that African apes are the closest relatives of humans. Not to scale.

Figure 4.20 Fully flanged adult male orangutan. Photo courtesy of Tim Laman. (A black and white version of this figure will appear in some formats. For the color version, please refer to the plate section.)

Figure 4.21 Adult male gorilla, *Gorilla gorilla*. Photo: Brocken Inaglory.

(MacKinnon, 1974; Rodman & Mitani, 1986). Orangutans are not strong walkers, and in fact their clumsy terrestrial gait is ineffectual enough that they sometimes give up on it altogether and resort to somersaults and rolls. Males are solitary with respect to other males. Dominant males defend a territory, excluding all other males from an area encompassing several females. Males are not terribly attentive to females either, unless they are in estrus.

Gorillas (*Gorilla gorilla*) loom so large in popular culture they hardly need an introduction (Figure 4.21). They are simply enormous; females tip the scales at nearly 80 kg (nearly 200 lb), and males weigh 170 kg (375 lb). If you have ever stood beside a six-and-a-half foot, 160 kg (350 lb) football lineman, you have some sense of their size but gorillas are still more imposing. I once stood behind a silverback at a zoo, thick glass separating me from the gorilla. It was as if a bulldozer had come to life. Despite their immense size, the ones who live in forests can and do spend considerable time in trees (Remis, 1995), because it is there they find their favorite food, ripe fruit. However, mountain gorillas (as opposed to lowland gorillas) are obliged to subsist entirely on piths and other foliage.

Gorillas live in one-male harems, in most cases led by an older male who has typically developed the white back that gives mature males their name. When **Dian Fossey** first worked with them she described them as gentle giants. They are, most of the time.

Other times they are among the most bloody-minded of primates (Fossey, 1984; Harcourt & Stewart, 2007). Males typically acquire females through violence, attacking another breeding group, defeating the resident male, and not infrequently killing one of the infants in the group. This is violent coercion of the most brutal kind, yet within their group males are gentle, affectionate, and tolerant fathers.

Chimpanzees are our closest relative, but the enigmatic bonobo is also our sister species. In other words, the chimpanzee's place in nature is as the closest relative of the bonobo. Their next closest relative is humans, then gorillas, then orangutans. Darwin and Huxley may seem to have intentionally ignored the bonobo or "pygmy chimpanzee," *Pan paniscus*, but in fact they were simply unaware that it was a species separate from the common chimpanzee. Much of this book would be the same if I had chosen to focus on our second sister, though we know a lot less about her than the common chimpanzee.

I still find it extraordinary that in *The Descent of Man* Charles Darwin so accurately foresaw what is now conventional wisdom: "Africa was formerly inhabited by extinct apes closely allied to the gorilla and chimpanzee; and as these two species are now man's

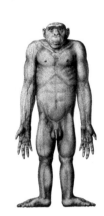
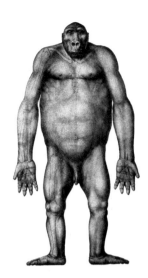

Figure 4.22 A.H. Schultz's comparison of the apes, with the flesh on, from *Life of Primates*.

nearest allies, it is somewhat more probable that our early progenitors lived on the African continent than elsewhere." This is an astonishing prescience, considering that as recently as 2009 some well-respected scientists were arguing that it is orangutans, not chimpanzees or bonobos, that are our sister species (Grehan & Schwartz, 2009). Gene sequencing has now answered this question definitively (well, every latest genetics study is presumed "definitive," as was the one before it, even if it came to a diametrically opposing conclusion). Still, it was Darwin's genius to recognize the truth nearly 150 years before the clinching argument had been put on the table.

It was Darwin's bulldog, T.H. Huxley (Huxley, 1863), who made Darwin's point so trenchantly by illustrating ape skeletons side by side in his brilliant publication, *Man's Place in Nature* (Figure 4.19). A century later, Adolf Schultz (1969) published his illustration of the hominoids with the skin on (but hair off) in a similar way (Figure 4.22). The two illustrations are among the most famous in evolutionary scholarship.

Let me tidy up my argument about the chimpanzee's place in nature by talking about humans. We are so closely related to the chimpanzee that calling ourselves anything other than "apes" is taxonomically wrong. We are, as are orangutans, gorillas, bonobos, and chimpanzees, a member of the family Hominidae, which has inspired authors to bestow their treatises on human nature titles like *The Third Chimpanzee* and *Last Ape Standing*. We are apes, though we are the human apes.

There you have it, our order, the primates. As fascinating as they are, though, we will now turn away from these non-*Pan* relatives and spend the rest of this book focusing on our sister species, the chimpanzee.

References

Bearder SK (1987) Lorises, bushbabies, and tarsiers: diverse societies in solitary foragers. In *Primate Societies* (eds. Smuts BB, Cheney DL, Seyfarth RM, Wrangham RW, Struhsaker TT), pp. 11–24. Chicago, IL: University of Chicago Press.

Bloch JI, Silcox MT, Boyer DM, Sargis EJ (2007) New Paleocene skeletons and the relationship of plesiadapiforms to crown-clade primates. *PNAS* **104**, 1159–1164.

Dammhahn M, Kappeler PM (2005) Social system of *Microcebus berthae*, the world's smallest primate. *Int J Primatol* 26, 407–435.

Dunbar RIM (1998) The social brain hypothesis. *Brain* 9, 178–190.

Erickson CJ (1991) Percussive foraging in the aye-aye, *Daubentonia madagascarensis*. *Anim Behav* 41, 793–801.

Fossey D (1984) Infanticide in mountain gorillas (*Gorilla gorilla beringei*) with comparative notes on chimpanzees. In *Infanticide: Comparative and Evolutionary Perspectives* (ed. Hausfater G, Hrdy SB), pp. 217–236. Hawthorne: Aldine.

Grehan JR, Schwartz JH (2009) Evolution of the second orangutan: phylogeny and biogeography of hominid origins. *J Biogeography* 36, 1823–1844.

Groves C (2001) *Primate Taxonomy*. Washington, DC: Smithsonian Institution Press.

Groves C, Shekelle M (2010) The genera and species of Tarsiidae. *Int J Primatol* 31, 1071–1082.

Harcourt AH, Stewart KJ (2007) *Gorilla Society: Conflict, Compromise, and Cooperation between the Sexes*. Chicago: University of Chicago Press.

Hartig G, Churakov G, Warren WC, et al. (2013) Retrophylogenomics place tarsiers on the evolutionary branch of anthropoids. *Sci Rep* 3, 1756. DOI: 10.1038/srep01756.

Hill WCO (1953–1970) *Primates: Comparative Anatomy and Taxonomy*, 8 vols. Edinburg: Edinburgh University Press.

Huxley TH (1863) *Evidence as to Man's Place in Nature*. London: Williams and Norgate.

Janečka JE, Miller W, Pringle TH, et al. (2007) Molecular and genomic data identify the closest living relative of primates. *Science*, 318, 792–794.

Jolly A (1966) *Lemur Behavior*. Chicago, IL: University of Chicago Press.

Larson SG (1998) Parallel evolution in the hominoid trunk and forelimb. *Evol Anthropol* 6, 87–99.

Linnaeus CV (1758) *Systema Naturae*, 10th ed., Vol. 1. Stockholm: L Salvii.

MacKinnon JR (1974) Behavior and ecology of wild orangutans (*Pongo pygmaeus*). *Animal Behavior* 22, 3–74.

MacKinnon JR, MacKinnon KS (1980) Niche differentiation in a primate community. In *Malayan Forest Primates* (ed. Chivers DJ), pp. 167–190. New York: Plenum Press.

Milton K (1981) Food choice and digestive strategies of two sympatric primate species. *Am Nat* 117, 495–505.

Mogicato G, Raharison F, Ravakarivelo M, Sautet J (2012) Normal nasal cavity and paranasal sinuses in brown lemurs *Eulemur fulvus*: computed tomography and cross-sectional anatomy. *J Med Primatol* 41, 256–265.

Murphy WJ, Eizirik E, Johnson WE, et al. (2001) Molecular phylogenetics and the origins of placental mammals. *Nature* 409, 614.

Napier JR (1980) *Hands*. London: George Allen & Unwin.

Napier JR, Napier PH (1967) *A Handbook of Living Primates*. London: Academic Press.

Napier JR, Walker AC (1967) Vertical clinging and leaping: A newly recognized category of primate locomotion. *Folia Primatologica* 6, 204–219.

Nash LT, Bearder SK, Olson TR (1989) Synopsis of *Galago* species characteristics. *Int J Primatol* 10, 57–80.

Nekaris NAI, Bearder SK (2011) The lorisiform primates of Asia and mainland Africa: diversity shrouded in darkness. In *Primates in Perspective*, 2nd ed. (eds. Campbell CJ, Fuentes A, MacKinnon KC, Bearder SK, Stumpf RM), pp. 24–45. Oxford: Oxford University Press.

Nieuwenhuys R, Ten Donkelaar HJ, Nicholson C (1998) *The Central Nervous System of Vertebrates*, Vol. 3. Berlin: Springer.

Perelman P, Johnson WE, Roos C, et al. (2011) A molecular phylogeny of living primates. *PLoS Genetics* 7. DOI: 10.1371/journal.pgen.1001342.

Perry S, Panger M, Rose LM, et al. (2003) Traditions in wild white-faced capuchin monkeys. In *The Biology of Traditions: Models and Evidence* (eds. Fragaszy D, Perry S), pp. 391–425. Cambridge: Cambridge University Press.

Pettigrew JD (1991) Wings or brain? Convergent evolution in the origins of bats. *Systematic Zoology* 40, 199–216.

Remis M (1995) The effects of body size and social context on the arboreal activities of lowland gorillas in the Central African Republic. *Am J Phys Anth* 97, 413–433.

Robinson JG, Janson JG (1987) Capuchins, squirrel monkeys, and atelines: socioecological convergence with Old World Primates. In *Primate Societies* (eds. Smuts BB, Cheney DL, Seyfarth RM, Wrangham RW, Struhsaker TT), pp. 69–82. Chicago, IL: University of Chicago Press.

Rodman PS, Mitani JC (1986) Social systems and sexual dimorphism of orangutans. In *Primate Societies* (eds Smuts BB, Cheney DL, Seyfarth RM, Wrangham RW, Struhsaker TT), pp. 146–154. Chicago, IL: University of Chicago Press.

Ross CF (1995) Muscular and osseous anatomy of the primate anterior temporal fossa and the functions of the postorbital septum. *Am J Phys Anthropol* 98, 275–306.

Schultz AH (1969) *The Life of Primates*. New York: Universe Books.

Shoshani J, Groves CP, Simons EL, Gunnell GF (1996) Primate phylogeny: morphological vs molecular results. *Molec Phylogenet Evol* 5, 102–154.

Smith RJ, Jungers WL (1997) Body mass in comparative primatology. *J Hum Evol* 32, 523–559.

Stephan H, Frahm H, Baron G (1981) New and revised data on volumes of brain structures in insectivores and primates. *Folia Primatol* 35, 1–29.

Sterling EJ (1994) Aye-ayes: specialists on structurally defended resources. *Folia Primatol* 62, 142–154.

Wilson DE, Reeder DM (eds) (2005) *Mammal Species of the World*, 3rd ed. Baltimore, MD: Johns Hopkins University Press.

5 Scratching Out a Living in an Unforgiving World
Habitat and Diet

Photo credit: Caro Deimel.

PART I: HABITAT

5.1 The Chimpanzee World

In captivity, chimpanzees seem … inadequate: confused; incompetent; to be pitied. Even when happy in captivity, by which I mean not driven to psychosis by solitary confinement or boredom, they seem out of their depth, lost in a human world that is complex beyond their imagining.

In the wild they are masters. They are purposeful, intent, and capable in the face of almost any challenge. They appear to be all-knowing when it comes to finding food and choosing the most efficient route between feeding trees. Rather than searching for food, they seem to know exactly where it is amid the thousands of species of trees, bushes, and ferns, most of which are useless to them as food.

We humans view the world as having depth and breadth: two-dimensional. Many experts believe we evolved in the savanna, where mental mapping calculations involved only horizontal distances but not the third dimension of elevation. So firmly are our feet (and minds) on the ground that our metaphor for any easily harvested resource is the proverbial low-hanging fruit. While chimpanzees are found in diverse habitats, including savannas (McGrew, 1974; McGrew et al., 1981; Hunt & McGrew, 2002), their most common habitat is forest, typically tropical rainforest, but they are also found in seasonal forests, higher-altitude forests, and forests confined to watercourses (riparian or riverine forests).

Chimpanzees cannot be two-dimensional, they must look up as much as into the distance; they live in a three-dimensional world where much of their food, and certainly their most important foods, are 50 meters or more in the air. They are forest-adapted creatures, at home in the sky, a world most humans know nothing about.

5.2 Habitat Types

Tropical rainforests are species-dense. While covering only 7 percent of the Earth's landmasses, they are home to half of the Earth's species (Whitmore, 1990). It is estimated that there are 170,000 species of flowering plants in the tropics, and 35,000 in mainland Africa alone. North America, in contrast, has only 171 species, one-tenth of a percent (0.1 percent) of the total in Africa. Europe is even less speciose, with one-third of the species of North America (Whitmore, 1990). In Southeast Asia, a single small slice of forest in Malaya, approximately the size of 50 football fields, was found to have more species than North America and Europe – 830 tree species.

Until recently, forests were often viewed as dark and dangerous jungles, a word derived from the Hindi word for dense foliage, "jangal." Jungles are now more typically referred to as rainforests. In the popular imagination, Africa is envisioned as coast-to-coast "jungle," but in fact **rainforest** covers only 6 percent of the African landmass, 1.8 million of 30.3 million square kilometers. Perhaps surprisingly, Africa is less forested than either the Americas or Southeast Asia.

The African forest is centered on Congo and peters out as you move from that center either to the west, beyond the Bay of Guinea (that 90° inflection linking the bulge of West Africa to Central Africa; Figure 5.1), or to the east, into East Africa. There are patchy forests in West Africa, or at least in the southern part, though there is a break called the Dahomey Gap, which leaves a patch of tropical rainforest isolated in westernmost Africa.

Rainforests are also referred to as **closed-canopy evergreen forests** (Table 5.1; the review below relies on Ellenberg & Mueller-Dombois, 1967; Greenway, 1973; White, 1983). That said, other than specialists, scholars are notoriously loose about terminology when it comes to forest types. When you read "rainforest," "evergreen forest," "closed-canopy forest" or sometimes just "forest," most academics mean *closed-canopy evergreen forest*. If you are familiar only with temperate climates, you might take evergreen to mean "trees bearing needle-like leaves," but in the tropics broadleaf trees seldom drop their leaves and are therefore evergreens. Such forests are "closed" because sunlight rarely penetrates to the forest floor. Viewed from the sky, the forest canopy covers 95 percent or more of the surface area, with a few patches here and there opened up by tree falls. Other areas are open because they are rocky or are created by animal activity. The tree crowns touch (are "interlocking"). These are rainforests because they are not dependent on river or lake water, but receive enough water from rain to maintain their substantial biomass.

Figure 5.1 African forests, in dark green. The chain of lakes in the east, where the dark green segues to lighter green, is the border of East Africa. West of this area, forest exists only in patches. Image from NASA. (A black and white version of this figure will appear in some formats. For the color version, please refer to the plate section.)

Tree crown height can exceed 50 m. The trunks of the larger trees in these forests are more massive than many of us from temperate areas would expect. In an old-growth forest – one that has been spared logging – tree trunk diameters of the largest trees average 4.5 m (15 ft) in diameter but the base of a truly large tree can be the size of a house. Tropical trees (Figure 5.2) often have **buttresses**, triangular, wing-like reinforcements that chimpanzees bang like drums during social displays. Draped from these towering trees are woody vines, also called **lianas** or **climbers**, some of them the diameter of trees. On the boles and in other nooks and crannies in the tree stem are found **epiphytes**, plants that live on trees but are not parasitic – they derive nutrients from air, rain, and detritus that accumulates around their roots.

Closed-canopy forests have a layered structure (Figure 5.2), traditionally divided into four elements, an **emergent layer** that consists of sporadically placed crowns at a height of 37 m and above, a denser, continuous **main canopy** from 24–36 m, a **lower canopy** layer consisting of shade-tolerant trees that rise to approximately 23 m above the forest floor, and the **ground layer** or forest floor.

The forest floor in a true, closed-canopy forest has sparse herbaceous vegetation (leafy plants, the stem of which is not woody), epiphytes, and ferns, but few trees, since too little light reaches the ground to

Table 5.1 **Biomes (vegetation) types used by chimpanzees**

	Canopy height (m)	Canopy structure	Forest cover (%)	Rainfall (mm)	Grass (%)
Closed-canopy rainforest	50	Layered	95–100	1800–3500	0
Riverine forest	5–50	Irregular	<95	Variable	0–5
Seasonal tropical forest	40	Less layered	70–95	1000–2500	5–30
Woodland	5–25	Unstructured	30–70	900–1200	30–70
Savanna (grassland)	0.5–15	Single trees	≤30	500–100	>70

Figure 5.2 Cross-section of a tropical rainforest. See Richards, 1952 and Whitmore, 1990. Image by the author.

nurture them. The ground is, of course, covered with dead vegetation: leaves, shed branches, decomposing vegetation, and other detritus. In forests with shorter and therefore less light-blocking canopies, there may be a significant herb layer, but grasses are virtually unknown. Although individual trees may drop leaves when distressed, the canopy is never completely without foliage. Forests on the taller, denser side have few vines and few palm trees, while the somewhat more open seasonal forests see a reduction in foliage density during the dry season and have more vines and palms.

Rainforests are consistently wet, warm, and unvarying. No individual month averages less than 18°C (64°F); average monthly temperatures vary by 5°C or less and rain is plentiful year-round at ≥ 150 mm per month and ≥ 1800 mm per year.

Trees get progressively shorter and species numbers decline as elevation increases; classic, tall-canopy rainforests are confined to altitudes of less than 1200 m (4000 ft).

Tropical forest trees typically flower about once per year, and so of course the fruit crop that develops from the fertilized blossoms is also annual. Figs, on the other hand, have no seasonal pattern and thus provide chimpanzees and other primates with a reliable nutrient-source year-round (Wrangham et al., 1993).

Among chimpanzee study sites familiar to you, this type of dense forest is found at Taï and Goulago. While Taï is typical chimpanzee habitat, similar to that covering all of Congo north of the Congo River and south of the Sahara, most forests where chimpanzees are well-studied are seasonal.

Riverine or *gallery* forests are like rainforests, but they depend on groundwater, most often watercourses, but occasionally water that pools in low-lying areas. They are closed – have interlocking crowns – but the canopy is less dense than closed-canopy evergreen forests. Trees range from 5 m to 50 m tall, substantially shorter than closed forests; species diversity is reduced. Trees often have buttresses. While the canopy is never without foliage, individual trees may shed leaves. Undergrowth is variable, but there are more vines and palms than in closed forests. This and other more open forests often have a more substantial ground layer than rainforest and may be dense with ferns and herbs, depending on light and soil moisture.

Seasonal tropical forest or **semi-evergreen** forests have either seasonal but still abundant rainfall or less rainfall than closed-canopy rainforests. The trees in these climatic zones are slightly shorter than the rainforest trees, and the canopy is slightly more open, with trees covering 70–95 percent of the ground's surface (Table 5.2). Species abundance is also slightly lower. Lianas are more abundant and ground cover is slightly thicker. The open canopy means that ground cover can be dense, particularly in low-lying areas that collect water. At Mahale, undergrowth is nearly impenetrable in some places, and visibility is about 2 m; at times there, I was nearly close enough to touch a chimpanzee before I saw it.

The seasonality of the rainfall in these forests means there is a wet season and a dry season. In India, the predictability of the rainy season (monsoon season) is much greater than in seasonal forests in Africa, but the difference between the seasons is still profound; some months the sun virtually never shines and it rains every day; other months there is no rain at all and near-constant sunshine. Variation in rainfall means that in seasonal forests there is also a rhythm to the abundance of fruit, with the rain eliciting flowering and then fruiting, yielding an increase in food supplies at the end of the rainy season or the beginning of the dry season. This forest type, seasonal forest, is also prime chimpanzee habitat, and characteristic of East African chimpanzee study sites such as Gombe and Mahale, though both are also interspersed with woodland and riverine forests.

Woodland forests are "open" compared to closed-canopy evergreen forests. Woody plants may cover as little as 30 percent or up to 95 percent of the ground surface area, but 70 percent coverage is a good mental image to hold onto. Crowns may touch but are not interlocking. Because the trees are shorter, predominantly ≥ 5 m to 25 m, the canopy layer is shallow and often irregular, so that even where canopy cover approaches 95 percent, enough light reaches the forest floor to nourish grass and other ground layer foliage. Drier woodlands often have succulents and forbs (plants with thick, moisture-retaining leaves); many trees also bear tough, thickened leaves. Trees may have thorns. Deciduous trees are variably common, and more common where rainfall is more seasonal and/or below 900 mm (White, 1983: 90). Epiphytes are rarer. Chimpanzees live in these forests, but more often when there are riverine or closed-canopy forests nearby to form part of their range, as at Gombe and Mahale.

In some ways, these are the most interesting forests, because in drier habitats you can stand near a water source and feel you are in the midst of a rainforest, yet as you walk away from the water, the forest dwindles first to a woodland-like habitat with more sky visible and a more substantial ground layer, then to a wooded savanna, then to a grassy savanna. Those of you who have worked in such a habitat know about this distinctive riverine forest feature: Just before you pass into the grassland you encounter a few meters of thorny plants that seem reluctant to allow you to leave the forest. This is the habitat where I work, at Semliki; at Gombe and Mahale some forest cover falls under this category.

Savanna or **grassland** is dominated by grasses, and woody plants cover less than 30 percent of the ground. Trees in savanna habitats may be deciduous. Tree crowns rarely or never touch. Trees are typically 2 m and up to 15 m, but they are variable in height and may be as short as 0.5 m. Trees are often fire-resistant, recovering in days from grass fires. Palms are variably common. Bare soil is uncommon. Tree leaves and stems may trend toward succulence in drier or more seasonal areas. Tree thorns are common.

No chimpanzee population lives in the savanna exclusively, but relies on nearby woodland areas for critical resources. Chimpanzees at Semliki range into this biome to feed on *Grewia* fruits, which they often harvest bipedally.

Scrubland, also known as **scrub forest** or **bush**, is a habitat with very short – waist-high – bush that are so dense their crowns often interdigitate. I have never seen chimpanzees in this biome even though it is common in the northern part of my field site, and I doubt any fossil hominin ever used it, either.

5.3 Woody Plants

Perhaps because we humans are such different life forms from plants, few consider the battle that trees fight to survive and reproduce. Trees are assaulted by enemies on all sides, and unlike animals they cannot flee to escape. Folivores – leaf-eating animals – ranging from ants to antelopes to colobus monkeys eat the main energy factories of trees, their leaves. Their reproductive organs, flowers, are also targeted. Suffering from these daily attacks occasionally things get much worse; a dense outbreak of predatory insects may attack a tree's fruits, perhaps just damaging a few, but at times infesting the entire fruit crop and leaving fruits inedible for animals that might typically eat it. At times a particular tree may be plagued by literally millions of caterpillars or other early-life-stage insects that can consume every leaf in their canopy. Vampires, at least bloodsuckers from the perspective of the trees, may gnaw through their "skin" and drink their sap, as many primate gummivores, such as marmosets, do. Microscopic fungi and bacteria can feast on just about every one of their body parts.

Trees have a defense. They strengthen themselves with woody armor or thicken their leaves with fibrous **cellulose**, an indigestible string of sugars, indigestible because the sugars are bonded so tightly that mammalian digestion cannot break them apart. They grow chemicals in their various body parts that deter predators – more about that in the next chapter. When their fruit crop – not a permanent part of the tree – is under assault, they are capable of canceling the whole operation, ceasing to grow fruits and dropping them; they abort the fruit crop. Richard Wrangham and I once stood in the shadowy darkness of an immense tree the fruits of which were infected, a crop we had estimated at more than one million fruits. We stood, mesmerized, listening to the din of the falling fruits, a thunderous cacophony that sounded like a torrential rain pounding on a tin roof.

Trees can be large enough in old-growth forests that even horizontal boughs are huge in human terms – once when I climbed a tree to get a photo, I found myself able to walk hands-free on a branch the size of the telephone pole. The world looks different from up there. We primatologists are used to staring up at chimpanzees, often with the bright sky as the background, leaving the chimpanzee you are trying to observe a mere featureless black silhouette. When perched high on a branch, the light is shining the other way, down onto the ground below, and the scene is lit up like a stage. No wonder chimpanzees always seem to see us coming before we see them.

The habitat classification in Table 5.1 is derived from perhaps the most cited work on the subject (White, 1983), but in fact scientists have never expressly agreed on a standardized terminology. Confusion abounds. I have seen the term "wooded savanna" used to mean savanna grading toward that upper limit of 30 percent tree cover. Experts would say either savanna or woodland, without the qualification. The term **woodland**, which I hope you will remember is a landscape with **tree cover of 30–70 percent**, seems particularly prone to misunderstandings; paleontologists are frustratingly fond of conflating woodland and forest. Thus, one might see the oxymoron "closed-canopy woodland," which is only slightly less misguided than "closed-canopy desert."

PART II: DIET AND FORAGING

5.4 Optimal Foraging

Whether forest, woodland or savanna, what chimpanzees most need to know is "where is the food." There was a time when quantitatively minded researchers held out hopes that **optimal foraging theory** (Emlen, 1966; MacArthur & Pianka, 1966; Schoener, 1971; Krebs, 1973; Charnov, 1976; Pyke et al., 1977) would provide us with a mathematically precise understanding of how animals in general and chimpanzees in particular go about finding food. Theorists wondered whether there might be some optimal search technique that would allow an animal to choose a foraging path that caused them to encounter the maximum number of food items. We might be able to quantitatively compare the behavior of a wild animal with an optimal strategy, one that might maximize "return" (calories acquired after subtracting the energy burned collecting it), after budgeting a little bit of time for social demands

(Schoener, 1971) and some few of life's other requirements.

To simplify the task, theorists thought we might start with the assumption that food items are distributed evenly and randomly. The simplest sort of search we can imagine might entail a forager walking in a relatively straight line, gathering whatever edible items it encountered, continuing until the edge of the gatherer's territory is encountered, then shifting an arm's length to one side, doing an about-face, and traversing the range again; repeat until the **home range** (an area in which an individual is most likely to be found) is surveyed. Or would a slightly more complicated zig-zag pattern be better? A chimpanzee-sized searcher might walk 10 meters, turn right and walk 10 meters, turn left and walk 10 meters, and so on, thoroughly covering ground in a wide swath. Or perhaps a completely random walk, a Brownian motion strategy, might be best? Mathematical modeling might predict an optimal strategy, which could then be tested against the behavior of real animals. As a general rule, animals are thought to forage optimally when they maximize their energy intake while minimizing the time they expend finding food (Schoener, 1971), so their foraging path should accomplish this for them.

The task theorists faced was to discover the maximum number of items encountered with the minimum distance traveled. Any one of many foraging paths might accomplish this. The best strategy would vary according to the distribution of food and factors inherent in the forager, such as reach, vision, olfactory acuity, body size, and so on. Of course, many of the simplifying assumptions made to make the modeling quantitative might be violated, but the idea was that we would start simple and add complexity as called for.

5.5 Chimpanzees Outsmart Optimal Foraging

The models built up from all these assumptions, so promising in prospect, turned out to have only equivocal success. Optimal foraging models do work for herbivores because grass is relatively evenly

Table 5.2 Ape versus monkey diets (percentage time feeding)

Site	Fruit	Piths, stems, roots	Leaf	Insects	Meat	Flowers	Bark	Other
Kibale chimpanzee[a]	79.0	16.9	2.6	0.0	0.9	~0.2	~0.2	0.4
Gombe chimpanzee[b]	63.3	16.5	16.6	3.4	0.3	0.0	0.2	0.0
Mahale chimpanzee[c]	56.7	19.1	10.6	5.9	0.9	0.1	0.2	6.0
Budongo chimpanzee[d]	64.5	3.2	19.7	0.0	–	8.8	–	–
Ngogo chimpanzee[e]	70.7	2.5	19.0	0.0	2.0[g]	2.6	0.0	3.6
Fongoli chimpanzee[f]	60.8	1.4	4.3	23.7	0.6	5.5	2.3	1.6
Chimpanzee average	65.8	9.9	12.1	5.5	0.8	2.9	0.8	2.3
Redtail monkey[h]	43.6	0.0	15.5	21.8	0.0	14.9	0.0	4.5
Blue monkey[i]	42.7	0.0	18.7	19.8	0.0	12.5	0.0	5.9
Mangabey[j]	58.8	0.0	5.3	10.9	0.0	5.9	0.0	4.0
Monkey average	48.4	0.0	13.2	17.5	0.0	11.1	0.0	4.8

[a] Kibale-Kanyawara; Wrangham et al., 1996; flowers and bark estimated. [b] Hunt, 1989. [c] Hunt, 1989. [d] Newton-Fisher, 1999. [e] Potts et al., 2011; Watts et al., 2012. [f] Estimated. [g] Bogart & Pruetz, 2011. [h] *Cercopithecus ascanius*; Struhsaker, 1978. [i] *Cercopithecus mitis*; Struhsaker, 1978. [j] *Lophocebus albigena*; Struhsaker, 1978.

distributed (Gross et al., 1995; Lima et al., 2001); but herbivores may be a special case because the foods of most animals are clumped. In short, after two decades of thoughtful, sometimes brilliant work, an influential review bemoaned the fact that only 6.5 percent of wild animal studies showed any sort of reasonable agreement with optimal foraging predictions (Stephens & Krebs, 1986). To rub salt in the wound, a highly cited work in the late 1980s was titled "eight reasons why optimal foraging theory is a complete waste of time" (Pierce & Ollason, 1987). Ouch.

By the late 1980s it had become apparent that the random aspect of the theory – the idea that animals might forage using a predetermined foraging path that maximized encounters with resources – hardly applies to mammals. Mammals have large brains and good memories, and therefore nearly complete information about the size and location resources (Perry & Pianka, 1997). In other words, rather than searching for food randomly, they walk directly from one feeding site to another. This is not "searching."

Some mammals work this omniscience gig better than others. Primates, of course, are in the "better" category.

5.6 Chimpanzee Diets

Chimpanzees are frugivores, a fancy word for fruit-eating. Two-thirds (66 percent; see Table 5.2) of the time chimpanzees spend gathering food is dedicated to getting fruit. They are efficient enough fruit harvesters that by weight fruit comprises over three-quarters of their intake. When fruit is ripe, chimpanzees eat it enthusiastically. They are gourmands. No hipster foodie obsesses over their food more than chimpanzees; as they feed on delectable foods they emit a feeding vocalization that at times sounds like a growlier version of the appreciative sound humans make to express their satisfaction with a particularly tasty food, "mmm, mmm."

Table 5.3 **Nutrients in fruit, by percentage of dry weight**

	Lipid	Protein	Sugars	Fiber
Ripe fruit	4.9	9.5	13.9	33.6
Pith	1.2	9.6	11.6	44.1
Domesticated fruit	2.6	5.1	42.8	10.6

Source: after Conklin-Brittain et al., 2002.

Like humans, who detect and adore digestible sugars from the moment of birth (Steiner, 1979; Rosenstein & Oster, 1988; Ventura & Mennella, 2011), chimpanzees are sugar fiends (Kellogg & Kellogg, 1933; Chapter 13). Chimpanzees are so fond of sugar that in captivity they can be manipulated into almost any painful medical procedure by offering them sugar as a reward. The evolution of this craving is easy to understand. Chimpanzees who craved sugar ate more calories, grew to breeding age faster, lived more often through droughts, were healthier, attained higher ranks, passed on more nutrients to their offspring via nursing, and had more surviving offspring than individuals who could take or leave sugar. They passed on their genes to future generations, including genes for detecting and craving sugar. While honey is the sweetest thing in the chimpanzee diet, on a daily basis ripe fruit is the food item with the most sugar (Table 5.3). As precious as the sugar molecule is, its constituents are utterly common: 24 atoms of carbon, hydrogen, and oxygen. Note that sugar is not the only important nutrient in wild fruits; while domesticated fruits have much more sugar, wild fruits have more fat and more protein. They also have more undesirable fiber.

While they may appreciate a particularly delightful food, chimpanzees rarely take time to savor it. When harvesting fruits, they are quick, efficient, and graceful. For foods that are relatively uniform in quality they are almost machine-like, planning their actions so that they reach for the next food item the moment the last has been delivered to the mouth, reaching out with a perfectly directed motion and without hesitation. In their economy of motion and precision they remind me of highly skilled, veteran factory workers, athletes in their own way who perform their complicated tasks with an elegant ease. What is only obvious to the experienced observer is that amid this elegant action, they are choosing the most nutritious, sugar-rich fruits as they feed, leaving behind more items than they pick.

For some species of fruits, often the larger ones, harvesting is slow, since gathering must be paced to the time it takes to chew and swallow a large item; here, chimpanzees are able to take the time to note the color or softness of individual fruits as they gather them, smelling or squeezing a fruit before plucking it (Goodall, 1986). Sometimes fruits are bitten into, dropped, and the bitten off part spat out. Presumably their keen sense of taste has told them these fruits are disappointing in sugar content.

The chimpanzee foraging strategy is memory-rich, thoughtful, flexible, and thorough, much more so than is the case for monkeys. They remember the location of food items; they consider time as they weigh whether a tree has enough ripe fruit to bother visiting; they change their plan if they discover a particular fruit is riper or more abundant than expected and they are very, very good at scouring the landscape for sugar-rich food items (Normand & Boesch, 2009; Normand et al., 2009). Timing is everything, and not only with fruit ripeness. Some seeds are edible only when the fruit is unripe, and thus fruits must be wrenched opened and the seed extracted.

Chimpanzees are particularly egregious foraging theory rule-breakers. Instead of obeying a rule-based searching strategy, hoping to encounter randomly dispersed resources as effectively as possible, chimpanzees were found to move in relatively straight lines and at high speeds (for chimpanzees), after which they spent long periods of time, an hour or two, feeding in one spot. Why? They know where the food is and they want to get there quickly. They were seen to travel farther when they intended to feed for a longer period of time; when they moved on to a new feeding site, they took minimum-distance paths between food resources (Menzel, 1973; Normand & Boesch, 2009; Janson, 2014).

The problem is that primates in general and ripe-fruit specialists in particular feed on items that are widely dispersed and highly concentrated (Boyer et al., 2006). In other words, as you have already figured out, chimpanzees "search" for food in a way that suggests they are not searching at all but have near-perfect knowledge of food resource locations (Wrangham, 1977; Janmaat et al., 2013).

Even when a mathematical model can be used to *describe* the movements of primates such as chimpanzees, such as the "thick tail" Levy walk model, which finds that once they set out humans and spider monkeys travel longer distances than expected at random (Ramos-Fernández et al., 2004; Lee et al., 2008; Smouse, 2010), the models are merely descriptive rather than predictive. They tell us what we already knew: Ripe-fruit specialists pass up most potential food items (e.g., leaves, flowers, unripe fruits) to feed on rare and nutritious ripe fruits.

We assume that chimpanzees have such excellent spatial memories to increase foraging efficiency. In fact, as we will learn in greater detail in Chapter 18, in trial after trial where individuals were rewarded for correctly recalling the location of objects in space, chimpanzees either equaled or exceeded humans (Tinklepaugh, 1932; Matsuzawa, 1985, 2003; Dolins et al., 2014). This laboratory-discovered cognitive ability plays out in the wild in the discovery that chimpanzees remember the location of large fruit trees precisely – many of us believe they remember particularly pleasant feeding experiences for years (Janmaat et al., 2013) – and they just go to known feeding sites rather than searching.

In short, chimpanzees almost never search for food, but instead their daily foraging path is one of an animal that knows exactly where the food is (Goodall, 1968; Wrangham, 1977; Normand & Boesch, 2009; Janson, 2014). Janmaat and colleagues, in fact, were able to demonstrate that chimpanzees anticipate their next meal long before they can actually see it (Janmaat et al., 2013, 2014). In truth, all of us chimpologists knew that because as chimpanzees approach a tree with particularly appealing fruits, they begin to food-grunt and even scream in happiness, sometimes long before they reach the feeding site, walking faster and sometimes breaking into a run as they near the food.

Chimpanzees plan their foraging not merely on a minute-to-minute or hour-to-hour basis, but in some cases *days* in advance. It was Richard Wrangham who noticed that chimpanzees sometimes approach a tree and spend a minute or so gazing into the crown – only to walk away without gathering a single fruit. They have a plan. A few days later when they return to the same tree they climb up without bothering to look to see how ripe the fruit is (Wrangham, 1977; Goodall, 1986). In their first look at the tree they judged the fruit would be ripe enough to harvest in only a few days and were so confident in their judgment that they felt no need to check the second time. Chimpanzees also feed from termite mounds month after month; females may feed from a particular mound many days in a row, approaching the mound from all directions (Goodall, 1986), suggesting they know exactly where it is. Their mental mapping is excellent. Not infrequently, they pause to prepare termiting tools even before the mound is in sight (Goodall, 1986), telling us they know exactly where the mound is. This is an ability few animals possess, and such skillful mental mapping is a challenge even for humans.

5.7 Chimpanzee Foraging Strategy

Chimpanzees combine their excellent memory for the location of food resources (Normand & Boesch, 2009; Normand et al., 2009) with outstanding skills as naturalists; they are excellent botanists (Wrangham, 1977), meaning they recognize an edible food when they see it. Identifying the proper food item is no small task, not even for humans; most readers would find it challenging to identify 50 or 100 species of trees, while also committing to memory which plant part is edible, as well – leaf, flower, seed, or fruit. For some species, blossoms are edible but nothing else; for others only the bark can be eaten; for still others new leaves and fruits, but nothing else. Their food list is surprisingly large. At Mahale they eat over 300 food items, at Gombe 200 (to be precise, 328 and 203; Nishida et al., 1983) and more than 100 at Kibale-Kanyawara (Wrangham et al., 1996) and Budongo (Newton-Fisher, 1999).

While chimpanzees have a long list of food items, this is not because they are not discriminating; they are very careful about exactly which items they eat, rejecting more than they consume. At Kibale-Kanyawara, chimpanzees fed from 35 species of trees, but the forest contains hundreds (Eilu et al., 2004). They pick and choose among the available items so that most of their food lacks two antifeedants that we will learn much more about in the next chapter – fiber and secondary compounds. After passing over two-thirds of the species of fruits in the forest, chimpanzees then pass over the majority of the fruit crop of species they do consume, attempting to gather fruits only when they are ripe.

Getting enough food is the principle worry of any chimpanzee on a typical day, and by far their most time-consuming activity. On an average day, between walking among food sources, climbing trees, maneuvering within trees, swinging between feeding sites, picking the actual fruits, scraping off the thick husks found on some fruits, peeling, stripping, opening fruits and chewing, six or seven hours is taken up in food-gathering – half of the chimpanzee active period (Goodall, 1968, 1986; Wrangham, 1977; Boesch & Boesch-Achermann, 2000). While we humans may linger over a meal for an hour or even two, few of us spend nearly an entire workday processing and eating food.

While the chimpanzee food list is long, on any one day it is rather monotonous. The bulk of their daily intake consists of two or three items (a flower, leaf, fruit, or seed of a single species would be considered four separate food items) and the total is only about 15 food items (Goodall, 1968, 1986; Wrangham, 1977). As days pass, new items are introduced and other items are dropped out as they fall out of season or are depleted.

In a summary of feeding at both Gombe and Mahale, a typical feeding bout lasted a little over 10 minutes, though occasionally a bout stretched to an hour and rarely to two hours (Nishida et al., 1983; Pruetz & McGrew, 2001). Males at Gombe have been found to break up their day into an average of 38 distinct feeding bouts, most of which (63 percent) last less than five minutes (Wrangham, 1977; Pruetz & McGrew, 2001).

I may have impressed you with chimpanzee efficiency so far, but I have hardly begun. They locate their sleeping nests having already planned out what they will eat for breakfast the next day (Janmaat et al., 2014); they plan their daily route to take advantage of fatty, high-calorie fruits (Ban et al., 2016); they keep track of the fruiting history of specific trees (Janmaat et al., 2013), and the better the fruit, the farther away they plan their foraging path to visit it (Ban et al., 2014). In short, chimpanzees use their intelligence to locate food efficiently and to buffer fluctuations in resources; by making optimal use of their time and the resources available, they experience the forest as more bountiful and more stable than it really is (Janmaat et al., 2016).

Nature gives of its bounty only reluctantly. Different food items require different food processing, and some are labor-intensive. *Myrianthus holstii* is a fruit that resembles a pineapple (and is in fact a relative). The individual fruit segments resemble a giant corn kernel, arrayed on a fat corncob. The individual segments must be wedged off with a finger rather than eaten like corn because only the interior side can be eaten. The outside is hairy, rough, and poisonous; I can say from experience that if it contacts your lips it is painful, itchy, and numbing.

Other fruits such as *Saba florida* have a centimeter-thick, tough husk that must be bitten into and torn off, a task I have never seen a human accomplish, but which chimpanzee do as a human might biting into an orange. Other fruits are grape-like; the whole fruit goes in mouth, but the skin is separated and spat out. As chimpanzees are manipulating a food item in their mouth they may push it onto their lower lip, extend the lip, and visually examine the partly processed item before sucking it back in the mouth to be finished off. Some seeds, such as *Diplorynchus condylocarpon*, are found in pods and attached to maple-seed-like wings. The pod must be opened and then the seeds bitten off.

Not infrequently at the end of a bout of fruit eating an individual will fill his or her mouth to the point of bursting and suck out the juice as they walk to their next feeding site; after extracting all the sugar they spit out a compressed, seven cm-long **wadge**, a sucked-dry chunk of seeds and fiber, looking like a little miniature brain. Many times I have found chimpanzees after discovering such a wadge and intensifying my search around it.

Much of the chimpanzee non-fruit protein intake comes from the 12 percent of their diet (Table 5.3) that is leaves (Wrangham et al., 1998). While pith is eaten only when fruit is scarce, and makes up for low sugar intake by substituting complex carbohydrates, leaves are eaten regardless of how much fruit is in the diet (Wrangham et al., 1998; Conklin-Brittain et al., 1999). They are eaten more commonly by pregnant and nursing females (Cook & Hunt, 1999) who need protein, but they are also eaten for their calorie content as a low-quality fallback food when ripe fruit is unavailable (Nishida, 1976; Goodall, 1986; Harrison & Marshall, 2011). Bark is another near-desperation food. The softer inside layer may contain significant concentrations of sugars, though it is just as often scarcely detectable; complex carbohydrates are a more dependable nutrient found in bark.

Chimpanzees are particular about which leaves they select because leaves are loaded with **antifeedants: cellulose, tannins,** and sometimes **alkaloids.** They carefully select leaves with the fewest of these antifeedants (e.g., *Baphia*; see Figure 5.3). They focus not only on species that are low in cellulose and tannin, but individual leaves with low antifeedants as well, probably by noting color, opacity, texture, and evidence that the leaves are new. Best are leaf flushes, the semi-transparent, emerald green young leaves; compared to mature leaves these are high in protein, have less cellulose, and are typically low in antifeedants, as well. Many of the leaves chimpanzees eat taste like spinach or kale, to me.

Chimpanzees eat a variety of insects, most famously termites (the 5 percent of their time devoted to feeding on insects is mostly termites), but they use tools to feed on ants as well, and when eating honey they also consume bees. Seeds are high in protein and fat, and make up 3 percent of the diet. While insects, and meat (0.8 percent of diet) are small parts of their diet, these foods are extremely important because they supply not only fat, but other essential nutrients. As living things, animals contain all the nutritional components necessary for life, and it is clear that meat (covered more completely in Chapter 17) is particularly prized. Meat, while less than 1 percent of their feeding time, provides 5 percent of their calories. Flowers (2.9 percent) often contain a little bit of sugar in the form of nectar, but also protein.

Chimpanzees consume some items that we seldom think of as food, among them termite soil, other clay-rich soils, dead wood, and minerals leeching out of rocky outcrops (Goodall, 1986; Mahaney et al., 1997, 2005; Reynolds et al., 2009). Termite mound soil is high in aluminum, iron, magnesium, and sodium; iron

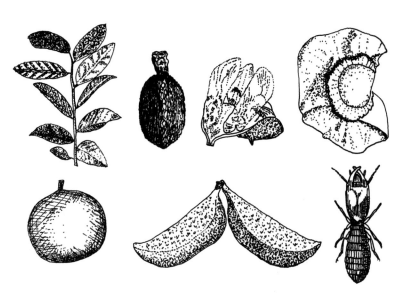

Figure 5.3 A sample of Tanzanian chimpanzee foods. Clockwise from top left, *Baphia capparidifolia* leaves, *Monanthotaxis poggei* fruit, *Diplorynchus condylocarpon* (note they are winged like maple-tree seeds), *Pterocarpus tinctorius* fruits (with surrounding paper-like wing), termite, *Diplorynchus condylocarpon* pod (an opened pod just above it) and *Dictyophleba lucida* fruit.

and sodium are important nutrients and aluminum and magnesium have an antacid action as well (Mahaney et al., 1997, 2005; Rothman et al., 2006). At Budongo, clay eaten by chimpanzees had high levels of metahalloysite and smectite, the combination of which is much like Kaopectate™, an antidiarrheal. Termite soils are eaten often when chimpanzees show evidence of gastrointestinal upset and thus are probably eaten to quell the symptoms of diarrhea. Dead wood and residue on rocky outcrops contain salt, potassium, calcium, and possibly magnesium – important nutrients (Wrangham, 1977; Goodall, 1986; Stanford & Nkurunungi, 2003; Reynolds et al., 2009, 2012, 2015; Watts et al., 2012).

5.8 Making Do When Fruit Fails

Much of the description so far emphasizes ripe fruit. The riper the fruit, the more the sugar. Even though chimpanzees pursue sugar, much of chimpanzee food has only a hint of sugar in it, nothing like the sugar found in domesticated fruits. Domesticated fruits have nearly four times the sugar of fruits chimpanzees typically eat (Table 5.3). When they have the opportunity to harvest truly sweet foods, they are manic. Honey, then, is a profoundly treasured food item that chimpanzees will endure much pain to obtain. They may labor tens of minutes to pry into a hive, using stout sticks to lever off bits of wood. As they work, disturbed bees swarm around them, stinging by the hundreds, yet chimpanzees persist. After a honey-harvesting bout their faces are often misshapen with swelling.

When ripe fruit is scarce, which is all too common from a chimpanzee's perspective, chimpanzees must rely on less palatable foods. Most fruits are consumed for their sugar content, but figs, at least in East Africa, are consumed with a different motivation. While cultivated figs can be quite sugar-rich, not so for wild figs. Instead, figs are rather like potatoes in their role in the chimpanzee diet, though perhaps with more protein (Conklin-Brittain & Wrangham, 1994). While they are not preferred, they are still an important source of nutrients when ripe fruits are unavailable. It is a testament to the scarcity of sugar-rich foods that figs, undesirable as they are, are one of the most important food items for chimpanzees – at least at the Kibale Forest (Wrangham et al., 1993) – partly because their density in the tree crown means they can be harvested quickly.

Figs, then, are a **fallback** food, a low-sugar, high-complex-carbohydrate food *turned to when preferred foods are not available*. Other such foods are piths such as *Pennisetum purpureum* (elephant grass), *Acanthus pubescens*, *Aframomum mala*, and *Marantochloa leucanth*. These celery-like foods make up 10 percent of the chimpanzee diet. I have tasted all these foods. Adding a *frisson* to the celery note in these foods are hints of pepper, traces of sugar, and an occasional suggestion of umami.

Chimpanzees have large incisors in part because opening these foods requires forceful incising. Elephant grass stems are the size of a broom handle and have a fibrous outer layer that must be removed before the celery-like interior can be eaten. The fronds of *Phoenix reclinata*, and at Gombe *Elaeis guineensis*, are perhaps the toughest things chimpanzees eat. They consume the central stem of the frond, not the leaves. The stems are 1–2 inches thick and 3–4 inches wide, round on one side, flat on the other. The surface on the top and bottom is smooth and waxy, but the sides sport ferocious-looking spikes. The exterior cortex is one-eighth of an inch thick and wood-like. This slippery, woody shell must be penetrated to reach the nutritious interior, a task beyond the ability of any human I have seen attempt it. The rounder surface of the frond must be pressed against the incisors with some force before the woody outer cortex can be cracked with a powerful bite, but the slippery outer coating of the frond makes it difficult to dig the incisors into while attempting this initial breach. Once the cortex is penetrated, a long strip of it can be pulled away to reveal the pithy interior. The task is not much easier than taking a bite out of a pine two-by-four building stud. The reward is a pith that, while digestible without dependence on coevolved microflora, is appetizing only to a fruit-deprived chimpanzee.

5.9 Unripe = Unpalatable

As impressive as the breadth and specificity of the chimpanzee diet is, the real surprise, and the secret to

Table 5.4 **Chimpanzee reliance on fruit (percentage of time feeding)**

	Chimpanzee diet		Monkey diet	
	Ripe fruit	Unripe fruit	Ripe fruit	Unripe fruit
Kibale-Kanyawara[a]	95	5	54	46
Kibale-Kanyawara[b]	97	3	–	–
Kibale-Ngogo[b]	98	2	–	–
Budongo[c]	85	15	–	–
Average	94	6	54	46

[a] Wrangham et al., 1998. [b] Potts et al., 2011. [c] Newton-Fisher, 1999.

nearly everything else that is unusual about chimpanzees in particular and apes in general, is discovered when we compare chimpanzee and monkey diets. Monkeys eat less fruit (Table 5.2), and rely much more on insects and flowers. The more important difference, however, is in the type of fruit (Table 5.4).

In the 1970s and 1980s, primatologists were hard put to detect a real difference in monkey and ape diets. While the insect and flower difference is real, nobody really believed that this difference could account for how distinctive apes are. The key turned out not to be in the item itself, but in its life-history. Richard Wrangham took note of this in the 1970s (Wrangham, 1977), but it was slow to sink in. Then, in several studies at Kibale, Wrangham and colleagues showed that 95 percent of the fruit chimpanzees eat is ripe (Wrangham et al., 1998; Conklin-Brittain et al., 1999; Potts et al., 2011; Watts et al., 2012) and at Budongo it is 85 percent (Newton-Fisher, 1999). In contrast, monkeys eat ripe and unripe fruits in nearly equal proportions (Wrangham et al., 1998). We will learn more detail in the next chapter about why ripe and unripe fruits are very different foods; for now, know that unripe fruits have fiber, poisons, and less sugar; ripe fruits are sugary.

Chimpanzees, in short, are *ripe-fruit specialists*, and the importance of ripe fruit can be seen in more ways than just its proportion in the diet; their whole gathering strategy is planned around maximizing ripe fruit. Chimpanzees not only prefer ripe fruit, they require it. The frugivorous monkeys in Table 5.2 eat only slightly more leaves than chimpanzees, but this simple statement belies the fact that they can make do on leaves alone for long periods of time. Chimpanzees cannot. Monkeys can eat foods that are poisonous to humans; chimpanzees cannot. Ripe fruit is critical to chimpanzees because their digestive physiology is unable to cope with the levels of fiber and poison monkeys take in and on which they thrive. They cannot process unripe fruits, and with their large brains they have come to depend on sugary foods.

As part of Wrangham's team in the 1990s, I helped gather data that showed the complex strategy chimpanzees employ to maximize sugar consumption and minimize fiber and secondary compound intake. We found that as more ripe fruit became available, chimpanzees increasingly ignored every other food (Figure 5.4) and the more ripe fruit there was, the more ripe fruit they ate (Figure 5.5). Monkeys were unaffected: As the proportion of trees with ripe fruit rose, monkeys maintained a steady 50–50 proportion of unripe to ripe fruit.

Chimpanzees need sugar in part because they burn calories at a surprisingly high rate. Chimpanzees consume 79 percent of the calories humans do, even though they weigh only 76 percent as much as humans (weights of hunter-gatherers; Chapter 7). Chimpanzees ingest 1934 calories per day, versus 2456 calories for humans (Pontzer et al., 2016).

In the chapters to follow I will show that the chimpanzee specialization on ripe fruits affects nearly

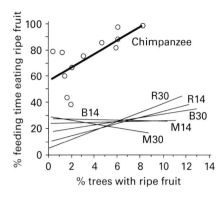

Figure 5.4 As the proportion of trees with ripe fruit rises, chimpanzees eat more and more fruit; monkeys (B = blue monkeys, R = redtails, M = mangabeys) are unaffected. From Wrangham et al., 1998. Used with permission of SpringerNature.

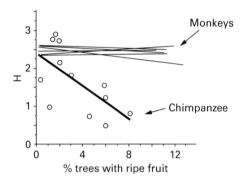

Figure 5.5 As the proportion of trees with ripe fruit rises, chimpanzee diet diversity (H) declines as chimpanzees focus more and more on ripe fruit; monkeys (B = blue monkeys, R = redtails, M = mangabeys) are unaffected. From Wrangham et al., 1998. Used with permission of SpringerNature.

every aspect of their behavior and biology. It affects their social system, because when food is truly scarce chimpanzee do something monkeys do not: They break into small parties, females alone with their young and males in groups of two or so. Dependence of this rare food item results in chimpanzees living at a low population density, forcing females away from the typical female-bonded social life most monkeys enjoy. Their anatomy and great intelligence are driven by their diet, as well.

In their daily scramble for nutrients, chimpanzees skirt a deadly precipice ringing the starvation abyss; we lose little in describing them as always hungry. In one of my classes I ask students to eat a chimpanzee diet for a day, or two days if they are resolute. Many cannot do it. Most complain of nearly irresistible cravings for oils (e.g., oil in fried foods or butter), protein, salt, and grain products. They suffer a feeling of emptiness or discontent even though they are permitted in the assignment to fill their stomachs with all the fruit and spinach they can eat.

When less food than normal is available, and chimpanzees, already lean, lose 10 percent of their body mass (Goodall, 1986) as they resort to the fallback foods we learned of earlier in the chapter.

Life is so precarious that females find it advantageous to reduce travel costs by focusing on a small core area in the community range and using it intensively (Pokempner, 2009). Nowadays we humans in industrialized societies mostly find ourselves in a dramatically different situation; food is superabundant and sugar is cheap and easy to acquire. This groaningly large cornucopia makes it difficult to relate to the wild primate in their daily struggle to ward off malnutrition. Those of us who have attempted to subsist in a forest without food provisions will have experienced how grudgingly a wild habitat gives of its plenty.

It is only on those rare and brief occasions when food is superabundant that chimpanzees can enjoy satiety. In 1992 at Kibale-Kanyawara there were massive numbers of the sugar-rich fruit *Mimusops bagshawei*. The Kibale chimpanzees were constantly near their favorite trees, even sleeping near them. They gathered in the largest groupings of the year; they groomed more, fought more, napped more, played more, and looked much healthier and happier. As the week wore on some females came into estrus.

It is during these times of plenty that chimpanzees are most likely to hunt (see Chapter 17). It was a surprise to chimpologists that the abundance of fruit was tied to meat eating (Mitani & Watts, 1999; Watts & Mitani, 2002; Gilby & Wrangham, 2007). Once explained, it makes sense, and it also says something about why meat is so valued among chimpanzees. Party sizes are large when food is abundant, and chimpanzees are most effective at hunting when their numbers are large (Mitani et al., 2002; Basabose, 2004; Mitani & Watts, 2005). Chimpanzee eat their fill, hunt and kill a monkey or two, and then when it seems there must have been excitement enough for one week, they top it all off by going on a **border patrol** (Mitani & Watts, 2005), an inspection of the boundaries of their territory, meant to expel or even kill extracommunity members. That chimpanzees hunt when they have more calories than they need hints that there are critical nutrients in meat beyond protein and fat.

PART III: NUTRITION AND OPTIMAL DIET

5.10 Optimal Diet: What to Eat and How Much

It may seem surprising that some foods that are eaten with great relish one year are passed over the next. Or items that are eaten at one site are ignored at another; only 60 percent of the food items available at both Gombe and Mahale are eaten in both areas (Nishida et al., 1983). Of course, some of this difference is cultural, and we cannot be certain yet how much. The explanation for this behavior comes from **Optimal Diet Theory**, the more successful sister of Optimal Foraging Theory. While chimpanzees have in essence outsmarted optimal foraging, their optimal diet list is held close to their hearts – and stomachs.

The important components of foods are (beginning with desirable parts) **proteins**, **lipids** (fats and oils), **sugars**, **complex carbohydrates** (or starch), **cellulose** (fiber), **vitamins**, **minerals**, and the undesirable **antinutrients** – negative-value food components. These will be discussed in greater detail in the next chapter. There is an optimal consumption of each of these food constituents and setting the costs and risks of obtaining each with its value (or its negative value) is the daily challenge of foraging.

We know what to feed chimpanzees to keep them healthy. Our century of experience caring for them in captivity suggests their nutritional needs are little different from those of humans (Bourne, 1971). Their daily energy expenditure is indistinguishable from that of humans, if you adjust for body size. We know what they like: With the exception of preferring slightly sour tastes, their food preferences are identical to ours (Chapter 13). They do suffer from heart disease more readily than humans, so some adjustment of sugars, fiber, and lipids is called for – they need a heart-healthy diet. Perhaps the similarity of human and chimpanzee nutritional needs is a surprise, but in fact such requirements are quite similar for all mammals (Robbins, 1983; Sizer et al., 2012) – it is the vastly different digestive physiologies of mammals that explain their dietary differences.

Chimpanzees obey several imperatives when it comes to optimizing their nutrient intake. Some nutrients, while absolutely essential, cease being valuable after intake has reached a certain **threshold**; beyond that they provide little or no benefit. In this case "optimal" consumption is different from "maximal" consumption. It is somewhat like fueling your automobile: Once the tank is full, continuing to pump gasoline is useless. Some such nutrients, like vitamin D, are even detrimental if overconsumed.

Protein is a threshold nutrient, even though in the popular mind maximizing protein intake is best. Protein requirements are elevated for nursing and pregnant females (though, perhaps surprisingly, less for pregnancy than nursing; Herrera & Heymann, 2004). While pregnant female chimpanzees do eat more leaves, presumably in pursuit of protein (Cook & Hunt, 1999), in general protein is not a nutrient that is pursued. "Protein," however, should not be confused with "meat." Chimpanzees deeply crave and pursue meat with a passion, presumably to obtain the things that come with animal protein such as vitamins, minerals, amino acids, and fat. Fat, in particular, is calorie-dense, and pursuit of calorie-maximization is a dietary imperative for chimpanzees. Insects and seeds also contain high proportions of protein, but are valuable for their fat and other nutrients, not their protein.

One may wonder whether protein is truly of no value at all once protein needs are met. What if one is starving? Protein *can* be broken down to provide energy, but only in the presence of fat, and even then it is almost worthless. In fact, in the complete absence of fat, protein is poisonous (Noli & Avery, 2010). Rabbits are notoriously lean animals, and consuming them provides no benefit unless supplemented with fats or oils, and can even sicken humans (Speth, 2010), a phenomenon known as **rabbit starvation.**

Vitamins and minerals are also threshold nutrients, and are little pursued by chimpanzees. Vitamins A, B-complex (including pantothenic acid [B_5], biotin [B_7], and folate), C, D, E, K, niacin, thiamin, and riboflavin all benefit humans and other animals until they have consumed some minimum amount, after which further consumption has no effect. Minerals required in threshold amounts include calcium, potassium, and salt (sodium chloride; USDA guidelines; www.nap.edu).

I have said these threshold nutrients seem to affect food choice little or not at all. If a threshold is necessary, why are they seemingly ignored? It is because most of these vitamins, minerals, and elements are found in small amounts in fruit and are consumed in adequate amounts as chimpanzees go about their daily task of sugar-seeking (Rode et al., 2003). As an example, studies show that howler monkeys, which consume no meat, still manage to consume the equivalent of the human recommended daily allowance of all minerals – with only the exception of salt – in the course of routine foraging (Milton, 1999).

Other nutrients, however, have no practical ceiling: These are nutrients that primates attempt to maximize. At levels available in the wild, high-calorie items continue to be beneficial, no matter how much has been consumed previously. Obviously, the stomach and gut have limited volume, but up to that point more is better. In this case "optimal" *is* "maximal." We have made much of sugar when discussing ripe fruits. These are the so-called simple sugars, fructose and glucose; we will learn more about sugar as a molecule and digestion in the next chapter. The sugar we are most familiar with –table sugar, or sucrose – is a combination of glucose and fructose. Complex carbohydrates, or starches, are easily digested and are consumed by chimpanzees in piths and fruits. While fats are not very abundant in nature, they can be found in a few fruits – you are familiar with avocadoes – though as mentioned above they are found mostly in meat, insects, and seeds.

As chimpanzees seek to maximize caloric intake while consuming a threshold amount of protein, vitamins, and minerals, they face a challenge that we humans rarely think about: minimizing antifeedants. For these unhealthful compounds, optimal is *minimal*.

Much of the decision-making in daily foraging activity consists of balancing the effort expended in maximizing calories with the cost of minimizing antifeedants and deciding when to leave a feeding tree of which the fruit crop is gradually depleted until it is best abandoned. Chimpanzees must feed themselves while competing with other animals and other chimpanzees. This competition determines where the food can be gathered and how chimpanzees must get it. This in turn affects social behavior both in the near term and evolutionarily; selective pressures on anatomy and cognition are affected as well. The constraints the chimpanzee diet places on these other aspects of their lives has resulted in the evolution of apes as we know them. We humans shifted our foraging strategy a little bit two or so million years ago when we expanded the role of meat in our diet. As with chimpanzees, the acquisition of that particular food item has sent filaments of change squirming into every other aspect of our existence. Food, glorious food. Also glorious is the anatomy that processes all that food, as we will see in the next chapter.

References

Ban SD, Boesch C, Janmaat KR (2014) Taï chimpanzees anticipate revisiting high-valued fruit trees from further distances. *Animal Cognition* 17, 1353-1364.

Ban SD, Boesch C, N'Guessan A, et al. (2016) Taï chimpanzees change their travel direction for rare feeding trees providing fatty fruits. *Anim Behav* 118, 135-147.

Basabose AK (2004) Fruit availability and chimpanzee party size at Kahuzi montane forest, Democratic Republic of Congo. *Primates* 45, 211–219.

Boesch C, Boesch-Achermann H (2000) *The Chimpanzees of the Taï Forest: Behavioural Ecology and Evolution*. Oxford: Oxford University Press.

Bogart SL, Pruetz JD (2011) Insectivory of savanna chimpanzees (*Pan troglodytes verus*) at Fongoli, Senegal. *Am J Phys Anthropol* 145, 11–20.

Bourne GH (1971) Nutrition and diet of chimpanzees. In *The Chimpanzee*, Vol. 4 (ed Bourne GH), pp. 373–400. Basel: Karger.

Boyer D, Ramos-Fernández G, Miramontes O, et al. (2006) Scale-free foraging by primates emerges from their interaction with a complex environment. *Proc R Soc B* 273, 1743–1750.

Charnov E (1976) Optimal foraging: the marginal value theorem. *Theoret Population Biol* 9, 129–136.

Conklin-Brittain NL, Wrangham RW (1994) The value of figs to a hind-gut fermenting frugivore: a nutritional analysis. *Biochem Ecol Systemat* 22, 137–151.

Conklin-Brittain NL, Wrangham RW, Hunt KD (1999) Dietary response of chimpanzees and cercopithecines to seasonal variation in fruit abundance: II. Macronutrients. *Int J Primatol* 20, 971–998.

Conklin-Brittain NL, Wrangham RW, Smith CC (2002) A two-stage model of increased dietary quality in early hominid evolution: the role of fiber. In *Human Diet: Its Origin and Evolution* (eds. Ungar PS, Teaford MF), pp. 61–76. Westport, CT: Bergin & Garvey.

Cook DC, Hunt KD (1999) Sex differences in trace elements: status or self-selection? In: *Gender in Palaeopathological Perspective* (eds. Grauer A, Stuart PL), pp. 64–78. Cambridge: Cambridge University Press.

Dolins FL, Klimowicz C, Kelley J, Menzel CR, et al. (2014) Using virtual reality to investigate comparative spatial cognitive abilities in chimpanzees and humans. *Am J Primatol* 76(5), 496–513.

Eilu G, Hafashimana DL, Kasenene JM (2004) Tree species distribution in forests of the Albertine Rift, western Uganda. *African J Ecology* 42, 100–110.

Ellenberg M, Mueller-Dombois D (1967) Tentative physiognomic-ecological classification of plant formations of the earth. *Berichte des Geobotanischen Institutes der Eidgenossischen Technische Hochschule Stiftung Rubel* 37, 21–46.

Emlen JM (1966) The role of time and energy in food preference. *Am Nat* 100, 611–617.

Gilby IC, Wrangham RW (2007) Risk-prone hunting by chimpanzees (*Pan troglodytes schweinfurthii*) increases during periods of high diet quality. *Behav Ecol Sociobiol* 61, 1771–1779.

Goodall, J van Lawick (1968) The behavior of free-living chimpanzees in the Gombe Stream Reserve. *Anim Beh Monogr* 1, 165–311.

Goodall J (1986) *The Chimpanzees of Gombe: Patterns of Behavior*. Cambridge, MA: Harvard University Press.

Greenway PJ (1973) A classification of the vegetation of East Africa. *Kirkia* 9, 1–68

Gross JE, Zank C, Thompson Hobbs N, St Spalinger D (1995) Movement rules for herbivore in spatially heterogeneous environments: responses to small scale patterns. *Landscape Ecol* 10, 209–217.

Harrison M, Marshall AJ (2011) Strategies for the use of fallback foods in apes. *J Primatol* 32, 531–565.

Herrera ERT, Heymann EW (2004) Does mom need more protein? Preliminary observations on differences in diet composition in a pair of red titi monkeys (*Callicebus cupreus*). *Folia Primatol* 75, 150–153.

Hunt KD (1989) *Positional behavior in Pan troglodytes at the Mahale Mountains and Gombe Stream National Parks, Tanzania*. PhD dissertation, University of Michigan. Ann Arbor, MI: University Microfilms.

Hunt KD, McGrew WC (2002) Chimpanzees in the dry habitats of Assirik, Senegal and Semliki Wildlife Reserve, Uganda. In *Behavioural Diversity in Chimpanzees and Bonobos* (eds. Boesch C, Hohmann G, Marchant LF), pp. 35–51. Cambridge: Cambridge University Press.

Janmaat KR, Simone D, Ban SD, Boesch C (2013) Chimpanzees use long-term spatial memory to monitor large fruit trees and remember feeding experiences across seasons. *Anim Behav* 86.6, 1183–1205.

Janmaat KR, Polansky L, Ban SD, Boesch C (2014) Wild chimpanzees plan their breakfast time, type, and location. *PNAS* 111, 16343–16348.

Janmaat KR, Boesch C, Byrne R, et al. (2016) Spatio-temporal complexity of chimpanzee food: how cognitive adaptations can counteract the ephemeral nature of ripe fruit. *Am J Primatol* 78, 626–645.

Janson C (2014) Death of the (traveling) salesman: primates do not show clear evidence of multi-step route planning. *Am J Primatol* 76.5, 410–420.

Kellogg WN, Kellogg, LA (1933) *The Ape and the Child: A Study of Environmental Influence upon Early Behavior*. New York: McGraw-Hill.

Krebs JR (1973) Behavioral aspects of predation. In *Perspectives in Ethology* (eds. Bateson PPG, Klopfer PH), pp. 73–111. New York: Plenum.

Lee K, Hong S, Kim SJ, Rhee I, Chong S (2008) Demystifying levy walk patterns in human walks. NCSU Technical Report.

Lima GF, Martinez AS, Kinouchi O (2001) Deterministic walks in random media. *Phys Rev Lett* 87, 010603.

MacArthur R, Pianka E (1966) On the optimal use of a patchy environment. *Am Nat* 100, 603–609.

Mahaney WC, Milner MW, Sanmugadas K, et al. (1997) Analysis of geophagy soils in Kibale Forest, Uganda. *Primates* 38, 159–176.

Mahaney WC, Milner MW, Aufreiter S, et al. (2005) Soils consumed by chimpanzees of the Kanyawara community in the Kibale Forest, Uganda. *Int J Primatol* 26, 1375–1398.

Matsuzawa T (1985) Use of numbers by a chimpanzee. *Nature* 315, 57–59.

Matsuzawa T (2003) The Ai project: historical and ecological contexts. *Anim Cogn* 6(4), 199–211.

McGrew WC (1974) Tool use by wild chimpanzees in feeding upon driver ants. *J Hum Evol* 3, 501–508.

McGrew WC, Baldwin PJ, Tutin CEG (1981) Chimpanzees in a hot, dry and open habitat: Mt. Assirik, Senegal, West Africa. *J Hum Evol* 10, 227–244.

Menzel EW Jr (1973) Chimpanzee spatial memory organization. *Science* 182, 943–945.

Milton K (1999) Nutritional characteristics of wild primate foods: do the diets of our closest living relatives have lessons for us? *Nutrition* 15, 488–498.

Mitani JC, Watts DP (1999) Demographic influences on the hunting behavior of chimpanzees. *Am J Phys Anth* 109, 439–454.

Mitani JC, Watts DP (2005) Correlates of territorial boundary patrol behaviour in wild chimpanzees. *Anim Behav* 70, 1079–1086.

Mitani JC, Watts DP, Lwanga JS (2002) Ecological and social correlates of chimpanzee party size and composition. In *Behavioral Diversity in Chimpanzees and Bonobos* (eds. Boesch C, Hohmann G, Marchant LF), pp. 102–111. Cambridge: Cambridge University Press.

Newton-Fisher NE (1999) The diet of chimpanzees in the Budongo Forest Reserve, Uganda. *Afr J Ecol* 37, 344–354.

Nishida T (1976) The bark-eating habits in primates, with special reference to their status in the diet of wild chimpanzees. *Folia Primatol* 25, 277–287.

Nishida T, Wrangham RW, Goodall J, Uehara S (1983) Local differences in plant-feeding habits of chimpanzees between the Mahale Mountains and Gombe National Park, Tanzania. *J Hum Evol* 12, 467–480.

Noli D, Avery G (1988) Protein poisoning and coastal subsistence. *J Archaeological Science* 15, 395–401.

Normand E, Boesch C (2009) Sophisticated Euclidean maps in forest chimpanzees. *Animal Behaviour* 77.5, 1195–1201.

Normand E, Boesch C, Ban SD (2009) Forest chimpanzees (*Pan troglodytes verus*) remember the location of numerous fruit trees. *Anim Cogn* 12, 797–807.

Perry G, Pianka ER (1997) Animal foraging: past, present and future. *Trends in Ecology & Evolution* 12, 360–364.

Pierce GJ, Ollason JG (1987) Eight reasons why optimal foraging theory is a complete waste of time. *Oikos* 49, 111–118.

Pokempner AA (2009) *Fission–fusion and foraging: sex differences in the behavioral ecology of chimpanzees (Pan troglodytes schweinfurthii)*. PhD dissertation. Stony Brook: SUNY Stony Brook.

Pontzer H, Brown MH, Raichlen DA, et al. (2016) Metabolic acceleration and the evolution of human brain size and life history. *Nature* 533, 390–392.

Potts KB, Watts DP, Wrangham RW (2011) Comparative feeding ecology of two communities of chimpanzees (*Pan troglodytes*) in Kibale National Park, Uganda. *Int J Primatol* 32, 669–690.

Pruetz JD, McGrew WC (2001) What does a chimpanzee need? Using natural behavior to guide the care and management of captive populations. In *Care and Management of Captive Chimpanzees* (ed. Brent L), pp.17–37. Chicago, IL: American Society of Primatologists.

Pyke GH, Pulliam H, Charnov EL (1977) Optimal foraging: a selective review of theory and tests. *Quart Rev Biol* 52, 137–153.

Ramos-Fernández G, Mateos JL, Miramontes O, et al. (2004) Lévy walk patterns in the foraging movements of spider monkeys (*Ateles geoffroyi*). *Behav Ecol Sociobiol* 55(3), 223–230.

Reynolds V, Lloyd AW, Babweteera F, English CJ (2009) Decaying *Raphia farinifera* palm trees provide a source of sodium for wild chimpanzees in the Budongo Forest, Uganda. *PLoS ONE* 4, e6194.

Reynolds V, Lloyd AW, English CJ (2012) Adaptation by Budongo Forest chimpanzees (*Pan troglodytes schweinfurthii*) to loss of a primary source of dietary sodium. *Afr Prim* 7, 156–162.

Reynolds V, Lloyd AW, English CJ, et al. (2015) Mineral acquisition from clay by Budongo Forest chimpanzees. *PLoS ONE* 10, e0134075.

Richards PW (1952) *The Tropical Rain Forest: An Ecological Study*. Cambridge: Cambridge University Press.

Robbins C (1983) *Wildlife Feeding and Nutrition*. New York: Academic Press.

Rode KD, Chapman CA, Chapman LJ, McDowell LR (2003) Mineral resource availability and consumption by colobus in Kibale National Park, Uganda. *Int J Primatol* 24, 541–573.

Rosenstein D, Oster H (1988) Differential facial responses to four basic tastes in newborns. *Child Dev* 59 1555–1568.

Rothman JM, Van Soest PJ, Pell AN (2006) Decaying wood is a Na source for mountain gorillas. *Biol Lett* 2, 321–324.

Schoener TW (1971) Theory of feeding strategies. *Ann Rev Ecol Syst* 2, 369–404.

Sizer FS, Piché LA, Whitney EN, Whitney E (2012) *Nutrition: Concepts and Controversies*. Boston, MA: Cengage Learning.

Smouse PE, Focardi S, Moorcroft PR, et al. (2010) Stochastic modelling of animal movement. *Phil Trans Biol Sci* 365, 2201–2211.

Speth JD (2010) The paleoanthropology and archaeology of big-game hunting: protein, fat, or politics? In *Interdisciplinary Contributions to Archaeology* (ed Eerkens J), pp. 149–116. New York: Springer.

Stanford CB, Nkurunungi JB (2003) Behavioral ecology of sympatric chimpanzees and gorillas in Bwindi Impenetrable National Park, Uganda: diet. *Int J Primatol* **24**, 901–918.

Steiner JE (1979) Human facial expression in response to taste and smell. *Adv Child Dev* **13**, 257–295.

Stephens DW, Krebs JR (1986) *Foraging Theory*. Princeton, NJ: Princeton University Press.

Struhsaker TT (1978) Food habits of five monkey species in the Kibale Forest, Uganda. In *Recent Advances in Primatology*, Vol. 1 (eds. Chivers DJ, Herbert J), pp. 225–248. New York: Academic Press.

Tinklepaugh OL (1932) Multiple delayed reaction with chimpanzees and monkeys. *J Comp Psychol* **13**, 207–243.

Ventura AK, Mennella JA (2011) Innate and learned preferences for sweet taste during childhood. *Curr Opin Clin Nutrition Metabol Care* **14**, 379–384.

Watts DP, Mitani JC (2002) Hunting behavior of chimpanzees at Ngogo, Kibale National Park, Uganda. *Int J Primatol* **23**, 1–28.

Watts DP, Potts KB, Lwanga JS, Mitani JC (2012) Diet of chimpanzees (*Pan troglodytes schweinfurthii*) at Ngogo, Kibale National Park, Uganda, 1. Diet composition and diversity. *Am J Primatol* **74**, 114–129.

White F (1983) *The Vegetation of Africa*. Paris: UNESCO.

Whitmore TC (1990) *An Introduction to Tropical Rain Forests*. Oxford: Oxford University Press.

Wrangham RW (1977) Feeding behaviour of chimpanzees in Gombe national park, Tanzania. In *Primate Ecology* (ed. Clutton-Brock TH) pp. 503–538. New York: Academic Press.

Wrangham RW, Conklin NL, Etot G, et al. (1993) The value of figs to chimpanzees. *Int J Primatol* **14**, 243–256.

Wrangham RW, Chapman CA, Clark Arcadi AP, Isibirye-Basuta G (1996) Social ecology of Kanyawara chimpanzees: Implications for understanding the costs of great ape groups. In *Great Ape Societies* (eds. McGrew WC, Marchant LF, Nishida T), pp. 45–57. Cambridge: Cambridge University Press.

Wrangham RW, Conklin-Brittain NL, Hunt KD (1998) Dietary response of chimpanzees and cercopithecines to seasonal variation in fruit abundance. I. Antifeedants. *Int J Primatol* **19**, 949–970.

6. Guts, Glorious Guts, Large Stomach, and Colon

Plant Chemistry, Fruit Ripening, Digestive Physiology, and Gut Anatomy

Photo by author

PART I: PLANT BIOLOGY

While the fondness chimpanzees hold in their hearts for sugar and their equally strong aversion to antifeedants is well understood today, as late as 1975, when Jane Goodall had already spent 15 years at Gombe, primatologists were baffled by the many differences between monkeys and apes. Given that both live in forests and eat fruits, why should they differ so much in social system, population density, body size, and cognition? The answer, it turns out, can be traced to differences in digestive physiology and how those differences dictate their diet and foraging behavior.

Before we can understand how chimpanzee digestive physiology differs from that of monkeys, we must first understand a little bit about the foods themselves. This means learning a little bit about the chemistry of flowers, leaves, bark, seeds, and fruit, and a little about their biological roles.

Much of what scientists have learned about plants was motivated by agricultural needs – our desire to improve our capacity to feed ourselves by improving the quality of cultivated plants. By *quality* I mean crop yields and nutrient content. While biologists down here at Indiana University are enamored with fruit flies (Raff & Kaufmann, 1983; Chapter 11), scientists up the road at Purdue University have occupied themselves with figuring out how plants function, not neglecting – this is Indiana, after all – corn (Schnable et al., 2009). As scientists unlocked the secrets of plant chemistry, they unwittingly uncovered evidence concerning the evolutionary relationship between plant-eating animals and their prey. The influence of plant chemistry on the daily search for food and the role of food in determining social structure has gradually been laid bare, one startling revelation after another, almost like a detective novel (Glander, 1982). Unexpectedly, many of us believe that food chemistry is the key to understanding why apes – including humans – have evolved the extraordinary characteristics that distinguish us from other animals, not least in terms of intelligence and violence (Wrangham, 1979, 1980; Wrangham et al., 1998; Hunt, 2016).

6.1 Primary Compounds

One might expect that a simple organism like a plant might need 10 or 100 building blocks. Instead, we discovered in the nineteenth century that plant chemistry involves a startlingly diverse array of organic chemicals. Instead of the hundreds of chemical components we expected, there are hundreds of thousands. There is some order in this knotted tangle of complexity, in that many of these chemicals are universally shared among plants; these, we have gradually come to know, are essential for life (Kossel, 1891; Berenbaum, 1995); chlorophyll, tryptophan, vitamin A, and sucrose are some such essential compounds. The giant of nineteenth-century plant chemistry, Albrecht Kossel, named them **primary compounds** (Kossel, 1891; Mothes, 1980). While these biochemicals were recognized as essential early on, there were many others for which it remained unclear for an embarrassingly long period of time whether they were necessary for life – primary compounds – or had some other, less-than-necessary, function. The importance of vitamins, for instance, was slow to be appreciated, since they were found to exist in very low concentrations compared to many other compounds; their critical role was so poorly appreciated that scientists only got around to coining the word "vitamin" in 1912 (Berenbaum, 1995).

Biologists who were interested in improving strains of domesticated plants, like those affiliated with Purdue, found some utility in tinkering with the proportion of primary compounds in crop species. They found that some predators, often insects, key on particular primary compounds, which meant that lower concentrations of these biochemicals resulted in less crop damage. This solution to improving crop health and pest resistance turned out to be a dead end. Plants bred to have lower concentrations of primary compounds waned, withered, and sometimes died (Berenbaum, 1995). Blocking the action of these compounds had the same result. Primary compounds, as Kossel taught us, are indeed necessary for life.

6.2 Secondary Compounds

Other compounds, even some that were quite abundant – so abundant that they were identified even before many primary compounds – were slowly discovered to be superfluous (Berenbaum, 1995). Such profligacy on the part of nature was and remains entirely unexpected; in the death-match arena that is the struggle for life, surely no competitor could afford to waste energy manufacturing useless biochemicals. And it was even worse than it seemed, because research slowly revealed that these superfluous compounds not only served no apparent function, they were found to be detrimental to the very plants that made them! Experiments showed that when these compounds were experimentally reduced or neutralized, not only was there no damage to the health of the plant, quite the opposite – plants grew faster, achieved a larger size, or produced greater fruit or seed crops (Fraenkel, 1969; Berenbaum, 1995). Thus, their presence was mysterious.

Why would plants produce compounds that made them unhealthy? Were they unavoidable waste – "excretory products" – that were necessarily shed during normal plant biochemical processes? In this view, if plants could they *would* eliminate these compounds from their "bodies." So why did they not? Perhaps competing demands, such as growth and maintenance, made expelling them a low priority; or perhaps selection pressure to grow quickly before predators or disease could kill a young, vulnerable plant was so great that diverting energy to such "house cleaning" would be fatal. Both of these explanations turned out to be wrong.

Compounds like these, those for which plant biologists could find no definite function, and which even seemed to encumber plants, came to be called **secondary compounds** to distinguish them from essential biochemicals (Kossel, 1891; Mothes, 1980). Even while scientists were scratching their heads over the function of these secondary compounds, they kept right on discovering them. In 1977 Tony Swain totted up the total number of known secondary compounds and got a figure of 12,000; he guessed there might be as many as 400,000 (Freeland & Janzen, 1974; Gartlan et al., 1980). Nothing we have learned since suggests he underestimated; more are being discovered all the time.

6.3 Secondary Compounds are Defensive

While most plant biologists were unable or unwilling to speculate on the function of secondary compounds, one person hit the nail on the head quite early. Stahl (1888) suggested that *secondary compounds had nothing directly to do with important physiological processes, but instead existed only to protect plants from their enemies; they were a type of chemical defense.* His was a voice in the wilderness. In retrospect, we might find it curious that Stahl's hypothesis failed to resolve the issue immediately, but keep in mind that not only was his hypothesis one of many competing for acceptance, those who assumed these compounds must be essential for life were unable to reverse course.

Plants, Stahl realized, have a profoundly important handicap that animals do not; they cannot flee from predators. The plant equivalents of skin, blood, bones, genitals, and energy-gathering organs (i.e., bark, sap, wood, flowers, and leaves) are all subject to the voracious attention of folivores. A common human nightmare, being unable to move our legs while some danger bears down on us, is an eternal reality for members of the plant kingdom. Humans sometimes protect themselves with armor. Plants already came up with that. With no avenue of escape, plants have evolved either armor-like thick bark or an alternative poisoning-the-well strategy; they interperse among their body parts chemicals that deter those who would eat them.

Tests of Stahl's defensive hypothesis kept coming up positive. One test showed that secondary compounds deterred plant-eaters (Fraenkel, 1969). Gradually the field of plant chemistry became less dismissive of the plant-defense hypothesis. At last, in 1976, for the first time, a scientist was emboldened to declare that the *only* function of secondary compounds was defense (Mothes, 1980). Clarity at last.

6.4 Juggling the Chemicals of Life

Incorporating the defensive hypothesis into a holistic view of plant life, we now view the need to grow, the need to reproduce, and the need to defend oneself as competing demands that a plant must balance, one against the other. As Herms and Mattson put it in the title of their excellent review article, "the dilemma of plants: to grow or defend" (Herms & Mattson, 1992). In this view, the cocktail of compounds coursing through the plant circulatory system and embedded in plant tissues is held in delicate balance: Too much of certain primary compounds makes the plant a juicier target for a plant predator; too little compromises growth, reproduction, and health; too much secondary compound slows those life functions; too little makes you lunch (Herms & Mattson, 1992). Plants are part of a community of living things, and as such, the most successful ones have balanced diverse threats exerted by plant predators, fungi, and bacteria, among other nuisances, and that balance long ago tilted toward protection by chemical means (Swain, 1977).

6.5 Types of Secondary Compounds

Secondary compounds can be divided into two principle classes, **antinutrients (digestion inhibitors)** and **poisons.** "Poison" and "toxin" are synonymous.

Tannins are among the most common secondary compounds. They are antinutrients, rather than poisons; they inhibit and slow digestion by binding with various amino acids and proteins, including digestive enzymes, and sometimes with nutrients themselves. They come in two types, **condensed tannins** and **hydrolyzable tannins.** Condensed tannins are indigestible and must be neutralized by attaching a molecule to them that makes them harmless – but making and attaching things takes energy. Hydrolyzable tannins are susceptible to digestion by enzymes that mammals can produce and, as a consequence, while they do affect growth and survival, most plant predators have evolved the ability to neutralize them (Feeny, 1970; Glander, 1982).

Antinutrients that bind with enzymes that digest protein are known as **proteinase inhibitors**; they can be broken down by high heat (boiling temperature), and while they are of little concern to humans – with our cooking technology – they deter other animals.

The second class of secondary compounds, **toxins,** are mostly a class of chemicals known as **alkaloids,** though alkaloids are not the only poisons primates encounter. Rhoades and Cates (1976) listed 23 further poisons, nearly all of which primates tolerate poorly (Glander, 1982). Most toxins are small molecules that are toxic in even very low concentrations; they work by passing through tissue barriers and into the bloodstream, where they cross cell walls and disrupt internal processes (Rhoades & Cates, 1976; Glander, 1982). Cyanide, found in apple seeds, binds with red blood cells and destroys their ability to carry oxygen. Other toxins interfere with nerve action. Nicotine, caffeine, cocaine, solanine, arsenic, and strychnine are all poisons evolved to deter plant predators.

Various types of terpenes are another type of chemical antifeedant that can interfere with neural function and other biological processes. They are volatile hydrocarbons that are often aromatic; turpentine is a terpene, and while turpentine smells foul, another terpene, limonene, found in citrus fruits, is rather pleasant.

6.6 Antifeedants

Antifeedant is a general term for anything in plants that discourages another living thing from eating them. Antinutrients and poisons are included in this category, but so are other things that most primates avoid, like cellulose. Thorns or armor-like structures such as thick, tough bark are **morphological defenses**, often obvious from a distance, but plants may also grow less obvious morphological defenses that are interspersed among the tissues they protect. **Opaline phytoliths**, or silica bodies, are tiny glass beads that are rather evenly distributed among some plant tissues, such as grass blades. The emerald green of new grass may promise a luscious meal to the naïve grazer, but when chewed (yes, I ate some) the blades produces a horrific scraping sound, as if your teeth are being ground down by sandpaper. Opaline silica bodies grate against the surface of the grazer's teeth and wear them down. These **inclusions** in plants deter plant predators by increasing the expense of eating plant tissues (Blackman, 1971). They have exerted such a strong selective force on mammalian grazers that many of them have evolved ever-growing teeth to replace tooth surfaces as they are worn down. Chimpanzees avoid grass blades (though one primate does eat new shoots – the gelada), but they do encounter inclusions when they break apart grass stems, puncture tough fruit rinds, and eat barks.

Some defenses are lucky accidents for plants – one such is cellulose, also known as fiber. Cellulose is the main physical structure of leaves and most other plant parts; it is the material that forms cell walls in plants. No mammal produces the enzyme **cellulase**, which is the chemical can-opener that slices open cell walls to get at the nutrients inside. Leaf-eating primates and other mammals digest cellulose only by relying on a work-around that has important consequences for physiology and anatomy (Chivers & Hladik, 1984). This work-around is a dependence on **coevolved microflora**, or bacteria, that live in the guts of mammals. We will examine this process in more detail later.

Even with such a microflora adaptation, cellulose-rich foods are a poor choice for chimpanzees. Their high activity levels burn calories so fast that the meager energy contents of fiber leave them energy-starved; their large, energy-hungry brain adds to the problem, requiring that they select foods that are easy to digest and contain few secondary compounds. Chimpanzees preferentially select new or immature leaves since, as leaves mature, their cell walls grow thicker and they manufacture more tannins. A mature leaf is a quite different food from a new leaf on the same tree. For chimpanzees, fiber is merely ballast that they must carry as they search for foods that truly benefit them: ripe fruits.

6.7 Fruits: "Eat Me, Please"

The seething brew of secondary compounds in plants illustrates how important defense is for them, but angiosperms – flowering plants – are deeply conflicted in this regard, because while they must set aside a large part of their energy budget to defending their body parts against predators, they must also expend another heaping measure of calories to building plant parts that they actually "want" to be eaten.

Trees and other sessile life forms face a unique reproductive challenge: Their offspring would be handicapped if they were to start life underneath the light-deflecting canopy of a parent, subsequently competing with that parent for nutrients, space, and even light. Yet parent and offspring cannot move away from one another. Trees find a way. They recruit a mobile organism, most often a vertebrate, to carry their offspring away from the parent and off to some spot (the parent "hopes") where every prospect pleases. **Seed dispersers** are in essence "paid" for this service. The best seed dispersers, from the tree's perspective, ingest a fruit, carry the seeds inside it some distance from the parent and deposit the embryonic tree undamaged in some congenial new home. The flesh of fruits is an enticement offered by the tree to the disperser for this service. Fruits "want" to be eaten, but, and here is the rub where it concerns dispersers, they want to be eaten *at the right time*, when the seed is ready to pass through its disperser's gut. Angiosperms are evolved to appeal most to the best dispersers but even then to appeal to dispersers only at the best time for the seed. Yet, they often fail to get their way.

6.8 Fruit Syndromes

There is thought to be a coevolutionary relationship between trees and their dispersers, one where fruit flesh nutrient composition, water content, color, skin thickness, and type and quantity of antifeedants carefully balanced to attract a particular class of dispersers (Tiffney, 1984) while repelling all others. Even the robustness of the attachment of fruits to trees is under selection; the fruit must come off easily if it is to appeal to a small, handless disperser (e.g., small birds), whereas a robust attachment will deflect the attention of birds and attract a disperser with hands. The cluster of fruit characteristics that appeals to a limited class of potential dispersers is known as a **fruit syndrome**; fruits that appeal to primates are thought to be red or orange, large, succulent, tough by virtue of possessing a thick skin (i.e., exocarp), and robustly attached to the tree (Janson, 1983; Terborgh, 1983; Gautier-Hion et al., 1985). Color vision is often cited as a feeding adaptation, evolved to detect the red or orange ripe fruits (Allen, 1879; Regan et al., 1998), though color vision also helps to detect the best leaves; red-tinted leaves are more nutritious than green leaves (Dominy & Lucas, 2001).

In addition to color, aroma is an important signal to frugivores. When chimpanzees are feeding they may sniff several fruits before choosing one, presumably seeking volatile chemicals that the fruit is evolved to produce when the seed is ready. One of these volatile chemicals is ethyl alcohol; alcohol content is correlated with sugar content, and primates are good at detecting alcohol, suggesting it is an honest signal of high sugar content (Dominy, 2004). The human fondness for alcohol may have a much deeper evolutionary origin than the relatively recent invention of brewing.

While long-term evolutionary trends (Eriksson et al., 2000) and the relationships among dispersers and tree species (Lambert & Garber, 1998) are complicated and not yet completely understood, most researchers maintain that the fruit syndrome/disperser relationship explains much of fruit morphology as well as disperser morphology and physiology. Dispersers and tree species each exert strong evolutionary pressures on one another.

PART II: FRUIT CHEMISTRY

As one of nature's most fascinating miracles, fruits deserve our attention. You may imagine a fruit as an envelope that the parent tree gradually loads up with nutrients. Not so. A fruit is more like a time bomb. Early in its life the parent tree has produced nothing but a colorless mass of cardboard-tasting cellulose and secondary compounds mixed with time-release enzymes. In other words, it is mostly antifeedants; the fruit "wants" to be left alone at this point. Over days it is converted into a colorful, sugar-rich, aromatic packet of nutrients, not by the mother plant pulling out these unpleasant antinutrients while pumping in sugar and delectable scents, but within the fruit in a process that almost seems like that of a caterpillar metamorphosing into a butterfly. The inedible mélange of fiber and chemicals in the unripe fruit is transmuted in-house into a sugary snack by photosynthesis, respiration, and enzyme-driven chemical processes (Giovannoni, 2001, 2004, 2007).

Antifeedants in unripe fruits may include tannins, alkaloids, or both. You already know that cellulose is a string of sugars. The right enzymes can digest cellulose and tannins into sugar and neutralize the alkaloids. Photosynthesis powers this and other processes that produce the alluring scent of a ripe peach or strawberry. If you thought the tree rather than the fruit was most active in this transformation, your own experience tells you otherwise. Most of us have brought home an unripe peach or banana from the grocery and left it out to ripen in our kitchen. No parent tree is in sight, yet in a matter of days antinutrients in the peach disappear; it becomes less fibrous, softer, more brightly colored, and more sugary; as it reaches full maturity it fills the room with an intoxicating scent we primates find alluring. On its own it changed from a lump of tissue hardly more nutritious than sawdust to a delicious food.

The chemical constituents of fruits are unique to each species, but we can still put together a basic outline of fertilization, growth, and ripening (Sacher, 1973; Coombe, 1976; Yang & Hoffman, 1984; Brady, 1987; Giovannoni, 2001, 2004, 2007; Seymour et al., 2013).

6.9 Pollination

Fruit begins life when a pollinator – we are most familiar with bees, but many other animals can serve as pollinators, including primates – deposits a pollen grain, half of the genetic material of the plant, on a flower. Trees have evolved features that attract pollinators; they may emit pleasing aromas, or decorate themselves with striking coloration to signal they harbor a reward – nectar. Other plants rely on deception to attract pollinators; some orchids smell as if they offer nutrients, but do not. Others smell like female insects, causing males to attempt to mate with them; the most bizarre of all is a type of dogbane that smells of decaying bees and attracts flies that normally steal dead bees from spiders (Heiduk et al., 2016).

Once a pollen grain is deposited, it builds a tunnel from the surface of the flower to the ovary, where sperm contained in the pollen granule fertilizes an **ovule**, which will eventually become a seed. Even at this early stage the ovule is a complicated structure. A layer of tissue known as the **endocarp** will become the seed shell, while the **mesocarp** will become the flesh of the fruit and the **exocarp** will be the outer skin or husk layer (Figure 6.1).

6.10 Unripe Fruit

After fertilization, the fruit grows by cell division, driven mostly by resources from the mother plant; this

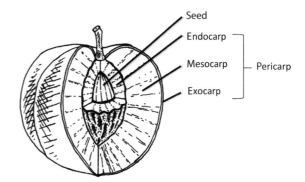

Figure 6.1 The fruit mesocarp is the nutritious part of the fruit; the exocarp may be thick to prevent small animals from penetrating it and thus exposing the mesocarp to fungi and bacteria. The endocarp protects the seed during fruit ingestion and as the seed passes through the gut. Image by the author.

may go on for a month. Similar to a developing embryo (Chapter 11), as the fruit grows it passes through phases of tissue differentiation and differential tissue growth. As with mammalian embryos, the local environment around a cell signals it to turn on or turn off certain genes, upregulating genes that encourage cell division. Significantly, the walls of the multiplying cells in the fruit, as is the case with other cells in plants, are made up of indigestible cellulose. Thus, in this first quarter of their lives (Figure 6.2), fruits are hard, fibrous, and non-nutritious. This will soon change. Although much fiber is indigestible to mammals without the help of coevolved bacteria, there are a number of types of fiber, and some are common in fruit, like **soluble fiber**.

As the fruit grows and cells produce first one then another biochemical, percolating through it are secondary compounds, meant to discourage consumption – for the time being – by both normal dispersers and parasites. If the fruit is eaten at this stage, before the seed is fully formed and capable of surviving passage through a digestive tract, the seed will die; these consumers are not dispersers but **seed predators**, killing the tree's offspring rather than helping to disperse the seed. Birds, insects, or mammals too small to swallow a tree's seed, but which may try to penetrate the exocarp to consume part or all of the mesocarp, are parasites.

The tree suffers not only by losing some fruit flesh to a non-disperser; once the exocarp of the parasitized fruit is

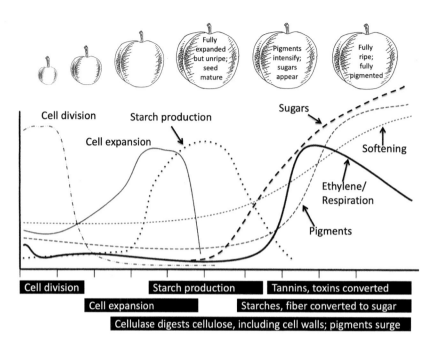

Figure 6.2 Fruits mature and ripen first by cell division and expansion, then enzymes convert cellulose, tannins, toxins, and starches into sugars. Image by the author.

penetrated, it is open to invasion by bacteria, fungi, or insects. Microscopic parasites are not only damaging the fruit, they are making it inedible; once installed in the now-ruined fruit, they emit their own secondary compounds as a means of preventing the fruit from being consumed by animals, which would kill the parasitic microorganisms. A fruit with its exocarp penetrated has a last-ditch strategy; it ripens rapidly in "hope" of attracting a disperser and serving its evolutionary purpose before it is inedible, but more likely it will fall to the ground in the parent's shadow and die. A thick, tough exocarp can protect the fruit from such interlopers, and for plants that rely on primates for dispersal this is a typical morphology. At this early stage of fruit development, the endocarp or shell is still poorly developed, which is why the seed cannot survive gut passage. You may have noticed incompletely fused seeds in unripe peaches or apples; they would not survive digestion.

6.11 Ripening

In the next quarter of development – the fruit now has all the cells it will ever have – genes that produce the enzyme **expansin** upregulate and this special enzyme weakens fibers in the cell walls; cellulase begins to digest fiber, provoking growth as individual cells expand and as their weakened walls are digested. The fruit becomes soft and more digestible.

Powered by light, chlorophyll produces starches, which are manufactured in great quantities in the middle stages of fruit development. In a very real sense, fruits breath: Respiration expels waste gases and takes in carbon dioxide as raw material to build carbohydrates. A mixture of amino acids and enzymes slowly produces the gas ethylene, which builds up in the fruit, stimulating a number of ripening processes. Sugar and organic acids are manufactured in the fruit and the higher concentration of sugar, in particular, creates an osmotic pressure that pulls in water from the parent tree, just like water climbing up an absorbent paper towel, increasing the fruit's succulence.

We know sunlight hastens all the chemical processes of ripening because fruits exposed to more sunlight ripen faster, have more sugars, and are less acidic (Houle et al., 2014), though we are uncertain how. Fruits in the sunny tops of trees are more sugary

Table 6.1 **Digestion**

Class of food	Digestive enzymes	Digestion site	Simplest product
Protein	Protease, pepsin	Stomach	Amino acids
Lipids (fats and oils)	Lipase	Small intestine	Fatty acids
Sugars	Disaccharidase	Small intestine	Single-sugar molecules
Complex carbohydrates	Amylase	Small intestine, mouth	Single-sugar molecules
Cellulose (fiber)	Cellulase in bacteria	Cecum, colon	Single-sugar molecules, fatty acids

that fruits in the shadowy lower strata (Houle et al., 2014). There is some controversy here because some analyses suggest that light only affects pigment production, not other aspects of ripening (Alba et al., 2000; Giovannoni, 2001).

In a rapid cascade of events in the last quarter of fruit ripening, enzymes finish the digestion of cellulose and digest complex carbohydrates into sugar. Enzymes neutralize acid and deactivate toxins. Tannins, made up of carbon, oxygen, and hydrogen, are digested and become sugar; thus an antifeedant is converted to a nutrient. Fruit flesh softening accelerates as cellulase and expansin finish their work. Other enzymes produce a plethora of volatile compounds – as many as 330 types are found in oranges alone – giving fruit its alluring aroma (Van Straten et al., 1983). At the last stages of ripening, chlorophyll is broken down, some of it morphing into pigments.

PART III: DIGESTIVE PHYSIOLOGY AND GASTROINTESTINAL ANATOMY

In the modern world, a poor diet – one that sickens us – is more likely to be one that contains too many rather than too few calories. How different it is for the chimpanzee, teetering as they do on the edge of malnutrition off and on for their entire lives. The abundance in which we wallow makes it difficult for us to fully comprehend the challenge chimpanzees face in the wild. While they are in constant pursuit of sugar, their ambitions are rarely satisfied; most of their foods have a lower concentration of sugar, salt, and fat than the human diet, even when we consider the human diet to be that of hunter gatherers. While monkeys hit trees hard while their fruits are still unripe, antifeedant-intolerant chimpanzees must fall back on bulky, low-nutrient-density foods such as piths and herbs when ripe fruits are not available. These foods have little sugar, but do contain starches and are low in tannins and toxins. Chimpanzees evolved large stomachs to accommodate the volume of these **fallback** foods (Chivers & Hladik, 1984).

6.12 Gut Parts

The gut is divided into several parts, each of which has a principal function, reviewed in Table 6.1. The larger stomachs of chimpanzees give them impressive consumption abilities. Old hands at Gombe recall that during the banana provisioning days one individual ate over 50 bananas in one sitting. The **small intestine** (Figures 6.3 and 6.4) digests and absorbs richer nutrients. Chimpanzees have larger **colons** (also known as the large intestine) and **cecums** (Chivers & Hladik, 1984; Milton, 1999) than humans, gut parts dedicated to nurturing the coevolved microbiome that digests cellulose. The cecum is a small extension to the colon that protrudes beyond its connection with the small intestine (Figure 6.4). In humans, this organ is shrunken to almost nothing, with the vestigial appendix in its place.

6.13 Digestion

Digestion begins in the mouth. Chimpanzees, like other primates, spill the digestive enzyme **amylase** in their saliva (Jacobsen, 1970; Murray, 1975), where it begins to break down starch as food is chewed. Chewing breaks food particles up into small bits, increasing the surface area of the digesta – gut contents – so that digestive enzymes can get at them. The smaller the food particles the better. Amylase functions best in a slightly alkaline environment (i.e., on the opposite end of the pH scale from acid) and thus saliva is mostly neutral or slightly alkaline. Chimpanzee teeth are slightly better than those of humans in slicing up leaves, piths, and fruits. As food is chewed, amylase begins to neutralize tannins, should there be any – and for chimpanzees there often are. Saliva also contains **lipase**, a fat-digesting enzyme, so that some few fats are broken down in the mouth. The remainder of digestion occurs in the stomach, small intestine, cecum, and colon (Southgate, 1995).

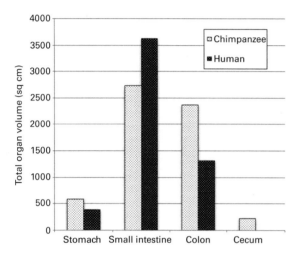

Figure 6.3 Though only two-thirds the weight of humans, chimpanzees have a larger stomach, a colon double the size of humans, and a cecum nearly as large as the human stomach (Image by the author; data: Chivers & Hladik, 1980; Rubin et al., 1988).

6.14 Stomach

In the stomach, foods are drenched in acid and exposed to further digestive enzymes. Acid plays perhaps less of a role in digestion than most imagine (it does sterilize food, and that can be important), but it is digestive enzymes that do most of the work. However, acid *is* important for the digestion of protein because protein-digesting enzymes (e.g., pepsin and protease; Table 6.1) are activated by acid, breaking proteins down into amino-acid chains; most protein digestion occurs in the acidic stomach (Lambert,

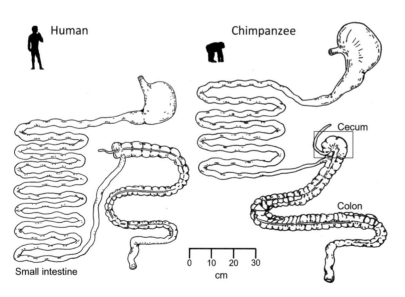

Figure 6.4 The chimpanzee gut, compared to the human, has a larger stomach, shorter small intestine, larger colon, and much larger cecum, reflecting the reliance on coevolved gut microflora to digest cellulose. After Stevens & Hume, 2004; with permission.

1998). The stomach is a muscular organ that churns, squashes, and mixes food particles and enzymes so that foods are broken down not just by the chemical action of enzymes, but also mechanically by the churning action of the heaving stomach.

6.15 Small Intestine

The stomach pushes the digesta into the small intestine, where the pancreas infuses a cocktail of enzymes (Table 6.1) into the mix to break it down into still smaller constituents; fats are broken down by the enzyme lipase into absorbable fatty acids, and complex carbohydrates are digested into sugar. Absorption is an equally important function for the small intestine. Here, most of the sugar content of food is absorbed and single amino acids pass across the small intestine tissue border into the bloodstream, where they are eventually used either to manufacture enzymes and hormones in the liver or to build the necessary proteins elsewhere in the body.

One unusual aspect of amino acids is that mammals have a difficult time storing them, which means that once the body has met its need for whichever amino acid is in demand, further protein intake is useless; this is why mammals only require a threshold amount of protein. Fats are the most calorie-rich nutrients of all, gram for gram, and unlike amino acids, lipids are nutrients that *can* be stored.

Humans diets are richer in fats, sugars, and complex carbohydrates than those of chimpanzees, which accounts for the longer chimpanzee small intestine.

6.16 Sugar Digestion in the Small Intestine

The digestion of sugar is slightly more complicated than it might appear above. The simplest molecules with the sweet taste we all know and love as "sugar" actually comes in three versions. These relatively simple molecules are called **monosaccharides**, single-sugar molecules: **glucose**, **fructose**, and **galactose**. Glucose (specialists would call it D-glucose), fructose, and galactose all have the same chemical formula – 6 carbons, 6 oxygens, and 12 hydrogens, $C_6H_{12}O_6$ – but are linked together slightly differently.

Two monosaccharides, glucose and galactose, do not simply osmose across the intestinal wall, but must pass through a sugar gate, the key for which is salt (Figure 6.5). Salt (actually, just the sodium part of salt) helps to push sugars across the tissue barrier by pushing it into this gate, which as it closes, pushes the sugars into the bloodstream on the other side

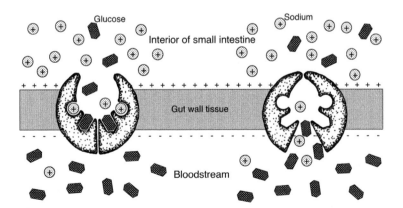

Figure 6.5 Glucose is pushed across the tissue barrier in the gut by an electrical charge. The gut side of the tissue barrier is positively charged because it has sodium (half of the salt molecule) ions. This pressure turns sodium into a key that, with glucose, opens the gate and ultimately spills glucose into the bloodstream, where it is shunted to the liver. Galactose uses the same mechanism. Image by the author.

(Figure 6.5; Southgate, 1995). A high concentration of sodium in the gut and a lower one in cells and the circulatory system creates the electrical potential or charge difference that pushes glucose through the gate and into the arteries surrounding the gut.

Fructose absorption is less well understood. It may be able to pass across the intestinal cell wall and into the bloodstream by passive diffusion (Southgate, 1995), just as water is absorbed by a sponge, but more likely it is pushed through the gut wall with a system similar to that of glucose and galactose, except we believe it is dependent on transporter proteins, rather than sodium.

Table sugar, or sucrose, is a **disaccharide**, as is lactose (the sugar found in milk) and maltose. Each of the three disaccharides is two single-sugar molecules linked together: lactose is glucose and galactose, and table sugar is glucose and fructose. Disaccharides are too big to pass through the sugar gate (Figure 6.5) but are easily broken down into single-sugar molecules by an enzyme secreted by the cells of the intestinal wall, predictably named **disaccharidase**. Starches, or complex carbohydrates (Table 6.2), I hope you have already learned, are chains of single-sugar molecules loosely linked together and readily digestible by amylase. It is an important enough nutrient that we, and presumably chimpanzees, perceive it as a unique taste, like "sweet" or "sour"; it is a pleasant "bready" or "ricey" taste (Lapis et al., 2016).

6.17 Colon and Cecum

When the gut contents reach the colon, they are largely devoid of nutrients and consist of a watery mixture of fiber, seeds, and other waste. As this slurry moves along the colon, bacteria digest some of the cellulose, after which water and some last nutrients are extracted. The secret of the cecum will be unsealed in the next section.

PART IV: FOREGUT VERSUS HINDGUT FERMENTERS

To our regret, we mammals cannot synthesize cellulase and so cannot digest fiber directly. Instead, fiber is digested for us by coevolved bacteria. These symbiotic microorganisms are so tightly coevolved with the grazers and browsers they inhabit that they are utterly dependent on one another.

Foliage specialists are, in essence, big fermenting vats; their guts, their physiology, and even their behavior are evolved to make their guts comfortable environments for their cellulose-digesting microbiome. Gut bacteria do their work by dissolving plant cell walls and consuming the plant cell contents, processing them into – not sugars, as you might expect – but fatty acids, which can then be absorbed by the mammal. As they split open plant cell walls, they spill cell contents into the slurry of digesta, after which both the cell contents and the bacteria themselves are digested into amino acids (Lambert, 1998).

The morphological and physiological length to which folivores go to digest fiber may seem extreme, but it pays off because fiber is superabundant in nature. The supermolecule cellulose is a structural molecule that forms both the supporting scaffold for plant cell walls and ultimately superstructures like stems. This tough, indigestible molecule is really just a long chain of the sweet, nearly perfect nutrient glucose (Table 6.2). As intimated above, when tied together loosely, glucose chains are starch, a wonderful food for us. When tied together in a much, much tighter configuration the chain is so different from starch – even though both are chains of sugars – that chemists label it a "non-starch polysaccharide." "Poly" for many and "saccharide" for sugar.

Mammals have gone in two directions with this coevolved system: In one system, their stomach has been evolved into an efficient fermenting organ, specializing in cellulose digestion above everything else; in the other, mammals have a mixed-use gut that, while less efficient, has its own ingenious advantages. Chimpanzees are as far away from foregut fermentation as possible, yet they can still digest cellulose, though not very well.

6.18 Foregut Fermenters

Mammals with the first system are **foregut fermenters** (Parra, 1978). Primate foregut fermenters

Table 6.2 **Carbohydrates in the diet**

Class of Carbohydrate			Source	Digestibility	Importance to human diet
1. Sugars					
	Monosaccharides ("simple sugars")				
		Glucose	Fruits, grains, beans, vegetables, nuts	Highest (no digestion)	Major
		Fructose	Fruits	High (converted in liver)	Major
		Galactose	Fermented milks such as yogurt	High (converted in liver)	If fermented milk is in diet
	Disaccharides				
		Sucrose (glucose + fructose)	Fruits, some vegetables (beets, corn)	High (requires sucrase)	Major
		Lactose (glucose + galactose)	Milk	High (requires lactase)	In proportion to milk in diet
		Maltose (two glucoses)	Syrups (small amounts in some grains)	High (requires maltase)	Minor (in syrups)
2. Oligosaccharides					
	Trisaccharides				
		Maltotriose compounds	Syrups	Somewhat high	Minor
		Raffinose	Seed legumes, vegetables	Somewhat high	Mostly minor but variable
		Tetrasaccharide	Vegetables	Somewhat high	Mostly minor but variable
		Pentasaccharide	Vegetables	Somewhat high	Mostly minor but variable

Table 6.2 (*cont.*)

Class of Carbohydrate			Source	Digestibility	Importance to human diet
3. Polysaccharides (polymers; ≥10 residues)					
	Starches (complex carbohydrates)				
		Amylose	Root vegetables, grains	Medium (requires amylase)	Major
		Amylopectin	Grains, potatoes, of fruits only bananas	Medium (requires amylase)	Major
	Non-starch polysaccharides				
		Dietary fiber			
		Cellulose	Plant cell walls e.g., leaves	Low (requires microflora)	Depends on fiber intake
		Noncellulosic fiber	Plant cell walls e.g., leaves	Low (requires microflora)	Depends on fiber intake
		Hemicelluloses, pectin	Plant cell walls e.g., leaves	Low (requires microflora)	Depends on fiber intake
		Isolated polysaccharides	Gums, mucilages	Low (requires microflora)	Minor

Source: after Southgate, 1995.

such as leaf monkeys and their relatives (the taxonomic subfamily Colobinae; see Appendix 1) have large stomachs that are segmented into four fermenting chambers so that they have more than one food/microbiome mixture brewing at any one time. Although this system vastly broadens the variety of plant parts that can be eaten, it carries with it harsh limitations (Parra, 1978; Lambert, 1998). Gut bacteria require either a neutral or alkaline pH, not an acidic environment. Eating acidic fruits disrupts the foregut fermenter's microbiome. Fats, because they are digested into amino *acids*, have a similar effect.

Foregut fermenters live in constant risk of lowering their pH below the critical level at which their gut bacteria begin to die: **acidosis**. Thus, while foregut fermenters have the seeming superpower of the ability to digest the indigestible, their need for a certain pH takes calorie-rich fruit or fat-rich food items off the table. Yet, humble as the festive board may seem for the foregut fermenter, their numbers have exploded over the last 20 million years.

6.19 Hindgut Fermenters

Hindgut fermenters such as chimpanzees have a much more ingenious system that patches a faux-foregut system onto the back end of the gut, thus allowing them to eat sugar- and fat-rich foods when they can, yet permitting them to digest cellulose well enough to see them through lean times. As digesting food is pushed through the small intestine and lipids, sugars, starches, and protein are digested out, the gut contents contain an ever-lower sugar and fat content until, as the digesta approach the colon (see Figure 6.4), only waste and indigestible cellulose remains. Most of the cellulose is then shunted into the cecum rather than pushed along into the colon, and in this back-end, supplementary digesting chamber, coevolved microflora do their work, digesting cellulose and yielding fatty acids, amino acids, a little sugar, and some waste gases such as methane and carbon dioxide. These cecal digestion products are then pushed into the colon where they continue to be absorbed. The fiber that fails to enter the cecum is not a total loss; some fermenting digestion also occurs in the colon.

When the chimpanzee diet is rich in high-sugar fruits there is little cellulose in the diet and the gut contents reaching the colon are just waste. When chimpanzees are forced to eat leaves, piths, and barks, the cellulose in these is digested in the cecum.

6.20 Digestion Is Not Free

We cannot lose sight of the fact that the complex process of digestion, while essential, is far from free. Food fuels us, but digestion is an expensive proposition: 10 percent of your caloric intake is consumed in the process (Secor & Diamond, 1995). The cost is not only the cost of manufacturing all those digestive enzymes, but the gut muscles also get a workout. The churning action of the stomach, the wave-like actions of the small intestine, and even the extraction of water from the liquid slurry toward the end of the process have costs.

PART V: SECONDARY COMPOUND PROCESSING

6.21 Tannins

I have been a little sketchy so far about how the body copes with secondary compounds. Chimpanzees employ the best defense against tannins at all times: avoid them; eat foods that lack them. In experiments, chimpanzees rejected bananas treated with tannins, while monkeys did not (Wrangham & Waterman, 1983). When the forest has no ripe fruit, which unfortunately for chimpanzees occurs all too often, they must fall back on leaves, piths, herbs, unripe fruits, and other low-quality foods. In these lean times higher loads of secondary compounds are unavoidable. They cope by producing excess saliva, sacrificing valuable saliva contents such as water, amylase, and lipase to the task of neutralizing toxins; similarly, enzymes in the gut are likewise sacrificed. While some tannins can be broken down by digestive enzymes, many cannot.

6.22 Toxins

Toxins are dealt with in a completely different manner. Digestion does little to alter them, even in the guts of the most powerful digesters such as grazers, but the weak digestive system of the chimpanzee does nothing. Kidneys cannot filter out unadulterated toxins, so they are captured in the liver and disabled by enzymes that fasten onto them a neutralizing molecule; the larger supermolecule is then captured by the kidneys and excreted. Whether a particular alkaloid is poisonous to an animal or not depends on whether the species has evolved the capacity to synthesize enough of the neutralizing enzyme specific to that alkaloid quickly enough to avoid death.

6.23 Chimpanzees and Humans are Poor Detoxifiers

Because chimpanzees are poor at detoxification – a lack that makes many of the foods eaten by monkeys inedible to chimpanzees – their best means of coping with alkaloids is to recognize them in foods and reject those foods. The bitter taste of alkaloids is no accident – we and chimpanzees both rely on this detection and rejection mechanism to avoid being poisoned.

6.24 Toxins Target Specific Enemies and Can be Medicines

Plants are threatened by a variety of enemies, among them seed predators, fungi, bacteria, and insects, but often one or two enemies are paramount. Thus, toxins are often evolved to target specific adversaries. There is a bright side to animal specificity. Some toxins that are relatively harmless to us are quite poisonous to other organisms, some of which are disease vectors. This specificity is the basis of pharmacology. Clever humans – and chimpanzees, to some extent (Chapter 14) – can consume these toxins to kill disease vectors. Penicillin, helpfully produced by the *Penicillium* fungus, kills many disease-causing bacteria. Other poisons attack cancer cells but are much less toxic to normal cells; taxol, an alkaloid found in the bark of the Pacific yew, is one of these. Quinine kills malaria parasites. Aspirin, morphine, novocaine, and codeine are (if used wisely) painkillers. Other poisons such as cocaine, nicotine, and caffeine have pleasant psychoactive effects when consumed in tiny doses. Others alkaloids are simply poisonous: solanine, arsenic, cyanide, and strychnine.

6.25 Few Plants are Chimpanzee Food

If you once viewed the forest as a green cornucopia of the delectable, you now know that plants are "aware" of your slavering hunger and invest heavily in antifeedants to make their body parts less gustatorily appealing. They need fruit dispersers, but they fine-tune the appeal of their fruits, "managing" their fruit dispersers by manipulating fruit size, nutrient content, and the timing of the disappearance of antifeedants.

We will discuss this in greater detail in Chapter 24, but as a preview: Digestive physiology has a profound effect on society. The monkey tolerance of antifeedants is ultimately responsible for social bonds among females. Antifeedant tolerance yields female-bondedness.

Chimpanzees, like other apes, eat a food that is so thin on the ground – and in the trees – that it cannot be defended. As a consequence, female chimpanzees are solitary and they permanently settle into a small range, a core area, where they can memorize the location of every important food resource. That settledness means that males can increase their reproductive success by forming bonds and by defending a territory, defending mating access to females.

The fission–fusion nature of chimpanzee society, then, is a result of the sparseness of their food and the variability in ripe fruit supply from place to place, a variability that afflicts monkeys much less because they can thrive on foods high in antifeedants, such as unripe fruit and leaves.

Fission–fusion society demands intelligence (Chapter 25). In an indirect way, when our ancestors failed to evolve antifeedant tolerance they set the stage for the evolution of sophisticated social awareness, which in turn resulted in increased intelligence. We and chimpanzees owe the pattern of our social relationships as well as our intelligence to our long-ago determined avoidance of antinutrients. As we hear so often, it all comes down to *Food, Glorious Food*.

References

Alba R, Cordonnier-Pratt MM, Pratt LH (2000) Fruit-localized phytochromes regulate lycopene accumulation independently of ethylene production in tomato. *Plant Physiol* 123, 363–370.

Allen G (1879) *The Colour Sense: Its Origin and Development*. London: Trübner.

Berenbaum MR (1995) Turnabout is fair play: secondary roles for primary compounds. *J Chemical Ecol* 21, 925–940.

Blackman E (1971) Opaline silica bodies in the range grasses of southern Alberta. *Canadian J Botany* 49, 769–781.

Brady CJ (1987) Fruit ripening. *Ann Rev Plant Physiol* 38, 155–178.

Chivers DJ, Hladik CM (1980) Morphology of the gastrointestinal tract in primates: comparisons with other mammals in relation to diet. *J Morph* 166, 37–86.

Chivers DJ, Hladik CM (1984) Diet and gut morphology in primates. In *Food Acquisition and Processing in Primates* (eds. Chivers DJ, Wood BA, Bilsborough A), pp. 213–230. New York: Plenum Press.

Coombe BG (1976) The development of fleshy fruits. *Ann Rev Plant Physiol* 27, 207–228.

Dominy NJ (2004) Fruits, fingers, and fermentation: the sensory cues available to foraging primates. *Int Comp Biol* 44, 295–303.

Dominy NJ, Lucas PW (2001) Ecological importance of trichromatic vision to primates. *Nature* 410, 363–366.

Eriksson O, Friis EM, Lofgren P (2000) Seed size, fruit size, and dispersal systems in angiosperms from the Early Cretaceous to the Late Tertiary. *Am Nat* 156, 47–58.

Feeny PP (1970) Seasonal changes in oak leaf tannins and nutrients as a cause of spring feeding by winter. *Ecology* 51, 565–581.

Fraenkel G (1969) Evaluation of our thoughts on secondary plant substances. *Entomol Exp Appl* 12, 473–486.

Freeland WJ, Janzen DH (1974) Strategies in herbivory by mammals: the role of plant secondary compounds. *Am Nat* 108, 269–289.

Gartlan JS, McKey DB, Waterman PG, Mbi CN, Struhsaker TT (1980) A comparative study of the phytochemistry of two African rain forests. *Biochem Syst and Ecol* 8, 401–422.

Gautier-Hion A, Duplantier JM, Quris R, et al. (1985) Fruit characters as a basis of fruit choice and seed dispersal in a tropical forest vertebrate community. *Oecologia* 65, 324–337.

Giovannoni J (2001) Molecular biology of fruit maturation and ripening. *Ann Rev Plant Physiol Plant Mol Biol* 52, 725–749.

Giovannoni JJ (2004) Genetic regulation of fruit development and ripening. *Plant Cell* 16, S170–S180.

Giovannoni JJ (2007) Fruit ripening mutants yield insights into ripening control. *Curr Opin Plant Biol* 10, 283–289.

Glander KE (1982) The impact of plant secondary compounds on primate feeding behavior. *Ybk Phys Anthropol* 25, 1–18.

Heiduk A, Brake I, von Tschirnhaus M, et al. (2016) *Ceropegia sandersonii* mimics attacked honeybees to attract kleptoparasitic flies for pollination. *Curr Biol* 26, 2787–2793.

Herms DA, Mattson WJ (1992) The dilemma of plants: to grow or defend. *Quart Rev Biol* 67, 283–335.

Houle A, Conklin-Brittain NL, Wrangham RW (2014) Vertical stratification of the nutritional value of fruit: macronutrients and condensed tannins. *Am J Primatol* 76, 1207–1232.

Hunt KD (2016) Why are there apes? Evidence for the co-evolution of ape and monkey ecomorphology *J Anat* 228, 630–685.

Jacobsen N (1970) Salivary amylase. II: alpha amylase in salivary glands of the *Macaca irus* monkey, the *Cercopithecus aethiops* monkey, and man. *Caries Res* 4, 200–205.

Janson CH (1983) Adaptation of fruit morphology to dispersal agents in a neotropical forest. *Science* 219, 187–189.

Kossel A (1891) Über die chemische Zusammensetzung der Zelle. *Archiv für physiologie* 4, 181–186.

Lambert JE (1998) Primate digestion: interactions among anatomy, physiology, and feeding ecology. *Evol Anthropol* 7, 8–20.

Lambert JE, Garber PA (1998) Evolutionary and ecological implications of primate seed dispersal. *Am J Primatol* 45, 9–28.

Lapis TJ, Penner MH, Lim J (2016) Humans can taste glucose oligomers independent of the hT1R2/hT1R3 sweet taste receptor. *Chemical Senses* 41, 755–762.

Milton K (1999) Nutritional characteristics of wild primate foods: do the diets of our closest living relatives have lessons for us? *Nutrition* 15, 488–498.

Mothes K (1980) Historical introduction. In *Secondary Plant Products* (eds. Bell EA, Charlwood BV), pp. 1–10. New York: Springer-Verlag.

Murray P (1975) The role of cheek pouches in cercopithecine monkey adaptive strategy. In *Primate Functional Morphology and Evolution* (ed. Tuttle RH), pp. 151–194. The Hague: Mouton.

Parra R (1978) Comparison of foregut and hindgut fermentation in herbivores. In *The Ecology of Arboreal Folivores* (ed. Montgomery GG), pp. 205–229. Washington, DC: Smithsonian Institution Press.

Raff RA, Kaufman TC (1983) *Embryos, Genes and Evolution*. New York: MacMillan.

Regan BC, Julliot C, Simmen B, et al. (1998) Frugivory and color vision in *Alouatta seniculus*, a trichromatic platyrrhine monkey. *Vision Res* 38, 3321–3327.

Rhoades DF, Cates RG (1976) A general theory of plant antiherbivore chemistry. *Rec Adv Phytochem* 10, 168–213.

Rubin J, Clawson M, Planch A, Jones Q (1988) Measurements of peritoneal surface area in man and rat. *Am J Med Sci* 295, 453–458.

Sacher JA (1973) Senescence and postharvest physiology. *Ann Rev Plant Physiol* 24, 197–224.

Schnable PS, Ware D, Fulton RS et al. (2009) The B73 maize genome: complexity, diversity, and dynamics. *Science* 326, 1112–1115.

Secor SM, Diamond J (1995) Adaptive responses to feeding in Burmese pythons: pay before pumping. *J Exp Biol* 198, 1313–1325.

Seymour GB, Østergaard L, Chapman NH, Knapp S, Martin C (2013) Fruit development and ripening. *Ann Rev Plant Biol* 64, 219–241.

Southgate DA (1995) Digestion and metabolism of sugars. *Am J Clin Nutrition* 62, 203S–210S.

Stahl E (1888) Pflanzen und Schnecken. *Jena Z Med Naturwiss* 22, 559–684.

Stevens CE, Hume ID (2004) *Comparative Physiology of the Vertebrate Digestive System*. Cambridge: Cambridge University Press.

Swain T (1977) Secondary compounds as protective agents. *Ann Rev Plant Physiol* 28, 479–501.

Terborgh J (1983) *Five New World Primates: A Study in Comparative Ecology*. Princeton, NJ: Princeton University Press.

Tiffney BH (1984) Seed size, dispersal syndromes, and the rise of the angiosperms: evidence and hypothesis. *Ann Mo Bot Garden* 71, 551–576.

Van Straten S, Maarse H, de Beauveser JC, Visscher CA (1983) *Volatile Compounds in Foods*, 5th ed. Zeist: Division for Nutrition and Food Research TNO.

Wrangham RW (1979) The evolution of ape social systems. *Soc Sci Infor* 18, 335–368.

Wrangham RW (1980) An ecological model of female-bonded primate groups. *Behaviour* 75, 262–299.

Wrangham RW, Waterman PG (1983) Condensed tannins in fruits eaten by chimpanzees. *Biotropica* 15, 217–233.

Wrangham RW, Conklin-Brittain NL, Hunt KD (1998) Dietary response of chimpanzees and cercopithecines to seasonal variation in fruit abundance. I. Antifeedants. *Int J Primatol* 19, 949–970.

Yang SF, Hoffman NE (1984) Ethylene biosynthesis and its regulation in higher plants. *Ann Rev Plant Physiol* 35, 155–189.

7 Thews, Sinews, and Bone
Chimpanzee Anatomy and Osteology

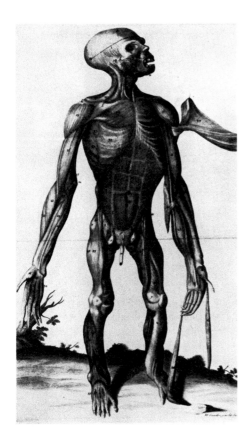

Image: Tyson, 1699.

Long ago, before the invention of many things that are so much a part of our daily lives that we take them for granted – before bicycles, matches, trains, the canning of food, and the invention of flight – we had in our hands a thorough record of chimpanzee anatomy (Tyson, 1699; Brockman, 2000). As improbable as it seems, Western science first confronted the astounding particulars of chimpanzee anatomy over 300 years ago with the publication of Edward Tyson's brilliant monograph documenting his dissections (see Figure 7.22 for an illustration from the book).

Superficial similarities between chimpanzees and monkeys makes calling chimpanzees monkeys an irresistible error; perhaps their hairiness, snoutiness, and lack of a forehead fosters that. However, an objective, scientific accounting of human, monkey, and chimpanzee characteristics reveals them to be nearly as different from monkeys as we are. In his monograph, as Tyson worked through detail after detail, he could scarcely contain his astonishment at the humanness of his specimen. The implication was startling, if slow to sink in: We humans are merely one of many apes, a reality with which we are still coming to grips. Perhaps I should repeat that in a different way: We are not just related to apes, we *are* apes.

Anatomy is the scholarly study of muscles, ligaments, tendons, internal organs, arteries, veins, and even microscopic and biochemical physical structures. In this chapter we will focus on muscles, **tendons** (fibrous tissue that connects muscles to bones, cartilage, or other muscles), **ligaments** (fibrous connective tissue that connects bones to one another), and the bones themselves. You may remember from an earlier chapter that the study of bones is called **osteology**. Chimpanzee biochemistry, neuroanatomy, hormones, and the form of the gastrointestinal tract will be covered in other chapters. When research relates function to ecology, the field of study is called **ecomorphology**. Study of the geometric qualities of some morphological feature is known as **geometric morphology**, sometimes shortened to **geomorphology**; the study of function is called **functional morphology**. When function is quantified by examining the forces acting on parts of the body as if the body were a mechanical device, this is **biomechanics**. We will examine the function of chimpanzee anatomy in Chapter 8.

7.1 Human and Chimpanzee Similarity

For those of you encountering comparative primate anatomy for the first time, the assertion that humans closely resemble chimpanzees may seem nothing more than a slippery verbal trick, an attempt to shock, ivory-tower gobbledy-gook that academics engage in all too often when they write for an audience outside their specialization. "Actually, from a mathematical perspective," a specialist might write, "rings and coffee cups are the same shape." Right; sure they are. This is not that.

Anatomists are not blind; they recognize that at first glance chimpanzees and other apes do look a lot like monkeys. They are covered with hair and lack an external nose; both have sloping foreheads, and both have hand-like feet with divergent and gripping great toes (Figure 7.1). Both have big mouths, receding chins, and protruding faces – more correctly, they have **prognathic** faces (Figure 7.2). Massive chewing muscles (Figure 7.3) give both apes and monkeys jaws with twice the power of humans, or more, compared to body mass (Eng et al., 2013).

Unlike monkeys, however, they lack a tail (Figure 7.4) and have a broad thorax, giving them shoulders. However, it is when you get really close up that the resemblance to humans is remarkable. The major internal organs of chimpanzees – the heart, lungs, liver, kidneys, and pancreas – differ in only the finest of details (Sonntag, 1923; Bourne, 1972). If not for the problem of tissue rejection, any one of these internal organs could be transplanted into humans with no adverse consequences whatsoever. The similarity is so profound that after reading a few pages of the twentieth century's most definitive chimpanzee gross anatomy exposition, Chaeles Sonntag's *On the Anatomy of the Chimpanzee*, one wonders if any part of chimpanzees differs from human anatomy. His description of the configuration of the orbicularis oculi muscle is simply "as in Man." He moves on to the muscles of the lips to note they are in two layers, "as in Man." The words "as in Man" appear again and again – in fact, Sonntag is a positive bore on the subject. He leaves off describing a feature to merely note that it is indistinguishable from humans 182 times in the 97-page work.

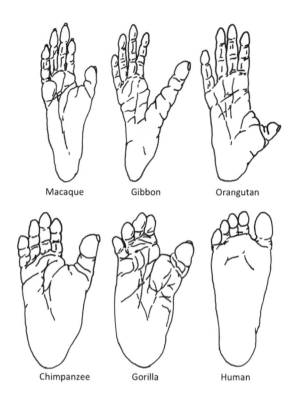

Figure 7.1 The chimpanzee foot has a divergent great toe that functions more like a thumb than a human big toe. The analogy to the hand extends to the lateral toes which are, like fingers, long and powerful. As in the hand, the lateral toes are much longer than the great toe. From Hunt, 2016; after Schultz, 1950.

Thus, while there are great differences in face shape, hairiness, and macro-scale body proportions, at a centimeter and millimeter scale there is a near identity. In other words, humans and chimpanzees resemble one another not in the way a pearl resembles a ball bearing, but more in the way a minivan resembles a MINI Cooper. At first glance the shape and size are dramatically different, but when you look closely, they have nearly identical steering wheels, tires, spark plugs, fuel tanks, disk brakes, seats, safety belts, and so on.

7.2 Body Size

Chimpanzees are smaller than humans (Table 7.1). However, documenting the exact difference is

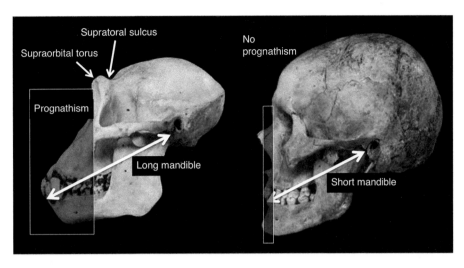

Figure 7.2 Chimpanzee and human skulls. Chimpanzee faces are prognathic; that is, much of the face projects in front a vertical line placed in front of the eye orbits. Chimpanzees have a prominent **supraorbital torus**, behind which there is a distinct depression, a **supratoral sulcus**. Chimpanzees lack a bony external nose.

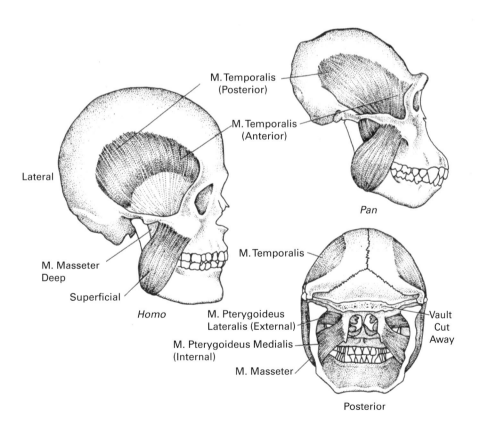

Figure 7.3 Muscles of mastication compared in humans (left) and the chimpanzee. With permission from Swindler & Wood, 1982.

Table 7.1 **Chimpanzee body mass (kg)**

Population	Species	Male mass	Female mass	Average
West Africa	*Pan troglodytes troglodytes*	57.5	45.0	51.3
West Africa	*Pan troglodytes troglodytes*	42.7	33.7	38.2
Gombe	*Pan troglodytes schweinfurthii*	39.9	31.3	35.5
Chimpanzee average		46.7	36.6	41.7
Ache humans				55.0

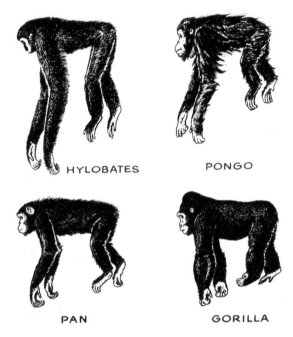

Figure 7.4 Body proportions among the apes. From Erikson, 1963. With permission.

environment similar to that in which we evolved. This principle is incredibly important, the idea that organisms are evolved to function well in a particular environment (their EEA), and when removed from that environment they may not even be able to survive. The Ache, a group of South American hunter-gatherers, are a good model (Walker & Hill, 2003). They average 55 kg (about 120 lb). Gombe chimpanzee males weigh ~40 kg and females 31 kg (Table 7.1; Wrangham, 1977; Pusey et al., 2005), though West African chimpanzees weigh a little more – 58 kg and 45 kg for wild individuals, according to Jungers and Susman (1984) and 43 kg and 34 kg according to Smith and Jungers, (1996). A good estimate, average males and females, is 42 kg (93 lb). Chimpanzees body weight exceeds that of any monkey.

It is easy to overlook how important body size is when we compare species, yet it is one of an animal's most important attributes. It determines how successful a species is when competing for food at a crowded feeding site, how quickly individuals mature, and how dangerous predators are. Larger species are less adept in the periphery of trees because branches are smaller. Not insignificantly, the larger the body mass the less costly it is to supply energy and oxygen to the brain; if gorillas had a brain the size of a walnut, doubling it in size would require very little change in its metabolic needs, percentage-wise.

challenging. Body weight is much easier to measure in captivity, but captive animals grow faster and reach larger body masses than wild animals. And we humans are also affected by "captivity"; with better medical care and a virtually unlimited diet, body weights are much greater among industrialized populations than they are in hunter-gatherers, purely as a developmental phenomenon, not a genetic one. And 15,000 years ago we were all hunter-gatherers. It might be best to compare wild chimpanzees to humans living closer to our **environment of evolutionary adaptedness** (EEA; Bowlby, 1958), an

7.3 The Skull and its Soft Tissue

Turning from size to the anatomy itself, and starting from the top of the head: Compared to humans, chimpanzees have a prominent brow ridge, a structure called a **supraorbital torus** by osteologists (Figure 7.2). Behind that brow ridge is a **supratoral sulcus**, a depression that separates the torus from the braincase; the face is in front of the brain in chimpanzees, not underneath it as in humans (Figures 7.2, 7.3, and 7.5). The chimpanzee face looks as if you gripped a monkey face on either side of the eyes and pulled it forward, pulling the face out in front of the braincase and leaving a gap between the top of the eyes and the braincase – creating a supratoral sulcus. Neither monkeys nor humans – modern ones at least (*Homo erectus* is chimpanzee-like) – have a torus or a sulcus. Chimpanzees are prognathic; underneath the eyes the face angles out at a 45° angle (Figures 7.2 and 7.3).

Chimpanzees lack a bony external nose, another feature that resembles monkeys. We are so used to the prominent human nose that it looks natural to us – but it is a very unusual feature among primates. The bony nose extends the nasal passage beyond the face, out into the world beyond, lengthening the functional nose to the size of chimpanzees; even though they have no external nose, the chimpanzee nasal passage is about the same size as ours because their face projects much farther forward than ours (Figure 7.5).

The anatomy of the eye is nearly indistinguishable in humans and chimpanzees (Hines, 1942; Prestrude, 1970), though chimpanzee eyes are slightly smaller than those of humans (Young & Farrer, 1964). Chimpanzees have longer rods and cones (Detwiler, 1943). The nuclear (or outer) layer of the retina is thinner in the fundus and at the periphery; Polyak (1941) found 8–10 rows of cells in the outer nuclear layer of the human fovea compared to 4–6 rows for chimpanzees. The function of this difference is unknown.

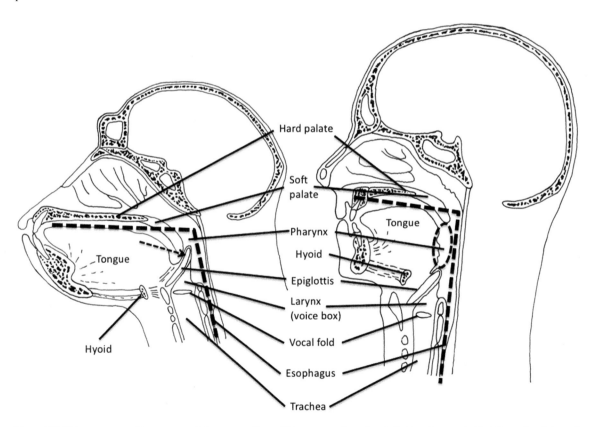

Figure 7.5 Chimpanzee and human cranial cross-sections. Chimpanzees have long, flat tongues; note that the soft palate and epiglottis overlap in chimpanzees (dotted arrow), whereas in humans there is a gap (dotted ellipse). Note the large tongue and less tightly angled esophagus. Modified from Mark Liberman; with permission.

The **temporalis** muscle, a fan-shaped muscle on either side of the head, is so large in chimpanzees that in some individuals the two muscles meet at the top of the head (Figure 7.3). When that happens, the bone responds by growing a fin-like extension called a sagittal crest. The **masseter** muscle is a muscle on the side of your jaw, again much larger in chimpanzees than in humans.

Chimpanzees can produce saliva in staggering abundance, due in part to salivary glands that are large compared to those of humans (Tyson, 1699; Sonntag, 1923). Drenching the chewed bolus in saliva prevents it from becoming a solid, unswallowable mass. Perhaps swallowing is easier because their esophagus is at a less dramatic angle with the oral cavity than is ours (Figure 7.5). Some fruits are so bitter, so sour, or so astringent that when you begin to chew the rush of saliva into your mouth is actually painful, and as you chew the tannins coagulate your saliva and the chewed mass becomes much like a sticky lump of clay, very difficult to swallow. Chimpanzees cope with these difficult foods with long, broad, flat, muscular tongues, three times the size of a human tongue (Takemoto, 2008) and strong enough to power even the coarsest food down the throat; the human tongue is rounded and short from front to back (Figure 7.5).

Greater prognathism gives chimpanzees a longer mandible than humans, yet they lack a chin. Their incisors protrude forward beyond their bony foundations (the mandible and the maxilla), whereas human incisors are more vertically implanted. We humans have teeth pulled back into the jaw, but the mandibular corpus, the lower margin of the jaw, protrudes enough to create a chin, of which we are inordinately proud. And rightly so – a chin is unknown in the rest of the primate order. Science, however, is still not sure why the chin is there; I offer an explanation in the next chapter. The back of the mandible at the midline sports two horizontal ridges at the midline (Figure 7.5), the **superior transverse torus** and the **inferior transverse torus**, neither of which humans have and upon which, again, science casts little light, and unfortunately neither can I.

Chimpanzee facial skin is wrinkly, even in juveniles. Gorillas are similar; orangutans are only

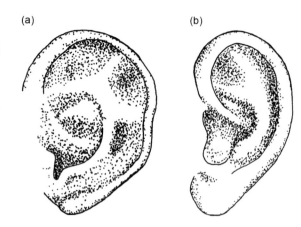

Figure 7.6 (a) Chimpanzee (after Schwalbe, 1916) and (b) human ears. Note the chimpanzee ear is wider and lacks an earlobe. Image by the author.

slightly more human-like. I have never seen an adaptive explanation for the smoothness of human skin, or even a comment on how curious this difference is.

Chimpanzees grow bald as they age, in patterns similar to those of humans.

The chimpanzee external ear or **pinna** is quite human-like in shape (Figure 7.6) compared to other animals, with similar infoldings and substructures, but with a smaller earlobe. Cartoonists are quick to catch onto the fact that chimpanzees have large ears, though is it the width that makes them appear larger than those of humans; they are approximately as broad as they are tall (Hershkovitz, 1977; Coleman, 2009). Compared to other apes, chimpanzee ears are larger in every dimension, nearly twice the size of gorilla ears, despite the gorilla's immense size (2117 mm^2 versus 1289 mm^2), and over four times the size of orangutans (2117 mm^2 versus 487 mm^2; calculations based on data from Schwalbe, 1916).

7.4 Teeth

Chimpanzees have the same number of teeth as humans: 32. Dividing the mouth into four quadrants, each has two **incisors**, a **canine**, two **premolars**, and three **molars**. Most of the teeth are human-like.

While similar in shape, chimpanzee incisors are much larger. Canines, on the other hand, differ from those of humans dramatically in both shape and size. While our canines (you may know them as eye teeth) are not much different from our incisors, the chimpanzee canine is a fang, cone-shaped and so large that the upper and lower canines overlap (Figure 7.3). To accommodate the overlap there is a space or **diastema** between the lower canine and the first premolar and between the upper canine the lateral incisor.

The front of the chimpanzee lower first premolar, the anterior-most of the teeth your dentist would call a bicuspid after its two (bi) cusps (cuspid), is shaped to properly occlude with the upper canine, which sits in front of it; that premolar is **sectorial**. The two rub together in such a way as to sharpen the back of the canine. The second premolar is quite similar in chimpanzees and humans, with two side-by-side cusps.

Chimpanzee molars differ little from those of humans, though the tooth enamel is a little thinner, perhaps another "less-is-more" adaptation; the thinner enamel wears through in some places early in life, leaving sharp edges that may serve to slice up foliage more effectively than a flat surface would.

7.5 The Neck

One particularly striking difference between humans and chimpanzees is the distance between the soft palate and the **epiglottis** (Figure 7.5). The epiglottis is a cartilaginous valve that closes over the air passage (the voice box or **larynx**, and the rest of the air passage, the **trachea**, below that) during swallowing so that food or liquids are pushed down the esophagus to the stomach, rather than sucked into the lungs. The vocal tract is made up of two spaces, a horizontal one called the **horizontal supralaryngeal vocal tract** (horizontal dotted line segments in Figure 7.5), the space that is the oral cavity or mouth; and the **vertical supralaryngeal vocal tract**, or **pharynx**, which is the space between the soft palate and the epiglottis. Between the shortening of the face and the elongation of the space between the soft palate and the epiglottis, humans have evolved to have these two spaces almost equal in size (reviewed in Lieberman et al., 2007). As you can see (Figures 7.5 and 7.7), the human esophagus makes a nearly 90° angle with the oral cavity (dotted lines), but the angle is opened up in chimpanzees, which may make it easier to swallow some items. In my career I have eaten just about every plant item chimpanzees eat, and some of them are extremely difficult to swallow; leaves that are coarse, almost like sandpaper, or covered with hairs come to mind. Chimpanzees can swallow these leaves whole (Wrangham & Nishida, 1983), but I have never managed it without chewing.

The larynx contains the vocal cords protected by a cartilaginous box, part of which you can feel as the Adam's apple. The larynx is larger in humans and there is some difference in muscle size and orientation between chimpanzees and humans, but otherwise vocal tract anatomy is remarkably similar. We make the sounds required for speech by varying the relative proportions of the horizontal supralaryngeal vocal tract and the vertical supralaryngeal vocal tract, by varying the shape of the tongue, by changing the shape of the lips and mouth, and by varying the passage of air through the trachea. These activities are finely controlled by the brain, and it is largely that neural control that enables speech, more than any differences in the shape of the vocal tract.

Chimpanzees have a **laryngeal sac**, a bag-like extension of the larynx, that fits into an opening in the hyoid (Figure 7.8; Schultz, 1969), the function of which is not completely clear. Most experts think the sac serves as an air reservoir that helps to extend the duration of vocalizations, but it may be that it prevents hyperventilation; during vocalizations the sac rather than the lungs can be filled with inspired air then used to vocalize. Gorillas and indeed all apes except gibbons have a laryngeal sac, as did australopiths (Perlman & Salmi, 2017), but among humans they are extremely rare.

Chimpanzees have a shorter, less rotatable neck than humans, causing them to tend to turn their whole torso to look behind them. Compared to humans, chimpanzees have immense **nuchal muscles** – the muscles of the back of the neck (Figure 7.9). In fact, the muscles are so large, in males in particular, that there is too little room on the braincase to accommodate them;

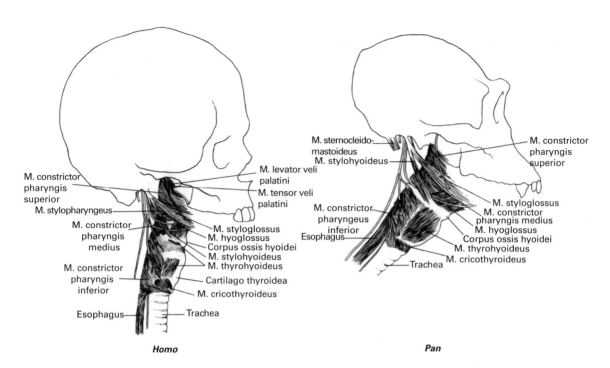

Figure 7.7 Larynx, esophagus, and associated anatomy. The larynx is enclosed by the cricoid cartilage (Cartilago thyroidea, in the more formal naming). Note the similarity of the overall morphology. From Swindler & Wood, 1982. Used with permission.

to increase the surface area for their attachment, a shelf of bone grows outward to ring the back and sides of the braincase (Figure 7.9).

Brain research is such an active field that rather than reviewing neuroscience here I gave it its own chapter, Chapter 19. Just as a preview, chimpanzee brains are so similar to human brains that little is lost in characterizing them as a human brain in miniature; they average 350 cm^3, compared to 1350 cm^3 for humans. Human brains are the size of a cantaloupe and chimpanzee brains are the size of a grapefruit.

7.6 The Hand

While some monkeys have human-like thumbs, the thumbs of chimpanzees are shorter, less mobile, and less well-muscled compared to humans (Napier, 1963, 1967). Chimpanzee fingers are long, thick, and powerful compared to those of humans (Figure 7.10), while monkey fingers are even shorter than human fingers. Chimpanzees have thick fingers in part because the skin is thick and tough, but the diameters of their **phalanges** (finger bones; the singular term is **phalanx**) are also large and the bones are dense and robust. When viewed from the side, chimpanzee fingers are curved (Figure 7.11) and the palm side of the phalanx is cupped, or concave (Figure 7.12). In life this concave area accommodates tendons of the **digital flexor** muscles, the muscles that flex the fingers; the flexor tendons are so thick they are nearly round in cross-section. In humans these tendons are flat slips of tissue (Figure 7.13) and the concavity seen in Figure 7.12 is rare. Large flexors give chimpanzees extremely powerful grips. Chimpanzee fingers are pointy (Figure 7.10) – that is, their terminal phalanges are narrow compared to those of humans and monkeys (Mittra et al., 2007).

Chimpanzees and gorillas have a ridge of bone on their metacarpals that is an adaptation to knucklewalking (Figures 7.11 and 7.14). While this feature may seem unimportant, australopiths and *Ardipithecus* lack it, assuring us that, unlike chimpanzees, they did not engage in knucklewalking.

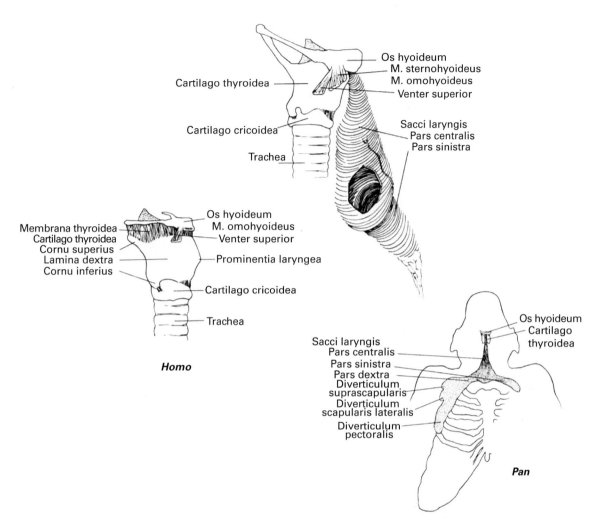

Figure 7.8 The laryngeal sac. From Swindler & Wood, 1982. Used with permission.

7.7 The Arm

Considering the forelimb in its entirety, a chimpanzee arm lying on a dissecting table, especially such a specimen without its skin, would hardly look any different to you than a human arm. The length would be very similar, though this similarity is a little deceptive since chimpanzees are somewhat smaller than humans, making their arms slightly longer than ours in proportion. Chimpanzee forelimb elements are progressively more elongated the farther they are from the shoulder: the forearm of chimpanzees (and other apes, too) is a little bit longer than that of humans, and the fingers are much longer (Figures 7.10 and 7.11; Schultz, 1930, 1936).

Chimpanzees share with humans a globular humeral head (solid lines in Figure 7.15). While the ball part of the ball-and-socket shoulder joint is only half a sphere in monkeys and can be wedge-shaped, it describes a large portion of a sphere in chimpanzees. When it is wedge-shaped and the **glenoid fossa** of the scapula, the complementary articulation for the humeral head, is trough-shaped, arm movement is limited to the sagittal plane – the arms move forward and back as in running – and shoulder mobility is poor (Figure 7.15). Muscle attachments for the shoulder girdle, the tubercles, are pulled away from the head so as to give it more clearance with the rim of the glenoid, allowing astonishing shoulder mobility.

In most mammals, including monkeys, the surface of the humeral head faces backward and its

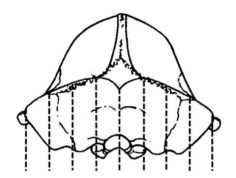

Figure 7.9 Rear view of a chimpanzee skull. Dotted vertical lines represent nuchal muscles. Note that the nuchal area of the skull is larger than the breadth of the braincase, so that the muscles attach to an enlarged plateau. After White et al., 1981. Used with permission.

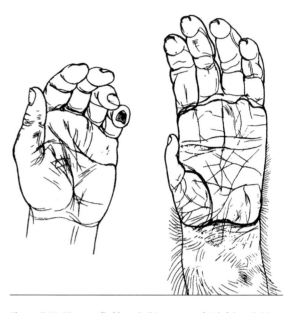

Figure 7.10 Human (left) and chimpanzee (right) hand. Note the small chimpanzee thumb, long fingers, and narrow terminal phalanges (fingertips). From Napier, 1960; with permission.

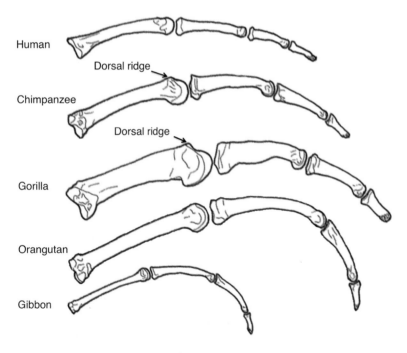

Figure 7.11 Side view of ape and human metacarpals and phalanges; apes, compared to humans, have long, curved digits. Knucklewalking apes, gorillas, and chimpanzees have a dorsal ridge to prevent hyperextension of the finger against the metacarpal (palm bone). After Etter, 1974; with permission.

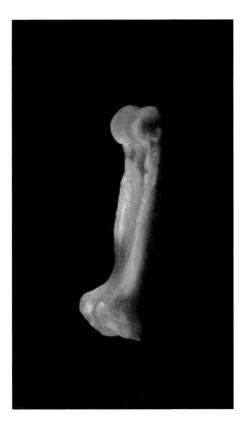

Figure 7.12 Chimpanzee phalanx; note the concave surface on the palmar side. Photo by the author.

complementary joint on the scapula, the glenoid fossa, faces forward. This orientation of the humeral head places the long axis of the elbow at a right angle to the direction the humeral head faces (Figure 7.15). In apes and humans, the glenoid fossa faces laterally, requiring the humeral head to face mostly medially, toward the center of the thorax. The monkey **humerus** (upper arm bone) is twisted, in other words, compared to that of humans and apes (Figure 7.15).

7.8 The Elbow

Osteologists break the humeral elbow joint down into three parts (Figure 7.16). The part that articulates with the ulna and provides most of the stability is called the **trochlea**; the deeper the groove in the trochlea, the greater the elbow stability. Viewed from the front, the chimpanzee trochlea is deeply notched and symmetrical, making it look a little like a bow tie; it is, in other words, spool-shaped (Figure 7.16). Just above the trochlea, apes often have an opening, or foramen, where, when the elbow is fully extended, the olecranon process of the ulna pokes through. In some sense the trochlea is the bottom of a stirrup and the ulna is hung from this stirrup like a clothes hanger hanging from a rod.

The trochlea is connected to the **capitulum**, the articulation for the radius, by a sloping area called the **zona conoidea**, the "cone-shaped zone." The capitulum is sphere-shaped in apes and humans and articulates with the radial head; the sphere shape allows the radius to rotate freely, facilitating the rotatable wrist characteristic of humans and chimpanzees. Monkeys are not very different from dogs, cats, and squirrels in having relatively little wrist mobility. Taken together, the elbow joint of chimpanzees, when viewed from the front, is somewhat saw-tooth shaped (Figure 7.17), whereas humans have a gentler undulation to this surface. The extreme topography of the ape elbow stabilizes the joint.

Apes and humans have a reduced **olecranon process** of the ulna that allows full elbow extension (Figure 7.17). A muscle on the back of the arm, the triceps, attaches here; the longer the olecranon process is, the greater the lever arm for this elbow extensor. A long olecranon process, though, prevents full elbow extension.

7.9 The Wrist

Monkeys have large ulnar styloid processes with an extensive complementary articulation in the wrist, giving the monkey wrist great stability but little mobility. Compared to monkeys, apes have a reduced articulation between the forearm and the wrist, allowing rotatory motion of the wrist and flexibility in other planes as well (Figure 7.18).

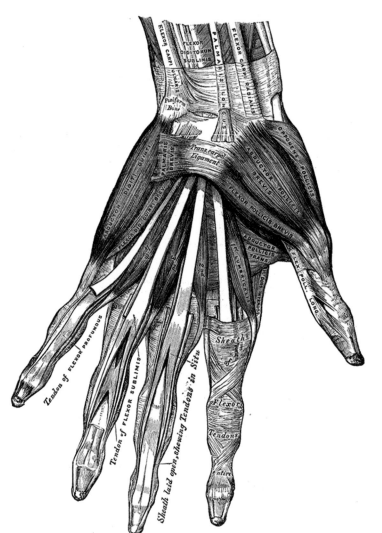

Figure 7.13 Human hand, palm side. The deep flexor tendons that power your hand when you flex your fingers run along the palm side of the phalanges.

7.10 The Thorax

When chimpanzees are sitting, their shorter leg length is hardly noticeable, giving them a human-like appearance, partly because they share a feature with humans that we notice almost subconsciously: they have shoulders. Viewed from the front, monkey shoulders are so sloping that they meld with their neck. You can almost draw a straight line from just below their ear to their elbow. Apes, in contrast, have a nearly 90° angle between their neck and shoulders (look back to Chapter 4, and see Figure 7.23), making them visually and anatomically distinctive. Put another way, ape and human torsos are broader (wider side to side) than they are deep (front to back; Figure 7.19). This orientation in combination with the position of the scapula results in ape and human shoulder joints that point to the side.

With the skin on, then, human and chimpanzee thoraxes are shaped very similarly. Underneath the skin and muscle, however, there are some important differences. Chimpanzees have an extraordinarily short back (Figure 7.20, A_2) compared to humans (A_1). In fact, they only have four **lumbar vertebrae** (lowest

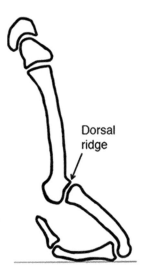

Figure 7.14 Side view of a chimpanzee hand in the knucklewalking position; note the dorsal ridge.

vertebrae, below vertebrae that have ribs attached), whereas we humans have five and monkeys have six or seven (Schultz, 1930; Ankel, 1972); the individual lumbar vertebrae are short as well (Ankel, 1972). The lumbar segment of the vertebral column is so short that the last or floating rib of chimpanzees nearly touches the crest of the **iliac blade** of the **os coxa** (Figure 7.20). The number, height, and shape of chimpanzee lumbar vertebrae make the lower back short and stiff. Monkeys, in contrast, have long backs, which means that in this respect humans are more similar to monkeys than they are to chimpanzees. The elongated pelvis of chimpanzees serves to stiffen the back; rather than having a back made up of segmented vertebrae, chimpanzees have reduced the number of their vertebrae and elongated the os coxae, which is, of course, completely inflexible. Viewed from the front, the **sacrum** is pushed down between the iliac blades so that only two of the lumbar

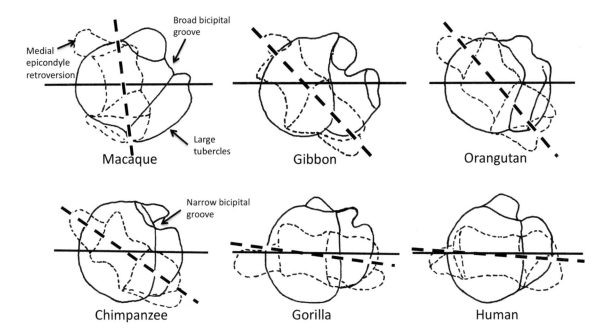

Figure 7.15 Top view of monkey and ape humeral heads (solid) and the long axis of the elbow joint (dotted line). Humeral torsion among macaques orients the long axis of the distal humeral articular surface (dotted outline) nearly at a right angle to the direction of the humeral head (solid line). Hominoids have humeral heads oriented medially and more nearly parallel to the long axis of the distal humerus (dotted line). Hominoids have large, globular humeral heads that more completely describe a complete sphere. Note the narrow bicipital groove in hominoids, a consequence of moving the muscle attachments at the shoulder (the tubercles) away from the humeral head to give it more clearance with the rim of the glenoid. Moving the tubercles pushes them together and narrows the bicipital groove. With permission of John Wiley & Sons.

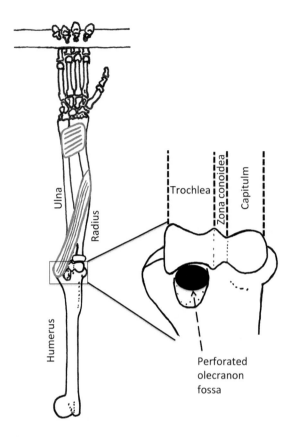

Figure 7.16 Ape elbow osteological terms. From Hunt, 2016, with permission of John Wiley & Sons.

vertebrae are free to move. A back with two vertebrae is less flexible than one with five.

Chimpanzees also have a **cone-shaped** (also known as "funnel-shaped") ribcage that narrows evenly from the bottom to the top of the ribcage (Figure 7.20, dotted lines); humans have a more barrel-shaped thorax. Humans have a much more clearly defined waist than chimpanzees.

Still deeper under the skin, the internal organs of chimpanzees are virtually indistinguishable from those of humans, with the exception of the gastrointestinal tract.

7.11 The Scapula

Compared to the human scapula (shoulder blade; plural: scapulae or less formally, scapulas) the chimpanzee scapula is narrower (Figure 7.21, dimension A_1). Humans have a triangular scapula, but chimpanzees, gibbons, and – can you guess the other primate? – spider monkeys have elongated, oblong scapulae. The orientation of the bowl-like articulation for the shoulder joint points upward (arrows B_1 and B_2). To be more anatomically precise, the chimpanzee **glenoid fossa** is **cranially oriented**. "Cranial" just means in the direction or on the side of the head.

We introduced the shoulder by noting the globular humeral head. This morphology gives all apes and humans the shoulder mobility to move their arms in practically any orientation, from pointing straight up to the sky (**full abduction** in anatomical terms), to pointing directly behind them, to crossing the arm over the chest. This mobility is partly due to the fact that the scapular glenoid fossa faces the side (Figure 7.21) and partly due to the ball-and-socket shape of the shoulder joint. The typical monkey (not surprisingly, spider monkeys, to name one, are an exception) have little more shoulder mobility than your pet dog or cat. If you try to pick up your dog by a front paw, she will probably bite you – it would be painful to them to completely abduct their humerus and their elbow cannot extend completely. Chimpanzees, in contrast, have even greater flexibility than we do, not only in their shoulders but in their wrists, too (Tuttle, 1969).

7.12 Upper Body Musculature: Elbow Flexors are Large

Many muscles differ in size between chimpanzees and monkeys – the relative size of these muscles differs less between humans and chimps; in this way, we are, again, apes. Among these muscles are the **biceps** (Figure 7.22), **brachialis**, and **brachioradialis** muscles, which are large; these are muscles used to flex the elbow, as you would to do a pull-up.

7.13 Arm-Raising Muscles are Large

The **deltoid** is another of the large ape muscles (Figure 7.22). We can contract each of its three different parts separately. They all raise the arm, but

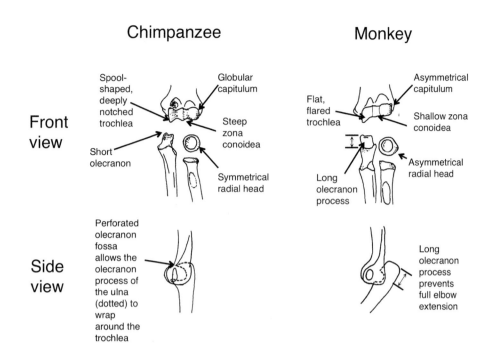

Figure 7.17 Ape elbows have a spool-shaped trochlea, a steep zona conoidea, a globular capitulum, a short olecranon process and a symmetrical radial head. Apes have a topographically distinct or saw-tooth shaped elbow joint. The olecranon process of the ulna is short enough to allow the process to fit in a depression on the back of the humerus, allowing the elbow to completely extend. The long olecranon process of the monkey elbow prevents complete extension, but gives elbow extensors a long lever arm. Images by the author.

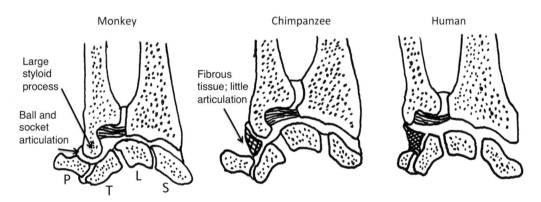

Figure 7.18 Wrists of a monkey, chimpanzee, and human. In monkeys the ulnar styloid process closely articulates with a cup-like surface formed by two carpal bones, the pisiform and triquetral, stabilizing the wrist but limiting wrist rotation. Apes have less articulation and the styloid process and the carpus have a flexible meniscus between them, allowing ulnar deviation and supination. Humans have an anatomy allowing still more flexibility. P, pisiform; T, triquetral; L, lunate; S, scaphoid. From Hunt, 2016, with permission of John Wiley & Sons; after Lewis, 1972.

the anterior deltoid raises the arm to the front, the lateral or middle part to the side, and posterior to the back. **Cranial trapezius**, your shrugging muscle, has its origin on the spine of the scapula (see Figure 7.21) and inserts on the spinous processes of the lower neck and upper thoracic vertebrae (Figure 7.23); supraspinatus (not shown) is also large. All of these muscles raise the arm.

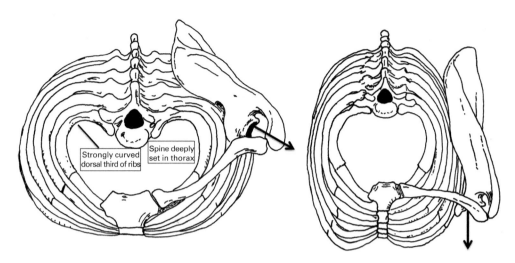

Figure 7.19 Cranial or top view of ribcage, scapula, clavicle, and manubrium and sternum of typical human (left) and monkey (right) torsos. Monkeys have narrow torsos with a glenoid fossa that faces forward (arrow), whereas humans have a shallow and broad torso and shoulder joints face to the side. From Hunt, 2016; after Schultz, 1950. Used with permission.

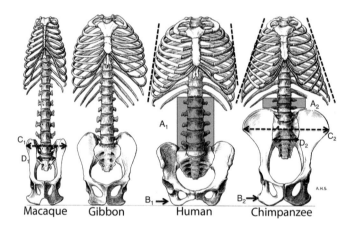

Figure 7.20 Torsos (ribcage and pelves) of monkeys and apes. Monkeys have narrow torsos and narrow pelves. Apes have broad torsos, giving them shoulders and an upper-body appearance very similar to that of humans. Humans have long backs (shaded A_1) compared to chimpanzees (A_2), and the chimpanzee torso is cone-shaped. Humans (B_1) lack the flaring ischia of chimpanzees (B_2) that serves as the foundation for the fibrous tissue that forms sitting pads. Apes have broad pelves compared to gibbons and macaques; note that the ribcage is broader than the pelvis in macaques and gibbons, slightly broader in humans, and exactly as broad as the pelvis in the chimpanzee. Apes have a broad pelvis (C_2) compared to monkeys (C_1), whereas monkeys have a broad sacrum (D_1) compared to the narrow sacrum of the chimpanzee (D_2). Modified from Schultz, 1969, with permission of the Adolph H. Schultz Foundation.

7.14 Arm Retractors are Large

Caudal (or lower) **pectoralis**, caudal trapezius, and latissimus dorsi (Figure 7.23) are also large compared to those of monkeys (Ashton & Oxnard, 1963; Napier, 1963). **Caudal** means *toward the tail* and we use the term even though we apes have no tails. These are muscles that pull the entire arm down, such as when

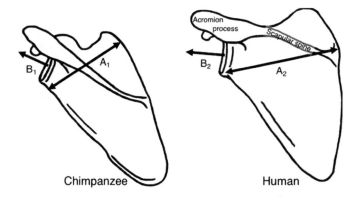

Figure 7.21 Chimpanzee scapulae are narrower (A_1 versus A_2) than in humans. The articulation for the humerus, the shallow bowl-shaped surface from which arrows B_1 and B_2 emerge, is the glenoid fossa. It faces upward in chimpanzees (B_1) compared to being side-facing in humans (B_2). Images by the author.

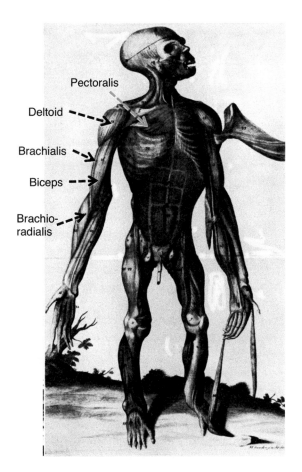

Figure 7.22 Compared to monkeys, chimpanzees have larger pectoralis major muscles, and larger deltoids, biceps, and brachioradialis muscles. From Tyson, 1699.

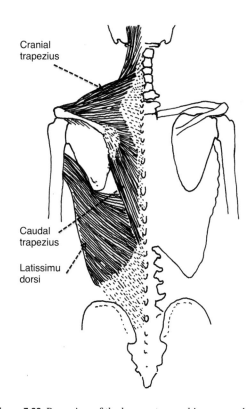

Figure 7.23 Rear view of the human torso; chimpanzees have a large cranial trapezius, latissimus dorsi, and caudal trapezius. After Aiello & Dean, 1990; with permission.

you climb a ladder or do a pull-up. **Latissimus dorsi** is a particularly interesting muscle. It has its origin attachment on the rim of the pelvic bone, the **os coxa**, and on a ligamentous sheath that attaches to the spinous processes of the thoracic and lumbar vertebrae (Figure 7.23). It takes a diagonal course up the back, passes under the armpit, and attaches to the top front of the humerus, close to the shoulder joint. This muscle pulls the humerus down as when you climb a ladder or do a pull-up. The elbow flexors and humerus retractors work together. Biceps and brachioradialis flex the elbow, and caudal pectoralis; caudal trapezius, posterior deltoid, and latissimus dorsi pull the upper arm down so that together these large muscles give chimpanzees incredible power pulling the arm down, as you would climbing a rope or doing a pull-up.

7.15 The Pelvis

At first glance, the bulging belly of chimpanzees suggests they have fat deposits there, just as an overweight human might. Not so. As we learned in Chapter 6, they have a much larger gastrointestinal system, so large that their belly swells to accommodate it. Chimpanzees are actually very lean.

Turning to the bony system supporting that large gut, the pelvis probably differs the most between apes and humans of any skeletal complex. It is made up of three bones, a pair of **os coxa** and a **sacrum**. Each os coxa has three parts, the **ilium** (or **iliac blade**) the **pubis** and the **ischium** (Figure 7.23). The iliac blade of the chimpanzee is flat and truly blade-like; the flat blade is parallel to the surface of the back. The pelvis is so broad that it and the ribcage are equal in breadth (Figure 7.20). This feature is found among all the great apes, but not in the gibbon; it is intermediate in the siamang (Hunt, 2016). This variation in breadth has yet to be adequately explained, but may be to accommodate larger **gluteals** to maintain something close to mechanical equivalency for extending the femur during climbing or leaping. Larger gluteals require a larger attachment area. Humans, in contrast, have a bowl-shaped pelvis (Figure 7.24). The crest of the os coxa curves around as one moves from the sacrum to bring the attachments for muscles that flex the hip to the front. Between the short human pelvis and the long human back, the lower back and pelvis of chimpanzees contrasts markedly with humans.

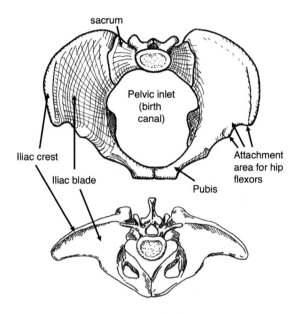

Figure 7.24 Top view of human (top) and chimpanzee (bottom) pelvis. When you lift your leg to take a step, muscles attached to the front of the pelvis are fired. Chimpanzee iliac blades are flat; human ilia are curved.

Monkeys have a narrow pelvis (Figure 7.20), part of their overall narrow body plan, with iliac blades the surface of which faces both to the side and the back, rather than backwards, as in apes. This narrow body plan makes sense for monkeys, with their high frequencies of walking, running, and leaping: It decreases torque on the hip joint and shoulder when only limbs on one side of the body support body weight. The monkey ribcage is broader than the pelvis, a condition also seen in humans and gibbons. Despite the narrow pelvis, monkeys have a broad sacrum compared to apes. In Chapter 8 we will see that the broad iliac blade, yielding a broad pelvis, and the narrow chimpanzee sacrum may be functionally related.

Although not as extensive as those on Old World monkeys such as the macaque, chimpanzees have *ischial pads*, fibrous, fat-infused, hairless pads that make sitting on hard surfaces a little more comfortable; these pads, when present, are anchored to an expanded ischial portion of the pelvis (Figure 7.20, B_1). In life the pads can be seen as bald spots on either side of the anus; having had a chimpanzee sit on my hand once when we were crowded into tight quarters, I can testify that the pads are similar in texture and firmness (and warmth) to the pads on your palm.

The unique constellation of characters around the pelvis in apes is, to my mind, the most enduring mystery of ape functional morphology. A tall pelvis and short vertebral lumbar segment, yielding a short back that is mostly taken up with the utterly rigid os coxae, and an extraordinarily broad pelvis are characters found in orangutans, gorillas, and chimpanzees – and no other primates. A clue to the function can be found in the siamangs, since they are intermediate in body weight between the great apes on the one hand and the gibbons on the other; their expression of these features is intermediate as well, suggesting that great ape body size exerts a strong selective force on the lower back and pelvis.

7.16 The Femur

Chimpanzee legs are much shorter than ours, making the length of the arms compared to the legs dramatically different. Anatomists compare arm to leg length using the intermembral index (IMI), the percentage of arm versus leg length. It is 105 in chimpanzees, versus about 85 in humans.

Below the waist, as we learned earlier, the chimpanzee hindlimb is a shorter version of the human leg, but there are further differences. Chimpanzees have a **varus** femur; that is, the skeletal elements of the leg look straight, though with the muscles in place they appear to be bow-legged. Humans, in contrast, have a **valgus** knee. With the flesh on, human legs look straight, but the skeletal leg appears knock-kneed (Figure 7.25).

The human femur, in side view (not pictured), has an elongated knee joint. More formally, the femoral condyles are expanded antero-posteriorly, a morphology that helps to lock the human knee when it is completely extended, a feature useful for stabilizing standing bipedal posture. Not surprisingly, chimpanzees lack this feature.

The chimpanzee femoral head differs from that of monkeys similarly to how the humeral head differs. In chimpanzees the femoral head describes a more

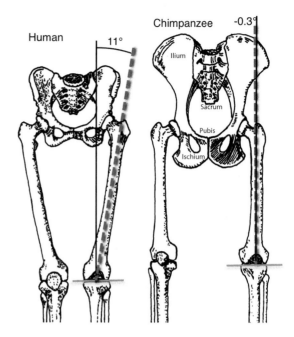

Figure 7.25 In this front view the chimpanzee femur is nearly straight, whereas the human femur is angled in at the knee; this *valgus* orientation gives the human skeleton a distinctly knock-kneed appearance. Image by the author.

complete sphere than in monkeys, which have more of a half-hemisphere femoral head. The femoral neck is angled upward in chimpanzees and the greater trochanter is distally displaced.

7.17 The Foot

The chimpanzee foot possesses a powerful, opposable big toe (Figure 7.1). The lateral toes are long, curved, and robust. Muscles that flex or extend the toes, both in the foot and in the leg, are large and powerful. One muscle that is particularly intriguing for its wandering course is the peroneus longus muscle. Its origin is just below the knee, on the outside of the tibia on the top of the smaller of the lower leg bones, the fibula, and along the shaft of the fibula (Figure 7.26). From there this long muscle strings along on the outside of the calf, and its tendon passes behind the distal fibula; that is, behind the knobby ankle bone on the outside of the ankle; finally, it passes underneath the foot and attaches to the first metatarsal that is the base of the

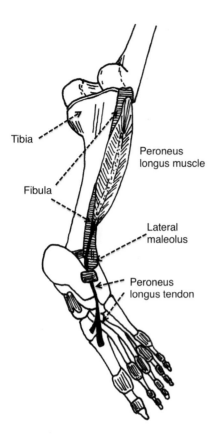

Figure 7.26 Lower leg and foot, knee joint. The peroneus longus muscle has its origin on the fibula (hatched bone); the tendon of the muscle passes behind the medial malleolus, that bump on the outside of your ankle, behind a bump at the midpoint of the outside of your foot (the base of the fifth metatarsal), underneath the arch of your foot, and at last attaches to the base of the big toe metatarsal. Image by the author.

big toe. When the toes are pointed, this muscle can be felt or seen tensing on the outside of the calf. Thus, we have an interesting phenomenon: A muscle on the outside of your lower leg powers the toe farthest to the inside of the foot, the big toe. The peroneus longus is bigger in chimpanzees than monkeys or humans, and the muscle has a different function – it causes the big toe to press against the sole of the foot or the other toes, as our thumb presses against our palm and other fingers when we grip a tennis racket. It is part of the gripping mechanism of the chimpanzee foot, useful during **vertical climbing**, which we will hear much more about in the next chapter.

7.18 Power

A persistent mystery in anatomy is the difference between humans and chimpanzees in raw power. Various studies have found that chimpanzees are, pound for pound, 2–3 times as powerful as humans (Bauman, 1923, 1926; Finch, 1943; Bozek et al., 2014). Bauman describes one spontaneous test in which the subject, the 135 lb Suzette, seemingly attempting to maliciously "pull [the testing apparatus] to pieces, sprang at the rope and, bracing both feet against the bars, pulled back with both hands upon the rope, making a pull [of] 1,260 ... the viciousness of this pull was something remarkable." Bauman tested 135 lb college students and found the maximum pull to be 332 lb. For students of all weights "one out of every hundred can ... pull 500 lb." In a later test of seven athletic young men weighing on average 148 lb, the mean two-handed pull was 406 lb. Endurance is another matter, but for this sort of task (not for walking, however) apes are still superior. In one test, an infant chimpanzee hung one-handed for five minutes, while the maximum for a human infant, two-handed, was 2.5 minutes (Edwards, 1970).

Recent studies have found chimpanzees to be only 1.3 (O'Neill et al., 2017) to 2.0 times (Bozek et al., 2014) as powerful as humans. Even so, the difference is still curious. While gross and microscopic muscle architecture is nearly identical, O'Neill and colleagues suggest that the greater chimpanzee power may be due to more fast-twitch muscle fibers (in technical terms, they have more MHC II isoforms [proteins]). Bozek and colleagues (2014) suggest that in chimpanzees metabalomes (energy storage and use mechanisms) deliver muscle twitching power more effectively. Primate metabolome form differs from that of mice about as much as would be expected, given their evolutionary relationship, while humans and chimpanzees were much *less* similar than their close genetic relationship would suggest. Bozek and colleagues detected rapid evolution of human metabolomes in the prefrontal cortex and the skeletal musculature, suggesting to them that the demands of the immense human brain have required diverting energy from their muscles to neural tissue. We sacrifice muscle power to ramp up brain power.

7.19 Anatomy Lesson

The lesson we learn from chimpanzee anatomy is that despite some points of resemblance, chimpanzees are not half-monkeys, nor half-humans for that matter. Their heads may resemble those of monkeys at first glance, but with their large supraorbital torus and expanded brain, the resemblance is superficial. From the neck to the waist, chimpanzees are human-like. Where they differ from humans – longer arms and fingers – they differ even more from monkeys. Monkeys tend to have long tails; chimpanzees have no tail at all. Chimpanzees have a short back, a feature found only in the great apes. While chimpanzees are more human- than monkey-like, we do best when we meet them on their own terms. They are chimpanzees.

References

Aiello LC, Dean MC (1990) *An Introduction to Human and Evolutionary Anatomy.* New York: Academic Press.

Ankel F (1972) Vertebral morphology of fossil and extant primates. In *Functional and Evolutionary Biology of Primates* (ed. Tuttle RH), pp. 223–240. Chicago, IL: Aldine Press.

Ashton EH, Oxnard CE (1963) The musculature of the primate shoulder. *Trans Zool Soc Lond* 29, 553–650.

Bauman JE (1923) The strength of the chimpanzee and orang. *Scientific Monthly* 16, 432–439.

Bauman JE (1926) Observations on the strength of the chimpanzee and its implications. *J Mammalogy* 7, 1–9.

Bourne GH, de Bourne MNG (1972) The histology and histochemistry of the chimpanzee tissues and organs. In *The Chimpanzee*, Vol. 6. (ed. Bourne GH), pp. 1–76. Basel: Karger.

Bowlby J (1958) The nature of the child's tie to his mother. *Int J Psycho-Analysis* 39, 350–373.

Bozek K, Wei Y, Yan Z, et al. (2014) Exceptional evolutionary divergence of human muscle and brain metabolomes parallels human cognitive and physical uniqueness. *PLoS Biol* 12. DOI: 10.1371/journal.pbio.1001871.

Brockman J (2000) *The Greatest Inventions of the Past 2,000 Years.* New York: Simon & Shuster.

Coleman MN (2009) What do primates hear? A meta-analysis of all known nonhuman primate behavioral audiograms. *Int J Primatol* 30, 55–91.

Detwiler S (1943) *Vertebrate Photoreceptors.* New York: Macmillan.

Edwards J (1970) Factors in the posture and grasping strength of monkeys, apes and man. *US Gov Res Dev Rep* 70, 45–61.

Eng CM, Lieberman DE, Zink KD, Peters MA (2013) Bite force and occlusal stress production in hominin evolution. *Am J Phys Anthropol* 151, 544–557.

Erikson GE (1963) Brachiation in the new world monkeys and in anthropoid apes. *Symp Zool Soc Lond* 10, 135–164.

Etter H-UF (1974) Morphologisch- und metrisch-vergleichende Untersuchung am Handskelett rezenter Primaten. *Gegenb Morph Jb* 120, 153–171.

Finch G (1943) The bodily strength of chimpanzees. *J Mammal* 24, 224–228.

Hershkovitz P (1977) *Living New World Monkeys (Platyrrhini)*, Vol. 1. Chicago, IL: University of Chicago Press.

Hines M (1942) Recent contributions to the localization of vision within the central nervous system. *Arch Opthalmol* 28, 913–937.

Hunt KD (2016) Why are there apes? Evidence for the co-evolution of ape and monkey ecomorphology *J Anat* 228, 630–685.

Jungers WL, Susman RL (1984) Body size and skeletal allometry in African apes. In *The Pygmy Chimpanzee* (ed. Susman RL), pp. 131–178. New York: Plenum Press.

Lieberman P, Fecteau S, Théoret H, et al. (2007) The evolution of human speech: its anatomical and neural bases. *Curr Anthropol* 48, 39–66.

Mittra ES, Smith HF, Lemelin P, Jungers WL (2007) Comparative morphometrics of the primate apical tuft. *Am J Phys Anthropol* 134, 449–459.

Napier JR (1960) Studies on the hand in living primates. *Proc Zool Soc Lond* 134, 647–657.

Napier JR (1963) Brachiation and brachiators. *Symp Zool Soc Lond* 10, 183–195.

Napier JR (1967) Evolutionary aspects of primate locomotion. *Am J Phys Anthropol* **27**, 333-342.

O'Neill MC, Umberger BR, Holowka NB, Larson SG, Reiser PJ (2017) Chimpanzee super strength and human skeletal muscle evolution. *PNAS* **114**, 7343-7348.

Perlman M, Salmi R (2017) Gorillas may use their laryngeal air sacs for whinny-type vocalizations and male display. *J Language Evolution* **2**, 126-140,

Polyak S (1941) *The Retina*. Chicago, IL: University of Chicago Press.

Prestrude AM (1970) Sensory capacities of the chimpanzee. *Psych Bull* **74**, 47-67.

Pusey AE, Oehlert GW, Williams JM, Goodall J (2005) Influence of ecological and social factors on body mass of wild chimpanzees. *Int J Primatol* **26**, 3-31.

Schultz AH (1930) The skeleton of the trunk and limbs of higher primates. *Hum Biol* **2**, 303-438.

Schultz AH (1936) Characters common to higher primates and characters specific for man. *Quant Rev Biol* **11**, 259-283, 425-455.

Schultz AH (1950) The physical distinctions of man. *Proc Am Philosoph Soc* **94**, 428-449.

Schultz AH (1969) The skeleton of the chimpanzee. In *The Chimpanzee* (ed. Bourne GH), pp. 50-103. Basel: Karger.

Schwalbe G (1916) Beiträge zur Kenntnis des äußeren Ohres der Primaten (Contributions to the knowledge of the outer ear of the primates). *Zeitschrift für Morphology und Anthropologie* **19**, 545-668.

Smith RJ, Jungers WL (1996) Body mass in comparative primatology. *J hum Evol* **32**, 523-559.

Sonntag, C.F. 1923. On the anatomy, physiology and pathology of the chimpanzee. *Proc Zool Soc Lond* **23**: 323-429.

Swindler DR, Wood CD (1982) *An Atlas of Primate Gross Anatomy: Baboon, Chimpanzee and Man*, 2nd ed. Seattle, WA: University of Washington Press.

Takemoto H (2008) Morphological analyses and 3D modeling of the tongue musculature of the chimpanzee (*Pan troglodytes*). *Am J Primatol* **70**, 966-975.

Tuttle RH (1969) Quantitative and functional studies on the hands of the Anthropoidea: the Hominoidea. *J Morphol* **128**, 309-363.

Tyson E (1699) *Orang-outang, sive Homo sylvestris: or The Anatomy of a Pygmie Compared with that of a Monkey, an Ape and a Man*. London: Osborne.

Walker R, Hill K (2003) Modeling growth and senescence in physical performance among the Ache of eastern Paraguay. *Am J Hum Biol* **15**, 196-208.

White TD, Johanson DC, Kimbel WH (1981) *Australopithecus africanus*: its phyletic position reconsidered. *S Afr J Sci* **77**, 445-470.

Wrangham RW (1977) Feeding behaviour of chimpanzees in Gombe national park, Tanzania. In *Primate Ecology* (ed. Clutton-Brock TH) pp. 503-538. New York: Academic Press.

Wrangham RW, Nishida T (1983) *Aspilia*: a puzzle in the feeding behaviour of chimpanzees. *Primates* **24**, 276-282.

Young F, Farrer D (1964) Refractive characteristics of chimpanzees. *Am J Optometry* **41**, 81-91.

8 Arboreal Gathering, Terrestrial Traveling
Locomotion and Posture

Image after Fleagle, 1980; with permission of the Japan Monkey Centre.

Fig trees are the best for spying on chimpanzees. Their leaves are sparse, which means an observer need not stagger around under the tree while looking up, searching for a clear view through the foliage, stepping in holes, stumbling on fallen branches, or barging into large stones all the while. You can see chimpanzees on the far side of the tree as easily as the near side. As a bonus, the trees are simply esthetically pleasing, their branches are elegantly curved and their bark is a pleasingly warm shade of brown.

Once, sitting on a smooth, water-worn stone in the stream bed of one of the watercourses at Mahale, I watched a party of half a dozen chimpanzees feed on figs, each individual going about his or her business in their own little sphere of influence, evenly dispersed around the tree crown. One male left his feeding sphere and picked his way through the branches, gradually making his way onto the farthest side of the tree where a high-ranking male was already feeding. As the distance between them lessened, the high-ranking male suddenly lunged at the interloper, who instantly turned and ran, *ran* through the tree crown, directly toward me, descending as he crossed from the far side of the tree to the near side. His pursuer was close on his heels. When they came to the lowest branch in the tree, nearly 10 feet from the ground, first the absconder and then the pursuer leapt to the ground. The pair raced headlong down the stream bed, intermittently splashing in the water or running on the smooth stones of the stream bed. This entire scenario, from the dominant male's first lunge to the pair's disappearance down the stream bed, played out in about five seconds.

As the echoing screams of the lower-ranking male receded into the distance, I looked down at the wet footprints on the stones, then up at the tree. The males had run through the canopy almost as fast as they can run on the ground; I studied the tree to try to find a solid path among the various branches and boughs on which they each might have placed their hands and feet as they ran. I could not, even after minutes of trying. Yet they did it on the fly, with fractions of a second to decide where to place each hand or foot.

Chimpanzees are born acrobats. They have incredible balance, and an incomprehensible capacity to pick out acceptable support points in a fraction of a second; they possess superhuman muscular power and the precise hand–eye coordination required to allow running on unevenly distributed branches 30 feet high in a tree. Watching these two males, their movement seemed no less alien to me than if they had been flying. It almost *looked* like flying. Without training, chimpanzees reared by humans display similar abilities.

For the year I walked around following chimpanzees for my dissertation research, focusing on a single target individual each day, a male one day, a female the next, the thing that struck me most powerfully was the stark contrast between their "normal" or most common movements, mostly slow walking, and their sudden incredible speed and athleticism. Most of the time – more so for females than males – chimpanzees move slowly, so slowly that as you walk along behind them you often have to stutter step or even stop to keep your pace slow enough not to bump into them. I learned that chimpanzees are lazy so-and-sos. They move almost as if they are ill or in pain. The most common chimpanzee positional mode – experts refer to a specific locomotor or postural behavior is a **positional mode** – is sitting (Figure 8.1). Chimpanzees spend about 64 percent of their time in this mode (see Table 8.1). Occasionally I will lapse into expert-speak with regard to locomotion and posture; the term **positional behavior** refers to both or either.

8.1 Are Chimpanzees Arboreal Armhangers or Terrestrial Knucklewalkers?

We binary-minded humans prefer things to be black or white, up or down. Chimpanzees are equally up as down, equally at home on the ground, where they walk and often rest or socialize, and in the trees, where they can be incredibly nimble. They are often

Figure 8.1 Sitting. Tissue over the ischia bear most of the body weight. After Hunt et al., 1996; with permission.

Table 8.1 Chimpanzee positional behavior

Study	Sit	Lie	Quad. stand	Squat	Cling	Bipedal stand	Arm-hang	Walk	Vertical climb	Quad. Run	Bipedal walk	Brachiate	Suspensory	Transfer	N
Hunt, 1989, 1992, 2016	62.4	12.1	2.5	0.7	0.3	0.3	4.4	16.1	0.9	0.3	0.1	0.1	0.1	0.2	16,303
Doran, 1996	65.1	14.5	5.0	0.0	0.0	0.1	1.3	12.1	~1.1	0.0	0.1	0.1	0.0	~0.3	10,077
Chimpanzee mean	63.8	13.3	3.8	0.4	0.2	0.2	2.9	14.1	1.0	0.2	0.1	0.1	0.1	0.2	

described as either terrestrial knucklewalkers or arboreal armhangers. This is a false dichotomy. If chimpanzees were constrained from using either trees or the ground, they could not survive. They are terrestrial travelers – nearly 99 percent of their travel is on the ground – but arboreal feeders: 85 percent of their feeding is in the trees (Hunt, 1989, 2016). Thus, the chapter title: arboreal gathering, terrestrial traveling. Their food is so dispersed that if they could not use the speed and efficiency of traveling on the ground, both their social system and their anatomy would be dramatically altered.

8.2 Walking

When chimpanzees travel, they typically do so by **knucklewalking**, as do gorillas and bonobos. Knucklewalking differs from the quadrupedal (or four-legged) walking of other primates principally in the orientation of the fingers. Rather than the palm, the outside (or dorsal aspect) of the index, middle, and ring fingers, the second, third, and fourth phalanges touch the ground (Figure 8.2). During both walking and standing the fingers are flexed almost into a fist, but with the first phalanx hyperextended against the palmar bones (Figures 8.2 and 8.3). Otherwise, the movement of the arms and legs is little different from the walking of any other primate. Chimpanzees knucklewalk even in trees when the support on which they are moving is large enough, and "large enough" is smaller than you might think, typically not much larger than a tennis racket grip. Walking makes up nearly 90 percent of chimpanzee locomotion (89.2 percent, to be exact; Table 8.1). Vertical climbing (see Figure 8.7) is the next most common at 6.3 percent of locomotion (though less than 1 percent of all

Figure 8.2 Knucklewalking. Image by the author.

Figure 8.3 Quadrupedal standing; note the knuckling positioning of the fingers. From Schultz, 1936; with permission.

positional behavior; Table 8.1). Monkeys specialize in walking, leaping, and running (see Figure 8.8).

Chimpanzees only run in films. I exaggerate. They do run, but it makes up only 0.2 percent of their behavior (I mean 0.2 percent of both locomotion and posture; it is 1.3 percent of locomotion), while walking constitutes 14.1 percent of their active day. Running makes up 10 times that for monkeys (Hunt, 2016). Put another way, chimpanzees knucklewalk about 1 hour and 50 minutes per day, while running takes up only a minute and a half. Chimpanzees travel ~5 km each day, but they move 2–3 times slower than expected for their body mass (Hunt, 1989; Pontzer & Wrangham, 2004). Long, uninterrupted walking bouts that are perfectly comfortable for a human may leave chimpanzees so fatigued they pause to rest and catch their breath (pers. obs.). Still, while they may be slow and inefficient walkers, and while they walk less than monkeys (see Figure 8.8), they walk long distances compared to other apes.

Chimpanzees are occasionally **bipedal**, but they appear awkward walking on two legs, with the exception of the honorary human Faben, who became quite a good biped when forced to: His right arm was paralyzed by polio (Bauer, 1977; Figure 8.4). Still, he tired so quickly he could not keep up with other chimpanzees. Chimpanzees are bipedal more often posturally than locomotorily (Figure 8.4); bipedal walking is a fraction – 1 percent – of all positional behavior; bipedal standing is 0.2 percent. Bipedalism is not a travel mode; of 1828 locomotor observations, three were display, two were feeding, none were travel.

After sitting and walking, the next most common chimpanzee positional behavior is lying. Chimpanzees are great liers. They take every opportunity to lounge supine, especially when on the ground, and particularly when it is hot and they can lie in a cool spot. It makes up over 13 percent of their behavior. Sometimes they lie because they are napping, but more often they are simply resting from what is, despite their typically slow movements, a very tiring day.

8.3 Quadrupedal Standing

Standing with the knees and elbows completely extended makes up 3.8 percent of the chimpanzee day. As they move on the ground they may pause in a quadrupedal stance while they wait on a companion, or they may look up into a tree to determine if the fruit there warrants a climb. If they anticipate stopping for more than a few seconds, they sit.

8.4 Suspensory Posture

The chimpanzee's suspensory postures are shown in Figure 8.5.

8.5 Unimanual Armhanging

This involves suspending the body from a single limb. It requires a strong grip and exerts powerful tensile forces on the shoulder; the body is, in essence, hanging from an arm, with only the ligamentous attachments of the humerus to the scapula supporting the body. While armhanging made be bimanual, or two-handed, most chimpanzee armhanging is unimanual, or one-handed (Figure 8.5(c)). Few humans can hang for more than a few seconds, but chimpanzees very casually spend five minutes at a time suspended unimanually.

8.6 Suspensory Locomotion

Several locomotor modes exert forces on the musculoskeletal system similar to armhanging, and therefore presumably requiring the same anatomical adaptations. These are pooled in a general category often labeled **suspensory behaviors**, and include

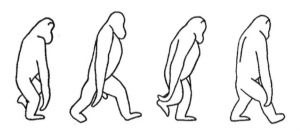

Figure 8.4 Bipedal walk, featuring Faben the chimpanzee. From Bauer, 1977; used with permission.

A. Assisted forelimb suspension

B. Arm-foot hang

C. Unimanual forelimb suspension; armhanging

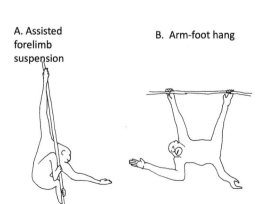

Figure 8.5 Common suspensory behaviors: (a) armhanging with support, (b) arm-foot hanging, (c) and unimanual suspension. Images by the author, used with permission of the Japan Monkey Centre.

armhanging, **brachiation**, **transferring**, **quadrumanous climbing**, and **arm swinging**. Sometimes an individual does something so contorted and odd that it fits into none of these categories, but they are doing it while suspended beneath a branch; these behaviors end up in the "other suspensory" basket (Table 8.1; Figure 8.5). Experts sometimes refer to suspensory modes as **"limbs in tension"** modes, contrasting these behaviors to positional modes in which the limbs are in net compression. In most instances the torso is vertical during suspensory behavior – perpendicular to the horizon – the specialist would call this "orthograde" positional behavior, contrasting it to "pronograde" modes where the back is horizontal.

Brachiation (Figure 8.6) is a hand-over-hand locomotor mode most perfectly expressed in gibbons; it is what humans do when they swing under horizontal ladder-like "monkey bars" in the playground. Transferring refers to moving from one tree to another, and among chimpanzees it is almost always suspensory; it is a behavior somewhere between armhanging and brachiation. The transferring individual begins in a suspensory armhanging mode, or perhaps hanging from one arm and one foot, and reaches out to grasp branches in a neighboring tree, gradually transferring weight onto them, thus moving between tree canopies. Sometimes the transferring individual sways his or her body weight back and forth to bend the supporting tree closer to the target. All suspensory modes together, armhanging, transferring, brachiation, and "other suspensory" make up nearly 4.4 percent of all behavior.

While less common than sitting, lying, standing, and walking, suspensory modes are vital behaviors because the most important food item for chimpanzees, fruit, tends to be found among the outermost branches in trees. In some species the fruit only grows in the outermost branches, but in others it might be found dispersed more evenly throughout the tree when the fruit is set, early in its fruiting cycle, but by the time the fruit is ripe fruits in the core of the tree have been eaten by monkeys, leaving a hollow sphere of fruit for chimpanzees.

Armhanging (often referred to as unimanual suspension) and other suspensory behaviors are *the* ape adaptation. They account for more of the unusual anatomy found in apes than any other behavior, and the behavior is rare among other primates, the suspensory spider monkeys and relatives (subfamily Atelinae) excepted. In both legend and science, they are suspensory; if you search for images of apes or chimpanzees, a good proportion of them will be depicted as hanging. Chimpanzees spend about 35 minutes a day hanging by one arm (Hunt, 1989, 1992a, 2016; Table 8.1).

8.7 Support Diameter Dictates Suspension

As foreshadowed above, armhanging and most other suspension are feeding behaviors: 90 percent of chimpanzee unimanual suspension occurs during food gathering. Chimpanzees armhang not because they enjoy it, but because they must. Suspensory behavior answers a specific challenge – negotiating small-diameter, flexible (or compliant) supports where fruit is more likely to be found. When branches are large enough to allow sitting, chimpanzees sit. Small branches, however, are too flexible for sitting – they bend so easily that the body weight of a chimpanzee

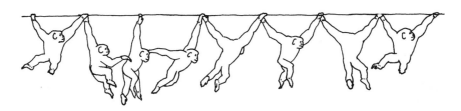

Figure 8.6 Gibbon brachiation is much like a locomotor version of armhanging; image by author, with permission of the Japan Monkey Centre.

easily deforms them until they are so steeply angled that a chimpanzee attempting to sit would be spilled off. In essence, the branch dumps the chimpanzee out of the tree by bending until the sitter is tossed forward, away from the support. What is required to negotiate these compliant supports is gripping the branch so it cannot get away. Nearly 80 percent of armhanging is engaged in among the terminal branches, the tiny twigs at the outermost periphery of the tree, and the smaller the support, the greater the proportion of suspensory behavior (Hunt, 1992b). Let me elaborate on this point.

8.8 Suspensory Behavior is 10–100 Times More Common among Compliant Supports

As an illustration of how tightly linked suspensory behavior is to canopy location, consider that while armhanging makes up only 2.5 percent of postural behavior when support diameter is greater than 10 cm; it increases to 8.3 percent among 3–10 cm branches and constitutes 24 percent of all posture when supports are less than 3 cm (Hunt, 1992b).

We see the same trend when we look at suspensory locomotion. It makes up only 0.2 percent of all chimpanzee locomotor behavior, but among the small branches in the outermost canopy layer it is over 100 times that, 29.4 percent of all behavior (Hunt, 1992b).

Comparing the behavior of dominant to subordinate individuals illustrates chimpanzee positional preferences. Where does a 60 kg, high-ranking chimpanzee male eat? Wherever he wants. He can monopolize the most preferred feeding sites where he can achieve a high rate of return – the most calories for the least effort. He does this by balancing the energetic demands of posture against the rate at which he can take calories on board. The ideal spot is one where the fruit is ripe and abundant and where sitting is possible. My research demonstrates that chimpanzees prefer feeding sites that seem most comfortable to us, sites where they need not engage in awkward contortions to reach food. Large branches that just happen to be near the tree edge are good: I found that although high-ranking individuals spent more time in the tree periphery than subordinates, they used larger supports, sat more often, and armhung less often (Hunt, 1992b).

8.9 Vertical Climbing

While it makes up only 1 percent of daily activity (though 6.3 percent of locomotion), vertical climbing is vital for chimpanzees. Because they tend to travel on the ground and because most of their food is in the trees, when they reach a feeding site they climb a vertical support with a locomotor mode similar to that of a human climbing a ladder. Rather than climbing ladder-like supports, though, chimpanzees climb what are essentially vertical poles, either vines or small-diameter trees, in hand-over-hand fashion (Figure 8.7). While chimpanzees are capable of climbing even barrel-sized boles, 20 cm or larger, they prefer smaller supports, approximately the diameter of a soda can. At Gombe and Mahale fully 50 percent of vertical climbing utilized supports 4 cm or smaller and only 5 percent of vertical climbing was engaged in on boles 20 cm or larger (Hunt, 1992a). I suspect chimpanzees pass over trees that can be accessed only via a large bole when they can.

Figure 8.7 Vertical climbing; images by the author, with permission of the Japan Monkey Centre.

8.10 Monkey Tree Ascension

You may be wondering how monkeys get into trees if they rarely vertical climb. They leap. They push off with their hindlimbs, land on a higher branch, and repeat until they arrive at their destination. They also climb less because they have less of a need to ascend; rather than climbing down to walk between trees, they simply leap between trees. Their smaller size makes such leaping easier for them because the same branch that would deform under the force required to launch a chimpanzee through the air will stay rigid when a 5 kg monkey pushes off from it.

One incident (pers. obs., Gombe, November 1986) demonstrates this difference. Once I watched a chimpanzee enter a tree crown first by vertical climbing on its approximately 15 cm vertical bole for a brief stretch, only a few meters, until she reached the first horizontal branch. She then vertical climbed farther by grasping horizontal branches as if climbing a ladder. A male baboon arrived a short time later and ascended the same tree by first making a prodigious leap to reach the first horizontal branch, then leaping from branch to branch in a spiral path around the bole, ascending to the same elevation as the chimpanzee.

Returning to the role of rank in feeding-site selection, with my interest in social rank and feeding, I had expected that high-ranking individuals would feed lower in trees. I was wrong, as we will see. My reasoning was that their dominance would allow them to monopolize feeding sites with the highest rate of return, and since climbing is expensive, the lowest sites would be the most desirable. It is apparent that vertical climbing is fatiguing from the observation that the positional mode that most often follows vertical climbing is sitting, while the most common contexts following climbing are resting and grooming (Hunt, 1989). After a brief rest on a large branch, a forager might then knucklewalk and (as smaller supports are encountered) palm-walk to the tree periphery, where harvesting might occur with a sitting posture if the support is large enough, or armhanging if the support is small-diameter and compliant.

8.11 Travel and Dispersed Ripe Fruit

The scarcity and dispersed nature of the preferred chimpanzee food, ripe fruit (Wrangham et al., 1998), determines travel choices. When an adjacent tree offers edible foods – a happy circumstance, when encountered – chimpanzees move between tree crowns by suspensory transferring, brachiation, or (rarely) leaping. It is often the case that an adjacent tree has nothing a chimpanzee would want to eat, and when the next tree after that also offers nothing very appetizing, feeding is followed by a descent and travel on the ground. Chimpanzees descend using hand-over-hand vertical climbing similar to descending on a ladder, often punctuating such descents with suspensory drops, a behavior that starts with hanging underneath in a bimanual suspensory posture and simply letting go. Occasionally, an individual might brachiate along the length of a branch before dropping to a lower branch or to the ground, but chimpanzees rarely brachiate more than two or three strides (Hunt, 2016). After descending, the forager will knucklewalk to a new food source and climb.

8.12 Midday Rest

Sometimes, after a long feeding bout or several shorter ones, especially as the middle of the day approaches, an individual will leave off feeding and move to a commonly used socializing location, a spot surprisingly often found in the midst of a vine tangle, after which individuals may groom or rest quietly while sitting or lying on the ground.

8.13 Monkey versus Ape Solutions to Arboreal Challenges

The contrast in monkey versus chimpanzee food-gathering helps to illuminate their contrasting evolutionary paths. The best way to learn what is distinctive about ape positional behavior (Figure 8.8) is to study an ape and a monkey in the same habitat. There, the same foods are available and they face the

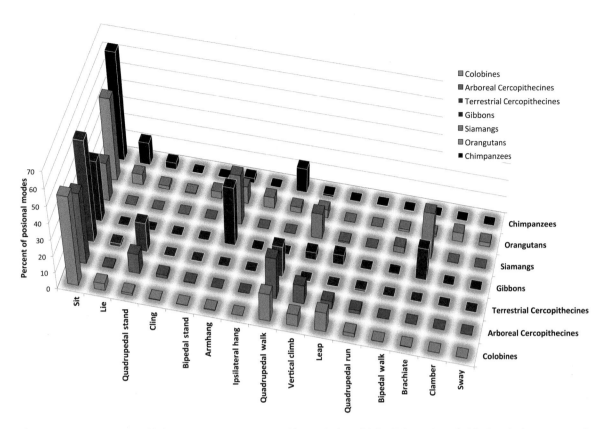

Figure 8.8 Primate positional behavior. From Hunt, 2016, with permission of John Wiley & Sons. (A black and white version of this figure will appear in some formats. For the color version, please refer to the plate section.)

same obstacle course as they travel and feed. This sort of same-site comparison, despite its utility, has been adopted in only a handful of studies.

In the 1980s I compared chimpanzees to savanna (or olive) baboons (Hunt, 1989, 1991, 1992a, 1992b). At the time we had still not clearly determined why chimpanzee anatomy was different from that of monkeys, but of course it had to be related to how they moved and positioned themselves in their habitat. Scientists had long suspected suspensory behavior was the distinctive ape behavior, but an important camp thought it was vertical climbing. My research suggested it was both.

Baboons and chimpanzees sit, walk, and lie in similar proportions. Armhanging, however, is quite different; it makes up 4.4 percent of all chimpanzee positional behavior, while in baboons it is zero. Armhanging is important not only because it is more common among chimpanzees, but because it has unique anatomical demands, requiring shoulder mobility and forelimb power. The contrast in vertical climbing is less striking, but still important. Chimpanzees vertical climb about twice as much as baboons, 1 percent versus 0.5 percent. Because these two comparisons are the only significant ones, it suggests that any explanation for difference in chimpanzee and baboon anatomy will be found in the stresses associated with these two modes. Figure 8.8 compares many primates (from Hunt, 2016) and confirms what we know from a chimpanzee/baboon comparison: Monkeys are leapers, walkers, and runners, while apes are suspensory primates that vertical climb.

In the most comprehensive assessment, Cant (1992) compared orangutans and gibbons to macaques and langurs. He devised an ingenious system of dividing trees into an upper third, a middle third, and a lower third, after which he documented which layer each primate used. He found that two primates used all levels: orangutans and langurs. Macaques fed in the middle and lower canopy layers and gibbons focused on middle and upper levels. This pattern illustrates one of the somewhat counter-intuitive discoveries about primate food-gathering: Primates that have the *least* similar diets are most likely to use the *same parts of the tree*. This suggests competition, to me, because the pair that are least likely to compete over food are found in the same places, and pairs that eat the same foods feed in different places. I speculate that this pattern arises out of the ability of orangutans to use their size and power to displace macaques from the more desirable higher strata, leaving them to feed in the less productive lower layers (Houle et al., 2006, 2014; Hunt, 2016). The tiny 7 kg gibbons are comfortable among smaller supports, enabling them to clean up orangutan leftovers when they feed in the same trees. Langurs are more folivorous (leaf-eaters) and so can feed in the same places as orangutans without competing with them – if they did compete, they would lose.

In another same-site comparison, Cannon and Leighton (1994) studied gibbon and macaque locomotion. They found that gibbons utilized the upper parts of the tree canopy more, could cross wider gaps, and were able to negotiate the small-diameter branches in the tree periphery better than the similar-sized macaque. Suspensory behavior is superior when moving among tiny supports.

8.14 How to Get Fruit

Tropical forests are bustling places, filled with an abundance of primates and other types of animals besides; you might well wonder how nature can cram so many species into one forest. I have argued that chimpanzees fit into this jigsaw puzzle by being specialized for foraging among the small-diameter, peripheral branches.

Not everyone sees it that way. Recently, Alain Houle and colleagues (Houle et al., 2006, 2007, 2010, 2014) found that when chimpanzees feed in the same trees as three monkeys – redtails, blues, and mangabeys – chimpanzees monopolize the higher, more productive portion of the tree canopy (much as Cant found for gibbons and orangutans versus macaques), displacing monkeys to lower feeding sites with fewer and less-ripe fruits (Houle et al., 2010). They showed that the upper half of trees produce nearly five times the number of fruits as the lower half, and the fruits in the upper half are larger, contain higher concentrations of sugars, and harbor fewer toxins. This is probably why I was wrong when I thought high-ranking males might remain low in trees.

In such a scenario, Houle would argue, suspensory behavior may have evolved to afford chimpanzees and other apes *equal* access to resources that their larger body mass might otherwise make inaccessible, while monkeys evolved antifeedant-tolerant digestive physiology as a competitive response to being displaced from the most preferred feeding sites. In Houle's view, monkeys evolved small body size to allow them to utilize small, low-density fruit crops that are below an ape's "giving-up" density (Brown et al., 1994), crops that are so small, in other words, they are not worth bothering with for apes.

My view is that apes evolved larger body size to displace monkeys and widen their diet breadth – larger body size allows them to process low-quality foods better (Chapter 6) – and then evolved suspensory behavior to reach the smallest branches.

The two hypotheses are: (1) apes have adaptations for feeding among the terminal branches in order to harvest what ripe fruits are left after monkeys have had earlier access; versus (2) apes compete by using suspensory capabilities to maintain equal access. In the scenario I judge more likely (Hypothesis 1), chimpanzees are expected to arrive at individual trees later in the fruiting cycle than monkeys, on average, and are expected to feed closer to the canopy edge (i.e., peripheral terminal branches); monkeys are expected to alter their movements to avoid chimpanzees. I found all of these to be true in a comparison of chimpanzees and baboons at Gombe (Hunt, 1992a).

In Houle's "giving-up-density" alternative, no difference in the timing of the arrival of monkeys and chimpanzees at fruit trees is expected; no difference in the proximity of monkeys and chimpanzees to the tree edge or in branch sizes is expected; and when feeding in the same trees, monkeys should remain in trees longer.

While there are differences between the two views that may seem boring blather to those who study other aspects of primate behavior, my hypothesis suggests that feeding competition explains why monkeys have become so numerous, whereas Houle's hypothesis sees the competition as more balanced and leaves the explanation for monkey dominance in numbers to some other aspect of evolution. Still, both scenarios are consistent with the hypothesis that suspensory behavior evolved as an adaptation for harvesting fruits among compliant, peripheral branches, and that apes evolved larger body size in part to displace monkeys, while monkeys evolved antifeedant tolerance to cope with ape physical dominance.

8.15 Monkeys Locomote Differently because They Tolerate Chemicals in their Diet

As a team member in Richard Wrangham's primate diet research program from 1989 to 1993, I studied differences in the feeding behaviors of monkeys and chimpanzees. Wrangham and his team gathered data to show definitively that monkeys tolerate the cocktail of chemicals found in unripe fruits much better than do chimpanzees. Monkeys, Wrangham found, eat both ripe and unripe fruit, in proportions – about half and half – that remain the same whether there is a little or a lot of fruit. Chimpanzees, in contrast, eat more ripe fruit as the supply increases, and when ripe fruit is unavailable they are likely to turn to low-antifeedant food items like piths and herbs rather than unripe fruit.

In other words, monkeys – tolerant of antifeedant chemicals such as terpenes, alkaloids, and tannins – are much less picky eaters than chimpanzees. So the forest offers more food items to them, and this perceived abundance plays out in their positional behavior. Because more things are edible, they are more likely to find food in adjacent trees.

They can stay in the trees rather than descending, walking on the ground, and re-ascending. In one tree they may eat leaves that have too much fiber and too much tannin for chimpanzees; in the next tree they may eat blossoms that contain nectar, but also poisonous alkaloids that apes avoid; in the next they may eat unripe fruit that is indigestible for chimpanzees; ripe fruit in the next. Thus, they stay high in the canopy, moving from tree canopy to tree canopy, finding something to eat in a high proportion of the trees they pass through. They sit to feed more often because they harvest food more often among the larger branches in the tree core. They walk and run within trees and leap between them. They have a "**feed-as-you-go**" foraging strategy, alternating rapidly between moving and eating (Wrangham et al., 1998), whereas among chimpanzee feeding and traveling bouts are more sharply defined.

8.16 Chimpanzees have Adapted to a Planet of the Monkeys

The world is very different for chimpanzees and monkeys. Chimpanzees are constrained to finding the few, widely dispersed patches of ripe fruits in the forest. There are even fewer such fruits than there might be because monkeys have eaten many of them before they ripened. In my view, monkeys eat the most accessible fruits, those in the tree core, leaving chimpanzees leftovers in the tree periphery. Chimpanzees are too heavy to bound between trees. Besides, they less often find food in the next tree, or even the next one after that, or five after that, making it unprofitable for them to stay in the tree canopy. Instead, they knucklewalk to a distant food source, vertical climb to enter the tree crown, walk quadrupedally to the edge of the canopy, and armhang among the tiny branches where the fruit is. When they have exhausted the food supply in that feeding tree, they descend, knucklewalk, and climb the next feeding site.

Monkeys have solved the primate food puzzle better than apes – this world is not a Planet of the Apes, but a Planet of the Monkeys. The necessity of moving and feeding among small branches makes severe demands on the chimpanzee anatomy that we have only alluded to in this chapter. Keeping in mind the locomotion and posture of chimpanzees, we will now examine how their bodies are "engineered" to accommodate this behavior. Of course, evolution, not engineering, created chimpanzee anatomy. Still, at cocktail parties and on planes when people ask what I do for a living, I often say "I reverse engineer chimpanzees."

References

Bauer HR (1977) Chimpanzee bipedal locomotion in the Gombe National Park, East Africa. *Primates* 18, 913–921.

Brown JS, Kotler BP, Mitchell WM (1994) Foraging theory, patch use, and the structure of a Negev Desert granivore community. *Ecology* 75, 2286–2300.

Cannon CH, Leighton M (1994) Comparative locomotor ecology of gibbons and macaques: selection of canopy elements for crossing gaps. *Am J Phys Anthropol* 93, 505–524.

Cant JGH (1992) Positional behavior and body size of arboreal primates: a theoretical framework for field studies and an illustration of its application. *Am J Phys Anthropol* 88, 273–283.

Doran DM (1996) Comparative positional behavior of the African apes. In *Great Ape Societies* (eds. McGrew WC, Marchant LF, Nishida T), pp. 213–224. Cambridge: Cambridge University Press.

Houle A, Chapman CA, Vickery WL (2007) Intratree variation in fruit production and implications for primate foraging. *Int J Primatol* 28, 1197–1217.

Houle A, Vickery WL, Chapman CA (2006) Testing mechanisms of coexistence among two species of frugivorous primates. *J Anim Ecol* 75, 1034–1044.

Houle A, Chapman CA, Vickery WL (2010) Intratree vertical variation of fruit density and the nature of contest competition in frugivores. *Behav Ecol Sociobiol* 64, 429–441.

Houle A, Conklin-Brittain NL, Wrangham RW (2014) Vertical stratification of the nutritional value of fruit: macronutrients and condensed tannins. *Am J Primatol* 76, 1207–1232.

Hunt KD (1989) *Positional behavior in Pan troglodytes at the Mahale Mountains and Gombe Stream National Parks, Tanzania*, PhD dissertation. Ann Arbor, MI: University of Michigan.

Hunt KD (1991) Mechanical implications of chimpanzee positional behavior. *Am J Phys Anthropol* 86, 521–536.

Hunt KD (1992a) Positional behavior of *Pan troglodytes* in the Mahale Mountains and Gombe Stream National Parks, Tanzania. *Am J Phys Anthropol* 87, 83–107.

Hunt KD (1992b) Social rank and body weight as determinants of positional behavior in *Pan troglodytes*. *Primates* 33, 347–357.

Hunt KD (2016) Why are there apes? Evidence for the co-evolution of ape and monkey ecomorphology. *J Anat* 228, 630–685.

Hunt KD, Cant JGH, Gebo DL, et al. (1996) Standardized descriptions of primate locomotor and postural modes. *Primates* 37, 363–387.

Pontzer H, Wrangham RW (2004) Climbing and the daily energy cost of locomotion in wild chimpanzees: implications for hominoid evolution. *J Hum Evol* 46, 315–333.

Schultz AH (1936) Characters common to higher primates and characters specific to man. *Q Rev Biol* 11, 259–283, 425–455.

Wrangham RW, Conklin-Brittain NL, Hunt KD (1998) Dietary response of chimpanzees and cercopithecines to seasonal variation in fruit abundance. I. Antifeedants. *Int J Primatol* 19, 949–970.

9 Forged in Nature's Cauldron
Engineering the Chimpanzee

Credit: Ali Majdfar / Moment Open / Getty Images

If we humans were a little less egocentric, we might celebrate chimpanzees rather than humans as nature's last word. Fast and powerful; capable of running up a flagpole; able to hang by one arm for 15 minutes at a time; at home on the ground and in the trees; endowed by nature almost from birth with the balance and agility of a circus acrobat – they are quite literally superhuman. And beyond these physical attributes they are astonishingly intelligent. Chimpanzees are a natural wonder.

We learned in Chapter 3 that the unique anatomy that drives these abilities has been known since Tyson's brilliant anatomical work in the late 1600s (Tyson, 1699). Unexpectedly, given the long history of ape anatomical study, it is only recently that we have begun to understand the engineering principles that explain their anatomy; in fact, experts are still arguing about it.

Functional morphologists often start their quest by determining which anatomical features are unusual or unique in the species under consideration, then attempting to link the morphology with a particular behavior that is thought to stress that part of the body. If a muscle is particularly large in only one species and is critically important for executing a behavior that is also unique to that species, we suspect a link between the two. For example, chimpanzees have extremely long fingers (Chapter 7), and their most distinctive positional behavior is unimanual suspension (Chapter 8). Long fingers are likely to have evolved to allow chimpanzees to completely encircle a branch with their fingers so as to armhang without fear of losing their grip. What we functional morphologists attempt to do, in essence, is reverse engineer animals. We look at animal bodies and speculate about how their various anatomical structures – things like squid tentacle suckers or ever-growing incisors in rodents – function in the everyday lives of these organisms.

Biomechanics applies more quantitative physical principles to these biological systems. "Physical" in this case means "physics principles." Biomechanists calculate exactly how much power or speed a muscle might generate, given its cross-sectional area and its lever (or moment) arm.

9.1 Body Weight Differences Muddy the Water: Allometry

Unfortunately, when we look for consistencies in a particular shape, robust manual phalanges, for example, we often find ourselves studying different-sized animals, and this introduces a vexing complication that, while not obvious at first glance, is probably intuitively familiar to you. The body proportions of large and therefore heavy animals are different from those of smaller animals. You have probably noticed that a tiny spider might have legs almost hair-like in their thinness, whereas the legs of their large relatives, tarantulas perhaps, are thick. Elephants have thicker legs than kangaroo mice; gorillas have thicker legs than marmosets (Figure 9.1).

Heavier animals must have disproportionally larger legs in orders to leap and run with the same agility as

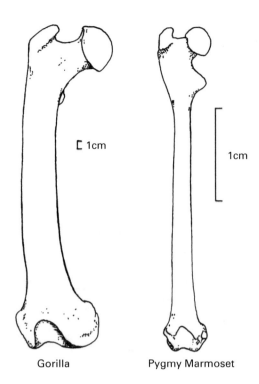

Figure 9.1 Larger animals are not simply geometrically expanded versions of smaller animals. Larger animals must have disproportionately larger cross-sectional areas. A gorilla femur (left) is much more heavily built than the femur of the tiny pygmy marmoset. After Fleagle, 2013; used with permission.

smaller animals. Although it might seem counter-intuitive, a mouse-sized animal blown-up to the size of a minivan would have very different – and diminished – capabilities, even if it kept the exact same body proportions. Not only would the house-sized ant of horror movie lore be unable to lift an object many times its body weight, it would not even be able to lift its own body. We will soon see exactly why.

The phenomenon of requiring a different shape as body size increases in order to retain the same function is called **allometry**. Agility deteriorates as the mass of an organism increases, according to a law that physics professors and engineers deal with on a daily basis, the **square-cube law**. This law applies to all objects, both living and inanimate. As a preview, the square-cube law takes account of the fact that as an animal increases in size, its volume (which is a *cubic* measure, length times width times depth)

increases at a cube of its length, while the cross-section (a *squared* measure) of its legs increases at only a square. Physics majors and engineers are yawning at the simplicity of this issue, but some of you are probably beginning to worry this is beyond your engineering abilities. It is not.

Look for a concrete example in a moment, but first try to grasp the concept from theory alone. As any physical object, inanimate or animate, increases in length, its *cross-section* increases in two dimensions, *width* and *depth*, while its *volume,* and thus its weight, increases in three dimensions, *length*, *width*, and *depth*. Because the cube of a series of numbers increases faster than its square, the weight of an animal increases disproportionately faster than a cross-section of its legs. This has profound consequences for function because the power of a muscle or the strength of a bone is proportional to its cross-sectional area, and while this cross-sectional area increases at a square of length, the thing that must be supported or moved (the body weight) increases at a *cube* of length – much faster. For a large primate to leap the same number of body lengths as a smaller primate, its muscles must be disproportionately large.

For example, imagine an animal with a cube-shaped body perhaps 4 cm in diameter and legs each 1 cm in diameter. I have put together such a body, with legs, by gluing together 1 cm cubes (Figure 9.2(a)). Let us say that each 1 cm cube weighs 1 g; if you count the cubes in Figure 9.2(a) you will see that there are 64; the animal's body, then, not considering its legs for the moment, weighs 64 g. We need not count them if we rely on our math; the body of the animal weighs 4 × 4 × 4 g, or 64 grams; $4^3 = 64$. In Figure 9.2, each of the animal's legs has a cross-sectional area of 1 cm². There are four legs, so the total weight-bearing support for the 64 g body is 4 cm². This means the body weight-to-cross-sectional leg area of the animal is 64:4, or 16:1. Another way of putting this is that they have one unit of power for every 16 units of weight.

God-like, let us endow this abstract 4 cm animal with the capability to leap four times its body length. A similar but larger version of this animal would have to retain that 16:1 body-weight-to-leg-muscle proportion in order to retain that capability. Table 9.1 summarizes this quantitative argument.

Now imagine an animal twice as long, not 4 cm but 8 cm (Figure 9.2(b)). The 4 cm and 8 cm forms of my schematic animal are called **geometrically scaled** versions of one another. Without the outlines of each cubic block to give scale (Figure 9.2), the two would look exactly the same. The number of cubes in the body of the blown-up 8 cm animal is 8^3, or $8 \times 8 \times 8$, which is 512 cubes. With each cube weighing 1 g, the total weight of the body (not including the legs) is 512 g. The legs double in diameter as well; they are 2 cm across (Figure 9.2(b,c)). This means the cross-section is now 4 cm² per leg or (summing the four legs) 16 cm² total. Whereas our 4 cm animal had a body to leg ratio of 16:1, an animal double its length, and therefore with a body weight of 512 g and legs of 16 cm², has a body-to-leg ratio of 512:16. This can be reduced to a 32:1 ratio, giving our 8 cm animal one unit of power for every 32 units of weight. Our smaller animal had a 16:1 ratio, which means the legs of our larger animal have only half as much power. This is always a surprise to those encountering this phenomenon for the first time. Even though they look *exactly* the same, the larger schematic organism has only half the leaping ability of the smaller animal. This is allometry.

Expressed in a verbal rather than mathematical way, the weight to leg-power proportion of the larger animal is twice that of the smaller animal. Because the power of a muscle is proportional to its cross-sectional area, the larger animal could not leap four times its body length, as the smaller animal could, but only twice. With less power, the larger version of the animal would appear ponderous and less agile than the smaller animal.

Now let us put the two versions of the animal in an imaginary world where natural selection is such that the same function is selected for, that both the small and large versions of the animal *must* be able to jump four times its body length. Let us say that each species has its own predator, and to escape being eaten the animal must jump four times its body length. To retain identical function, the larger animal will have to evolve *disproportionately* large legs. To retain the 16:1 weight to leg-strength ratio, the 512 g animal, the larger one, would need leg cross-sections of 8 cm² each (Figure 9.2(d)), or a total of 32 cm² for all four legs.

Table 9.1 **Scaling for a hypothetical abstract animal**

Animal	A	B
Length	4 cm	8 cm
Weight	64 g	512 g
Leg cross-section (all four legs)	4 cm	16 cm
Weight to leg cross-section	16:1 (or 64:4)	32:1 (or 512:16)

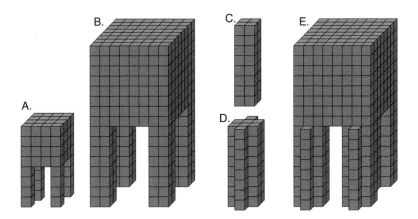

Figure 9.2 Abstract 4 cm long animal (a) and its geometrically scaled-up twin, double its size (b). The legs of the original animal (a) have a cross-section of 1 cm². The leg of the scaled-up animal is 2 cm in breadth, and its cross-section is 4 cm² (c). However, to achieve functional equivalency, the leg of the larger animal must be 8 cm² (d), not 4 cm². The functionally (or mechanically) equivalent scaled-up animal must have disproportionately large legs (e). This allometric principle accounts for differences in body proportions between small and large organisms. Image by the author.

I have mocked up such an animal in Figure 9.2(e). While the abstract animal in 2B is a **geometrically scaled** version of the first animal (Figure 9.2(a)), it cannot function as does the smaller animal. The *mechanically* or *functionally* equivalent upscaled version of our abstract animal is version Figure 9.2(e), not Figure 9.2(b).

9.2 Nature Often Compromises

Note that in the real world it seems that as animals grow larger the cost of retaining precisely the same function is simply not worth it. In other words, natural selection favors larger animals that have given up a little bit of their functionality rather than retaining perfect mechanical equivalency; they have compromised abilities. Elephants and giraffes cannot leap. Yet, they have disproportionately larger legs than a tiny Thompson's gazelle; they have compensated partly for their larger body size, but not completely. To put in another way, real-life large animals are less ponderous than if they were *geometrically scaled* versions of their smaller relatives, but their legs are not so large that they are *mechanically scaled* versions.

Because chimpanzees are large animals compared to monkeys and lesser apes, a functional morphologist must consider allometric effects when comparing them to smaller relatives.

9.3 Nature is Stingy

Before the middle of the twentieth century, much of our understanding of muscle function came from anatomists who were principally interested in humans; they were medical school professors who taught doctors how the body worked. They looked at muscle attachment sites and the shape of the joints the muscles acted on to make their best guesses about what the muscle did. Mostly they were right, or approximately right. While their understanding of muscle function was intuitive, rather than experimental, their work was not done in a vacuum; they tested their theories against clinical experience.

Rather than getting themselves arrested as mad scientists by experimenting on humans, they studied patients who had lost function of a particular muscle, either due to accident or paralysis. This allowed anatomists to test their functional hypotheses against reality by looking at functional deficits. When theoretical speculation about the function of a certain muscle jibed with the lost movement in paralysis victims, functional anatomy inched forward a notch.

After World War II, a new field of anatomical research emerged that took some of the guesswork out of muscle function research. John Basmajian and others (Joseph & Nightingale, 1952; Basmajian, 1957, 1965; MacConnaill & Basmajian, 1969) studied function by inserting fine wires into muscles and registering the electrical signals when the muscles contracted. Researchers could even send their own signal along the wire and into the muscle to see how a body part moved when they artificially stimulated a muscle. I have participated in such experiments, as both a researcher and a subject. The wires are inserted with a needle, and I can testify it is unavoidably painful. The subject might be asked to perform a specific task, such as hammering a nail or a specific movement such as flexing the fingers; the electrical activity in nerves and muscles during this movement is recorded. Since Basmajian's early work, EMG (electromyographic) research has been extended to primates and many other animals, with spectacular advances in our understanding of functional morphology and evolution (Tuttle & Basmajian, 1974, 1977; Stern et al., 1980).

Based on their observations of muscle function, Basmajian and colleagues established a tenet that is difficult to test, but has come to be accepted as valid: Anatomy evolves to reduce muscle action and strain on bones and ligaments during common behaviors, thereby conserving energy, reducing fatigue, and preventing wear (Basmajian, 1965; MacConnaill & Basmajian, 1969; Cartmill et al., 1987). This principle is called the **muscle-sparing principle**. The body evolves to be sparing with the work muscles must do. Arising from this principle is the rather counter-intuitive notion that the animal best adapted to performing a certain behavior will have the *least* muscle activity during that behavior. A wildebeest migrating 1000 miles while running from lions benefits by evolving to be efficient, thus using less fuel and arriving at breeding grounds healthier; a chimpanzee that is highly evolved for climbing has an extra few calories to pass on to her nursing, growing infant.

Functional morphologists try to understand the peculiar adaptations of different species by looking for unusually efficient actions, particularly large muscles, strangely shaped joints, or odd body proportions, and then relating this unusual anatomy to an unusual behavior in the species.

9.4 Ape Allometry

Returning to allometry, in Chapter 7 we learned that great apes have a number of distinctive features, a short back with only four lumbar vertebrae, broader, shorter lumbar vertebrae, a broad pelvis, a narrow sacrum, a more cone-shaped ribcage, and a tall, broad pelvis. Gibbons have a long back, a short, narrow pelvis, and a barrel-shaped thorax. We have already learned that long backs are related to leaping. Can we make that inference here? It makes sense, but here allometry rears its ugly head. For every one of the features that distinguish gibbons and chimpanzees, siamangs – intermediate in body weight between great apes and gibbons – are also intermediate in form (Figure 9.3). Siamangs are halfway between gibbons and chimpanzees in pelvic height, back length, pelvic breadth, and ribcage cone-shaped-ness. I will speculate more on this below, but I think allometry is partly responsible; however, it will take a dissertation-level research effort to resolve the issue.

9.5 The Relationship of Behavior to Anatomy

The first step in explaining anatomy, then, is to study what the animal does with its body day in and day

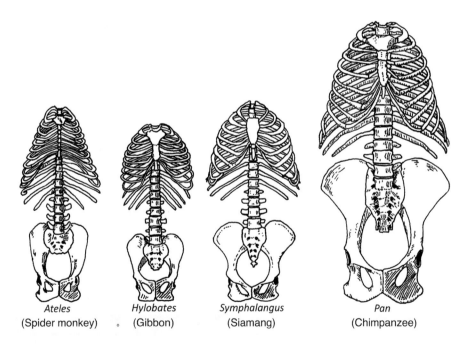

Figure 9.3 Great apes have broad pelves, a short back so that the floating ribs nearly touch the crest of the iliac blade, and a tall pelvis, taking up most of the space between the hip joint and the lower ribs, making the back inflexible. Siamangs, intermediate in size (7 kg for the gibbon, 15 kg for the siamang, and 40 kg for the chimpanzee), are also intermediate in form. Gibbons have five lumbar vertebrae, chimpanzees have four, and siamangs can have either four or five, averaging 5.5 (see Hunt, 2016 for review); siamangs are intermediate in sacrum breadth, lumbar vertebral height, pelvic breadth, and how cone-shaped their ribcage is. Allometry may explain these differences. From Hunt, 2016, Wiley Publishers; used with permission.

out. At one time, "they seem to leap a lot," was considered adequate; nowadays we record data from dawn to dusk, across seasons, and in many individuals, and document the frequency of locomotor and postural modes down to a fraction of a percent. As we consider how natural selection acts to cause a particular morphology to evolve, we can identify several pressure-points where natural selection might focus. We apply the muscle-sparing principles to wild animals. If a behavior is common, the animal must be able to perform it without injury and while avoiding extreme wear and tear on the body. We expect adaptations to prevent such injuries. We expect efficiency to evolve because, as we learned in Chapter 5, a chimpanzee must get up and go to work nearly every day, and starvation is a constant threat; a calorie saved is a calorie earned. The more efficient an animal is at any one task, the less energy it needs to find. As they gather food they must avoid or escape predators, and in the case of chimpanzees, murderous strangers. Individuals who gather their nutrients efficiently also leave time for other important activities, such as childcare and strengthening social bonds.

While efficiency is important, sometimes agility is even more important. Joint flexibility may be required to reach an inaccessible part of a feeding tree to harvest an essential resource. At the same time, flexibility reduces efficiency; yet, the individual has to have that food. Animals are compromise machines, trading economy for flexibility when agility in trees is advantageous, or trading speed for power when the species has to dig in the hard ground, or power for economy when long migrations are required.

Using this sort of reasoning, we have made tremendous progress in explaining how many of the traits that are distinctive in chimpanzees evolved, though even among experts quite a few are still the subject of disagreement.

9.6 Gape

Compared to humans and many monkeys, the chimpanzee's face projects far in front of the braincase (Figures 9.4, 9.5); they are prognathic. Their

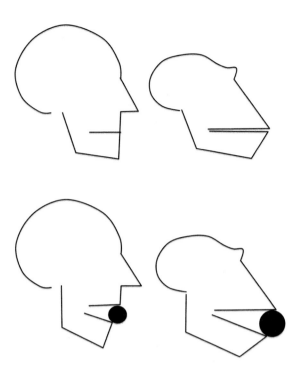

Figure 9.4 When the mouth is opened by rotating the mandible through the same angle, say 5°, a longer jaw results in a wider gape. Because chimpanzees (right) have longer jaws than humans (left), they can bite into larger-diameter objects. Images 9.4, 9.5 by the author.

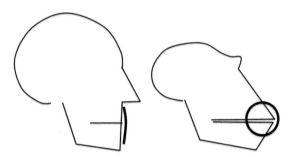

Figure 9.5 The chimpanzee prognathic face and receding chin allow them to scrape inside highly concave objects.

greater prognathism gives them a very long mandible, and there are important functional implications to long jaws. As the distance between the hinge of the mandible (the TMJ or **temporomandibular joint**) and the tip of the incisors increases, so does **gape**, or how wide the mouth can be opened. When a human opens her mouth 5°, the teeth are about an inch apart, but because the chimpanzee jaw is longer, a 5° opening separates the incisors by nearly twice as much (Figure 9.4). In other words, chimpanzees have huge gapes compared to humans.

Large gapes are not an accident of facial geometry, gape is under selective pressure and has evolved to provide important functional advantages, although it may serve slightly different functions in males and females. Bill Hylander (2013) has shown that among the monkeys of Asia and Africa, the cercopithecoids, males have much larger gapes than females, and male gape is scaled to the length of the canines: Species with longer canines have larger gapes. This is because long canines are of little use unless an individual can open his mouth wide enough to get something in between the tips of the canines. In species where males engage in intense physical battles – and in particular where those battles are fought one-on-one – large canines and the large gapes that accompany them give individuals an advantage during fights by allowing them to encompass and then puncture large-diameter objects – like an opponent's throat. Females are (at least among the monkeys and apes) less adapted to interpersonal violence and have both smaller canines and smaller gapes.

Gape is no use without bite force. Hylander showed that species with long canines and large gapes also have muscular adaptations that improve bite forces. When the mouth is opened as wide as possible, the muscle fibers of the jaw-closing muscles – the masseter, the pterygoids, and the temporalis muscles – can become so stretched that it is impossible to bite down with any force. A practical experiment demonstrates the effect of gape on bite force; biting into an apple seems an easy task, not one that might require a jaw-straining, vein-popping application of muscular force. Yet, if you place the entire apple between your incisors you will find yourself unable even to pierce the skin. Hylander showed that, in contrast to the feeble human bite force, species with large gapes have evolved much longer muscle fibers than humans, enabling them to generate great force even when the mouth is wide open.

There is another selective force acting on gape, however. Compared to humans, chimpanzees have immense incisors. Years ago, Hylander and Rich Kay (Hylander, 1975; Kay & Hylander, 1978) showed that incisor size is correlated with fruit diameter among primates. If you eat large-diameter fruits, you have to have incisors large enough to get a good bite out of the fruit and large enough that they can last through a lifetime of daily wear. Female gapes must still be great enough to avoid overstretching their muscle fibers when biting through such large fruits (Figure 9.4), though this requires a much smaller gape than does long canines. I think fruit processing is as responsible for chimpanzee faces as canine use. Chimpanzees have shorter canines than baboons, yet they have long jaws made even longer by the geometry of the supratoral sulcus (Figure 9.6).

9.7 Chin Function

The functional length of the chimpanzee jaw is greater still because the business end of the jaws, the incisors,

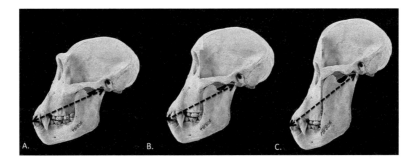

Figure 9.6 When we envision a normal chimpanzee skull (a) altered so that the face is pulled back toward the braincase (b) we find that skull length, in particular prognathism, is decreased (dotted line). The same jaw length could be maintained if the face were taller (c) but this introduces its own complications. Image by the author. (A black and white version of this figure will appear in some formats. For the color version, please refer to the plate section.)

protrude forward from their bony foundations, leaving the snouty chimpanzee without the distinctive jutting chin of humans. These angled incisors give them a sloping chin quite unlike the square, protruding chin of humans (Figures 9.4 and 9.5). It is humans who are odd in this feature, not chimpanzees. Every other primate has the same chinless morphology. Our own human ancestors were chimpanzee-like in this feature up until about 50,000 years ago (Daegling, 1993). Rather than the human chin having some important function, it seems more likely that the lack of a chin and protruding incisors function in chimpanzees to allow them to push their jaws into tight places where a chin would interfere (Figure 9.5). They can scrape the inside of even extremely convex items. When chimpanzees break open a hard-husked fruit the size of an orange (e.g., the divinely scented but wooden-husked *Oncoba spinosa*), they can scrape the inside of the shell with their protruding (procumbent, to be technical) incisors. Humans, with their orthognathic (the opposite of prognathic), vertically implanted incisors, prominent chin, and prominent nose, are incapable of scraping a surface that is only slightly concave.

Because the lack of a chin is such an obvious chimp–human difference, it is worth examining this unusual human feature in more detail. While the protruding incisors and receding chin of chimpanzees may have a clear feeding function, the origin of the human chin is not as clear. Early on (Robinson, 1914) it was thought that a projecting chin (formally known as the "mentum osseum" [Daegling, 1993]) aided speech somehow, but the argument was weak since it boiled down to "only humans have chins, only humans have speech." Others speculated that the chin functions only as an indicator of health, vigor, and masculinity – it might be a sexually selected character that females were evolved to find sexy only because it signaled vigor, while males were selected to display this badge of manhood as a mating strategy (Hershkovitz, 1970; Barber, 1995). If so, it is curious that the sexy chin made so little headway for the first six million years of our evolution. And even if this explanation is true, the feature may have evolved for some other reason and only later been coopted into a sexually selected character.

Another hypothesis holds that the short jaw of humans creates unique stresses on the front of the mandible that the prominent chin resolves. In this view, while the human mandible has decreased in length, the human brain has been increasing, pushing apart the jaw joints, preventing a matching reduction in its breadth, changing the stresses that act on it. Even with no experimental evidence to counter it, we might wonder why Neanderthals lack a protruding chin even though they have large brains. They do have longer jaws than modern humans, but it seems more likely to me that, like chimpanzees, they were evolved to work their incisors into recessed places.

Yet another hypothesis for chin origin takes the position that the chin is present in all primates, but most primates have large incisors that are pitched outward, leaving the bony structure that is apparent in humans less apparent in them (Lewontin, 1978). Humans have reduced their incisors, leaving the chin out in front of a pulled-back and atrophied dental arch. This hypothesis is mostly consistent with the getting-incisors-into-tight-places theory.

Others (DuBrul & Sicher, 1954; Riesenfeld, 1969; Daegling, 1993) dismiss this explanation, reasoning that the reduced power and size of human jaws requires less bone to support chewing stresses, so the lower margin of the jaw should be equally reduced to the part holding the teeth. Why should the teeth and the bone just below it recede, while the chin retains its ancestral robust dimensions?

Possibly the best-accepted hypothesis is that as human teeth reduced, the jaw became shorter and reduced the amount of stress concentrating on the chin when chewing muscles contract. The long mandible of other primates gives these muscles long lever arms that pull the mandible apart like breaking a wishbone. The shorter mandible does nothing, however, to reduce twisting stresses from the masseters that results in tensile stresses on the lower part of the chin, so it remains large (Hylander, 1979, 1984, 1985; Hylander et al., 1987; Daegling, 1993). Humans have a chin because they no longer have a need for the ability to push their incisors into tight places, but they still create stress in the chin area that requires that this bone remain.

I side with the reduced-incisors, reduced need for poking teeth into things hypothesis. Humans use tools to poke into things, allowing the incisors to reduce. The lower margin of the chin cannot reduce because it anchors and accommodates the larynx and other structures of speech.

9.8 Supraorbital Torus and Supratoral Sulcus

The prominent chimpanzee supraorbital torus or brow ridge and the depression or supratoral sulcus behind it have been the object of intense scientific curiosity, in part because we humans had a large supraorbital torus until very recently in our evolutionary history. It is a more difficult problem than it might seem. Many have speculated that it is reinforcement to resist large stresses created by the powerful jaws of chimpanzees and ancient humans. The problem with this is that measures of stress in monkeys show every little is concentrated in the brow ridge (Endo, 1970); the brow ridge is much larger than needed to counter these minor masticatory forces (Hylander et al., 1991).

My view is that chimpanzees have a face pulled far forward, dragged forward by selection to increase jaw length and gape. Large brow ridges set the eyes deep in a protective ring, shielding them from the slashing canines of their enemies. The eyes could be tilted backwards, rather than having a prominent brow ridge, but this would also take up space where muscle fibers are. Chimpanzee eyeballs are recessed in the face, with a bony ring encircling the eye. The brow ridge is part of this ring. The edges of the bowl protect the eye, in part because the eye is too recessed to be damaged by the teeth of an attacker, and as we know, chimpanzee intercommunity attacks result in horrific wounds; 28.6 percent of Gombe males had fractures or punctures on the face (Jurmain 1989, 1997). The orangutan has no brow ridge, but the large, violent males have raised fatty pads, flanges that I once thought may serve a similar purpose. Now we know they direct sound, so that males communicate more effectively with females.

If the face were pulled back (Figure 9.6(b)) so that there was no sulcus and the brow ridge was molded into the forehead – orangutans are built that way – the jaw would be shortened and the gape decreased (dotted line). It would be possible to keep the same mandibular length if the face were taller (Figure 9.6(c)), but the additional mandibular height would reduce gape: When the jaw opened the mandible would move backward rather than gaping.

Why not keep the condition in Figure 9.6(b) but simply elongate the jaw? Why is it that the eyes are pulled forward? When the eyes are pushed forward, well in front of the brain, the temporalis muscles are oriented more posteriorly, making them oriented better to pull the jaw backwards, rather than powerfully closing the mouth.

9.9 Superjaws

Chimpanzees have extremely powerful jaws, and they need them. Opening piths requires a substantial bite force – palm fronds, in particular, are tough; no human can open them. Imagine a grass stem the size of a broomstick, with a thick skin more like wood than you might expect. Chimpanzees push cylindrical piths like these into their incisors and rip out long strips of the tough outer layer using their powerful jaw muscles. Inside is a tasty pith the taste and consistency of celery. Palm fronds are even larger and tougher than grass stems; the centimeter-thick outer cortex is waxy and slippery, requiring considerable power to open. Chimpanzees grip these plant items with their incisors and rock their heads back and forth using their large nuchal muscles. Again, using the nuchals, they pull back on the tough fibers while pushing the stem away with their arms. The maneuver requires powerful jaws, necks, and arms. The posterior orientation of the muscle fibers is shown in Figure 9.7 When pushing a palm frond away from the mouth, the mandible is pulled forward by the pushing of the hands. Muscle fibers must counteract this force or the jaw will be dislocated forward. In 1987 over one-quarter of my feeding observations were on the broomstick-sized stems of *Pennisetum*. Ripping open these woody stems, containing as they do sandpaper-like opaline phytolith, wears the incisors; chimpanzees evolved huge incisors in response. As immense as their incisors are, they are often worn to the gums in older chimpanzees.

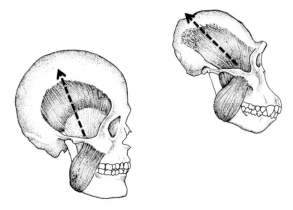

Figure 9.7 The average orientation of the temporalis muscle fibers is posterior (arrows), providing a retraction force during stripping and pulling actions. Image by Charles Wood; used with permission.

9.10 Other Dental Adaptations

Chimpanzee canines are long, cone-shaped, pointed, and overlapping. While some primates use canines to open fruits, chimpanzee canines are weapons of war. They are used to injure enemies in battle. Chimpanzees attacked in intercommunity battles endure many tens of puncture wounds, the sum of which are fatal. Their sharp canines, driven by the powerful chimpanzee jaw muscles, can penetrate skin and bone. As the teeth are pulled away, the sharp back edge of the canine cuts through tissue, opening huge, gaping wounds.

The chimpanzee lower first premolar, one of the teeth your dentist calls a bicuspid, is **sectorial**; the front surface has a long, sloping face that is shaped to rub against the back of the upper canine as the mouth closes (Figure 9.8). This rubbing sharpens the back of the canine.

The other teeth, the nonsectorial premolars and the molars, differ little from those of humans in external morphology, but the enamel caps are thinner. This adaptation may be a "less-is-more" adaptation; the thinner enamel wears through in some places early in life, leaving sharp edges that may serve to slice up foliage, more common in the chimpanzee diet than the human diet, more effectively than a flat surface could.

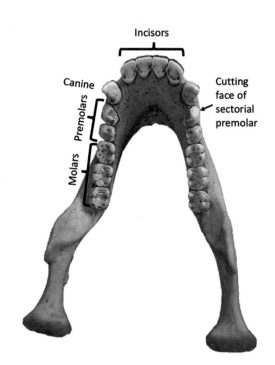

Figure 9.8 Chimpanzee mandible and teeth. Note the asymmetric first premolar; the light portion of the tooth in this image rubs against the back of the upper canine, sharpening it. Photo by the author. (A black and white version of this figure will appear in some formats. For the color version, please refer to the plate section.)

9.11 Pinnae

In Chapter 11 we will learn more about chimpanzee hearing, but the short story is that it differs little from that of humans; here we focus our interest on the external ear, the pinna (pl.: pinnae). The various protrusions and recesses of the ear reflect some frequencies and absorb others, fine-tuning the sound that reaches the internal ear. Some structures create microechoes that are thought to increase clarity. The combination of reflection, absorption, and echoes are thought to increase the clarity and intelligibility of words (Freedman & Gerstman, 1972) and aid in identifying the direction of the source of sounds (Blauert, 1997). The close similarity of human and chimpanzee ears suggests that chimpanzees and humans are evolved to detect similar sorts of sounds. Sounds recorded on normal microphones are less intelligible than sounds fitted with molds of ears.

I know of no publication that speculates on the function of the larger chimpanzee ears, but I believe they improve detection of sounds as an adaptation to detecting long-distance calls. Humans who have had parts of their ears removed because of disease have poorer hearing, possibly due to shape change, but just as likely due to their smaller size (Goode et al., 1977). Larger ears collect more sound waves and increase the volume of sounds reaching the ear canal. The chimpanzee external auditory meatus (the skeletal ear hole) is small in diameter (Figure 9.6) compared to that of humans. I hypothesize that the smaller ear concentrates sound better, increasing the ability to detect faint sounds in the distance, presumably at the expense of clarity. Before electronics, humans with poor hearing took advantage of this larger-is-better factor and the concentration of sound by using an ear trumpet, a large artificial ear that served the same function as large chimpanzee ears.

Chimpanzees need acute hearing because they break into small groups that may be separated by miles, and they rely on their excellent hearing to detect the calls of their compatriots (and enemies as well) at these great distances. Chimpanzees use their pant-hoots to tell their friends where they are – they recognize one another's voices – and to call other community males to large food sources. Gathering larger parties protects them from deadly attacks from neighboring males. The ability to hear enemies in the distance helps to locate extracommunity males to target them for attack. Gorillas have smaller ears and do not have these long-distance calls. Instead, they live in small groups that maintain close proximity throughout their daily foraging.

9.12 The Nose

Chimpanzees lack a bony external nose, a feature that is among the most noticeable of the human–chimpanzee differences. Most primates that have long noses house their extensive olfactory tissue inside their large nasal cavities, and they have a keen sense of smell. Not humans. The large human bony nose extends the nasal passage beyond the face, but not to increase the sense of smell. Inside the external nose is well-vascularized mucous tissue, specialized skin that secretes moisture, but it does not contain olfactory tissue. Instead, it seems to function to warm or humidify air or both, improving oxygen uptake in the lungs and shielding sensitive lung tissue from damage (Churchill et al., 2004; Wolf et al., 2004). Among human populations, those who live in drier climates and/or colder climates have larger noses (Carey & Steegmann, 1981). This adaptation gives humans the ability to tolerate much drier and colder conditions than many primates. Humans adapted to dry environments in particular require larger nasal mucous membrane surface areas to provide enough moisture to properly humidify inspired air; they have larger noses to accomplish this (Franciscus & Long, 1991). Only a few breaths of dry or cold air stimulate mucus production in the nose. Chimpanzees tend to live in warm, moist forests, leaving them with no need to extend their nasal passages.

9.13 The Mouth, Lips, and Tongue

The mobile lips and broad, flat, muscular chimpanzee tongues of chimpanzees – three times the size of a human tongue (Takemoto, 2008) – aid in processing food items. Chimpanzees can produce saliva in staggering abundance, and their salivary glands are likewise large compared to those of humans (Tyson, 1699; Sonntag, 1923). Large volumes of saliva are needed to neutralize tannins or toxins in some foods and to lubricate dry, fibrous food items to make them easier to swallow. There is an immediate, subconscious physiological response when bitter, sour, or astringent foods hit the tissue in the mouth. I find that some fruits provoke such a sudden and profound response that there is a painful cramping of the submandibular salivary glands, followed by a rush of saliva into the mouth.

9.14 Postcranial Adaptations

Much of the functional morphology of the head concerns feeding, and most of the rest is related to sensory perception. When we shift our attention south of the muscular neck of the chimpanzee, however, we focus more on locomotion and posture. Our food fixation will remain intact, though, because much of chimpanzee movement is about walking to feeding sites, climbing trees to get at food, and negotiating the difficult, small-branch milieu where fruits are found. As we consider the functional morphology of the body, a little history will help to show why science has not yet settled on a complete consensus on chimpanzee morphology, and where future research may move.

9.15 Keith's Brachiation Hypothesis Fails to Explain Chimpanzees

For decades, nearly a century, long forelimbs, mobile shoulders, and powerful upper body musculature were thought to be adaptations to brachiation. It should be no surprise that it was the study of gibbons that motivated Sir Arthur Keith to formulate this hypothesis; gibbons are brachiation exemplars *par excellence* (Keith, 1891). Keith speculated that among thin, pliable branches in the periphery of trees primates could not walk but instead must hang underneath the flexible supports found there (Keith, 1891; Avis, 1962; Napier, 1967). Long arms, long fingers, rotatable wrists, flexible shoulders, and powerful musculature were hypothesized to be adaptations to propelling the body during brachiation. Short legs reflected their lesser importance.

While this hypothesis is arguably still valid for the lesser apes, the great apes – despite the wishes of many researchers – stubbornly refused to conform to it. Jane Goodall's many hours of chimpanzee observation and Vernon Reynolds early work (Reynolds & Reynolds, 1965; Goodall, 1968) found almost no brachiation among chimpanzees; orangutans brachiate only slightly more often. By the 1970s, as the weight of accumulating ape locomotor data impressed itself on functional morphologists, they largely abandoned the brachiation hypothesis. Some avoided a wholesale paradigm shift by modifying Keith's hypothesis, merging brachiation with generalized slow suspensory movement. Meanwhile, a completely new hypothesis emerged.

9.16 The Vertical Climbing Hypothesis

While brachiation is uncommon among larger apes, all apes climb vertical supports. A number of workers speculated that the high intermembral indices (IMI) of apes – long arms and short legs – work together, improving the ability of apes to vertically climb tree boles as large as a meter in diameter. When arms are long and legs are short, the arms can reach far around even large tree trunks. When one or both forelimbs grasp the far side of a bole, the weight of the body presses the sole of the foot against the trunk, providing the friction required to push the body upwards. The longer the arms, the greater the friction. Long arms allow the knees to be extended, giving the muscles of the leg a better mechanical advantage (Kortlandt, 1968; Cartmill, 1974; Jungers, 1976; Sarmiento, 1987). Kortlandt compared the physics of the chimpanzee high IMI to a human climbing a large bole using a rope or belt to extend the arms (Kortlandt, 1968). Telephone- and power-line workers use a version of this technique to climb telephone poles, often increasing the purchase of the foot with spiked boots (Figure 9.9).

Consistent with the vertical climbing hypothesis, climbing recruits muscles that are distinctively large

Figure 9.9 A pole strap fastened to a body belt allows a lineman to increase the distance between the center of gravity and the telephone pole, increasing friction on the sole and allowing the knee and hip to be extended in a more natural position. Photo by Eric Willett; with permission.

in apes (biceps, brachialis, brachioradialis), muscles that flex the elbow (biceps, brachialis, brachioradialis), and muscles that pull the arm down (latissimus dorsi, lower pectoralis, posterior deltoid; Tuttle & Basmajian, 1974), as when doing a pull-up. Other muscles that are large in apes (parts of deltoid, upper pectoralis) raise the arm, as is required when reaching up for a new handhold.

The mobility of the shoulder found in apes (but few other animals) was hypothesized to have evolved to allow reaching up for a higher handhold while vertical climbing. The gripping great toe functioned to propel the body upwards when climbing smaller supports. The wrist would need to rotate as the body moved upwards and the elbow would have to be completely extended to reach around the tree trunk (Stern et al., 1980; Fleagle et al., 1981).

Isler and colleagues (Isler & Thorpe, 2003; Isler, 2005) offered support for the vertical-climbing-is-linked-to-elbow-extension-and-shoulder-mobility hypothesis from a different perspective. She analyzed 3D kinematics of apes during vertical climbing and demonstrated that three apes – orangutans, gorillas, and gibbons – raised their arms almost completely above the head during vertical climbing (Isler & Thorpe, 2003; Isler, 2005) – this call-on-me-teacher posture is called **full abduction**.

Isler and colleagues' experiments, unfortunately, were conducted on a flexible rope; in the wild, chimpanzees vertical climb trees or woody vines as thick as your arm, supports much stiffer than rope. These allow the torso to be tilted away from the vertical support, thus increasing friction on the sole (Jungers, 1976; Sarmiento, 1987, 1989). The leaning-back kinematic of climbing on rigid supports closes the angle the humerus makes with the thorax, precluding the need for full flexion or full abduction. Leaning back increases the lever arms of muscles that extend the hip and knee, reducing muscle action. My prediction is that when Isler's experiments are repeated on supports more like those on which wild animals actually locomote, joint angles for orangutans and gibbons will come into line with my chimpanzee observations.

9.17 Chimpanzees Seldom Climb Large Trunks

I was convinced of the vertical climbing hypothesis when I went into the field to study chimpanzee locomotion and posture in 1986, and in fact I was still convinced as I began to analyze my data. My data, however, stubbornly refused to fall in line with the high IMI equals vertical climbing hypothesis. I found that chimpanzees rarely climbed the large tree trunks (Hunt, 1991a, 1991b, 1992a, 1992b), a foundational aspect of the vertical climbing hypothesis, evoked to explain the function of long arms and completely extensible elbows. The average size of the stems on which chimpanzees vertical climbed was 7.6 cm (Hunt, 1992a), the size of a soda can; of all vertical climbing only 2.5 percent was on trunks larger than 10 cm, the diameter of a large-ish coffee mug. When they climbed large boles, 20 cm and more in diameter, the arms *were* completely extended; however, such extension left their large biceps and other elbow flexors, supposedly adaptations to large-diameter vertical climbing, useless. Vertical climbing on large boles does not seem to require shoulder flexibility, since the arm was almost never extended all the way above the head, never fully abducted.

Chimpanzees tended to ascend vines, saplings, and small trees, even when climbing into large trees. Climbing on these small supports *did* require a gripping big toe, and also required powerful flexion of the elbow. It did not require, however, complete elbow extension or abduction.

My colleagues who still adhere to the large-trunk vertical climbing hypothesis remind me that even though chimpanzees rarely climb such large trunks, they are astonishingly comfortable and casual climbing up a tree the size of a smoke stack – such climbing might be extremely important even if it is rare. The results of the rope-climbing experiments, they say, may turn out to be right. And perhaps the habitats in Mahale and Gombe required climbing smaller trunks that in the forests where chimpanzees evolved. All this is possible, but I prefer another explanation for rotatable wrists, shoulder mobility, and completely extensible elbows, a hypothesis a little closer to Keith's brachiation hypothesis.

9.18 Unique Chimpanzee Behaviors

My data suggest that chimpanzee elbow flexors, gripping feet, and (as we will see below) short lumbar segments are adaptations to climbing, but most other chimpanzee specializations are adaptations to suspension. We seem to have come to a hybrid theory that suggests that both vertical climbing and a sort of postural and slow-motion brachiation – suspensory behavior – are the two unique ape behaviors that we can link to unique ape anatomy.

9.19 Long, Curved Fingers with Large Flexor Sheath Ridges

Chimpanzees have disproportionately large **digital flexors**. When a digital flexor contracts, a finger flexes. The functional anatomist then asks: "Is there an unusual action that might involve this unusual anatomy, and might the two be linked?" Do chimpanzees do something unusual that requires flexing their fingers? They do indeed. They hang by one arm more than their relatives, the Old World monkeys. You might think, "What about walking? Is flexing the wrist not important during push-off? Or some part of the motion during walking? What about climbing? They must grip the bole as they ascend." It is a surprise, but the digital flexors are largely inactive when standing or walking. The hand is supported entirely by ligaments and bones, not muscles (Susman, 1979). Digital flexors were active during climbing, but not nearly as active as during armhanging. Maximum activity was seen during manual suspension.

The length of the fingers matters not at all when moving on the ground – animals specialized for efficient terrestrial walking often lack fingers altogether, instead having hooves. When moving among large-diameter branches, chimpanzees knucklewalk, or walk on their palms (Hunt, 1992a). The body weight presses the hand and foot onto the branch, and there is no need for gripping.

When hanging underneath a branch, however, long, powerful fingers are a necessity, better explaining both the powerful digital flexors of chimpanzees and their long, curved fingers. Curved fingers are preformed, and therefore prestressed to the shape of the cylindrical branches they are evolved to grip. The cup-shaped surface of the palmar side of the finger bones of chimpanzees is an accommodation for the thick tendon that runs along the inside of the fingers and powers their gripping ability, and robust flexor sheath ridges are the anchors that hold the ligaments that power that flexion.

9.20 Long Arms and Short Legs May Be Independent Adaptations

Bill Jungers pointed out that among primates, the larger the species, the shorter the hindlimbs in relation to the arms (Jungers, 1984). He suggested that short hindlimbs serve to lower the center of gravity, making the animal more stable on flexible supports. In fact, when supports are unstable, primates fake short legs; they make the effective length of their legs shorter by crouching closer to the branch on which they are walking (Schmitt, 1994). Perhaps short legs improve stability on unstable surfaces like tree branches, while long arms evolved for a different reason entirely.

9.21 Long Arms Aid in Terminal Branch Feeding

Suspensory behavior solves a specific challenge, negotiating small-diameter, flexible (or compliant) twigs like those among the outermost branches of trees. In the previous chapter I noted that among chimpanzees, suspensory locomotion makes up only 0.2 percent of all locomotion, yet its frequency rises to 29.4 percent in the outermost meter of the tree canopy. Among the smallest category of supports, 8.8 percent of locomotion is brachiation and 20.6 percent is a very similar behavior, "transferring," which is a sort of lunging armhanging used when the individual reaches out to grab branches in a tree into which it is transferring its weight.

We see the same story for posture, except that when armhanging, chimpanzees really only do one thing: eat. And they eat in places that encourage suspensory behavior; as the diameter of supports decreases, armhanging increases. Long arms increase the diameter of what has been called the "feeding sphere" (Grand, 1972). They improve feeding efficiency by extending the reach of the chimpanzees, a helpful feature when collecting fruits in difficult-to-reach areas. There is less need to shift feeding sites, a difficult and expensive proposition among small branches in the tree-edges.

It may be that long arms are an adaptation to both hang-feeding and climbing, while short legs lower the center of gravity, easing movement among unstable supports.

9.22 Armhanging Requires Complete Abduction, Full Elbow Extension

In summary, while a high IMI might have multiple selective advantages, armhanging is the only behavior in which the arm is fully extended above the head, suggesting that it is armhanging for which mobile shoulders in apes are evolved (Hunt, 1992a). It is also the position that most often requires elbow extension, and is also exhibited during quadrupedal standing.

9.23 The Cone-Shaped Thorax

While vertical climbing has never been linked to narrow scapulae or cone-shaped ribcages, armhanging and other suspensory behavior may explain them. To see why, we will resort to some "thought experiments" that consider stresses on the thorax.

The torso of the typical primate is rather barrel-shaped; arms attach at the sides. One-armed hanging causes stresses on the thorax like those in a barrel that is lifted from a single point on its rim (Figure 9.10). Imagine a heavy but flexible barrel, made of rubber perhaps, and filled with something malleable – like water or (if you prefer) internal organs like lungs and intestines. If you picked it up gripping the rim, what

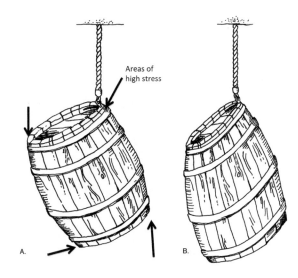

Figure 9.10 Stresses acting on a barrel lifted by the rim (a) can be visualized in this thought experiment. If the barrel were flexible it would deform into a shape such that the areas with the highest stress would deform the most (b). These are areas (arrows) that have the sharpest angles, the areas that have the farthest to deform before stress is evened out over the entire object. Image by the author.

might happen? When the barrel is lifted this way, it would sag; the bottom would bulge as in Figure 9.11(c). Stresses at the corners where the top and bottom meet the sides would be particularly high, creating hotspots where failure might be more likely (Figure 9.11).

An animal hanging by one arm generates hotspots in places where bone and ligament are repeatedly overstretched; not just muscle and ligament, but even bone fatigues; microfractures appear, much as bending a piece of metal over and over again eventually breaks it. If some areas are overstressed while others are understressed, the overstressed areas fatigue and fracture, while the understressed areas are being underutilized. Hotspots, the overstressed areas, could be eliminated if we "preformed" the barrel so that the corners were less sharp. If the barrel were shaped as in Figure 9.10(b), the hotspots would be a little cooler and the remainder of the barrel would bear slightly more stress, reducing the maximum stress at any one point.

There is a problem with this preforming plan, however. Primates have two arms, and preforming the

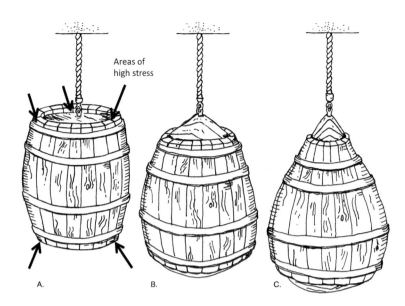

Figure 9.11 Stresses acting on a barrel lifted by the rim (a) can be visualized in this thought experiment. If the barrel were flexible it would deform into a shape such that the areas with the highest stress would deform the most (b). These are areas (arrows) that have the sharpest angles, the areas that have the farthest to deform before stress is evened out over the entire object (c). Image by the author.

torso so that it would be less stressed when hanging by one arm would actually make the hotspots even *more* stressed when hanging by the other arm. Mammals tend to be more or less symmetrical; Figure 9.10(b) is a poor evolutionary solution.

To get closer to a solution we can imagine that instead of hanging by an edge, the rubber barrel was suspended from a rope attached to the *center* of its top (Figure 9.11(a)). The barrel would deform to resemble Figure 9.11(b). Stresses would be reduced in Figure 9.11(b), but there would still be strains at the edges where the top meets the sides. Note that the bottom bulges out, straining the corners where the bottom meets the sides. If we extend this thought experiment and imagine that the barrel were more flexible still and that all of the stress is distributed perfectly evenly, we might get the form in Figure 9.11(c). Nature provides us with an example of this in a water droplet, which settles into its shape by a balance between surface tension and gravity: Stress is exactly equal over the entire surface area of a droplet; the result is a distinctive teardrop shape mimicked in Figure 9.11(c). In this case, because stress is distributed over the entire surface, there are no hotspots. The maximum stress is much lower because there are no hotspots, and there are no underutilized areas that are barely deformed, which would leave the thorax in those places essentially over-built. The hunched appearance of chimpanzees is a result of the narrower upper thorax.

9.24 The Scapula

If an armhanging primate were to optimize its shape, then it might be the shape of a teardrop. There is still a problem. In this model the barrel is suspended from a single point in its top. This is great for single-armed animals, but primates have two arms, and a midline that is already occupied by a head, an essential body part.

The only way a primate could assume this shape would be to have shoulders that swing up until they reach the midline, and on both the left and right side. The head is there, but perhaps the shoulder could move behind the head. This is exactly the function of chimpanzee shoulder blades (scapulas).

In humans, the scapula is triangularly shaped so that an inside edge runs into the spine before it can rotate far enough to allow the shoulder joint to move to the midline (Figure 9.12), but the narrow scapula of the chimpanzee allows it to swing far up, so far that when an armhanging chimpanzee is viewed from the

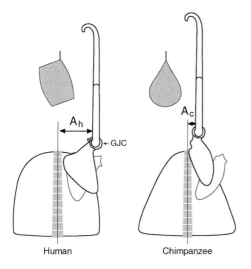

Figure 9.12 Rear view of the ribcage; vertebrae at the centerline, scapula presented in an arm-raised position (solid lines) and normal arms-down posture (shaded image). The eccentrically placed human shoulder joint creates stress hotspots in highly angled anatomical fields. The narrow scapula of the chimpanzee allows the shoulder to swing up to the centerline, reducing the distance between the centerline and the shoulder joint (a) and allowing the thorax to assume a stress-relieving teardrop shape. The tilted-up shoulder joint allows ligaments around the entire circumference of the joint to bear weight, whereas in humans these ligaments (GJC) are loose on the top and overstretched on the bottom. Image by the author.

front, the shoulder has all but disappeared behind the neck (Figure 9.12).

To make armhanging still more comfortable – in other words, to reduce the overstretching of ligaments – the chimpanzee shoulder joint is tilted up. The **glenohumeral joint capsule** (GJC in Figure 9.12), the ligaments that anchor the shoulder, are stretched nearly equally around their entire circumference of the glenoid. In humans, the sideways facing glenoid fossa leaves the ligaments on the top of the GJC loose and underutilized, while the ligaments underneath bear all the weight and are overstretched.

9.25 Muscle Sparing

In the 1970s, Russ Tuttle and colleagues examined muscle activity during chimpanzee armhanging and found something astounding. During armhanging, digital flexors were virtually the only active muscles. That bears repeating. A chimpanzee hanging underneath a branch by one arm, a stressful posture, has evolved so that muscles that allow the fingers to grip are active, *but almost no other muscle* is. Rather than wasting calories flexing muscles to support armhanging, chimpanzees bear weight on the skeleton, ligaments, intramuscular septa, and by passive muscular tension (Tuttle & Basmajian, 1974, 1977).

9.26 Broad Thorax

Chimpanzees, then, have narrow scapulas that can swing up to allow the shoulder joint to approach the midline, and their ribcage is cone-shaped. Add a bulging belly hanging underneath and the thorax begins to resemble a teardrop, where stress is evenly distributed both on the ligaments, bones, and muscles of the thorax and on the ligaments of the GJC.

During armhanging, the thorax is in essence hung from the arm, attached in the front via the scapula and clavicle to the sternum (breast-bone) and via muscles attached to the scapula in the back. This front-and-back attachment compresses the torso as if a heavy weight were placed on a person lying on their back on the floor.

However, unlike this lying-on-the-floor analogy, less stress would be exerted if the torso were smaller in diameter from front to back. That is, when the thorax is broader than it is deep, the thorax is pre-shaped to reduce stress pressing the front and back of the thorax together. Imagine picking up a bag with a loaf of bread in the top, placed so that the long axis was between the handles. It would be squashed. Turn it 90° so that the length is side-to-side and it would be unharmed.

Prestressing the thorax so that it is narrow front-to-back and wider side-to-side gives an animal shoulders, as all apes and humans have, but which your pet cat or dog does not.

9.27 Short Back

While armhanging has shaped the top half of the thorax, fashioning the scapula, cone-shaped thorax, and gripping forelimb, the lower part of the thorax, the lumbar vertebrae, are under a different selective pressure – adaptation to vertical climbing.

The reduced lumbar length and vertebral number (four lumbar vertebrae, and sometimes only three) and the bizarrely elongated os coxa (pelvis; see Figure 7.20) mean that much of the chimpanzee lower back is taken up by a completely inflexible pelvis, rather than the segmented vertebrae that are engineered for flexibility (Schultz 1930, 1936; Erikson, 1963). The short ape back is utterly different from that of the monkey; monkeys have six, seven, or even eight lumbar vertebrae (Schultz, 1961; Erikson, 1963; Rose, 1975) and their backs are heavily muscled. The function of the long, powerful monkey back is well understood as a leaping adaptation (Hatt, 1932; Currey, 1984). Leapers flex their backs, hips, and hindlimbs, gathering their legs underneath them and scrunching their backs down so that their shoulders are near their knees. With explosive power, they extend their legs, hips, and backs, accelerating their bodies and propelling them many body lengths. The longer the back, the more deeply the monkey can flex it and the longer force is applied to the platform; the longer the feet can push against the branch the greater the acceleration and the longer the leap (Napier, 1967; Ripley, 1967). Leaping selects for long, powerful backs.

You may ask, why not simply increase the length of the legs to increase the length of time the muscular force accelerating the body can be applied to the take-off point, rather than elongating the complicated and troublesome back? Good question. Many animals, for example ambush predators (e.g., cats) and primates, require a lower center of gravity, for mobility in the first case and to increase stability on compliant supports in the second (Schmitt, 1994). A long back retains a lower center of gravity while allowing rapid acceleration and long-distance leaping.

While the function of a tall lumbar region is a settled issue, the function and adaptive value of the short, stiff back among apes is poorly understood. Early explanations focused on hand-over-hand brachiation (Keith, 1891), reasoning that as the torso twisted first one way and then the other, the lower body had to be whipped around. Hypothetically, the lumbar vertebrae bear most of the stress as the powerful upper body twists the weight of the hips and legs. This theory is clearly wrong: Not only do larger apes rarely brachiate, the best brachiator, the gibbon, has the longest back among apes.

Others noted that even if locomotor suspensory behavior is rare among the larger apes, suspensory posture – armhanging – *is* common (Rose, 1975; Cartmill & Milton, 1977; Hunt, 1992a), and perhaps the back was stressed during repositioning or by the pull-up movement used to transition from armhanging to walking on top of branches. The slow movement typical during these activities makes this speculation unlikely; or at least makes it incomplete.

An alternative might be that the backs of apes are merely an accommodation to their large body size; the short back might be an allometric effect. Because, as we learned in Section 9.1, weight increases as a cube of length (the allometric square-cube issue explained in Figure 9.1); the back becomes less and less stable as body weight increases. I would argue that this cannot be the full explanation. Monkey spines withstand huge forces during leaping. It seems unlikely that a bizarrely reinforced configuration is required in apes to accommodate static postures and slow movement. Humans are even heavier than chimpanzees, and they bear stress during bipedal carrying, yet their backs are longer than those of chimpanzees.

9.28 A Model for the Spine

We can think about the vertebral column best by reminding ourselves that it is a column. If you imagine jostling a stack of cylindrical objects, like soup cans, you already have an intuitive understanding of its stability from real-world experiences. If you stack the cans too high and then place a weight on top, they topple easily. Similarly, a vertebral column with tall, narrow, and numerous vertebrae is in greater danger of either **buckling** (Figure 9.13) or collapsing. During climbing, the back is stressed. A hindlimb is pushing the body up first on one side then the other, which bends the spine back and forth, risking buckling. As first one leg then the other propels the heavy upper body of apes upward

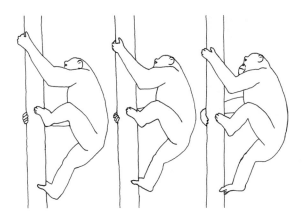

Figure 9.13 Chimpanzee vertical climbing. Note the gripping great toe. Image by the author.

(see also Cartmill & Milton, 1977), the spine is bent first one way then the other. A long, narrow spine risks buckling (Figure 9.14) under this stress (Jungers, 1984), whereas a short, broad lumbar vertebral segment is more stable and less stressed (Ward, 1993).

The latissimus dorsi muscle has its origin on the crest of the ilium and attaches to the humerus. When it contracts it pulls the arm down, or if the arm is holding onto a vertical pole, it pulls the hip upward. As the hip is drawn up toward the pulling arm, the spine is flexed, first one way, then the other.

Given that trees sway, grips fail, trunks are uneven, and obstacles like horizontal branches present themselves, the forces acting on the spine are erratic and eccentric. With a long, narrow back there is the constant issue of the middle of the lumbar section passing outside the compressive stress from the upper body on the one hand and the force of the pushing leg (Figure 9.14). Danger is severe when some part of the column flexes so that it falls outside the line of force acting on the column (Figure 9.14(a)), and this happens easily when the point of pressure is outside the edge of a column. As the column flexes, the farther the middle section passes outside the line of force compressing the spine, the greater the leverage of the body weight on the ligaments supporting the spine and the greater the danger of complete collapse. The more segments the column has, the greater the flexibility and the greater the buckling risk; the narrower the column, the less flexing is required to initiate buckling.

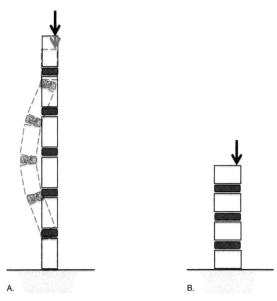

Figure 9.14 Schematic representation of a vertebral column of a monkey (left) and an ape (right). Open rectangles represent vertebrae, stippled ones represent intervertebral disks. When force acts asymmetrically on a column (arrows), a narrow column with many segments (left) has a greater tendency to buckle (dotted image) than a column with broad, short vertebrae. Image by the author.

If a pulling arm is on one side and a pushing leg on the other, the two forces balance out and there is little bending, only compressive stress. However, if the arm and leg act together, left arm pulling, left leg pushing, bending stress on the spine is enormous. Chimpanzees do just this. They tend to propel the body for a substantial part of their climbing by pulling up with an arm and pushing up with a leg on the same side. The combination of the latissimus dorsi pulling the hip up and the shoulder down and the power of the pushing leg also pushing that side of the hip upward risks such dramatic bending that buckling is a risk. A short back with broad lumbar vertebrae is a solution to this stress.

Instead of bearing weight on six or eight lumbar vertebrae, apes have taken up most of their backs with the rigid os coxae. The short, squat form of the lumbar vertebrae makes them more like disks than cylinders. They are less likely to over-flex and buckle, just as a stack of short, squat cans is less likely to topple (Figure 9.14(b)).

Bill Jungers speculated that vertical climbing – but not leaping – introduces destabilizing forces on the narrow monkey spine. During leaping, if the primate pushes off with both legs, the spine is compressed along the midline and the powerful back muscles and symmetrical forces keep the stack of soup cans straight.

9.29 The Broad Pelvis

The combination of larger body size and the demands of climbing are the best explanation for the short, stiff back of chimpanzees, but what about the broad ilia? One hypothesis is that the pelvis must be as broad as the ribcage. The large gut of great apes demands a broad ribcage and therefore a broad pelvis. The broad thorax and narrow pelvises of gibbons (Figure 9.3) and spider monkeys (Figure 7.20; Schultz 1930, 1961) disproves this speculation. Another explanation seems more likely, and it too relates to allometry. Because chimpanzees are heavy compared to smaller primates, they require disproportionately large muscles that extend the hindlimb – the gluteus minimus and medius, to name two. The iliac blades might be expanded to provide a larger muscular attachment for the disproportionately large gluteal muscles, large in order provide this functional equivalency (Stern, 1971). Gorillas leap during fights (Figure 9.15) and their immense weight requires equally immense gluteals and quadriceps.

9.30 Allometry Redux

Recall the discussion concerning the gibbon–siamang–chimpanzee series. It provides us with something of a test for allometry hypotheses; siamangs are twice as heavy – about 13.5 kg versus 7 kg for gibbons – but the locomotion of the two is very similar. They both engage in brachiation, armhanging, and vertical climbing. If we just compare gibbons to great apes, there is a clear distinction in locomotor behavior; gibbons leap more and have a longer back, seeming convergent on monkeys. Just as we begin to feel comfortable with this link, however, allometry raises its ugly head. For every one of the

Figure 9.15 Gorillas leap not in trees but during one-on-one fights, where their large, powerful hindlimb and gluteal musculature propels them in a period of free-flight, despite their immense body size. Photo by Nicholas Godsell; used with permission.

features listed above, siamangs – intermediate in body weight between great apes and gibbons – are convergent on chimpanzees (Figure 9.3). The os coxa of the smaller gibbon, specifically the ilia, are narrow, offering a small surface area for hip extensors; the pelvis is narrower than the ribcage (Figure 9.3). Siamang pelves resemble those of the great apes in that they are as broad as the ribcage, in accord with the square-cube argument (Stern, 1971) that larger animals require larger gluteal muscles. Gibbons have five lumbar vertebrae, more than the four typical of great apes, while the larger siamang falls right in between the two: half of them have four vertebrae, half have five, yielding an average of 4.5 (Schultz, 1961). Excuse me now re-muddying water that may have seemed to have cleared. While the difference is not great, siamangs leap less than gibbons, which suggests we might explain back length without resorting to allometry. But look closely at the ribcage and pelvis. Siamangs are halfway between gibbons and chimpanzees in pelvic breadth and ribcage cone-shaped-ness. How exactly could a cone-shaped thorax and a broad iliac blade be related to lower frequencies of leaping? Or does the larger size of siamangs allow them to cross caps in the forest canopy by brachiating rather than leaping? They brachiate more than gibbons, so this might be the case. I will speculate more on this below, but I can only speculate. It will take a dissertation-level research effort to resolve this issue.

9.31 Vertical Climbing

To summarize all this, we have the gripping hind foot, powerful muscles that aid in the pull-up action, a short lumbar spinal segment, and tall os coxae. Chimpanzees must extend their hip to climb, so a broad pelvis may be an adaptation to vertical climbing, though leaping may also be a factor.

9.32 Taillessness

The lack of a tail in chimpanzees may be due to allometry. We have long known that the tail plays a critical role in maintaining balance on narrow, unstable supports. In an early study, rodents with their tails amputated were much less competent on narrow supports. Their center of gravity wobbled first one way and then the other as they attempted to stay balanced, compared to subjects with tails (Buck et al., 1925). Intact subjects whip their tails back and forth to make minor adjustments to balance and, as a consequence, they can better maintain their center of gravity directly above the support (Buck et al., 1925; Kelley, 1997; Larson & Stern, 1998; Russo & Shapiro, 2011). Even when arboreal supports are medium-sized (e.g., the size of your wrist), they can be unstable for a chimpanzee because another individual lands or takes off from them, because the wind oscillates them, because the point of support is far from a stabilizing branch-point with the trunk, or simply because the body weight of the primate walking on them deforms them – all situations where a tail would come in handy.

As animals grow larger the same square-cube law that affects leg size acts on tails. Because body weight increases at a cube to body length, for a tail to retain the same function in large versus small monkeys, it would have to be disproportionately long. We can imagine a gorilla with a tail three times its body length, but such an organ would have to be so massive it would be a hindrance rather than a help. If this were not reason enough to forego a tail, for the same reason that long backs are stressed more in larger animals, the tail vertebrae would be under impossibly large stresses as the tail whipped back and forth to stabilize the center of gravity. The trade-off is apparently not worth it. Some monkeys have very long tails, but in accord with this explanation the largest monkeys – baboons and mandrills – have quite small tails. Consider also that long tails are not "free"; there is an energetic cost of carrying an extra body part and they increase predation risk; chimpanzees often get their first grip on monkey prey by seizing their tails (Stanford & Wrangham, 1998).

9.33 Narrow Terminal Phalanges

While much of the functional morphology you have just reveled in is messy and complicated by allometry, let us finish postcranial functional morphology with a feature that is not very controversial, largely because few researchers have paid it any attention. Chimpanzee terminal phalanges are narrow – their fingers are pointy compared to the broad fingertips of humans, and we have no idea why. Likewise, we understand the small thumb of chimpanzees – a trait shared by gorillas and orangutans – only poorly, if at all. The typical explanation is a negative one: Chimpanzees use their thumbs for tool use and other manipulative tasks far less often than do humans (Ziegler, 1964), so there is no need for a long, powerful thumb. This explanation seems inadequate since in monkeys such as macaques and capuchins the thumb is more human-like (Napier, 1967). I hope to read your dissertation on one of these topics someday.

9.34 The Miracle That Is the Chimpanzee

As incredible as it seems, we have at last a reasonable explanation of chimpanzee (and great ape) anatomy. Their short backs, tall pelves, gripping feet, powerful body-lifting muscles (reviewed in Figure 9.16 and Table 9.2), and long arms (partly) are adaptations to

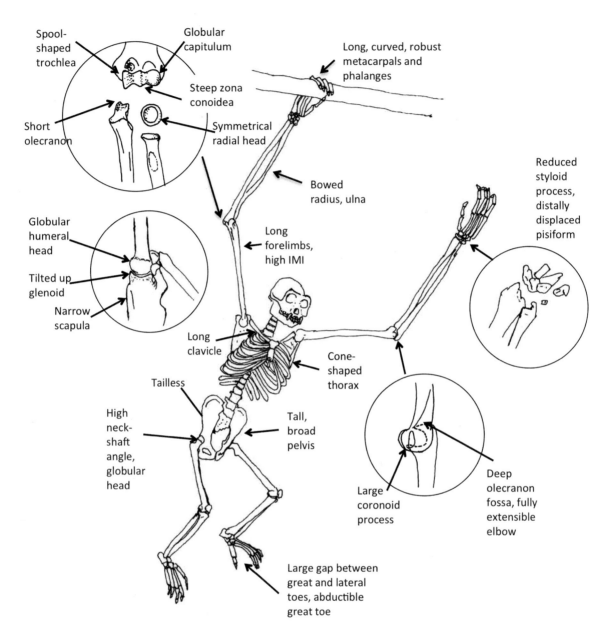

Figure 9.16 Ape specializations. A globular capitulum, round symmetrical radial head, reduced ulnar styloid process, and distally displaced pisiform allow free pronation and supination. A bowed radius and ulna increase the moment arms of pronators and supinators. A cranially oriented glenoid fossa reduces strain on the glenohumeral joint capsule during armhanging. A globular humeral head allows wide shoulder joint excursion through all three planes. A narrow scapula allows the shoulder joint and the vertebral border to approach the midline during full abduction, reducing strain on the thorax during armhanging. A short olecranon process allows complete elbow extension during armhanging, and a humeral articular surface for the ulna that wraps completely around the trochlea (dotted line in lateral view of humerus and ulna) allows the body weight to be borne by bone during suspension, rather than muscle, tendon, or ligament. A large coronoid process increases the moment arm for brachialis to initiate elbow flexion when the elbow is completely extended. Robust, curved manual rays reduce muscle action during suspensory gripping and reduce stress on the hand. Long forelimbs increase the gathering sphere and aid in vertical climbing. A gripping great toe allows the hindlimb to exert powerful vertical force, propelling the body weight upward during climbing. A high neck-shaft angle displaces the femoral head cranially, away from the greater trochanter, to increase hip flexibility. After Fleagle (2013) and Hunt (2016), with permission of John Wiley & Sons.

Table 9.2 **Chimpanzee functional anatomy**

Anatomical Feature	Function
Skull	
Prognathism	Increase gape to broaden capacity to open large food items
Receding chin, procumbent incisors	Increase capacity to incise concave food items
Large brow ridge, sulcus	Intraspecific violence: eye protection
Large brains	Adaptation to fission–fusion; adaptation to ripe fruit
Large temporalis, masseter	Stripping, ripping, and gripping, esp. piths, fronds
Large nuchal muscles	Stripping, ripping, and gripping, esp. piths, frond
Dentition	
Thin enamel	Fallback diet of piths and leaves
Y-5 cusp pattern	Diet of ripe fruit
Large canines	Intraspecific violence: offensive weapon
Sectorial premolar	Intraspecific violence: hone upper canine
Large incisors	Stripping, ripping, and gripping, esp. piths, frond
Upper body musculature	
Large elbow flexion muscles	**VC (vertical climbing)**: lift body mass
Large humerus retracting muscles	VC: lift body mass
Robust flexors	**Armhanging (AH)**: grip support, esp. during unimanual AH
Body proportions	
Long arms	VC (increase friction on sole of foot) or AH (reaching out to harvest food)
Short legs	VC; arboreality: increase friction on sole of foot, lower center of gravity during arboreal walking
Arms (forelimb)	
Distinct radiocarpal articulation	AH: allows pronation/supination during armhanging
Reduced ulnar styloid	AH: wrist rotation, ulnar deviation during food gathering
Globular humeral head	AH: shoulder mobility
Humeral torsion	AH: adaptation to broad torso, which in turn reduces stress
Reduced ulnar olecranon process	AH: allows full elbow extension
Humeral olecranon foramen	AH: decrease stress on elbow ligaments, increase stability
Globular articulation for radius	AH: rotatory wrist increases harvesting sphere

Table 9.2 (*cont.*)

Anatomical Feature	Function
Spool-shaped humeral trochlea	AH, VC: elbow stability/integrity during AH, forelimb-driven arboreal movement including suspensory and vertical climbing
Hands (technically: the manus)	
Long fingers	AH: circumduct support
Curved fingers	AH: decrease stress on fingers, better circumduct support
Dorsal ridges	**Knucklewalking (KW):** reinforce hand to prevent fingers hyperextending
Small thumbs	**Leaping:** gripping landing support
Legs (hindlimb)	
Distally displaced greater trochanter	AH: allows mobile hop for gripping support
Open femoral neck angle	AH: allows mobile hop for gripping support
Feet (technically: the pes)	
Gripping great toe	VC: grip support during hand-over-hand vertical climbing
Long, curved toes, digits II–IV	AH: allows gripping support
Thorax	
Cone-shaped ribcage	AH: reduce stress torso
Broad torso = sharply curved ribs	AH: reduce stress torso
Narrow scapula	AH: reduce stress on GHJC, and torso
Cranially oriented glenoid	AH: reduce stress on GHJC
Short lumbar segment	VC: reduces buckling
Round or oval glenoid fossa of scapula	AH: shoulder flexibility with stability; prevent dislocation during powerful movement
Pelvis	
Broad pelvis	Unknown. Large gut? If so, why not the same in leaf monkeys?
Tall pelvis	VC: reduces vertebral buckling
No tail	AH not above branch? Large body size? Terrestrial ancestry?

vertical climbing. Their long fingers, powerful flexors, mobile shoulders, extensible elbows, narrow scapulae, tilted-up shoulder joints, cone-shaped thoraxes, and (partly) long arms give them their amazing suspensory abilities. Knuckling during walking accommodates long fingers to allow long-distance travel.

Chimpanzees are sometimes characterized as terrestrial animals, but such a label is deceptive. They take advantage of the decreased travel costs the ground provides when they need to move long distances, but they gather their food and sleep in trees, where their extraordinary physical abilities allow them a competence and comfort in the canopy that might seem like flying to us. You may have heard you cannot have your cake and eat it, too, but chimpanzees come close; they have their superb adaptations to the trees, but they have the benefits of terrestrial travel, too.

References

Avis V (1962) Brachiation: the crucial issue for man's ancestry. *Southwestern J Anthropol* 18, 119–148.

Barber N (1995) The evolutionary psychology of physical attractiveness: sexual selection and human morphology. *Ethol Sociobiol* 16, 395–424.

Basmajian JV (1957) Electromyography of 2-joint muscles. *Anat Rec* 129, 371–380.

Basmajian JV (1965/1977) *Muscles Alive: Their Functions Revealed by Electromyography*. Baltimore, MD: Wilkins & Wilkins.

Blauert J (1997) *Spatial Hearing: The Psychophysics of Human Sound Localization*. Hong Kong: MIT Press.

Buck C, Tolman N, Tolman W (1925) The tail as a balancing organ in mice. *J Mammal* 6, 267–271.

Carey JW, Steegmann AT (1981) Human nasal protrusion, latitude, and climate. *Am J Phys Anthropol* 56, 313–319.

Cartmill M (1974) Pads and claws in arboreal locomotion. In *Primate Locomotion* (ed. Jenkins FA, Jr.), pp. 45–83. New York: Academic Press.

Cartmill M, Milton K (1977) The lorisiform wrist joint and the evolution of "brachiating" adaptations in the Hominoidea. *Am J Phys Anthropol* 47, 249–272.

Cartmill M, Hylander WL, Shafland J (1987) *Human Structure*. Cambridge, MA: Harvard University Press.

Churchill SE, Shackelford LL, Georgi JN, Black, MT (2004) Morphological variation and airflow dynamics in the human nose. *Am J Hum Biol* 16, 625–638.

Currey JD (1984) *The Mechanical Adaptations of Bones*. Princeton, NJ: Princeton University Press.

Daegling DJ (1993) Functional morphology of the human chin. *Evol Anthropol* 1, 170–177.

DuBrul EL, Sicher H (1954) *The Adaptive Chin*. Springfield, IL: Charles C. Thomas.

Endo B (1970) Analysis of stresses around the orbit due to masseter and temporalis muscles respectively. *J Anthrop Soc Nippon*, 78, 251–266.

Erikson GE (1963) Brachiation in the new world monkeys and in anthropoid apes. *Symp Zool Soc Lond* 10, 135–164.

Franciscus RG, Long JC (1991) Variation in human nasal height and breadth. *Am J Phys Anthropol* 85, 419–427.

Fleagle JG (2013) *Primate Adaptation and Evolution*, 3rd ed. San Diego, CA: Academic Press.

Fleagle JG, Stern JT, Jr., Jungers WL, Vangor AK, Wells JP (1981) Climbing: a biomechanical link with brachiation and with bipedalism. In *Vertebrate Locomotion* (ed. Day MH), pp. 359–375. New York: Academic Press.

Freedman SJ, Gerstman HL (1972) The role of pinnae in speech intelligibility. *J Commun Disorders* 5, 286–292.

Goodall J van Lawick (1968) The behaviour of free-living chimpanzees in the Gombe Stream Reserve. *Anim Behav Monogr* 1, 161–311.

Goode RL, Friedrichs R, Falk S (1977) Effect on hearing thresholds of surgical modification of the external ear. *Ann Otol Rhinol Laryngol* 86 441–450,

Grand TI (1972) A mechanical interpretation of terminal branch feeding. *J Mammal* 53, 198–201.

Hatt R (1932) The vertebral columns of ricochetal rodents. *Bull Am Mus Nat Hist* 63, 599–738.

Hershkovitz P (1970) The decorative chin. *Bull Field Museum Nat Hist* 41, 6–10.

Hunt KD (1991a) Mechanical implications of chimpanzee positional behavior. *Am J Phys Anthropol* 86, 521–536.

Hunt KD (1991b) Positional behavior in the Hominoidea. *Int J Primatol* 12, 95–118.

Hunt KD (1992a) Positional behavior of *Pan troglodytes* in the Mahale Mountains and Gombe Stream National Parks, Tanzania. *Am J Phys Anthropol* 87, 83–107.

Hunt KD (1992b) Social rank and body weight as determinants of positional behavior in *Pan troglodytes*. *Primates* 33, 347–357.

Hylander WL (1975) Incisor size and diet in anthropoids with special reference to Cercopithecidae. *Science*, 189, 1095–1098.

Hylander WL (1979) The functional significance of primate mandibular form. *J Morph* 160, 223–240.

Hylander WL (1984) Stress and strain in the mandibular symphysis of primates: a test of competing hypotheses. *Am J Phys Anthropol* 64, 1–46.

Hylander WL (1985) Mandibular function and biomechanical stress and scaling. *Am Zool* 25, 315–330.

Hylander WL (2013) Functional links between canine height and jaw gape in catarrhines with special reference to early hominins. *Am J Phys Anthropol* 150, 247–259.

Hylander WL, Johnson KR, Crompton AW (1987) Loading patterns and jaw movements during mastication in *Macaca fascicularis*: a bone strain, electromyographic, and cineradiographic analysis. *Am J Phys Anthropol* 72, 287–314.

Hylander WL, Picq, PG, Johnson, KR (1991) Function of the supraorbital region of primates. *Arch Oral Biol*, 36(4), 273–281.

Kay RF, Hylander WL (1978) The dental structure of mammalian folivores with special reference to Primates and Phalangeroidea (Marsupialia). In *The Biology of Arboreal Folivores* (ed. Montgomery GG), pp. 173–191. Washington, DC: Smithsonian Institute Press.

Kelley J (1997) Paleobiological and phylogenetic significance of life history in Miocene hominoids. In *Function, Phylogeny, and Fossils* (eds. Begun DR, Ward CV, Rose MD) pp. 173–208. Boston, MA: Springer.

Isler K (2005) 3-D kinematics of vertical climbing in hominoids. *Am J Phys Anthropol* 126, 66–81.

Isler K, Thorpe SKS (2003) Gait parameters in vertical climbing of captive, rehabilitant and wild Sumatran orang-utans (*Pongo pygmaeus abelii*). *J Exp Biol* 206, 4081–4096.

Joseph J, Nightingale A (1952) Electromyography of muscles of posture: leg muscles in males. *J Physiology* 117, 484–491.

Jungers WL (1976) Hindlimb and pelvic adaptations to vertical climbing and clinging in *Megaladapis*, a giant subfossil prosimian from Madagascar. *Ybk Phys Anthropol* 20, 508–524.

Jungers WL (1984) Scaling of the hominoid locomotor skeleton with special reference to lesser apes. In *The Lesser Apes: Evolutionary and Behavioral Biology* (ed. Preuschoft H, Chivers D, Brockelman W, Creel N), pp. 146–169. Edinburgh: Edinburgh University Press.

Jurmain R (1989) Trauma, degenerative disease, and other pathologies among the Gombe chimpanzees. *Am J Phys Anthropol* 80, 229–237.

Jurmain R (1997) Skeletal evidence of trauma in African apes, with special reference to the Gombe chimpanzees. *Primates* 38, 1–14.

Keith A (1891) Anatomical notes on Malay apes. *J Straits Br Asiat Soc* 23, 77–94.

Kortlandt A (1968) Handgebrauch bei freilebenden Schimpansen. In *Handgebrauch und Verstandigung bei Affen und Fruhmenschen* (ed. Rensch B), pp. 59–102. Bern: Hans Huber.

Larson SG, Stern JT, Jr (1998) Maintenance of above-branch balance during primate arboreal quadrupedalism: coordinated use of forearm rotators and tail motion. *Am J Phys Anthropol* 129, 71–81.

Lewontin RC (1978) Adaptation. *Sci Am* 239, 212–230.

MacConnaill MA, Basmajian JV (1969) *Muscles and Movements: A Basis for Human Kinesiology*. Baltimore, MD: Williams and Wilkins.

Napier JR (1967) Evolutionary aspects of primate locomotion. *Am J Phys Anthropol* 27, 333–342.

Reynolds VF, Reynolds F (1965) Chimpanzees in the Budongo Forest. In *Primate Behaviour* (ed. De Vore I), pp. 368–424. New York: Holt, Rinehart & Winston.

Riesenfeld A (1969) The adaptive mandible: an experimental study. *Cells Tissues Organs* 72, 246–262.

Ripley S (1967) The leaping of langurs, a problem in the study of locomotor adaptation. *Am J Phys Anthropol* 26, 149–170.

Robinson L (1914) The story of the chin. *Ann Rep Smithsonian Inst*, 1914, 599–610.

Rose M (1975) Functional proportions of primate lumbar vertebral bodies. *J Hum Evol* 4, 21–38.

Russo GA, Shapiro LJ (2011) Morphological correlates of tail length in the catarrhine sacrum. *J Hum Evol* 61, 223–232.

Sarmiento EE (1987) The phylogenetic position of *Oreopithecus* and its significance in the origin of the Hominoidea. *Am Mus Novit* 2881, 1–44.

Sarmiento EE (1989) A mechanical model of ape and human climbing and its bearing on body proportions. *Am J Phys Anthropol* 78, 296.

Schmitt D (1994) Compliant walking in primates. *J Zool Soc Lond* 248, 149–160.

Schultz AH (1930) The skeleton of the trunk and limbs of higher primates. *Hum Biol* 2, 303–438.

Schultz AH (1936) Characters common to higher primates and characters specific for man. *Quant Rev Biol* 11, 259–283, 425–455.

Schultz AH (1961) Vertebral column and thorax. *Primatologia* 4, 1–66.

Sonntag, CF (1923) On the anatomy, physiology and pathology of the chimpanzee. *Proc Zool Soc Lond* 23, 323–429.

Stanford CB, Wrangham RW (1998) *Chimpanzee and Red Colobus: The Ecology of Predator and Prey*. Cambridge, MA: Harvard University Press.

Stern JT Jr. (1971) *Functional Myology of the Hip and Thigh of Cebid Monkeys and Its Implications for the Evolution of Erect Posture*. Basel: Karger.

Stern, JT Jr., Wells JP, Jungers WL, Vangor AK (1980) An electromyographic study of serratus anterior in atelines and *Alouatta*: implications for hominoid evolution. *Am J Phys Anthropol* 52, 323–334.

Susman RL (1979) Comparative and functional morphology of hominoid fingers. *Am J Phys Anthropol* **50**, 215–236.

Takemoto H (2008) Morphological analyses and 3D modeling of the tongue musculature of the chimpanzee (*Pan troglodytes*). *Am J Primatol* **70**, 966–975.

Tuttle RH, Basmajian JV (1974) Electromyography of the forearm musculature in the gorilla and problems related to knuckle-walking. In *Primate Locomotion* (ed. Jenkins FA Jr.), pp. 293–345. New York: Academic Press.

Tuttle RH, Basmajian JV (1977) Electromyography of pongid shoulder muscles and hominoid evolution: I. Retractors of the humerus and rotators of the scapula. *Ybk Phys Anthropol* **20**, 491–497.

Tyson E (1699) *Orang-outang, sive Homo sylvestris: or The Anatomy of a Pygmie Compared with that of a Monkey, an Ape and a Man.* London: Osborne.

Ward CV (1993) Torso morphology and locomotion in *Proconsul nyanzae*. *Am J Phys Anthropol* **92**, 291–328.

Wolf M, Naftali S, Schroter RC, Elad D (2004) Air-conditioning characteristics of the human nose. *J Laryngology & Otology* **118**, 87–92.

Ziegler A (1964) Brachiating adaptations of chimpanzee upper limb musculature. *Am J Phys Anthropol* **22**, 15–32.

10 Up from the Protoape
The Evolution of the Chimpanzee

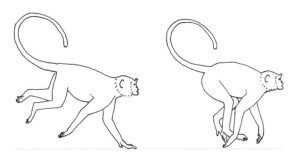

From Kevin D. Hunt, "Why Are There Apes?" *Journal of Anatomy* John Wiley & Sons, © 2016 Anatomical Society.

PART I: PLANET OF THE MONKEYS

Paleontologists face challenges seldom encountered in the scientific world. Their task is to put flesh – and behavior, if possible – on the bones of long-dead animals. Yet rarely do they have more than a handful of fossilized scraps. True, now and then a joyous cry echoes through the halls of academe as a half-skeleton of some important primate species is announced, but such rapturous events are rare. Instead, a typical haul is a chipped-up elbow or an inch or two of jaw with a tooth or two. Nature seems almost to be plotting against us.

Imagine looking at a 1000-piece jigsaw puzzle depicting a sailboat floating on a shimmery blue lake; puffy clouds float in the sky above it and a row of white cottages can be seen in the background, behind them a forested hill. If you had only *20 pieces* of this puzzle, the smallest bit of bad luck might leave you with all sky. Or mostly boat. Or entirely hills and cottages. Even if we were to draw a representative cross-section of puzzle pieces, 20 people might come up with 20 different theories of what the image depicted. This is where we are with primate paleontology.

Despite this impediment, primate paleontologists have made a brave beginning in uncovering the origin of chimpanzees some 25–30 million years ago (Mya). And it is an origin that is almost precisely the opposite of what we expected. We expected the smaller-brained leaf-eating monkeys to appear first; they might have wised up over a few million years and evolved into larger-brained fruit-eating monkeys; then some fruit-eating monkey evolved into an ape. Instead, our tale features tailless but nevertheless monkey-like apes, ape-like monkeys, and dozens of unrelated species evolving specific bits of anatomy in preposterously parallel fashion.

10.1 What Is an Ape?

Old Tom Fuller said it is darkest before the dawn, and the apes heard him; for years it seemed the more we searched for the dawn of the apes the darker it kept getting. As the fossil record filled in, the very concept of "ape" became murkier. What *are* apes? Perhaps we could define them as *any* species that falls on the ape side of the Y after the divergence of apes and monkeys. If we take that view, we face the discouraging fact that none of the early apes look much like living apes.

If we reject the ape-side-of-Y method, we can instead define "first ape" as "the first fossil species with the long arms, long fingers, gripping great toes and human-like shoulders that are so familiar to us among living apes." So familiar, in fact, that it is tempting to think that the ape body plan must have existed far into the evolutionary past as sort of primeval archetype. Not so. Instead, it seems quite likely, if you can believe it, that there never was such a thing. In other words, it is quite possible that the lesser apes (gibbons and siamangs), orangutans, gorillas, and the genus *Pan* (bonobos and chimpanzees) each evolved ape traits separately. If that fails to confound, it even may be that the first species to have ape-like bodies is not an ancestor of living apes at all, but an extinct ape cousin!

Let us return to the "ape-side-of-the-Y" criterion. Using that definition, ancient apes had only one truly ape-like feature, their teeth. These earliest apes had no brow ridges, no long arms, no long fingers, and no ape-like shoulders. With this in mind, let us call these early ape-toothed fossils *dental apes*. Our best guess is that this first ape lived about 25–30 Mya and was a member of or closely related to the family Proconsulidae, or less formally, a proconsulid; we will meet this primate soon. Once upon a time, many paleontologists cavalierly referred to all of the species I am calling "dental apes" – propliopithecoids, pliopithecoids, dendropithecoids, and proconsulids – as "Miocene apes." It took some time for the fossil-studying world to realize that while a primate may have ape-like teeth (that is to say, they were not bilophodont; see Figure 10.1), not all of the dental apes were close relatives to living apes. It might even be that some of these dental apes are in fact monkeys masquerading as apes. Short story: ape teeth ≠ ape.

10.2 The Dominance of Monkeys

You may not have considered how much more common monkeys are compared to apes: Worldwide,

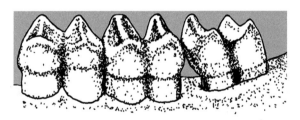

Figure 10.1 Monkey bilophodont cheek teeth (top) versus the low-crowned, rounded teeth of chimpanzees.

extant monkey species outnumber extant apes 136 to 20, a seven-to-one difference (Chapter 4; Appendix 1). If we bump it up a taxonomic notch to the level of genus, the comparison is 21 monkey genera to 5 ape genera. In Africa there are only four species of ape (I am counting lowland and mountain gorillas as separate species) compared to 71 species of monkey, an 18–1 difference. Drilling down to a specific example, the inventory of primates at Campo-Ma'an, a Cameroon forest, is rather typical (Matthews & Matthews, 2002); 11 species of monkeys inhabit the same area as 2 apes (gorillas and chimpanzees).

Monkeys dominate not only in species and genus numbers, but even more so in terms of individuals, both in Africa (Matthews & Matthews, 2002) and worldwide (Kelley, 1994). In Uganda, a survey of eight conservation areas reported monkey densities of 86 per square kilometer, versus $1.4/km^2$ for chimpanzees (Plumptre & Cox, 2006), a 60–1 difference. Surely this world of ours is a Planet of the Monkeys rather than a Planet of the Apes.

Perhaps this monkey dominance strikes you as unremarkable, just as it may seem normal that there are more rats than buffaloes and more buffaloes than blue whales. Perhaps it has always been that way. Quite the contrary. Not only is monkey dominance not preordained, many experts believe it was once exactly the opposite: Monkeys were once a tiny minority of primates, while apes, dental apes at least, were abundant. One expert has estimated that there were *hundreds* of species of apes in the Miocene (Begun, 2003; though others offer a more restrained count). As late as 9 Mya, which is about one-seventh the history of the primates (Figure 10.2), the most definitive assessment found apes were at an 80/20 advantage (Figure 10.2). Alas, the Planet of the Apes was ephemeral; over the next 10 million years monkey numbers exploded. Many of us think we know why: It will be no surprise to you that it concerned food.

10.3 Monkey Foods

In January 1987 I stood on the edge of the famed artificial feeding area at Gombe, a rectangle of closely trimmed foliage situated smack dab in the middle of

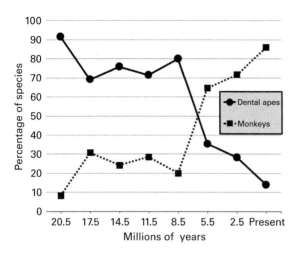

Figure 10.2 The relative abundance of apes and monkeys. Monkeys gradually increased in number for some millions of years until approximately 9 Mya, when there was a dramatic turnaround. One possible explanation is that it was then that monkeys evolved the physiological ability to deal with tannins, terpenes, and alkaloids, an ability apes lack. The rapid expansion of monkey species must have exerted a powerful selective pressure on apes. After Harrison, 2010b, from Hunt, 2016.

the 35 km² Gombe National Park. Good sight lines made it a wonderful place to observe primate behavior, which is why over the years much of the Gombe chimpanzee video footage has been shot there. I was following a chimpanzee who wandered in and lingered there for a frustratingly long time, perhaps considering whether she might luck into a banana or two if she hung around long enough. I had little to do while she dithered, since by convention Gombe researchers only take data outside the artificial feeding area.

My attention was drawn from her by the clamor of a dozen noisy baboons across the way, feeding from a tree just outside the edge of the trimmed area. My locomotion and posture study also encompassed baboons, so I felt no guilt in abandoning my chimpanzee target and wandering over to see what the commotion was all about. They were feeding on a grape-sized yellow fruit, the skins of which littered the ground under the tree, interspersed with a few whole fruits. This particular fruit pops out of its skin much like a grape. I picked up a glistening, deskinned piece from the ground. The fruit flesh was gooey, as if

it were sugary, but the scent that wafted up to my nose was not a fruity bouquet, but the smell of some volatile petroleum compound like turpentine or paint thinner. I tasted it. I thought I detected sugar in the few seconds before my lips and tongue went numb. I experienced a dry (more precisely, astringent) taste that I was quite familiar with: tannins. This fruit was a lesson in plant chemistry. It was also nothing a human would ever in a million years consider eating. When it comes to things you and I – or any chimpanzee – would prefer *not* to be in our fruit, this one was rich. While sugar was almost certainly the nutritional component that drew the baboons to this smallish tree, the fruit also contained two of four of the compounds that monkeys tolerate only grudgingly and apes hardly at all: terpenes (presumably providing the turpentine smell) and tannins (the same compound that makes unripe bananas taste dry). It mostly lacked the two other things primates prefer to avoid, fiber (or cellulose) and alkaloids (poisons such as arsenic and strychnine).

My chimpanzee research subject was ignoring this fruit for a very good reason; as we learned in Chapter 5, tannins and terpenes (and fiber and alkaloids as well) are **antifeedants**. They make the food inedible to apes, or nearly so, even if monkeys are not so picky.

Although I knew much of this at the time, I was still astounded that baboons were eating this noxious thing. I had no idea that I would later spend three years of my life gathering data to show that the key difference between monkeys and apes, and (in my view) the key to the divergence between apes and monkeys around 25 Mya, was the edibility of fruits like this.

10.4 The Primate Diet Project

That three years of work was a two-year postdoc and a further research stint as part of Richard Wrangham's newly constituted Primate Diet Project. Wrangham and his team were investigating a hypothesis he had developed in the 1970s but not yet tested: that antifeedants were critical monkey–ape diet differences. My role was to supervise teams of local assistants observing five species of monkey as well as the chimpanzees. We also monitored fruiting trees to see how much fruit they produced and when it ripened. The experience profoundly affected my understanding of primate feeding behavior. As a supervisor, I walked from team to team, making sure the staff were out doing their jobs. I would pause, asking what the monkeys were eating that day and whether they had seen chimpanzees; if the monkeys were doing something particularly interesting I stayed a while. Then I would move on to the next team and the next species. The experience gave me a good feel for how different primates moved in the trees, their feeding style, their interactions as they fed, and how long they spent in each tree. Some days I stayed with a particular team all day, helping to gather data. These data beautifully supported Wrangham's hypothesis that it was the inability of apes to tolerate noxious compounds that was the most important factor in the evolution of the African apes (Wrangham & Waterman, 1983; Wrangham et al., 1998).

10.5 The Superiority of Monkeys

While they may have given little thought to it before, primate paleontologists were forced to confront the cause of the turnover in ape/monkey numbers when Peter Andrews published a famous paper in 1981. From a contemporary perspective, he inappropriately lumped all the dental apes with proconsulids, somewhat deceptively referring to all of them as "apes." As we learned earlier, most primate paleontologists now believe this pooling is wrong. But whether it was right or not, he was asking an extremely important question: Why have monkeys become so dominant? Working along similar lines, Alis Temerin and John Cant (1983) stepped in to fold Wrangham's contention that it is dependence on ripe fruits that distinguishes apes from monkeys (Wrangham, 1979, 1980) into Andrews' fossil observations. Thus, Temerin and Cant identified the ecological driver that spurred the 25 million-year-long turnover in "ape" and monkey numbers: It was because apes were constrained to subsist on high-quality food items that are rare, dispersed in space,

and patchy in distribution (Wrangham, 1979). Given that the most ape-like feature of those early Miocene "apes" was their teeth, Temerin and Cant's dietary observations still resound.

In the 1970s Suzanne Ripley (1970, 1979) posited that ape under-branch, suspensory locomotion was a specialization for feeding among small, compliant branches in tree edges, reasoning that Old World monkey-style above-branch walking and leaping might be adapted to moving high in trees where the canopy would be discontinuous. Later research confirmed the terminal branch speculation; chimpanzees spend more time in the outer edges of trees among the tiny terminal branches than baboons, even though baboons are lighter (Hunt, 1992a). She was wrong about monkeys feeding higher; on average, apes feed higher (Cant, 1992; Houle et al., 2010), presumably because there are more and better fruits there (Houle et al., 2014). I think an advantage of larger ape body size is that it allows them to exclude the smaller monkeys from preferred sites; in fact, chimpanzees sometimes go out of their way to chase baboons and other monkeys not just from a particular feeding site, but from the entire tree.

Apes not only feed in peripheral, terminal branches; their suspensory adaptations allow them to move between tree canopies among tiny tree-edge branches using a locomotor mode called "transferring," a slow brachiation (Hunt, 1992a, 2016; Houle et al., 2014). Monkeys often cross gaps both within and between trees by leaping (Rose, 1973; Cannon & Leighton, 1994; Chatani, 2003); 34 percent of gap crossings (Cannon & Leighton, 1994) are by leaping. They resort to an awkward quadrupedal "bridging" at other times.

10.6 "Ape" Morphology Has Evolved Repeatedly as a Solution to Feeding Challenges

In 2016 I published a review article that updated the monkey–ape contrast discussed by Wrangham, Ripley, Andrews, and Temerin and Cant (Hunt, 2016). Many chimpanzee traits, I argued, evolved as work-arounds to make up for their inability to consume antifeedants; monkeys took the alternative route, evolving physiology to cope with poisons. In other words, competition with monkeys was the selective pressure that led to the evolution of the chimpanzee body plan.

It seems that when primates compete with one another over millions of years, evolution seems to just keep coming up with something like a chimpanzee. It happened among the early apes such as *Oreopithecus* and *Hispanopithecus*, but also among South American monkeys. South American monkeys split from the ape/Old World monkey branch 40 Mya, when primates were small-brained, short-legged squirrel-like animals. Yet South America produced spider monkeys (see Chapter 4), a chimpanzee-like New World primate that has mobile shoulders, long, curved fingers, a large brain, and a slow life history (Harvey & Clutton-Brock, 1985). They share with great apes 34 distinctive anatomical features, each of which must have evolved independently (Larson, 1998). They also possess a social system shockingly similar to that of chimpanzees, featuring male territoriality, male bonds, a fission–fusion grouping pattern, and border patrolling (Aureli et al., 2006). In Asia, orangutans appear to have had rather monkey-like bodies until true monkeys – macaques – appeared; today their bodies look like those of chimpanzees.

Nature seems to have a soft spot in its evolutionary heart for a long-armed, mobile-shouldered, armhanging primate.

10.7 Refining Concepts of Ape Locomotor Adaptation

Returning to Ripley's comparison of ape and Old World monkey feeding adaptations (Ripley, 1970; Grand, 1972; Rose, 1973; Temerin & Cant, 1983), she theorized that the ape under-branch locomotor adaptation was not simply *useful* among small, flexible supports, but was *superior* to the above-branch monkey adaptation – apes could take shorter, quicker routes between the best feeding sites among peripheral branches, whereas monkey incompetence in the small-branch milieu would force them to return to the core of the tree on a main bough, even when moving only a few meters to an adjacent branch. In other words, apes moved near the surface of an imaginary sphere that made up the outermost edges of the tree canopy, easily negotiating twigs by brachiating, while monkeys inefficiently zig-zagged back and forth between the tree bole and the periphery.

Superior ape competence among small branches is now well accepted (Hunt, 1992a; Cannon & Leighton, 1994), but the idea that apes move more efficiently or more quickly at the tree-edge than monkeys has fallen into disfavor. The shorter-route benefit, if there is one, is surely offset by the greater energetic cost of suspensory locomotion compared to walking (Parsons & Taylor, 1977). This inefficiency is expressed behaviorally by the fact that while chimpanzees are comfortable among tree-edge twigs, they avoid them when they can (Chapter 8; Hunt, 1992b); they enter these precarious areas not because they prefer them, but because they must – this is where their food is. While apes are more capable among small branches, they have but a slight advantage, since the smaller body size of monkeys allows them creep out into many small-branch feeding sites. Their leaping ability gives them a means of moving between feeding sites on adjacent boughs without zig-zagging back to the trunk.

10.8 Apes as Superior Beings (Not!)

Greater ape intelligence once led many to assume, perhaps subconsciously, that ape-ness was a glorious consummation of evolution's wisdom. The fine-grinding mills of natural selection had evolved an energetically efficient and intellectually superior being from a lowly monkey-like first draft. Instead, the insights of Andrews, Wrangham, and Temerin and Cant say that while apes may have superior intellects, they are inferior at the evolution game. They teeter now on the edge of the ultimate short end of the evolution stick – extinction – while monkeys thrive. I see ape locomotor adaptations and large bodies as a sort of work-around, compensating for the inability to tolerate antifeedants (Hunt, 2016). It is the monkeys that have evolved the superior adaptation: robust digestive physiology, rapid propagation, and a resilience in the face of food fluctuations, born of their ability to eat a wider variety of food types, all of which led to more species and greater population densities.

But precisely why do apes have larger bodies?

10.9 The Advantage of Being a Heavyweight

I have argued (Hunt, 2016) that apes evolved large body size to bully monkeys away from the best feeding sites. But larger body mass confers two further advantages. Larger animals have lower pound-for-pound dietary needs (Kleiber, 1932; Sailer et al., 1983), helping apes to weather food shortages. Second, without the robust digestive system of monkeys, apes are more dependent on coevolved bacteria to digest fiber, and larger animals have slower gut passage rates, giving gut microflora more time to process plant structural materials (Bell, 1971; Geist, 1974; ; Clutton-Brock & Harvey, 1977a, 1977b; Jarman, 1974; Gaulin & Konner, 1977; Gaulin, 1979; Wheatley, 1982; Sailer et al., 1983). This is a bug, or at least a work-around for a bug, not a feature; chimpanzees rely on their microbiome because it is the only way they can get any food value at all out of fiber. Ape suspensory adaptations are in turn an accommodation to high body weight. Monkeys, unable to feed in trees when apes occupy them, instead evolved the ability to get to the fruit earlier via antifeedant tolerance, thus nullifying ape superiority in one-on-one contests.

10.10 Ape Molar Morphology is Primitive

The transition was not so fast that we cannot see it. A group of paleontologists led by Nancy Stevens believes it has found the earliest monkey, a jaw fragment with a single molar, the 25 million-year-old *Nsungwepithecus* from Tanzania. It has five cusps arranged in a slightly skewed but still ape-like Y-5 pattern (Stevens et al., 2013), rather than the tightly interlocking bilophodont modern arrangement (Figure 10.2). A later, slightly better-preserved monkey is the 19 million-year-old *Prohylobates* (Benefit, 1999; Benefit & McCrossin, 2015); it bears in its name a relic of our early misapprehension of primate evolution, a time when monkeys were viewed as primitive, with apes evolving from them: *Prohylobates* means "gibbon ancestor" – *Prohylobates* is a monkey ancestor masquerading as a gibbon progenitor. In other words, while dozens of ape-toothed primates frolicked in the early 30-million-year-old forests, monkey-toothed primates had not yet evolved. While they may not be proud of it, monkeys descend from dental apes. Monkeys must have changed their diet to one where bilophodont molars were advantageous, while the unchanged ape teeth suggest their diet changed little.

The remarkable monkey bilophodont teeth work like an industrial plastic shredder, slicing up unripe fruit, cellulose-rich mature leaves, and other foliage into a finely comminuted digestible slurry that resembles green sand (Hladik, 1978). Chewing relatively indigestible plant parts until they are reduced to tiny bits increases their surface area and breaks open some individual cells, increasing digestibility and allowing monkeys to exploit lower-quality food items. Accompanying this new dental shape was a shift in monkey physiology to allow the tolerance of antifeedants; surely this transformation has left some trace in the monkey genome, but this work is yet to be completed – though some have tried (Nemeth, 2007). All this came slowly, though; not only at 25 Mya, but for another 10 million years monkeys were gradually abandoning a ripe-fruit diet.

A 15 Mya nearly complete skull, dentition, and postcrania of the fossil *Victoriapithecus* is gloriously helpful. Even at this late date its molar morphology (Figure 10.3) most resembles species with diets dominated by fruits, estimated as 79 percent of their diet (Benefit, 1999), a proportion of fruit dramatically higher than that seen not only in living Old World monkeys but even living apes (Wrangham et al., 1998). Benefit calls Victoriapithecus a *superfrugivore*. Not just a frugivore, she and her colleagues interpret *Victoriapithecus* molar morphology as adapted for a diet of *ripe fruit*. Put another way, "The dietary category of ancestral catarrhines ["dental apes"] and

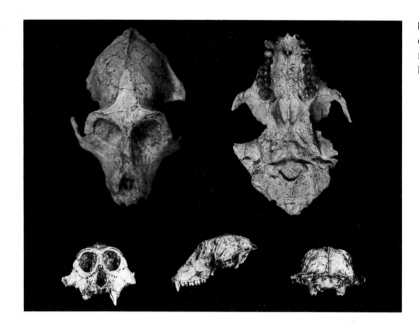

Figure 10.3 *Victoriapithecus*, the earliest monkey skull. The first, best monkey skull. Image courtesy of Brenda Benefit.

earliest hominoids is widely interpreted as soft-fruit frugivory" (Andrews & Martin, 1991: 206).

10.11 The Monkey Body is Primitive

Even more ripe fruit than apes? Would that mean an ape-like body, too? Not on your life. *Victoriapithecus* (Figure 10.3) has not only a beautifully preserved skull but substantial postcrania (body skeleton elements) as well, and its body is quite monkey-like (Benefit, 1999; Figure 10.4). Whereas living apes have adaptations for joint mobility that allow them to assume the contortionist-like postures necessary to feed among small branches in tree-edge microhabitats, monkeys are quite different. *Victoriapithecus* had a narrow thorax, as evidenced by its backward-facing humeral head, a bend in the humerus typical of monkeys, and a tail. In combination with other early fossils we know early monkeys had long backs and legs and arms of almost equal length – again, monkey-like. *Victoriapithecus* fingers are monkey-like in length and the great toe is less abducted than that of apes; that is, the space between the big and lateral toes is smaller (Harrison, 1989; Benefit, 1999). Ischial morphology indicates the presence of sitting pads, or ischial callosities, something much more common among contemporary monkeys than apes, an adaptation to extended periods of sitting during feeding and even while sleeping. Muscles that are active during running and that propel leaping have long lever arms (Figure 10.4; for you experts: high humeral tubercles, a long olecranon process, and a posteriorly reflected medial epicondyle; extended greater trochanters at the hip and a long ischium gave the hamstrings a longer moment arm). Brains are modern-monkey-like in shape, but the size matches that of the decidedly un-brainy lemur (Gonzales et al., 2015).

In short, the earliest monkeys had a monkey body, ape teeth, and a lemur brain.

PART II: ANCIENT APES

10.12 The First Apes

Genetics hint that apes and Old World monkeys (cercopithecoids) split roughly 28 Mya, but the fossil record from the Oligocene epoch is sparse. If the

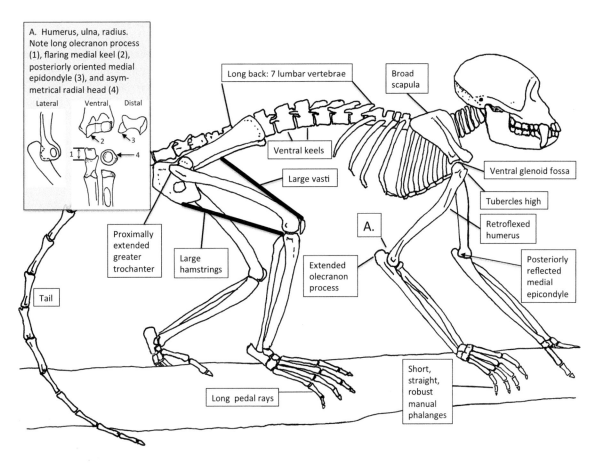

Figure 10.4 Monkey specializations. From Hunt, 2016, with permission of John Wiley & Sons.

genetics is right, Nancy Stevens' 25 million-year-old fragmentary monkey fossil *Nsungwepithecus gunnelli* (Stevens et al., 2013) was just after this split. While the site of Rukwa, where *Nsungwepithecus* was found, was stingy about providing us with body parts, it was generous in species diversity. Another fossil from the same site, *Rukwapithecus fleaglei*, may well be the earliest known member of the ape lineage. *Rukwapithecus* is only a fragment of a mandible with a premolar and two molars, but the tooth cusps, despite the early date, more closely resemble later apes than other contemporaneous fossils; even though it is only a jaw, body weight could be estimated from it: 12 kg (26 lb), over twice the 5 kg weight of the contemporaneous *Nsungwepithecus*. While their placement in any lineage is tentative, if these two fossils do represent members of the ape and monkey lineages, ape-like body masses emerged almost the moment apes evolved.

Our first solid evidence of the *bodies* of the early apes comes from the early Miocene (about 22 Mya) proconsulids (Walker & Pickford, 1983; Walker & Teaford, 1989; Harrison, 2010a), a taxon recently divided into two genera, *Proconsul* and *Ekembo* (McNulty et al., 2015). Proconsul is found earlier, from about 23 Mya (Meswa Bridge; Harrison & Andrews, 2009) to 20 Mya; Ekembo follows at 19 to 17 Mya. Experts disagree on how many species are represented among the proconsulids, but for our purposes we can parse them into four, a larger and smaller pair early and another pair later: (1) *Proconsul africanus*, weighing 10–12 kg (about 25 lb) and (2) *Proconsul major*, weighing nearly 90 kg (almost 200 lb); the later pair are (3) *Ekembo* (previously *Proconsul*) *heseloni*, similar in size to *P. africanus* and probably its descendent, though dentally (McNulty et al., 2015) and postcranially a little more ape-like, and finally (4) *Ekembo nyanzae*, much larger at 30–40 kg (65 lb and up).

While one or another of these East African proconsulids has been long considered the earliest well-preserved ape ancestor, many paleontologists question that assignment. In fact, of all the fossils available as candidates for "first ape," the fossil species that paleontologists see as most likely to be the earliest ancestor of living apes, is … well … none of them. For instance, in a recent review of fossil primates (Figure 10.5), a figure shows that as we go back in time the known ancestors of the great apes (thick lines) trickle out at about 12 Mya. We not only have nothing (thin lines) suggesting the origin of the apes, there is nothing from the origin of the ape + monkey lineage (the Catarrhini). Proconsulids, on the other hand, went extinct – note the timing – at the exact point that the first fossils in the great ape lineage appear (the thick lines).

This situation strikes me as quite similar to the state of affairs in human paleontology once upon a time. Before about 1960, diagrams tracing human ancestry back to the split with apes featured a human lineage entirely devoid of fossils. All the specimens we now put in the human lineage, Javan *Homo erectus*, Peking man, Neanderthals, australopiths, were hanging out on dead-end twigs – none was thought to be a human ancestor. Where were the human fossils? We gradually realized the fossils we had relegated to side branches were ancestors. Part of the problem was that paleontologists wanted *fossil* humans to look just like *living* humans;

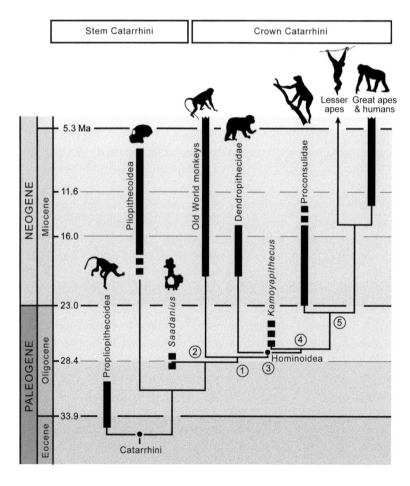

Figure 10.5 A recent phylogeny of fossil and living monkeys and apes. Note that the fossil record of the apes from 23 to 12 Mya is blank – the thin line – while the Proconsulidae were an evolutionary dead-end. For want of a better model for the morphology of the first ape, proconsulids will have to do. From Zalmout et al., 2010, with permission. (A black and white version of this figure will appear in some formats. For the color version, please refer to the plate section.)

my mentor wondered whether these deniers were Platonic essentialists, unreasonably expecting to see the first ειδος of humanity, a perspective he characterized as antievolutionary (Brace, 1964). I worry that primate fossils experts are doing the same thing, expecting early apes to look like living apes.

While I consider proconsulids our best prospect for an early Miocene ancestor, even if that happens to be wrong and some yet-undiscovered early Miocene species proves to be the true ancestor of the living apes, the proconsulids are so close to the ape–monkey split that their anatomy will closely resemble the ghost fossil (Harrison, 2010a). Recall that the earliest fragment of ape we have, Nancy Stevens' *Rukwapithecus*, precedes *Proconsul* by only two million years.

10.13 Proconsulids Resembled Monkeys More than Apes

The perception of *Proconsul*'s ape-ness was influenced by the fact that, as so often happens, teeth and fragments were discovered first (Hopwood, 1933); it was sold as an ancestor of the chimpanzee – it was named after a popular bicycle-riding zoo chimpanzee, Consul – with little dissent ("Pro-Consul," then, would be "Consul's ancestor"). In 1948 the famous Mary Leakey skull made its spectacular debut. Maybe it was a bit less ape-like than expected, its lack of a brow ridge perhaps the most discordant note, and it had a bit of a snout, too – but it had a definitely ape-ish vibe. If you squinted a bit you could see it as an ape (Figure 10.6). In 1951, bits and pieces of the postcrania began to trickle in (Walker & Teaford, 1989) and the head-scratching began. These parts were not really ape-like at all. By the 1980s, Mike Rose made no bones about the fact that proconsulids not only differed from apes, they were patently unlike *any* living primate (Rose, 1983). Rose wrote in 1993 that while proconsulids "share some morphological features with some living primates, there are no close overall similarities between them and particular living taxa … [making them] rather puzzling chimeras" (Rose, 1993: 252). His analyses placed them closest to a monkey, but unexpectedly not an African monkey; they resembled the New World howler, genus *Alouatta* (Rose, 1983,

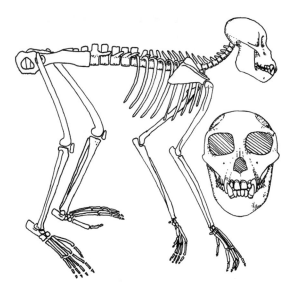

Figure 10.6 *Proconsul africanus*. An ape-like head and teeth are paired with a monkey-like body, with a narrow thorax, arms and legs of equal length, a monkey-like rear-facing humeral head, short fingers, monkey-like vertebrae, and a broad, monkey-shaped scapula, but the presence of a more ape-like divergent great toe. After Walker and Teaford, 1989, with permission.

1993, 1996) – with the important distinction that proconsulids lacked tails, while howlers have a powerful one. Proconsulids resembled capuchin monkeys next most closely, also a New World monkey. This is puzzling. The common ancestor of the New World monkeys and catarrhines (Old World monkeys plus apes) is nothing like the proconsulids, and instead is possibly the quite primitive *Aegyptopithecus*. Our best understanding of this unexpected resemblance is that proconsulids and primitive monkeys were less specialized than living catarrhines. Two paths diverged in the African woods. Old World monkeys evolved specializations for more refined walking, running, and leaping, while apes evolved away from their generalized ancestor to become better armhangers and vertical climbers.

10.14 Monkey-Like Body Proportions

Two things about arm and leg length are important, length compared to body weight and interlimb proportions. Monkeys are long-limbed compared to, say, squirrels (Figure 10.6) and proconsulids are the same – not squirrel-like but monkey-like (Rose, 1993). Proconsulid interlimb proportions are also monkey-like, with arm length 96.4 percent of leg length (Rose, 1993). Fingers are again monkey-like both in length and form; they are straight, rather than curved as they are in apes (Walker & Pickford, 1983; Rose, 1993). Fragmentary ribs and a rather complete pelvis give us a sense that the thorax was narrow – monkey-like (Figure 10.7; Rose, 1993). As is the case with New World monkeys, the proconsulid pelvis lacks the flared ischium (lower pelvis) that anchor sitting pads in monkeys, often interpreted as a sleep-sitting adaptation (Washburn, 1957).

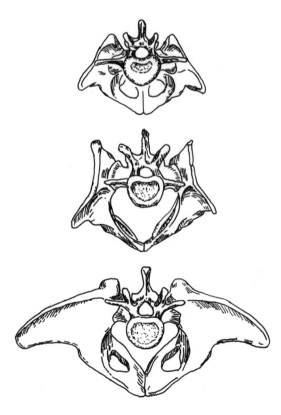

Figure 10.7 The pelvis of *Ekembo/Proconsul* (middle) is narrow and monkey-like, resembling the baboon (top) rather than the chimpanzee (bottom). After Ward, 1993; used with permission.

10.15 Monkey-Like Humerus and Scapula

The humeral head (shoulder joint) faces backwards, as it does in animals with narrow, monkey-like thoraxes, confirming the rib data. The upper humerus is "retroflexed," bent in side view (Figure 10.6; Rose, 1993). In contrast to these monkey-like features, when viewed from the front the articular surfaces of the elbow are more deeply notched than in monkeys, a necessary stabilizing reinforcement to withstand the powerful but sometimes erratic forces that act on the elbow to maintain balance among small, flexible supports. Imagine holding your body weight stable while walking on all four limbs on a rope-bridge; considerable power must be exerted to counter the tendency of the rope to sway out from under your hands. The elbow is also under great strain during vertical climbing, which requires lifting the body with a pull-up like action.

Monkeys have less need for a stabilized elbow because they move on larger, more stable supports; they vertical climb less.

The elbow joint also has a somewhat posteriorly oriented medial epicondyle (Rose, 1993), a monkey-like feature that increases the moment arm (lever arm) of muscles that powerfully extend the elbow during running or leaping (Hunt, 2016). Proconsulids have a long (i.e., proximally extended) ulnar olecranon process (Figure 10.4, label A) that increases the lever arm for the triceps but would have prevented full extension of the elbow, necessary for armhanging (Rose, 1993). There is an extensive articulation between the ulna and the wrist bones, suggesting monkey-like stability rather than ape-like mobility.

The proconsulid scapula resembles that of colobine monkeys or howler monkeys (Rose, 1993; MacLatchy & Bossert, 1996). Remember those colobines.

At the other end of the body, the presence of the vertical climbing adaptation – a gripping big toe in proconsulids – tells us that hand-over-hand vertical climbing (versus the leaping-like climbing of monkeys) was more common than it is among living monkeys.

10.16 Long, Muscular Back and Powerful Legs

The lumbar spine had six (or possibly seven) vertebrae, as do monkeys. Vertebral bodies were tall, narrow, and wedge-shaped (Figure 10.8) – that is, monkey-like.

When vertebrae are viewed from the top, the positioning of the **transverse processes**, the spikes protruding from vertebrae (Figure 10.9), are positioned forward on the vertebral body to accommodate large back muscles, as in monkeys. This is important because when we find anteriorly placed transverse processes among living primates, the back is long; among living monkeys these features are always accompanied by a narrow thorax, a narrow pelvis, medium leg length, and powerful leg muscles.

A monkey leaping bout begins with spine, hip, knee, and often ankle strongly flexed, after which each is violently extended to accelerate into take-off. In a comparison of two leaf monkeys, one of which leapt more than the other (Stern, 1971; Fleagle, 1976, 1977), leapers had longer hindlimbs, longer lumbar vertebral segments, powerful erector spinae muscles, and long feet. Both a longer back and longer hindlimb segments increase the time over which force can be applied to the take-off support, thus increasing take-off velocity, which in turn increases the length of the leap (Fleagle, 1976, 1977). *Proconsul/Ekembo* also has a long ischium that increases the lever arm of the hamstrings, a monkey-like running/walking/leaping adaptation (Rose, 1993).

10.17 Proconsulids were Tailless

Ekembo heseloni, and presumably other proconsulids, was tailless, as are living apes (Ward et al., 1991). The lack of the tail might seem a minor issue, but that loss holds critical information about early ape adaptations. Ancient apes were arboreal primates, so the loss of this very valuable locomotor organ is unexpected. As we saw in Chapter 9, tails are important for balance – but they become less effective as body weight increases. The heaviest Old World monkeys have reduced or absent tails (Fleagle, 2013). Miocene apes are big. Even the smallest proconsulid species outweighs a typical monkey. *Ekembo heseloni* females weighed 9.8 kg (22 lb), more than the *tailless* Celebes macaque, *Macaca nigra* (female body mass

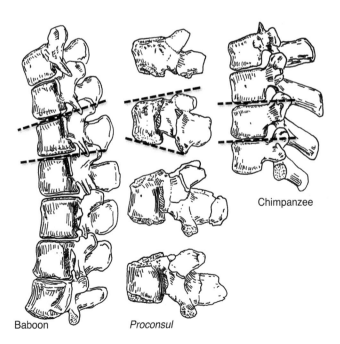

Figure 10.8 Proconsulid and monkey (baboon in this case) vertebrae are wedge-shaped (dotted lines), an adaptation to the extreme flexion of the back in preparation for leaps. There are six or seven lumbar vertebrae, compared to four for living chimpanzees. Redrawn and modified from Ward et al., 1993.

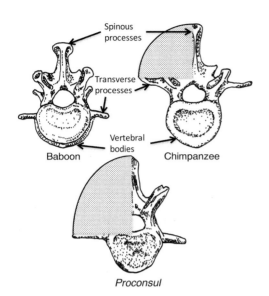

Figure 10.9 Top view of lumbar vertebrae of a baboon, chimpanzee, and *Proconsul*. Compared to chimpanzees, the transverse processes of proconsulids have their base at a more forward or anterior position, onto the vertebral body. This placement leaves a larger compartment (hatched area) for the back extensor muscles, allowing for a more powerful extension of the back during leaping. Redrawn and modified from Ward et al., 1993.

5.5 kg or 12 lb [Fleagle, 2013]). *Ekembo* is double the size of contemporary monkeys and equal in weight to female savanna baboons (*Papio* spp.: 12 kg or ~25 lb; Table 9.2), which have small, almost vestigial tails. Female body masses for other proconsulid species are even larger, estimated as ranging from 30.0 to 63.4 kg (66–140 lb).

While large bodies may explain tailless-ness, their manner of locomotion may also figure in. Their joint flexibility (see Section 10.18; Rose, 1993), gripping thumbs and toes (Kelley, 1997), and semicircular canal morphology (Ryan et al., 2012) all suggest a slow, cautious locomotion that did not require the help of a tail.

10.18 Long Femoral Necks and Globular Femoral Heads Allow Hip Flexibility

While the back suggests strong leaping, the femur and hip joint are more ape-like, allowing flexibility. The femur of leapers has a proximally extended greater trochanter that increases the lever arm of the upper-leg-extending gluteal muscles. The acetabulum, the concave part of the hip joint, is deep and bucket-shaped to accommodate a cylindrical femoral head. Proconsulids have a larger, more globular femoral head than monkeys, and it fits into a shallow, bowl-shaped (i.e., ape-like) hip joint, giving them greater hip mobility (MacLatchy & Bossert, 1996). Their femoral neck is long and tilted up to move the femoral head away from the greater trochanter. This is the morphology that provides the flexibility required to negotiate unstable, erratically placed branches in the outer edges of trees, and may make up for the lack of a balance-enhancing tail (Buck et al., 1925; Kelley, 1997). Establishing contact with as many supports as possible, regardless of their random placement, becomes more important as body size increases.

The lack of a tail, however, cannot fully explain joint shape because proconsulids strongly resemble howler monkeys, which have among the largest tails in relation to body weight in the animal world. Hip joint flexibility is likely also an adaptation to foot-hanging in the terminal branches. Howler monkeys (Figure 10.10) often assume a hindlimb suspensory posture.

The proconsulid knee (the distal femoral articular surface) is broader than that of monkeys (Rose, 1993), though not fully ape-like, and the shape of the tibial surface of the ankle joint is more ape-like than monkey-like. Knees may bear greater torques because apes vertical climb using alternate feet for propulsion, while monkeys push off with both feet.

10.19 Ape-Like Foot and Ankle

Proconsulid lateral toes are monkey-like, but there is a long, robust great toe indicating a more ape-like than monkey-like gripping ability, suggesting a compromise sort of hindlimb use, with less refined adaptations to running, walking, and leaping than monkeys, but effective adaptations to vertical climbing – though less highly evolved than seen among apes.

Proconsulids have a slightly snoutier, more primitive face, with a bit of a protruding nose (Figure 10.6).

10.21 Teeth

While proconsulids had howler monkey-like bodies (albeit without tails), their low, blunt molar cusps (Figure 10.2) suggest a "superfrugivore" diet (Kay, 1977), an even greater dependence on ripe fruit than is found among living apes. This is no surprise since monkeys at the time were the same; everyone was going for ripe fruit. Lack of shearing crests suggests that proconsulids were less often forced to fall back on piths, leaves, unripe fruits, and herbaceous items, as chimpanzees must. Chimpanzees and gorillas have thin enamel, often argued to be an adaptation to chewing piths and leaves: the enamel wears through to the dentin, leaving sharp edges at the enamel–dentine border that finely shred fibrous foods like leaves, herbs, and piths. Herbs and piths are eaten principally on the ground; perhaps these mostly terrestrial food items became part of the chimpanzee diet when apes were forced into terrestrial travel as ripe fruit became increasingly dispersed.

Proconsulid incisors are large (Andrews & Martin, 1991), but not as large as those of chimpanzees. We know that primate incisors are large among species that must break into large fruits, but piths and barks select for large incisors, too. Apes may have added these resources as they became more terrestrial.

Figure 10.10 Proconsulids differ from all living primates, but in statistical analyses they often fall close to the New World howler monkeys. Like howlers, proconsulids had arms slightly shorter than legs, adaptations to leaping such as a long back, a narrow body plan, including a narrow ribcage and pelvis, short fingers and toes, and other adaptations for walking and leaping, but few suspensory adaptations. Photo: Paulo B. Chaves.

We will put all this anatomical information into a behavioral reconstruction shortly.

10.20 The Cranium

The proconsulid brain is slightly larger than that of a monkey, 167 cm^3, but not ape-sized. Proconsulids lack a brow ridge and sulcus, giving them a skull that, while presenting an overall impression of ape-ness, is in many ways more monkey-like. The ape face is prognathic, but it is only the jaw – the teeth and supporting maxilla – that protrudes, not the nose.

10.22 A Mosaic Adaptation

Proconsulids possessed less highly refined versions of important living Old World monkey adaptations, with ape-like compromises here and there. Put another way, proconsulids have a mostly monkey-like anatomy, but where their anatomy diverges from living Old World monkeys it varies in the direction of chimpanzees or other apes. These ape-like features provide hip flexibility, elbow stability, and a gripping great toe, part of an adaptation for hand-over-hand vertical climbing, as opposed to the two-footed, pulse-like, quasi-leaping method of ascension that monkeys use (Hunt, 1992a).

Taken together, these characters suggest a primate that is better adapted than monkeys for foraging among the erratically placed, small-diameter, compliant branches in tree edges. Perhaps hip mobility means they engaged in more hindlimb suspension than is seen among extant apes. Movement among small branches requires powerful stabilization of the limbs to cope with the wobbly twigs there; this instability would require relatively slow, careful movement compared to monkey-like walking and running. Their long backs suggest they often engaged in leaping, perhaps to cross gaps within and between tree crowns, though their ape-like hips suggest they could never push off as powerfully, as do Old World monkeys.

Dental evidence of a ripe-fruit diet in concert with their monkey-like bodies suggests that proconsulids collected much of their fruits as monkeys do, in the central core of the tree canopy, but more than living OWM, they must have spent time among thin branches in the tree periphery. Their blunt-cusp tooth morphology suggests they did not supplement their ripe-fruit diet to any great extent with low-calorie terrestrial herbaceous plants or with leaves.

Spun out into a complete daily routine, proconsulids probably had an active, feed-as-you-go, above-branch walking, leap-between-trees-and-bridge-when-you-must foraging regime, much like that of monkeys, finding food in every few trees they visited and rarely climbing down to the ground to travel. Their gripping feet suggest that they engaged in hand-over-hand vertical climbing more than do monkeys – monkeys would leap up, rather than climb up. Their mobile hips would help them to forage in the tree periphery and to bridge between tree crowns, reaching out with the arms to grasp a hold in the new tree and transferring weight carefully. By analogy with howlers and in accord with their hip mobility, they probably spent more time hanging by their hindlimbs than do living apes. Table 10.1 summarizes proconsulid functional morphology.

10.23 The Long and Winding Road to Ape-dom

10.23.1 Trends within the Proconsulids

Over the next few million years most of the Miocene apes evolved still larger body masses and features that are often attributed to suspensory behavior (ovoid glenoid fossae, longer arms, more mobile shoulders, longer fingers, flexor sheath ridges). However, they *lacked* bowing in the ulna and radius (armhanging features), and they retained leaping features such as tall vertebrae and developed (assuming the proconsulid condition is primitive) tall greater trochanters. Their faces are very prognathic and large compared to the braincase, probably merely a consequence of larger body size. They lack a brow ridge and sulcus, so that their eye orbits are sloped backwards. *Morotopithecus* exhibits much of this cranial morphology. There is only a snout, really, but it is astonishingly similar to the cranium of the later proconsulid-like *Afropithecus* (Gebo et al., 1997; Nakatsukasa & Kunimatsu, 2009). It has an ovoid glenoid fossa, suggesting some suspensory behavior (MacLatchy et al., 2000; MacLatchy & Bossert, 1996); *Proconsul major* also bears this feature (Nakatsukasa & Kunimatsu, 2009). While the lumbar vertebrae are tall, the transverse processes are ape-like, attaching to the back of the vertebral body (Figure 10.9; Walker & Rose, 1968). This feature is not a definitive one, because I have seen transverse processes from different chimpanzees originating from very different spots. The femur is more monkey-like, with a tall greater trochanter, suggesting leaping. There is no evidence for thorax breadth, but later hominoids, 14–15 Mya *Kenyapithecus* and *Nacholapithecus*, have backward-facing humeral heads, suggesting a narrow thorax.

My interpretation of this morphology is that *Morotopithecus* and later apes were leapers, but so heavy that the suspensory final phase of the leaps required quasi-armhanging; if this seems obscure, it will be explained in detail below. Surely, though, *Morotopithecus* was also capable of incipient suspensory behavior. Its 20.6 Mya date (Gebo et al., 1997) is surprising, making it contemporaneous with

Table 10.1 **Proconsulid functional morphology**

Feature	Function
Short, straight fingers	Monkey-like quadrupedalism; little suspension
No flexor sheath ridges	Monkey-like quadrupedalism; little suspension
IMI 95	Monkey-like quadrupedalism, leaping
Humerus with torsion	Narrow torso: walking, running, leaping adaptation
Retroflexed humerus	Monkey-like walking
Incipient spool shaped trochlea	Robust elbow: cautious climbing, vertical climbing
Long olecranon process	Monkey-like quadrupedalism, leaping
Straight radius, ulna	Monkey-like quadrupedalism, leaping; no armhanging
Large ulnar styloid	No rotatory wrist; quadrupedalism
Long lumbar region	Leaping
Tall lumbar vertebrae	Leaping
Wedge-shaped lumbar vertebrae	Leaping
Transverse processes ventral	Leaping
Tailess-ness	Slow, cautious movement
Semicircular canal shape	Slow, cautious movement
Monkey-like scapula	Quadrupedalism
Monkey-like glenoid	Quadrupedalism
Gripping great toe	Vertical climbing
Somewhat ape-like ankle	Vertical and small support climbing; hindlimb suspension?
Narrow pelvis	Leaping, walking, running
Angled femoral neck	Some hip flexibility: slow small-branch locomotion
Shallow hip joint, globular fem. head	Some hip flexibility: slow small-branch locomotion
Ape-like ischial moment arms	Vertical climbing; hindlimb suspension?
Medium body size (~22 kg)	Cautious locomotion
Small brain size	Monkey-like cognition
Ape-like (non-bilophodont) molars	Ripe-fruit food processing
Thicker enamel than apes	Less herbaceous and leaf-based diet than living apes

other, more primitive proconsulids, but other dates have been offered from 12–14 Mya to 15–17 Mya to 17.5 Ma (Young & MacLatchy, 2004). My friend Laura MacLatchy is very confident of her 20.6 Mya date, but ironclad dates have been wrong before. *Morotopithecus* would be perfectly at home among later large-bodied proconsulids.

10.23.2 Mobile Shoulders, Long-Armed Leapers

What can we make of an ovoid glenoid, long arms, and long fingers, but no bowing among forearm bones? What use are long arms and shoulder mobility other than for suspension? The best interpretation I can offer is a retention of leaping adaptations as body size increased, with long arms increasing reach and more mobile shoulders evolving as a consequence of the landing phase of leaping.

Colobus monkeys are leapers *par excellence*. Their spectacular gap-crossings bounds can be up to 10 times their body length, and the distance covered in *descent* can be greater than the horizontal distance covered. Colobus monkeys land using their powerful fingers to grasp a horizontal branch with both hands, though the legs sometimes aid as well. This means that as the leaper lands the forelimbs must bear much of the deceleration effort, working to prevent the body's downward inertia. Imagine how difficult it would be to keep your elbows bent and your arms from extending all the way above your head if you leapt onto a horizontal bar from above it. You could not. Apes have large bodies – remember that muscle power cannot keep up with body size, so resisting elbow extension at the end of leaps is more difficult – their elbows must extend more. The same thing happens at the shoulder; at its extreme, the arms can be forced into a two-handed armhanging posture. This arm-extension argument has a long history. The anatomy associated with colobus-like arm extension during landing led to colobines being described as "semibrachiators," not because they were thought to brachiate, but because their anatomy seemed like that of a partly committed brachiator (Ashton & Oxnard, 1963, 1964a, 1964b; Napier, 1963; Oxnard, 1963, 1967).

After landing in what becomes a two-handed, semi-armhanging posture, a "pull-up" or hoist is required to pull the body up into a sitting or standing position. If the landing substrate is compliant, which is common, the support deforms to near-vertical under the weight of the leaper and instead of a pull-up, the landing is followed immediately by either angled-ascent walking, an ascending leap, or hand-over-hand vertical climbing. Whichever it is, the leaper often has to climb up a vertical stem.

Colobus monkeys have long fingers, a hook-like hand, and *no thumbs* – I find it intriguing that chimpanzees have small thumbs (Figure 10.11). Leaping, then, is a pre-adaptation to both armhanging (shoulder mobility, elbow extension, long fingers) and vertical climbing (pull-up action, requiring strong elbow flexors and a large latissimus

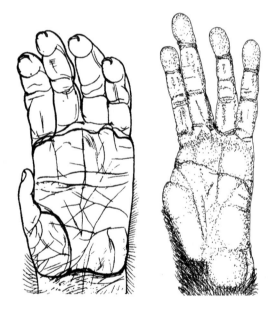

Figure 10.11 Chimpanzee hands (left) are reminiscent of those of colobus monkeys in that their thumb is quite reduced. Colobus monkeys lack a thumb altogether. Thumb reduction may be a leaping adaptation. The bulbous pad at the base of the thumb, the thenar eminence, and the thumb itself may interfere with grasping branches at high velocity. The thumb may bounce the landing support away from the hand. The tiny thumb of the chimpanzee suggests that they are more adapted to leaping than behavioral studies (Hunt, 1992a) suggest. Left image after Napier, 1960, with permission of John Wiley & Sons; right image by the author.

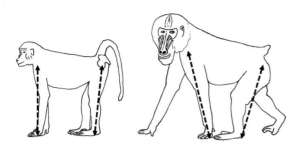

Figure 10.12 While much of baboon and mandrill morphology and locomotor behavior (to the extent we know it) is similar, the larger mandrill has disproportionately long forelimbs. It may be that long forelimbs extend reach to help compensate for the loss of arboreal competence that comes with greater body mass. Greater reach could help with feeding, leaping, and transferring. Image by the author.

dorsi). I have suggested a corollary to this long-distance leaping adaptation (Hunt, 2016). Compared to other cercopithecoids, colobines have wide interorbital distances, which may improve depth perception, an advantage in judging landing spots at the end of leaps, when velocities are high. Chimpanzees also have wide interorbital distances, perhaps a retention from a leaping ancestry.

But what about the long arms? It turns out there is a tendency among primates for larger species to have disproportionately longer arms (Jungers, 1979). Perhaps this is an adaptation that neutralizes the burden that greater body weight imposes on vertical climbing and movement among small, flexible supports; species with greater body mass but also greater reach could still harvest foods from tiny supports. Greater reach might also be advantageous during gap-crossing leaps or tree-transfer, where larger body size limits agility. Mandrills, for instance, have longer arms in relation to their legs than savanna baboons, though the rest of their anatomy is quite similar (Figure 10.12).

In my colobus- and mandrill-inspired evolutionary scenario, mobile shoulders and full elbow extension that originated as part of a leaping adaptation and long arms that resulted from the larger body size eased the transition to armhanging for food gathering as competition from monkeys pushed ancestral apes into ever-smaller, ever more peripheral feeding sites.

10.23.3 Robust Faces and Still Smaller Hind- versus Forelimbs

Over the next few million years ape species showed a tendency toward more robust faces (more like australopiths than living apes) while retaining bodies vaguely like those of a long-armed proconsulid (Kunimatsu et al., 2007) but with body weights of up to 40 kg (90 lb). Among the better-known are *Kenyapithecus* and *Nacholapithecus*. *Kenyapithecus* has a posteriorly facing humeral head (= narrow thorax), muscle attachments that intrude on the humeral head (= little shoulder flexibility), and humeral retroflexion, like monkeys. Bizarrely, the hindlimbs seem to be much smaller in relation to the forelimbs than even in chimpanzees (Kunimatsu et al., 2007); the mandrill model is the only explanation I have for this peculiarity.

10.24 Ape Evolution 10-12 Mya

This takes us to a little after 15 Mya, a time of few ape fossils in Africa. Paris, Vienna, Barcelona, and Berlin, on the other hand, were ape-country 12 million years ago. Possible ancestors for which we have more than fragments are *Pierolapithecus*, *Hispanopithcus*, *Rudapithecus*, and the venerable genus *Dryopithecus*, named in 1856; these four are similar enough that David Begun (2010) lumps them together in the single genus *Dryopithecus*. The oldest fossil, 12.5–13 Mya *Pierolapithecus* from Spain, has phalanges similar to but shorter than chimpanzees – that is, curved and bearing flexor sheath ridges (*Dryopithecus* fingers are longer, chimpanzee-sized; Moyà-Solà et al., 2004). A long clavicle and ribs curved similarly to chimpanzees suggest a broad thorax (Moyà-Solà et al., 2004), and chimpanzee-like carpal bones suggest rotatable wrists. Vertebrae have transverse processes nearly as displaced from the body as chimpanzees. Unlike chimpanzees, though, the joints where the fingers join the palm are shaped in such a way as to suggest *Pierolapithecus* walked on its palms, hyperextending the fingers, as do monkeys. *Pierolapithecus* lacks knucklewalking features, then. Furthermore, the pelvis is primitive – that is, proconsulid-like (Hammond et al., 2013). The lack of terrestrial knucklewalking and vertical climbing (tall pelvis) features in the presence of armhanging upper-body anatomy suggest a way of life much like that of chimpanzees – when in the trees – but the chimpanzee foraging pattern of a long feeding bout followed by terrestrial travel followed by ascension and another feeding bout was absent. I conclude that ape feeding sites were not dispersed enough to require long terrestrial jaunts, making late Miocene apes principally arborealists. *Dryopithecus* and *Hispanopithecus* are a little later in time (10–12 Mya) and had longer fingers, suggesting that armhanging was evolving in this period.

The face of these dryopiths in profile shows some slight divergence from the eye-orbits-tilted-back look of *Afropithecus/Morotopithecus*; there is the slightest beginning of a brow ridge. A relatively complete *Rudapithecus* skull exhibits an expanded brain size and vertical orbits.

10.25 The Last Common Ancestor of Chimpanzees and Humans

We now approach the time period of the last common ancestor (LCA) of humans and chimpanzees. Gorillas split from the human–chimpanzee line around 10 Mya and humans and chimpanzees parted ways about 6 Mya. Was the LCA australopith-like, perhaps even bipedal? Some would say so (Crompton, 2016). If so, a robust-faced fossil is appealing as an LCA. Others would say the common ancestor is quite similar to a chimpanzee (Wrangham & Pilbeam, 2002), which would leave hominins to do all the evolving. Unfortunately, not only can we not be confident in one or the other of these choices, diverse experts offer dozens more possibilities. Given this lack of consensus, I resolved to take a democratic approach. I asked 30 or so prominent paleontologists – those who had written the most highly cited scholarly works on the issue – to identify their best LCA candidate. Not quite two-thirds responded. If I parsed those 20 answers as finely as possible, I would have 20 different responses.

10.26 The Argument for a Chimpanzee-Like LCA

The most popular expert opinion, a perspective most closely associated with David Pilbeam, has it that the LCA is a yet-undiscovered fossil that strongly resembled a chimpanzee. Let us call this the **chimpanzee-like LCA hypothesis.** Pilbeam cautioned against lapsing into pedanticism. Perhaps you can hear the faint chatter of voices off-stage nit-picking the meaning of "chimpanzee-like" to death. This is not helpful. Pilbeam says he means "chimpanzee-like in the sense that baboons are macaque-like." Baboons look much like but not exactly like overgrown macaques. For someone unfamiliar with primates, we might say "in the sense that a wolf is German shepherd-like." A number of experts largely agreed with this chimpanzee-like LCA hypothesis, though nearly all viewed the ancestor as differing from chimpanzees in specific ways.

One expert suggested the LCA had "a greater amount of arboreal bipedalism than seen in living African apes" (perhaps along the lines of gibbons, which would be something like the fossils *Hispanopithecus* and *Dryopithecus*). Another expert suggested the LCA varied in the direction of bonobos (longer legs and a less robust face than chimpanzees). *Pierolapithecus* was named by one commentator as a chimpanzee-like LCA candidate but this stretches the concept of "chimpanzee-like" a bit since *Pierolapithecus* lacks knucklewalking dorsal ridges and the pelvis is monkey-like. I view this as meaning *Pierolapithecus* was not a knucklewalker or vertical climber, and therefore engaged in little terrestrial behavior, which seems not very chimpanzee-like to me. Others suggested the LCA was chimpanzee-like "but with shorter fingers." I have a hard time envisioning a short-fingered chimpanzee, since a short-fingered species suggests either suspension from small branches or no suspensory adaptations, in which case it would probably lack other chimpanzees armhanging traits such as a broad thorax. Figure 10.13 envisions an *Afropithecus/ Morotopithecus/Pierolapithecus* evolutionary sequence. Many experts dismiss all the species named

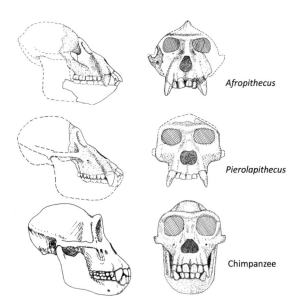

Figure 10.13 This evolutionary scenario envisions the early proconsulid *Afropithecus* as the earliest ape. *Afropitehcus* shares much in common with the later, more derived (Young & MacLatchy, 2004) Miocene ape, *Morotopithecus* – so much so that some see them as synonymous (Patel & Grossman, 2006). *Pierolapithecus* approaches the time range (12.5–13 Mya) of the gorilla–chimpanzee–human split (9 Mya), though it is not gorilla-like. The gorilla-like *Ouranopithecus* is dated to 9.6 Mya; in my phylogeny it would be interposed between *Pierolapithecus* and chimpanzees.

in this paragraph – European apes – as an LCA because they see them as on the orangutan branch.

10.27 A Robust-Faced Australopith-Like Ancestor

A close second to the chimpanzee-like LCA hypothesis imagines the LCA as possessing thick-enameled molars, a heavy mandible, and tall, heavy zygomatics – and perhaps a more bipedal locomotor adaptation than living apes. Some listed these features with no mention of australopiths, but others literally described the LCA as "australopith-like in these characteristics." In one case, *Australopithecus* itself was named as the LCA. I consider australopiths to be partly adapted to suspensory behavior (Chapters 28 and 30), but one correspondent explicitly stated "not suspensory."

Others tweaked "australopith-like" in various ways. One statement was, paraphrasing a little bit, "australopith-like face and quite human-like – or more precisely, African *Homo erectus*-like – in postcrania, though without some of the more highly refined adaptations for running and walking efficiency."

10.28 The Robust-Faced "Proconsulid" Evolutionary Path

Other experts suggested fossils with robust faces, as above, but bodies vaguely like those of a long-armed proconsulid (Kunimatsu et al., 2007), resembling or closely related to *Kenyapithecus* and *Nacholapithecus*; while these fossils are a little too early to be the LCA, we can apply Pilbeam's phraseology: "the LCA was *Kenyapithecus*-like in the sense baboons are macaque-like." These fossils share (where the right bones are present) large, thickly enameled premolars and molars, tall zygomatics, and thick mandibles, and include *Kenyapithecus* (and close relatives; Schwartz & Conroy, 1996; McCrossin & Benefit, 1997; Ward et al., 1999), *Nacholapithecus* (Ishida et al., 2004), and two similar to the European fossil *Ouranopithecus*, discussed below (*Nakalipithecus* – Kunimatsu et al., 2007; *Samburupithecus* – Ishida & Pickford, 1997), and *Ouranopithecus* itself (de Bonis & Koufos, 1994; Gülec et al., 2007). Measuring canine size is a little difficult, but in most cases these same fossils had smallish canines. To my mind, the consistently robust faces of post-10 Mya fossils increases the odds that this condition was primitive for the great ape–human clade; remember that this makes them australopith-like (Andrews & Martin, 1991; Alba et al., 2010).

Peter Andrews was rather specific, and gave me permission to quote him:

Large upper central incisors, thick molar enamel (but molar size unknown), parallel tooth rows, moderate midfacial prognathism, low alveolar prognathism; zygomatic shifted forward, low facial buttressing; robust jaws, inferior transverse torus; gripping foot; hand not elongated, short, curved fingers, long thumbs; broad pelvis but ischium long, long back with 5–6 lumbar vertebrae; not suspensory, semi-terrestrial palmigrade quadrupedal locomotion, seasonal woodland environment.

Of the fossils we have in hand, to me this falls closest to *Ouranopithecus*.

As you know, there are no apes in Europe now. By 5.5 Mya the European climate had shifted, skewing cooler, drier, and more seasonal. Grasslands expanded (Barry et al., 2002; Badgley et al., 2008). As the winter waxed cooler, ripe fruits were no longer available year-round. All these things stifle apes, and competition with monkeys possibly did as well. Perhaps as early as 11 Mya – but more confidently by 8.5 Mya – monkeys appeared in Europe (Delson, 1994; Harrison, 2005). The broader, more eclectic diet of monkeys (Merceron et al., 2009) may have allowed them to weather the climatic changes seizing Europe. Monkeys survived shortages by eating antifeedant-rich leaves; apes died. The European age of apes had passed.

10.29 *Ouranopithecus* as LCA

Just a little later (9.6 Mya) than the *Dryopithecus* group above, an intriguing robust-faced, late Miocene ape emerged, *Ouranopithecus macedoniensis*. It was nominated as LCA by its describers (de Bonis & Koufos, 1994, 2014; Koufos & de Bonis, 2005) and others (Cameron, 1997) as falling somewhere in or near the human lineage. A Greek first hominin! In its favor as LCA, two African fossils are so similar, *Nakalipithecus* (Kunimatsu et al., 2007) and *Samburupithecus*, that some experts consider all three to be geographical variants of the same species, a critical point since early hominins are found in Africa. *Ouranopithecus* (Figure 10.14) resembles gorillas in that they both share primitive characters with late Miocene apes: large nasal apertures, large molars, robust and tall zygomatics, smaller incisors than chimpanzees, square eye orbits, and a low zygomatic origin. This heavy-faced adaptation is accompanied, not unexpectedly, by immense premolars and molars, even larger than those of gorillas and australopiths (Kunimatsu et al., 2007; Güleç et al., 2007). This resemblance is interesting because there has been a recent florescence of gorilla-like-LCA advocates. Among the traits some have suggested as shared between gorillas and the LCA are life history (Duda & Zrzavy, 2013), thumb morphology (Almecija et al., 2015), hand and forelimb proportions (Schultz, 1930), foot morphology (Schultz, 1930, 1950; Crompton,

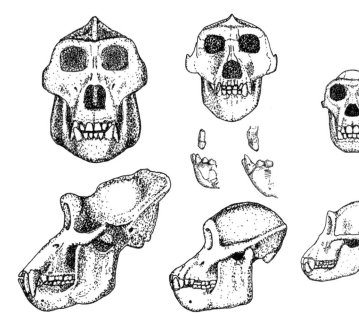

Figure 10.14 *Ouranopithecus* (middle) compared to the gorilla (left) and chimpanzee (right). *Ouranopithecus* has a robust face and cheek teeth the size of those of gorillas. Incisors, canines, and brow ridges are smaller than those of chimpanzees and gorillas. Not only are male canines small (inset, middle), female canines are diamond shaped, smaller than those of any living ape. Images by the author.

2016), foot mechanics (Wang & Crompton, 2004), ecology (Crompton, 2016), climbing behavior (Crompton, 2016), scapular morphology (Alemeseged et al., 2006), adaptation to bipedalism (Crompton, 2016), and kinematics of load carrying (Watson et al., 2009).

It has a body mass close to that of chimpanzees but a face more like that of a gorilla (Figure 10.14). In fact, it has been called a "mini-gorilla," though how "mini" is unclear; it may weigh as little as 48 kg (105 lb) for males (Kelley, 2001), with females perhaps half that, or as much as 72 kg for males (160 lb; de Bonis & Koufos, 1994). Molars larger than those of gorillas argue for the greater body mass (de Bonis & Koufos, 1994).

Ouranopithecus postcrania are poorly known, but two well-preserved phalanges are shorter than would be expected for a chimpanzee (given body size differences) and lack the robust flexor sheath ridges characteristic of chimpanzees (de Bonis & Koufos, 2014), resembling instead the fingers of terrestrial quadrupeds (de Bonis & Koufos, 2014). They are curved, however, approaching the curvature seen in chimpanzees and suggesting gripping habits. If this species is indeed the common ancestor of *Pan*, *Gorilla*, and hominins, knucklewalking *and* fully evolved armhanging must have evolved after the divergence.

I have a slight preference for a robust-faced LCA, in part because most ape fossils from this time are robust-faced, as are australopiths.

10.30 The Ancient African Ape Social System

Ouranopithecus males are much larger than females (Figure 10.14) and (where it is possible to say) the same holds for other apes of the time, a sex difference that may have exceeded (Scott et al., 2009) even gorilla body size dimorphism (note that the latter article discussed teeth only; but see Figure 10.14).

Among living primates, this size dimorphism is seen only when males engage in violent, one-on-one physical battles, which often result in gaping wounds, broken bones, bitten-off body parts, and occasional death. Chimpanzees, in contrast, have a much lower level of body size dimorphism; this is because males fight as a group – two 50 kg chimpanzees appear to be a superior fighting force to one 100 kg chimpanzee (I provide more information on this phenomenon in Chapters 24, 25, and 26).

Great body size dimorphism suggests that the *Ouranopithecus* social system included this one-on-one aggression, and this in turn suggests their society was one in which several females were bonded to a breeding male. This is a gorilla-like social system. Many female-bonded species such as savanna baboons also have great body weight sexual dimorphism, so it is fair to ask, why not a savanna baboon type of social system? Or why not a gelada-like system, in which females are bonded to a male, as in gorillas, but the females are also both related to and bonded to one another? A baboon-like – that is, female-bonded – system is unlikely because no living ape shares with baboons a female-kin-based social system. I suggested recently (Hunt, 2016) that while it is likely that proconsulids were female-bonded, over time, as monkeys proliferated, consuming an ever-greater proportion of the fruit crop, ape foods would have become more dispersed, and dispersed foods result in a loss of female bonding.

One hypothesis that seems sensible to me suggests that single-male units gathered in larger groups like hamadryas baboons (Swedell & Plummer, 2012). A hamadryas-like bonding among males with a chimpanzee-like fission–fusion social system conveniently shares many features with either chimpanzees, humans, or both.

Australopiths retained a gorilla- or *Ouranopithecus*-like dimorphism, as did many late Miocene apes (Kelley, 1995a, 1995b; Scott et al., 2009), suggesting a hamadryas/gorilla type of society might have been common among Miocene apes.

10.31 The Evolution of Monkey Antifeedant Tolerance

The short story for ape and monkey evolution is this: Monkeys changed their teeth and kept their bodies, whereas apes changed their bodies and kept their teeth.

Monkeys gather food and move through the forest in the same way the ape/monkey common ancestor did, even though they have a different diet – monkeys eat unripe fruit rather than ripe fruit (Hunt, 2016). While some fruits, such as the one I saw baboons eating at Gombe, are inedible to apes, we learned in Chapter 6 that fruits chimpanzees eat *start* their existence as tough, non-nutritious packets of fiber, loaded with one or more of toxins, tannins, terpenes, and acids, but as they mature these chemicals degrade or are neutralized and the fruit becomes ape food. Monkeys have evolved to harvest fruits early in the fruiting cycle, before larger, dominant apes can displace them. While monkey adaptations originated to help them compete for fruit, tolerance of antifeedants preadapted monkeys to include leaves, blossoms, and other foods that are packed with antifeedants. Colobines – the leaf monkeys – took the unripe-fruit adaptation one step further and evolved to take advantage of these foods.

As monkeys evolved to eat ever less-ripe fruit and as they refined their unripe fruit adaptation, their species numbers and population density increased. Apes found that by the time fruit was edible to them, as many as 10 species of monkeys had already depleted the fruit crop in the tree core, utilizing their ancient sit-to-feed adaptation. What remained was a

Figure 10.15 Hypothetical change in ripe fruit resources available to apes from the early Miocene (A–D) to the present (I–L). In the early Miocene, monkeys and apes alike were ripe-fruit specialists and were unable to harvest fruit until late in the fruiting cycle, both appearing at trees at time C or D. At an intermediate time, monkeys harvested fruits earlier and earlier in the fruiting cycle, and the remaining fruits formed more of a hollow sphere, forcing apes into the compliant terminal branches (H). Currently as many as 10 species of monkey harvest unripe fruit and partly ripe fruit (J) still earlier in the fruiting cycle, separating their harvesting time further from apes and leaving not only a hollow sphere (K, L). At some tipping point ripe fruit became so sparse (L) that a threshold was reached where traveling arboreally was inefficient because it required traversing numerous tree crowns, at which point descending, traveling terrestrially, and re-ascending evolved. Image from Hunt, 2016, with permission of John Wiley & Sons.

hollow sphere (Figure 10.15) with the ripe fruit their ancestors could eat while sitting in the core of the tree available only among small-diameter peripheral branches, where non-stereotyped postures and armhanging were required to feed (Hunt, 2016).

Inevitably, the large number of monkeys depleting the fruit crop leaves the ape food supply even sparser and more dispersed, in turn making walking on the ground, as opposed traveling arboreally, a better way to reach food sources. But that requires vertical climbing to re-ascend when they reach the new feeding tree.

Monkeys have two advantages, then. They can arrive at fruiting trees earlier and they can harvest a higher proportion of the fruit crop, since they can eat both ripe and unripe fruit (Wrangham et al., 1998). Looking back into the Miocene epoch 25 Mya and more, monkeys must have begun with only a slight advantage initially – perhaps they could feed on fruit just a day or two earlier than apes. By 8–9 Mya, however, the advantage must have grown to many days (Hunt, 2016). In other words, the monkey ability to tolerate antifeedants is responsible for ape and monkey differences (Wrangham, 1979, 1980; Wrangham & Waterman, 1983). The research that will show exactly how big the arrival-time difference is between coexisting monkeys and apes has not yet been done. The future holds much excitement.

10.32 Parallelism: Nature Keeps Making Suspensory Primates

Before wrapping up chimpanzee evolution, perhaps it is worth commenting on how dramatically our view of ape evolution has changed in the last 50 years. Observing the long arms, short legs, and suspensory behavior of apes, we once thought that the common ancestor of all living apes must have possessed these features. It seems more likely now that ape-like anatomy, once so closely associated with brachiation, evolved independently in many lineages, perhaps almost all of them.

Dainton and Macho (1999), Inouye and Shea (2004), and Kivell and colleagues (2009) argue that quadrupedal knucklewalking evolved independently in gorillas and chimpanzees. These authors show that most features hypothesized to be adaptations to knucklewalking (beaked scaphoid, dorsal concavity of the scaphoid, waisting of the capitate, distal concavities on the hamate and capitate, dorsal ridges on the hamate and capitate) are much more variable among the African apes than previously appreciated, and those that are consistent appear at different developmental stages (Kivell et al., 2009), suggesting independent origin. What about human ancestors? Corruccini and McHenry (2001; *contra* Richmond & Strait, 2000) maintain that australopith hand morphology shows hominins did not have a knucklewalking ancestry, which in turn suggests that knucklewalking evolved in chimpanzees after the ape–human split. Knucklewalking evolved when a long-fingered ape needed to move efficiently on the ground. If it evolved independently in gorillas and chimpanzees, the level of monkey dominance that could leave chimpanzee foods so dispersed they were forced to come to the ground to travel must have occurred after the 8.1 Mya chimpanzee-gorilla split (Raaum et al., 2005).

Orangutans are the most superbly arboreally adapted great ape – clumsy on the ground but contortionists in the trees – but their Miocene ancestor, *Sivapithecus* (8–13 Mya), had monkey-like muscle attachments on the humerus (see Figure 10.4), proximal humeral retroflexion, and a narrow pelvis – and therefore probably a narrow, monkey-like thorax (Morgan et al., 2015). Although *Sivapithecus* had a somewhat ape-like elbow (Rose, 1989; Pilbeam et al., 1990), hand bones (carpals) and a fragmentary radius are largely *Proconsul* or monkey-like, suggesting little wrist flexibility (Rose, 1993). Femoral fragments, a wrist bone (navicular), and several phalanges have been interpreted as consistent with vertical climbing but not suspensory behavior (Madar et al., 2002). As with *Proconsul*, there is some indication of hip mobility (Rose, 1993). These details suggest a semi-terrestrial positional repertoire with vertical climbing and (given hip mobility) low levels of suspensory terminal-branch food harvesting, on the order of that reconstructed for *Proconsul*. In other words, as late as 8 Mya, long after the gorilla-chimpanzee-human clade split off from the ancestors of orangutans,

orangutans remained rather monkey-like, or more precisely, *Proconsul*-like. If each of the great apes were not suspensory at their origin, hylobatids as well must have evolved suspensory behavior on their own.

Nor is this the end of the list of potentially independently evolved suspensory species. Rose (1993, 1996) argued that suspensory behavior emerged separately, not just in the apes, but in as many as a half-dozen further lineages unrelated to extant apes, including the gibbon-sized ~11 Mya Miocene *Pliopithecus*, several other small Miocene apes, *Dryopithecus*, and *Oreopithecus* (7–9 Mya).

If this is not enough to convince you, remember Susan Larson's documentation of the 34 suspensory characters shared by spider monkeys and apes (Larson, 1998). Yet another lineage, the odd-nosed monkeys, has also evolved some suspensory features independently (Covert et al., 2006; Su & Jablonski, 2009), which brings us to an argument often evoked to refute my contention that competition between monkeys and apes resulted in chimpanzee food-gathering adaptations. Some lineages evolve suspensory behavior without monkey competitors. What draws these species out into the terminal branches? Leaves. There is much yet to be studied about suspensory leaf-eaters, but it may be that new leaves are more common in the terminal branches; or it could be that some monkeys gather leaves sitting, forcing other monkeys into the terminal branches.

10.33 A Tentative Evolutionary Path

It may be disappointing that there is no consensus concerning chimpanzee origins. Have faith; a plausible evolutionary scenario can be drawn from the evidence we have at hand. Here is a five-step scenario that is as likely as any.

1. **Proconsulid phase (25 Mya).** The earliest ancestor on the ape arm of the ape-monkey Y was a rather slow-moving generalized quadruped whose habits included walking, leaping, sitting, and a little bit of climbing; it engaged in more non-stereotyped, acrobatic positioning among tiny branches than living Old World monkeys. It may have weighed around 30 kg (65 lb), similar to *Ekembo nyanzae*. We can visualize it as a tailless howler monkey; leave out the howling as well. The cranium might have been more like the prognathic *Afropithecus* than the smaller, more famous Mary Leakey *Proconsul* face (Figure 10.6). Compared to later apes it had only moderate facial robusticity and more tilted-back eye orbits. It had a monkey-sized brain. My best guess is that it was female-bonded.

2. **Forelimb dominance, robust-faced phase (20 to 15 Mya).** A second phase might be exemplified by *Kenyapithecus* and *Nacholapithecus*. With body weights nudging up toward 45 kg (100 lb), the trend for larger primates to have disproportionately long forelimbs pushes the IMI into ape-like ranges. Longer arms compared to legs and a taller greater trochanter suggests less hip mobility and a dependence on long arms to reach out into the tree edge to collect fruits, rather than placing the whole body out there. Walking and leaping are still part of the repertoire, but greater body weight demands shoulder mobility during the landing phase of leaping. Long olecranon processes and straight forearm bones, however, preclude ape-like suspensory behavior. Faces were more robust and molars larger than among proconsulids. The social system is still female-bonded.

3. **Nascent suspensory phase (15 to 10 Mya).** Curved fingers are approaching or even equaling chimpanzee lengths, with flexor sheath ridges. Combined with rotatable wrists and a broad thorax, there is little doubt they engaged in suspensory behavior. They still had, however, a proconsulid-like pelvis, suggesting they still engaged in leaping for crossing gaps. They engaged in less vertical climbing than chimpanzees. Lack of knucklewalking and vertical climbing adaptations suggest a largely or even completely arboreal lifeway; if so, the social system still may have been female-bonded.

4. **Gorilla-like phase (<10 Mya).** At some point food resources became so dispersed the all-arboreal lifeway of ape ancestors, where there was food in every few trees they passed through, came to an end, heralding the semi-terrestrial adaptation we see in living chimpanzees. Perhaps the *Nakalipithecus/Samburupithecus/Ouranopithecus* robust-faced, gorilla-like fossils represent this turning point. As the fruit supply declines, selection for female-bonded

fruit tree defense dwindles and an ape-like social system would have emerged, perhaps a single-male, multi-female system like that of gorillas, possibly with also hamadryas-baboon-like male clan alliances.

  5. **Proto- and true chimpanzees (6 Mya to present).** Dispersed food encouraged females to settle into stationary core areas, a distribution that encouraged males to defend a territory. Evidence of a gorilla-like human–chimpanzee LCA (both neontological and paleontological) and genetics suggest that demonic violence (Wrangham & Peterson, 1997) among males evolved *after* the human–chimpanzee split, and suggest that female sedentism and consequent increased bonding among males, with its increased terrestrial travel and territorialism, evolved in this fifth phase. As male bonds solidified, body size sexual dimorphism decreased. Hamadryas baboons are at this intermediate stage; they have no female bonds and medium-strength male bonds. Refinement of vertical climbing and suspensory adaptations, knucklewalking, and patrolling may have evolved as a package.

10.34 Planet of the Apes Monkeys

The overlap in chimpanzee and Old World monkey food item lists, the ability of monkeys to tolerate high levels of cellulose, tannins, terpenes, and alkaloids, the reversal in the abundance of dental apes versus monkeys, the monkey-like bodies of proconsulids, the gradual emergence of monkey bilophodonty, and the multiple origins of ape suspensory anatomy together suggest to many of us there was a 25-million-year history of competition and coevolution among frugivorous Old World monkeys and apes. Without this competition, evidence suggests, monkeys would not have their distinctive bilophodonty and digestive physiology, and apes would not have their large body size, advanced cognition, and anatomical specializations.

Evolution never sleeps. Evidence from genetics, paleontology, and primate ecomorphology suggests that the "chimpanzees" of 6 Mya were not the chimpanzees of today, but rather a proto-chimpanzee that continued to evolve rapidly after the human–chimpanzee split – and are evolving still.

Personally, I mourn the passing of the great age of apes. Of course, some of my colleagues would argue that I am overdoing it in labeling this section "Planet of the Monkeys," perhaps seeing this world as a Planet of the Naked Apes. Humans are apes, after all. But had an alien race with sporting blood visited the planet a million years ago, they probably would have bet on monkeys eliminating the apes. We are one of those stunning, bracket-busting, last-second turnarounds, coming out of nowhere to dominate the planet. Myself, I am hoping for a resurgence of the hairy apes, including our sister species, the chimpanzee.

References

Alba DM, Fortuny J, Moya-Sola S (2010) Enamel thickness in the Middle Miocene great apes *Anoiapithecus*, *Pierolapithecus* and *Dryopithecus*. *Proc Roy Soc B* **277**, 2237–2245.

Alemseged Z, Spoor F, Kimbel WH, et al. (2006) A juvenile early hominin skeleton from Dikika, Ethiopia. *Nature* **443**, 296–301.

Almecija S, Smaers JB, Jungers WL (2015) The evolution of human and ape hand proportions. *Nat Comm* **6**, 7717.

Andrews PJ (1981) Species diversity and diet in monkeys and apes during the Miocene. In *Aspects of Human Evolution* (ed. Stringer CB), pp. 25–41. London: Taylor and Francis.

Andrews PJ, Martin L (1991) Hominoid dietary evolution. *Phil Trans Roy Soc B* **334**, 199–209.

Ashton EH, Oxnard CE (1963) The musculature of the primate shoulder. *Trans Zool Soc Lond* **29**, 553–650.

Ashton EH, Oxnard CE (1964a) Locomotor pattern in primates. *Proc Zool Soc Lond* **142**, 1–28.

Ashton EH, Oxnard CE (1964b) Functional adaptations in the primate shoulder girdle. *Proc Zool Soc Lond* **142**, 49–66.

Aureli F, Schaffner CM, Verpooten J, et al. (2006) Raiding parties of male spider monkeys: insights into human warfare? *Am J Phys Anthropol* **131**, 486–497.

Badgley C, Barry JC, Morgan ME, et al. (2008) Ecological changes in Miocene mammalian record show impact of prolonged climatic forcing. *PNAS* **105**, 12145–12149.

Barry JC, Morgan MLE, Flynn LJ, et al. (2002) Faunal and environmental change in the Late Miocene Siwaliks of Northern Pakistan. *Paleobiology* **28**, 1–71.

Begun DR (2003) Planet of the apes. *Sci Am* **289**, 74–83.

Begun DR (2010) Miocene hominids and the origins of the African apes and humans. *Ann Rev Anthropol* **39**, 67–84.

Bell RH (1971) A grazing ecosystem in the Serengeti. *Sci Am* **225**, 86–93.

Benefit BR (1999) *Victoriapithecus*: the key to Old World monkey and catarrhine origins. *Evol Anthropol* **8**, 155–174.

Benefit BR, McCrossin ML (2015) A window into ape evolution. *Science* **350**, 515–516.

Brace CL (1964) The fate of the "Classic" neanderthals: a consideration of hominid catastrophism. *Curr Anthropol* **5**, 3–43.

Buck C, Tolman N, Tolman W (1925) The tail as a balancing organ in mice. *J Mammal* **6**, 267–271.

Cameron DW (1997) A revised systematic scheme for the Eurasian Miocene fossil Hominidae. *J Hum Evol* **33**, 449–477.

Cannon CH, Leighton M (1994) Comparative locomotor ecology of gibbons and macaques: selection of canopy elements for crossing gaps. *Am J Phys Anthropol* **93**, 505–524.

Cant JGH (1992) Positional behavior and body size of arboreal primates: a theoretical framework for field studies and an illustration of its application. *Am J Phys Anthropol* **88**, 273–283.

Chatani K (2003) Positional behavior of free-ranging Japanese macaques (*Macaca fuscata*). *Primates* **44**, 13–23.

Clutton-Brock TH, Harvey PH (1977a) Primate ecology and social organisation. *J Zool Lond* **183**, 1–39.

Clutton-Brock TH, Harvey PH (1977b) Species differences in feeding and ranging behavior in primates. In *Primate Ecology* (ed. Clutton-Brock TH), pp. 557–588. London: Academic Press.

Corruccini RS, McHenry HM (2001) Knuckle-walking hominid ancestors? *J Hum Evol* **40**, 507–511.

Covert HH, Quyet LK, Wright BW (2006) The positional behavior of the Tonkin snub-nosed monkey (*Rhinopithecus avunculus*) at Du Gia Nature Reserve, Ha Giang Province, Vietnam. *Am J Phys Anthropol* **42**, 78.

Crompton RH (2016) The hominins: a very conservative tribe? Last common ancestors, plasticity and ecomorphology in Hominidae. Or, What's in a name? *J Anat* **228**, 686–699.

Dainton M, Macho GA (1999) Did knucklewalking evolve twice? *J Hum Evol* **36**, 171–194.

de Bonis L, Koufos GD (1994) Our ancestors' ancestor: *Ouranopithecus* is a Greek link in human ancestry. *Evol Anthropol* **3**, 75–83.

de Bonis L, Koufos GD (2014) First discovery of postcranial bones of *Ouranopithecus macedoniensis* (Primates, Hominoidea) from the late Miocene of Macedonia (Greece). *J Hum Evol* **74**, 21–36.

Delson E (1994) Evolutionary history of the colobine monkeys in paleoenvironmental perspective. In *Colobine Monkeys: Their Ecology, Behaviour and Evolution* (eds. Davies AG, Oates JF) pp. 11–43. Cambridge: Cambridge University Press.

Duda P, Zrzavy D (2013) Evolution of life history and behavior in Hominidae: towards phylogenetic reconstruction of the chimpanzee human last common ancestor. *J Hum Evol* **65**, 424–446.

Fleagle JG (1976) Locomotor behavior and skeletal anatomy of sympatric Malaysian leaf monkeys (*Presbytis obscura* and *Presbytis melalophos*). *Ybk Phys Anthropol* **20**, 440–453.

Fleagle JG (1977) Locomotor behavior and muscular anatomy of sympatric Malaysian leaf-monkeys (*Presbytis obscura* and *Presbytis melalophos*). *Am J Phys Anthropol* **46**, 297–308.

Fleagle JG (2013) *Primate Adaptation and Evolution*, 3rd ed. San Diego, CA: Academic Press.

Gaulin SJC (1979) A Jarman–Bell model of primate feeding niches. *Hum Ecol* **7**, 1–20.

Gaulin SJC, Konner M (1977) On the natural diet of primates, including humans. In *Nutrition and the Brain*, Vol. 1. (eds. Wurtman R, Wurtman J), pp. 1–86. New York: Raven Press.

Gebo DL, MacLatchy L, Kityo R, et al. (1997) A hominoid genus from the early Miocene of Uganda. *Science* **276**, 401–404.

Geist V (1974) On the relationship of social organization and ecology in ungulates. *Am Zool* **14**, 205–220.

Gonzales LA, Benefit BR, McCrossin ML, Spoor F (2015) Cerebral complexity preceded enlarged brain size and reduced olfactory bulbs in Old World monkeys. *Nature Comm* **6**, 7580.

Grand TI (1972) A mechanical interpretation of terminal branch feeding. *J Mammal* **53**, 198–201.□

Güleç ES, Sevim A, Pehlevan C, Kaya F (2007) A new great ape from the late Miocene of Turkey. *Anthropol Sci* **115**, 153–158.

Hammond AS, Alba DM, Almécija S, Moyà-Solà S (2013) Middle Miocene *Pierolapithecus* provides a first glimpse into early hominid pelvic morphology. *J Hum Evol* 64, 658–666.

Harrison T (1989) New postcranial remains of *Victoriapithecus* from the middle Miocene of Kenya. *J Hum Evol* 18, 3–54.

Harrison T (2005) The zoogeographic and phylogenetic relationships of early catarrhine primates in Asia. *Anthropol Sci* 113, 43–51.

Harrison T (2010a) Apes among the tangled branches of human origins. *Science* 327, 532–534.

Harrison T (2010b) New estimates of hominoid taxonomic diversity in Africa during the Neogene and its implications for understanding catarrhine community structure. *J Vert Paleontol* 30, 102A.

Harrison T, Andrews P (2009) The anatomy and systematic position of the early Miocene proconsulid from Meswa Bridge, Kenya. *J Hum Evol* 56, 479–496.

Harvey PH, Clutton-Brock TH (1985) Life history variation in primates. *Evolution* 39, 559–581.

Hladik CM (1978) Adaptive strategies of primates in relation to leaf-eating. In *The Ecology of Arboreal Folivores* (ed. Montgomery GG), pp. 373–395. Washington, DC: Smithsonian Institution Press.

Hopwood AT (1933) Miocene primates from Kenya. *J Linn Soc Zool* 38, 437–464.

Houle A, Chapman CA, Vickery WL (2010) Intratree vertical variation of fruit density and the nature of contest competition in frugivores. *Behav Ecol Sociobiol* 64, 429–441. ☐

Houle A, Conklin-Brittain NL, Wrangham RW (2014) Vertical stratification of the nutritional value of fruit: macronutrients and condensed tannins. *Am J Primatol* 76, 1207–1232.

Hunt KD (1992a) Positional behavior of *Pan troglodytes* in the Mahale Mountains and Gombe Stream National Parks, Tanzania. *Am J Phys Anthropol* 87, 83–107.

Hunt KD (1992b) Social rank and body weight as determinants of positional behavior in *Pan troglodytes*. *Primates* 33, 347–357.

Hunt KD (2016) Why are there apes? Evidence for the co-evolution of ape and monkey ecomorphology *J Anat* 228, 630–685.

Inouye SE, Shea BT (2004) The implications of variation in knuckle-walking features for models of African hominoid locomotor evolution. *J Anthropol Sci* 82, 67–88.

Ishida H, Pickford M (1997) A new Late Miocene hominoid from Kenya: *Samburupithecus kiptalami* gen. et sp. nov. *CR Acad Sci II* 325, 823–829.

Ishida H, Kunimatsu Y, Takano T, Nakano Y, Nakatsukasa M (2004) *Nacholapithecus* skeleton from the Middle Miocene of Kenya. *J Hum Evol* 46, 69–103.

Jarman PJ (1974) The social organization of antelope in relation to their ecology. *Behaviour* 58, 215–267.

Jungers WL (1979) Locomotion, limb proportions, and skeletal allometry in lemurs and lorises. *Folia Primatologica* 32, 8–28.

Kay RF (1977) The evolution of molar occlusion in the Cercopithecidae and early catarrhines. *Am J Phys Anthropol* 46, 327–352.

Kelley J (1994) A biological hypothesis of ape species density. In *Current Primatology*, Vol. 1: *Ecology and Evolution* (eds. Thierry B, Anderson JR, Roeder JJ, Herrenschmidt N), pp. 11–18. Strasbourg: Louis Pasteur University.

Kelley J (1995a) Sex determination in Miocene catarrhine primates. *Am J Phys Anthropol* 96, 391–411.

Kelley J (1995b) Sexual dimorphism in canine shape among extant great apes. *Am J Phys Anthropol* 96, 365–389.

Kelley J (1997) Paleobiological and phylogenetic significance of life history in Miocene hominoids. In *Function, Phylogeny, and Fossils* (eds. Begun DR, Ward CV, Rose MD) pp. 173–208. Boston, MA: Springer.

Kelley J (2001) Phylogeny and sexually dimorphic characters: canine reduction in *Ouranopithecus*. In *Phylogeny of the Neogene Hominoid Primates of Eurasia* (eds. de Bonis L, Koufos GD, Andrews P) pp. 269–283. Cambridge: Cambridge University Press.

Kivell TL, Schmitt D, Walker A (2009) Independent evolution of knuckle-walking in African apes shows that humans did not evolve from a knuckle-walking ancestor. *PNAS* 106, 241–246.

Kleiber M (1932) Body size and metabolism. *Hilgardia* 6, 315–353.

Koufos GD, de Bonis GD (2005) The late Miocene homininoids *Ouranopithecus* and *Graecopithecus*: implications about their relationship and taxonomy. *Annales de Paleontologie* 91, 227–240.

Kunimatsu Y, Nakatsukasa M, Sawada Y, et al. (2007) A new late Miocene great ape from Kenya and its implications for the origins of African great apes and humans. *Proc Nat Acad Sci* 104, 220–225.

Larson SG (1998) Parallel evolution in the hominoid trunk and forelimb. *Evol Anthropol* 6, 87–99.

MacLatchy LM, Bossert WH (1996) An analysis of the articular surface distribution of the femoral head and acetabulum in anthropoids, with implications for hip function in Miocene hominoids. *J Hum Evol* 31, 425–453.

MacLatchy L, Gebo D, Kityo R, Pilbeam D (2000) Postcranial functional morphology of *Morotopithecus bishopi*, with implications for the evolution of modern ape locomotion. *J Hum Evol* 39, 159–183.

Madar SI, Rose MD, Kelley J, et al. (2002) New *Sivapithecus* postcranial specimens from the Siwaliks of Pakistan. *J Hum Evol* 42, 705–752.

Matthews A, Matthews A (2002) Distribution, population density, and status of sympatric cercopithecids in the Campo-Ma'an area, Southwestern Cameroon. *Primates* **43**, 155–168.

McCrossin ML, Benefit BR (1997) On the relationships and adaptations of *Kenyapithecus*, a large-bodied hominoid from the middle Miocene of eastern Africa. In *Function, Phylogeny, and Fossils* (eds. Begun DR, Ward CV, Rose MD) pp. 241–267. Boston, MA: Springer.

McNulty KP, Begun DR, Kelley J, Manthi FK, Mbua EN (2015) A systematic revision of Proconsul with the description of a new genus of early Miocene hominoid. *J hum Evol* **84**, 42–61.

Merceron G, Koufos GD, Valentin X (2009) Feeding habits of the first European colobine, *Mesopithecus* (Mammalia, Primates): evidence from a comparative dental microwear analysis with modern cercopithecids. *Geodiversitas* **31**, 865–878.

Morgan ME, Lewton KL, Kelley J, et al. (2015) A partial hominoid innominate from the Miocene of Pakistan: description and preliminary analyses. *Proc Nat Acad Sci* **112**, 82–87.

Moyà-Solà S, Köhler M, Alba DM, Casanovas-Vilar I, Galindo J (2004) *Pierolapithecus catalaunicus*, a new Middle Miocene great ape from Spain. *Science* **306**, 1339–1344.

Nakatsukasa M, Kunimatsu Y (2009) Nacholapithecus and its importance for understanding hominoid evolution *Evol Anthrol* **18**, 103–119.

Napier JR (1963) The locomotor functions of hominids. In *Classification and Human Evolution* (ed. Washburn SL), pp. 178–189. Chicago, IL: Aldine.

Nemeth LA (2007) *Cercopithecoid versus ceboid monkeys as models for human drug metabolism*. MS dissertation. Dublin: Hibernia College.

Oxnard CE (1963) Locomotor adaptations in the primate forelimb. *Symp Zool Soc Lond* **10**, 165–182.

Oxnard CE (1967) The functional morphology of the primate shoulder as revealed by comparative anatomical osteometric and discriminant function techniques. *Am J Phys Anthropol* **26**, 219–240.

Parsons PE, Taylor CR (1977) Energetics of brachiation versus walking: a comparison of a suspended and an inverted pendulum mechanism. *Physiol Zool* **50**, 182–188☐

Patel BA, Grossman A (2006) Dental metric comparisons of *Morotopithecus* and *Afropithecus*: implications for the validity of the genus *Morotopithecus*. *J Hum Evol* **51**, 506–512.

Pilbeam DR, Rose MD, Barry JC, et al. (1990) New *Sivapithecus* humeri from Pakistan and the relationship of *Sivapithecus* and *Pongo*. *Nature* **348**, 237–239.

Plumptre AJ, Cox D (2006) Counting primates for conservation: primate surveys in Uganda. *Primates* **47**, 65–73.

Raaum RL, Sterner KN, Noviello CM, Stewart CB, Disotell TR (2005) Catarrhine primate divergence dates estimated from complete mitochondrial genomes: concordance with fossil and nuclear DNA evidence. *J Hum Evol* **48**, 237–257.

Richmond BG, Strait DS (2000) Evidence that humans evolved from a knuckle-walking ancestor. *Nature* **404**, 382–385.

Ripley S (1970) Leaves and leaf-monkeys: the social organization of foraging gray langurs (*Presbytes entellus thersites*). In *Old World Monkeys: Evolution, Systematics, and Behavior* (eds. Napier JR, Napier PH), pp. 481–509. New York: Academic Press.

Ripley S (1979) Environmental grain, niche diversification, and positional behavior in Neogene primates: an evolutionary hypothesis. In *Environment, Behavior and Morphology: Dynamic Interactions in Primates* (eds. Morbeck ME, Preuschoft H, Gomberg N), pp. 91–104. Washington, DC: Smithsonian Institution Press.

Rose MD (1973) Quadrupedalism in primates. *Primates* **14**, 337–357.

Rose MD (1983) Miocene hominoid postcranial morphology: monkey-like, ape-like, neither, or both? In *New Interpretations of Ape and Human Ancestry* (eds. Ciochon RL, Corruccini RS), pp. 405–417. New York: Plenum Press.

Rose MD (1989) New postcranial specimens of catarrhines from the Middle Miocene Chinji Formation, Pakistan: descriptions and a discussion of proximal humeral functional morphology in anthropoids. *J Hum Evol* **18**, 131–162.

Rose MD (1993) Locomotor anatomy of Miocene hominoids. In *Postcranial Adaptation in Nonhuman Primates* (ed. Gebo DL), pp. 252–272. DeKalb, IL: Northern Illinois University Press.

Rose MD (1996) Functional morphological similarities in the locomotor skeleton of Miocene catarrhines and platyrrhine monkeys. *Folia Primatol* **66**, 7–14.

Ryan TM, Silcox MT, Walker A, et al. (2012) Evolution of locomotion in Anthropoidea: the semicircular canal evidence. *Proc Roy Soc London B* **279**, 3467–3475.

Sailer LD, Gaulin SJC, Boster JS, Kurland JA (1983) Measuring the relationship between dietary quality and body size in primates. *Primates* **26**, 14–27.

Schultz AH (1930) The skeleton of the trunk and limbs of higher primates. *Hum Biol* **2**, 303–438.

Schultz AH (1950) The physical distinctions of man. *Proc Am Philos Soc* **94**, 428–449.

Schwartz GT, Conroy GC (1996) Cross-sectional geometric properties of the *Otavipithecus* mandible. *Am J Phys Anthropol* **99**, 613–623.

Scott JE, Schrein CM, Kelley J (2009) Beyond *Gorilla* and *Pongo*: alternative models for evaluating variation and sexual dimorphism in fossil hominoid samples. *Am J Phys Anthropol* **140**, 253–264.

Stern JT, Jr. (1971) *Functional Myology of the Hip and Thigh of Cebid Monkeys and Its Implications for the Evolution of Erect Posture.* Basel: Karger.

Stevens N, Seiffert ER, O'Connor PM, et al. (2013) Palaeontological evidence for an Oligocene divergence between Old World monkeys and apes. *Nature* **497**, 611–614.

Su DF, Jablonski NG (2009) Locomotor behavior and skeletal morphology of the odd-nosed monkeys. *Folia Primatol* **80**, 189–219.

Swedell L, Plummer T (2012) A Papionin multilevel society as a model for hominin social evolution. *Int J Primatol* **33**, 1165–1193.

Temerin LA, Cant JGH (1983) The evolutionary divergence of old world monkeys and apes. *Am Nat* **122**, 335–351.

Walker AC, Pickford M (1983) New post-cranial fossils of *Proconsul africanus* and *Proconsul nyanzae*. In *New Interpretations of Ape and Human Ancestry* (eds. Ciochon RL, Corruccion RS), pp. 325–351. New York: Plenum Press.

Walker AC, Rose M (1968) Fossil hominoid vertebra from the Miocene of Uganda. *Nature* **217**, 980–981.

Walker AC, Teaford M (1989) The hunt for *Proconsul*. *Sci Am* **260**, 76–82.

Wang WJ, Crompton RH (2004) Analysis of the human and ape foot during bipedal standing with implications for the evolution of the foot. *J Biomech* **37**, 1831–1836.

Ward CV (1993) Torso morphology and locomotion in *Proconsul nyanzae*. *Am J Phys Anthropol* **92**, 291–328.

Ward CV, Walker A, Teaford MF (1991) *Proconsul* did not have a tail. *J Hum Evol* **21**, 215–220.

Ward CV, Walker A, Teaford MF, Odhiambo I (1993) Partial skeleton of *Proconsul nyanzae* from Mfangano Island, Kenya. *Am J Phys Anthropol* **90**, 77–111.

Ward S, Brown B, Hill A, Kelley J, Downs W (1999) *Equatorius*: a new hominoid genus from the middle Miocene of Kenya. *Science* **285**, 1382–1386.

Washburn SL (1957) Ischial callosities as sleeping adaptations. *Am J Phys Anthropol* **15**, 269–276.

Watson J, Payne R, Chamberlain A, et al. (2009) Carrying in humans and great apes: implications for the evolution of human bipedalism. *Folia Primatol* **80**, 309–328.

Wheatley BP (1982) Energetics of foraging in *Macaca fascicularis* and *Pongo pygmaeus* and a selective advantage of large body size in the orangutan. *Primates* **23**, 348–363.

Wrangham RW (1979) The evolution of ape social systems. *Soc Sci Infor* **18**, 335–368.

Wrangham RW (1980) An ecological model of female-bonded primate groups. *Behaviour* **75**, 262–299.

Wrangham RW, Peterson D (1997) *Demonic Males: Apes and the Origins of Violence.* New York: Houghton Mifflin.

Wrangham R, Pilbeam D (2002) African apes as time machines. In *All Apes Great and Small: Developments in Primatology – Progress and Prospects* (eds. Galdikas BMF, Briggs NE, Sheeran LK, Shapiro GL, Goodall J), pp. 5–17. Boston, MA: Springer.

Wrangham RW, Waterman PG (1983) Condensed tannins in fruits eaten by chimpanzees. *Biotropica* **15**, 217–233.

Wrangham RW, Conklin-Brittain NL, Hunt KD (1998) Dietary response of chimpanzees and cercopithecines to seasonal variation in fruit abundance. I. Antifeedants. *Int J Primatol* **19**, 949–970.

Young NM, MacLatchy L (2004) The phylogenetic position of *Morotopithecus*. *J Hum Evol* **46**, 163–184.

Zalmout IS, Sanders WJ, MacLatchy LM, et al. (2010) New Oligocene primate from Saudi Arabia and the divergence of apes and Old World monkeys. *Nature* **466**, 360–364.

11 Building a Natural Wonder
Growth, Development, and Life History

Photo credit: Adolf Naef, 1926.

The structural similarity of chimpanzees and humans corresponds so closely that an early work on chimpanzee anatomy found it necessary to repeat the words "as in Man" 182 times in a 97-page work (see Chapter 7). Yet, this tale of similarity has a twist: To the casual observer the many micro-scale similarities between *Homo* and *Pan* are overwhelmed by undeniable macro-scale differences. Whether an expert or neophyte, one cannot fail to notice that chimpanzees have longer, thicker body hair, a jutting face, a smaller braincase, long arms – longer the closer one moves to the tips of their extraordinarily long fingers – short legs, and a very unhuman-like nose, a nose that is little more than two holes in the face. Anatomists who emphasize similarities have not overlooked these differences, but they more often drill down to the millimeter- or microscopic-scale, and here neurons, pulmonary alveoli, muscle fibers, tendons, and the arterial networks are nearly identical in chimpanzees and humans.

To the neophyte, though, the macro-scale human–ape differences can make you question anatomists' eyesight: Plainly, humans and chimpanzees are nothing alike. If this is your view, perhaps the image of the young chimpanzee heading this chapter will give you pause. If you are surprised at his human-like appearance, you will have learned something about development. Chimpanzee faces change more than those of humans during growth. Humans are, in other words, *neotenous* (Shea, 1984, 1989; Penin et al., 2002) or juvenilized. In this chapter, we will consider how differences in the speed and pattern of growth, both whole-body and body part by body part, lead to human and chimpanzee differences. The profound chimpanzee and human differences emerge from *differential growth*, even though both are made of the same things at the microscale. Using a masonic analogy, just as the instructions "lay down a row of 200 hundred bricks and repeat until you have 20 courses" yields a different house than "lay down 100 bricks in 30 courses," so does rearranging the same raw materials yield a chimpanzee rather than a human.

11.1 Life History

One parallel between humans and chimpanzees that no amount of staring at any one individual can reveal is the similarity of our **life histories**, the schedule and duration of key events in the life of an organism; we both have a **slow life history**. If you have a pet, the typical mammalian fast life history is familiar; cats grow quickly, humans slowly. In fact, you probably already have a sense that *there is an allometry (a size* *effect) to growth and development*; all other things being equal, small animals mature more quickly than larger ones – they have a fast life history. A mouse can produce pups at two months of age; dogs, in contrast, are not reproductively mature until six months of age. Body size, however, is only a small part of the chimpanzee story, because they are off the charts compared to similar-sized mammals. A 40 kg female chimpanzee has her first pregnancy around the age of 13 (Nissen & Yerkes, 1943; Nishida et al., 2003), 10 times the age of a similar-sized goat. A racehorse, tipping the scales at a colossal 545 kg (1200 lb), is fully grown at the age of four. The Kentucky Derby is for three-year-olds – yet a three-year-old chimpanzee has reached only one-fifth of her adult body weight and will need another dozen years of growth to hope to win an all-comers chimpanzee foot race.

11.2 Evolution and Development

Details of growth will help you to understand how chimpanzees and humans came to differ, but will also reveal their separate evolutionary histories. As we trace our body shape back from adulthood to childhood, from fetus to embryo, we find that humans and chimpanzees resemble one another ever more. Adults, in other words, are not just enlarged versions of infants; we humans have legs that grow disproportionately fast. Put another way, as infants, humans and chimpanzees have arms and legs equal in length; adult humans have longer legs, adult chimpanzees longer arms.

PART I: EMBRYOLOGY

Gestation is, to my mind, the single most incredible phenomenon in nature. In a shockingly brief time a chimpanzee (or a human) grows from a single cell to a breathing, crying, sensate being. As stunning as it is that the 37.2 trillion cells in the human body (Bianconi et al., 2013) spring from a single cell in a mere 38 weeks, it is even more astounding that the nearly-as-complex dog accomplishes this feat in one-fifth of that time.

11.3 Fertilization and Preimplantation

Among humans this incredible journey begins when the ovum (also known as the oocyte) emerges from the ovary, surrounded by a gel-like cloud of tissue known as the **zona pellucida** (Figure 11.1). This egg-coat is made up of proteins and tissue that both protect the egg and interact with sperm to allow fertilization. Once the DNA of the sperm and egg unite, the single-cell fertilized egg, referred to for this brief single-

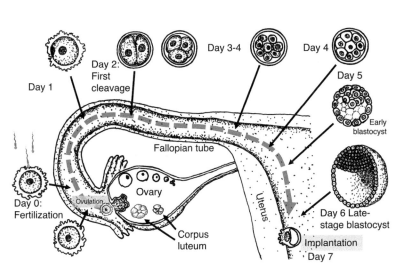

Figure 11.1 The first seven days of embryonic development. Fertilization takes place in the fallopian tube. Even before implantation, at day 7 the fertilized egg has developed into a multicellular blastocyst. Image by the author, after Sandra & Coons, 1997; with permission.

celled period as the **zygote**, begins to float down the fallopian tube toward the uterus. Early chimpanzee embryology is presumed to be virtually identical to that of humans.

Thus begins a process of duplication and diversification that will yield hundreds of cell types, including skin, muscle, liver, pancreas, bone, brain, and so on, all appearing quite different under a microscope, but miraculously all (except a scant few) with the same DNA and all having descended from that single cell. The process by which one set of DNA and one cell can produce this diversity is much better understood today than it was only a few decades ago. Solving this mystery is one of biology's greatest triumphs.

We start slow. Humans take an entire day and more (30 hours) just going from the single-cell zygote to two cells (Figure 11.1; this review is abstracted from Sandra & Coons [1997], Schoenwolf et al. [2014], and Yamada et al. [2015]). Astonishingly, even at this first division something profound has happened; while the two cells look identical under a microscope, the head side of the two-cell embryo – the "west" end or "pole," following the convention that developmental biologists use – has already been determined. Specialists refer to this phenomenon as **polarity**, the determination of the head versus tail end of the embryo.

How does one cell know it is the head end when the two cells appear identical and have identical genes? Although there are other explanations (Motosugi et al., 2005), most developmental biologists believe that polarity is established even before fertilization. It results from an uneven distribution of biomolecules inside the ovum, a difference established before the ovum emerges from the ovary. The differing biochemical environment in the west-end versus east-end cells, when the embryo is only two cells, determines which genes are switched on in each cell, beginning a whole cascade of cell differentiation.

11.4 Homeobox (Hox) Genes and Evo-Devo

The process of turning on the right genes at the right time is orchestrated by a relatively newly discovered class of genes called **homeobox genes,** or **Hox** genes. Few discoveries in science are as mind-bending as the discovery that Hox genes are nearly identical across all animal life. Our Hox genes evolved, in essence, over 500 million years ago; they evolved before worms, flies, and mammals embarked on their individual evolutionary paths (Raff & Kauffman, 1983; Raff, 1996, 2000; Carroll, 2005, 2008). For example, the Hox gene that directs the formation of the eye is nearly identical in flies and mice, even though the structure of the eyes themselves have little in common, other than that both have cells that detect light. Legs ditto. Transplant proto-leg tissue from a frog embryo to a mouse and a mouse leg grows.

The discovery of homeoboxes has profoundly altered the way we view evolution. Before Hox genes were discovered it was thought that the hundreds of million years of evolution that separated starfish, flies, frogs, and chimpanzees meant that there would be virtually no overlap in the respective gene sequences that directed their development. Bodies were thought to be evolved by duplicating genes, then coopting one of the duplicates for the task of directing the development of a new body part. This led to the expectation that organisms, even sister species, would have quite different genes.

11.5 Discovery of Homeobox (Hox) Genes

The discovery of Hox genes started – as so many discoveries do – when the curiosity of a genius inspired him to study a phenomenon few had even noticed. Over 100 years ago, William Bateson (Bateson, 1894) took an interest in several bizarre deformities in fruit flies: body parts that grow in the wrong places, like legs in place of antennae. He called coined the term *homeosis* to describe the phenomenon of misplaced or duplicated body parts, giving us the *homeo* part of what later became the term homeobox genes.

Bateson's work could be said to have been a dead-end. He documented a number of cases of homeosis, but he was unable to explain it; the discovery of its genetic basis was still a lifetime away. That discovery started in a small way with the one-celled organism *Escherichia coli* (Monod, 1971; Jacob, 1974; for more information refer to Sean Carroll's delightful 2005 book, *Endless Forms Most Beautiful*).

E. coli is a tiny gut bacterium that eats sugar, reproduces every 20 minutes and does just about nothing else, seemingly having found the meaning of life in these simple pleasures. Well, some of its evil strains give us diarrhea, but I hope it is not proud of it. It thrives on the single-molecule sugar *glucose*. We learned in Chapter 6 that sugar molecules come in several types, some with two sugar molecules fused together; **lactose**, for instance, is made up of the two sugars, glucose and galactose, neither of which can be utilized until they have been separated by the digestive enzyme **lactase**. *E. coli* eats glucose and normally completely ignores lactose, which means it need not and does not produce the enzyme to digest it, lactase. However, when things are not normal, when there is no glucose but much lactose, it shifts gears and begins making lactase. If asked to guess how this happens, you might guess there has been a mutation, but that would be wrong. It turns out that *E. coli* had the gene to make lactase all along, but it was turned off; why waste energy producing an unnecessary enzyme?

If you like a challenge, try this one: Imagine a mechanism by which this simple, brainless bacterium might detect lactose and then turn on its lactase gene. You will fail. Your tens of billions of neurons will never come up with the incredibly efficient mechanism humble *E. coli* arrived at. It uses the *lactose itself* as the key to turn on the lactase gene.

Normally a tiny blob of protein called a **DNA binding repressor** grips the beginning part of the DNA that holds the code for lactase. This repressor blob is clamped on so tight that DNA cannot be unzipped to start the first step of making a protein. The DNA binding repressor sticks to any length of DNA that has the base sequence that happens to be found at the beginning of the lactase gene. In this case the only place with the sequence the repressor sticks to is found in *E. coli* at the beginning of the lactase gene. We call the particular stretch of base pairs that is sticky to the repressor-blob the **lactase repressor attachment site**.

So this blob sticks to the attachment site and *E. coli* merrily digests glucose, ignoring lactose. Spill lactose into the petri dish, however, and things change. While the lactase repressor blob seems very committed to the lactase repressor attachment site, it is not, in fact, so monogamous. It has a secret lover: lactose. When a lactose molecule floats near it, the repressor blob folds it in its loving arms and is transformed: The two merge to form a larger lactose + lactase-repressor-blob, and the new couple, enamored with one another, have no use for the lactase repressor attachment site. The lactose + lactase-repressor pair detaches and drifts off without looking back. Freed from the repressor blob, the lactase gene is unleashed and begins doing what all genes do if they have no repressor blob – produce protein, in this case lactase. Now *E. coli* can digest lactose.

But now imagine what happens when lactose dwindles and disappears. Now lactose repressor blobs find themselves single again; when they encounter their old flames the attraction is rekindled and they latch onto the lactase repressor attachment sites, shutting down lactase-making genes. The section of DNA that is sticky to the lactase repressor is in essence a switch that activates the gene in the presence of lactose and switches off the gene when no lactose is present.

This was an incredible discovery, the sort of unambiguous triumph that is rare in science. The discoverers might have sat on their laurels, basking in the admiring glances of other bacteriologists, being stood to beers in return for retelling their thrilling narrative. The discoverers did no such thing. They excitedly realized that all living things might have such gene switches and they wasted no time in beginning to search for them (Monod, 1971; Jacob, 1974). They found them; in abundance.

Meanwhile, a completely different group of scientists were trying to better understand Bateson's homeosis by studying fruit flies. Many of these fruit fly folks were here at my home base, Bloomington, Indiana (Raff & Kauffman, 1983; Hughes & Kaufman, 2002), one of the world's great fruit fly centers. They soon found that the key to Bateson's misplaced body parts in fruit flies was a mutation that affected a very interesting gene normally active in the fly's head.

Normally. When this gene becomes active elsewhere, no matter where it becomes active, the tissue at this site forms an antenna. This gene was one of only eight genes geneticist Sean Carroll came to call **master genes** (Carroll, 2005), which are genes that govern development in flies. Drawing on Bateson's term homeosis, master genes have come to be better known as **homeobox genes,** or **Hox** genes for short.

The year 1983 turned out to be the year of homeobox synchronicity, because Hox genes were independently discovered that year (published in 1984) in two labs, that of Walter Jakob Gehring in Switzerland (McGinnis et al., 1984) and Thomas Kaufman's lab here at Indiana University, long a locus of *Drosophila* research (Wakimoto & Kaufman, 1981; Scott & Weiner, 1984). Both labs had managed to isolate genes that controlled the construction of antennae, and, as their work progressed, seven other Hox genes. They found that each of the eight was turned on in specific parts of the fly, in strips of tissue that form bands from the head end of the developing fly (Figure 11.2) to the tail. In other words, as each Hox gene switches on it spurs a different kind of development in each of eight strips, differentiating the nascent fly into the eight segments: (1) the mouthparts (directed by the homeobox gene "lab"; Figure 11.2); (2) the nose parts (pb); (3) the head (Dfd); (4) the first of three parts of the thorax, including a leg (Scr); (5) the middle part of the thorax, including a pair of wings (Antp); (6) the back part of the thorax, including a leg and vestigial wings (UBx); and (7, 8) two abdominal Hox genes (Abd-A and Abd-B).

You might feel I left you hanging when I jumped from lactase repressor DNA in *E. coli* to Hox genes in flies. Let me go back and link the two. To their astonishment, as fly researchers sequenced fruit fly Hox genes, they found that each Hox gene started with a nearly identical 180 base pair sequence, a sequence they found to their surprise was already known to science. It was similar to the now-familiar lactose switch in *E. coli*, the lactase repressor attachment site. We might need to pause to let that sink in: A gene in a microscopic, one-celled organism – a simple protozoan with only 5000 or so genes – turned out to be a version of a gene found in the decidedly macroscopic fruit fly, possessing over 15,000 genes. I will wait while you gape.

These researchers gradually realized that this 180 base pair beginning part of each Hox gene was a switch that repressed gene activity. They realized that when the right protein – let us call one of these "Formula X" for the moment – is present, the fly Hox gene Dfd starts to make proteins that in turn start a cascade of events that results in a head being made. Inject Formula X into any part of a fly embryo, a thorax, an abdomen, a leg, and those cells will begin trying to make a head. "Formula X" type proteins are called **signaling proteins** because their purpose is to send a signal that activates particular genes. Each of the eight Hox genes in the fly is activated by a different signaling protein, just as lactose signals the *E. coli* lactase gene to fire up. Among animals, from fly to gecko to horse, it is differences in the genes that contain the instructions for the details of the head, not

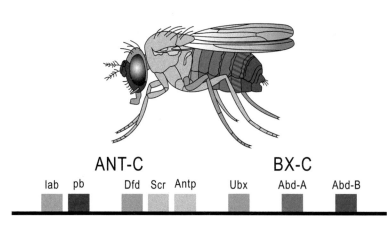

Figure 11.2 Fruit fly Hox genes; these genes appear on fly chromosome 3; intriguingly, they are lined up on the chromosome in the same order they appear in the fly's body. Image courtesy of WikiMedia Creative Commons. (A black and white version of this figure will appear in some formats. For the color version, please refer to the plate section.)

differences in the Hox genes that say "make a head," that cause the organisms to be different. If you were to say "build a toilet" (my analogy to a Hox gene signaling other genes to start making a body part) to my kinfolk in Kentucky during my father's youth, you would end up with a little house over a pit, inside of which you would find a bench with a hole cut in it. If you gave those same directions in Bloomington today, you would get a ceramic bowl connected to a sewage pipe; and in Uganda you would get a different shaped ceramic bowl, below floor level, centered over two footprint shapes meant to tell you where to place your feet. Context is everything.

Research on chimpanzees and humans actually anticipated the discovery of Hox genes in some way. In 1975, Mary-Claire King and Allan Wilson (King & Wilson, 1975) published a paper showing that chimpanzee and human proteins, and thus the genes that coded for them, were 99 percent similar. This 99 percent factoid has been so widely disseminated that you are undoubtedly already familiar with it, but at the time it was shocking. Chimpanzee and human genes nearly identical? Preposterous. King and Wilson tried to make sense of this surprising genetic identity with a remarkably prescient observation: They reasoned that rather than genes for making a leg differing dramatically between humans and chimpanzees, it must be that **regulatory genes**, the genes that turn on and off the genes for making a leg, must differ. We now know that altering the timing of homeobox gene activity, turning one on earlier and another later, can make for remarkably diverse organisms, even if the genes that are being turned off and on are virtually the same. By changing development only a little bit you can get dramatic differences in brain size, hairiness, and forelimb length, for example. Everything we have learned since the discovery of Hox genes in 1984 has shown us that most of evolution occurs by making subtle developmental changes rather than by evolving a new gene for every little change.

In the new Hox (post-Hox?) world we now live in, we know that body parts differentiate and develop by switching genes off and on; if we speeded up the development video it would look like thousands of blinking lights on a Christmas tree, each light eliciting growth in the right place and at the right time. We have learned further that many genes serve double- and triple-duty; they are activated in different places at different times. For example, in mammals a gene known at *MMP5* guides the development of not only the ribs, but also cartilage in the larynx and in the outer ear. It is turned on at different times in these different places. By switching genes off and on at different times and in different places, just a few genes – we have fewer than 25,000 – are needed to build the human (and chimpanzee) body.

By now you have probably guessed why I suddenly veered from discussing the fact that even a two-celled embryo has a head end and a rump end to discussing Hox genes in *E. coli* and fruit flies. It is homeobox genes that govern embryonic development in humans and chimpanzees, just as they do in flies. The asymmetric distribution of molecules in the first two cells activates different genes in each cell, thus producing different "Formula X" proteins in each cell. Let us give the Formula X proteins that activate genes their real name. These signaling proteins are called **morphogens.** It is thought (Carroll, 2005, 2008) that as the embryo continues to grow to tens then hundreds of cells, morphogens produced at one pole (hence, **polarity**) of the embryo travel part of the way across it by soaking through nearby cells, much like water soaks through a sponge. The morphogens become more diluted as they migrate across the developing embryo, eventually trickling out altogether.

Imagine a protein we can call morphogen A (I am making this up to keep it a little simpler than reality) being produced at the head end of the embryo and then diffusing quarter of the way across the embryo. High concentrations of morphogen A activate genes that produce a second morphogen, B, which happens to be better at diffusing and diffuses halfway across the embryo (Figure 11.3). The presence of both A and B elicits a third morphogen, C, which soaks three-quarters of the way across. The first quarter of the embryo has now been exposed to morphogens A, B, and C, the second quarter to morphogens B and C (but not A), and from halfway across to three-quarters of the way cells have been soaked only in morphogen C. The last quarter of the embryo has no morphogens.

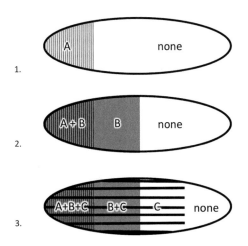

Figure 11.3 In this imaginary early embryo, imbalances in proteins in the left end of the embryo stimulate the production of morphogen A (vertical hatching), which soaks quarter of the way across the embryo (1). Morphogen A stimulates the production of morphogen B, which soaks halfway across the embryo (2). The presence of A + B stimulates the production of C, which soaks three-quarters of the way across the embryo, leaving four distinct bands. A + B + C turns on "head" Hox genes; B + C turns on "thorax" Hox genes, C turns on abdomen genes, and no morphogens causes tail genes to activate. Image by the author.

Imagine an embryo, then, with four stripes. The first stripe has a mix of morphogens that cause it to develop into one specific body segment, say the head; the second stripe becomes the thorax, the third the abdomen, and the last, with no morphogens, becomes the tail.

Differentiation can be made still more fine-grained by adding still other morphogens that switch off genes, rather than switching them on. Some morphogens activate yet other morphogens that switch on still other genes.

This is happening not just in one direction – head to tail – but also from back to belly. There are, in other words, two axes, two points of initial diffusion of morphogens, one from a pole where the spine will be and one from the head end. This means different morphogens are crossing at right angles, some switching on genes, other switching off genes. This interaction can create not just stripes, but spots. One of those spots will become a forelimb and one will become a hindlimb.

11.6 The Developing Embryo

Now that we know that Hox genes are running things, let us return to our two-celled embryo. Ten hours after the first division (40 hours in) there is a second division to create four cells. As these duplications are happening, the fertilized ovum is still traveling down the fallopian tube, on its way to the uterus (Figure 11.1). By day 4 there are eight cells, but now the rate of cell division increases dramatically. By day 6 the developing embryo, now called a **blastocyst**, has 200 or more cells and begins to take on a distinct form – though not that of an animal. Starting with that single cell – about the diameter of a human hair – each division halves the size of the cells, just as if you were dividing up a ball of bread dough. Thus, the hundreds of cells in the blastocyst are tiny compared to the original fertilized ovum, yet each has a full complement of DNA.

At first the embryo is just a roundish blob of outwardly identical cells – outwardly only, though; inside there have already been changes that will determine each cell's future. By the sixth day the roundish blastocyst has developed a cavity inside so that it now resembles a hollow ball. Most of the ball is only one cell thick, but there is a lump of cells on one side (Figure 11.1, days 5 and 6; Figure 11.4). The cells in the lump are called the **inner cell mass**, while the cells of the single-cell wall are called **trophoblast cells**. By now there is some differentiation; the cells of the inner cell mass and the trophoblast cells differ. The segregation into these two basic cell types is the foundation for all future cell differentiation. This inner cell mass/trophoblast distinction is very important; cells in the inner cell mass will form the embryo while those in the single-cell wall will become the placenta, the amniotic sac, and all other structures outside the embryo proper.

Besides dividing, up to now (day 6) the cells themselves have had little physiological work to do other than deciding for which part of the body they are destined, but at about day 7 they take on their first job: The trophoblast cells pump liquid into the hollow inside of the ball. They do this by first pushing sodium ions into the cavity, which causes water to enter by osmosis. This fluid-filled cavity is called the **blastocoel** (pronounced "BLAST-oh-seal").

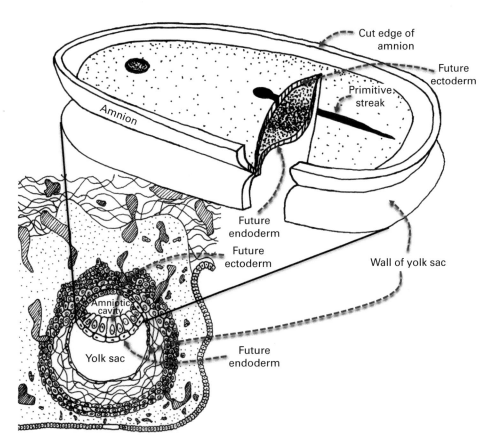

Figure 11.4 Week 2–3: the inner cell mass is now a two-layered embryonic disk, later to develop into the embryo. The cell layer with the primitive streak, the amnion side, will become the ectoderm and the yolk-sac side layer will become the endoderm. A depression or cleft called the primitive streak forms and cells from the ectoderm flow through the streak and into the space between the two layers, forming a third layer, to be called the mesoderm. Image by the author, after Sandra & Coons, 1997; with permission.

The trophoblast cells are also busy working with the uterus, building the placenta, an incredible organ that I think gets too little publicity. It generates hormones required for fetal growth and – later – birth; it transfers both nutrients and oxygen to the fetus throughout gestation; going the other way, it transmits waste from the growing fetus to the mother's bloodstream. I wrote "working with the mother" because half of the placenta is tissue from the fetus and half is from the mother. The mother–fetus connection is so intimate that more than waste products pass back to the mother; some of the fetal stem cells enter the mother's bloodstream and lodge in her tissues, where they remain for the rest of her life. Not only are you more your mother than father –

since you inherited *all* of your mitochondrial DNA from your mother – a small proportion of your mother actually *is* you. If your mother has ever said to you "you will always be a part of me," she knew what she was talking about. You should send her a card.

11.7 Implantation: Embryonic Disk

At the end of the first week, the blastocyst has at last implanted itself in the wall of the uterus, tunneling into the uterine lining – the endometrium – "inner cell mass" side first. It will soon be completely incorporated into the tissue of the uterus (Figure 11.4). Before implantation, the blastocyst was a free-floating blob, unnoticed by the mother's physiology; after implantation, the blastocyst becomes in some sense a parasite, a resource-draining foreign body (though, of course, with half of the mother's DNA) that will, for nine months while it goes from a single cell to a little person, rely on the mother's metabolic processes to provide all its nutrition and to dispose of all its waste. You should be amazed. Perhaps you should send flowers with that card.

After implantation, the fluid-filled blastocoel will develop into the yolk sac, a bag attached to one side of the flattening inner cell mass. Attached to the other side of the disk-like inner cell mass is another membrane, the **amnion**, which will become the amniotic sac. The yolk sac (Figures 11.4 and 11.5) is in some way a remnant from deep in our evolutionary history, when our ancestors hatched from eggs; it is important, though. It will change shape and move around as development progresses, and while most of it will gradually disappear as other cells around it divide and develop, it will serve some important functions before some of it morphs into the digestive system.

The inner cell mass continues to grow and by the second week the cells of the disk have organized themselves into two layers and the disk is oval-shaped (Figure 11.4). Blood cells begin to be produced in the yolk sac. A furrow forms on the oval-shaped disk, bifurcating the disk up to about one-third of its length; this furrow is called the **primitive streak**. At the end of the furrow, nearest the center of the disk, a small knot of cells forms a circular structure that causes the disk to resemble an oval with an exclamation point in its top third (Figure 11.4). You learned in Chapter 7 that "cranial" means "toward the head" and "caudal" means "toward the tail." The side of the disk where the knot forms will become the cranial end of the embryo, while the open end of the furrow will be the caudal end. As development proceeds, the primitive streak will be displaced farther and farther caudally (Figure 11.5).

11.8 Week Three: Gastrulation

As the third week begins, some ectoderm cells in the disk begin to migrate via an astonishing process called **gastrulation**, a process so bizarre it is difficult to fathom. It is only a slight exaggeration to say that at this point the embryo turns itself inside out. Some of the cells on the amnion (i.e., ectoderm) side, the top side in Figure 11.4, move toward the primitive streak furrow and, like water falling over a waterfall, flow over the side of the furrow, penetrate the floor of the groove, and flow in between the two layers, expanding to form a new, middle layer. With this added layer *we now have three layers: the ectoderm, the endoderm, and the new third layer which will be called the mesoderm.* Just to make sure you are oriented properly, the layer on the primitive streak/amnion side we now call the primitive ectoderm (from Latin: "ecto" = outside; "derm" = skin), the middle layer is the mesoderm ("middle skin"), and the other side, the yolk sac side, is the endoderm. These cells will take on vastly different roles as the embryo

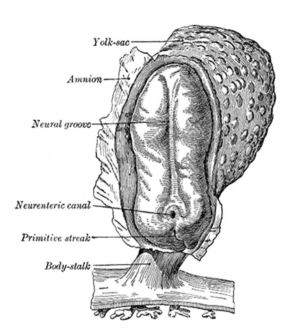

Figure 11.5 Day 20: the 2 mm long, now-elongated two-layered embryonic disk, later to develop into the embryo. The embryo is seen from the amnion side in this perspective, which means we are seeing the ectoderm; the amnion will become the amniotic sac. Here, it is torn away, leaving a ragged edge that rings the embryo. The top of the disk will become the head. From Gray, 1918.

develops (Table 11.1). The ectoderm will become the skin, nervous system, and brain; the mesoderm will become the circulatory system, bones, muscles, and reproductive tissues; and the endoderm will become the gut, lungs, and other organs.

11.9 Neural Tube and Somite Development

Toward the end of the third week, important action now moves away from the primitive streak to the cranial end. As the cranial side of the disk grows and the primitive streak is shifted lower, a fold forms, called in its shallow form the **neural fold** and later the **neural groove** (see Figure 11.5), under which we find the developing **notochord** (Figure 11.6). Now at day 20, cells are increasingly differentiating and beginning to form distinct structures. In the mesoderm, cells along the midline, just under the neural fold, coalesce into the rod-shaped notochord; this will eventually become cartilage and later bone – the backbone, in fact. The notochord stretches to halfway down the oval disk, approaching the dot-shaped end of the primitive streak. Over the rod-like notochord the neural fold deepens and its edges become more distinct; these edges we now call the **neural crests** (Figure 11.6); the neural fold is now deep enough to be called a groove (Figure 11.6). In the mesoderm, to either side of the notochord and just underneath each side of the neural groove, several knot-like structures form, like peas in a pod; these **somites** will later become the vertebrae (Figure 11.7).

The sides of the neural groove, the crests, start out as mere ridges of tissue, but they have differentiated from cells in the groove and are now called **neural crest cells**; they are destined for something extremely important. As the neural crests push up and the groove deepens, the neural crest sides arc over the groove, touch, and merge to form a separate tissue underneath the now closed ectoderm. Likewise, the differentiated tissue in the groove fuses, going from a U to an O, creating a tube we will now call the **neural tube** (Figure 11.6).

In cross-section, we now have a layer of ectoderm, then in the mesoderm layer an elongated strip of neural crest cells, then a hollow tube we now know as the neural tube, and finally even deeper in the mesoderm, directly under the hollow tube, the rod-shaped notochord.

The neural crest folding begins at the top (cranial end) of the disk and moves toward the bottom, closing like an upside-down zipper. This is only important because sometimes the folding ends prematurely; if your neural tube failed to completely close, you have a condition known as **spina bifida**.

Meanwhile, differentiating cells throughout the mesoderm have begun to form tubes that will become the circulatory system and heart. While the circulatory system is still quite simple at this point, at day 22 something miraculous occurs: A simple structure at the center of this tubular network – the primitive heart – begins to beat.

Table 11.1: **Fates of embryonic disk tissues**

Ectoderm	Skin, nervous system, brain
Mesoderm	Circulatory system, including heart, bones, muscles, reproductive system
Endoderm	Gut, lungs, other organs
Yolk sac	Gut

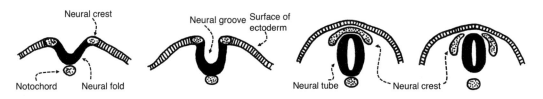

Figure 11.6 Cross-section of the embryonic disk showing the folding of the neural groove into the neural tube. Image by the author, after Sandra & Coons, 1997; with permission.

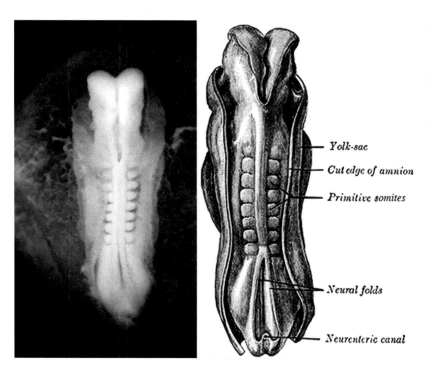

Figure 11.7 Carnegie Stage 10, at 23 days. Although only two are noted, the eight pairs of pea-shaped structures are somites. Left image from the Kyoto Collection (reproduced with the permission of Professors Shiota and Yamada, Kyoto University); right image Gray, 1918.

11.10 Week 4 and Carnegie Stages

At this point, day 22 or 23, the neural tube is complete, there are veins and arteries, and the primitive heart is beating, yet the embryo now looks a little like a segmented worm, as in Figure 11.7 and as "CS10" in Figure 11.8. In fact, at this stage the embryo has more in common with a flatworm than a human; this will soon change. "CS10" stands for **Carnegie Stage 10**, so named because much of our understanding of the stages of embryonic development comes to us by way of a group of embryologists who worked at the Carnegie Institution. They recognized that the vertebrate embryo – I mean all vertebrates from frogs to rats to humans – share the same 23 stages of development, each stage linked to the development of specific structures (O'Rahilly & Müller, 1987). This tagging of stages to specific developmental events, rather than to the length of time since conception, allows comparisons across many species, including those with vastly different gestation times (Streeter, 1942; O'Rahilly & Müller, 1987).

While the embryo still resembles a flatworm on day 23 (CS 10), this ends on days 24 and 25 when a second truly incredible event occurs, equal in unlikelihood to gastrulation; the flat embryo folds itself into a tube or zeppelin-like structure, ending up looking like a fat sausage.

To envision this folding, we will imagine looking at the embryo from the opposite side from the view we have taken up to now. Rather than viewing the embryo from the ectoderm side as we have in Figures 11.3, 11.4, and 11.6, let us mentally flip over the flatworm-like shape in Figure 11.7 and look at it from the yolk or endoderm side. In this yolk-sac-side view we are looking at what will become the interior of the body – it is, after all, the *endo*derm. Imagine the left and right sides of this oval disk curling toward you until they meet to form a tube. As the two sides approach one another they squeeze part of the yolk sac into the inside of the tube. This part of the yolk sac will become a tube itself, and it will form part of the gut (CS11; top row, right in Figure 11.8). This transverse folding is accompanied by a sort of

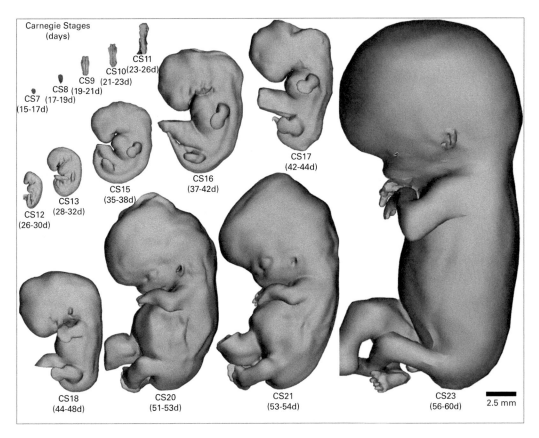

Figure 11.8 Carnegie stages. See the text for an explanation. From the 3D Atlas of Human Embryology (www.3datlasofhumanembryology.com) and de Bakker et al (2016). Used with permission. (A black and white version of this figure will appear in some formats. For the color version, please refer to the plate section.)

hunching, creating the chin-in-the chest posture visible in CS12–18.

The neural crest may have sounded like nothing special when it was introduced – just a ridge next to the forming neural tube – but those cells have already begun to change into neural tissue; neural crest cells will ultimately become the brain, the spinal cord, and the eyes – yes, your eyes are actually part of your brain.

On day 26, CS12, neural crest cells on the cranial end develop into eye spots, but nothing more than spots for the moment. Just below the developing head on the left and right side, a pair of small buds emerge: future arms. Even with these buds, the embryo still looks more like a seahorse than a primate. A tail is still visible on the caudal side. Just after the arm buds start, leg buds begin to form just above the "tail."

In CS13, days 28–32, a primitive mouth and slits that will become nostrils form. The embryo is still tiny, the size of a pea, yet in less than one month this tiny being has already gone through the incredible events of gastrulation and folding; the heart is beating.

Beginning slowly at about CS14, the tissue in the mesoderm begins to differentiate and coalesce into the cartilaginous precursors of bones. In CS14 to 15 a tail is still prominent, but between CS16 and 17 the rest of the thorax is beginning to catch up with it and surround it.

11.11 Week 5 and Beyond

By day 45 (CS18), ears are apparent and finger and toe buds have appeared on what had looked more like paddles than hands and feet. The fingers form in these disk-like paddles by cell death (apoptosis); the tissue between the fingers commits suicide and bones coalesce in the fingers. The embryo now has, at CS18, the distinct appearance of a mammal, though the head is still hunched over at a grotesque angle.

Each of these processes of differentiation continues incrementally until in another two weeks, by day 60, CS23, the embryo has developed all the hallmarks of humanity: hands, a human-like as opposed to a generalized primate foot, and an immense human-like head nearly as big as the rest of the body. In two short months the human embryo has gone from a single cell to a form we can recognize as human, even if it is still tiny and primitive – it is only an inch long. Not only growth, but differentiation will continue for the next seven months, but mostly growth; all the drama is in the first two months. Beyond CS23 the developing embryo is usually referred to as a fetus.

11.12 Evolution and Developmental Biology

You learned above that the fetus and newborn hold important information about the evolutionary history of humans and chimpanzees. At four months of fetal development, apes and humans resemble one another more than they will at birth, and they resemble one another more at birth than they will as adults (Figure 11.9; Richardson, 1999). The tendency of embryos and fetuses to resemble ancestors was immortalized by Ernst Haeckel (1866) in the phrase "ontogeny recapitulates phylogeny." If that flew past you on first pass, he meant that we see the evolutionary history (phylogeny) of an organism repeated (recapitulated) in its fetal development (ontogeny). This rule drew more controversy than it deserved in the 1980s and 1990s, mostly because some of Haeckel's advocates pushed the theory beyond what Haeckel had envisioned. Haeckel's rule (or law if you feel strongly enough about it) has partly regained its place in the biologist's toolkit with the emergence of evo-devo. However, even though embryos famously look quite like one another at some point, especially close relatives, the pattern is not a

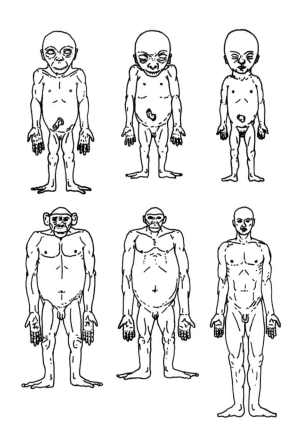

Figure 11.9 Four-month-old fetuses (top row) and adults (bottom), from left to right: chimpanzee, gorilla, and human. To my eye, the human fetus resembles the gorilla more than the chimpanzee. After Schultz, 1926; with permission of the Adolph H. Schultz Foundation.

simple one. If we look broadly across animals, early in development embryos may show a wide range of shapes (Sander, 1983; Slack et al., 1993; Richardson, 1995; Raff, 1996; Hall, 1997); however, midway through embryonic development morphology converges on a rather common form – a highly conserved stage, strongly resembling ancestral forms – then specializations begin to appear. Perhaps the most celebrated expression of this is that early in embryonic development, as we have already seen, humans have tails, though the growth of the rest of the body very soon outstrips this structure and it is shifted into the body to become the coccyx (tail bone).

Compared to other apes, adult humans and gorillas have short forearms and short fingers. During fetal development, gorilla and human fingers are never disproportionately long, while chimpanzee fingers are slightly elongated at four months and get proportionately longer as gestation proceeds (Figures 11.9 and 11.10). Humans display the common condition of ape embryos; as Schultz (1925) notes, "since an extreme length of the forearm, exceeding that of the upper arm [is found only in the chimpanzee] ... it can be concluded that man and gorilla are conservative," or more like the common ancestor of apes and humans, while the long chimpanzee fingers evolved after the human–chimpanzee split.

Consider the implications of this for the course of chimpanzee evolution. If gorillas and humans have arms and legs of roughly equal length and shorter fingers than chimpanzees, this suggests that the common ancestor of gorillas/humans/chimpanzees was less suspensory than chimpanzees, and chimpanzees evolved some adaptations for armhanging after they branched from humans, late in their evolution.

To my mind this strengthens the case for the common ancestors of chimpanzees and humans being *Ouranopithecus*-like (see Chapter 10), since this fossil has a gorilla-like face and a gorilla-like level of male–female body weight difference. It also fits well with genetic evidence that the human/chimpanzee ancestor slept on the ground, as do gorillas. If the idea that genetic evidence could suggest whether an ancestor slept on the ground seems surprising, your curiosity will be rewarded in the next chapter.

If you look at your hand, you will notice that the third digit, the middle finger, is longest. In primates that have a gripping foot, including chimpanzees, toe proportions are more similar to our hand proportions: the middle (third) toe is longest. This is so in human embryos as well, though only for a brief time (Schultz, 1925). Early in fetal life, after 2–3 months of gestation, the human fetus still has a chimpanzee- or gorilla-like gripping great toe (Figure 11.11). The root of the big toe, the branching point between the gripping great toe and the other toes, is quite deep in apes, giving them the capacity to grip large branches.

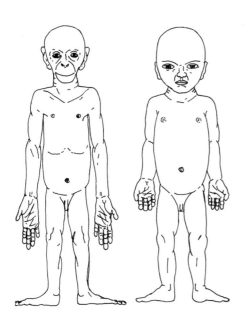

Figure 11.10 Newborn chimpanzee and human. Note that late in gestation (compare to Figure 11.9) chimpanzee arms and fingers grow disproportionately. After Schultz, 1925; with permission of the Adolph H. Schultz Foundation.

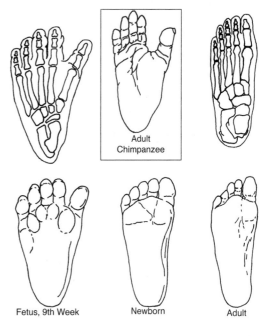

Figure 11.11 Fetal human foot at 9 weeks of gestation (top left: view of skeleton; bottom left external morphology) versus newborn (center bottom) and adult (right). The fetal divergent big toe and narrow heel resembles the foot of the adult chimpanzee (top center). After Schultz, 1925; with permission of the Adolph H. Schultz Foundation.

This deep branching point is acquired relatively late in embryonic development, suggesting that the ape ancestor had less of a gripping toe than living apes (Schultz, 1925); vertical climbing evolved late. Compared to the rest of the foot, the lateral toes, digits 2–5, are relatively short in all primate early embryos, suggesting that long toes are a primate specialization, not found in the pre-primate ancestor. In humans, relative toe length shortens as the embryo develops, while it lengthens in other primates.

The human brain follows a similar primitive-to-advanced path during gestation, from ape-like to human-like. A part of the frontal lobe of the brain, the orbitofrontal cortex, differs dramatically in size between adult humans and chimpanzees. In humans, the lateral orbitofrontal cortex is so large that it overlaps and obscures a part of the brain called the **insula** (highlighted with vertical white lines in Figure 11.12). In chimpanzees the insula is visible even in adults because the orbitofrontal cortex is smaller. Until the 28th week of gestation the frontal lobes of humans and chimpanzees appear quite similar, with the insula visible in both (Cunningham, 1892; Hurst, 2016); the orbitofrontal cortex (enclosed in white in Figure 11.12) is still small, chimpanzee-sized. Between 30 and 40 weeks, however, the human orbitofrontal cortex dramatically expands, gradually overlapping (for you experts, **operculating**) the insula until at 42 weeks the insula is completely obscured. While most of the growth that causes this overlapping is from frontal expansion, there is some contribution from parietal and temporal lobes, too (Figure 11.12).

PART II: LIFE HISTORY

Chimpanzee gestation lasts 33 weeks (228 days in Nissen & Yerkes [1943] and 232 in Dahl [1999]), compared to the 38 weeks (270 days) of humans, 85 percent as long. Our life histories are clearly quite similar, but keep in mind that this particularly close similarity is a little deceptive. It belies a huge difference in maturity at birth. Humans are born prematurely compared to the rest of the primates. If we were born at the same developmental stage as chimpanzees, based on something like brain size or dental development, human gestation "should" last 18 months. Evolution has altered one of the most important life history stages, gestation length, to accommodate perhaps the most distinctive human characteristic: our large brains. Passing the

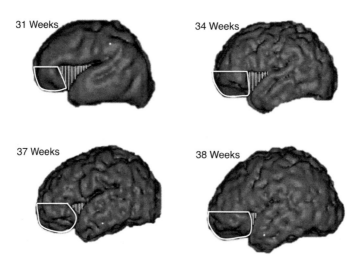

Figure 11.12 Side view of the left surface of the brain, front side on the left. In chimpanzees, a part of the brain called the insula (hatched area) is visible on the surface of the brain. This area is hidden in humans by the overlapping orbitofrontal cortex and, less so, the parietal cortex. In the course of human fetal brain development, the orbitofrontal cortex (outlined) expands to cover the insula (hatched). After Ellen Grant; reproduced with permission.

human neonatal head through the birth canal (see Figure 7.24) is so difficult that birth is a dangerous proposition for humans. In the poorest parts of the world one in six women dies in birth (Ronsmans & Graham, 2006), and birth is still difficult and sometimes fatal even with modern medical knowledge and technology.

That our gestation should "really" last much longer can be seen by the fact that our brain continues to grow at fetal rates for an entire year after birth, doubling in size by the end of that year (Leutenegger, 1973), whereas chimpanzee brain growth slows immediately after birth. Chimpanzees possess 40.5 percent of their adult brain size at birth; humans only 23 percent.

This gestational shift is important, but it is not the only phenomenon that makes infant chimpanzees seem more human-like than adults. As we learned early in this chapter, humans are, to some extent, neotenized; we will discuss this in more detail below.

The effect of early human birth on life history is apparent when we compare landmarks throughout the lifespans of humans and chimpanzees. After their similar gestation length of 84 percent that of humans, chimpanzees eventually settle into a pattern of reaching life stages in about 75 percent of the time of humans (Table 11.2). Considering that chimpanzees body weight is 65 percent that of humans, and smaller animals mature more quickly, the similarity is great, even if less than the gestational comparison.

Pregnancy is a burden, but more so for humans than for chimpanzees. Newborn chimpanzees weigh 4.3 percent of their mother's weight (Gavan, 1953) versus 5.5 percent for humans (Schultz, 1940). At birth, a chimpanzee weighs 1.8 kg (4 lb) on average, versus 3 kg (6.5 lb) for humans. This proportion differs mostly because human babies are quite fat at birth, possibly because humans require an energy buffer to protect our larger, faster-growing brains (Cunnane & Crawford, 2003) – another burden the mother must bear. However, chimpanzees grow faster (Figures 11.13), so that by the time chimpanzees are four years old they weigh more than a four-year-old human (Gavan, 1953). This reflects both the faster chimpanzee life history and the loss of baby fat in humans. Chimpanzee body weight changes little after the age of 13 (females) or 14 (males; Gavan, 1953; Leigh & Shea, 1995), whereas humans put on weight

Table 11.2 **Chimpanzee versus human life history**[a]

	Time (years)	
Life event	Chimpanzee	Human
Gestation	0.63	0.75
Begin walking	0.42	1.2
Gestation + walk	1.05	1.95
M1 eruption	3.3	6.2
M3 eruption	11.5	18
Adult weight	13.5	18.5
Reproduction	14.5	19
Senescence	33	45

[a] For wild chimpanzees and human hunter-gatherers. Abstracted from Schultz, 1925, 1926; Gavan, 1953; Graham, 1979; Tutin & McGinnis, 1981; Goodall, 1986; Smith, 1989, 1991; Leigh & Shea, 1995; Hamada et al., 1996; Nishida et al., 2003; Zihlman et al., 2004; Nishida, 2012; Smith et al., 2013 for chimpanzees. Taranger & Hägg, 1980; Forssberg, 1985; Bogin & Smith, 1996; Hill & Kaplan, 1999; Leigh, 2004; Mitteroecker et al., 2004; Walker et al., 2006 for humans.

for several more years, with hunter-gatherers reaching adult body weights at 18 and 19 (Walker et al., 2006). First reproduction is at 14.5 for chimpanzees versus 19 for humans, or 76.3 percent as long (Tutin & McGinnis, 1981; Hill & Kaplan, 1999). Perhaps first reproduction and attainment of adult body weight are the best life history markers; note that both take about 75 percent of the time in chimpanzees as in humans.

PART III: GROWTH AND DEVELOPMENT

11.13 Chimpanzee Infancy (Birth to Five Years)

Baby chimpanzees are playful, curious, and for the most part carefree. And infancy lasts a long

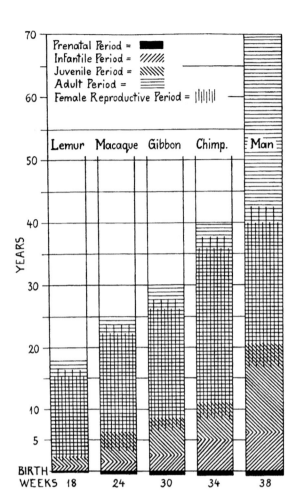

Figure 11.13 Primate life history events. Unlike other primates, humans live long after their reproductive career is over. After Schultz, 1969; infancy for chimpanzee updated following Goodall, 1986.

time – five years (Goodall, 1986). During this extended babyhood, the mother (see Chapter 16 for more on mothering) takes care of most of a baby's needs, including their nutritional and "housing" requirements; the baby sleeps in the mother's nest each night, a bed in the sky.

Infancy ends at weaning, and sometimes not placidly – infants often protest when the mother cuts off the milk supply.

As with humans, chimpanzee infants are relatively helpless at birth and seem hardly aware of their surroundings (Goodall, 1986), requiring the mother's steadying hand for some time, perhaps a month or two. Alertness and coordination, however, increase much more quickly for chimpanzees than for humans (Kellogg & Kellogg, 1933; Goodall, 1986). In little more than a week, some precocious infants cling to their mother on their own.

Chimpanzees begin to walk without the mother's help at five months (Doran, 1992), versus an average of a year and two months (Forssberg, 1985) for humans; a 35 percent proportion. They start climbing and armhanging early in infancy, though much of their early movement consists of climbing on the mother (Doran, 1992). Brain growth is rapid. By two years old, chimpanzee brains are 85 percent of their adult size, and by four years old the proportion is 94 percent; by six years old brain growth is complete (Schultz, 1940).

Typically, chimpanzee infants begin to ride on their mother's back, jockey-style, between one and two years old, perhaps an expression of their curiosity about the world at large, since they have an impressive view from high on mom's back.

By the age of three, chimpanzee infants have begun to engage in some adult behaviors. They provide an important part of their own energy requirements, even though nursing will continue for two more years. They gather food almost as a play behavior, copying what mother does. Three-year-olds begin to interact more with other individuals in the social group, even remaining in a feeding tree with playmates as their mother moves on, following her later. This relative independence means that even though the infant is not yet weaned, a three-year-old orphan has some prospect of survival. This was the age of the Gombe infant Mel when he was orphaned (see Chapter 16), the youngest known chimpanzee orphan to survive. From three to five years old independence increases, but the infant will continue to ride on the mother's back until weaning, particularly when travel distances are great.

11.14 First Molar Eruption

For both humans and chimpanzees, the first permanent tooth to erupt is the first molar (Table 11.2). The eruption of this tooth is viewed as a pivotal life history milestone by growth specialists, not only because it is the first permanent tooth, but because its emergence falls close to the completion of brain growth, coming at 90 percent of adult brain size. Chimpanzee M1s erupt at the age of 3.3, or 53.2 percent of the 6.2 years it takes humans (Nissen & Riesen, 1964; Smith, 1989; Bogin & Smith, 1996; Zihlman et al., 2004; Smith et al., 2013). Although mammals are typically weaned right around the time of M1 eruption (Smith, 1991), chimpanzees will continue to nurse for over a year and a half.

Consider what the extended period of nursing among chimpanzees means for humans; we humans would continue to nurse up to the age of eight and a half if we weaned at the same developmental stage as chimpanzees (Dettwyler, 1995). Some theorists see early weaning among humans as perhaps our most significant adaptation. **Lactational amenorrhea** is the term for infertility due to the stresses of nursing. By weaning babies early, humans shorten this infertile stage, which means the next pregnancy and birth occur before the first infant is independent, an adaptation that some believe gives humans a great advantage over the other apes (Dettwyler, 1995; Hawkes et al., 1998; Hill & Kaplan, 1999; Lancaster & Lancaster, 2010).

Animals grow, as we have seen, at vastly different rates. We started off by noting that horses mature at four, while chimpanzees take four times as long. The brain seems to be the key. While some consider life history variables like attainment of adult body mass or first birth as relatively free to vary over evolutionary time, the last few decades have seen a consensus coalescing around the idea that the caloric expense of the brain imposes very strict limits on growth rate, and this limitation in turn determines the set points for other life history variables. The brain, in other words, is the pacemaker of growth (Sacher & Staffeldt, 1974; Harvey & Clutton-Brock, 1985; Smith, 1989; Ulijaszek, 2002). Chimpanzees achieve their adult brain size almost exactly at the age of weaning; this is no coincidence.

11.15 Juvenile Period (5–7 Years Old)

As nursing winds down and weaning progresses, a female typically becomes fertile again. After spending years having mother to herself, the now-weaned infant finds herself tossed into a frenetic throng of large, aggressive males, jostling for the chance to mate with her mother. It is no wonder that juveniles sometimes attempt to interfere, leaping on a copulating male's back or pulling at the mother (Goodall, 1986); after five years of attentiveness, the mother has now directed her attention elsewhere. Childhood's end.

The juvenile period marks a watershed, and its hallmark is independence. While humans remain dependent on their parents and others for nutrition more or less right up until adulthood, once infancy has passed the juvenile chimpanzee meets all her own nutritional needs. She must now walk, gather food, and make her sleeping nest on her own. Soon her mother will be showering her affection on a new infant. Still, the well-adjusted brother or sister seems to enjoy the younger sibling, playing with the new sibling and playing at the role of mother.

While weaning marks physical independence, juveniles are still quite emotionally attached to the mother, leaving her only for brief periods of time, and mostly when reconnecting is easy. The association will persist for years to come.

11.16 Female Early Adolescence (8–10 Years Old)

As we reach the stage of early adolescence, males and females begin to follow different developmental tracks. Adolescent females continue to travel with their mothers, even when the daughter begins to manifest a sexual swelling. Estrus swellings start small, but sexual interest in males increases rapidly, though that interest is only rarely reciprocated. Adult males, in other words, show little sexual interest in adolescents. Indeed, it will be well into adulthood, after she has already borne and reared an infant, before most adult males will find her truly sexually attractive (Muller et al., 2006).

11.17 Male Early Adolescence (8–12 Years Old)

In early adolescence, males begin to show what some have described as fascination or even hero-worship of adult males. In early adolescence, males watch their hero's every move, mimicking his behavior and following him for hours at a time, sometimes leaving mother for a day or more. Gombe regulars quipped that adolescent Goblin was studying at the "university of Figan," the dominant male at the time.

In early adolescence, males embark on what will be a lifetime of obsession with social dominance. While they are far too young to compete with adult males, they satisfy their need to dominate by focusing on an easier target; they begin to dominate females at this age. An ambitious young male will succeed in dominating some of the lowest-ranking, most timid females even in early adolescence, at eight or nine, even though they are still quite small.

11.18 Female Late Adolescence (11–14 Years Old)

Sexual swellings increase to adult size in late adolescence, and to the delight of the maturing female she will find that some adult males begin to show sexual interest in her.

Whereas in early adolescence females stay close enough to their mother that males are encountered only accidentally, when they arrive at the same feeding tree or cross paths on heavily traveled animal tracks, things change as she reaches late adolescence. She becomes bolder when in estrus and leaves mother to actively seek out male parties, and then rather than engaging in a quick greeting and an equally brief goodbye, as would have been the case in early adolescence, she may tag along behind males for days, making her nest near them at night.

11.19 Group Transfer in Late Adolescence

As late adolescence progresses, a fateful and life-altering decision becomes more and more urgent (Pusey, 1979; Goodall, 1986), and females will follow one of two paths. A minority of females will continue to follow community males until she gets pregnant, around the age of 13 or so (Nishida et al., 2003). Most females, however, will leave their birth community, abandoning everything and everyone they have ever known. It must be a harrowing time.

Females who stay in their birth community will settle into a core area close to that of their mother, especially if she is high-ranking; these females have the benefit of their high-ranking mother's help as they compete for their core areas, increasing the likelihood of claiming a core area near the more desirable center of the community range. The downside for these stay-at-home females is that as they look around for an attractive mate they find a surplus of unappealing uncles, cousins, brothers, and fathers. Females resist copulatory overtures by close relatives (Pusey, 1980, 1990; Walker et al., 2017), and one case study suggests there is a cost if they fail – an inbred, sickly, scrawny offspring (Chapter 27). Because the costs of inbreeding are high, a female who cannot hope to inherit a desirable core area from her mother is likely to transfer.

A typical female will begin looking toward the horizon around the age of 11 (Nishida et al., 2003), long before she will become pregnant. She may hear the pant-hoots of strange males off in the hinterlands and feel a strange compulsion to travel toward them. Or she may obey the stirring of wanderlust (and other lust) one day when in estrus and strike out on her own toward the edge of her community territory, searching for stranger males. Such males will welcome her, treating her – we may imagine, reading her adolescent mind – like a grown woman of means; what a relief, after being dismissed as inconsequential and unattractive by resident males. She may not make the jump permanently the first time she meets these extracommunity males, instead returning to her birth community as her estrus swelling deflates so as to reunite with her mother. Eventually, though, she will follow the males into their community territory, an undiscovered country where she will spend the rest of her life.

After transferring she will remain with her new male compatriots month after month, in and out of estrus, traveling the length and breadth of her new community territory, all the time learning feeding-tree localities, memorizing geography, and encountering new females. And what of those new-met females? She will not find a welcoming sisterhood in her new home – far from it. She will be ignored, at best, and sometimes physically attacked. When this happens, she will rely on the male friendships she has established as she has wandered her new community; they will take her side in conflicts (Nishida & Hiraiwa-Hasegawa, 1985; Nishida, 1990).

For a year or two after she has transferred into her new community she will continue her estrous cycles. Here we see the advantage of the long, 2.8-year (Nishida et al., 2003) period of adolescent infertility. This long period of traveling with males is important. As she roams her new home, she will use these many months as a vagabond to get a sense of the relative rank of different females, the locations of the best food sources, and chimpanzee trails before she is saddled with the task of eating for two. She will gradually begin to identify a likely spot for her core area. As she gets to know the community females, she will begin to establish her social rank, for good (high) or ill (low).

When she at last becomes pregnant she will abandon the males and settle into as advantageous a core area as she can manage, close to the center of the community range if she is up to it. Without males to defend her she must now fall back on her own skills as a fighter. It will ease the strain if, as often happens, she finds friend among the females in her new community, though strong social bonds are not typical among females. Surely some very few females have the great fortune of encountering a long-lost sister in their new community, or a playmate from childhood. We can only imagine how uplifting this might be.

As foreshadowed above, females typically have their first pregnancy at the end of their late adolescence (Tutin et al., 1979, 1980; Pusey et al., 1997), but so difficult are the physical and cognitive demands of caring for an infant in the wild that many females, most females in fact, will lose their first infant. Half of all infants die before weaning (Nishida et al., 2003) and the risk of loss is proportionally greater for low-ranking females (Pusey et al., 1997). The famous Fifi at Gombe had her first infant at nine; he survived and she had a truly extraordinarily successful reproductive history after that (Pusey et al., 1997). The high mortality rate of first-born infants in particular prompts me to list "first birth" as 14.5 in Table 11.2 because that is the age at which the first infant is likely to live, rather than 13 when the first birth might be expected.

The last permanent tooth, the third molar or wisdom tooth, erupts in only 63.9 percent of the time it takes humans, at 11.5 years for chimpanzees versus 18 in humans (Smith, 1991; Table 11.2). This does not mark adulthood, however. If full adulthood is considered achieving maximum body weight – wild chimpanzees never become fat – adult body mass is realized as early as 14 (Leigh & Shea, 1995) or as late as 16–17 (Goodall, 1986).

11.20 Male Late Adolescence (12–15 years)

As late adolescence proceeds, males slip more and more into adult roles, traveling constantly with other adult males. They complete their mission of dominating all the females and soon begin to contest dominance with the very lowest ranking of the adult males (Goodall, 1986; Nishida, 1990). They will return only now and then to visit their mother and siblings, though if they suffer an injury or setback in their quest for dominance they are likely to seek solace in the bosom of their family.

11.21 Female Adulthood (15–33 Years Old)

For females, adulthood is dominated by infant-rearing (Goodall, 1986). A chimpanzee female may give birth to six or seven infants during her life. While many will achieve adulthood if she is high-ranking (Goodall, 1986; Nishida, 1990; Pusey et al., 1997), the average female loses two infants. For the rest of her life she will come into estrus only when she weans an infant, or when an infant dies. If she is an excellent mother, like Fifi of Gombe, and few or no infants die, she will be in estrus for about four months every five years. Macaques, by comparison, wean infants in a year and will come into estrus in less than two years (Fooden, 2000).

11.22 Male Young Adulthood (16–20 Years Old)

Male chimpanzees may continue to put on muscle until 16 or 17 years old (Goodall, 1986: 84), versus perhaps 20 in humans. In young adulthood males learn important social roles and begin to contest for social dominance seriously with other adult males (Goodall, 1986; Nishida, 1990).

Because courtship is a complicated affair for male chimpanzees (Tutin, 1979; Goodall, 1986; Nishida, 1990), involving multiple strategies (see Chapter 25) and often requiring the ability to stand your ground against highly motivated challengers, young males become more adept at the social graces required to succeed in courting females as young adulthood progresses.

11.23 Male Prime Adulthood (21–26 Years Old)

Males are at the height of their physical strength and stamina in their prime adulthood, though often not the height of their social status.

11.24 Male Middle Age (27–33 Years Old)

Although there is little decline in vigor at this early stage of senescence, by 27 chimpanzees have passed the first flush of youth. However, chimpanzee society is complicated enough that it is usually after their physical prime that males achieve their highest social status. At this life stage they have learned how to cement important alliances, how to counter other male strategies – important tactics when in actual physical combat – when to retreat to fight another day, and how to take advantage of rivals' weaknesses. Typically, the middle-aged male has established a wide network of relationships among his male community members and with a number of adult females as well.

11.25 Old Age (>33 Years Old)

The accumulation of a lifetime of injury and disease takes its toll, so that by age 33 chimpanzees begin to appear old. This life stage is achieved in about 73 percent of the time of humans (33 versus 45 years). Teeth begin to show wear in old age, and not unusually they are worn right down to the gum-line; many begin to fall out. Activity levels decline, including male social display. Older individuals show visible fatigue during long walking bouts. Unable to keep up the pace necessary to remain with the main male feeding party, old males become less social and may begin to avoid the conflicts inherent in social interaction altogether, traveling alone or with a similarly aged friend (Nishida, 1990). Hair begins to appear dull and often turns gray or brown; eventually bald patches appear on the body. As senescence advances, weight loss ensues. The mortality rate spikes after 43, according to one study (Figure 11.14).

While the signs of age are similar in males and females, the effects are different. Solitary males are easy targets for extracommunity males, and a life of competing for dominance and interpersonal combat take a toll; few males make it into their late forties.

Among females, as vigor declines keeping up the demanding foraging schedule necessary to recover from a pregnancy is more difficult. Interbirth intervals increase. It is not unusual for the stresses of motherhood to prove to be too great a challenge for an aged female. A female's last infant is often born after a particularly long interbirth interval, but even so between age and condition she may be unable to nurse and care for the infant and both may die.

11.26 The Role of Age in Competence

By the age of 14 both females and males have achieved an adult competence in foraging skills (Figure 11.14) and they sustain this competence into old age, with productivity gradually dropping off after

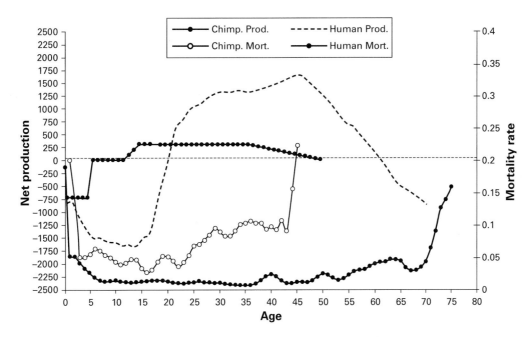

Figure 11.14 Human and chimpanzee foraging productivity. From Kaplan & Robson, 2002. Copyright (2002) National Academy of Sciences, USA. Used with permission.

age 35. Older individuals make up for declining vigor with know-how; they know where to find food and they have mastered the skill of prioritizing resources so that they move efficiently from one feeding spot to another.

Humans, in contrast, have such a complicated foraging regime, with its larger number of food items and its wider-ranging requirements, that they continue to increase their productivity well into adulthood and into early senescence (Figure 11.14), as learning continues to benefit them. Hunter-gathering is a physically challenging job, yet a 50-year-old woman is at her most productive life stage, and at 55 she is as productive as a 20-year-old. Impressive, we oldsters. It is this competence into old age that drives the "grandmother hypothesis" (Hawkes et al., 1998), the idea that humans have lifespans that stretch to long after our reproductive career is over because we confer such great benefits on our offspring and even our offspring's offspring (Jamison et al., 2002).

For years we had not studied chimpanzees enough to say whether they had menopause or not (Graham, 1979), but a recent study suggests they do, and it occurs at about the same time as human menopause, age 50 (Herndon et al., 2012). Fifty, however, falls very, very late in a chimpanzee's life, meaning few individuals will experience menopause, whereas among humans, because of our greater lifespan, menopause is common. This suggests that there is something to the grandmother effect.

I can still remember my first encounter with 50-year-old Wagamuma at Mahale (Nishida, 2012). She showed all the signs of extreme age we see in humans. She was frail, hunched over, and her voice was quieted to a mere whisper. She moved slowly, as if she were in pain; she had large patches of missing hair and often failed to respond to noises when others would turn to investigate. The skin around her watery, bloodshot eyes, whitened with cataracts, was puffy and wrinkled. I was impressed that she was still making a living for herself, but Figure 11.14 shows that she was typical.

11.27 The Pattern of Growth

After growing at a relatively steady rate for our childhood, we humans grow faster just before we reach adulthood; some researchers see this growth spurt as unique in humans (Bogin, 1997), while others suggest it is similar in pattern but merely shorter and less important for chimpanzees (Hamada et al., 1996; Leigh, 2004). Among humans the growth spurt begins at 10 for girls and 12 for boys (Taranger & Hägg, 1980). For captive chimpanzees in Kumamoto Primates Park, males had a growth spurt that peaked at about eight years of age, but females had no spurt at all (Hamada et al., 1996). The spurt is small enough that I have ignored it in Figure 11.15.

The African apes are all sexually dimorphic in body mass: Males are bigger than females. This is achieved differently in different species (Figure 11.15). In gorillas, the lines describing male and female weight gain completely overlap until the age of 9, at which point females cease growing while males continue to do so until 17 or 18; males simply continue to grow along the female growth trajectory. Bonobos do it differently. Males grow faster than females, and although they do continue to grow for a brief time after females are finished, for the most part the difference in body size is achieved by males growing faster (Figure 11.15). Chimpanzee males are different still; they grow faster *and* longer (Figure 11.15);

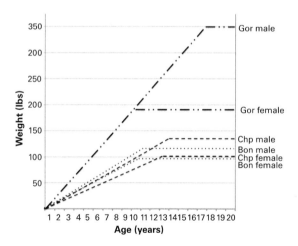

Figure 11.15 Growth in African apes. Gorilla males become large by continuing for eight more years the same growth curve followed by both males and females up to adolescence. Bonobo males are larger mostly because they grow faster. Chimpanzee males grow both faster and for longer. Figure by the author, but see Leigh & Shea, 1995.

females reach adult weight close to 13 years old. Males are already larger at that age, having grown faster, and then they also continue growing for another year.

11.28 Allometric Considerations

As we discussed in Chapter 7, larger organisms are often – almost always, in fact – shaped differently than their smaller relatives, and chimpanzees are no exception. This allometry means that as chimpanzees mature some body parts increase in size disproportionately. In the face, the principal difference is in **face height**, the distance from the incisors to a point between the eyes; this measure increases much more in chimpanzees than in humans. D'arcy Thompson expressed this by showing that you need not deform an adult human from the infant shape as much as you must deform an adult chimpanzee (Figure 11.16). Another way of saying this is that the chimpanzee face grows much more than a human face does. As the face grows, the eye orbits remain the same size (Figure 11.17) even as face height and, to a lesser extent, brow ridge thickness increases. As a result, very large individuals have disproportionately tall faces, thick brow ridges, and small eyes (Figure 11.17; Krogman, 1930; Mitteroecker et al., 2004). The same trend is seen in humans, though with less dramatic differences (Bastir & Rosas, 2004).

11.29 Neoteny

It is widely agreed that the human face is a consequence at least *in part* (Shea, 1989) of neotenization – not changing shape as much as chimpanzees do, but merely growing larger. Because humans are born "prematurely" and the brain continues to grow for an entire year (Deacon 1992, 1997; Leigh, 2004), the large size of the human braincase makes the adult human face more similar to that of an infant (Figure 11.16). Humans, in this view, resemble larger versions of juveniles.

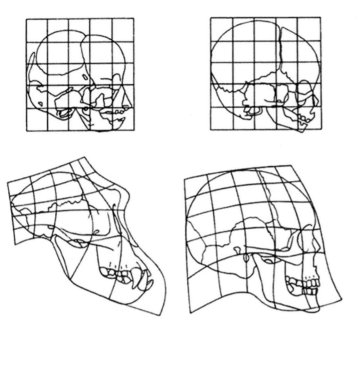

Figure 11.16 Adult chimpanzee deformed from an infant (left) versus infant and adult human. Humans are an infantilized or neotenized species. From Thompson & Wentworth, 1917/1970.

Figure 11.17 The principal allometric shape differences between small and large chimpanzees is height of the face and thickness of the brow ridges. Figure by the author, after Mitteroecker et al., 2004; used with permission.

Even some structures not apparent to the naked eye, such as synaptic spines in the prefrontal cortex, are neotenized (Petanjek et al., 2011). The idea that human morphology is infantilized is corroborated – at least it is considered so by some – by the presence of juvenile behaviors in adult humans, such as the adult human tendency to engage in play (Bjorklund, 1997). Bonobo faces resemble those of juvenile chimpanzees, suggesting that perhaps bonobos, too, evolved in part by neoteny (Wrangham & Pilbeam, 2002).

Infant chimpanzees are hauntingly human-like. Adolf Naef (1926) published a widely reproduced image of a young chimpanzee that had, to some extent, assumed an unnaturally human-like pose, but most of the infant's similarity to humans is real (Figure 11.18).

11.30 Domestication

Evolutionary theorists have long-studied domesticated animals to better understand evolution, and with good reason. Domestication and its consequences are fascinating, and one of them seems to be a tendency to become neotenized, which has led some to characterize human evolution as self-

Figure 11.18 Adolf Naef's (1926) young chimpanzee.

domestication. Domesticated animals are highly evolved versions of their wild relatives; Darwin used varieties of pigeons as an example of how human selection (versus natural selection) could alter animal shape. A 40-year fox-breeding project by Dmitry Belyaev showed us exactly what happens when wild animals are domesticated (Belyaev et al., 1985; Trut et al., 2009). Belyaev had a simple goal: He wanted to breed foxes for their pelts in order to avoid the trouble of trapping wild animals. Caged wild foxes, he found, become so highly stressed it affects their health and ultimately the quality of their pelts; caged animals often self-bite or lose their fur. Wild foxes were also aggressive, making them dangerous to care for. Belyaev selected foxes that seemed tamest for breeding, but what he got was a whole suite of physical changes (Trut et al., 2009), including rolled tails, floppy ears, and spotted or depigmented fur. Other domesticated animals exhibit smaller faces, smaller brains, lower stress responses, and less aggression compared to their wild counterparts. Some scholars have proposed that bonobos look juvenilized because they have evolved under the same selective pressures as those that caused the evolution of domesticated animals (Wrangham & Pilbeam, 2002; Hare et al., 2012). I will relate bonobo evolution to neoteny in greater detail in Chapter 29.

11.31 Summary and Conclusion

While it is common for us humans to place ourselves in one category and all animals in another, chimpanzees challenge that dichotomy. It is not just tool use, complex cognition, language, and politics that challenge the placement of chimpanzees with other animals, but the very pace of life itself. Chimpanzees reach milestones at about 75 percent of the time of that of humans. Some of this difference is due to body size, since chimpanzees weigh only 65 percent of what humans weigh. Brain size plays a role as well. The onerous caloric demands of the brain make growing it slowly so as not to outstrip food resources the best strategy for apes, theoretically causing the pace of growth and development to be set by the demands of building the complicated and expensive brain (Kuzawa et al., 2014). The body paces itself to brain growth.

This tardiness plays out socially as well. Among chimpanzees, adult roles are not achieved until about age 15 – very late compared to most mammals, some of which mature in a couple of months; even some large animals take only several years. While about 15 years of high-quality adult life would be a delight for a rat, it is hardly longer than the length of time it takes to become an adult chimpanzee. Declines associated with aging begin to set in during the early thirties.

This slow life history has dire implications for chimpanzee survival (Dobson & Oli, 2007). Chimpanzees and other apes are sensitive to population crashes; when disease or disaster decimates their numbers, it takes 15 years for an infant born into the now more thinly populated world to begin to have infants of her own. Because population growth is slow, several insults in a row can lead to local extinction. A recent estimate suggests the number of surviving wild chimpanzees is only 300,000 (Oates, 2006), the population of Toledo, Ohio. Perhaps enough to avoid immediate panic, but a worry for all of us in the long term.

Chimpanzee life history, more than any other feature, shows us why their population is only 0.0004 percent of ours. The human ecological niche is an extremely complex one. Chimpanzees reach adult gathering competence even before they stop growing. Humans only begin to approach our maximum productivity 10 years after reaching adulthood, but we keep going, ramping up our gathering haul until we reach 50. Chimpanzees rarely share food with others, but it is a human virtue; we share our wealth with our children, grandchildren, collateral relatives, and friends. It was once fashionable to think that the extended juvenile period of humans was an accommodation to the amount we had to learn. Chimpanzee life history is evidence against the learning hypothesis; chimpanzees grow slowly even though they learn how to gather effectively even before they are grown. We now believe it is more likely that we grow slowly because it takes a long time to build the immense human brain.

References

Bastir M, Rosas A (2004) Facial heights: evolutionary relevance of postnatal ontogeny for facial orientation and skull morphology in humans and chimpanzees. *J Hum Evol* **47**, 359–381.

Bateson W (1894) *Materials for the Study of Variation Treated with Especial Regards to Discontinuity in the Origin of Species*. London: Macmillan.

Belyaev DK, Plyusnina IZ, Trut LN (1985) Domestication in the silver fox (*Vulpes fulvus* Desm): changes in physiological boundaries of the sensitive period of primary socialization. *Appl Animal Behav Sci* **13**, 359–370.

Bianconi E, Piovesan A, Facchin F, et al. (2013) An estimation of the number of cells in the human body. *Ann Hum Biol* **40**, 463–471.

Bjorklund DF (1997) The role of immaturity in human development. *Psychol Bull* **122**, 153.

Bogin B (1997) Evolutionary hypotheses for human childhood. *Ybk Phys Anthropol* **40**, 63–90.

Bogin B, Smith BH (1996) Evolution of the human life cycle. *Am J Hum Biol* **8**, 703–716.

Carroll SB (2005) *Endless Forms Most Beautiful: The New Science of Evo-Devo*. New York: W.W. Norton & Co.

Carroll SB (2008) Evo-devo and an expanding evolutionary synthesis: a genetic theory of morphological evolution. *Cell* **134**, 25–36.

Cunnane SC, Crawford MA (2003) Survival of the fattest: fat babies were the key to evolution of the large human brain. *Comp Biochem Physiol A Mol Integr Physiol* **136**, 17–26.

Cunningham DJ (1892) *Contribution to the Surface Anatomy of the Cerebral Hemispheres*. Dublin: Royal Irish Academy of Science.

Dahl JF (1999) Perineal swelling during pregnancy in common chimpanzees and puerperal pathology. *J Med Primatol* **28**, 129–141.

de Bakker BS, de Jong KH, Hagoort J, et al. (2016) An interactive three-dimensional digital atlas and quantitative database of human development. *Science* **354**, aag0053.

Deacon T (1992) Impressions of ancestral brains. In *The Cambridge Encyclopedia of Human Evolution* (eds. Jones S, Martin R, Pilbeam D), pp. 116–117. Cambridge: Cambridge University Press.

Deacon T (1997) What makes the human brain different? *Ann Rev Anthropol* **26**, 337–357.

Dettwyler KA (1995) A time to wean: the hominid blueprint for the natural age of weaning in modern human populations. In *Breastfeeding: Biocultural Perspectives* (ed. Stuart-Macadam P), pp. 39–73. New York: Routledge.

Dobson FS, Oli MK (2007) Fast and slow life histories of mammals. *Ecoscience* **14**, 292–297.

Doran DM (1992) The ontogeny of chimpanzee and pygmy chimpanzee locomotor behavior: a case study of paedomorphism and its behavioral correlates. *J Hum Evol* **23**, 139–157.

Fooden J (2000) Systematic review of the rhesus macaque, *Macaca mulatta* (Zimmermann, 1780). *Field Zool* **96**, 1–180.

Forssberg HH (1985) Ontogeny of human locomotor control: I. Infant stepping, supported locomotion and transition to independent locomotion. *J Exp Brain Res* 57, 480–493.

Gavan JA (1953) Growth and development of the chimpanzee; a longitudinal and comparative study. *Human Biol* 25, 93–143.

Goodall J (1986) *The Chimpanzees of Gombe: Patterns of Behavior*. Cambridge, MA: Harvard University Press.

Graham CE (1979) Reproductive function in aged female chimpanzees. *Am J Phys Anthropol* 50, 291–300.

Haeckel E (1866) *Generelle Morphologie der Organismen*. Berlin: Georg Reimer.

Hall BK (1997) Phylotypic stage or phantom: is there a highly conserved embryonic stage in vertebrates? *Trends Ecol Evol* 12, 461–463.

Hamada Y, Udono T, Teramoto M, Sugawara T (1996) The growth pattern of chimpanzees: somatic growth and reproductive maturation in *Pan troglodytes*. *Primates* 37, 279–295.

Hare B, Wobber V, Wrangham RW (2012) The self-domestication hypothesis: evolution of bonobo psychology is due to selection against aggression. *Animal Behav* 83, 573–585.

Harvey PH, Clutton-Brock TH (1985) Life history variation in primates. *Evolution* 39, 559–581.

Hawkes K, O'Connell JF, Jones NB, Alvarez H, Charnov EL (1998) Grandmothering, menopause, and the evolution of human life histories. *PNAS* 95, 1336–1339.

Herndon JG, Paredes J, Wilson ME, et al. (2012) Menopause occurs late in life in the captive chimpanzee (*Pan troglodytes*) *Age* 34, 1145–1156.

Hill K, Kaplan H (1999) Life history traits in humans: theory and empirical studies. *Ann Rev Anthropol* 28, 397–430.

Hughes CL, Kaufman TC (2002) Exploring the myriapod body plan: expression patterns of the ten Hox genes in a centipede. *Development* 129, 1225–1238.

Hurst SD (2016) *Emotional evolution in the frontal lobes: social affect and lateral orbitofrontal cortex morphology in hominoids*. PhD dissertation. Bloomington, IN: Indiana University.

Jacob F (1974) *The Logic of Life: A History of Heredity*. New York: Pantheon.

Jamison CS, Cornell LL, Jamison PL (2002) Are all grandmothers equal? A review and a preliminary test of the "Grandmother hypothesis" in Tokugawa Japan. *Am J Phys Anthropol* 119, 67–76.

Kaplan HS, Robson AJ (2002) The emergence of humans: the coevolution of intelligence and longevity with intergenerational transfers. *PNAS* 99, 10221–10226.

Kellogg WN, Kellogg, LA (1933) *The Ape and the Child: A Study of Environmental Influence upon Early Behavior*. New York: McGraw-Hill.

King M-C, Wilson AC (1975) Evolution at two levels in humans and chimpanzees. *Science* 188, 107–116.

Krogman W (1930) Studies of growth changes in the skull and face of anthropoids. *Am J Anat* 46, 303–313.

Kuzawa CW, Chugani HT, Grossman LI, et al. (2014) Metabolic costs and evolutionary implications of human brain development. *PNAS* 111, 13010–13015.

Lancaster JB, Lancaster CS (2010) The watershed: change in parental-investment and family-formation strategies in the course of human evolution. In *Parenting Across the Life Span: Biosocial Dimensions* (eds. Lancaster JB, Altmann J, Sherrod LR, Rossi A), pp. 187–205. New York: Aldine-Transaction.

Leigh SR (2004) Brain growth, life history, and cognition in primate and human evolution. *Am J Primatol* 62, 139–164.

Leigh SR, Shea BT (1995) Ontogeny and the evolution of adult body size dimorphism in apes. *Am J Primatol* 36, 37–60.

Leutenegger W (1973) Maternal–fetal weight relationships in primates. *Folia Primatol* 20, 280–293.

McGinnis W, Levine MS, Hafen E, Kuroiwa A, Gehring WJ (1984) A conserved DNA sequence in homoeotic genes of the *Drosophila* Antennapedia and bithorax complexes. *Nature* 308, 428–433.

Mitteroecker P, Gunz P, Bernhard M, Schaefer K, Bookstein FL (2004) Comparison of cranial ontogenetic trajectories among great apes and humans. *J Hum Evol* 46, 679–698.

Monod J (1971) *Chance and Necessity: An Essay on the Natural Philosophy of Modern Biology*. New York: Alfred A Knopf.

Motosugi N, Bauer T, Polanski Z, Solter D, Hiiragi T (2005) Polarity of the mouse embryo is established at blastocyst and is not prepatterned. *Genes Dev* 19, 1081–1092.

Muller MN, Thompson ME, Wrangham RW (2006) Male chimpanzees prefer mating with old females. *Curr Biol* 16, 2234–2238.

Naef A (1926) The prototype of anthropomorphen and the phylogeny of human impairment. *Natürwissenschaften* 14, 472–477.

Nishida T (1990) *Chimpanzees of the Mahale Mountains*. Tokyo: University of Tokyo Press.

Nishida T (2012) *Chimpanzees of the Lakeshore: Natural History and Culture at Mahale*. Cambridge: Cambridge University Press.

Nishida T, Hiraiwa-Hasegawa M (1985) Responses to a stranger mother–son pair in the wild chimpanzee: a case report. *Primates* 26, 1–13.

Nishida T, Corp N, Hamai M, et al. (2003) Demography, female life history, and reproductive profiles among the chimpanzees of Mahale. *Am J Primatol* 59, 99–121.

Nissen HW, Riesen AH (1964) The eruption of the permanent dentition of chimpanzees. *Am J Phys Anth* 22, 285–294.

Nissen HW, Yerkes RM (1943) Reproduction in the chimpanzee: report on forty-nine births. *Anat Rec* 86, 567–578.

Oates JF (2006) Is the chimpanzee, *Pan troglodytes*, an endangered species? It depends on what "Endangered" means. *Primates* **47**, 102–112.

O'Rahilly R, Müller F (1987) *Developmental Stages in Human Embryos*. Washington, DC: Carnegie Institution.

Penin X, Berge C, Baylac M (2002) Ontogenetic study of the skull in modern humans and the common chimpanzees: neotenic hypothesis reconsidered with a tridimensional Procrustes analysis. *Am J Phys Anthropol* **118**, 50–62.

Petanjek Z, Judaš M, Šimić G, et al. (2011) Extraordinary neoteny of synaptic spines in the human prefrontal cortex. *PNAS* **108**, 13281–13286.

Pusey A (1979) Intercommunity transfer of chimpanzees in Gombe National Park. In *The Great Apes* (eds. Hamburg DA, McCown ER), pp. 465–479. Menlo Park, CA: Benjamin Cummings.

Pusey A (1980) Inbreeding avoidance in chimpanzees. *Anim Behav* **28**, 543–582.

Pusey A (1990) Mechanisms of inbreeding avoidance in nonhuman primates. In *Pedophilia* (ed. Feierman JR), pp. 201–220. New York: Springer.

Pusey A, Williams J, Goodall J (1997) The influence of dominance rank on the reproductive success of female chimpanzees. *Science* **277**, 828–831.

Raff RA (1996) *The Shape of Life: Genes, Development, and the Evolution of Animal Form*. Chicago, IL: University of Chicago Press.

Raff RA (2000) Evo-devo: the evolution of a new discipline. *Nature Rev Genet* **1**, 74–79.

Raff RA, Kauffman TC (1983) *Embryos, Genes and Evolution*. New York: MacMillan.

Richardson MK (1995) Heterochrony and the phylotypic period. *Dev Biol* **172**, 412–421.

Richardson MK (1999) Vertebrate evolution: the developmental origins of adult variation. *Bioessays* **21**, 604–613.

Ronsmans C, Graham WJ (2006) Maternal mortality: who, when, where, and why. *Lancet* **368**, 1189–1200.

Sacher GA, Staffeldt EF (1974) Relation of gestation time to brain weight for placental mammals. *Am Nat* **108**, 593–616.

Sander K (1983) The evolution of patterning mechanisms: gleanings from insect embryogenesis and spermatogenesis. In *Development and Evolution* (eds. Goodwin BC, Holder N, Wylie CC), pp. 124–137. Cambridge: Cambridge University Press.

Sandra A, Coons WJ (1997) *Core Concepts in Embryology*. Philadelphia, PA: Lippincott-Raven.

Schoenwolf GC, Bleyl SB, Brauer PR, Francis-West PH (2014) *Larsen's Human Embryology*, 5th ed. Amsterdam: Elsevier Health Sciences.

Schultz AH (1925) Embryological evidence of the evolution of man. *J Washington Academy of Sciences* **15**, 247–263.

Schultz AH (1926) Fetal growth of man and other primates *Quarterly Rev Biol* **1**, 465–521.

Schultz AH (1940) Growth and development of the orang-utan. *Contrib Embryol* **28**, 57–110.

Schultz AH (1969) *The Life of Primates*. London: Weidenfeld & Nicolson.

Scott MP, Weiner AJ (1984) Structural relationships among genes that control development: sequence homology between the Antennapedia, Ultrabithorax, and fushi tarazu loci of *Drosophila*. *PNAS* **81**, 4115–4119.

Shea BT (1984) An allometric perspective on the morphological and evolutionary relationships between pygmy (*Pan paniscus*) and common (*Pan troglodytes*) chimpanzees. In *The Pygmy Chimpanzee: Evolutionary Biology and Behavior* (ed. Susman R), pp. 89–130. New York: Plenum.

Shea BT (1989) Heterochrony in human evolution: the case for neoteny reconsidered. *Am J Phys Anthropol* **32**, 69–101.

Slack JM, Holland PW, Graham CF (1993) The zootype and the phylotypic stage. *Nature* **361**, 490–492.

Smith BH (1989) Dental development as a measure of life history in primates. *Evolution* **43**, 683–688.

Smith BH (1991) Dental development and the evolution of life history in Hominidae. *Am J Phys Anthropol* **86**, 157–174.

Smith TM, Machanda Z, Bernard AB, et al. (2013) First molar eruption, weaning, and life history in living wild chimpanzees. *PNAS* **110**, 2787–2791.

Streeter GL (1942) *Developmental Horizons in Human Embryos: Description of Age Group XI, 13 to 20 Somites, and Age Group XII, 21 to 29 Somites*. Washington, DC: Carnegie Institution.

Taranger J, Hägg U (1980) The timing and duration of adolescent growth. *Acta Odontologica Scandinavica* **38**, 57–67.

Thompson DW, Wentworth A (1917/1970) *On Growth and Form*. Cambridge: Cambridge University Press.

Trut L, Oskina I, Kharlamova A (2009) Animal evolution during domestication: the domesticated fox as a model. *Bioessays* **31**, 349–360.

Tutin CEG (1979) Mating patterns and reproductive strategies in a community of wild chimpanzees (*Pan troglodytes schweinfurthii*). *Behav Ecol Sociobiol* **6**, 29–38.

Tutin CEG (1980) Reproductive behavior of wild chimpanzees in the Gombe National Park, Tanzania. *J Reprod Fert* (Suppl) **28**, 43–57.

Tutin CEG, McGinnis PR (1981) Chimpanzee reproduction in the wild. In *Reproductive Biology of the Great Apes* (ed. Graham CE), pp. 239–264. New York: Academic Press.

Ulijaszek, SJ (2002) Comparative energetics of primate fetal growth. *Am J Hum Biol* **14**, 603–608.

Wakimoto BT, Kaufman TC (1981) Analysis of larval segmentation in lethal genotypes associated with the Antennapedia gene complex in *Drosophila melanogaster*. *Dev Biol* **81**, 51–64.

Walker KK, Rudicell RS, Li Y, et al. (2017) Chimpanzees breed with genetically dissimilar mates. *R Soc Open Sci* **4**, 160422.

Walker R, Gurven M, Hill K, et al. (2006) Growth rates and life histories in twenty-two small-scale societies. *Am J Hum Biol* **18**, 295–311.

Wrangham R, Pilbeam D (2002) African apes as time machines. In *All Apes Great and Small: Developments in Primatology – Progress and Prospects* (eds. Galdikas BMF, Briggs NE, Sheeran LK, Shapiro GL, Goodall J), pp. 5–17. Boston, MA: Springer.

Yamada S, Hill M, Takakuwa T (2015) *Human Embryology*. London: InTech.

Zihlman A, Bolter D, Boesch C (2004) Wild chimpanzee dentition and its implications for assessing life history in immature hominin fossils. *PNAS* **101**, 10541–10543.

12 The Source of Similarity
Chimpanzee Genetics

Credit: Andrew Brookes / Cultura / Getty Images

With the possible exception of natural selection, the discovery of DNA and its function must be the greatest breakthrough in the biological sciences. Genetics is increasingly a part of all our lives as DNA sequencing allows more targeted medical interventions and as we anticipate gene therapy permanently eliminating some genetic diseases. Bringing it closer to home, some diseases are the result of the afflicted person possessing a chimpanzee version of a gene. Humans and chimpanzees differ because their DNA differs. We have known the basics for over half a century, and now at last we can point to specific genes when we discuss human/chimpanzee differences. While your grandparents can probably remember the very first time they heard the term "deoxyribonucleic acid" and puzzled over the mysterious workings of this newly discovered molecule, you probably learned that a **gene** is a length of DNA that codes for a specific protein before you got out of primary school. How long before we have a short list of the genes that make us human, rather than chimpanzee?

The fields of study that explain the microscopic processes that turn a sequence of base pairs in DNA into a living thing are called **genetics** or **molecular biology**. Molecular biology focuses on the function of a wide variety of biological molecules, such as proteins, enzymes, and RNA and the variation in these functions both within and among species, whereas genetics is more limited to the study of DNA itself. The entirety of an organism's hereditary information is called its **genome**, and the comparison of large segments of DNA among different living beings is called **genomics**. In this chapter we will explore the genetic differences between chimpanzees and humans and consider what these differences mean.

If you are like me, when you first heard about DNA you thought of it as a big filing cabinet, bulging with the instructions for making all the proteins, enzymes, and hormones needed to keep the human body purring along for an entire lifetime. There is more to the story. DNA is more a factory than a filing cabinet. It both (1) stores instructions for how to build a chimpanzee (or human) starting from a single cell; and (2) transmits the instructions for how to keep that creature alive and functioning on a second-by-second basis.

We might be inclined to feel grateful to genes for keeping track of so much important information for us, and for their admirable tenacity in keeping copying errors – mutations – in check. They do that, true, but that perspective misses an important aspect of genetics. In fact, DNA has another secret agenda that is more than a little bit sinister. It was Richard Dawkins' (1976) *The Selfish Gene* that brought this home to many us. He pointed out that you, mind and body, are a device built by your genes solely for the purpose of protecting and reproducing gene copies. Your wants, desires, talents, quirks, and your very consciousness came to be as genes caused resources to be sucked in and used to build your mind and body, and the sole purpose of your existence is to protect those genes and help them to duplicate themselves.

Genes themselves are tiny, non-conscious little molecules that, tough as they are in their own way, are incapable of protecting themselves against the many dangers in their environment, from the weather to predators to starvation. To do that, they create an immense (compared to the DNA itself), intelligent robot programmed to shelter them from danger and to make as many copies of themselves as possible. You are that robot, built up from a single cell for that one mission. Somehow it seems a betrayal that your genes are completely unconcerned whether "you" are frustrated, unhappy, racked with pain, unfulfilled in your career, or indeed even dead, so long as any of these conditions serves to protect your genes or help replicate them. Your genes and the structures they caused to be built, such as your brain, tell you to eat, avoid predators, find a loving partner, and shower affection on your children. They prevent you from feeling positive emotions like euphoria until you have finished some huge, consequential life project; you, of course, would rather feel euphoric all the time; indeed, many humans induce it artificially and too many die as a consequence.

Your genes are, in fact, toying with your emotions in the crassest way in order to motivate you – *manipulate you* – into replicating them. Perhaps I have dwelt too long on this, DNA's creepiest secret.

After all, most of the time you and your genes are in rough agreement about how you should live your life.

There are those who (misguided as I think they are) disagree with this selfish gene view, but whether genes are selfish or not, "we" are their handiwork. I will start this chapter with a look at how genes function – if you are familiar with genetics and protein synthesis, you can probably skip to Part II, though you might miss a few interesting tidbits of genetic trivia if you do. But if you could benefit from an introduction or a brush-up, read on.

PART I: HOW DNA WORKS

12.1 DNA

Your body, you learned in grammar school, is largely made up of cells, the smallest living unit that contains your DNA. You are undoubtedly aware that a cell is a little bag of fat and sugar with important things that help the cell do its job suspended inside, such as ribosomes and mitochondria (Figure 12.1). The

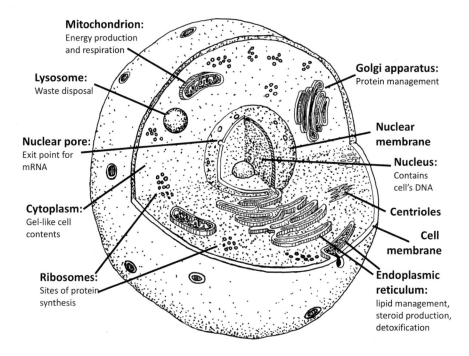

Figure 12.1 The cell. Discussed in the text are the cell membrane, nucleus, mitochondria, ribosomes, and cytoplasm. Image by the author.

outside cell envelope is a sturdy bag called the **cell membrane**; sturdy, but made up of organic material that would be no more substantial than cooking oil in our macro world; in the microscopic world of the cell, though, the fat and protein cell membrane is resilient and rugged. It interacts with its environment by keeping certain things inside (nutrients) and letting other things out (wastes, proteins). It demands a password of sorts from molecules that "want" to get inside and in this way it gathers resources it needs and repels disease vectors.

Floating in the middle of the jelly-like **cytoplasm** that fills the cell, suspended like the yolk in an egg, is the **nucleus** (Figure 12.1), containing your chromosomes. Chromosomes are merely lengths of DNA, or deoxyribonucleic acid, and genes are short pieces of chromosomes. You probably know that chromosomes are found in the nucleus of every cell in your body (well, nearly every cell; not erythrocytes – red blood cells lack a nucleus). To make an analogy with a book, the chromosome is a book; DNA is the paper that makes up the book, and a gene is a segment of DNA that makes a protein or part of a protein, much like a chapter is a segment of a book.

Outside the nucleus, hundreds of **ribosomes** are sprinkled throughout the cytoplasm; these are protein-factories. Nuclear DNA is not your only genetic material; you also have DNA in **mitochondria**, hundreds or thousands of which are floating inside the cell. You inherited your mitochondria from your mother; it is inside the ovum that became you even before it was fertilized by a sperm. Sperm contribute no mitochondria, so none comes from your father, which means you have just a tiny bit more DNA from your mother than your father. I hope that helps you remember Mother's Day.

You might think your DNA is delicate, considering how much we hear about mutations, but in fact DNA is wiry and tough, and it has to be, considering how much these slender strands are pulled and tugged during their lifetime. As strong as it is, DNA can be wounded by free radicals, radiation, caustic chemicals, and many other enemies – even heat. Your DNA suffers as many as one million mutations per day (Lodish & Berk, 2004; Clancy, 2008), many of which, however, are repaired. When researchers began to identify specific genes that affect longevity, they found that it was the genes that direct DNA repair that were most responsible for extending life (Browner et al., 2004). The cell machinery for repairing DNA is constantly at work. Thank goodness. Inevitably, though, sometimes the repair mechanism fails and in that case DNA suffers a mutation.

DNA is a long, ladder-like chain (Figure 12.2) that is twisted. *Sugars and phosphates make up the sides of the ladder, while each rung is made up of a pair of the four chemicals adenine, thymine, guanine, and cytosine; these rungs are often referred to as* **bases**. Imagine a ladder on which each of the rungs is sawn in two but held together by extremely strong magnets. You could split such a ladder into two halves, each looking a little like a comb. When the half-rung that is adenine (A) encounters a thymine half-rung (T), they

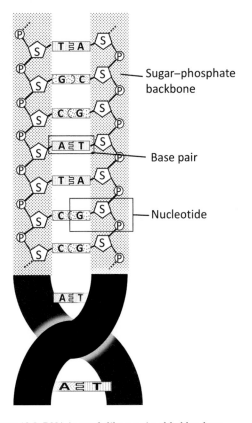

Figure 12.2 DNA is much like a twisted ladder, here untwisted to better see its components. The sides of the ladder are made of sugar and phosphates; the rungs are the bases adenine, thymine, guanine, and cytosine. Illustration by the author.

snap together. Guanine and cytosine snap together, too, but neither is attracted to A nor T. You will have to memorize these pairings, but they are easy: the straight letters A and T go together; the round letters G and C go together.

When, as occasionally happens, an A ends up with a G, it is a mutation. There are other types of mutations; several bases and the ladder-sides that go with them can be lost, or a section might get sliced out and may reattach upside down, or it may attach in a completely different place. Such major errors are rare.

The sugar and phosphate sides with the base pair rungs, a structure that can be explained in a single paragraph, is nothing less profound than the secret of life: it is the sequence of these bases that determines whether you are a mosquito, a monitor lizard, or a human.

12.1.1 Protein Structure

DNA is best understood from the perspective of its working life, its role in producing the biological molecules that run the body. DNA, in other words, is a working molecule, not just a filing cabinet with growth directions; it is the fundamental template in the process that is the source of all the molecules that make your body work. DNA makes these molecules almost without cease, and at speeds incomprehensible to us. We give these molecules names according to their role in the body, but their generic name is **protein**.

Proteins are chemicals that constitute the body and run it. The skin is made of protein, so is muscle and hair. Other proteins gather minerals and deposit them in regular patterns to add a rigid structure to bones and teeth. **Hormones** are a type of protein. Adrenalin and HGH (human growth hormone) are hormones that stimulate the metabolism in ways with which you are already familiar. Other proteins called **enzymes** speed up chemical reactions. Digestive enzymes are proteins that break down nutrients into simpler molecules we can use as fuel; without digestive enzymes food would pass through our bodies little altered (although not *un*altered, since your microbiome would be at work even if you had no digestive juices; see Chapter 6). Lactose is a type of very nutritious sugar, but it cannot be used as fuel by the body without digestion. An enzyme – conveniently named lactase – cleaves lactose into two single-molecule sugars called glucose and galactose which pass through the gut wall – as lactose cannot – where they can be used to power the body. Some other digestive enzymes are **lipase**, which breaks up fat, and **amylase**, which with maltase turns the starch you eat in your potatoes into the more useful glucose. **Elastase** helps you digest meat by breaking down elastin, a fibrous connective material in tissues like arteries and skin. **Hemoglobin** is a protein in blood cells that grabs oxygen and holds onto it tightly in high-oxygen environments, like your lungs, but releases it when it encounters a low-oxygen environment.

Whether protein, enzyme, or hormone, these molecules are made by reading off base pairs in DNA and turning that information into biomolecules. DNA, through these molecules, regulates our growth and determines the function of our body on a second-by-second basis.

Proteins are made up of smaller molecules called **amino acids**, which are strung together like pearls on a string, or maybe we should say like lengths of Brio™ wooden railroad track. Proteins work because they have a particular shape, a shape determined by the way amino acids bind to one another, just as the shape of a Brio wooden railroad track is determined by the shape of its track segments.

Chimpanzees and humans have a mere 20 amino acids, and all of our proteins, including hormones, enzymes, and the rest, are built from these 20 amino acids. Perhaps it is even more amazing that the instructions for building those proteins are spelled out using only four base pairs. It is the order of the bases in DNA, the order of the As, Ts, Gs, and Cs, that determines which amino acids go into proteins and in what order.

12.2 Protein Synthesis

Proteins are constructed by a process called, not surprisingly, **protein synthesis**. The process is simple if you take it step by step. We start with DNA in the nucleus of the cell. As I reminded you earlier, a length of DNA that codes for a protein is a gene. Geneticists say that such a stretch of DNA is **expressed** and so, taking the *ex* from this word, we call protein-coding parts of DNA **exons**. Parts that are noncoding, or not expressed, are **introns**. Only 1.5 percent of our DNA carries the codes for proteins. DNA that is not expressed, the intron part, was once known as "junk DNA," and most geneticists thought it did nothing, or at least nothing very important. Some thought junk DNA was just another expression of "selfish genes" – DNA that selfishly managed to insinuate itself into our genome without doing anything useful for us. Nowadays, we give junk DNA more respect and call it **noncoding** DNA. Among the noncoding bits of DNA are the switches that determine when and how much protein is made; it is also known as **regulatory** DNA.

Protein synthesis begins in the nucleus when the nuclear DNA receives some sort of chemical signal that a protein is needed. DNA has a kind of resting state, a state where it is locked down and not making proteins or enzymes. You will remember from the last chapter that a gene is resting, locked down, really, when a blob of protein called a **repressor** is gripping the beginning part of the gene. Up to now I have neglected to mention an eager enzyme clinging to the DNA, pressing up against the repressor, impatient to start its job but held back by the repressor. Left on its own, it would zip down the DNA, doing something amazing – making mRNA. It is left, instead, straining at its leash, blocked from doing its work by the repressor. That eager enzyme is called RNA **polymerase**. When the right signaling chemical is present, that signal molecule melds with the repressor, making it unattracted to the repressor attachment site. The signal-molecule–repressor twosome drifts off, clearing the way for RNA polymerase to begin construction of **messenger** RNA or **mRNA** for short (Figure 12.3); the process of producing mRNA is known as **transcription** – Latin for copying a text or writing something out: "script" for "writing," "trans" for "across." Different tissues have differing parts of the DNA working to make different mRNA. You may know that the pancreas makes insulin, whereas stomach tissue produces digestive enzymes; this is because the length of DNA that codes for insulin is active in the pancreas, whereas in the stomach cells DNA that makes digestive enzymes is active.

The bases in RNA (ribonucleic acid) are not exactly like DNA, in that RNA lacks the base **thymine** and instead has a base called **uracil** (U). Where there is a C on the exposed DNA, an RNA with an exposed G attaches; where there is a G, a C attaches; and where there is a T an A attaches. So far this is the same as in DNA. However, where there is an A on the DNA, the matching letter is U (rather than T) on the growing mRNA molecule.

As the transcription begins, RNA polymerase pulls the bases apart, leaving each A, T, G, or C exposed,

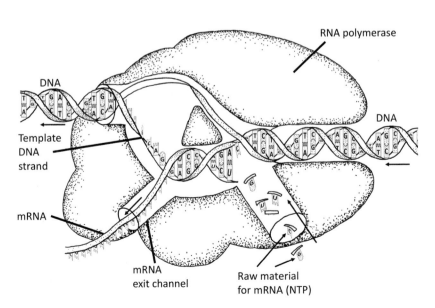

Figure 12.3 Construction of an mRNA strand on the template strand half of DNA. See the text for an explanation. Illustration by the author.

unattached to its partner (Figure 12.3). The raw material to make mRNA is floating around in the nucleus in the form of the t-shaped molecules consisting of a base with a phosphate and sugar attached. Using one side of the pulled-apart (or unzipped) DNA strand as a template, RNA polymerase crawls along the DNA strand, guiding a complementary base and its short piece of mRNA backbone onto each of the raw bases, one by one. The magnetic-like attraction of A to U and G to C helps RNA polymerase to pull the mRNA fragment onto the DNA template, where it sticks.

As As are matched to Us and Gs to Cs, a long strip of mRNA is built up. When the mRNA is complete, it detaches from the DNA and is drawn out of the nucleus through a nuclear pore, into the cytoplasm where it will rendezvous with a little biomachine called a **ribosome**. Ribosomes are the site of protein synthesis (Figure 12.1). I will mention only briefly that the mRNA chain is worked on a bit by enzymes before it leaves the nucleus, editing out some "junk" segments and placing protective caps on the ends.

It is in the ribosomes that protein is actually made, using the mRNA transcript. The ribosome grips the mRNA strand and pulls it along, attaching to each three-letter sequence another molecule called **transfer RNA, or tRNA** (Figure 12.4). The tRNAs are short RNA backbones with three exposed bases sticking out one side, C, U, and A (Figure 12.4); the bases are in essence a little three-pronged plug. An amino acid is attached to the non-plug side of the tRNA. Each individual three-base set can have only one particular amino acid on its back, and can have no other: in the case of "CUA" (Figure 12.4) the amino acid is **aspartic acid**; when the "plug" on tRNA has as its bases ACC, it will be tryptophan. Proteins are built as the mRNA is pulled through the ribosome and its correct tRNA is plugged onto the mRNA strand (Figure 12.5). As each tRNA attaches to the mRNA, the amino acid on its back is juxtaposed to another amino acid. The amino acids stick together and at the same time they become less sticky to the tRNA backbone, allowing a chain of amino acids to peel off. The resulting string of amino acids is a protein. The empty tRNAs are now junk to be recycled. Turning the sequence of bases on mRNA to a protein, the placing of the amino acids, is called **translation**.

As an example, let us trace a code back to the DNA and move forward to the protein. If there is a sequence of ACC on the exposed half of the DNA ladder (the template strand), the mRNA that complements those bases will be UGG (remember than RNA has U, not T). The mRNA then carries the message "UGG" to a ribosome, where a tRNA with the plug ACC is pulled into place to match the UGG on the mRNA. An "ACC" tRNA will have tryptophan riding on its back. You can try it with another sequence. When DNA has the three-letter sequence of TTT, the matching mRNA is AAA. In the ribosomes, tRNA with the plug UUU sticks to the mRNA AAA; this tRNA has on its back lysine.

If you have become a genetics nerd, you will find this amazing. I am going to build a real protein for you. The DNA sequence of TTA AAT ATA TAA CTT ACC AAT TTT CTA CCA CCA GGA AGA AGA CCA GCA GGA GGA GGA AGA yields (by now you can do this yourself) an mRNA strand with the code AAU UUA UAU AUU GAA UGG UUA AAA GAU GGU GGU CCU UCU UCU GGU CGU CCU CCU CCU UCU (by the way, if you decide to major in genetics you must know that the three-letter codes on mRNA are called *codons*).

This mRNA sequence will match up with these "plugs" on the tRNAs (these are called the *anticodons*) UUA AAU AUA UAA CUU ACC AAU UUU CUA CCA

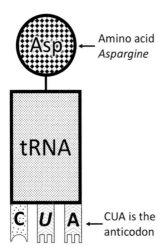

Figure 12.4 Transfer RNA, or tRNA, molecule. Base pair prongs of cytosine, uracil, and adenine are at the bottom and an attached amino acid, aspartic acid, on top. Image by the author.

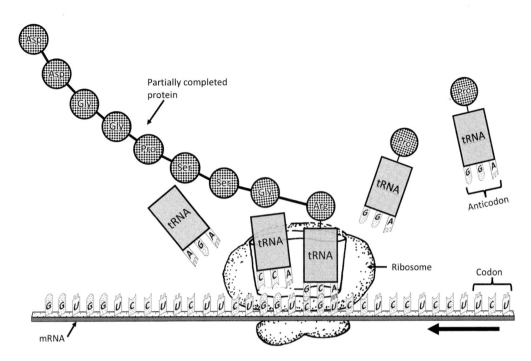

Figure 12.5 Protein synthesis. The ribosome pulls the mRNA strand along, attaching tRNAs, while the growing, partially completed protein (still called a peptide chain until it is finished) peels off. Illustration by the author.

CCA GGA AGA AGA CCA GCA GGA GGA GGA AGA. These particular tRNAs will have on their backs, in this order, asparagine, leucine, tyrosine, isoleucine, glutamine, tryptophan, leucine, lysine, aspartic acid, glycine, glycine, proline, serine, serine, glycine, arginine, proline, proline, proline, and serine – 20 in total.

We can summarize it this way:

DNA	TTA AAT ATA TAA CTT ACC AAT TTT CTA CCA CCA GGA AGA AGA CCA GCA GGA GGA GGA AGA
mRNA	AAU UUA UAU AUU GAA UGG UUA AAA GAU GGU GGU CCU UCU UCU GGU CGU CCU CCU CCU UCU
tRNA	UUA AAU AUA UAA CUU ACC AAU UUU CUA CCA CCA GGA AGA AGA CCA GCA GGA GGA GGA AGA
Amino acid	Asn Leu Tyr Ile Gln Trp Leu Lys Asp Gly Gly Pro Ser Ser Gly Arg Pro Pro Pro Ser

This sequence of amino acids is the protein **TRP-Cage**, a protein found in the saliva of the Gila monster. This

Figure 12.6 The protein TRP-Cage. From www.pdb.org. (A black and white version of this figure will appear in some formats. For the color version, please refer to the plate section.)

protein, as it happens, is the smallest protein known to science, only 20 amino acids in length.

Once extruded from the ribosome, the TRP-Cage protein pops into its peculiar shape (Figure 12.6) in an astonishing four nanoseconds (Qiu et al., 2002) – you could sequentially fold a *quarter of a million* TRP-

Cage proteins in a single second. Figure 12.6 is an illustration of TRP-Cage; you can see it in 3D at www.pdb.org. Most of the functionality of proteins comes from the way the strip of amino acids is folded; the rest comes from chemical bonds. The two together, shape and bonds, make them like magnetized square pegs, floating around in your bodily fluids, ready to stick to any square hole they encounter. There are also round pegs, triangular pegs, hexagonal pegs, and so on. Your blood is a stew teaming with these proteins, like letters in an alphabet soup, all ready to slot into the right-shaped receptacle to do their work.

For the sake of completeness, there is a complication when it comes to translation among vertebrates. While repressors alone run the show for simple organisms like *E. coli*, vertebrates have other molecules that come into play. Just as in *E. coli* a certain base sequence, called a **repressor attachment site**, attracts a specific repressor, in more complicated animals a noncoding stretch of DNA called a **binding site** attracts a protein folded in such a way that it has a little clamp on one side; this protein blob is a **transcription factor**. Some people describe them as little Pac Mans™. So far they may sound no different than repressors, but transcription factors are more multitalented than repressors. When a transcription factor clamps onto a binding site it causes the regulatory DNA near it to bend and warp so that it touches a gene that it regulates. This deformation causes some genes to stop making protein, doing the same job as repressors, and other genes to become even more active, dramatically increasing protein synthesis. This is why shape is so important. Later you will see I have a reason for mentioning these; keep in mind that **binding sites** are in the intron or nonexpressed part of DNA.

Transcription factors are so important because they allow our bodies to have an evolved-in flexibility, an ability to respond to specific conditions not only by turning genes on or off, but making them, while still "on," more or less active. The study of this modulation and the conditions that cause it is called **epigenetics**. For instance, studies found that people who are starved during growth and development turn on genes that cause fat to be retained particularly well. This is the body's way of preparing the individual for the next famine. The Dutch famine of 1944–1945, caused by a Nazi embargo and other unfortunate events, caused a severe food shortage (Roseboom et al., 2001). Those born just after this embargo, people whose mothers were malnourished during pregnancy, retain fat better than other people from the area. Unfortunately, in our current environment of plenty, this trait is not helpful; these people suffer from higher rates of obesity and heart disease.

PART II: CHIMPANZEE GENETICS

In the past, we humans looked out over the living world and found ourselves impressed not only with our own beauty and perfection, but also with how different we are from every other mammal. We were next to the angels, made in the image of God. We placed ourselves in one category, *humans*, and we lumped all the other vertebrates into one bulging group, "animals," despite the fact that the animal "chimpanzee" is far more closely related to us than they are to monkeys. From this somewhat arrogant traditional view, the history of the scientific study of humanity's place in nature has been a long series of humiliating come-downs as science has pecked away at the distinctiveness of one human quality after another. Genetics is no exception.

The history of chimpanzee molecular biology is one of a gradual awakening to the realization that human and chimpanzee DNA are far more similar than anatomy would suggest. To be more precise, our DNA is as similar as the *micro* similarities of humans and chimpanzees, rather than the *macro* differences in arm length and hairiness, for instance. This is a clue that humans and chimpanzees differ not so much because our genes differ, but because our regulatory DNA differs.

12.3 Counting Chromosomes

While as early as the mid-1800s the towering figures Charles Darwin and T.H. Huxley had speculated that humans and apes were closely related (Huxley, 1863), they were in the minority. Richard Owen placed humans in a different taxonomic *subclass*, almost as different from chimpanzees as whales are from mice. As we first began to understand the biological basis for the shape of animal bodies, Owen seemed closer. We have 23 pairs of chromosomes, whereas chimpanzees have 24 (Figure 12.7), as do other apes (De Grouchy, 1987), a difference of 4 percent. All apes together, humans separate. Of course! This suggested the ancestor of all living apes split off from humans far in the past, then differentiated into the various living species.

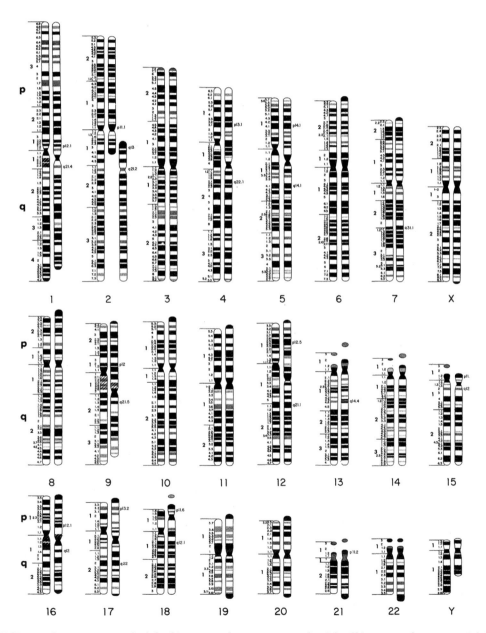

Figure 12.7 Human chromosomes on the left, chimpanzee chromosomes on the right. Chimpanzee chromosome 2 is broken into two chromosomes, giving chimpanzees 24 pairs of chromosomes compared to 23 for humans. From Yunis et al., 1980, with permission.

On closer inspection the difference in chromosome number is deceptive. Specialists now contend that the common ancestor of apes and humans had 24 pairs of chromosomes (Yunis & Prakash, 1982) and sometime after we separated from chimpanzees what had been two chromosomes in our common ancestor fused to form the human chromosome 2. It is a bit egocentric that we call the two separate chromosomes in the other apes chromosomes 2A and 2B (Figure 12.7). But fusing two chromosomes barely changes their base pair sequences, just as Shakespeare's *Collected Works* have the same word order whether presented in two volumes or one big one. In fact, the sequence of bases in our chromosome 2 are nearly identical; you can see in Figure 12.8 that stained ape chromosomes 2A and 2B line up perfectly with human chromosome 2 (Yunis et al., 1980).

12.4 Protein Similarity Revealed by Electrophoresis

While chromosomes were discovered in the late 1800s and gradually became associated with inheritance, the field of genetics had its true birth when Watson and Crick – with critical help from Rosalind Franklin and Maurice Wilkins – unraveled the structure of DNA; over the next few decades genetics were gradually brought to bear on the human/chimpanzee issue, and Huxley and Darwin were proved right.

In the 1950s and 1960s, before gene sequencing was possible, scientists like the late Morris Goodman began to get at ape genetics indirectly by using a technique known as **electrophoresis** to compare proteins in the blood in different primates. Electrophoresis took advantage of the fact that different proteins have different weights and different electric charges. In the technique of electrophoresis, blood serum proteins are dropped onto one end of a gel medium and an electric charge is applied (Figure 12.9). Because proteins – fragmented DNA and RNA have also been analyzed – differ in weight and electric charge, they are pushed across the gel at different speeds and end up separated into bands. Goodman pioneered two-dimensional gel electrophoresis, separating proteins in two dimensions, one 90° from the other. The resulting patterns (Figure 12.10) provides a visually arresting

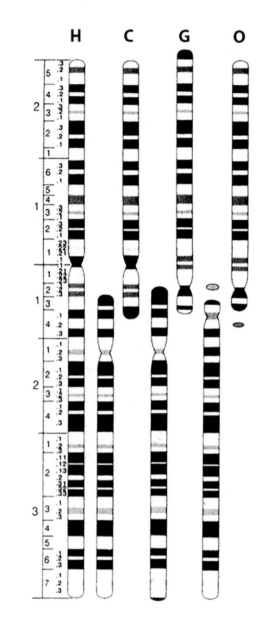

Figure 12.8 Human chromosome 2 (left) compared to chromosomes 2A and 2B of the chimpanzee (C), gorilla (G), and orangutan (O). From Yunis & Prakesh, 1982, with permission.

demonstration of how similar ape and human proteins are – extremely. Goodman and others came up with various schemes to turn differences in the shape and distribution of the blotches into some quantitative expression of the relationship among species, but despite some success there remained as much art as science in interpreting their results. Although

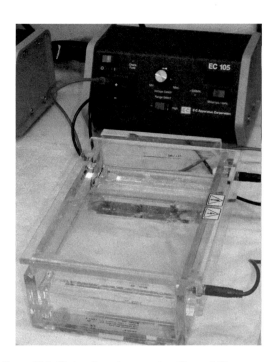

Figure 12.9 Electrophoresis apparatus. Photo: Jeffrey M. Vinocur. Used with permission.

Goodman was unable to determine just which ape was our closest relative, by 1963 he made the startling declaration – considering how different apes and humans look – that many proteins were *identical* in humans, gorillas, chimpanzees, and bonobos ("pigmy chimpanzee" in Figure 12.10).

While molecular biologists were marveling at the near-identical human and ape proteins, anatomists continued examining macro features such as limb lengths, brain size, and skin morphology, and were finding contrary results with pronounced differences between apes and humans (Biegert, 1963; Napier, 1963). Some anatomical contrasts were an order of magnitude greater than the contrasts Goodman was finding with his electrophoresis work. How could such dramatic differences in anatomy *not* require similarly dramatic differences in proteins?

12.5 Nailing Down 99 Percent

Science was about to get the definitive quantitative comparison between human and chimpanzee genes (indirectly, still, but boldly stated) that Goodman's

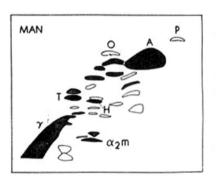
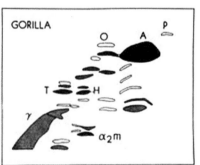
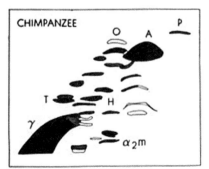
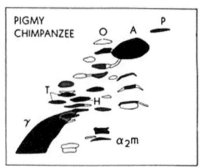

Figure 12.10 Electrophoresis results. From Goodman, 1963, with permission.

methods could not supply. Even softened up by Goodman's electrophoresis work, it was still a shock to biologists when, in 1975, a masterful compilation of our burgeoning knowledge about protein similarities laid out the facts (King & Wilson, 1975): For one amino acid sequence after another, human and ape proteins were identical. While most scientists offered squishy conclusions, preferring phrasing like "we don't know enough yet," or "well, it depends on what you mean by 'similar,'" King and Wilson announced their startling findings with numerical precision: Chimpanzee and human genes were 99 percent identical. Their elegant visual summary is reproduced here as Figure 12.11.

We are not just as similar as might be expected of two species; our two species teeter on the figurative razor's edge between species that are closely related enough to be in the same genus (congeneric species), like lions and panthers, and species that are not-quite-separate-species ("semispecies"). Keep in mind that science still uses separate genera, *Homo* and *Pan*, to classify humans and chimpanzees. Instead, Mary-Claire King and Alan Wilson told us, we are sister species, and if we were only the tiniest bit more closely related, we could be considered the same species. King and Wilson's stark statement hit the scientific world like a thunderclap. Some scientists grumbled that King and Wilson had overshot their data. Surely, some believed, given ape–human anatomical differences King and Wilson must have missed something. At the risk of stating a thrice-told factoid, in actuality the anatomy of chimpanzees and humans is nearly identical at the level of fine details; the dramatic differences in hairiness, body proportion, and brain size are to some extent deceptive.

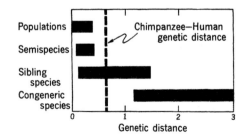

Figure 12.11 King and Wilson's (1975) estimate of genetic differences between humans and chimpanzees. Used with permission.

12.6 DNA–DNA Hybridization

In the late 1970s and 1980s geneticists moved beyond examining proteins and a little closer to sequencing bases. Charles Sibley and Jon Ahlquist did this using a new technique called DNA–DNA hybridization (Sibley & Ahlquist, 1984). Think again about the structure of DNA, and imagine splitting DNA down the middle. In this technique, DNA was separated down the middle for the two animals to be compared. Now imagine lining up the left side of zebra DNA with the right side of donkey DNA and allowing them to stick together. Some base pairs will be mismatches – a G on one side and, say, a T on the other, but most would match. The greater the number of base pair matches, the tighter the two sides will be bonded. This is the principle in **DNA–DNA hybridization**.

The process works like this. DNA is chopped up into short segments by a special enzyme. One organism's DNA is then irradiated so that a Geiger counter can detect it. The DNA of the two species to be compared is then mixed together and allowed to combine, forming hybrid DNA. The resulting mixture is slowly heated until the DNA "melts," freeing the two sides so that they split apart. As the DNA is heated, the mixture is rinsed, sweeping any loose DNA away. The rinse water is shunted past a radiation detector to measures how much radioactively marked DNA emerges in the wash. A base level measure, the melting temperature of the DNA of a species, is determined by melting human–human DNA; the temperature when radiation reaches a peak for human–human DNA yields a "same-species" value to compare to the mixed-species DNA. The more similar the DNA of the two species, the higher the temperature has to be to melt the hybrid DNA, and the higher the temperature will be when the radiomarked DNA reaches a peak in the wash. Experiments suggested that when the genomes of two species are 1 percent different, they melt at one degree lower than human–human DNA. Chimpanzee and human DNA melted at about that one-degree difference, which translated to the same 1 percent King and Wilson had found when analyzing proteins. The technique was not without its detractors, but it was an advance over electrophoresis. The technique did yield something more, though. Using this method, Sibley and Ahlquist were emboldened to say something out loud that would have been unthinkable a generation before: Humans are more closely related to chimpanzees than chimpanzees are to gorillas.

12.7 Sequencing

Gene sequencing came of age in the 1980s. Sequencing was slow, difficult, and so expensive that only small sections of DNA were examined – segments of only hundreds or thousands of base pairs long, a tiny fraction of the three billion base pairs in the human genome. Often this limited sequencing yielded a human/chimpanzee difference of around 1 percent. But it was like trying to determine how forested the whole USA is by counting trees along a randomly chosen 50-mile stretch of highway. If your random stretch of highway happened to be in Utah or Oklahoma your answer would be very different than if it was in Oregon or Maine.

Still, we progressed. In 1987 the late Morris Goodman and colleagues (Miyamoto et al., 1987) compared 7000 or so base pairs that code for hemoglobin, a blood protein, and in support of Sibley and Ahlquist found the human–chimpanzee pair was more similar than the chimpanzee–gorilla one.

12.8 "Functional" Genes

Gradually, the number of genes sequenced mounted up and the 1 percent figure held up. As confidence in this similarity grew, Goodman and colleagues examined 97 "functional" genes – that is, genes that code for proteins. They found that humans and chimpanzees were identical at 99.4 percent of the sites. Humans and chimpanzees were so closely related that if we were any other pair of species – if we were two species of antelope or two species of rodents – we would be placed in the same genus (Wildman et al., 2003). Goodman and company were inspired to propose something implicit in the King and Wilson work 25 years earlier: Chimpanzees (and bonobos) should be in the genus *Homo*.

Table 12.1 **Gene copies gained and lost**

	Number of copies gained	Number of copies lost
Humans	689	86
Chimpanzees	26	729

From Hahn et al., 2007.

12.9 Chimpanzee Genome Sequenced

In 2005, after a massive effort, the Chimpanzee Sequencing and Analysis (CSaA) Consortium published their rough draft of the chimpanzee genome (CSaA Consortium, 2005). Their results showed that of the nearly three billion base pairs in the human and chimpanzee genomes, there is a 1.23 percent difference. But this 1.23 percent difference, scholars inside and outside the project noted, is deceptive because it measured how many base pairs differed in the genes we shared in common. Some genes – a tiny fraction of the three billion – have been lost entirely in one species or another. How can we fairly state that difference? Furthermore, some genes are exactly the same in both humans and chimpanzees, but they appear once in one species and four or five times – even ten times – in the other (Table 12.1). How do you count that? Ten times as many genes might mean 10 times as many proteins doing whatever those proteins do. Or it might not. We had finally gotten to the point where we had such a precise grip on human and chimpanzee base pair sequences that we were forced to pause and ask "what do we mean when we say 'the same' or 'different?'" At last we were close enough to the forest to see the trees, and that meant the argument now shifted to whether the forest or the trees was most important.

12.10 Counting Gene Copies Yields a Different Result

Matthew Hahn and colleagues (2007) took a different approach. They purposely looked for genes that had multiple copies. They found that since humans and chimpanzees went their separate ways around six million years ago, humans have gained 689 copies of genes and lost 86 copies; chimps gained only 26 but lost 729, about the same number that humans have picked up. If you just look at these gene copies, the difference between humans and chimps is much larger than if you just compare all the base pairs (which counts both gene and non-gene segments, but ignores how many copies there are). The difference in the number of copies of genes is 6.4 percent. Keep in mind, though, that this is just the difference in the number of copies. As a way of thinking about these results, consider the song "I want to hold your hand." It has 187 words and 691 letters (not counting spaces). If this song were our genome, looking now at sequences in the genes, it would differ by only two words. You probably would hardly notice the difference. In fact, you probably never even noticed that 44 seconds into the song the Beatles do not sing "You'll let me hold your hand" as the published lyrics say they do, but "I let me hold your hand." For shame, George Martin!

12.11 Rapidly Diverging Genes

The rough draft of the chimpanzee genome (CSaA Consortium, 2005) provided some intriguing hints about human and chimpanzee evolution by identifying genes that have rapidly diverged since the two species went their separate ways (Table 12.2). This list was intriguing, but not shocking. It was when we began to identify more confidently *which* genes had changed in which species that we began to find surprises.

The biggest difference was in skin development, a gene complex that is in part responsible for helping the skin limit water lost; it also influences snout and trunk development. We walk in the sun without the protection of fur, whereas chimpanzees live in forests and avoid the sun. Our snouts are much smaller, so these differences make sense. Cystatins do a number of things, among them fighting viruses; we must face different diseases than do chimpanzees. Our hair is different, so keratin differences are no surprise. Chimpanzees are sugar-detecting machines; perhaps human emphasis on animal fats left us with less need for sugar-seeking genes than chimpanzees. Other rapidly diverging genes have to do with immune function and brain function. All quite sensible. However, I fail to understand why genes for sense of smell diverged so much.

12.12 Positively Selected Genes

Soon after the publication of the rough draft of the chimpanzee genome, a new comparison was made that focused on **positively selected genes** (Nielsen et al., 2005). Positively selected genes are new genes that go from 0 percent (at first appearance) to some much higher percentage, not unusually 100 percent; the frequency of that gene increases because the gene is advantageous, or "positively selected." The initial analysis (Table 12.3) of these genes contained surprises, though nothing like what was to come. Nielsen and colleagues found that unexpectedly few positively selected human genes were associated with brain development. Instead, human genes for the immune system, sperm production, and olfaction were those that proliferated. Genes related to programmed cell death (apoptosis) were also positively selected, but there were more genes for *suppressing* cell death than anything else; humans weakened their ability to kill off rogue cells, hypothetically because they interfered with spermatogenesis – but of course chimps need to make sperm too. Sense of smell and sperm production are not on any human evolutionists' list of expected evolutionary changes among early humans.

We will see that for the most part the genes that were positively selected made sense; however, the difference in the *number* of positively selected genes made no sense at all (Figure 12.12). Humans are distinct from the apes because we have big brains, naked skin, organs of speech, and so on. When we consider the myriad barriers to evolving a big brain, it would seem that we need to *add* something to the last common ancestor (LCA) to get a human: We need more neurons; we need extra calories during growth and development to grow that big brain; we need more interconnections to wire up the additional neurons; we need more oxygen for that larger brain. We need *more*. You would think, would you not, that

Table 12.2 **Rapidly diverging gene clusters in human and chimpanzee**

Cluster	Median K_A/K_I	Function
Epidermal differentiation complex	1.46	Skin development; mutation in IKK in mice yields shiny skin, death by dehydration. Snout and trunk development?
Olfactory receptors and HLA-A	0.96	Sense of smell
Cystatins	0.94	Protease inhibitors = antiviral; Alzheimer's; kidney function; heart disease; suppresses oral microorganisms; cancer tumor suppression; skin inflammation; maturation of dendritic cells
Pregnancy-specific glycoproteins	0.94	"Immunomodulator to protect the growing fetus;" prevents mother's immune system from attacking; when lacking: low birth weight, hypoxia, retardation, abortion
Keratins and keratin-associated proteins	0.93	Hair
CD33-related Siglecs	0.90	Sugar detection
WAP domain protease inhibitors	0.90	Antiviral
Immunoglobulin-l/breakpoint critical region	0.85	Immune function
Taste receptors, type 2	0.81	Sugar detection
Chemokine (C-C motif) ligands	0.81	Inflammatory response
Leukocyte-associated immunoglobulin-like receptor	0.80	Immune function
Protocadherin-b	0.77	Schizophrenia
Complement component 4-binding proteins	0.76	Antibody action
Keratin-associated proteins and uncharacterized ORFs	0.76	Hair
CD1 antigens	0.72	"Presentation of lipid antigens to T cells"
Chemokine (C-X-C motif) ligands	0.70	Brain function

From CSaA Consortium, 2005.

the way we evolved these new features was by preserving mutations that created new genes that coded for these new features, while our ape sisters, you might guess, were left with the "old" genes and would share more genes with the LCA?

For instance, a mutation might cause a gene that influences brain size to code for a slightly larger brain, after which that gene would be highly positively selected. Among apes, such a gene might not be advantageous; the benefits of being slightly more

Table 12.3 **Positively selected genes or gene clusters in humans and chimpanzees**[a]

Positively selected in chimpanzees	Positively selected in humans
Proteolysis (protein recycling)	Fatty acid metabolism (fat storage, use)
Protein metabolism	Phosphate transport (brain function)
Stress response (social system changes?)	Phosphatase activity (brain function)
mRNA transcription (protein synthesis)	Ectoderm development (skin)
Nucleic acid binding (protein synthesis)	G-protein mediated signaling (multiple; see text)
Nuclease (protein synthesis)	Anion transport (energy regulation)
Transferase (protein synthesis)	Lyase production (energy regulation)
mRNA transcription regulation (protein synthesis)	Ion transport (energy regulation; neural?)
	ApoE3 (lipid transport: brain function)[b]

[a] Bakewell et al., 2007.
[b] Finch, 2010.

More genes underwent positive selection in chimpanzee evolution than in human evolution

Margaret A. Bakewell, Peng Shi, and Jianzhi Zhang*

Department of Ecology and Evolutionary Biology, University of Michigan, Ann Arbor, MI 48109

Figure 12.12 In 2007 Bakewell, Shi and Zhang announced the surprising result that chimpanzee genes have evolved more than human genes.

intelligent might be outweighed by the difficulty of obtaining the extra nutrients needed to grow and power that larger brain. Among humans, however, such a mutation would be advantageous, perhaps because our diet was richer, making the costs of brain expansion lower; or perhaps we lived in larger social groups, making greater intelligence so much more valuable that it outweighed its costs. In short, we expected that humans would have a whole multiplicity of new, positively selected genes, while our cousins the great apes, more similar to the LCA, would be left mostly with old genes.

That was wrong. In an analysis published soon after the rough draft of the chimpanzee genome appeared (Figure 12.12), Bakewell and colleagues (Bakewell et al., 2007) showed that not only did humans *not* have more positively selected genes, it was exactly the opposite; chimpanzees had more positively selected genes than humans (Figure 12.12). In other words, since the chimpanzee–human split, chimpanzees genes have changed more. The difference was not huge, 233 genes positively selected for chimpanzees versus 154 for humans, but we were expecting the difference to be huge the other way. And at first glance the genes that were selected made little sense (see Table 12.3). Many new chimpanzee genes had to do with protein synthesis; for humans they were related to energy control and fat. You would think that our lineage would be the one to change the way we metabolize protein, since we need protein for our large brain. We eat a lot more protein-rich foods – like meat – than do chimpanzees. Yet, among humans, nine genes associated with protein metabolism were positively selected, whereas chimpanzees had *40*.

Many genes that were strongly selected in chimpanzees had to do with protein synthesis: mRNA transcription, nucleic acid binding, nuclease function, transferase function. Why would chimpanzees require fundamental changes in these cellular functions? We have no idea, but it may be that since our lineages split, chimpanzees have changed the way they eat, fight, and form social groups even more than we expected.

I have focused on results that seem not to make sense so far, but there are plenty of genetic changes that we can make sense of. One group of genes related to proteins that proved valuable to chimpanzees as they evolved from our common ancestor were genes for proteolysis: 16 such genes have been positively selected in chimpanzees versus only 2 for humans. Proteolysis is the "*digestion*" of protein – I have placed special emphasis on this because this means not only digestion in the gut, but protein turnover in the body, such as when the protein in dead cells is recycled. In dogs, infection causes an upregulation of proteolysis-related genes (Hagman et al., 2009) and in humans appears to have something to do with wound healing (Lauer et al., 2000; Eming et al., 2014; Leoni et al., 2015), perhaps speeding up the transition from inflammation to healing (Landén et al., 2016). This emphasis on wound healing suggests that the extreme violence observed among living chimpanzees evolved *after the split* between our two species.

Chimpanzees had 14 positively selected genes for stress response versus 2 for us. Getting wounded, fighting for your life, and facing mortal enemies on all sides – those are stressful things and, as we will see, they define chimpanzees. Given the high body size dimorphism of late Miocene apes, the LCA had a gorilla-like social system, one less murderous than the current chimpanzee social system. Chimpanzees took on this greater stress after our two lineages split.

There is more. There has also been rapid evolution of the *MRG* gene family in humans, genes responsible for pain reception (Choi & Lahn, 2003). Chimpanzees seem to have a higher pain threshold than humans; somehow it was advantageous for humans to be more sensitive to pain. Perhaps our greater sociality meant we could turn to others for aid when we were sick and injured, and expressing pain was a way to gain such assistance.

While there were many fewer than expected fast-evolving genes related to brain development in humans (Gu & Gu, 2003; Dorus et al., 2004), there were some, and two of these genes, *ASPM* and *MCPH1*, have been studied in detail. They regulate cell division early in brain growth. When *ASPM* is mutated, neuron duplication is slowed and brains are smaller. Incredibly, large-brained cetaceans – dolphins and some whales – have independently evolved the same *ASPM* gene (Xu et al., 2012).

A number of positively selected human genes direct the storage and metabolism of fat. The brain is a calorie-hungry, fatty organ – it is no surprise that humans might need to store fat differently than chimpanzees. Phosphatase regulation and phosphate transport both have to do with regulating cell activity, but probably more importantly with regulating neuronal and synaptic brain activity. Other positively selected genes among humans have broad functions in energy regulation and production, some possibly to do with neural function; the brain is an energy hog, so these genes may be related to brain development.

Humans have evolved a new apolipoprotein since the split. Apolipoproteins bind with lipids to shepherd them through the circulatory system so they reach the right place. Humans have a new form, *ApoE3*, that may be only about 225,000 years old (Finch & Stanford, 2004; Finch, 2010). The human form helps direct fat to the brain. Some unfortunate people have the chimpanzee form (*ApoE4*); they suffer from higher cholesterol, greater risk of arterial diseases, higher rates of heart disease, earlier onset of Alzheimer's, and chronic inflammation. For unknown reasons, *ApoE3* also helps in recovery from brain injury. The human version of this gene is almost certainly an adaptation to cope with the bad consequences of eating more meat, and thus getting more fat in the diet.

Also evolving more quickly in humans, not surprisingly, were genes that regulate ectoderm development – skin development. The gene *KRT* is responsible for the development of the tough exterior of human skin cells; without hair to protect us, we would certainly need different skin than chimpanzees (Varki & Altheide, 2005).

G-protein mediated signaling may be the most intriguing yet unsatisfying of the positively selected human gene groups – unsatisfying because this group of genes is involved in so many physiological processes (Vassilatis et al., 2003; Wettschureck & Offermanns, 2005) that nailing down why they differ so much between humans and chimpanzees is difficult. By one estimate, 34 percent of marketed drugs target G-protein-coupled receptors, including drugs for a baffling array of medical conditions, including schizophrenia, anxiety, depression, asthma, seasonal allergies, heartburn, migraines, high blood pressure, and Parkinson's disease (Hauser et al., 2018).

12.13 Lost Genes among Humans

Our genes hold yet another revelation. Not only have we not added many new genes during our evolution, we have shut down more genes than have chimpanzees. It has been estimated that 80 or more genes (not *copies* of genes, as we were talking about earlier, but the gene itself) have been lost in humans since we split from the other apes (Wang et al., 2006; Varki et al., 2008).

Maynard Olson explained this loss-of-function trend among humans as a "less-is-more" type of evolution (Olson, 1999). He speculated that shutting down genes that regulate different aspects of growth was the simplest way to accomplish the major make-over necessary to turn a chimpanzee-like organism into a human. But how can *deleting* genes *add* features? You may remember an example of a lost gene discussed earlier that increases digestion, the gene that *turns off* lactase production. By losing that gene we can drink milk as adults. There are many mutations similar to this one, mutations where shutting down a gene causes something to happen rather than stopping something from happening. One gives people an immunity to a kind of malaria. In domesticated mice – lab mice – a gene is disabled that normally suppresses reproduction during times of low light, the winter, when wild mice need to hunker down and avoid unnecessary energetic investments. It allows lab mice to breed year-round. Get a mutation, produce more offspring. This type of genetic change is common when a species is domesticated.

We humans have evolved, in other words, like domesticated livestock. Why? Domesticated animals have evolved under severe selective pressure – human selection; we choose which animals breed and which become dead-end lineages – and we evolve new animals very quickly. Shutting down genes rather than evolving new ones is a sort of evolutionary shortcut. Only two million years ago, when macaques and giraffes and hippos were hardly different from now, human ancestors were small-brained, long-armed, fruit-eating apes. We have changed profoundly and quickly in that short period of time, which has necessitated evolving by a less-is-more shortcut. We are, in Olson's words, "hastily made-over apes" (Olson, 1999).

12.14 Gain Brain, Lose Elsewhere

Our most unique feature, our large brain, might be expected to have required an across-the-board genetic transformation to effect such rapid increase. Brains have a *lot* of cells – by one estimate 86 billion neurons (Azevedo et al., 2009), three times as many as chimpanzees. Brains grow during fetal development through a series of cell doublings; with 86 billion neurons (Herculano-Houzel, 2014), human brains must require two more doublings than chimpanzees. Growth, however, is a double-edged sword. Extremely rapid growth might be cancer; attacking runaway cell growth benefits most species. Many cancer therapies specifically target fast-growing cells. Humans, however, must grow the brain more rapidly than other animals to attain that huge size, which means growth-checking genes could repress brain expansion. One way that humans have evolved bigger brains is by shutting down genes that limit rapid cell growth.

John McDonald and his team at Georgia Tech have suggested that deleting these governor-genes not only allowed more rapid brain growth; at the same time, it slammed humans with a disease that seems to be a signature affliction for our species – cancer (Arora et al., 2009, 2012). At its simplest, we get cancer when a single cell begins to divide uncontrollably. McDonald's team has shown that humans have modulated programmed cell death compared to chimpanzees (Arora et al., 2012); in one comparison they found a 10-fold difference. Some studies found that by one measure cancer kills around 2 percent of chimpanzees versus more than 20 percent of humans (Puente et al., 2006; Weis et al., 2008). In McDonald's view, we shut down genes to allow our big brains to grow, but we got cancer in the bargain.

12.15 Muscles

Yet another example of deletions shaping humans has to do with our reduced chewing muscles. Chimpanzees have enormously powerful jaws compared to humans. As captive ape caretakers know, a chimp can snip off a finger with a single bite. The gene *MYH16* is responsible for muscle development in the head. Small chewing muscles in humans are partly the result of a mutation that crippled this gene by a deletion of only two base pairs (Stedman et al., 2004). Humans have replaced teeth with tools, so the reduction of less-needed temporalis and masseter muscles is an advantage in that it frees up calories for other more important purposes – like growing and maintaining a large, energy-hungry brain.

12.16 Endurance

Humans lost function of the *CMAH* gene when a 92 base pair chunk of our DNA was deleted, and we think we have a very good idea of when this happened – 3.2 million years ago (Altheide et al., 2006). This deletion leaves humans without a sialic acid molecule on the cell surface (Kehrer-Sawatzki & Cooper, 2007). The disabled gene spread rapidly through the entire species, eliminating all alternative genes. The fixation of this inactivated gene is particularly curious because its effect is that when the diet is too rich in fats, it shuts down the ability of the pancreas to produce insulin, afflicting us with type 2 diabetes and all its health consequences. It also makes us more susceptible to obesity (Kavaler et al., 2011). It could not have swept through our species so rapidly unless it had some very valuable function, but what? It turns out losing this gene improves endurance by improving muscle capacity for oxygen utilization (Okerblom et al., 2018). Perhaps this adaptation is helpful for the long-distance pursuit of game. Lose a gene, become a marathoner.

12.17 Hair

The gene *KRTHAP1* functions in keratin production and, we think, makes hair thicker and stiffer (Table 12.3). It is yet another gene that is disabled in humans, leaving us with thinner hair. Humans are not "hairless," of course, but much of our hair is so thin we appear hairless. It is thought to have been shut down relatively recently, about 240,000 years ago (Winter et al., 2001).

12.18 Immune Function

Humans have an inactivated caspase-12 (*CASP12*) gene, a functional gene in other mammals (Saleh et al., 2004). Caspase-12 is thought to play an important role in apoptosis and the processing of inflammatory cytokines (Nakagawa et al., 2000; Lamkanfi et al., 2002; Kehrer-Sawatzki & Cooper, 2007). People with the inactivated version have a reduced inflammatory response, which in turn decreases the likelihood of dying from sepsis. The mutated *CASP12* spread 60,000 years ago, perhaps as population densities were increasing and making infectious diseases a greater burden (Xue et al., 2006).

12.19 Lost Genes among Chimpanzees

While, as we have learned, humans have lost more genes than chimpanzees, those few that have been lost by our ape cousins are still interesting. Among those lost genes are three genes that regulate inflammatory response through the protein interleukin, *IL1F7*, *IL1F8*, and *ICEBERG*. The action of some of these genes is to raise the body temperature and regulate inflammation. We are all familiar with the signs of inflammation – redness, swelling, and often warmth at the site of inflammation.

These symptoms are the result of increased blood flow to the site and an influx of healing white blood cells (Herwald & Egesten, 2011). While uncomfortable, inflammation speeds healing; it also leaves more scar tissue. As a consequence, the body appears to have a regulatory system to pump up inflammation to speed healing but not to overdo it. *ICEBERG* and *IL1F7* dial down inflammatory response, while *IL1F8* ramps up inflammation. The deletion of *ICEBERG* and *IL1F7* would speed healing. The genes have not been lost in humans, gorillas, and orangutans (CSaA Consortium, 2005), suggesting it was chimpanzees that took the unusual evolutionary path rather than humans or any other ape. This family of genes is also involved in pain sensitivity; chimpanzees may not have the luxury of limiting activity when injured. It must be that extreme violence, and the biological enablers of it – rapid healing and pain-dampening – evolved in chimpanzees after the human–chimpanzee split. Lastly, and perhaps this is most important of all, mice with *IL1* genes knocked out have decreased anxiety (Koo & Duman, 2014; Wohleb et al., 2014); when your life is a constant struggle for dominance, interspersed with fatal confrontations with murderous enemies, high anxiety is not helpful.

One very odd loss among chimpanzees are the genes *APOL1* and *APOL4*, genes that offer a modicum of resistance to sleeping sickness (trypanosomiasis). Now, why would chimpanzees *not* need resistance to this disease? Wild chimpanzees do show evidence of having contracted trypanosomiasis (Jirku et al., 2015); perhaps it is uncommon or was in the recent past. The vector for sleeping sickness is the tsetse fly (*Glossina moristans*), and we know a little bit about tsetse fly habits (Pilson & Pilson, 1967). They tend to stay close to the ground – in some descriptions as close as 2 m from the ground – and they are diurnal. Chimpanzees do spend time on the ground, but perhaps not as much as humans and gorillas – yet certainly more than orangutans. This loss is a puzzle.

12.20 Chimpanzee Y-Chromosomes Are Bizarre, and so Is Male Behavior

The so-called "**autosomal**" chromosomes, all the chromosomes except the sex chromosomes, are evidence enough that the chimpanzee social system has diverged dramatically from the LCA. The change in the non-sex chimpanzee chromosomes, however, is nothing compared to the chimpanzee **Y-chromosome** evolution. The chimpanzee and human Y-chromosomes differ from one another as much as human autosomal chromosomes differ from those of chickens (Hughes et al., 2010). Think for a moment about how much change that is. In six million years the chimpanzee Y-chromosome has evolved as much as would be expected in non-sex chromosomes in 320 million years (Figure 12.13). Something about chimpanzee and human maleness is very, very different. Most of the genes on the Y-chromosome control testis development and sperm production, though some genes make other proteins, the function of which we have not yet discovered. Chimpanzees

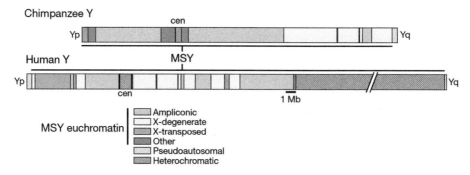

Figure 12.13 Human and chimpanzee Y-chromosomes compared. It is immediately obvious that the two chromosomes are very different, but they are more different than anyone would have expected, as different as humans are from chickens. From Hughes et al., 2010, with permission. (A black and white version of this figure will appear in some formats. For the color version, please refer to the plate section.)

have huge testicles, great sperm production (Harcourt et al., 1981), and they copulate on average six times a day, day after day. The chimpanzee male is dramatically different from nearly every other primate male; bonded like Spartan warriors, they are murderously violent and fiercely protective of their territory. They are, in short, males built on an entirely different blueprint than any other primate male. From this perspective, perhaps it is not so odd that the bit of chromosome that governs maleness has diverged dramatically from that of other primates (Hughes et al., 2010).

Your mind may have gone exactly where mine went when I read of these results. I have offered evidence elsewhere in this volume that the common ancestor of humans and chimpanzees was more gorilla-like than chimpanzee-like. If this is true, should we not expect that the Y-chromosome of gorillas is more similar to that of humans than chimpanzees, presuming that the mating system of the LCA of chimpanzees and humans was gorilla-like? In 2016 this was found to be true (Tomaszkiewicz, et al., 2016).

In Chapter 10 I discussed fossil apes, and you learned that *Ouranopithecus macedoniensis*, possibly the *Homo-Pan* LCA, was extremely sexually dimorphic in body size: Males were twice the size of females, or more. Chimpanzee females, in contrast, are 85 percent of the size of males, not 50 percent. *Ouranopithecus* also had smaller canines than living chimpanzees. Early hominins, *Australopithecus afarensis*, were more like *Ouranopithecus* than living chimps. This and other evidence (Chapter 29) suggests an australopith social grouping more like that of gorillas or hamadryas baboons than chimpanzees, a group with a single male mated to several females (Swedell & Plummer, 2012).

Such a nearly incomprehensible evolutionary rate in the Y-chromosome suggests that the chimpanzee mating behavior, and thus social system, must have taken its current form after the human–chimpanzee split. In other words, it was only recently that chimpanzees evolved their current social system, one in which males copulate frequently (gorillas do not) and form large, highly integrated paramilitary groups for the purpose of stalking and killing their neighbors (Wrangham, 1999). They not only dish it out, they must take it as well. They suffer from much higher rates of violence and wounding than typical primates (Wilson et al., 2014), and thus they need advanced stress response, ramped up defense against infection, and accelerated wound healing. Genetics shows they have these traits.

12.21 Noncoding or Regulatory DNA Evolution

Genetic studies show that humans have evolved via rapid changes in regulatory DNA. I mentioned these stretches of DNA earlier. Once called "junk DNA," we know that some parts of these noncoding areas interact with genes. Katherine Pollard and her colleagues have identified over 200 noncoding areas in the human genome that have evolved rapidly since the chimpanzee–human split (Pollard, 2009). When Pollard and her team found the first of these, they named it Human Accelerated Region 1 (HAR1). HAR1 is tiny, containing only 118 base pairs, and so seemingly an unlikely candidate for an evolutionary hotspot; however, while HAR1 is almost identical in rodents and chimpanzees – differing by only two base pairs – humans and chimpanzees differ by 18 pairs! It is probably no surprise that HAR1 is involved in brain development. Incredibly, half of the HARs Pollard and her colleagues have found are involved in brain development. The second most accelerated region in the human genome, HAR2, is involved in wrist, thumb, hand, and ankle development, anatomical areas that differ dramatically in humans and chimpanzees. Most HARs are like HAR1 – they do not code for proteins, but instead regulate genes that do.

12.22 *FOXP2*

It was long after the human-ape split, but I cannot resist commenting on one gene that seems to be the superstar of human-specific genes: *FOXP2*. During development, *FOXP2* encourages neuron production in areas that govern speech; people who have a mutation in this gene have speech abnormalities. By reverse logic, having the gene may have endowed our ancestors with superior speech. *FOXP2* supposedly became more or less a human universal about 200,000 years ago and its appearance has become one of the most widely discussed of human evolutionary events – suggesting that human speech dates to 200,000 years ago, suggesting truly modern speech emerged after humans and Neanderthals diverged at about 500,000 years ago. Perhaps, some have suggested, the lack of language accounts for the extinction of Neanderthals. Not quite. Much of the Neanderthal genome has been sequenced, and, lo and behold, it turns out Neanderthals had this gene, too (Krause et al., 2007). Perhaps the 200,000 year date is wrong.

12.23 Lessons

Powerful, fast-healing, smart, endowed with fearsome offensive weaponry, and murderously aggressive: The genes of chimpanzees show them to be highly evolved war-making machines; significantly, many genes linked to these traits appeared after the *Homo–Pan* split. This suggests that the LCA had a social system more similar to that of one of the other apes rather than chimpanzees. Gorillas and orangutans attempt to secure mating rights to a number of females, but for the most part they engage in that defense individually, one on one. It is our sister species that evolved a group-based make-war social system that benefited from rapid wound healing and stress management. But we knew that from behavior alone, Wrangham and Peterson (1997) called chimpanzee males "demonic."

Gorilla and chimpanzee violence are different. Among gorillas, once an opponent is defeated there is little to be gained from pursuing the defeated male to extend the attack, to inflict further damage. Furthermore, chasing after the defeated male leaves a silverback's infants vulnerable to attack by a third male or an all-male group; it risks further injury and exhaustion by extending the violence, and it accomplishes little. It is difficult for a single individual to kill another, and the killing is of little advantage, since the male is, short-term at least, no longer a danger.

Chimpanzees do gain an advantage by continuing the confrontation after the opponent is defeated. There are no infants nearby at risk, and at the same time killing an opponent weakens the future fighting force of the enemy. While it is very difficult for a single individual to kill another, a group can do it safely. Rather than depending on sheer body weight, chimpanzees draw strength from outnumbering their foes. When they attack as a group some attackers immobilize the victim while others inflict damage. This is why they have less body weight sexual dimorphism than gorillas; chimpanzees rely on an imbalance of power to win intercommunity battles, the side that has strength in numbers can kill their rivals with very little risk (Wrangham, 1999).

Sequencing the human and chimpanzee genomes has revealed two huge surprises. First, in many ways chimpanzees have evolved faster than humans since the split. Much of this evolution probably has to do with "demonic" behavior among male chimpanzees (*sensu* Wrangham & Peterson, 1997). Chimpanzee evolution is a more typical kind of evolution; selection increases the frequency of some genes, while new genes appear and become more common. Human evolution has proceeded by copying genes, by shutting down genes, and by the rapid evolution of regulatory regions (HARs).

The irascible geneticist Frank Livingstone, one of my professors during my doctoral work, was prone to dramatic proclamations and quixotic scholarly posturing. In the early 1980s, long before the chimpanzee genome had been sequenced, he pronounced humans and chimpanzees to be so closely related that, "Give me 500 years and 50 chimpanzees and I'll breed you a human." A human from a chimp in 25 generations? It seemed so preposterous that it always generated laughs when he said it. Think about it, though. Many human and chimpanzee genetic differences are either merely gene duplications – the genes are virtually identical, but are repeated a different number of times – or came about by turning off genes. These are changes that created domesticated animals, and they evolve quickly. Evolutionarily, these are "easy" changes, far more common that advantageous mutations. Inserting copies of genes happens on a generational scale – very quickly.

Put another way, most of the differences between humans and chimpanzees are to be found regulatory genes, in which genes are turned off and on, and in copies of genes, not the base pair sequences of the genes. As Maynard Olson of the less-is-more genetic hypothesis put it, humans are "hastily made-over apes" (Olson, 1999). All, or nearly all, of the raw material you need to make a human is there in the chimpanzee genome. We could do it by knocking out a few hundred genes, adding copies of a few more, and tweaking regulatory genes. Well put, Professor Livingstone.

References

Altheide TK, Hayakawa T, Mikkelsen TS, et al. (2006) System-wide genomic and biochemical comparisons of sialic acid biology among primates and rodents. *J Biol Chem* 281, 25689–25702.

Arora G, Polavarapu N, McDonald JF (2009) Did natural selection for increased cognitive ability in humans lead to an elevated risk of cancer? *Med Hypoth* 73, 453–456.

Arora G, Mezencev R, McDonald JF (2012) Human cells display reduced apoptotic function relative to chimpanzee cells. *PLoS ONE* 7, e46182. DOI: 10.1371/journal.pone.0046182.

Azevedo FA, Carvalho LR, Grinberg LT, et al. (2009) Equal numbers of neuronal and nonneuronal cells make the human brain an isometrically scaled-up primate brain. *J Comparative Neurol* 513, 532–541.

Bakewell MA, Shi P, Zhang J (2007) More genes underwent positive selection in chimpanzee evolution than in human evolution. *PNAS* 104, 7489–7494.

Biegert J (1963) Evaluation of characteristics of skull, hands and feet for primate taxonomy. In *Classification and Human Evolution* (ed. Washburn SL), pp. 116–145. Chicago, IL: Aldine.

Browner WS, Kahn AJ, Ziv E, et al. (2004) The genetics of human longevity. *Am J Med* 117, 851–860.

Choi SS, Lahn BT (2003) Adaptive evolution of MRG, a neuron-specific gene family implicated in nociception. *Genome Res* 13(10), 2252–2259.

Clancy S (2008) DNA damage & repair: mechanisms for maintaining DNA integrity. *Nature Ed* 1, 103.

CSaA Consortium (2005) Initial sequence of the chimpanzee genome and comparison with the human genome. *Nature* 437, 69–87.

Dawkins R (1976) *The Selfish Gene*. Oxford: Oxford University Press.

De Grouchy J (1987) Chromosome phylogenies of man, great apes, and Old World monkeys. *Genetica* 73, 7–52.

Dorus S, Vallender EJ, Evans PD, et al. (2004) Accelerated evolution of nervous system genes in the origin of Homo sapiens. *Cell* 119, 1027–1040.

Eming SA, Martin P, Tomic-Canic M (2014) Wound repair and regeneration: mechanisms, signaling, and translation. *Sci Transl Med* 6, 265sr6.

Finch CE (2010) Evolution of the human lifespan and diseases of aging: roles of infection, inflammation, and nutrition. *PNAS* 107, 1718–1724.

Finch CE, Stanford CB (2004) Meat-adaptive genes and the evolution of slower aging in humans. *Q Rev Biol* 79, 3–50.

Goodman M (1963) Serological analysis of the systematics of recent hominoids. *Human Biol* 35, 377–436.

Gu J, Gu X (2003) Induced gene expression in human brain after the split from chimpanzee. *Trends Genet* 19, 63–65.

Hagman R, Rönnberg E, Pejler G (2009) Canine uterine bacterial infection induces upregulation of proteolysis-related genes and downregulation of homeobox and zinc finger factors. *PLoS One* 26, e8039.

Hahn MW, Demuth JP, Han SG (2007) Accelerated rate of gene gain and loss in primates. *Genetics* 177, 1941–1949.

Harcourt AH, Harvey PH, Larson SG, Short RV (1981) Testis weight, body weight and breeding system in primates. *Nature* 293, 55–57.

Hauser AS, Chavali S, Masuho I, et al. (2018) Pharmacogenomics of GPCR drug targets. *Cell* 172, 41–54.

Herculano-Houzel S (2014) The glia/neuron ratio: how it varies uniformly across brain structures and species and what that means for brain physiology and evolution. *Glia* 62, 1377–13791.

Herwald H, Egesten A (eds.) (2011) *Sepsis: Pro-Inflammatory and Anti-Inflammatory Responses*. Basel: Karger.

Hughes JF, Skaletsky H, Pyntikova T, et al. (2010) Chimpanzee and human Y chromosomes are remarkably divergent in structure and gene content. *Nature* 463, 536.

Huxley TH (1863) *Evidence as to Man's Place in Nature*. London: Williams & Norgate.

Jirku M, Votýpka J, Petrželková KJ, et al. (2015) Wild chimpanzees are infected by *Trypanosoma brucei*. *Int J Parasitol* 4, 277–282.

Kavaler S, Morinaga H, Jih A, et al. (2011) Pancreatic β-cell failure in obese mice with human-like CMP-Neu5Ac hydroxylase deficiency. *FASEB J* 25(6), 1887–1893.

Kehrer-Sawatzki H, Cooper DN (2007) Understanding the recent evolution of the human genome: insights from human–chimpanzee genome comparisons. *Hum Mutat* 28, 99–130.

King MC, Wilson AC (1975) Evolution at two levels in humans and chimpanzees. *Science* 188, 107–116.

Koo JW, Duman RS (2014) Interleukin-1 receptor null mutant mice show decreased anxiety-like behavior and enhanced fear memory. *Neurosci Lett* 456, 39–43.

Krause J, Lalueza-Fox C, Orlando L, et al. (2007) The derived FOXP2 variant of modern humans was shared with Neandertals. *Curr Biol* 17, 1908–1912.

Lamkanfi M, Declercq W, Kalai M, Saelens X, Vandenabeele P (2002) Alice in caspase land: a phylogenetic analysis of caspases from worm to man. *Cell Death Differ* 9, 358–361.

Landén NX, Li D, Ståhle M (2016) Transition from inflammation to proliferation: a critical step during wound healing. *Cell Mol Life Sci* 73, 3861–3885.

Lauer G, Sollberg S, Cole M, et al. (2000) Expression and proteolysis of vascular endothelial growth factor is increased in chronic wounds. *J Invest Dermatol* 115, 12–18.

Leoni G, Neumann PA, Sumagin R, Denning TL, Nusrat A (2015) Wound repair: role of immune–epithelial interactions. *Mucosal Immunol* 8, 959.

Lodish H, Berk A (2004) *Molecular Biology of the Cell*, 5th ed. New York: Freeman.

Miyamoto MM, Slightom JL, Goodman M (1987) Phylogenetic relations of humans and African apes from DNA sequences in the Globin region. *Science* 238, 369–373.

Nakagawa T, Zhu H, Morishima N, et al. (2000) Caspase-12 mediates endoplasmic-reticulum-specific apoptosis and cytotoxicity by amyloid-beta. *Nature* 403, 98–103.

Napier JR (1963) The locomotor functions of hominids. In *Classification and Human Evolution* (ed. Washburn SL) pp. 178–189. Chicago, IL: Aldine.

Nielsen R, Bustamante C, Clark AG, et al. (2005) A scan for positively selected genes in the genomes of humans and chimpanzees. *PLoS Biology* 3, e170.

Okerblom J, Fletes W, Patel HH, et al. (2018) Human-like Cmah inactivation in mice increases running endurance and decreases muscle fatigability: implications for human evolution. *Proc R Soc B* 285. DOI: 10.1098/rspb.2018.1656.

Olson MV (1999) When less is more: gene loss as an engine of evolutionary change. *Am J Hum Gen* 64, 18–23.

Pilson RD, Pilson BM (1967) Behaviour studies of *Glossina morsitans* Westw. in the field. *Bull Entomol Res* 57, 227–257.

Pollard KS (2009) What makes us human? *Sci Am* 300, 44–49.

Puente XS, Velasco G, Gutiérrez-Fernández A, et al. (2006) Comparative analysis of cancer genes in the human and chimpanzee genomes. *BMC Genomics*, 7, 15.

Qiu L, Pabit SA, Roitberg AE, Hagen SJ (2002) Smaller and faster: the 20-residue Trp-cage protein folds in 4 μs. *J Am Chem Soc* 124, 12952–12953.

Roseboom TJ, Van Der Meulen JH, Ravelli AC, et al. (2001) Effects of prenatal exposure to the Dutch famine on adult disease in later life: an overview. *Molec Cell Endocrinol* 185, 93–98.

Saleh M, Vaillancourt JP, Graham RK, et al. (2004) Differential modulation of endotoxin responsiveness by human caspase-12 polymorphisms. *Nature* 429, 75–79.

Sibley CG, Ahlquist JE (1984) The phylogeny of the hominoid primates, as indicated by DNA–DNA hybridization. *J Mol Evol* 20, 2–15.

Stedman HH, Kozyak BW, Nelson A, et al. (2004) Myosin gene mutation correlates with anatomical changes in the human lineage. *Nature* 428, 415–418.

Swedell L, Plummer T (2012) A Papionin multilevel society as a model for hominin social evolution. *Int J Primatol* 33, 1165–1193.

Tomaszkiewicz M, Rangavittal S, Cechova M, et al. (2016) A time-and cost-effective strategy to sequence mammalian Y chromosomes: an application to the de novo assembly of gorilla Y. *Genome Res* 26, 1–11.

Varki A, Altheide TK (2005) Comparing the human and chimpanzee genomes: searching for needles in a haystack. *Genome Res* 15, 1746–1758.

Varki A, Geschwind DH, Eichler EE (2008) Human uniqueness: genome interactions with environment, behaviour and culture. *Nature Rev Genet* 9, 749–763.

Vassilatis DK, Hohmann JG, Zeng H, et al. (2003) The G protein-coupled receptor repertoires of human and mouse. *PNAS* 100, 4903–4908.

Wang X, Grus WE, Zhang J (2006) Gene losses during human origins. *PLoS Biol* 4, e52.

Weis E, Galetzka D, Herlyn H, Schneider E, Haaf T (2008) Humans and chimpanzees differ in their cellular response to DNA damage and non-coding sequence elements of DNA repair-associated genes. *Cytogenet Genome Res* 122, 92–102.

Wettschureck N, Offermanns S (2005) Mammalian G proteins and their cell type specific functions. *Physiol Rev* 85, 1159–1204.

Wildman DE, Uddin M, Liu G, Grossman LI, Goodman M (2003) Implications of natural selection in shaping 99.4%

nonsynonymous DNA identity between humans and chimpanzees: enlarging genus *Homo*. *PNAS* **100**, 7181–7188.

Wilson ML, Boesch C, Fruth B, et al. (2014) Lethal aggression in *Pan* is better explained by adaptive strategies than human impacts. *Nature* **513**, 414–417.

Winter H, Langbein L, Krawczak M, et al. (2001) Human type I hair keratin pseudogene φ hHaA has functional orthologs in the chimpanzee and gorilla: evidence for recent inactivation of the human gene after the *Pan–Homo* divergence. *Hum Genet* **108**, 37–42.

Wohleb ES, Patterson JM, Sharma V, et al. (2014) Knockdown of interleukin-1 receptor type-1 on endothelial cells attenuated stress-induced neuroinflammation and prevented anxiety-like behavior. *J Neurosci* **34**, 2583–2591.

Wrangham RW (1999) Evolution of coalitionary killing. *Am J Phyl Anthropol* **110**, 1–30.

Wrangham RW, Peterson D (1997) *Demonic Males: Apes and the Origins of Violence*. New York: Houghton Mifflin.

Xu S, Chen Y, Cheng Y, et al. (2012) Positive selection at the ASPM gene coincides with brain size enlargements in cetaceans. *Proc Roy Soc Lond B* **279**, 4433–4440.

Xue Y, Daly A, Yngvadottir B, et al. (2006) Spread of an inactive form of caspase-12 in humans is due to recent positive selection. *Am J Hum Gen* **78**, 659–670.

Yunis JJ, Prakash O (1982) The origin of man: a chromosomal pictorial legacy. *Science* **215**, 1525–1530.

Yunis JJ, Sawyer JR, Dunham K (1980) The striking resemblance of high-resolution G-banded chromosomes of man and chimpanzee. *Science* **208**, 1145–1148.

13 Making Your Way in the Great Wild World
Chimpanzee Senses

Photo: Shara Johnson, with permission

If we could transplant your brain into the skull of a chimpanzee so as to allow you to experience the world through their sensory system, you would see things... just as you do now. Chimpanzee and human senses are so similar that your vision, touch, taste, smell, and hearing would be so little different you would hardly notice. As unalike as our bodies appear at first sight, the data the chimpanzee body sends to the brain are virtually identical to ours.

13.1 Vision

Among the primates, monkeys and apes in particular rely on vision for communication, rather than smell; not surprisingly, then, the ape/monkey lineage has seen an increase in eye size (but not shape) that has enhanced visual acuity (Ross, 2000; Dominy et al., 2004; Kirk, 2004; Ross & Martin, 2007). Drawing on our superior vision, we monkeys and apes communicate with facial gestures and find our food with our eyes and memory, not with our nose.

Apes and humans receive visual information from a part of the electromagnetic spectrum that we rather anthropocentrically call "visible light," ranging from the long-wavelength red side of the spectrum right through to what we perceive as violet light, on the shorter side, petering as we reach the ultraviolet range. Many other animals have a broader range, seeing farther into either the infrared or ultraviolent ends of our range – or both – and thus perceiving images and patterns invisible to us (Detwiler, 1943; Nilsson, 2009; Fain et al., 2010).

Light passes through our (and the chimpanzee's) clear cornea, through the opening called the pupil, through a lens that muscles can stretch, thinning the lens to allow us to focus on things at different depths, through a clear, jelly-like substance called the vitreous fluid and finally striking light-perceiving tissue in the back of the eye, the retina.

The retina has *two basic kinds of light-detecting cells, "rods" that detect only light and dark and "cones," which allow us to detect colors*. Humans and chimpanzees perceive light in three different color ranges, giving us what vision specialists know as "trichromatic" vision. There are three slightly different type of cones: S-type cones detect wavelengths of light centered around blue, at wavelengths of 400–500 nm; M-type cones detect bluish-green light with wavelengths of 450–630 nm, which gives them some overlap with the S-type cones on one side and L-type cones on the other, which detect red light in the 500–700 nm range (Grether, 1940, 1941; Prestrude, 1970).

Impulses from the rods and cones pass via the optic nerve to the visual cortex of the brain, where the information, both where on the retina the light fell and which of the rods and the three cones were stimulated, is organized into what we perceive as images. The location of the visual cortex is one of those anatomical details that testify that we are the product of evolution; once located conveniently close to the eyes in our distant ancestors, it has been displaced by the expansion of the neocortex, the part of the brain responsible for higher thought, to the back of the brain in humans and other primates. This means humans take a fraction of a second longer to perceive visual images than, say, a rat.

We are fortunate to detect all three colors, otherwise van Gogh paintings would lose some of their charm, though of course we would never know what we were missing if we saw only two, as most diurnal mammals do – including, surprisingly most primates from South and Central America; they are dichromats, having only two-color receptors (Detwiler, 1943; Jacobs, 1993). While chimpanzees are trichromats (Grether, 1940, 1941; Prestrude, 1970), their perception appears to be displaced 5 nm toward the short wavelength or violet end of the spectrum (Grether, 1940, 1941; Jacobs et al., 1996). With a visual spectrum spanning 300 nm, a displacement of 5 nm, or less than 2 percent, is minor enough that Maier and Schneirla (1935: 515) characterized the color discrimination abilities of chimpanzees and humans as "identical, to all intents and purposes."

Our eyes, and those of chimpanzees, are the eyes of frugivores (fruit-eaters). In experiments in which fruits were placed against a uniform green background, after which an objective measure of how distinguishable they are from the background was made using a spectroradiometer, fruits the color of those typical in primate diets – yellow or orange (Gautier-Hion, 1988) – were more discernible than similar objects of different colors (Osorio & Vorobyev, 1996; Sumner & Mollon, 2000). Interestingly, nutritious leaves are also reddish in the tropics, making yellow-red color detection valuable for many primates, even those that are specialized for leaf-eating (Dominy & Lucas, 2001).

There is much direct evidence that chimpanzees see what we see, but their eye anatomy should lead us to expect that; it is virtually identical to that of humans, though their eyeball is slightly smaller, probably only because their entire bodies are slightly smaller (Prestrude, 1970; Young & Farrer, 1964; Young, 1971). While color perception may be slightly different in humans and chimpanzees, visual acuity is so similar that it is not statistically different (Spence, 1934), even though the smaller eye of chimpanzees should give them less acuity (Prestrude, 1970). Tests of other visual capacities – discrimination among different levels of brightness (not color brightness, but intensity of white light) or the ability to detect movement – showed no detectable differences. Vision is in the brain, however, rather than just the eye; different animals process the information from the eye differently.

13.2 Color Perception

We humans tend to emphasize edges, but other animals are more acutely aware of motion. It might be, then, that brain adaptations cause chimpanzees to see shapes differently than us. However, it turns out not: When asked to sort objects on the basis of color, chimpanzees test out identical to humans (Ladygina-Kohts, 1935; Jarvik, 1956; Matsuzawa, 1985; Rumbaugh, 1977). Some of these experiments were by Nadia Ladygina-Kohts and are worthy a quick admiring comment. They are extraordinary in part because they occurred so early, 100 years ago. But remarkable also because Kohts performed them with almost no institutional support and at a time when women faced intense discrimination in academia (Yerkes & Petrunkevitch, 1925). She sat her five-year-old chimpanzee study subject, Ioni (let us Anglicize it to "Johnny"), across from her at a table and showed him a test object. His task was to choose an exact match from among a sample of similar objects arrayed in front of him ("matching to sample"). By making the difference in color between the test object and the comparison sample closer and closer, she was able to determine just how similar colors must be for Johnny to recognize that they match, and how similar this ability was to humans.

13.3 Shape Perception

Kohts tested shape perception in the same way. Johnny discriminated at human-like levels on pictures of circles, ovals, rectangles, rhombuses, trapezoids, and pie-shaped segments, as well as polygons with five, six, eight, ten, and twelve sides. He could distinguish one from another of spheres, cylinders, cones, cubes, pyramids, and prisms with varying numbers of surfaces. Recall that Johnny was only five years old.

Her experiments say something about the chimpanzee personality, in addition to their cognitive ability. When Johnny failed and Kohts verbally announced "wrong," the announcement was said to be met with an "expression of sadness sometimes accompanied by protrusion of the lips and crying." When he got it right he was rewarded with what he desired most, a play session with Kohts in the form of "chasing, catching, and wrestling" (Ladygina-Kohts, 1935; Yerkes & Petrunkevitch, 1925).

If the test object was withdrawn, Johnny quickly forgot it so that if the presentation of the test sample was much more than 15 seconds after the object with which he was meant to match it, he failed, but this was a matter of shape-memory rather than perception. Similarly, while Johnny was excellent at discriminating among colors, Kohts noted that the vividness of color seemed to matter little to him (Ladygina-Kohts, 1935), possibly as a matter of motivation rather than perception. Kohts did not detect the slight blue-shift that others did later.

13.4 Size Perception

The chimpanzee ability to detect differences in the size of objects was also indistinguishable from that of humans; when two objects differ by 5 percent we (and chimpanzees) can tell (Menzel, 1961; Prestrude, 1970).

13.5 Night Vision; Aging

Chimpanzees have the same night vision acuity as humans: very little (Prestrude, 1970). Perhaps surprisingly, given how vital good eyesight is for an animal that must judge distances during leaps and must detect predators and enemies from a distance, chimpanzees (at least in captivity), like humans, become near-sighted as they grow older, to the same extent and at approximately the same ages as humans (Prestrude, 1970).

13.6 Edge Detection

It is sometimes reported that chimpanzees are baffled by line drawings, but the Hayeses (1953) reported that one chimpanzee, Vicki, famous as their main language-study chimpanzee, was able to match an objects to black and white line drawings even if there was a time delay. Perhaps Kohts' Johnny's failure was an individual quirk.

13.7 Facial Discrimination

Face recognition is more of a cognitive processing task than a visual task, but it is worth noting that chimpanzees can discriminate among different faces – whether human or chimpanzee – as well humans, and they do so using the same neural mechanisms (Boysen & Bernston, 1989; Parr et al., 2000; Nelson, 2001; Martin-Malivel & Okada, 2007).

13.8 Hearing (Audition)

Chimpanzee hearing is nearly identical to that of humans, but only "nearly"; they can detect higher-frequency sounds than humans (Elder, 1934, 1935; Spector, 1956; Farrer & Prim, 1965; Kojima, 1990). Human detection generally tops out near 24 kHz, and this is for young test subjects; some middle-age subjects without diagnosed hearing impairment failed at 14 kHz. One chimpanzee, in contrast, got nearly 50 percent of the trials right at 34 kHz. Other studies report similar results, with chimpanzees about 10 kHz better than humans (Spector, 1956; Farrer & Prim, 1965; Gilbert & Kaplan, 1965). Elder (1934) felt comfortable characterizing the human upper range as 20 kHz and chimpanzees as 30 kHz. On the other end of the scale, chimpanzees are slightly less sensitive than humans to low-frequency sound, below 250 Hz (Kojima, 1990).

Interestingly, chimpanzee perception of higher frequencies improves as they reach adulthood and this sensitivity persists into middle age (Elder, 1935), perhaps because detecting distant pant-hoots is more important for adults. From a distance, the lower register of chimpanzee pant-hoots is lost and the hoots sound like high-pitched chirps. The difference between close-up and distance was brought home to me one day when I was in the field with one of my assistants. She was very experienced with chimpanzees in captivity and thus intimately familiar with the sound of pant-hoots. Yet, as we were searching for chimpanzees that day I heard a distant pant-hoot, probably more than a mile away, and without comment began to move quickly toward the hoots. She was baffled. "Those are pant-hoots?" she asked. "I've heard that a lot – I thought it was bird calls." Pant-hoots call community members to abundant resources so that they can gather for mutual protection; other times they notify each other of one another's location. Hearing these high-pitched tones could be the difference between life and death for a chimpanzee.

Beyond detecting high-pitched sounds, it is often stated that chimpanzees can detect sounds at lower volumes than humans, at all frequencies, not just higher ones, but support for this assumption is a little squishy. Fieldworkers often comment on how astute chimpanzee hearing is. Many times, I have been sitting with chimpanzees in a group when every head in the group – except mine – swivels to look in a particular direction. Some individuals stand up and orient their bodies in the direction of the sound, and sometimes they run off in the direction they are facing. If the group walks very far, soon I can begin to hear the distinctive high-pitched hoo-hoo-hoos of a distant chimpanzee pant-hoot. You might worry about my hearing since I have borne decades of hearing-damaging antimalarials, worked with power tools and attended rock concerts, but even my young colleagues and assistants tell me they hear nothing in situations like this. Experiences like mine, and reports of humans who rear chimpanzees in their homes (Kellogg & Kellogg, 1933), seem to support the keener-hearing hypothesis.

13.9 Ear Size

I am aware of no lab tests that have attempted to examine the function of the larger ears (pinnae), especially ear breadth (see Chapter 6) or the narrower ear canals of chimpanzees (Nevell & Wood, 2008). These might give chimpanzees better hearing in a practical sense, even though it has not been detected in the lab with headphones on. The larger external ear and narrower canal may mean that the chimpanzee ear both gathers more energy from sound waves, as you can by cupping a hand over your ear, and concentrates it more by narrowing the ear opening, like an old-fashioned mechanical hearing aid, the ear trumpet. Such an adaptation may compromise subtle aspects of sound quality, but its advantage is that it allows chimpanzees to hear the distant pant-hoots that may warn them of a murderous intruder or call them to a critical food resource.

The Kelloggs found another advantage that has not been tested in the lab, the ability of chimpanzees to locate the direction from which the sound is coming. They found their son Donald succeeded in locating the exact source of a sound half as often as their chimpanzee subject.

13.10 Taste (Gustation)

Sense of taste is referred to scientifically as **gustation**. Mammalian gustation differentiates five tastes: salty, sweet, umami, sour, and bitter (Table 13.1). Chimpanzees detect all five tastes at concentrations very like those of humans (Kalmus, 1970; Glaser, 1986; Simmen & Hladik, 1998; Hellekant & Ninomiya, 1991, 1994; Hellekant et al., 1996, 1997, 1998; Hladik et al., 2002). Each of these tastes has its purpose.

13.10.1 Salty

Humans and chimpanzees are evolved to love salt because it makes digestion more efficient (Chapter 6), but it is also essential for maintaining the proper balance between extracellular and intracellular fluids. When the amount of fluids outside of cells is too low it leads to apathy, weakness, fainting, anorexia, and

Table 13.1 **Chimpanzee gustation**

Taste	Comparison with humans	Function	Role in diet
Sweet	Both highly prefer	Detect sugar	Maximize caloric consumption
Salty	Both highly prefer	Detect sodium	Improve digestion
Umami	Unknown, but likely the same	Detect fat, protein	Maximize fat, detect protein
Sour	Chimpanzees prefer	Detect acid	Unexplained
Bitter	Both avoid	Detect alkaloids	Avoid poisons

low blood pressure. When the condition is extreme, the circulatory system collapses and death soon follows (Nadal et al., 1941).

13.10.2 Sweet

Our fondness for sugar is an evolved preference that motivates us to maximize caloric consumption. Humans and chimpanzees both love sugar. Chimpanzee and human sensitivity to sugar, combined with our intense motivation to consume it, makes us sugar maximizers. Combine these two things – taste and motivation – with intelligence and you have a very effective sugar collector. As much as we humans like sugar, my impression is that chimpanzees like it more. In 1986/1987, when I was in Tanzania, it was nearly impossible for me to get sugar, and I often went months without any at all. I can testify that this deprivation intensified my desire for sugar, but it still seems to me that it is innately more intense among chimpanzees.

13.10.3 Umami

Umami is associated with protein and fat, both important nutrients. It is rare for chimpanzees to need to seek out protein; they get enough while pursuing sugar, though sometimes needs can be high, such as when nursing. The taste for and desire to consume fat serves to maximize caloric intake.

13.10.4 Bitter

Bitter compounds are often poisonous; alkaloids are the biggest danger. Humans and chimpanzees both taste bitterness, giving them the ability to detect poisons so as to avoid them. Both humans and chimpanzees respond to bitter tastes with negative facial expressions such as grimaces, eye squinches, nose wrinkles, and frowns – at *birth* (Ueno et al., 2004; Chapter 6); it is clear this is a biological, not a cultural preference. Chimpanzees have less of a negative reaction to bitter tastes than humans, but the difference between the two species is slight. Briefly (see Chapter 6 for details), compounds such as strychnine, arsenic, cyanide, and other alkaloids interfere with internal physiological processes, and can be fatal. Humans may be slightly more variable, as a species, in their ability to detect bitter tastes; the *TAS2R* receptor gene, the gene responsible for the ability to detect bitter tastes, appears to be under relaxed selection in humans (Wang et al., 2004). In other words, we rely on culture to avoid poisons even more than do chimpanzees – and the chimpanzee diet is *very* culturally determined.

13.10.5 Sour

There is one real difference between our two species: Chimpanzees show a preference for sour things. The Kelloggs first documented this preference by testing chimpanzee Gua against their son Donald. They fed each a 5 percent citric acid mixture, to which Gua reacted positively and Donald not (Kellogg & Kellogg, 1933). This is a concoction we would experience as sour, a taste also found in many unripe fruits (Behrens & Meyerhof, 2011). Gua was completely undeterred by the addition of the acid mixture, whereas Donald turned his nose up at it. Kohts had noted the same preference, though without experimenting (Ladygina-Kohts, 1935/2002: 82–83). One might think we evolved the ability to detect sour tastes so as to allow us to modulate our intake of unripe fruits, inspiring us to focus on more ripe fruits, but this cannot explain chimpanzee preference since they strongly prefer ripe fruits. Likewise, vitamin C, an essential nutrient, is sour – it is an acid – but is so ubiquitous in fruits that chimpanzees would have little need to seek it out. Furthermore, many fruits are sour even when ripe, partly as a means of extending the healthy life of the fruit – acid is in essence an antibiotic. It is easy to understand why it might fail to deter chimpanzees, but why they prefer it is another matter. We know of no desirable nutrient that is associated with acids that is not also associated with sugar. The mystery remains.

13.11 Smell (Olfaction)

As with other senses, the chimpanzee olfactory sense is quite similar to that of humans, but there is genetic evidence that smell, or olfaction, has evolved to be less acute in humans (Fischer et al., 2004; Gimelbrant et al., 2004) or perhaps just different (Nielsen et al., 2005). This information jibes with observations of wild chimpanzees; they sometimes pause to sniff at the ground, paying close attention to a spot that is odorless to me. Captive chimpanzees seem to detect odors better than humans (Köhler, 1925; Yerkes & Yerkes, 1929; Ladygina-Kohts, 1935/2002: 78), though the difference seems minor. Chimpanzees, like humans, are sensitive to the smell of ethanol, presumably to help detect ripe fruits (Dudelly, 2004). Definitive lab tests are lacking.

13.12 Touch (Somatosensation)

All we know of the chimpanzee sense of touch, or of skin sensitivity, comes from the Kelloggs' study – and there appears to be no difference between humans and chimpanzees. Nociception, or pain perception, is quite a different matter.

13.13 Pain Perception (Nociception)

The Kelloggs, Ladygina-Kohts, the Yerkeses, and many others who have lived among chimpanzees have noted how impervious they are to pain. An anecdote about the language-experiment chimpanzee Washoe illustrates this. Washoe and a caretaker had been playing outside when it came time to go back inside the facility. The door of the facility was constructed of heavy-duty steel and mounted with a powerful spring to ensure it could not be left ajar. Returning from outside, the caretaker entered first and held the door for Washoe, who rushed forward as if she intended cross the threshold without stopping. The caretaker, seeing Washoe was nearly inside, turned away and released the door. At the last second Washoe paused outside but then attempted to leap in ahead of the rapidly closing door; she failed. The door slammed on her ankle. The caretaker turned just in time to see this horrific sight, and her facial expression in response was one of pain and horror. Washoe, seeing this expression, eagerly approached the caretaker, asking her in sign language if she were okay. The pain of the injured ankle was not severe enough that Washoe judged it to be important, compared to her empathy for her friend, even though the ankle was later found to be broken.

13.14 Temperature Perception (Thermoception)

Little has been done to document chimpanzee temperature perception, but in captivity chimpanzees appear to suffer from heat stress similarly to humans (Köhler, 1925; Fouts, 1974). In the wild they avoid direct sun at Gombe (Goodall, 1986) and Budongo, where chimpanzees also have been shown to avoid the sun more as the temperature rises, to curtail activities in the heat, and to move to the ground into the shade when it is hot (Kosheleff & Anderson, 2009). My experience is in line with some of the earliest observations of chimpanzees, which is that, compared to humans, they avoid exposure to direct sunlight (Nissen, 1931). In the wild they begin to shiver at about the same temperature as humans (Goodall, 1986).

13.15 Body Sense (Proprioception)

Kinesthetic sense, the ability to perceive the relationship of body parts to one another and to the environment – often called proprioception – has been studied scientifically only quite superficially; no difference was found between humans and chimpanzees (McCulloch, 1941; Prestrude, 1970).

13.16 Balance (Equilibroception)

Chimpanzees have extraordinary balance (Figure 13.1), a sense that is principally governed by the semicircular canals. Chimpanzees are born gymnasts. Those who have reared chimpanzees as humans find them to acquire climbing, leaping, and most particularly high-wire balance skills with no training or encouragement whatsoever; if anything, they are discouraged. The sight of a chimpanzee running in a tree among irregularly placed and unstable branches leaves first-time observers awestruck. We see similar skills among humans only among highly trained Olympic athletes and circus acrobats, and I believe even these adepts do not approach chimpanzee agility.

We might expect the superior chimpanzee skills to be reflected in semicircular canal shape, but function has not been clearly linked to morphology for the canals (Spoor et al., 2007; Gunz et al., 2013). We do know that an extremely important function of the canals is stabilizing the visual field during locomotion, and animals with great agility and/or who commonly utilize locomotor movements that jostle the head have larger semicircular canals (Spoor et al., 2007). When mice are bred for competence during quick locomotor movements, their semicircular canal size increases (Schutz et al., 2014). Chimpanzee canal size is large in relation to body size, suggesting they have superior agility and, for their body size, quite rapid locomotion (Ryan et al., 2012). Their canals differ from those of humans, but only slightly (Figure 13.2; Gunz et al., 2013).

13.17 Summary

To sum up, chimpanzee vision is little different than that of humans; their perception of shape, color, and size, and their edge detection, are very similar.

Figure 13.1 Chimpanzees have an extraordinary sense of balance, and spontaneously perform astonishing feats of acrobatics without training, as the climbing and leaping behavior of chimpanzees reared by humans demonstrates. Photo Shara K. Johnson.

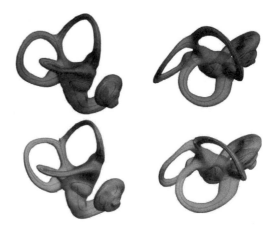

Figure 13.2 Semicircular canal shape of humans (top) and chimpanzees differ, but only slightly, seemingly at odds with evidence of superior chimpanzee balance and body sense. From Gunz et al., 2013, with permission.

Compared to humans, chimpanzees register sounds better in higher frequencies, and their external ear and auditory canal shape may make their hearing superior to that of humans. The ability to hear higher frequencies is unexplained, but my speculation is that the large external pinnae (external ears) and ability to hear higher registers are adaptations that help them hear pant-hoots in the distance. Their sense of taste is the same as ours, except that they prefer sour tastes more. They probably have slightly more sensitive noses. There is no evidence their sense of touch or thermoception is different from ours. They seem to register pain less intensely. Their agility suggests that their sense of balance is superior to that of humans, but similar to that of monkeys.

All of these differences – perhaps with the exception of pain – are so small that, as I wrote to open this chapter, if we had the same senses as chimpanzees, we would not even notice. The similarity of our senses reflects a long shared evolutionary history, the fact that we require the same nutrients, and our shared need to avoid poisons. When it comes to senses, we are indeed as alike as sisters.

References

Behrens M, Meyerhof W (2011) Gustatory and extragustatory functions of mammalian taste receptors. *Physiology & Behavior* **105**, 4–13.

Boysen ST, Bernston GG (1989) Conspecific recognition in the chimpanzee (*Pan troglodytes*): cardiac responses to significant others. *J Comp Psych* **103**, 215–220.

Detwiler S (1943) *Vertebrate Photoreceptors*. New York: Macmillan.

Dominy NJ, Lucas PW (2001) Ecological importance of trichromatic vision to primates. *Nature* **410**, 363–366.

Dominy NJ, Ross CF, Smith TD (2004) Evolution of the special senses in primates: past, present, and future. *Anat Rec* **281**, 1078–1082.

Dudelly R (2004) Ethanol, fruit ripening, and the historical origins of human alcoholism in primate frugivory. *Integr Comp Biol* **44**, 315–323.

Elder JH (1934) Auditory acuity of the chimpanzee. *Comp Psychol* **17**, 157–183.

Elder JH (1935) The upper limit of hearing in chimpanzee. *Am J Physiol* **112**, 109–115.

Fain GL, Hardie R, Laughlin SB (2010) Phototransduction and the evolution of photoreceptors. *Curr Biol* **20**, R114–R124.

Farrer DN, Prim MM (1965) A preliminary report on auditory frequency threshold comparisons of humans and pre-adolescent chimpanzees. Technical Report No. 65-6, 6571st Aeromedical Research Laboratory, Holloman Air Force Base.

Fischer A, Gilad Y, Man O, Pääbo S (2004) Evolution of bitter taste receptors in humans and apes. *Molec Biol Evol* **22**, 432–436.

Fouts RF (1974) Capacities for language in great apes. In *Proceedings of the 18th International Congress on Anthropological and Ethnological Sciences*, pp. 1–20. The Hague: Mouton.

Gautier-Hion A (1988) The diet and dietary habits of the forest guenon. In *A Primate Radiation: Evolutionary Biology of the African Guenons* (eds. Gautier-Hion A, Bourliere F, Gautier, J-P, Kingdon J), pp. 257–283. Cambridge: Cambridge University Press.

Gilbert G, Kaplan G (1965) A wide band frequency auditory test system for use with primates. Technical Report No. ARL-TR-6S-5) 6571st Aeromedical Research Laboratory, Holloman Air Force Base.

Gimelbrant AA, Skaletsky H, Chess, A (2004) Selective pressures on the olfactory receptor repertoire since the human–chimpanzee divergence. *PNAS* **101**, 9019–9022.

Glaser D (1986) Geschmacksforschung bei Primaten. *Vjschr Naturf Ges Zürich* **131**(2), 92–110.

Goodall J (1986) *The Chimpanzees of Gombe: Patterns of Behavior*. Cambridge: Harvard University Press.

Grether WF (1940) A comparison of human and chimpanzee hue discrimination curves. *J Exp Psychology* **26**, 394–403.

Grether WF (1941) Comparative visual acuity thresholds in terms of retinal image widths. *J Comp Psychology* **31**, 23–33.

Gunz P, Ramsier M, Kuhrig M, Hublin JJ, Spoor F (2013) The mammalian bony labyrinth reconsidered, introducing a comprehensive geometric morphometric approach. *J Anat* **220**, 529–543.

Hayes K, Hayes C (1953) Picture perception in a home-raised chimpanzee. *J Comp Physiol Psychol* **46**, 470–474.

Hellekant G, Ninomiya Y (1991) On the taste of umami in chimpanzee. *Physiol Behav* 49, 927–934.

Hellekant G, Ninomiya Y (1994) Bitter taste in single chorda tympani taste fibers from chimpanzee. *Physiol Behav* 56, 1185–1188.

Hellekant G, Ninomiya Y, Danilova V (1997) Taste in chimpanzees: II. Single chorda tympani fibers. *Physiol Behav* 61, 829–841.

Hellekant G, Ninomiya Y, DuBois GE, Danilova V, Roberts TW (1996) Taste in chimpanzee: I. The summated response to sweeteners and the effect of gymnemic acid. *Physiol Behav* 60, 469–479.

Hellekant G, Ninomiya Y, Danilova V (1998) Taste in chimpanzees: III. Labeled-line coding in sweet taste. *Physiol Behav* 65, 191–200.

Hladik CM, Pasquet P, Simmen B (2002) New perspectives on taste and primate evolution: the dichotomy in gustatory coding for perception of beneficent versus noxious substances as supported by correlations among human thresholds. *Am J Phys Anthropol* 117, 342–348.

Jacobs GH (1993) The distribution and nature of colour vision among the mammals. *Biol Rev* 68, 413–471.

Jacobs GH, Deegan JF, Moran JL (1996) ERG measurements of the spectral sensitivity of common chimpanzee (*Pan troglodytes*). *Vision Res* 36, 2587–2594.

Jarvik ME (1956) Simple color discrimination in chimpanzees: effect of varying contiguity between cue and incentive. *J Comp Physiol Psychol* 46, 390–392.

Kalmus H (1970) The sense of taste of chimpanzees and other primates. *The Chimpanzee* 2, 130–141.

Kellogg WN, Kellogg LA (1933) *The Ape and the Child: A Study of Environmental Influence upon Early Behavior.* New York: McGraw-Hill.

Kirk EC (2004) Comparative morphology of the eye in primates. *Anat Rec A* 281A, 1095e1103.

Köhler W (1925/1959) *The Mentality of Apes*, 2nd ed. New York: Viking.

Kojima S (1990) Comparison of auditory functions in the chimpanzee and human. *Folia Primatol* 55, 62–72.

Kosheleff VP, Anderson CNK (2009) Temperature's influence on the activity budget, terrestriality, and sun exposure of chimpanzees in the Budongo Forest, Uganda. *Am J Phys Anth* 139, 172–181.

Ladygina-Kohts NN (1935/2002) *Infant Ape and Human Child*. Moscow: Museum Darwinianum [taDP].

Maier N, Schneirla T (1935) *Principles of Animal Psychology*. New York: McGraw-Hill.

Martin-Malivel J, Okada K (2007) Human and chimpanzee face recognition in chimpanzees (*Pan troglodytes*): role of exposure and impact on categorical perception. *Behav Neurosci* 121(6), 1145.

Matsuzawa T (1985) Colour naming and classification in a chimpanzee (*Pan troglodytes*). *J Hum Evol* 14, 283–291.

McCulloch T (1941) Discrimination of lifted weights by chimpanzees. *J Comp Psych* 32, 507–519.

Menzel EW, Jr. (1961) Perception of food size in the chimpanzee. *J Comp Physiol Psychol* 54, 588–591.

Nadal JW, Pedersen S, Maddock WG (1941) A comparison between dehydration from salt loss and from water deprivation. *J Clin Invest* 20, 691–703.

Nelson CA (2001) The development and neural bases of face recognition. *Infant Child Dev* 10(1), 3–18.

Nevell L, Wood B (2008) Cranial base evolution within the hominin clade. *J Anat* 212, 455–468.

Nielsen R, Bustamante C, Clark AG, et al. (2005) A scan for positively selected genes in the genomes of humans and chimpanzees. *PLoS Biol* 3, e170.

Nilsson DE (2009) The evolution of eyes and visually guided behaviour. *Phil Trans Roy Soc B* 364, 2833–2847.

Nissen HW (1931) A field study of the chimpanzee: observations of chimpanzee behavior and environment in western French Guinea. *Comp Psychol Monogr* 8, 122.

Osorio D, Vorobyev M (1996) Colour vision as an adaptation to frugivory in primates. *Proc Biol Sci* 263, 593–599.

Parr LA, Winslow JT, Hopkins WD, de Waal F (2000) Recognizing facial cues: individual discrimination by chimpanzees (*Pan troglodytes*) and rhesus monkeys (*Macaca mulatta*). *J Comp Psychol* 114(1), 47.

Prestrude AM (1970) Sensory capacities of the chimpanzee. *Psychol Bull* 74, 47–67.

Ross CF (2000) Into the light: the origin of Anthropoidea. *Ann Rev Anthropol* 29, 147–194.

Ross CF, Martin RD (2007) The role of vision in the origin and evolution of primates. In *Evolution of Nervous Systems*, Vol. 5: *The Evolution of Primate Nervous Systems* (eds. Preuss TM, Kaas JH), pp. 59–78. Oxford: Elsevier.

Rumbaugh D (1977) *Language Learning by a Chimpanzee: The Lana Project*. New York: Academic Press.

Ryan TM. Silcox MT, Walker A, et al. (2012) Evolution of locomotion in Anthropoidea: the semicircular canal evidence. *Proc Roy Soc Lond B* 279, 3467–3475.

Schutz H, Jamniczky HA, Hallgrimsson B, et al. (2014) Shape-shift: semicircular canal morphology responds to selective breeding for increased locomotor activity. *Evolution* 68, 3184–3198.

Simmen B, Hladik CM (1998) Sweet and bitter taste discrimination in primates: scaling effects across species. *Folia Primatol* 69, 129–138.

Spector W (1956) *Handbook of Biological Data*. Philadelphia, PA: WB Saunders.

Spence KW (1934) Visual acuity and its relation to brightness in chimpanzee and man. *J Comp Psychol* 18, 333–361.

Spoor F, Garland T, Krovitz G, et al. (2007) The primate semicircular canal system and locomotion. *PNAS* 104, 10808–10812.

Sumner P, Mollon JD (2000) Catarrhine photopigments are optimized for detecting targets against a foliage background. *J Exp Biol* 203, 1963–1986.

Ueno A, Ueno Y, Tomonaga M (2004) Facial responses to four basic tastes in newborn rhesus macaques (*Macaca mulatta*) and chimpanzees (*Pan troglodytes*). *Behav Brain Res* 154, 261–271.

Wang X, Thomas SD, Zhang J (2004) Relaxation of selective constraint and loss of function in the evolution of human bitter taste receptor gene. *Hum Molec Genet* 13, 2671–2678.

Yerkes RM, Petrunkevitch A (1925) Studies of chimpanzee vision by Ladygin-Kohts. *J Comp Psychol* 5, 99–108.

Yerkes RM, Yerkes DN (1929) Concerning memory in the chimpanzee. *J Comp Psychol* 8, 237–271.

Young F (1971) Visual similarities of nonhuman and human primates. In *Medical Primatology* (eds. Goldsmith EI, Moor-Jankowski J), p. 316. Basel: Karger.

Young F, Farrer D (1964) Refractive characteristics of chimpanzees. *Am J Optometry* 41, 81–91.

14

The Grim Reaper in the Forest Primeval
Wild Chimpanzee Diseases and Lessons for Healthy Living

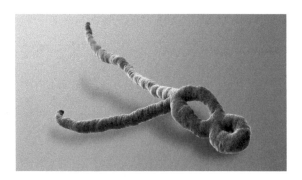

Ebola image by Scientific Animations (www.scientificanimations.com), licensed under CC BY-SA 4.0

We humans suffer from a multitude of maladies. No matter our age, the ever-watchful Grim Reaper may spring upon us from any direction. One of the lessons chimpanzees teach us is that many of our ailments are self-inflicted – we kill ourselves in car crashes, shootings, and suicides, yes, but rather than external forces I mean internal afflictions that sap our vitality and tip us into an early grave. Comparing rates of heart disease, arthritis, and cancer among chimpanzees and different human populations tells us these ailments are the product of the mismatch between our current mode of life and the living conditions in which our ancestors evolved. Listening to these wellness lessons could yield profound health benefits, if only our cognitive selves could overpower instincts bequeathed us by our hunter-gatherer ancestors.

PART I: PARASITES AND MEDICINE

March 9, 1987, Mahale, Tanzania: In the predawn darkness my colleague Hiroyuki Takasaki and I stood under a 40-foot-tall tree on a hilltop, waiting for the female chimpanzee Chausiku and her son Chopin (Figure 14.1) to wake up and start their day. I had followed her and her friend Gwekula to their night nests the day before. Hiroyuki and I looked forward to what might be, with a little luck, a 13-hour-long, all-day follow of this charismatic, high-ranking female. We chatted quietly as the diffuse light of the growing dawn filtering down through a cloudy sky. Just after dawn Gwekula emerged from her nearby nest, climbed down, and departed. I did not consider following her, even though my protocol called for me to switch targets often, and I had followed Chausiku the day before; an injury to a leg some years ago had left Gwekula lame and unable to engage in the full range of locomotor and postural behavior, the subject of my study. Minutes passed. We grew impatient. Finally, we decided that she must have moved her nest in the middle of the night, an unusual but not unheard-of event, especially when there is a full moon. I was leaning against the tree as we discussed whether last night's waxing-gibbous half-moon under cloudy skies precluded a nest-switch, when I felt a shudder in the tree. We looked up and saw Chopin peering at us over the edge of the nest. A minute later Chausiku's rump edged over the lip of the nest and we were forced to leap out of the way to avoid the following cascade of urine and feces.

Still, she did not emerge. 7:00 came, very late for a chimpanzee to start her day, then 7:30. It was 7:35 before she exited her nest, climbed down, and walked – ever so gingerly – down a winding path in the low foliage of the hilltop, toward Lake Tanganyika to the east. As we watched, she paused to peer into a knot of foliage for some time, like a diner dithering over the wine list, finally plucking a plant and sitting to eat it. Holding the stem of the plant in her hand she gently closed her mouth over a leaf, opening and closing her mouth several times, finally pulling off the lanceolate-shaped leaf. For perhaps half a minute she seemed to roll the leaf around in her mouth and then, with a visible effort, swallowed it. She repeated the process with a dozen leaves. As she

Richard Wrangham and Toshisada Nishida, discovered the use of medicinal plants by chimpanzees a few years before (Wrangham & Nishida, 1983).

Chimpanzees eat a wide variety of plants with medicinal qualities, often to help rid themselves of intestinal parasites (Huffman & Wrangham, 1997). When consumed, they are eaten slowly, are not chewed, are eaten only when chimpanzees seem ill, and are often the only thing consumed in the course of the day; they often emerge in the dung whole (Huffman & Wrangham, 1997). Some species in the genus *Aspilia* have a compound that is effective in suppressing the parasites (Rodriguez et al., 1985); the same goes for *Lippia* (Takasaki & Hunt, 1987). The compound in the leaves has both an antibiotic effect and an anti-parasite effect, or more precisely an "antihelminthic" effect.

In our modern world we get an infection, whether parasitical, fungal, or bacterial, and we go to the doctor to get it cleaned up. Chimpanzees give us a whiff of our ancestral past. You live with it. Nearly all chimpanzees are plagued by one or another intestinal parasite (Myers & Kuntz, 1972; McGrew et al., 1989; Ashford et al., 2000; Gillespie et al., 2010). Nematodes *Oesophagostomum* spp. and *Strongyloides fulleborni* are ubiquitous. In a 2007 survey, 100 percent of individuals sampled at Gombe were infected with the latter (Gillespie et al., 2010). You will not avoid it by being the king of the jungle, either: Higher-ranking males at Kibale have more rather than fewer parasites (Muehlenbein, 2006; Muehlenbein & Watts, 2010).

Intestinal parasites, it turns out, are less of a worry for chimpanzees than malaria, since a chimpanzee can take medicine for worms – *Aspilia*, I mean, or some other medicinal plant. Even so, parasites almost certainly cause the same symptoms in chimpanzees as in humans: lethargy, pain, and discomfort. We know they cause diarrhea (Huffman, 1997). *Aspilia* leaves are rough-surfaced – perhaps the texture helps to scour the parasites off the intestinal walls or opens cuts on them (Huffman & Wrangham, 1994; Huffman et al., 1996), but the antihelminthics certainly help, too (Takasaki & Hunt, 1987). Many medicinal plants that chimpanzees consume induce the production of copious diarrheal fluids, helping to flush the parasites

Figure 14.1 Chausiku and her son Chopin pause before entering a low-lying stand of *Pennisetum purpureum*; Mahale Mountains are in the background. Chausiku was a socially dominant female who has been observed to have used more medicinal plants than any other Mahale individual (Takasaki & Hunt, 1987; Huffman & Wrangham, 19941997. Photo by the author.

ate, Takasaki whispered to me: "It's a medicinal plant."

After sitting unmoving for some minutes, Chausiku gimped over to a short tree, climbed up, made a nest, and went back to bed.

The plant she ate was *Lippia plicata*, never seen to be eaten by chimpanzees previously (Takasaki & Hunt, 1987). Thank goodness Takasaki was there. I would have recorded this as a normal feeding bout, never thinking anything of it. Probably Hiroyuki would have, too, had not my more astute colleagues, out. Shock them, scrape them, and flush them out – all three may be at work (Huffman, 1997).

PART II: WHAT CAN KILL YOU

14.1 Types of Diseases

There are three distinct classes of disease, each with a different cause, and each calling for a different remedy; we humans have nearly solved the first, though it afflicts chimpanzees as perniciously as ever, while the other two prey on humans much more than on chimpanzees. The three types of diseases are:

1. **Infectious diseases.** Maladies caused by an external source, usually a virus, bacterium protozoa, or fungus.
2. **Noninfectious diseases.** Disorders caused by internal malfunctions such as cancer, autoimmune afflictions, and system collapse, such as heart failure.
3. **Degenerative diseases**, or "lifestyle" diseases such as arthritis, heart disease, stroke, and type 2 diabetes.

14.2 Human Infectious Diseases

Infectious diseases are pervasive among chimpanzees (Pusey et al., 2008). While the lush and inviting forest seems a place where every prospect pleases, this green paradise harbors many hidden killers; in a survey by Goodall and colleagues, 58 percent of chimpanzees died from infectious disease. Another 28 percent were killed by other chimpanzees (Williams et al., 2008).

Not so very long ago it was the same for humans (Table 14.1; Figure 14.2). Before the discovery of antibiotics, concern about plagues and other infectious diseases were a wearying menace casting a dark shadow over everyday life (Armstrong et al., 1999). Small pox was every parent's nightmare, a common disease and one that killed a heart-breakingly high proportion of those afflicted, and it was not simply the scourge of the lower classes, but elites as well; Ben Franklin's son Benny died of small

Table 14.1 **Leading causes of death in the United States, 1900 versus 1999**

	1900	1999
1.	Pneumonia	Heart disease
2.	Tuberculosis	Cancer
3.	Diarrhea, enteritis	*Stroke*
4.	Heart disease	*Lung disease (smoking)*
5.	Stroke	Pneumonia/influenza
6.	Liver disease	*Diabetes*
7.	Injuries	*Obesity*
8.	Cancer	HIV
9.	Senility	*Suicide*
10.	Diphtheria	*Chronic liver disease*

pox. In 1900, 30 percent of childhood deaths were attributed to infectious diseases; how grateful we should all be that the figure stands at 1.4 percent now (Centers for Disease Control [CDC], 1999; Kochanek et al., 2016). In the last 100 years or so, public health programs have given us better waste disposal, water purification, and dissemination of information about disease transmission, and these have steadily driven down death rates. The first antibiotic, sulphonamide, was introduced to clinical use in 1935, followed by penicillin (1941) – between them these two almost eliminated bacterial infection as a cause of death. Tuberculosis (TB) was the second leading cause of death in 1900 – only pneumonia was a worse killer – and until the introduction of streptomycin in 1943 there was no treatment (Table 14.1; Armstrong et al., 1999; CDC, 1999). Nowadays, diseases that are treatable with antibiotics such at TB, diarrhea, and diphtheria have been knocked right out of the top 10 causes of death in the USA. Flu and pneumonia – and COVID-19 – are still dangers, but most of our killers (Table 14.1, bold print) "lifestyle diseases," which we will return to later. The dramatic change in our threat from disease is reflected in lifespan data; life expectancy in the USA was 47 in 1900; it is 79 now.

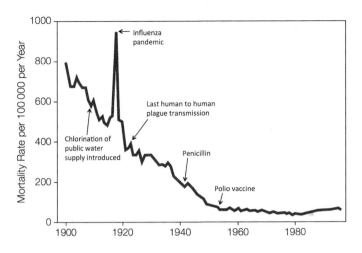

Figure 14.2 Causes of death in the USA, 1900–1996. Better sanitation, government health initiatives, antibiotics, and medical advances have completely altered human life in the last 100 years; chimpanzees are not so lucky. After Armstrong et al., 1999 and CDC, 1999, with permission.

14.3 Chimpanzee Infectious Diseases

Not so for chimpanzees. And so far as we know, there is no infectious disease that afflicts humans but not chimpanzees – including pneumonia, polio, influenza, malaria, and anthrax (Leendertz et al., 2006) – and they must cope with very little medicine, though (as we have seen) not *no* medicine.

On the other hand, we are not identical (Table 14.2). The progression of several diseases, including HIV (and other retroviruses), hepatitis C, falciparum malaria, and influenza, afflict humans and chimpanzees differently (Bailey, 2011; Strong et al., 2016). Chimpanzees die from these scourges (O'Neil et al., 2000; Keele et al., 2009), but their immune response ramps up faster and works better to slow the progression of the disease (Barreiro et al., 2010).

The strain of HIV found in both captive and wild chimpanzees is formally known at SIVcpz (Keele et al., 2009), and it is clear humans came to be infected with HIV by contracting the chimpanzee version from chimpanzees, perhaps in Congo (Gao et al., 1999; Yuan et al., 2016). In fact, it was passed between our two species three separate times, almost certainly when chimpanzees were being butchered for meat and not, as some folklore suggests, through sexual intercourse. Exposure to the blood of chimpanzees is not the only risk factor; monkeys can pass on a version of HIV as well. Thirteen species harbor SIV (Peeters et al., 2002). I despise poachers for their effect on wildlife, but we can blame them for this deadly malady as well. Several outbreaks of Ebola, the terrifying, often fatal type of hemorrhagic fever made famous in Richard Preston's *The Hot Zone*, have been attributed to poaching as well (Leroy et al., 2004).

Chimpanzee resistance to these diseases, slight as it is, is of extraordinary interest to the medical community, since chimpanzees may serve as an **animal model** for the progression of and cure of human diseases; that is, at least some medical researchers are interested in laboratory experiments in which chimpanzees are infected with diseases to see whether new treatments are effective. In 1967 a group proudly announced that they had successfully infected chimpanzees with kuru by injecting afflicted human brain matter into the unfortunate experimental subjects (Gajdusek et al., 1967). The number of chimpanzees in captivity in the USA was driven up by a plan to use them to study AIDS. It now stands at over 1000 individuals (Grimm, 2017).

I consider this experimentation to be unethical. In a discussion about this issue in an online forum, an immunologist once retorted to my objections: "Be honest, if a close relative could be saved by the death of a chimpanzee, would you not accept that exchange?" This criticism hit home, since my father had just died. I responded: "I might well accept it, but

Table 14.2 **Differences between humans and apes in incidence or severity of medical conditions**

Medical condition	Humans	Great apes
Definite		
HIV progression to AIDS	Common	Very rare
Hepatitis B/C late complications	Moderate to severe	Mild
P. falciparum malaria	Susceptible	Resistant
Endemic infectious retroviruses	Rare	Common
Influenza A symptomatology	Moderate to severe	Mild
Probable		
Menopause	Universal	Rare?
Alzheimer's disease pathology tangles	Complete	No neurofibrillary
Epithelial cancers	Common	Rare?
Hydatiform molar pregnancy	Common	Rare?
Possible		
Early fetal wastage (aneuploidy)	Common	Rare?
Differences likely attributable to current human evolutionary mismatches		
Myocardial infarction	Common	Very rare
Atherosclerotic strokes	Common	Rare?
Rheumatoid arthritis	Common	Rare?
Endometriosis	Common	Rare?
Pre-eclampsia/toxemia in pregnancy	Common	Rare?
Bronchial asthma	Common	Rare?
Autoimmune diseases	Common	Rare?
Major psychoses	Common	Rare?

From Varki & Altheide, 2005.

that is no argument that chimpanzee experimentation is ethical, since I would probably agree to trade *your* life for that of a loved one, too." Fortunately (in my view at least), the declaration that chimpanzees are endangered means they can no longer be used in such experiments (Grimm, 2017).

Some worrisome human diseases that pass easily across the species barrier to afflict chimpanzees, and to which they have no special immunity, are Ebola, measles, pneumonia, scabies, and anthrax (Leendertz et al., 2006), as well as gonorrhea (Brown et al., 1972), typhoid (Edsall et al., 1960), foamy virus (Heneine et al., 1998), leprosy (Donham, 1977), polio, viral leukemia, and mange (Williams et al., 2008). We may even have originally contracted a particularly lethal strain of malaria, falciparum malaria, from bonobos (Krief et al., 2010). This should not be a shock, given that malaria is quite common among chimpanzees.

Polio is a horrific affliction for both humans and chimpanzees, and one that Goodall was forced to confront early in her work at Gombe, famously recorded in Hugo van Lawick's *National Geographic* films. Deaths were heart-breaking, and many of the survivors were crippled as well. Faben, son of Flo, was paralyzed in one arm and spent the rest of his life walking bipedally (Figure 14.3). He lived nine years in this state, though he was clearly handicapped, falling behind other males on long treks – and a lone chimpanzee is a vulnerable one. He was ultimately killed by neighboring males (Goodall, 1983), an indirect victim of the polio.

The respiratory diseases pneumonia and bronchitis are particularly lethal to chimpanzees and are responsible for a large proportion of chimpanzee deaths in captivity (Strong et al., 2016; see the chapter heading image); many chimpanzees at Taï, Ivory Coast have also died from herpes and influenza, diseases that rarely kill humans, and which were probably contracted from humans (Boesch, 2008). In fact, pneumonia and other respiratory diseases are so easily passed from humans to chimpanzees that researchers and tourists are urged to take special steps to prevent infections, even if they never come into physical contact (Lukasik-Braum & Spelman, 2008).

In the wild, chimpanzees often suffer from colds during the rainy season (Goodall, 1986), with the same unpleasant runny nose we humans get. Once at Gombe, while I was following chimpanzees, I heard a rasping, human-sounding cough nearby that went on so long I left off observation and went to try to talk what sounded like a very sick field assistant into going back to camp and resting. It was a chimpanzee, looking extremely miserable and, of course, unwilling to risk starvation by taking a day off from foraging; the wet season is a time when fruit is not very abundant in the forest and every calorie is valued. The toll infectious disease takes on chimpanzees is compounded during times of food scarcity, when malnutrition weakens the immune system. For much of the year wild chimpanzees skirt the bounds of malnutrition, making do on bark, leaves, and grass stems that are no tastier – and hardly more nutritious – to them than to us. Unlike humans, they cannot call in sick. Taking off one or two days might be possible if they have built up fat reserves, but most days they must go to work sick – missing three, four, or more days of foraging would leave them still more susceptible to disease or malnutrition, perhaps provoking a downward spiral from which they could not recover.

Between colds, flu, and bronchial infections it is common in the wet season to hear labored, raspy breathing, coughing, and wheezing from one individual or another (Hanamura et al., 2008). Many is the time I have lain in my dry tent under the thatch listening to rain lash the forest and thought of the

Figure 14.3 Faben was stricken with polio and his right arm permanently paralyzed. Unable to walk in the normal fashion, for the remainder of his life he often walked bipedally. Photo by David Bygott.

chimpanzees in their nests trying to catch a little bit of shut eye in a wet world. When I see chimpanzees the day after a soaking rain they often seem tired and slow to get going.

Given chimpanzee intelligence, one might expect that when individuals are seriously ill, friends or loved ones might give them a day off by offering provisions. This is virtually unknown. If you have ever been hospitalized with an illness, you know the first thing the medical staff do is start an IV to give you fluids. Chimpanzees have no access to such aids to ward off dehydration and starvation, much less access to life-saving surgery. Perhaps this is one of the biggest differences between our species; we humans take care of one another.

14.4 Noninfectious Diseases: Cancer

While chimpanzee suffer from all our infectious diseases, they are spared some noninfectious ones. Cancer is exceedingly rare in chimpanzees, especially the types of cancers that are fatal among humans (McClure, 1973; Seibold & Wolf, 1973; Beniashvili, 1989; Waters et al., 1998; Puente et al., 2006; Varki & Varki, 2015). "Chimpanzee tumors are both rare and biologically different from human cancers" (Bailey, 2009: 408). Breast, prostate, and lung cancers are responsible for 20 percent of human deaths, but fewer than 2 percent of chimpanzees die from these (Puente et al., 2006). Humans and apes have lost the enzyme uricase, leaving us with higher levels of uric acid; while this gives us gout, some believe it helps to ward off cancer (Ames et al., 1981). Apparently it is quite an effective defense for chimpanzees, but not for humans. As we learned in Chapter 10, human susceptibility to cancer is hypothesized to be a compromise that allows the rapid growth of brain tissue; genes that act to destroy fast-growing tissues in chimpanzees have been disabled in humans (Arora et al., 2009, 2012).

When captive chimpanzees do get cancer, the most likely affliction is uterine or ovarian cancers (49 percent of all chimpanzee cancers; Brown et al., 2009), possibly because chimpanzees have so many more estrous cycles in captivity, an eventuality for which they are poorly evolved (more on this below), a *mismatch*. Skin, liver, pituitary, adrenal, pancreatic, kidney, stomach, colon, rectal, gall bladder, gut, brain, and bone marrow cancers (leukemia) are also known (Brown et al., 2009).

14.5 The Toll of Injury and Aging on Chimpanzees

Injury and afflictions of old age are a familiar sight in any chimpanzee population. Missing fingers and ear wounds are common fight injuries; puncture wounds and internal injuries are sustained in intercommunity battles and are thought to be the cause of death in many cases (such injuries are discussed in more detail in Chapter 25).

A detailed record of the toll of age and injury on chimpanzees is written in the skeletons that Goodall and others have collected at Gombe over the years (Jurmain, 1977a, 1977b, 1980). Despite the patchy availability of sugar-rich foods, now and then chimpanzees gorge on sweet, ripe fruits, and this, in combination with their worn teeth, yields dental caries; 10 percent of chimpanzee teeth have cavities. Extreme wear is universal in chimpanzees in their thirties and above, progressing not uncommonly to the exposure of the tooth pulp cavity, which often results in infection and abscesses. Tooth loss, often following extreme wear, is found in one in four individuals. Despite the large amount of dental material in the big chimpanzee incisors, they are typically worn to the gum line by the time an individual approaches old age. Tooth wear is made worse by enamel defects: As many as half of all chimpanzees suffer from *enamel hypoplasia*, the failure to fully form enamel on the tooth crown during growth. These imperfections may be caused by illness or malnutrition that struck just as that bit of tooth with missing enamel was being constructed. The thinner enamel makes the tooth more susceptible both to dental caries and to breakage – and without cooking or food preparation implements, chimpanzees need their teeth.

The Gombe skeletons also reveal just how tough chimpanzees are. Hugo, a chimpanzee of Gombe, had nine different bone fractures (Jurmain, 1977a, 1977b, 1980); needless to say, not one of those breaks was properly set by a doctor; bones heal in whatever position they end up after a break, sometimes totally displaced. While Hugo was unusual, at Gombe 30.8 percent of the individuals had healed fractures; males had more breaks than females. Fractures were found most often in the hands, feet, arms, and clavicles. Injuries to the hands and feet are most often sustained during brutal intercommunity conflicts; 28.6 percent of the Gombe males also had fractures or punctures to the face and skull (Jurmain, 1977a, 1977b, 1980), sustained the same way. Breaks to the long bones, including the clavicle, are typically caused by falls. Goodall (1986) recorded 51 falls of more than 35 feet, or 10 m. While I have seen chimpanzees walk away seemingly unharmed from similar falls, they have great pain tolerance, so they might be walking away injured. And such falls are sometimes fatal; Goodall saw an infant break his neck in a fall (Goodall, 1986). Fractures require grit on the part of chimpanzees because the individuals who suffer them must simply carry on with their lives, foraging for food despite the pain. Occasionally a chimpanzee will forego eating for a day or two when suffering from a serious infectious disease, but so far as I know no chimpanzee has ever been known to lie in bed for a day because of a fracture.

Flesh wounds less often leave traces on the bones, but they are common. When a single individual is caught in an intercommunity attack, he is lucky to escape with his life, and if he does, he will likely be left to cope with punctures and gaping wounds. Occasionally a predation episode leads to a serious wound; a Gombe male who had a run in with a bush pig sustained a wound so deep the putrefaction stank and maggots were visible (Goodall, 1971), yet he survived. Deep wounds that would leave disfiguring scars on humans heal quickly and with little scarring in chimpanzees; such rapid healing is partly due to the high metabolic rate of chimpanzees, stimulated by their near-constant movement, but some of it may be due to genetic adaptations to wound healing that have evolved since humans and chimpanzees diverged.

14.6 "Lifestyle" Diseases

Four of the current top-10 human killers, those in bold in Table 14.1, are "lifestyle" diseases, maladies that we *could* avoid with a change in habits. Three further diseases (those in italic font) are entirely or almost entirely self-inflicted: Lung disease is mostly due to smoking and most liver disease is due to excessive alcohol consumption; the third cause is suicide.

These diseases are recent problems for humans. The idea of someone dying of too much food (obesity, number 7) would be laughable to a hunter-gatherer, who must work hard for nearly every calorie. The Tsimane, Bolivian hunter-gatherers (though they do engage in some farming), have the lowest levels of heart disease ever reported (Kaplan et al., 2017). Yet, when hunter-gatherers are drawn into a more Western lifestyle they quickly acquire all the precursors of lifestyle diseases and soon thereafter they develop the diseases themselves (Dounias & Froment, 2006). Chimpanzees face the same challenges. In the wild, none of these diseases afflict chimpanzees; of the 11 necropsies conducted on Gombe chimpanzees and considering also the inferred causes of death for tens of other individuals over the years, not one could be attributed to "lifestyle" (Goodall, 1986; Williams et al., 2008; Terio et al., 2011). Yet lifestyle diseases are common among captive chimpanzees (Seiler et al., 2009; Strong et al., 2016).

Like hunter-gatherers, in the wild chimpanzees have an extremely low frequency of heart disease (Terio et al., 2011), in contrast to its high frequency in captivity (Hubbard et al., 1991). It appears that chimpanzees are even more susceptible to heart disease than are humans; evidence suggests that under exactly the same dietary and exercise regimes chimpanzees would suffer significantly greater disease than humans. Several studies have demonstrated that an astonishing 38 percent of deaths in zoos were wholly or partly attributable to heart disease (Seiler et al., 2009; McManamon & Lowenstine, 2012; Lowenstine et al., 2016).

Heart disease is not the only malady chimpanzees experience when they have a sedentary lifestyle.

Just as with humans, putting on weight, eating too much fat, exercising too little – and just plain getting old – lead to high blood pressure, insulin resistance, and high cholesterol levels (Rosseneu et al., 1979; Andrade et al., 2011; Ely et al., 2013). Chimpanzee blood pressure increases when salt increases in the diet (Denton et al., 1995; Elliott et al., 2007).

14.7 Mismatch Theory

The consequences of the disturbance of an organism's equilibrium when it finds itself in a new environment – in our case an environment of superabundant food and freedom from gathering and chasing prey – is the subject of **mismatch theory**. When a species finds itself in an environment sufficiently different from that in which it evolved, its anatomy and biochemistry are often unable to cope with the new environment. The body of the organism does not match its needs in the new environment, hence *mismatch*.

Surely we humans live in circumstances our ancestors never faced, with our low workloads and high caloric consumption. It is widely acknowledged that the contrast between the diet and activity levels of our ancestors and our current behavior is the reason we suffer from lifestyle diseases (O'Keefe et al., 2011; Kaplan et al., 2017). Mismatch theory is a special case of classic Darwinian natural selection: An animal faces a new environment; some individuals are better adapted to that new environment, while the losers (and their genes) are weeded out. The difference is that mismatch theory tends to examine cases where the old and new environments are dramatically different and the contrast between them appeared extremely rapidly, often due to human innovation (Lloyd et al., 2011). Earlier I mentioned that John Bowlby (1969) wrote about the original conditions, those in which a species lived and evolved for millennia, as an animal's **environment of evolutionary adaptedness** (EEA). Another way of saying this is "the conditions under which the animal evolved." When an animal is outside its EEA, it is typically in stress and struggling to cope – or it may fail to cope and die.

In the last few pages I have begun to lay out the well-accepted evidence that it is this mismatch between the presumed ancestral environment of evolutionary adaptedness and our newly established, industrialized EEA that has generated the current pandemic of diabetes, obesity, heart disease, osteopenia, osteoporosis, osteoarthritis, and other lifestyle diseases – even certain cancers (Olshansky et al., 2005; Malina & Little, 2008). This evidence comes from chimpanzees (as well as other apes), living hunter-gatherers, and ancient humans. Let us examine these each in turn in more detail.

14.8 Consequences of Mismatches: Evidence from Ape Behavior

In classic nature films, chimpanzees run, make daring leaps among trees, and grapple with one another. This is not normal. The pressing need to gather food every day, regardless of health, social status, injury, or infant care obligations means that wild chimpanzees are nearly uniformly (in)active throughout the day, with only slight activity peaks at dawn and dusk (Hunt, 1992).

I wrote "inactive" because chimpanzees engage in only three or so bouts of brisk walking, lasting perhaps 10–20 minutes at a time, and the rest of the day they move at a pace much slower than normal human walking. While they spend 15.7 percent of their day – over two hours – walking, only 0.3 percent (several minutes) is spent running (Hunt, 1991, 1992).

While the activity level is low, it is near-constant; they indulge in half a dozen bouts of relaxed behavior such as grooming or playing; a further 19 percent resting (Hunt, 1992). That 2 hours and 45 minutes of resting is typically dispersed throughout the day in short episodes, though there is one long bout, perhaps an hour, at midday.

While they are going about their daily business, they engage in somewhat acrobatic behaviors that require extensive joint excursion for 8 percent of their active period (Hunt, 1991, 1992), but they do this in a manner more like stretching before a workout or light

Table 14.3 **Day range length and non-trauma induced degenerative joint disease (DJD)**

Species	Site	Sex	Day range (km)	Knee DJD	Spine DJD
Pan troglodytes	Kanyawara	Male	2.4[a]	0%[b]	0%[b]
		Female	1.9[a]	40%[b]	20%[b]
	Gombe[c]	Male	4.6[c]	0%[i]	20%[i]
		Female	3.2[c]	40%[i]	40%[i]
	Taï	Male	3.7[d]		
		Female	3.6[d]		
	Mahale[e]	Pooled	4.8[e]		
P. troglodytes species mean			3.46	Low[i]	Low[i]
Gorilla g. beringei	Karisoke		0.5[f]		
Gorilla g. gorilla	Lope Gabon		1.1[g]		
Gorilla g. graueri	Itebero Congo		0.5[h]		
Gorilla gorilla mean			0.70	High[j]	High[j]

[a] Pontzer & Wrangham, 2006; [b] Carter et al., 2008; [c] Wrangham, 1977; [d] Herbinger et al., 2001; [e] Matsumoto-Oda, 2002; Pontzer & Wrangham, 2004; [f] MacNeilage, 2001; [g] Tutin, 1996; [h] Yamagiwa et al., 1994; [i] Jurmain, 1989, Rothschild & Ruehli, 2005a; [j] Lovell, 1990, Rothschild & Ruehli, 2005b.

yoga than propelling their bodies through space like an Olympic gymnast. While such behaviors require extensive joint mobility, they rarely require the application of forces great enough to generate high joint stresses (Hunt, 1991). In a typical day, chimpanzees will engage in perhaps half a dozen bouts of energetically demanding climbing.

14.9 Consequences of Mismatches: Evidence from Ape Skeletal Collections

All this description of chimpanzees would mean nothing if there were no evidence that wild behavior is healthful, and sedentary zoo behavior is not. Let me be more quantitative about joint problems in wild chimpanzees (Table 14.3). In the same study I cited earlier, 344 vertebral joint surfaces – the surfaces your vertebral disks contact – were examined in eight wild chimpanzees. None had pathological changes. Of the 748 apophyseal synovial surfaces (the contact points on the various protuberances on the vertebrae), only 8 had degenerative changes (Jurmain, 1989). Of 186 surfaces in the shoulder, elbow, hip, knee, and TMJ, only 2 surfaces showed even moderate arthritis, even though the sample included the skeletons of two individuals in their forties, thought to be near the limit of the wild chimpanzee lifespan. At Kibale, Carter et al. (2008) reported on joint health in the skeletons of 12 wild chimpanzees. If cases of trauma are excluded – when a joint is crushed it will inevitably have pathology – two cases of moderate or severe osteoarthritis were observed, both in old females. Arthritis, in other words, is quite rare among wild chimpanzees.

In stark contrast to their wild conspecifics, captive chimpanzees have a frequency of arthritis that is thought to approach 100 percent (Videan et al., 2011), though precise data have not yet been published. Several explanations for the state of captive chimpanzee joint health have been offered. It might be due to hard substrates in captivity (Videan et al.,

2011), greater body weights increasing the stress on joints (Nunn et al., 2007; Videan et al., 2011), or possibly the fact that they may live longer (Videan et al., 2011). I disagree. If advanced age in captive populations could explain arthritis, the two quite old, wild Gombe female skeletons might be expected to have had advanced disease, but they did not. If greater body weight among captive individuals explained higher rates of arthritis, males might be expected to have higher rates of arthritis than females, but in fact, they have lower rates (Table 14.3).

If high activity levels contributed to arthritis, males should have more arthritis than females, since they have greater day ranges, but they do not. In cross-species comparisons, chimpanzees have the *longest* daily ranging distance of any living ape, yet they also have the *lowest* rates of arthritis (Table 14.3); instead, it is gorillas that have small ranges and high rates of arthritis. Note also, if you are still wondering about body weight, that female gorillas have body masses not dramatically greater than those of chimpanzee males.

14.10 Consequences of Mismatches: Arthritis in Industrialized Countries

Degenerative joint disease, or arthritis, is an affliction that blights the lives of too many people in industrialized countries. Joint pain troubles 29.2 percent of US adults, and the disease that most often accompanies it, osteoarthritis, is found in one or more joints in 22.6 percent of adults (US Department of HHS et al., 2012). Keep in mind this survey includes young and old alike. Not surprisingly, the statistics are even worse among older Americans: Among individuals 45 and older, 40.6 percent report joint pain and 37.1 percent report arthritis (US Department of HHS et al., 2012). By the age of 75, 80 percent of Americans exhibit radiographic evidence of arthritis (Dekker et al., 1992; Arden & Nevitt, 2006; Jensen, 2008), and you know if you can see it on an x-ray the pain came long before.

The consequences to health and mobility are dire: Osteoarthritis is second only to ischemic heart disease as a cause of disability in men over 50 (Arden & Nevitt, 2006). Radical joint surgery is increasingly common; in recent years the USA has averaged over half a million knee replacements annually (Sharma et al., 2011). Reduced physical activity from joint discomfort contributes to the onset of lifestyle diseases such as heart disease, obesity, and diabetes, afflictions that often lead to premature death (Malina & Little, 2008; Chodzko-Zajko et al., 2009). So our "captive" humans have joint problems. What about humans living a more natural lifestyle, one more similar to their EEA?

14.11 Evidence from Hunter-Gatherers

A few years ago, a colleague and I applied for an NIH grant to study locomotion and posture (to remind you, the two together are commonly referred to by the general term *positional behavior*) in hunter-gatherers. These people come closest to living in our EEA and thus my colleague and I argued that their behavior would give us the best idea of what is healthful for us now. We failed to convince NIH of this, so the study remains a dream, and one unlikely to be realized since the number of hunter-gatherers who could be studied declines year by year. I can still make a convincing argument about healthful living from the evidence we do have, combining what we know of primate behavior, clinical evidence, and what we already know of hunter gatherers.

While we do not yet have positional behavior data on hunter-gatherers, we know that their activity levels are high compared to those of Westerners (Malina & Little, 2008; Pontzer et al., 2012). Contemporary hunter-gatherers sustain levels of exertion common in affluent societies only among high-level athletes in training. Hunter-gatherers are rarely immobile for long, often walking long distances and engaging in what we would call physical labor such as processing food or building structures (Hill et al., 1985; Hurtado et al., 1985). There is an unexpected contrast on the *other* end of the activity spectrum as well. They also rarely engage in activities that drive them to exhaustion, or require heavy lifting or involve rapid acceleration.

Human ancestors that were ape-like – australopiths – probably had a similar activity profile,

Table 14.4 **Ache hunter-gatherer activity budgets**

	Male (minutes)	Female (minutes)	Average	Percent
Walk	237	114	175.5	25
Pursue game	110	3	56.5	8
Search for plant foods	18	55	36.5	5
Search for insect foods	0	15	7.5	1
Search for honey	9	6	7.5	1
Process food	18	32	25	4
Eat	86	64	75	11
Work on tools	36	71	53.5	8
Childcare	10	16	13	2
Camp maintenance	8	42	25	4
Groom	18	21	19.5	3
Inactively waiting for a companion	4	0	2	0
Waiting while collecting	11	21	16	2
Rest/sleep/talk (daytime)	122	245	183.5	26
Undetermined	0	17	8.5	1

After Hill et al., 1985; Hurtado et al., 1985.

but when hunting became more important the frequency of running and walking must have increased, whereas the time committed to feeding and food processing may have decreased, if living hunter-gatherers are a guide (Hill et al., 1985; Hurtado et al., 1985; Leonard & Robertson, 1997; Leonard, 2008; Malina & Little, 2008). The running of hunter-gatherers, however, is not a sprint, but a comfortable jog. For our EEA, then, the average male/female activity budget was about 3 hours of walking, 2 hours of food gathering (more walking, mostly) and 1 hour and 15 minutes of eating – time commitments similar to those of chimpanzees – plus 3 hours of resting (just a little more than chimpanzees) and perhaps a half-hour of running versus just a few minutes for chimpanzees (Table 14.4).

Contrast this hunter-gather activity budget to that of people in industrialized countries. Static postures are a near-universal requisite of contemporary job tasks. Rather than walking to gather food, "work" in affluent societies often consists of extended periods of immobility; we sit at our workstations and stare at screens. And how do we try to make up for that inactivity? All too often we follow extreme inactivity with brief periods of intense, high-stress physical activity – the "workout" – involving high-impact movements, heavy lifting, and sprinting. We were not evolved to do this.

14.12 Evidence from Ancient Human Skeletons

Not so long ago all humans were hunter-gatherers, so skeletal remains from archeological sites provides further information on the relationship between joint health and cultural practices. Around 11,000 years ago, humans began the transition from a hunter-gatherer lifeway to more sedentary lifestyles – though still active, compared to you and me – involving agriculture (Kelly, 1992). Archeologists recognize the impact of that shift on the body. Agricultural lifeways involve more repetitive activities such as grinding flour, paddling, chopping, and earth-moving. Just as in contemporary societies, the ancient increase in these high-stress, repetitive activities was accompanied by an increase in arthritis in the joints that were over-used for such tasks (Jurmain, 1977a, 1977b, 1980; Martin et al., 1979; Pickering, 1979, 1984; Goodman et al., 1984; Bridges, 1992). Contrast this to frequent, low joint-stress movements common among hunter-gatherers.

14.13 Arthritis and the "Golden Mean"

Although attitudes are changing in response to new discoveries, medical authorities have often blamed exercise for joint deterioration. This perspective views aging as a sort of wearing-out process and joint pain as a normal part of senescence (Chodzko-Zajko et al., 2009). Obesity, injury, and autoimmune involvement have long been recognized as significant contributors to arthritis, to be sure, but the fact that even many lean, injury-free, disease-free individuals suffer from joint problems was considered evidence that joints inevitably must wear out. Reinforcing this perspective was evidence of a **dose effect**: joints wear out more quickly the more one engages in repetitive, prolonged, or forceful movements (Kumar, 2001; Vignon et al., 2006). High rates of arthritis among individuals in professions where joints are highly stressed by heavy lifting or that focus weight-bearing on small surface areas – professions such as foundry workers, cotton mill workers, farmers, ballet dancers, and athletes – supported the wearing-out perspective. It was noticed also that joint problems arose among individuals forced to hold their joints in awkward positions, even when stress was not high, and among individuals who had poor joint mechanics (reviewed in Larsen, 1995). Because muscle fatigue distorts movement and negatively affects joint mechanics, severe fatigue has

also been linked to arthritis (Kumar, 2001; Kendall et al., 2005; Delahunt et al., 2006). Quick, forceful movements are known to induce injury, low back pain in particular (Punnett et al., 1991; Kumar, 2001), a ubiquitous and costly health complaint in affluent nations.

If arthritis is a wearing out, and if cartilage cannot regenerate, inevitably wear must be irreversible and progressive (Anderson & Felson, 1988). Budget your exercise, in other words, because you only have so much cartilage to use up, and you should spend it wisely. In accord with this view, the most common physician-recommended prescription for patients presenting with joint pain was to reduce activity; stop running or weight-lifting; conserve what you have left (Spector et al., 1992; Felson et al., 1997). With rest and time, it was thought, perhaps we might hope for some small repair of chondral tissue (Waller et al., 2011), but such was a remote possibility.

At the same time, it has been long recognized that at the other end of the activity spectrum, sedentism leads to degenerative diseases such as muscle atrophy, obesity, and ultimately heart disease. This is why you have to hit that **Golden Mean**; balance joint wear against the benefits of exercise (Haskell et al., 2007). Even so, this view suggests a no-win perspective on exercise: too much and get arthritis; too little and get heart disease; hit the Golden Mean and make your decline shallowest. The best we can do is to find some low-level exercise that is healthful because it improves cardiac function, cholesterol balance, muscle tone, and connective tissue health, yet is not so vigorous as to wear out joints quickly (Haskell et al., 2007; Chodzko-Zajko et al., 2009). We will find out the best path to joint health is different in a few paragraphs.

14.14 Exceptions that Challenge the Golden Mean Rule

This "wear and tear" perspective ignores a decisive bit of data: 60 percent of older Americans maintain healthy, pain-free joint function, even though some of them have had very active lives. Why are their joints not wearing out? Why do some individuals require joint replacement at 50, while other individuals remain pain-free into their eighth decade? This is where we should look for insight. When we treat non-symptomatic active older adults – pain-free 80-year-olds – as noise rather than critical information, we risk missing an opportunity to discover important preventive prescriptions.

An increasing body of knowledge suggests that high levels of physical activity can be a positive for joint health (Arthritis Foundation et al., 1999), even though the relationship is complicated. Perhaps the evidence from foundry workers and ballet dancers, in combination with information on the lifestyles of pain-free older adults, means that overdoing it is bad, but low-impact exercise is nearly always positive.

We can go further. There is some evidence, in stark contrast to once-prevailing Golden Mean wisdom, that exercise is good not only as a preventive measure, but is neutral and perhaps even healthful even for *damaged* joints (Chodzko-Zajko et al., 2009; Hunter & Eckstein, 2009). In one study, patients with severe arthritis who happened to ignore their doctor's orders and exercised more than prescribed had some improvement in joint pain (Alaaeddine et al., 2012). Perhaps low-level activity such as jogging regenerates tissue in the knee joint, in stark contrast to high-stress activities like sprinting or leaping, which likely damage it further (Urquhart et al., 2011).

Åstrand (1992) was an early proponent of the perspective that the proper "manner" or type of activity improves the functional components of the joints, essentially providing a conditioning effect and reducing pain and wear. Rather than high activity levels alone, it may be that the combination of repetitive movements during intense, high-stress exercise followed by long periods of immobility (Riddoch et al., 2004) causes arthritis. Such an exercise regime is exactly what hunter-gatherers *do not* do.

14.15 Clinical Studies

While evidence from ancient humans, hunter-gatherers, chimpanzees, and active older adults offers powerful testimony that sustained but low-level activity is healthful, it is clinical studies that show why: Cartilage degeneration is linked not to cartilage wear *per se*, but to a local deterioration of underlying bone that in turn leads to cartilage damage (Zhang et al., 2000, 2011; Burr, 2004; Neogi et al., 2011; Cox et al., 2012). When a small crater forms in the bony matrix that serves as the rigid support for the joint cartilage, the sharp edges of the damaged bone then wear the cartilage. It is lack of exercise that leads to collapse of joint bone. Long walks strengthen bone; sitting on the couch for hours weakens it. Our bodies try to warn us. Kumar (2001) found that tendons, ligaments, and muscles respond to occupational biomechanical hazards first, becoming painfully swollen and inflamed; only later does bone and cartilage damage follow.

Something subtle is going on here. Both sporadic exercise (as opposed to consistent activity) and inactivity lead to arthritis, not necessarily because cartilage wears, but because underlying bone collapses. Bone mineralization and density are known to be higher among active people (Lane et al., 1986). For example, the first lumbar vertebrae of long-distance runners have 40 percent more bone mineral than matched controls, but despite their extremely high activity level they have no more arthritis (Lane et al., 1986).

These results suggest that high levels of physical activity are a preventive prescription for arthritis, so long as the level of activity is not high-stress or excessive.

It has now become a well-accepted maxim that static posture that involves extended periods of muscle contraction can result in musculoskeletal damage (Sjogaard, 1986; Li & Buckle, 1999; Kumar, 2001). Sitting postures, surprisingly, can concentrate stress on a few joints, produce unnaturally high joint loads, decrease circulation, and increase bone decalcification (Magnusson & Pope, 1998; Vieira & Kumar, 2004), leading to low-back joint pain, discomfort, and musculoskeletal damage. Pope et al. (1985) identified the predictors of low-back pain: weakened abdominal strength, atrophied hamstring musculature, and increased lumbar lordosis.

PART III: FIGHTING LIFESTYLE DISEASES

14.16 Activity: The APBR

In review, evidence from apes, hunter-gatherers, archeological skeletal collections, and clinical studies of people in industrialized countries suggests that a regime similar to, but perhaps not exactly the same as, that of chimpanzees is healthful. Chimpanzees run their joints through their entire range of motion while under light stress, spending more than half the day moving, engaging in aerobic – but not high-stress or exhausting – exercise. Hunter-gatherer data are less thorough, but their activity budget is quite similar with the exception that they spend about 30 minutes jogging. This (Table 14.4) is likely our EEA, the activity profile for which our bodies evolved. We might call this the **Ancestral Positional Behavior Regime (APBR)**, the pattern of movement or exercise that approximates that of our ancient hunter-gatherer ancestors. An APBR would involve walking 15–25 percent of the day – though hardly running at all, perhaps jogging for half-hour a day – bearing weight for some minutes on all our joints, running them through the limits of their excursion, and avoiding sitting for long periods. When we exercise, the jostling not only strengthens our muscles and bones, but our arteries as well. Weakened arteries and high blood pressure lead to blow-outs – strokes. While walking for four hours a day is impractical for most people, increasing activity levels as near to that as possible would help avoid arthritis as well as granting freedom from heart disease, diabetes, high blood pressure, stroke, and obesity – with an appropriate diet.

Recent research suggests that the benefits of exercise extend even to boosting the immune system and preventing cancer. In one 11-year study involving 1.4 million people, healthful exercise was found to result in a 7 percent lower cancer rate; high-activity-level participants had lower frequencies of 26 different kinds of cancer (Moore et al., 2016). Some cancers were more strongly affected: Esophageal cancer was reduced 42 percent while lung, kidney, and stomach cancers were reduced 20 percent or more. Even blood cancers and breast cancer were lower. There is a dose effect; the more exercise you get, the lower the cancer risk. In a study of cyclists over 55, those who regularly biked long distances had more muscle mass, less fat, better memory, and a more robust immune system – comparable to inactive participants half their age (Duggal et al., 2018).

14.17 Diet and Mismatch

Perhaps I have convinced you that long-duration, low-impact activity is good for your health. I have neglected to mention something every bit as important: diet. Hunter-gatherers consume a low-fat, low-calorie, high-fiber diet – the healthful nutritional regimen you may know as the "paleodiet" (Eaton et al., 1997; Jenike, 2001). The idea that an ancestral-type diet is healthful is no longer the least bit controversial. This is our EEA.

The hunter-gatherer diet contrasts just as sharply with that of twenty-first century industrialized nations as their activity level. In Chapter 5 we learned that sugar and fat are precious resources for chimpanzees, resources they get less of than they wish. At the same time, they cannot avoid fiber. In the course of getting the sugar and other nutrients they need, they get more fiber than they prefer. We humans in affluent societies have a different problem. We eat potato chips packed with salt, fat, and carbohydrates, but little fiber. Our most sophisticated cuisine is rich in sauces packed with fat. Our meat, bread, and dairy diets are nearly devoid of fiber. We celebrate our birthdays with a food that should come with a biohazard warning, consisting of practically pure sugar and fat.

This diet, like our inactive lifestyle, is an evolutionary mismatch. Our bodies are not adapted to deal with it. Our hunter-gatherer instincts urge us to avoid fiber and hoard calories, but our bodies are unable to cope with this abundance. The constant presence of sugar in our blood floods our system with insulin. After a while, our bodies begin to ignore what it "sees" as an impossibly persistent supply of sugar and begins to ignore the signal to produce insulin, which causes elevated blood glucose. High blood sugar leads to heart disease, eye problems, kidney failure, and nerve damage. Our blood is also packed with bad cholesterol (low-density lipoproteins; LDL), and has too little high-density lipoprotein (HDL); heart disease and high blood pressure are a consequence.

Your intestines have muscles that push digesta along; just as muscles in your arms and legs would atrophy if not used, so does your gut require exercise. Bulky fiber, bulk our ancestors could not avoid, must be moved through our gut by muscle action, and this workout strengthens the walls of our intestines, thickening the tissue. When we get too little fiber, our intestinal walls atrophy and undesirable gut microflora proliferate. Through means not completely understood, but surely related to the atrophy of the gut wall, a common consequence of a low-fiber diet is diverticulosis (Tursi, 2016). This is a condition in which the thinned and weakened intestinal wall gives way and a bubble forms, bulging into the abdomen and becoming a kind of backwater where decomposing food and toxins accumulate (Figure 14.4). The constant juxtaposition of poisons on the tissue of the intestinal wall is in itself dangerous – it increases the likelihood of colon cancer. If the thin-walled tissue that bubbles out from the intestinal wall bursts, toxins will be spilled into the abdominal cavity, with possibly fatal consequences. Maybe that low-sugar, low-fat, high-fiber, woody-tasting breakfast cereal is looking better to you now.

For reasons we understand poorly, a high-fiber diet not only strengthens the gut walls, it alters your blood chemistry, even when fat content stays the same. It lowers our LDLs (Brown et al., 1999), raises our HDLs (Nicolosi et al., 1999), and improves insulin function (Weickert et al., 2006). These in turn lower blood pressure, reduce the risk of stroke (Chen et al., 2013), and help avoid type II diabetes (Salmeron et al., 1997).

Lastly, and perhaps unexpectedly, if I must add one more virtue to a high-fiber diet. I have written "high-fiber diet" often enough that you are probably ready to slam me over the head with a high-fiber two by four – but note this: It not only helps with cancer and cholesterol, it even reduces knee arthritis (Dai et al., 2017).

Looking back at Table 14.2, you can see why I consider heart attacks (myocardial infarction), strokes, and arthritis to be **mismatch diseases**. Exercise and diet are mismatches, be we also suffer from a plethora of other mismatches. Endometriosis is a migration of uterine lining tissue outside the uterus into the fallopian tubes, and even into the abdominal cavity; it is likely a result of high body mass, low pregnancy rates, and lower rates of nursing. Our ancestors often went months and years without menstruating because nursing and low-calorie diets suppress ovulation. Women in industrialized countries, unlikely to nurse, rarely pregnant, ovulate many, many times more often than our ancestors, and in some people with each repeated cycle the uterine lining creeps farther out of the uterus. Toxemia is caused by a rich diet, and obesity, hypertension, older age, and diabetes all contribute. All are products of our current lifestyle. Asthma, allergies in general, and perhaps even autoimmune diseases have been blamed on lack of exposure to allergens and pathogens our ancestors were unable to avoid; we are too clean for our own good, and too disease-free (Yazdanbakhsh et al., 2002). Even gut parasites have their good side. Intestinal parasite infection during childhood, unavoidable for our ancestors, refines the maturing

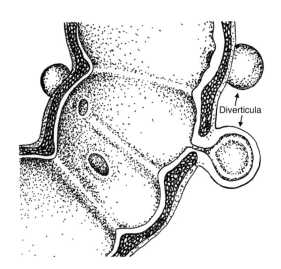

Figure 14.4 Diverticulosis, a bubbling out of the atrophied intestinal wall of the large colon that can lead to serious infection and even death. The diverticula may contain pus, harmful bacteria, or both. Image by the author.

immune system's ability to tell real health threats from harmless allergens in the environment. This confusion might even cause the immune system to attack our own bodies, leading to autoimmune diseases.

Lastly, in our EEA we were constantly surrounded by close relatives and rarely, very rarely, completely alone. We were virtually never in the midst of strangers. Although we crave privacy and many of us enjoy the anonymity of living among strangers, such isolation and disconnection from loved ones is unnatural. It may even lead to psychoses (Reininghaus et al., 2008). Having nosy, ever-present aunt Edith living in the spare bedroom may cramp a teenager's style, but it may also be a prescription for mental health.

14.18 Review and Lessons

A lesson from the chimpanzee is that an activity budget dominated by low-stress exercise and a high-fiber, low-sugar diet makes us healthier. Sugar and slothfulness are killing us. Exercise and dietary fiber delay death and decay. Mundane as it seems, walking is healthful; it is what we were born to do. The vibration caused by your heel striking the ground travels up the entire length of your body (Smeathers, 1989), and the heel-strike vibration is at a frequency that stimulates bone and tissue growth that strengthens arteries, veins, ligaments, tendons, and bones (Verschueren et al., 2004; Morgado Ramirez et al., 2017). The APBR was likely similar to that of living hunter-gatherers who, like chimpanzees, engage in low-stress activity for prolonged periods of time, spending little time in activities similar to the joint-pounding stress of basketball or weight-lifting. Hunter-gatherers spend very little time not moving at all, as we might when watching TV or working on a computer.

Obesity, stroke, arthritis, colon cancer, diverticulosis, high blood pressure, heart disease, and premature aging are avoidable. Take up an APBR and a paleodiet, and your prospects for a long healthful life improve dramatically.

References

Alaaeddine N, Okais J, Ballane L, Baddoura R (2012) Use of complementary and alternative therapy among patients with rheumatoid arthritis and osteoarthritis. *J Clinical Nursing* 21, 3198–3204.

Ames BN, Cathcart R, Schwiers E, Hochstein P (1981) Uric acid provides an antioxidant defense in humans against oxidant-and radical-caused aging and cancer: a hypothesis. *Proc Nat Acad Sci* 78, 6858–6862.

Anderson JJ, Felson DT (1988) Factors associated with osteo-arthritis of the knee in the first national-health and nutrition examination survey (HANES-I): evidence for an association with overweight, race, and physical demands of work. *Am J Epidemiol* 128, 179–189.

Andrade MC, Higgins PB, Mattern VL, et al. (2011) Morphometric variables related to metabolic profile in captive chimpanzees (*Pan troglodytes*). *Comp Med* 61, 457–461.

Arden N, Nevitt MC (2006) Osteoarthritis: epidemiology. *Best Pract Res Clin Rheumatol* 20, 3–25.

Armstrong GL, Conn LA, Pinner RW (1999) Trends in infectious disease mortality in the United States during the 20th century. *JAMA* 281, 61–66.

Arora G, Polavarapu N, McDonald JF (2009) Did natural selection for increased cognitive ability in humans lead to an elevated risk of cancer? *Med Hypoth* 73, 453–456.

Arora G, Mezencev R, McDonald JF (2012) Human cells display reduced apoptotic function relative to chimpanzee cells. *PLoS ONE* 7, e46182. DOI: 10.1371/journal.pone.0046182.

Arthritis Foundation, Association of State and Territorial Health Officials, Centers for Disease Control and Prevention (1999) *National Arthritis Action Plan: A Public Health Strategy*. Atlanta, GA: Arthritis Foundation. www.arthritis.org/media/Delia/NAAP_full_plan.pdf.

Ashford RW, Reid GDF, Wrangham RW (2000) Intestinal parasites of the chimpanzee *Pan troglodytes* in Kibale Forest, Uganda. *Ann Trop Med Parasit* 94, 173–179.

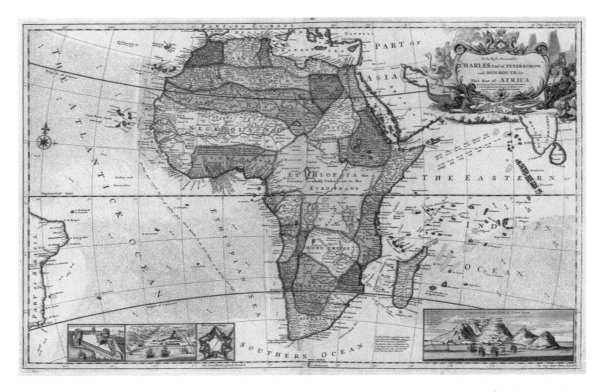

Figure 3.2 A 1710 map of Africa. The immense country taking up most of the continent is "Aethiopia." (A black and white version of this figure will appear in some formats.)

Figure 3.7 Essentialist Greek philosopher Plato (~428–348 BC). (A black and white version of this figure will appear in some formats.)

Figure 3.10 St. Thomas Aquinas surrounded by the doctors of the Catholic Church. *The Apotheosis of St. Thomas Aquinas* 1631, Franccisco De Zurbaran. With permission from The Museum of Fine Arts (Museo Provincial de Bellas Artes), Seville. (A black and white version of this figure will appear in some formats.)

Figure 4.14 (a) Cotton-top tamarin (*Saguinus oedipus*; photo by Terry Waters), (b) common marmoset (*Callithrix jacchus*; photo by Raimond Spekking) and (c) the emperor tamarin (*Saguinus imperator;* photo by TheBrockenInaGlory). (A black and white version of this figure will appear in some formats.)

Figure 4.18 (a) A Gombe olive baboon female with a kidnapped infant that later died of dehydration (photo by author), (b) the golden snub-nosed monkey (*Rhinopithecus roxellana*), (c) the Tonkin snub-nose monkey (*Rhinopithecus avunculus*; photo by Tito Nadler), (d) the proboscis monkey (*Nasalis larvatus*; photo by Charlesjsharp, and (e) the black and white colobus (*Colobus guereza*; photo by author). (A black and white version of this figure will appear in some formats.)

Figure 4.20 Fully flanged adult male orangutan. Photo courtesy of Tim Laman. (A black and white version of this figure will appear in some formats.)

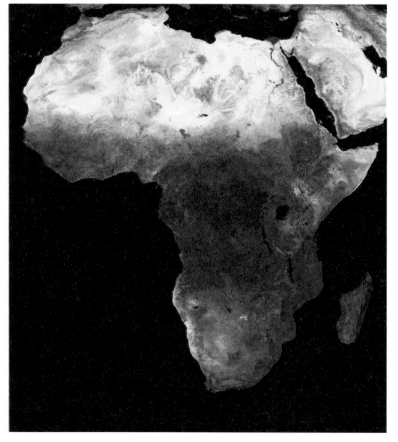

Figure 5.1 African forests, in dark green. The chain of lakes in the east, where the dark green segues to lighter green, is the border of East Africa. West of this area, forest exists only in patches. Image from NASA. (A black and white version of this figure will appear in some formats.)

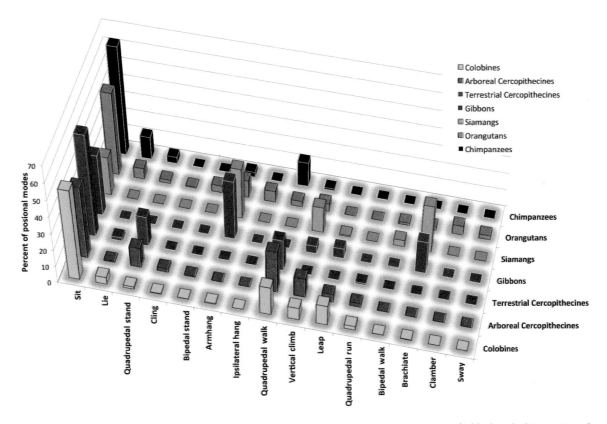

Figure 8.8 Primate positional behavior. From Hunt, 2016, with permission of John Wiley & Sons. (A black and white version of this figure will appear in some formats.)

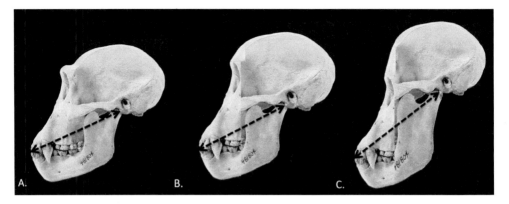

Figure 9.6 When we envision a normal chimpanzee skull (a) altered so that the face is pulled back toward the braincase (b) we find that skull length, in particular prognathism, is decreased (dotted line). The same jaw length could be maintained if the face were taller (c) but this introduced its own complications. Image by the author. (A black and white version of this figure will appear in some formats.)

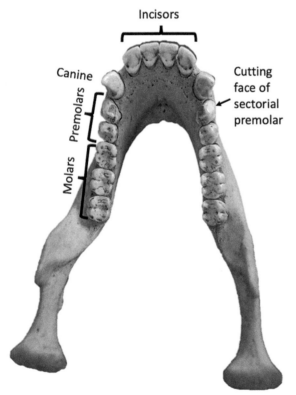

Figure 9.8 Chimpanzee mandible and teeth. Note the asymmetric first premolar; the light portion of the tooth in this image rubs against the back of the upper canine, sharpening it. Photo by the author. (A black and white version of this figure will appear in some formats.)

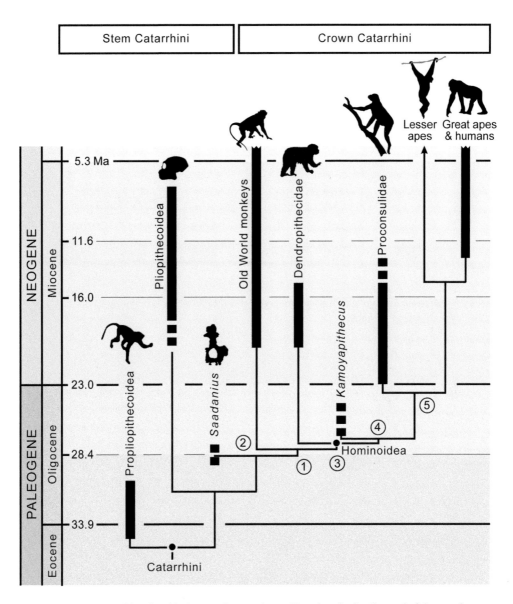

Figure 10.5 A recent phylogeny of fossil and living monkeys and apes. Note that the fossil record of the apes from 23 to 12 Mya is blank – the thin line – while the Proconsulidae were an evolutionary dead-end. For want of a better model for the morphology of the first ape, proconsulids will have to do. From Zalmout et al., 2010, with permission. (A black and white version of this figure will appear in some formats.)

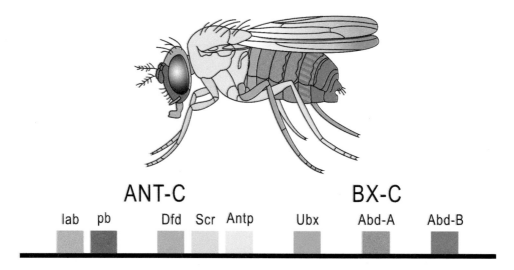

Figure 11.2 Fruit fly Hox genes; these genes appear on fly chromosome 3; intriguingly, they are lined up on the chromosome in the same order they appear in the fly's body. Image courtesy of WikiMedia Creative Commons. (A black and white version of this figure will appear in some formats.)

Figure 12.6 The protein TRP-Cage. From www.pdb.org. (A black and white version of this figure will appear in some formats.)

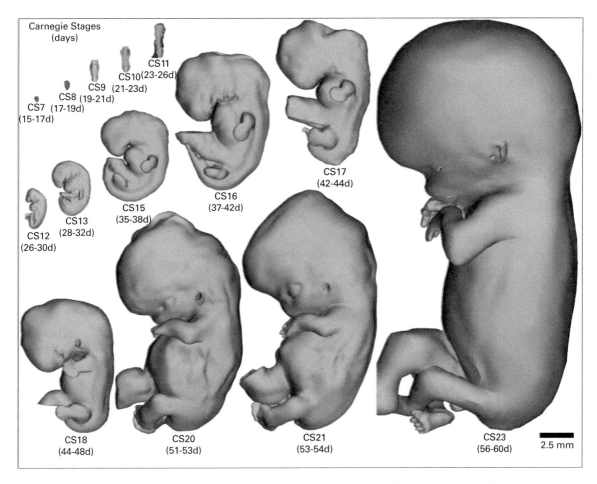

Figure 11.8 Carnegie stages. See the text for an explanation. From the 3D Atlas of Human Embryology (www.3datlasofhumanembryology.com) and de Bakker et al (2016). Used with permission. (A black and white version of this figure will appear in some formats.)

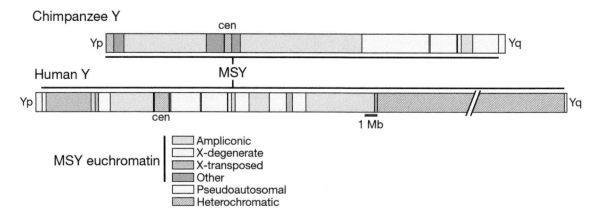

Figure 12.13 Human and chimpanzee Y-chromosomes compared. It is immediately obvious that the two chromosomes are very different, but they are more different than anyone would have expected, as different as humans are from chickens. From Hughes et al., 2010, with permission. (A black and white version of this figure will appear in some formats.)

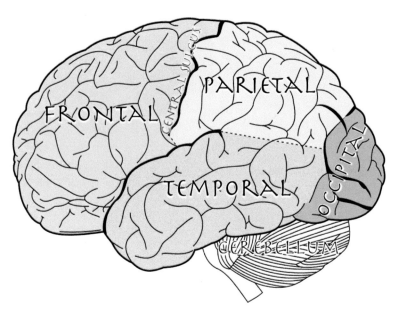

Figure 19.1 View of left side of the brain, with the four lobes of the cerebrum color-coded. Frontal lobe (blue), parietal lobe (yellow), temporal lobe (green), and occipital lobe (pink). The hatched globule at the base is the cerebellum and the mostly hidden white structure is the brain stem. By Henry Vandyke Carter; after Gray (1918). (A black and white version of this figure will appear in some formats.)

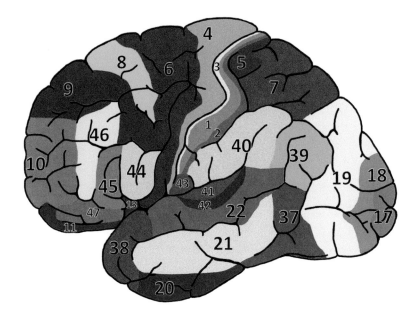

Figure 19.4 Brodmann's areas visible on the surface of the brain; many areas are hidden in the interior of the brain. Areas 44 and 45 are Broca's area and area 39 is Wernicke's area. The premotor cortex is area 6 and the primary motor cortex is area 4; the visual cortex is made up of areas 17, 18 and 19. The orbitofrontal cortex areas visible on this image are Brodmann area 11 and 47. Area 13, not visible here, is just on the other side of 11. After Brodmann, 1909. Illustration by the author. (A black and white version of this figure will appear in some formats.)

Figure 20.1 Semliki chimpanzee Hunter relaxes in a day nest. Photo: Caro Deimel. (A black and white version of this figure will appear in some formats.)

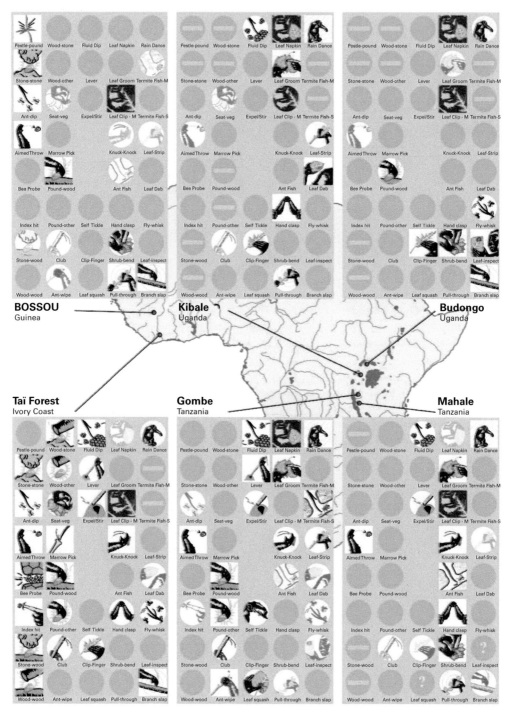

Figure 23.2 Behaviors at the six sites with the longest research history. Square color icons signify behaviors that are customary at that site, meaning it has been seen in nearly all able-bodied members of at least one age–sex class (such as adult males); circular color icons indicate a habitual behavior, one that has been seen often and in multiple individuals, but not as pervasively as "customary"; monochrome icons signify a behavior that is present, meaning it is present in the community; a gray circle means absent; a gray circle with a horizontal bar means absent with an ecological explanation; a question mark signifies uncertain; no symbol at all indicates unknown. From Whiten et al., 1999; with permission. (A black and white version of this figure will appear in some formats.)

Figure 26.3 Female lion approaching a dark-maned dummy male. Courtesy of Craig Packer. (A black and white version of this figure will appear in some formats.)

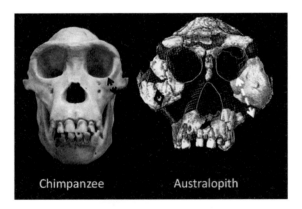

Figure 28.1 Australopith faces are heavy and reinforced compared to chimpanzees. Zygomatic height (dotted lines) is much smaller in chimpanzees than australopiths. Image by the author. (A black and white version of this figure will appear in some formats.)

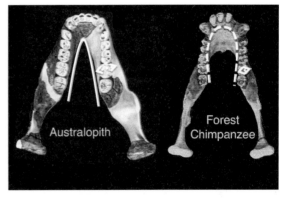

Figure 28.2 The australopith's mandibular body is so thick that when viewed from above the inner surface of the mandible makes a V shape; forest chimpanzees have a thick body that is hardly thicker than the width of the molars, even though the molars are small so that the inner surface describes a U. Image courtesy of Tim White; used with permission. (A black and white version of this figure will appear in some formats.)

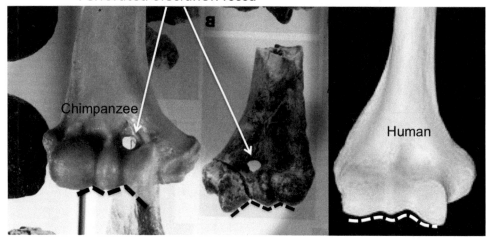

Figure 28.6 Chimpanzees (left) and australopiths (center) share a perforated olecranon fossa, allowing the olecranon process of the ulna to completely wrap around the humeral trochlea; the articular surface is M-shaped (black dotted line). Human humeri have a curved, undulating articular surface (white dotted line). Photos by the author. (A black and white version of this figure will appear in some formats.)

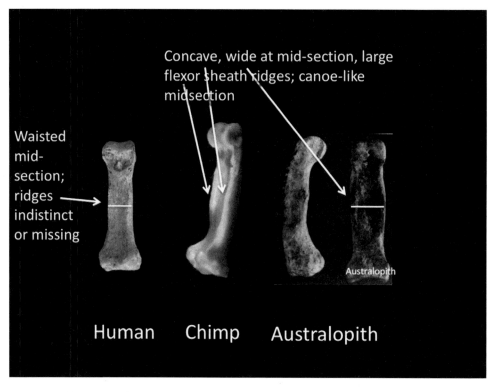

Figure 28.8 Phalanges (finger bones) of a human (palmar view), chimpanzee (three-quarter view), and australopith (side and palmar views) fingers have large flexor sheath ridges, and are curved and expanded side to side at mid-section. Australopith fingers are slightly shorter versions of chimpanzee fingers. Human, chimp photos by the author; australopith images courtesy of Wiley; with permission. (A black and white version of this figure will appear in some formats.)

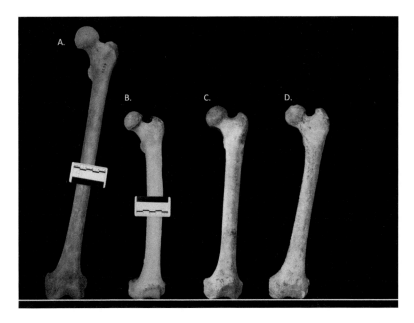

Figure 28.16 Human femora (A) are angled so that the shaft is 13° off vertical, compared to the articular surface of the knee (white line). Forest chimpanzees (B) have femora nearly perfectly vertical, varying by less than one-third of a degree from vertical. The average Semliki femur (C) falls in between humans and forest chimps. One of the Semliki femora (D) is right at the human average. No other chimpanzee population has this human-like angle. Image by the author. (A black and white version of this figure will appear in some formats.)

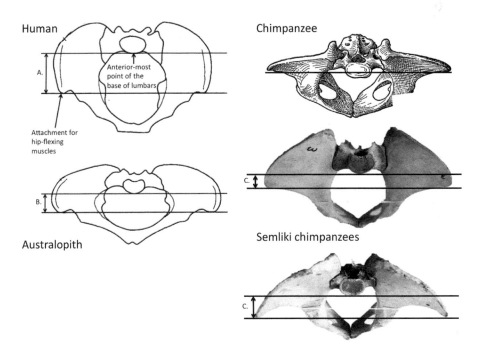

Figure 28.17 Top view of human pelvis, australopith (Lucy, bottom left), forest chimpanzee (top right), and two Semliki pelves (right, second from top, bottom). Humans have a bowl-shaped pelvis with a large distance between the attachment for hip flexors (muscles in the front of the thigh that raise the knee) and the forward-most point of the pelvic base for the spine (distance A). Australopiths (distance B) vary in the direction of chimpanzees, which have no distance between the spine and the hip flexors (top right). Semliki chimpanzees (distances C) vary from forest chimpanzees in the direction of humans. Forest chimpanzee after Schultz, 1936. (A black and white version of this figure will appear in some formats.)

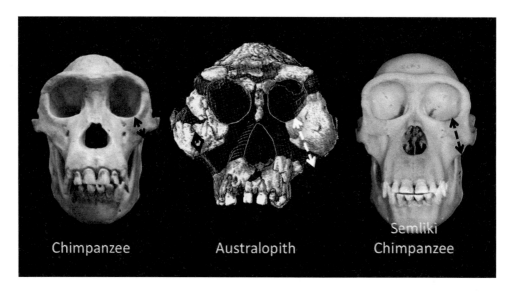

Figure 28.18 Forest chimpanzees (left) have short (top-to-bottom) zygomatics (black line, left) compared to australopith zygomatic height (white dotted line, middle). Semliki chimpanzees have australopith-like zygomatics (black dotted line, right); Semliki photo by the author. (A black and white version of this figure will appear in some formats.)

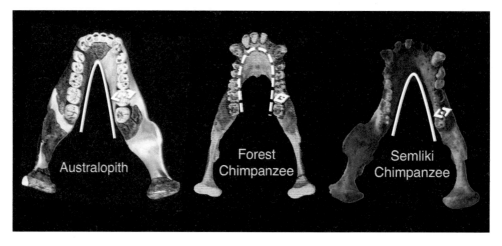

Figure 28.19 Australopiths (left) have robust mandibles with mandibular bodies so thick the mandible has a V-shape. Chimpanzee mandibles are thinner, giving them a round inner surface compared to australopiths. Semliki mandibles are thick and australopith-like. Semliki mandible by the author. (A black and white version of this figure will appear in some formats.)

Åstrand PO (1992) Why exercise? *Med Sci Sport Exerc* 24, 153–161.

Bailey J (2009) An examination of chimpanzee use in human cancer research. *Alternatives to Laboratory Animals-ATLA* 37, 399–416.

Bailey J (2011) Lessons from chimpanzee-based research on human disease: the implications of genetic differences. *Alternat Lab Animals*, 39, 527.

Barreiro L, Marioni J, Blekhman R, Stephens M, Gilad Y (2010) Functional comparison of innate immune signaling pathways in primates *PLoS Genet* 6 DOI: 10.1371/journal.pgen.1001249.

Beniashvili DS (1989) An overview of the world literature on spontaneous tumors in nonhuman primates. *J Med Primatol* 18, 423–437.

Boesch C (2008) Why do chimpanzees die in the forest? The challenges of understanding and controlling for wild ape health. *Am J Primatol* 70, 722–726.

Bowlby J (1969) *Attachment and Loss*, Vol. 1: *Attachment*. New York: Basic Books.

Bridges PS (1992) Prehistoric arthritis in the Americas. *Ann Rev Anthropol* 21, 67–91.

Brown L, Rosner B, Willett WW, Sacks FM (1999) Cholesterol-lowering effects of dietary fiber: a meta-analysis. *American J Clin Nutr* 69, 30–42.

Brown SL, Anderson DC, Dick EJ, Jr., et al. (2009) Neoplasia in the chimpanzee (*Pan* spp.). *J Med Primatol* 38, 137–144.

Brown WJ, Lucas CT, Kuhn US (1972) Gonorrhoea in the chimpanzee: infection with laboratory-passed gonococci and by natural transmission. *Br J Vener Dis* 48, 177.

Burr DB (2004) The importance of subchondral bone in the progression of osteoarthritis. *J Rheumatol* 31, 77–80.

Carter ML, Pontzer H, Wrangham RW, Peterhans JK (2008) Skeletal pathology in *Pan troglodytes* schweinfurthii in Kibale National Park, Uganda. *Am J Phys Anthropol* 135, 389–403.

Centers for Disease Control (1999) Achievements in public health, 1900–1999: control of infectious diseases. *Morbid Mortal Weekly Rep* 48, 621–629.

Chen GC, Lv DB, Pang Z, Dong JY, Liu QF (2013) Dietary fiber intake and stroke risk: a meta-analysis of prospective cohort studies. *Eur J Clin Nutr* 67, 96–100.

Chodzko-Zajko WJ, Proctor DN, Singh MAF, et al. (2009) Exercise and physical activity for older adults. *Med Sci Sport Exerc* 41, 1510–1530.

Cox LGE, van Donkelaar CC, van Rietbergen B, Emans PJ, Ito K (2012) Decreased bone tissue mineralization can partly explain subchondral sclerosis observed in osteoarthritis. *Bone* 50 1152–1161.

Dai Z, Niu J, Zhang Y, Jacques P, Felson DT (2017) Dietary intake of fibre and risk of knee osteoarthritis in two US prospective cohorts. *Ann Rheum Dis* 76, 1411–1419.

Dekker J, Boot B, van der Woude LHV, Bijlsma JWJ (1992) Pain and disability in osteoarthritis: a review of biobehavioral mechanisms. *J Behav Med* 15, 189–214.

Delahunt E, Monaghan K, Caulfield B (2006) Altered neuromuscular control and ankle joint kinematics during walking in subjects with functional instability of the ankle joint. *Am J Sports Med* 34, 1970–1976.

Denton D, Weisinger R, Mund NI, Wickings EJ (1995) Blood pressure of chimpanzees. *Nature Med* 1, 1009–1016.

Donham KJ, Leininger JR (1977) Spontaneous leprosy-like disease in a chimpanzee. *J Infect Dis* 136, 132–136.

Dounias E, Froment A (2006) When forest-based hunter-gatherers become sedentary: consequences for diet and health. *UNASYLVA-FAO* 57, 26–33.

Duggal NA, Pollock RD, Lazarus NR, Harridge S, Lord JM (2018) Major features of immunesenescence, including reduced thymic output, are ameliorated by high levels of physical activity in adulthood. *Aging Cell* 17. DOI: 10.1111/acel.12750.

Eaton SB, Eaton SB, Konner MJ (1997) Paleolithic nutrition revisited: a twelve-year retrospective on its nature and implications. *Eur J Clin Nutr* 51, 207–216.

Edsall G, Gaines S, Landy M, et al. (1960) Studies on infection and immunity in experimental typhoid fever: I. Typhoid fever in chimpanzees orally infected with *Salmonella typhosa*. *J Exp Med* 112, 143.

Elliott P, Walker LL, Little MP, et al. (2007) Change in salt intake affects blood pressure of chimpanzees: implications for human populations. *Circulation* 116, 1563–1568.

Ely, JJ, Zavaskis T, Lammey ML (2013) Hypertension increases with aging and obesity in chimpanzees (*Pan troglodytes*). *Zoo Biol* 32, 79–87.

Felson DT, Zhang Y, Hannan MT, et al. (1997) Risk factors for incident radiographic knee osteoarthritis in the elderly: the Framingham Study. *Arthritis Rheum* 40, 728–733.

Gajdusek DC, Gibbs CJ, Alpers M (1967) Transmission and passage of experimental "kuru" to chimpanzees. *Science* 155, 212–214.

Gao F, Bailes E, Robertson DL, et al. (1999) Origin of HIV-1 in the chimpanzee *Pan troglodytes troglodytes*. *Nature* 397, 436–441.

Gillespie TR, Lonsdorf EV, Canfield EP, et al. (2010) Demographic and ecological effects on patterns of parasitism in eastern chimpanzees (*Pan troglodytes schweinfurthii*) in Gombe National Park, Tanzania. *Am J Phys Anthropol* 143, 534–544.

Goodall J (1983) Population dynamics during a fifteen year period in one community of free-living chimpanzees in the Gombe National Park, Tanzania. *Zeitschrift fur Tierpsychologie* 61, 1–60.

Goodall J (1986) *The Chimpanzees of Gombe: Patterns of Behavior*. Cambridge, MA: Harvard University Press.

Goodall J van Lawick (1971) *In the Shadow of Man*. London: Collins.

Goodman AH, Lallo J, Armelagos GJ, Rose JC (1984) Health changes at Dickson Mounds Illinois (A.D. 950–1300). In *Paleopathology at the Origins of Agriculture* (eds. Cohen MN, Armelagos GJ), pp. 271–305. Orlando, FL: Academic Press.

Grimm D (2017) Research on lab chimps is over: why have so few been retired to sanctuaries? *Science*. DOI: 10.1126/science.aan6956.

Hanamura S, Kiyono M, Lukasik-Braum M, et al. (2008) Chimpanzee deaths at Mahale caused by a flu-like disease. *Primates* 49, 77–80.

Haskell WL, Lee IM, Pate RR, et al. (2007) Physical activity and public health: updated recommendation for adults from the American College of Sports Medicine and the American Heart Association. *Med Sci Sports Exerc* 39, 1423–1434.

Heneine W, Switzer WM, Sandstrom P, et al. (1998) Identification of a human population infected with simian foamy viruses. *Nature Med* 4, 403–407.

Herbinger I, Boesch C, Rothe H (2001) Territory characteristics among three neighbouring chimpanzee communities in the Taï National Park, Ivory Coast. *Int J Primatol* 32, 143–167.

Hill K, Kaplan H, Hawkes K, Hurdato AM (1985) Men's time allocation to subsistence work among the Ache of eastern Paraguay. *Hum Ecol* 13, 29–47.

Hubbard GB, Lee DR, Eichberg JW (1991) Diseases and pathology of chimpanzees at the Southwest Foundation for Biomedical Research. *Am J Primatol* 24, 273–282.

Huffman MA (1997) Current evidence for self medication in primates: a multidisciplinary perspective. *Am J Phys Anthropol* 104(S25), 171–200.

Huffman MA, Wrangham RW (1997) Diversity of medicinal plant use in wild chimpanzees. In *Chimpanzee Cultures* (eds. Wrangham RW, McGrew WC, de Waal FB, Heltne PG), pp. 129–148. Cambridge, MA: Harvard University Press.

Huffman MA, Page JE, Sukhdeo MV, et al. (1996) Leaf-swallowing by chimpanzees: a behavioral adaptation for the control of strongyle nematode infections. *Int J Primatol* 17, 475–503.

Hunt KD (1991) Mechanical implications of chimpanzee positional behavior. *Am J Phys Anthropol* 86, 521–536.

Hunt KD (1992) Positional behavior of *Pan troglodytes* in the Mahale Mountains and Gombe Stream National Parks, Tanzania. *Am J Phys Anthropol* 87, 83–107.

Hunter DJ, Eckstein F (2009) Exercise and osteoarthritis. *J Anat* 214, 197–207.

Hurtado AM, Hawkes K, Hill K, Kaplan H (1985) Female subsistence strategies among Ache hunter-gatherers of Eastern Paraguay. *Hum Ecol* 13, 1–28.

Jenike MR (2001) Nutritional ecology: diet, physical activity and body size. In *Hunter-Gatherers: An Interdisciplinary Perspective* (eds. Panter-Brick C, Layton RH, Rowley-Conwy P), pp. 205–238. Cambridge: Cambridge University Press.

Jensen LK (2008) Knee osteoarthritis: influence of work involving heavy lifting, kneeling, climbing stairs or ladders, or kneeling/squatting combined with heavy lifting. *Occup Environ Med* 65, 72–89.

Jurmain RD (1977a) Paleoepidemiology of degenerative knee disease. *Med Anthropol* 1, 1–14.

Jurmain RD (1977b) Stress and etiology of osteoarthritis. *Am J Phys Anthropol* 46: 353–366.

Jurmain RD (1980) The pattern of involvement of appendicular degenerative joint disease. *Am J Phys Anthropol* 53, 143–150.

Jurmain RD (1989) Trauma, degenerative disease, and other pathologies among the Gombe chimpanzees. *Am J Phys Anthropol* 80, 229–237.

Kaplan H, Thompson RC, Trumble BC, et al. (2017) Coronary atherosclerosis in indigenous South American Tsimane: a cross-sectional cohort study. *The Lancet* 389, 1730–1739.

Keele BF, Jones JH, Terio KA, et al. (2009) Increased mortality and AIDS-like immunopathology in wild chimpanzees infected with SIVcpz. *Nature*, 460, 515–519.

Kelly RL (1992) Mobility/sedentism: concepts, archaeological measures and effects. *Ann Rev Anthropol* 21, 43–66.

Kendall FP, Kendall McCreary F, Provance PG, Rodgers M, Romani W (2005) *Muscle Testing and Function with Posture and Pain*, 5th ed. Baltimore, MD: Lippincourt, Williams & Wilkins.

Kochanek KD, Murphy SL, Jiaquan Xu J, Tejada-Vera B (2016) Deaths: final data for 2014. *National Vital Stat Rep* 65, 1–121.

Krief S, Escalante AA, Pacheco MA, et al. (2010) On the diversity of malaria parasites in African apes and the origin of *Plasmodium falciparum* from Bonobos. *PLoS Pathog* 6. DOI: 10.1371/journal.ppat.1000765.

Kumar S (2001) Theories of musculoskeletal injury causation. *Ergonomics* 44, 17–47.

Lane NE, Bloch DA, Jones HH, et al. (1986) Long-distance running, bone density and osteoarthritis. *J Am Med Assoc* 255, 1147–1151.

Larsen CS (1995) Biological changes in human populations with agriculture. *Ann Rev Anthropol* 24, 185–213.

Leendertz FH, Pauli G, Mätz-Rensing K, et al. (2006) Pathogens as drivers of population declines: the importance of systematic monitoring in great apes and other threatened mammals. *Biol Conserv* 131, 325–337.

Leonard WR (2008) Lifestyle, diet, and disease: comparative perspectives on the determinants of chronic health risks. In *Evolution in Health and Disease*, 2nd ed. (eds. Sterns

SC, Koella JC), pp. 265-276. New York: Oxford University Press.

Leonard WR, Robertson ML (1997) Comparative primate energetics and hominid evolution. *Am J Phys Anthropol* 102, 265-281.

Leroy EM, Rouquet P, Formenty P, et al. (2004) Multiple Ebola virus transmission events and rapid decline of central African wildlife. *Science* 303, 387-390.

Li G, Buckle P (1999) Current techniques for assessing physical exposure to work-related musculoskeletal risks with emphasis on posture-based methods. *Ergonomics* 42, 674-695.

Lloyd E, Wilson, DS, Sober E (2011) Evolutionary mismatch and what to do about it: a basic tutorial. *Evol Appl*, 2-4.

Lovell NC (1990) Skeletal and dental pathology of free-ranging mountain gorillas. *Am J Phys Anthropol* 81, 399-412.

Lowenstine LJ, McManamon R, Terio KA (2016) Comparative pathology of aging great apes bonobos, chimpanzees, gorillas, and orangutans. *Vet Pathol* 53, 250-276.

Lukasik-Braum M, Spelman L (2008) Chimpanzee respiratory disease and visitation rules at Mahale and Gombe National Parks in Tanzania. *Am J Primatol* 70, 734-737.

MacNeilage A (2001) Diet and habitat use of two mountain gorilla groups in contrasting habitats in the Virungas. In *Mountain Gorillas: Three Decades of Research at Karisoke* (eds. Robbins MM, Sicotte P, Stewart KJ), pp. 265-92. Cambridge: Cambridge University Press.

Magnusson M, Pope MH (1998) A review of the biomechanics and epidemiology of working postures (it isn't always vibration which is to blame). *J Sound Vibr* 215, 965-976.

Malina RM, Little BB (2008) Physical activity: the present in the context of the past. *Am J Hum Biol* 20, 373-391.

Martin DL, Armelagos GJ, King JR (1979) Degenerative joint disease of the long bones in Dickson Mounds. *Henry Ford Hosp Med J* 27, 60-63.

Matsumoto-Oda A (2002) Behavioral seasonality in male chimpanzees. *Primates* 43, 103-117.

McClure HM (1973) Tumors in nonhuman primates: observations during a six-year period in the Yerkes primate center colony. *Am J Phys Anthropol* 38, 425-429.

McGrew WC, Tutin CEG, Collins DA, File SK (1989) Intestinal parasites of sympatric *Pan troglodytes* and *Papio* spp. at two sites: Gombe (Tanzania) and Mt. Assirik (Senegal). *Am J Primatol* 17, 147-155.

McManamon R, Lowenstine LJ (2012) Cardiovascular disease in great apes. *Zoo Wild Animal Med Curr Ther* 7, 408-415.

Moore SC, Lee IM, Weiderpass E, et al. (2016) Association of leisure-time physical activity with risk of 26 types of cancer in 1.44 million adults. *JAMA Intern Med* 176, 816-825.

Morgado Ramirez DZ, Strike S, Lee R (2017) Vibration transmission of the spine during walking is different between the lumbar and thoracic regions in older adults. *Age Aging* 46, 982-987.

Muehlenbein MP (2006) Intestinal parasite infections and fecal steroid levels in wild chimpanzees. *Am J Phys Anthropol* 130, 546-550.

Muehlenbein MP, Watts DP (2010) The costs of dominance: testosterone, cortisol and intestinal parasites in wild male chimpanzees. *Biopsychosocial Med* 4, 21.

Myers BJ, Kuntz RE (1972) A checklist of parasites and commensals reported for the chimpanzee. *Primates* 13, 433-471.

Neogi T, Felson D, Niu J, et al. (2011) Cartilage loss occurs in the same subregions as subchondral bone attrition: a within-knee subregion-matched approach from the multicenter osteoarthritis study. *Arthritis Rheum Arthritis Care Res* 61, 1539-1544.

Nicolosi R, Bell SJ, Bistrian BR, et al. (1999) Plasma lipid changes after supplementation with β-glucan fiber from yeast. *Am J Clin Nutr* 70, 208-212.

Nunn CL, Rothschild B, Gittleman JL (2007) Why are some species more commonly afflicted by arthritis than others? A comparative study of spondyloarthropathy in primates and carnivores. *J Evol Biol* 20, 460-470.

O'Keefe JH, Vogel R, Lavie CJ, Cordain L (2011) Exercise like a hunter-gatherer: a prescription for organic physical fitness. *Prog Cardiovasc Dis* 53, 471-479.

Olshansky SJ, Passaro DJ, Hershow RC, et al. (2005) A potential decline in life expectancy in the United States in the 21st century. *New Engl J Med* 352, 1138-1145.

O'Neil SP, Novembre FJ, Hill AB, et al. (2000) Progressive infection in a subset of HIV-1-positive chimpanzees. *J Infect Dis* 182, 1051-1062.

Peeters M, Courgnaud V, Abela B, et al. (2002) Risk to human health from a plethora of simian immunodeficiency viruses in primate bushmeat. *Emerg Infect Dis* 8, 451-457.

Pickering RB (1979) Hunter-gatherer/agriculturalist arthritic patterns: a preliminary investigation. *Henry Ford Hosp Med J* 27, 50-53.

Pickering RB (1984) *Patterns of degenerative joint disease in Middle Woodland, Late Woodland, and Mississippian skeletal series from the Lower Illinois Valley*. PhD thesis. Evanston, IL: Northwestern University.

Pontzer H, Wrangham RW (2004) Climbing and the daily energy cost of locomotion in wild chimpanzees: implications for hominoid evolution. *J Hum Evol* 46, 315-333.

Pontzer H, Wrangham RW (2006) Ontogeny of ranging in wild chimpanzees. *Int J Primatol* 27, 295-309.

Pontzer H, Raichlen DA, Wood BM, et al. (2012) Hunter-gatherer energetics and human obesity. *PLoS One* **7**, p. e40503.

Pope MH, Bevins T, Wilder DG, et al. (1985) The relationship between anthropometric, postural, muscular, and mobility characteristics of males ages 18–55. *Spine* **10**, 644–648.

Puente XS, Velasco G, Gutiérrez-Fernández A, et al. (2006) Comparative analysis of cancer genes in the human and chimpanzee genomes. *BMC Genomics* **7**, 15.

Punnett L, Fine LJ, Keyserling WM, Herrin GD, Chaffin DB (1991) Back disorders and non-neutral trunk postures of automobile assembly workers. *Scand J Work Environ Health* **17**, 337–346.

Pusey AE, Wilson ML, Anthony Collins DA (2008) Human impacts, disease risk, and population dynamics in the chimpanzees of Gombe National Park, Tanzania. *Am J Primatol* **70**, 738–744.

Reininghaus UA, Morgan C, Simpson J, et al. (2008) Unemployment, social isolation, achievement-expectation mismatch and psychosis: findings from the ÆSOP study. *Soc Psych Psychiatr Epidemiol* **43**, 743–751.

Riddoch CJ, Anderen LB, Wedderkopp N, et al. (2004) Physical activity levels and patterns of 9- and 15-yr-old European children. *Med Sci Sports Exerc* **36**, 86–92.

Rodriguez E, Aregullin M, Nishida T, et al. (1985) Thiarubrine A, a bioactive constituent of Aspilia (Asteraceae) consumed by wild chimpanzees. *Experientia* **41**, 419–420.

Rosseneu M, Declercq B, Vandamme D, et al. (1979) Influence of oral polyunsaturated and saturated phospholipid treatment on the lipid composition and fatty acid profile of chimpanzee lipoproteins. *Atherosclerosis* **32**, 141–153.

Rothschild BM, Ruehli FJ (2005a) Comparison of arthritis characteristics in lowland *Gorilla gorilla* and mountain *Gorilla beringei*. *Am J Primatol* **66**, 205–218.

Rothschild BM, Ruehli FJ (2005b) Etiology of reactive arthritis in *Pan paniscus, P. troglodytes troglodytes*, and *P. troglodytes schweinfurthii*. *Am J Primatol* **66**, 219–231.

Salmeron J, Manson JE, Stampfer MJ, et al. (1997) Dietary fiber, glycemic load, and risk of non-insulin-dependent diabetes mellitus in women. *JAMA* **277**, 472–477.

Seibold H, Wolf RH (1973) Neoplasms and proliferative lesions in 1065 nonhuman primate necropsies. *Lab Animal Sci* **23**, 533–539.

Seiler BM, Dick EJ, Jr., Guardado-Mendoza R, et al. (2009) Spontaneous heart disease in the adult chimpanzee (*Pan troglodytes*). *J Med Primatol* **38**, 51–58.

Sharma R, Vannabouathong C, Bains S, et al. (2011) Meta-analyses in joint arthroplasty: a review of quantity, quality, and impact. *J Bone Joint Surg* **93**, 2304–2309.

Sjogaard G (1986) Intramuscular changes during long-term contraction. In *The Ergonomics of Working Postures* (eds. Corlett N, Wilson J, Manenica I), pp. 136–143. London: Taylor and Francis.

Smeathers JE (1989) Transient vibrations caused by heel strike. *J Eng Med* **203**, 181–186.

Spector TD, Cooper C, Cushnaghan J, Hart DJ, Dieppe P (1992) *A Radiographic Atlas of Knee Osteoarthritis*. London: Springer-Verlag.

Strong VJ, Grindlay D, Redrobe S, Cobb M, White K (2016) A systematic review of the literature relating to captive great ape morbidity and mortality. *J Zoo Wildlife Medicine* **47**, 697–710.

Takasaki H, Hunt KD (1987) Further medicinal plant consumption in wild chimpanzees? *Afr Study Monog* **8**, 125–128.

Terio KA, Kinsel MJ, Raphael J, et al. (2011) Pathologic lesions in chimpanzees (*Pan trogylodytes schweinfurthii*) from Gombe National Park, Tanzania, 2004–2010). *J Zoo Wildlife Medicine*, **42**, 597–607.

Tursi A (2016) Diverticulosis today: unfashionable and still under-researched. *Therapeut Adv Gastroenterol* **9**, 213–228.

Tutin CG (1996) Ranging and social structure of lowland gorillas in the Lopé Reserve, Gabon. In *Great Ape Societies* (eds. McGrew WC, Marchant LF, Nishida T), pp. 58–70. Cambridge: Cambridge University Press.

Urquhart DM, Tobing J, Hanna FS, et al. (2011) What is the effect of physical activity on the knee joint? A systematic review. *Med Sci Sports Exerc* **43**, 432–442.

US Department of Health and Human Services, Centers for Disease Control and Prevention, National Center for Health Statistics (2012) The National Health Interview Survey. DHHS Publication No. (PHS) 2012-1580.

Varki A, Altheide TK (2005) Comparing the human and chimpanzee genomes: searching for needles in a haystack. *Genome Res* **15**, 1746–1758.

Varki NM, Varki A (2015) On the apparent rarity of epithelial cancers in captive chimpanzees. *Phil Trans R Soc B* **370**, 214–225.

Verschueren SM, Roelants M, Delecluse C, et al. (2004) Effect of 6-month whole body vibration training on hip density, muscle strength, and postural control in postmenopausal women: a randomized controlled pilot study. *J Bone Mineral Res* **19**, 352–359.

Videan EN, Lammey ML, Lee DR (2011) Diagnosis and treatment of degenerative joint disease in a captive male chimpanzee (*Pan troglodytes*). *Am Assoc Lab Anim Sci* **50**, 263–266.

Vieira ER, Kumar S (2004) Working postures: a literature review. *J Occup Rehab* **14**, 143–159.

Vignon E, Valat J, Rossignol M, et al. (2006) Osteoarthritis of the knee and hip and activity: a systematic international review and synthesis (OASIS). *Joint Bone Spine* **73**, 442–455.

Waller C, Hayes D, Block JE, London NJ (2011) Unload it: the key to the treatment of knee osteoarthritis. *Knee Surg Sports Traumatol Arthroscop* 19, 1823–1829.

Waters DJ, Sakr WA, Hayden DW, et al. (1998) Workgroup 4: spontaneous prostate carcinoma in dogs and nonhuman primates. *Prostate* 36, 64–67.

Weickert MO, Möhlig M, Schöfl C, et al. (2006) Cereal fiber improves whole-body insulin sensitivity in overweight and obese women. *Diabet Care* 29, 775–780.

Williams JM, Lonsdorf EV, Wilson ML, et al. (2008) Causes of death in the Kasekela chimpanzees of Gombe National Park, Tanzania. *Am J Primatol* 70, 766–777.

Wrangham RW (1977) Feeding behaviors of chimpanzees in Gombe National Park, Tanzania. In *Primate Ecology* (ed. Clutton-Brock TH), pp. 503–538. London: Academic Press.

Wrangham RW, Nishida T (1983) *Aspilia*: a puzzle in the feeding behaviour of chimpanzees. *Primates* 24, 276–282.

Yamagiwa J, Mwanza N, Yumoto T, Maruhashi T (1994) Seasonal change in the composition of the diet of eastern lowland gorillas. *Primates* 35, 1.

Yazdanbakhsh M, Kremsner PG, Van Ree R (2002) Allergy, parasites, and the hygiene hypothesis. *Science*, 296, 490–494.

Yuan Z, Kang G, Ma F, et al. (2016) Recapitulating cross-species transmission of simian immunodeficiency virus SIVcpz to humans by using humanized BLT mice. *J Virol* 90, 7728–7739.

Zhang Y (2011) Cartilage loss occurs in the same subregions as subchondral bone attrition: a within-knee subregion-matched approach from the multicenter osteoarthritis study. *Arthritis Rheumat-Arthritis Care Res* 61, 1539–1544.

Zhang YQ, Hannan MT, Chaisson CE, et al. (2000) Bone mineral density and risk of incident and progressive radiographic knee osteoarthritis in women: the Framingham Study. *J Rheumatol* 27, 1032–1037.

15 Powering Life
Endocrinology and Physiology

Photo by Anne Frohlich, licensed under CC BY-ND 2.0

PART I: HORMONES

Imagine the workings of a factory (Figure 15.1). The X-Cell-O corporation produces building materials, everything from steel beams to iron roofing to windows – the same sorts of material, in fact, of which the factory building itself is made; in a way we could almost say it reproduces itself. Iron ore and scrap metal are trucked in and shunted to the blast furnace, where they are melted. Impurities are skimmed off before the molten iron is poured into molds; the still-soft molded steel is then pressed between rollers to produce a thin sheet, much as dough is rolled out with a rolling pin. A series of rollers might shape some of the still-glowing, malleable steel into I-beams. Metal sheets are sent to stamping machines to shape them into window frames or iron roofing sheets. Finished products are shuttled to the dock for pick-up.

As manufacturing proceeds, workers place rejected parts, packing materials, and other waste products in waste containers outside. Worn out or broken machinery is either recycled or discarded.

Fresh air is constantly pumped in, some of it refined into pure oxygen for the manufacturing process; the rest of diverted to the open shop floor so that workers can breathe clean air. The prodigious heat generated by the work is dispersed by fans or cooled by air conditioners. In the winter – even with the blast furnaces generating heat – some parts of the factory need to be warmed by a central heating system.

Business is good and at times finished products are used to build new structures that then become working parts in the now-expanded factory. The roof, walls, and other structures are regularly scoured and repainted to prevent corrosion. Broken windows and roof panels are replaced and other damaged or worn out structural parts are fixed.

While this pleasing image of manufacturing efficiency might remind us of a Diego Rivera mural, we can imagine a less felicitous scenario. An inept manager or an overwhelmed foreman might communicate with his team poorly, or he might make poor decisions. Maybe too little raw material arrives at the dock, causing production to shut down; or the foreman might order too much rolled steel made so that it begins to pile up in a staging area, impeding the movement of workers and equipment.

Now imagine something even more disastrous. Imagine that none of these processes are regulated at all. Instead of being modulated according to need, each process is running full bore. Scrap is delivered at a constant rate; sometimes it piles up outside the

Figure 15.1 Volkswagen assembly line. To function properly a factory must acquire raw materials, turn effort into product, and remove waste, all while keeping the environment habitable for the workers. Photo by Alden Jewell. Used with permission.

factory and other times it is used up before the next delivery, slowing production. Constant forging produces more steel than can be rolled and stamped and the excess quickly accumulates on the factory floor. Finished product is created faster than transportation can carry it away and it jams the dock outside. Sanitation workers constantly patrol the factory floor, looking to gather up debris even when there is nothing to gather, adding to the confusion and congestion. The heating system is going full blast, fighting with the air conditioning also running at full capacity. Repairs are being made without regard to need so that energy and materials are wasted as perfectly good windows and roof panels are unnecessarily replaced.

Under these conditions the whole operation might lurch forward for a time, but with incredible waste and inefficiency. But only for a time; within hours congestion and confusion would bring everything to a complete standstill. If this were a living organism, it would be dead. And if even this factory functioned for a little while, the marketplace would quickly prune this inefficient business from its list of ongoing concerns.

A well-functioning factory must be regulated so that processes are turned on when needed; information on what raw materials are needed or which product is needed must make it to the right worker. An instruction book on how many screws to order per month and what date to switch on the air conditioning would make things run more smoothly. Some prediction or anticipation of future needs might be helpful so that the right raw materials could be ordered in advance, rather than ordering them only after they have run out. New-product transport could arrive to carry away product as it is completed, before the docks are actually jammed and at a standstill.

A good manager would use past experience to anticipate needs. She would pass that information on to the foreman effectively and he would then run the stamping machines only when needed. A supervisor should be assigned to inspect the building to discover where repairs are needed, rather than sending out repair staff willy-nilly. Refuse would be removed only when needed.

An exceptional X-Cell-O CEO would closely monitor economic news so as to anticipate demand and set production accordingly, suspending repair efforts when other more urgent matters press. We can imagine something even more sophisticated, a wise manager who can trade off demands when resources are scarce, directing effort to tasks that will pay off best in the future. An efficient factory such as this would deliver more product to the marketplace, run some other businesses into bankruptcy, and expand while other businesses languished. Soon we would find the X-Cell-O logo everywhere on the cultural landscape.

15.1 The Organism as Factory

A sharp reader such as yourself has already recognized the analogy with living organisms. While I may have gone overboard with my factory description, our time has not been wasted; each of the processes discussed as a manufacturing process also must be accomplished in a living organism. The factory that is the chimpanzee body will supply nourishment to her fetus, manufacture mother's milk to nurse her infant only when she actually has an infant, and will maintain her body so as to have the vigor to protect and support her offspring. Her health, vigor, power, athleticism, and ability to heal will allow her to rise in social rank and her success will lead to healthy prodigy, guaranteeing success of her "brand." Decisions about where to allocate resources within her body will take into consideration which trade-off is best for long-term success, ramping up her metabolism when she is sick, suspending reproduction when times are hard, and storing excess energy as fat when times are good. The efficient chimpanzee (and human) body assesses needs accurately, making sure the right raw materials are present when needed, and effectively disposing of waste.

Most of the communication that makes this model of a successful organism possible – by analogy the boss telling the foreman to shift from making windows to rolling out roof sheets – is in the form of hormones. The study of hormones and their function is known as **endocrinology**.

15.2 What Are Hormones?

Hormones are molecules that, once released in the bloodstream, communicate to one or – much more often – many tissues and organ systems – mostly instructing them to turn off or turn on a particular process. With incredible efficiency, some hormones turn off one system at the same time they turn another one on, just as it might benefit a factory to always turn off the air conditioning when turning on the heat. Many hormones pass easily across cell membranes so that they affect the internal processes of cells – the equivalent of a comprehensive internal memo that goes to every worker.

Rather than burning glucose at maximum capacity to create warmth, eating without let up and so on, hormones turn up the metabolism when needed – or even when it anticipates needing more energy. Hormones signal cellular processes to ramp up during extreme exertion. Some hormones work on the brain to create motivation, leaving it to the brain itself to figure out ways to satisfy the desire intelligently: Hormones make you hungry, but it is advanced cognition that figures out how to get food.

Evolution has sculpted the hormonal system so that the body responds almost as if it has consciously anticipated dangers and future needs, like that wise corporate executive. Furthermore, the endocrine system has evolved to shift energy allocation over both the short term, something you already knew, but also over the entire lifespan, focusing energy on the most urgent metabolic processes and those that will pay off best later.

At least, it does these wonderful things when conditions are similar to those under which it evolved, but when "market conditions" change unexpectedly, the mismatch can find the organism ill-prepared. More on that as we go along. Refer back to Chapter 14 for a more thorough discussion of evolutionary mismatch theory.

15.3 Chimpanzee Hormones

Hormones are not fast-evolving molecules; they hardly differ among organisms as diverse as marmosets, wolves, and humans. We know for a fact that human and chimpanzee hormone structure and function are nearly identical for dozens of hormones and we know of no hormone that is substantially different (Haynes et al., 1982; Forsyth, 1986; Fan et al., 1989; Kenter et al., 1992; Fisher, 1997; Birken et al., 2001; Morgan et al., 2003; Gombart et al., 2009; Ronke et al., 2015). Perhaps surprisingly, given how different chimpanzee estrus is from the human sex cycle, this near-identity extends even to sex hormones (Winter et al., 1975, 1976; Copeland et al., 1985; Agnew et al., 2016). In other words, we have every reason to expect that discoveries about hormone function in humans can be applied to chimpanzees and *vice versa*; I cannot go far wrong when I am forced to use human research as I describe chimpanzee endocrinological processes we suspect are the same. Now and then we will range further afield to draw insights on chimpanzee endocrinology from species such as mice and penguins.

15.4 What Initiates Hormonal Action?

We might start by asking what causes a hormone to be released. Often it is another hormone. And that one? Sometimes yet another hormone. Much of hormonal communication is bafflingly complicated, so much so that we are still in the dark about even some important processes. What we do thoroughly understand tells us that nature has often accomplished its ends by quite complicated, multistep processes that seem unnecessarily convoluted. Such complications in nature inspired the King of Galicia, Alfonso X (1221–1284), to have uttered the much-paraphrased quip "If the Lord Almighty had consulted me before embarking on creation thus, I should have recommended something simpler." The utterance fits the endocrine system to a T. One only moderately complicated system has been called "miserably confusing" by the prominent researcher who has made it his specialization (Sapolsky, 1998). The simplest systems are those where one hormone turns a system on and a different hormone turns it off; these pairs are known as antagonists (see Table 15.1) because their actions oppose one another. For example, angiosinogen, endothelin, renin, and thromboxane all induce vasoconstriction (see Table 15.1), while antagonists atriopeptin, vasoactive intestinal peptide, and vasopressin cause vasodilation. Ghrelin stimulates the appetite, while leptin suppresses it.

Most hormonal signals begin with brain activity in the **hypothalamus**, which then often signals a sort of "master gland," **the pituitary** (noting, of course, that most hormonal action begins in the brain, making it more of the "master" than the pituitary [Sapolsky, 1998]); then, in a cascade of hormone release, hormone-production, and hormone interactions, the proper response eventually emerges.

15.5 Adrenaline: Coping with a Bad Moment

Let us consider the action of a hormone at the center of one of the simpler systems. In the hormone hall of fame, **adrenaline** (also known as **epinephrine**) must be the most illustrious, after estrogen and testosterone; all of us are familiar with its effects. Let an elephant who was calmly grazing turn and charge your Land Rover and adrenaline is felt immediately; in you, I mean, not the elephant.

Adrenaline is an emergency hormone that tells your body to abandon all long-term plans and prepare itself for immediate action (Sapolsky, [1998] reviews this subject wonderfully). Your heart races, your breathing quickens, your air passages expand, and your senses become sharper. Your pupils dilate to improve visual acuity. Your blood pressure spikes to more effectively push oxygen into muscles and organs. Forget digesting food; blood is directed away from your digestive organs and toward organs of action, muscles, heart, and lungs. Quite the opposite of storing nutrients, stress hormones instead cause fat cells to dump energy into the bloodstream in the form of glucose, providing extra fuel to power anticipated muscle activity. Heat generated by muscle activity is expected any fraction of a second; sweating is activated. Pain perception is suppressed. Growth, immune response, and in fact all long-term activities are suspended. Even repair of tissues is abandoned until danger has passed. This is the body's version of suspending manufacturing when a hurricane is approaching. No executive would order construction on a new wing or initiate routine maintenance at such a time. Instead, she will close down production, send workers home, close the hurricane shutters, top-up fuel in the emergency generators, and make sure essential systems can keep running if the power goes out.

Table 15.1 **Human hormones: the complete list**

Name	Abbreviation	Producing gland or tissue	Target tissue	Effect
Adiponectin	Acrp30	Adipose tissue	Many cells	Regulates fat metabolism
Adrenaline (or epinephrine)	EPI	Adrenal	Many tissues	Stimulates processes supporting extreme physical activity
Amylin	IAPP	Pancreas	Digestive system	Slows gastric emptying, secretion
Androstenedione		Adrenals, gonads	Diverse cells	Estrogen and testosterone precursor
Aldosterone		Adrenals	Kidneys, salivary and sweat glands, colon	Sodium retention, thus increasing blood volume by water retention
Angiotensinogen, angiotensin	AGT	Liver	Kidney, salivary, and sweat glands, colon	Vasoconstriction, salt retention
Anti-Müllerian hormone	AMH		Reproductive tissues	Inhibits prolactin, prevents growth of vagina, uterus
Atriopeptin	ANP	Heart	Kidney, adrenals, blood vessels, fat cells	Reduces aldosterone, vasodilator, lowers blood pressure, ups sodium loss, releases fatty acids
Calcitonin	CT	Thyroid	Bone	Bone building, reduces blood calcium, inhibits osteoclasts
Calcitriol		Kidney	Gut, kidney	Increases absorption of calcium and phosphate in gut
Cholecystokinin	CCK	Duodenum	Pancreas, gallbladder	Release digestive enzymes
Corticotropin	ACTH	Anterior pituitary	Adrenal gland	Stimulates production of glucocorticoids

Table 15.1 (*cont.*)

Name	Abbreviation	Producing gland or tissue	Target tissue	Effect
Cortisol		Adrenals	Diverse cells	"Stress" hormone; inhibits fat storage, increases blood sugar, suppresses the immune system, metabolizes fat
Corticotropin releasing hormone	CRH	Hypothalamus	Pituitary	Releases ACTH (corticotropin)
Cortistatin	CORT	Cerebral cortex	Brain	Induces sleep, immobilizes body
Dehydroepiandrosterone	DHEA	Testes, ovary, kidney	Kidney, testes, ovaries	Testosterone, estrogen precursor, increases muscle mass, induces sex characters
Dihydrotestosterone	DHT	Converted from T in diverse cells	Prostate, sex organs, skin, hair	Male embryonic development, induce male characters
Enkephalin		Kidney	Brain, nerves	Opioid-like peptide, reduces pain
Endothelin	EDN	Blood vessels	Blood vessels	Vasoconstriction
Erythropoietin	EPO	Kidney	Marrow	Stimulates blood cell production
Estrogen	E	Ovary, diverse tissues	Diverse cells, ovary or testes	Primary female sex hormone
Follicle-stimulating hormone	FSH	Pituitary	Reproductive tissue	Ovarian follicles in females, sperm production in males
Galanin	GAL	Gut, brain, spinal cord	Brain, nerves	Modulates neuronal activity
Gastric inhibitory polypeptide	GIP	Gut	Pancreas	Induces insulin secretion
Gastrin	GAS	Gut	Gut (esp. stomach)	Induces gastric secretion
Ghrelin	GHRL	Stomach	Brain	Stimulates appetite, releases pituitary growth hormone

Table 15.1 (*cont.*)

Name	Abbreviation	Producing gland or tissue	Target tissue	Effect
Glucagon	GCG	Pancreas	Liver	Converts glycogen to glucose
Glucagon-like peptide-1	GLP1	Gut	Pancreas	Synthesis and release of insulin
Gonadotropin-releasing hormone	GnRH	Hypothalamus	Pituitary	Releases FSH, luteinizing hormone
Growth hormone	GH, hGH	Pituitary	Diverse cells, liver	Stimulates growth, release of IGF-1, strengthens bone
Growth hormone-releasing hormone	GHRH	Hypothalamus	Pituitary	Releases growth hormone
Guanylin	GN	Gut	Gut, kidney	Stimulates diarrhea
Hepcidin	HAMP	Liver	Gut, diverse cells	Retains iron, both in cells and gut
Human chorionic gonadotropin	hCG	Placenta	Corpus luteum	Maintains corpus luteum, inhibits immune response, prevents immune system rejecting embryo
Human placental lactogen	HPL	Placenta	Pancreas	Increases insulin, growth factors (IGF-1), reduces insulin sensitivity thus increasing glucose levels
Inhibin		Testes, ovary, fetus	Pituitary	Inhibits production of FSH
Insulin	INS	Pancreas	Pancreas, liver, cells	Directs glucose intake, production, storage
Insulin-like growth factor	IGF	Liver	Diverse cells, liver	Regulates cell growth, development
Leptin	LEP	Adipose tissue	Hypothalamus	Opposes ghrelin; kills appetite, increases metabolism
Leukotrienes	LT	White blood cells	Blood vessels, air passages	Controls vascular permeability, bronchial restriction, inflammation

Table 15.1 (*cont.*)

Name	Abbreviation	Producing gland or tissue	Target tissue	Effect
Lipotropin	LPH	Pituitary	Brain, diverse cells	Unclear; lipolysis, steroidogenesis?
Luteinizing hormone	LH	Pituitary	Gonads	Estrogen production in females, testosterone production in males
Melanocyte stimulating hormone	MSH or α-MSH	Pituitary	Melanocytes	Stimulates production of pigment
Melatonin	MT	Pineal	Brain	Regulates sleep
Motilin	MLN	Small intestine	Gut	Stimulates gut activity, pepsin
Neuropeptin	BNP	Heart	As with ANPNPR	As with ANP
Orexin		Hypothalamus	Brain	Promotes wakefulness, increase energy expenditure and appetite
Osteocalcin	OCN	Skeleton	Muscle, brain, pancreas, testes, bone	Increases muscle function, bone growth, testosterone production
Oxytocin	OXT	Pituitary	Breast, female sex organs, brain	Releases breast milk, stimulates uterine contractions; bonding
Pancreatic polypeptide	PP	Pancreas	Pancreas, liver PP cells	Stimulates production of digestive enzymes and glycogen
Parathyroid hormone	PTH	Parathyroid	Bone, kidney	Increases blood Ca^{2+} by dissolving Ca from bone; activates vitamin D
Pituitary adenylate cyclase-activating peptide	PACAP	Multiple cells	Stomach, pituitary	Stimulates gastric activity
Progesterone	PR	Ovary, adrenals, placenta	Diverse cells, reproductive tissues	Supports pregnancy in multiple ways

Table 15.1 (cont.)

Name	Abbreviation	Producing gland or tissue	Target tissue	Effect
Prolactin	PRL	Pituitary, uterus	Mammary glands	Milk production, sexual gratification
Prolactin releasing hormone	PRH	Hypothalamus	Pituitary	Releases prolactin from pituitary
Prostaglandins	PG	Seminal vesicles	Blood vessels	Vasodilation, erection
Prostacyclin	PGI_2	Endothelium	Blood vessels	Vasodilation, clotting inhibition
Relaxin	RLN	Corpus luteum, ovaries, placenta	Blood vessels, ligaments	Allow larger blood volume, increase elasticity of ligaments to allow childbirth
Renin		Kidney	Blood vessels	Vasoconstriction
Secretin	SCT	Duodenum	Liver, pancreas	Stops production of gastric juice
Somatostatin GH-inhibiting hormone	SRIF GHIH	Hypothalamus, pancreas, gut	Pituitary, stomach, smooth muscles, pancreas	Inhibits release of growth hormone, reduce stomach acid production, gut (including stomach) activity
Testosterone	T	Testes	Diverse cells but particularly genitals	Raises libido, builds muscle, increases bone density, induces male sex characters
Thrombopoietin	TPO	Liver, kidney, striated muscle	Marrow	Increases red blood cell production
Thromboxane	TXA_2	Platelets	Blood vessels	Vasoconstriction
Thyroid-stimulating hormone	TSH	Pituitary	Thyroid and via thyroid hormones diverse cells	Stimulates the secretion of thyroid, hormones, increases metabolism
Thyrotropin-releasing hormone	TRH	Hypothalamus	Thyroid	Releases TSH thus increasing metabolism, releases prolactin

Table 15.1 (cont.)

Name	Abbreviation	Producing gland or tissue	Target tissue	Effect
Triiodothyronine	T3	Thyroid	Most tissues	Increases metabolism
Thyroxine	T4	Thyroid	Most tissues	Increases metabolism
Vasoactive intestinal peptide	VIP	Gut, pancreas, hypothalamus	Circulatory system, trachea, stomach, gall bladder	Stimulates contractility in the heart, vasodilation lowering blood pressure
Vasopressin	ADH	Posterior pituitary	Hypothalamus, pituitary, kidneys	Water retention, vasoconstriction, corticotropin release
Uroguanylin	UGN	Gut	Gut, kidney	Induces feeling of "fullness"

Have you noticed that adrenaline (or epinephrine) is one of those multipurpose hormones mentioned above? It turns on some systems – heart, respiration rate, sweating, and so on – but shuts down others: digestion, growth, and immune function. At first this system shows the promise of being relatively simple: You get scared, adrenaline is released, and your capacity to flee or fight is augmented. The process is, in fact, not so simple – but simple enough to explain here and use as a model for many endocrine systems.

Adrenaline production and release is a double-pronged cascade of neural and hormonal responses. Your senses register and relay information on the external world to the brain – registering a threat, in this case. In response to the electrical and chemical activity in the brain that constitutes the realization of imminent danger, a part of the brain known as the hypothalamus sends a nerve impulse to the inner layer of your adrenal glands, the adrenal medulla, small pyramidal glands that sit on top of your kidneys. The signals from the hypothalamus stimulate a near-instantaneous spilling of adrenaline into the bloodstream. In a heartbeat, literally, adrenaline begins a rapid dispersal through your entire body.

15.6 Cortisol: Coping with a Bad Week

While all this is going on under the influence of the short-term hormone epinephrine, the body is also taking a longer-term strategy. The hypothalamus squirts a small dose of **corticotropin releasing hormone (CRH**; so now you see why the hypothalamus is more the "master gland" than the pituitary) into the nearby pituitary gland via a miniscule interconnecting tube. In short order the pituitary releases **corticotropin** – which explains why the first hormone is called CRH. Note that **specialists often refer to corticotropin as ACTH.** Corticotropin (or ACTH) in turn activates a different layer of the adrenal gland, the outer part, known as the adrenal cortex, to release longer-acting hormones – glucocorticoids, hormones that suppress inflammation, accelerate the production of glucose in the liver, and stimulate the breakdown of fat reserves into something your body "anticipates" needing, which is yet more glucose. **Cortisol**, perhaps the most widely known glucocorticoid, is more or less a longer-acting version of adrenalin, and one that provokes many of the same responses: mobilize sugar, shut down the immune system, stop storing fat, leave off making repairs to the body; get ready to dodge that charging rhino or hungry leopard. It even breaks down muscle to use as energy.

Cortisol also stimulates the pancreas to release the hormone **glucagon**. Glucagon is a miraculous hormone that stimulates diverse cells throughout the body to offer up support for the hard-working muscles by converting their stored glucose (stored as **glycogen**) into raw glucose, dumping this fuel into the blood.

The body evolved to produce cortisol as a longer-term response to threats than adrenaline. Cortisol sustains the stress response to hours or perhaps several days, rather than just minutes. It is the brain that juggles data from many sources to determine that long-term trouble is on the horizon, which in turn stimulates the hypothalamus, starting the cascade of events we discussed above. *Cortisol is commonly referred to as the "stress hormone,"* but it is only one of several glucocorticoids, some of which you may know of because your doctor might prescribe them: cortisone, hydrocortisone, and prednisone, the latter widely used clinically to quell inflammation.

Imagine our factory going on 24-hour-a-day production to fill an emergency order, shutting down research and development, postponing that planned expansion of work space, putting off replacing machinery that will break down soon. At the same smelting furnaces are cranked up, air cleaners run on high, emergency fuel stores are drawn down, and workers are shifted from roof repair to the production line. All hands on deck until the emergency has passed.

15.7 Longer-Term Planning: Growth

The all-hands-on-deck effort allocation above is extreme, but subtler trade-offs occur constantly. The body is adapted to cope not only with minute-to-minute (adrenalin) or week-to-week stresses (cortisol), it juggles trade-offs over much longer time scales as well, some of them over the entire lifespan. Growth and development were reviewed in Chapter 8, but let us note here that it is hormones that determine where energy will be allocated during this critical period. When times are tough, energy is reserved for emergencies and cortisol shuts down growth by turning off growth-hormone production (Sävendahl, 2012). At the age of five (in humans) hormones dictate a life history trade-off strategy that allocates most energy (or at least all it can spare after minimal body-running needs are met) to brain growth, anticipating that, as risky as it is to allocate so much energy to one organ, it will pay off over the long term (Kuzawa et al., 2014). Beginning only a year later, from ages 6–10, hormones guide a further shift, once brain growth is taken care of, dedicating all available energy to body growth. Next, adolescence arrives and the body shifts energy to completing reproductive maturation. Once reproductive maturation is complete, in the early adult years there is a shift toward bodily processes related to reproduction. Hormones even fine-tune bodily functions as the organism begins to show the effects of age (Finch & Rose, 1995; Bribiescas & Muehlenbein, 2010).

15.8 Energy Management and Storage

Much of hormone activity involves less dramatic pastimes than fleeing lions or pumping up brain growth. Normally, instead of withdrawing energy from storage, we make plans to help us avoid the lion, more like the foreman switching production from I-beams to sheet metal, or from roof repair to window replacement. In planning for the future, whenever it can the body stores supplies for hard times. Some glucose, fats (often called **lipids**), and deconstituted proteins (i.e., amino acids) that enter our bloodstream via digestion are used immediately, but many others are stored for future use, either shunted into fat cells as a long-term investment or stuck together to form glycogen, those long glucose chains we encountered earlier, which is then stored either in the muscles themselves or in the liver. The very important hormone **insulin** is the hormone that stimulates all this storage, affecting a diverse array of cells and tissues. When the body detects glucose in the bloodstream, insulin is instantly spilled from the pancreas and proceeds to mop it all up and stockpile this energy for hard times down the road, provoking cells to store it as either fat or glycogen. Insulin also acts on other nutrients that digestion provides, the constituents of fat, called fatty acids; these are strung together and stored in the fat cells. Other body cells absorb amino acids, the building blocks of proteins.

These and many, many other rather mundane functions are directed by various of the 71 human hormones. You may wonder what those hormones are, and what their functions are; they are listed in Table 15.1 along with their actions; I also list the organs that produce them and their target tissues. Readers often skip tables, but, truly, if you have gotten this far in the book, you will find this list fascinating.

PART II: HORMONES GONE BAD

15.9 Modern Life: Eating Yourself to Death

In the last chapter we talked about evolutionary mismatches. Organisms are adapted to cope with circumstances their ancestors faced; if their ancestors rarely encountered a challenge, natural selection could not select those who dealt with it best; the population will have no adaptations to deal with it. The widespread availability of sugars in food such as cultivated fruits (typically more sugary than their wild counterparts), sugary snacks, sugared drinks, and many other foods – sugar is packed into deli meats, ketchup, bread, and soup – means that the amount of sugar we consume is unprecedented in our evolutionary history. Our blood is constantly awash in glucose, which mean the pancreas is constantly releasing insulin to do something with it. Our bodies, unevolved for this abundance, cannot cope. Individual fat cells, filled to bursting, begin to ignore insulin and no longer take up glucose; its level in the blood rises. Or the pancreas begins to ignore signals to produce insulin, "thinking" glucose could never be so constant a presence. In either case, without insulin stimulating its storage, glucose languishes in the blood.

At first glance, having our principal energy source, glucose, present at high levels in the blood, right where it can be used, might seem like a good thing, or at the very least nothing to worry about. Our factory manager would never worry about having the factory's fuel tank constantly full. It turns out it *is* a big problem because elevated levels of blood glucose interfere with numerous physiological processes, leading to heart disease, eye problems, kidney failure, and nerve damage. Sugar is killing us.

If only there were some way to suppress our tendency to overconsume sugar, one that draws on the body's natural processes rather than inserting into our bloodstream some exotic chemical our species has never encountered and may not be able to cope with. We are only just now beginning to understand the function – and the wonders – of a hormone that regulates hunger, ghrelin. **Ghrelin** stimulates the appetite. When the stomach is full and its tissues stretched, this wonder hormone is inactive; but when the stomach is empty and slack, ghrelin is secreted, stimulating the appetite. Just like the wise factory supervisor, your body refrains from ordering supplies when it is still processing earlier deliveries of raw materials. And the moment the body's factory floor is clear – when the stomach is empty – it sends a message that more raw material is needed. Incredibly, ghrelin also stimulates memory (Cahill et al., 2014) – a hungry person learns better. You will hardly believe it, but in mice an injection of ghrelin increases *hoarding* behavior!

By now you have undoubtedly realized that a "ghrelin blocker" would be a best-seller. Recall that many hormones have an antagonist, and ghrelin is no exception; its antagonist is **leptin**. Leptin is produced in fat cells when they get full, a fact that we were completely unaware of until 1994 (Bribiescas & Anestis, 2010). It acts to suppress appetite and increase metabolism, an ideal combination for weight loss. Leptin levels are higher in chimpanzee females (Bribiescas & Anestis, 2010), perhaps because they need to mobilize energy during pregnancy. Ghrelin and leptin are not the only appetite hormones; uroguanylin, which is released by the gut during digestion and induces a feeling of fullness, is an important signal that tells us to stop eating before we burst.

You may be eager to try out appetite-suppressing hormones. Short-circuiting the body's desire for food is not simple. Yes, you can buy leptin, but side-effects such as inflammation and autoimmune responses have been reported. Because ghrelin also has something to do with such diverse phenomena as concentration, depression, PTSD, psychological stress response, and task persistence (Lutter et al., 2008; Meyer et al., 2014), suppressing it is likely to affect some other important aspect of biology, producing the dread of all drug companies: the side-effect. Better to diet and get more exercise.

Fat storage is a little less aggressive in apes. We humans have a hormone response that tends to lead to fat storage, and while apes can get fat in zoos, they tend to be lean animals. Among bonobos, females have body fat similar to that of extremely lean humans, such as male marathon runners (3.7 percent), but, incredibly, males are even leaner, with body fat percentages that are essentially zero (0.005 percent; Zihlman & Bolter, 2015).

15.10 Modern Life: Sweating the Small Stuff

It is only relatively recently that endocrinologists have come to realize that our lifestyle creates yet another mismatch, a psychological environment where we continuously face "crises" that mimic confronting some imminent physical danger. Being stuck in an unexpected traffic jam just when an important presentation looms, opening a notice that a big late fee has been slapped on your credit card account because your payment was late, or hearing gossip that a big layoff is threatening at work – these all evoke in your endocrine system the same response the proverbial zebra experiences when chased by a lion (Sapolsky, 1998). Responding this way to what are not serious threats is a mismatch.

Our misperception of danger exposes us to high cortisol levels. In response, for weeks and months our bodies leave off repairs, forget about protecting the body from disease, and even abandon reproduction – sometimes for years. Chronic stress, by causing the abandonment of repair and protection from disease, makes you sick. Cortisol evolved to deal with emergencies; having it float around your body constantly is a recipe for an early grave.

If all this has so far failed to motivate you to take action to better cope with your cortisol-inducing lifestyle, maybe this last thought will. Glucocorticoids like cortisol are so debilitating that some species use them as suicide hormones. In both salmon and marsupial mice, programmed death arrives via overproduction of glucocorticoids (Sapolsky, 1998), a phenomenon we have verified by experimentally removing the glucocorticoid-producing adrenal glands in these animals. After adrenal gland removal, individuals in these species live months after reproducing, rather than hours. Suicide seems like the ultimate in extreme strategies, so you may be thinking there must be some evolutionary advantage to it. It turns out salmon are very unlikely to live long enough to reach a second breeding season, which means that from an evolutionary perspective once they lay or fertilize their eggs, their reproductive lives are over. Given that, is there *anything* a salmon can do to help its still defenseless fertilized eggs? Dying and leaving their bodies floating in the same stream as their maturing offspring might actually help by providing so much food for predators and scavengers that the eggs are left alone.

PART III: REPRODUCTIVE HORMONES

15.11 Prolactin: The Bonding Hormone

Estrogen gets all the publicity, and it will get its due below, but the wonders of **prolactin** are perhaps even more impressive, given that it has such diverse effects. Its principal function is to induce lactation, but it has been coopted to play a role in contraception and bonding with offspring. The pituitary gland, and to a lesser extent the uterus, release prolactin when a mother nurses, or in fact when the nipple is stimulated in any way, encouraging milk production and the release of the bonding hormone oxytocin. As a good endocrinologist you might be thinking "hmm; maybe an adoptive mother could begin to produce milk if she offers her breast to a nursing infant." Indeed, this is true (Thearle & Weissenberger, 1984).

Among other functions, prolactin interferes with a hormone called **progesterone**, which works to sustain pregnancy; it also acts directly on the ovaries to discourage ovulation. If a mother nurses often enough, high prolactin levels serve as a kind of birth control. This is, however, unlikely to become a fashion in your neighborhood because "often enough" is about every 15 minutes (Konner & Worthman, 1980), a schedule to which hunter-gatherers often adhere, but one impractical for lawyers and teachers. Prolactin also increases interest in and affection toward infants in rats, marmosets, and in fact any mammal tested so far; non-mothers injected with prolactin exhibit maternal care, while untreated females ignore infants (Samuels & Bridges, 1983). Sexual activity also releases prolactin, producing feelings of well-being and sexual gratification, after which it acts as a temporary libido suppressant (Crenshaw & Goldberg, 1996; Saks & Gillespie, 2002).

Prolactin is not just a female hormone; it plays no small role when it comes to male parenting as well. In numerous species where males help care for offspring, including humans, prolactin levels are highly elevated in fathers (Figure 15.2; Schradin & Anzenberger, 1999, 2003; Storey et al., 2000; Fleming et al., 2002; Gray et al., 2007; Gettler et al., 2012). If prolactin is injected into unmated males they engage in more parental behavior (Figure 15.3; Schradin & Anzenberger, 1999). Among primates it was the monogamous male marmosets that were first found to have this elevated level of prolactin (Schradin & Anzenberger, 1999, 2003; Storey et al., 2000; Fleming et al., 2002). While humans are not strictly monogamous, both pair-bonding and male involvement in child-rearing are species-typical, so it is not surprising that human fathers show elevated prolactin levels (Fleming et al., 2002; Gray et al., 2007; Gettler et al., 2012). Men with elevated levels of prolactin respond more profoundly to a crying infant (Fleming et al., 2002). The unexpected intensity of feelings many new fathers feel reminds us that hormones can affect us powerfully, and prolactin is certainly one of those hormones (Curtis et al., 1997).

15.12 Oxytocin: Multiple Functions, Including More Bonding

Oxytocin is another multi-duty bonding hormone. Its most important function is to induce uterine contractions to begin labor. If you have children, you may be familiar with the drug Pitocin™, an artificial version of oxytocin commonly used to induce labor. It does so by mimicking the function of oxytocin,

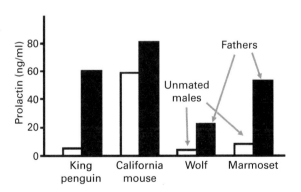

Figure 15.2 Among animals as diverse as penguins and mice, prolactin is higher in fathers than in unmated males. Prolactin induces paternal behavior. After Schradin & Anzenberger, 1999, with permission.

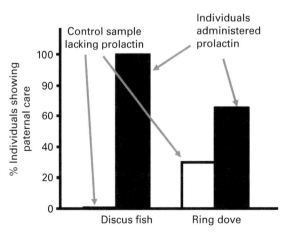

Figure 15.3 When prolactin is artificially administered to animals as varied as fish and birds, parenting behaviors increase. From data in Schradin & Anzenberger, 1999; Buntin, 1996, with permission.

inducing and strengthening uterine contractions. One risk in the administration of Pitocin is uterine tears, an illustration of the principal that most physiological processes involve multiple responses to multiple hormones: If the cervix has not been stimulated to dilate, birth is impossible and the powerful contractions tear the uterus instead.

As birth proceeds, oxytocin has another role; it encourages blood vessel constriction and blood coagulation, preventing hemorrhage at the site of the expelled placenta. Nor is oxytocin finished even when the baby is in the mother's arms. After birth it stimulates the release of breast milk, the "let down" phenomenon with which nursing mothers will be familiar. The release of oxytocin during nursing increases the bonds the mother feels with baby (Donaldson & Young, 2008).

Initiating birth, preventing hemorrhages, and promoting milk delivery are an impressive enough list, but this wonder-hormone is still far from finished. This "love hormone" is responsible not only for infant bonding, but also induces pair bonding, or to be less clinical, it is responsible for perhaps one of the most transcendent of human experiences: falling in love. Oxytocin is released not only during nursing, but during any breast stimulation. New bonding functions for this hormone are still being discovered. Some research suggests that it inspires social connections to the extent that it improves "mind reading" in humans (Domes et al., 2007). It increases the tendency to make eye contact (Guastella et al., 2008) and bolsters trust (Kosfeld et al., 2005) among social partners, so much so that in experiments in which test subjects are given injections of oxytocin they continue to invest money with a financial advisor even after the advisor is exposed as a crook (Donaldson & Young, 2008). If Bernie Madoff had weaponized oxytocin we would all be broke. To top it all off, oxytocin suppresses the hormone I portrayed as a plague on modern life earlier, the stress hormone cortisol (Heinrichs et al., 2003).

15.13 Sex Hormones

Given the frequency of multiple-effect responses to hormones, it is no surprise that sex hormones undertake many tasks: spermatogenesis, ovulation, and sexual maturation. They initiate and later maintain sex differences.

Sex hormones start their lives as boring old fat, more or less – both male and female hormones are modified cholesterol (Figure 15.4). The most important hormone in the **estrogen** family, **estradiol**, goes through many transitions before finally being converted to estrogen, the penultimate step being testosterone. The notorious cortisol is also a steroid, as you can see in Figure 15.4.

Shortly we will learn that estrogen is an important hormone in the **menstrual cycle**, but many of us are more familiar with its role during fetal and other development. Estrogen diffuses throughout the body, but it affects tissues rich in estrogen receptors in particular, principally in the ovaries, uterus, breasts, and brain. As estrogen levels increase during puberty they not only stimulate breast development, they increase body fat and affect its distribution. Estrogen sharpens cognitive function (Hara et al., 2015), reduces susceptibility to heart disease (Rosano and Panina, 1999), reduces inflammation (Nadkarni et al., 2011), helps to sustain healthy bone density (Weiss et al., 1980), and maintains skin health by preventing collagen loss and skin thinning (Shah & Maibach, 2001).

15.14 Reproduction

I will refrain from extending my factory analogy to reproduction, with this one exception: Just as there are propitious times to expand factory capacity and times when it is best to wait, there is a time for reproduction and a time to hunker down, and hormones govern that timing. Prolactin suppresses ovulation, and it is thought that a mix of hormones associated with low body fat also aids in signaling the body to stop ovulating and hold back supplies for later reproduction. The familiar villain cortisol plays a role, too, in suppressing ovulation; the body has evolved to put off reproducing when emergencies threaten. In the past, cortisol might have been stimulated by news of approaching war or knowledge of the loss of critical food stores. Having a baby would

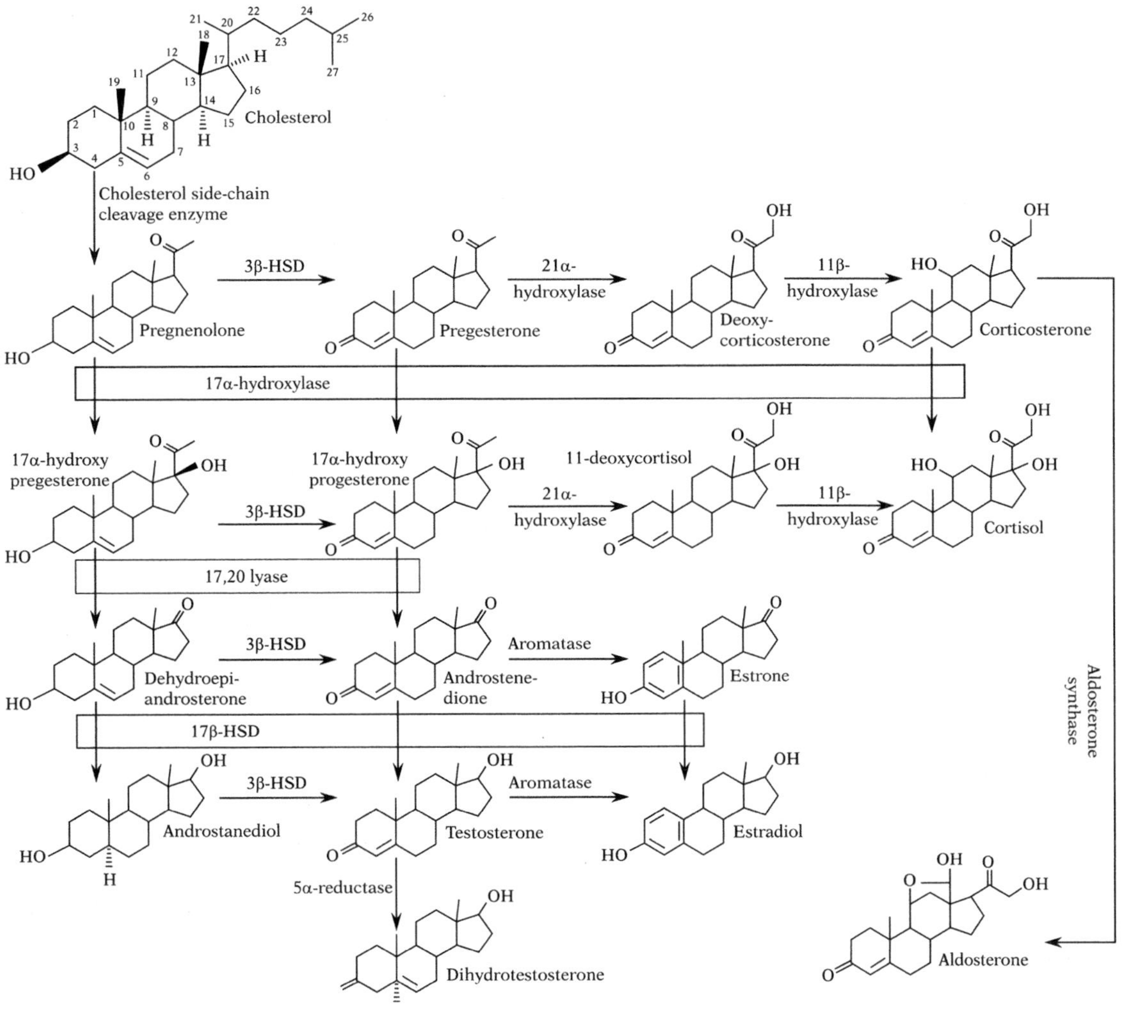

Figure 15.4 Steroid hormone synthesis and structure. After Bribiescas & Muehlenbein, 2010, with permission.

be risky for both mother and infant in such bad times. Worrying that you failed to get a promotion, however – treating it as if your sole winter food supply was wiped out – is a mismatch.

15.15 The Menstrual Cycle: Ovulation

As we might expect, ovulation and pregnancy arrive at the behest of a wide assortment of hormones and processes (see recent reviews by Bribiescas & Muehlenbein [2010] and Muehlenbein & Bribiescas [2010]). The menstrual cycle, or more appropriately the **sex cycle**, is considered to have two halves, each of approximately 14 days, so 28 days in total (Figure 15.5). In the first 14 days, action is driven by a structure known as the **follicle**, to which you will soon be introduced, while the second half is directed by the equally important **corpus luteum**. The first 14-day period is the **follicular phase**, the part of the cycle before ovulation; the second half of the sexual cycle, that after ovulation, is the **luteal phase** (Figure 15.5).

As with most hormone stories, this one begins with the hypothalamus, which releases the aptly named

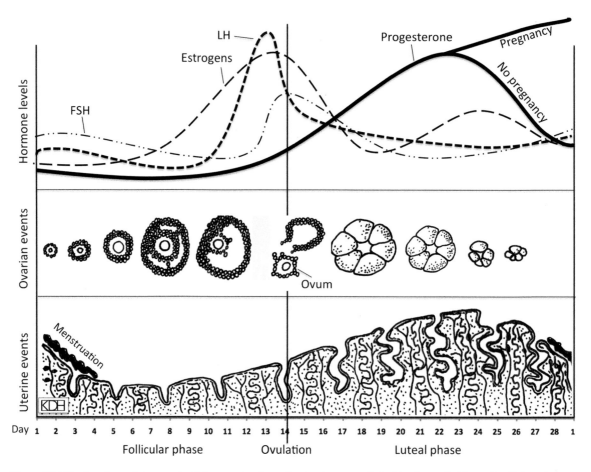

Figure 15.5 The female ovulatory cycle. Estrogen stimulates the development of the blood vessel-rich uterine lining, the endometrium, while rising levels of progesterone prepare the surface of the endometrium for the implantation of the fertilized egg. A spike in luteinizing hormone (LH) and follicle-stimulating hormone (FSH) spurs the developing follicle to release the ovum. If a fertilized egg implants, the flower-shaped corpus luteum helps to keep progesterone levels high and sustains the highly vascularized uterine lining; if there is no implantation, progesterone levels fall and the uterine lining is shed in menstruation. Illustration by the author.

gonadotropin-releasing hormone, which in turn stimulates the pituitary to begin to release two of the four hormones involved in ovulation, the gonadotroph pair **follicle-stimulating hormone** (FSH) and **luteinizing hormone** (LH). Follicle-stimulating hormone acts directly on the ovum receptacle in the ovary, the follicle (hence "follicle stimulating hormone"), causing it to grow and develop (Figure 15.5). The follicle helps out in the ovulation process by producing the third of the four ovulation hormones, estrogen (the follicle actually releases estrogen precursors which are then converted to estrogen in the fat cells), which in turn stimulates growth of the endometrial lining of the uterus (more on this below). Driven by the ever-more-active follicle, estrogen levels rise until they peak in the middle of the ovulatory cycle, just before ovulation. This peak does not precipitate ovulation in and of itself, but provokes the pituitary to release a burst of LH and FSH, the combination of which stimulates the follicle, now swollen to almost an inch in diameter, to release the egg. Luteinizing hormone and FSH then tail off and play little role in what is to follow.

After the release of the egg, the follicle is far from a spent force; instead it undergoes a transformation, folding in on itself to form a flower-like structure called the **corpus luteum** that will soon serve a critical function, should there be a pregnancy (see "ovarian events" in Figure 15.5).

We have not yet met the fourth of the ovulatory hormones, **progesterone**, which is produced in the adrenal glands. It begins rising late in the first half of the cycle, indirectly controlled by estrogen, which stimulates the adrenals to release it. While it is estrogen that promotes the development of the uterine lining via cell multiplication and tissue growth, progesterone helps in uterine preparation by stimulating the **endometrium**, the uterine lining, to change in texture and thickness, yielding a tissue more conducive to the implantation of the fertilized ovum. This blood vessel-rich lining will nurture the fertilized ovum if a baby is to be made – or be shed as menstrual blood if there is no pregnancy. After ovulation, the corpus luteum begins to release progesterone, adding its production to that already emitted by the adrenals. For the time being, progesterone levels will remain high (Figure 15.5).

15.16 Pregnancy

If the ovum is fertilized it will implant in the uterine wall about 7–9 days after ovulation, beginning a physiological relationship between the zygote, i.e., the fertilized egg, and the mother. As implantation proceeds, the zygote will insinuate itself ever more deeply into the endometrial lining of the uterus and then begin to grow a placenta, the organ that passes nutrients and oxygen to the developing embryo and waste products into the bloodstream of the mother. You will recall details about embryology from Chapter 8.

At this point the implanted zygote begins to contribute something to the pregnancy – it releases **human chorionic gonadotropin** (hCG), a pivotal pregnancy hormone. This hormone stimulates the corpus luteum not only to sustain its progesterone production, but to increase it. This is critical, because without a sustained high level of progesterone the uterine lining will be shed – and with it the zygote. As the embryo and fetus develop, the placenta will also begin to bear some of the burden of progesterone production.

If there is no implanted zygote, hCG levels do not rise and the corpus luteum will leave off producing progesterone. The fall of progesterone precipitates the collapse of the uterine lining and menstruation guarantees that there is no pregnancy. Pregnancy tests, you will be interested to know, work by detecting levels of hCG, a hormone that will not be present in any abundance unless there is a thriving zygote.

During pregnancy, progesterone not only signals the body to sustain the pregnancy, it reduces the mother's immune response so that the developing embryo is not rejected. It also shuts down two other activities that are either unneeded or counterproductive at this point, uterine contractions and lactation. Small wonder that progesterone is often called the pregnancy hormone.

Up to now all the hormones we have discussed, other than a brief mention of oxytocin, are involved in events leading up to and shortly following ovulation. As the pregnancy progresses, several organs or tissues – the corpus luteum, ovaries, and placenta – begin to produce a fascinating new hormone, **relaxin**. Relaxin affects both the cervix and ligaments, allowing the ligamental stretching in the pelvis that is necessary to give birth. Late in pregnancy many women experience this ligamentous loosening, and some can even feel the various bones of the pelvis shifting as they walk. Relaxin serves yet another purpose, too, acting as a vasodilator to accommodate the increase in blood volume necessary to sustain the developing fetus.

As pregnancy progresses, the pituitary ramps up production of prolactin, stimulating breast tissue development; later, as birth nears, it will stimulate actual lactation, but only when and if progesterone levels drop, as it does near birth.

And how are chimpanzees different? Any differences are said to be minor (Lin & Halbert, 1978; Graham, 1981; Hrdy & Whitten, 1987; Smith et al., 1999; Shimizu et al., 2003), and a typical assessment of chimpanzee reproductive endocrinology is that hormonal patterns during pregnancy in the chimpanzee are remarkably similar to those in humans (Reyes et al., 1975).

15.17 Testosterone and Male Reproduction

Compared to female reproductive endocrinology, males are simple. **Testosterone** (T) appears at various times during fetal and embryonic development to produce male sexual characteristics, then surges at puberty to spur the development of secondary sexual characteristics. Most of us are familiar with the muscle-building qualities of T, because it is so often abused by athletes to gain a competitive advantage. Testosterone affects not only muscle, but bone development (though perhaps less so than for muscle; van Beld et al., 2000), metabolic rate, and endurance (Fukagawa et al., 1990). It may surprise you to learn that as males mature, the same hormone that stimulates follicle development in females, FSH, provokes sperm production in males.

The increased muscle mass and physiological effects of testosterone improve a male's chance of winning male–male competitions involved in guarding mates, establishing territories, guarding resources, and competing for status (Wingfield et al. 1990).

The effect T has on the body, however, is only one of its roles; the surprise comes in T's effect on behavior and its capacity to adapt to diverse reproductive challenges throughout the male lifespan. The higher the T level, the more likely males are to engage in violence (Batrinos 2012), in part because T lowers the amount of stimulus required for an aggressive response (Pope et al., 2000). Testosterone also acts to increase confidence and can even lead to mistakes in cognitive tasks due to *over*confidence (Nave et al., 2017).

In popular media, the effects of testosterone are often boiled down to their role in producing the "dominant male," but here things are more complicated than motorcycle and perfume ads might lead us to believe. Assuredly big muscles and aggressive behavior are helpful when two males meet in combat. And numerous studies of animals and people show that males who become dominant have more testosterone, but perhaps not when and why you might think. They are winners in competitions not because they start with higher testosterone, necessarily – more reliably, testosterone surges *after* a male wins a competition, both in mice (Oyegbile & Marler, 2005) and in humans (Booth et al., 1989; Gladue et al., 1989; Mazur & Booth, 1998). Testosterone increases not only for the winners in an actual contest, but even for a fan of a winning athletic team (Bernhardt et al., 1998).

So T has something to do with contests. But note that its appearance *after* a victory suggests its role is to *sustain* social rank more than it is to achieve it. In support of this observation, we have long known that injecting T into a male has little or no effect on his rank; instead, when (for example) a middle-ranking monkey has his T level augmented, he just becomes a more anxious, more aggressive middle-ranking individual (Dixson and Herbert, 1977), not a high-ranking male. While he may be more aggressive, his aggression will be directed toward individuals to which he is already dominant.

Testosterone provides an advantage not only in **intrasexual**, i.e., male–male, competition, but in ***inter*sexual competition** as well. Among humans, males with high levels of T during adolescence have more prominent cheekbones, larger chins, larger brow ridges, and bigger faces (Symons, 1995; Thornhill & Møller, 1997), traits attractive to females, perhaps because these features are an honest signal that males are healthy, and further they show they have high enough T to win male–male competitions that are required to sustain their status (Thornhill & Gangestad, 1999).

Whatever its advantages, there are trade-offs to high levels of T (Muehlenbein & Bribiescas, 2005). There is evidence that T reduces lifespan (Ketterson & Nolan, 1999) and its downregulating effect on immune function is well-known (Wingfield et al., 1990). Its effect of increasing metabolism means high-T males must forage longer or more effectively than low-T males. Higher rates of aggression require more energy as well (Wingfield et al., 1990), and they increase the risk of injury. The advantages of *low* T are apparent in the fact that eunuchs live longer than intact males – a lot longer; in one study, eunuchs lived 14–19 years longer than three comparison groups. Perhaps T is the reason men live much shorter lives than women (Buettner, 1994).

We find much the same costs and the same benefits in the chimpanzee. High-ranking males have higher T levels than lower-ranking males (Muehlenbein et al., 2004; Muller & Wrangham, 2004), and the difference is even greater than in baboons, perhaps because the constant shuffling of individuals in chimpanzee parties makes the status of a chimpanzee alpha less certain (Mitani et al., 2002). Incredibly, when T is measured on an hourly basis we find that the higher the T level, the more the chimpanzee pant-hoots (Fedurek et al., 2016). Not only that, males with high T levels hit higher notes during the pant-hoot, perhaps because they reflect better physical condition, since higher notes require more energy to produce (Fedurek et al., 2016). The louder the pant-hoot the better, because pant-hooters call allies to the side of the vocalizer, giving him safety in numbers, should an attack come from another community. Mike Wilson and I have been debating this issue for some time. I maintain that the higher notes of the alpha male evolved principally because they carry farther in the forest, allowing him to gather more males to his side. The alpha male benefits most when territory expands. He maintains, with Richard Wrangham and colleagues, that it is an honest signal of health and vigor, functioning to aid the alpha in dominating competitors.

In a test of chimpanzee cognitive skills, individuals with higher T levels performed better on tasks requiring spatial cognition (Wobber & Herrmann, 2015). Similar sex differences in cognition are found in humans. Perhaps spatial skills help facilitate ranging over larger geographical areas; both chimpanzee and human hunter-gatherer males range farther than females (Goodall, 1986; Hill & Hurtado, 1996). In contrast to chimpanzees, among male bonobos there is no correlations between testosterone and performance on cognitive tasks (Wobber & Herrmann, 2015). Bonobo males and females travel together so that there is no difference in ranging.

15.18 Rank and Repose

There was a time when we thought life was easy for top-ranking males, but hormonal studies have made us more aware of the complexities of this equation. We now know that alpha males must work hard to sustain status, and they suffer for it. High-ranking males have higher cortisol levels (Muller & Wrangham, 2004); their immune system, as we would expect, is compromised, giving them higher **parasite loads** (Muehlenbein, 2005); and the greater the T level, the greater the parasite load (Muehlenbein, 2006; Muehlenbein & Watts, 2010). The life of a male chimpanzee is a violent one. Territorial defense, territorial expansion, and competition for social rank all demand aggressive responses to threats, but males pay a heavy price for this competence.

15.19 Body Maintenance Hormones

Many hormones are indispensable even if they have less dramatic roles than sex and bonding. We tend to think of **growth hormone** (GH) and **insulin-like growth factor** (IGF-1) in their roles during childhood growth and development, but GH and IGF also play roles in body maintenance, stimulating the repair of tissues and maintaining bone strength. In fact, they are so effective at stimulating muscle growth, repair, and recovery that they are subject to abuse by athletes and are one of the substances banned by Olympic and other athletic organizations. Calcitonin builds bone as well. **Erythropoietin** turns on red blood cell production. **Thyroid-stimulating hormone** (TSH) ramps up the metabolism and is the hormone most implicated, when too low, in extreme weight gain.

15.20 Chimpanzee Fathers

Chimpanzee adult males, unlike human males, participate very little in infant care. Among humans, prolactin levels rise in fathers and T decreases (Archer, 2006; Gray et al., 2006). It would be interesting to know whether chimpanzee males who might be experimentally forced into infant care would have the same response. Certainly everything we know about chimpanzee endocrinology reinforces what we know about anatomy, behavior, and genetics: Humans and chimpanzees are extremely similar.

PART IV: PHYSIOLOGY

While hormones are central to the operation of an organism, they are merely 71 of hundreds of biochemicals that mingle in a heaving stew in the body, most of which are not hormones even though we find them acting on instructions from hormones. The study of the chemical and physical functions that power the body is called **physiology**. Digestive physiology was covered in Chapter 6, and to be sure it is critical for understanding chimpanzees, but here we will confine our study to chimpanzee metabolism, the inner fire of chimpanzees.

Not surprisingly, chimpanzee body temperature is virtually indistinguishable from that of humans (98.7°F; Morrison, 1962). Due to differences in body size, though, we might expect different metabolic rates and different energy consumption. Instead, they are more similar than expected. **Basal metabolic rate (BMR)** is the unavoidable energetic costs an organism bears, the amount of energy required just to sustain the body without considering things like movement, temperature stress, or the expense of digestion. Of equal interest is the actual amount of energy expended during normal activity, which would include not just BMR, but energy expended on foraging and other activities. This we call the **total daily energy expenditure** (TEE; Frohle & Schoeninger, 2006; Pontzer et al., 2014, 2016).

Differences in body size can be important to TEE, so we might expect a difference between males and females; but there is none – neither in humans nor in chimpanzees (Frohle & Schoeninger, 2006). Up until quite recently, reports suggested a dramatic difference in TEE between humans and chimpanzees, 1409 calories for chimpanzees and 2293 calories for humans (Frohle & Schoeninger, 2006), yielding a chimpanzee figure only 60 percent that of humans – a dramatic difference exceeding expectations based on body size difference between the two species. This would suggest that chimpanzees had found a foraging strategy that is somehow less demanding than that of humans. However, a recent report using the more reliable doubly labeled water method revised TEE for chimpanzees upward: 1934 calories versus 2456 calories for humans (Pontzer et al., 2016). In this latter study, chimpanzees burned 79 percent of the calories of humans. Given body size differences (chimpanzees body weights are 76 percent that of human hunter-gatherers; see Chapter 4), the two values are indistinguishable.

Chimpanzees, along with their closest relatives, including us, have lost the urate oxidase gene expression (Friedman et al., 1985). The consequence is that we humans (and chimpanzees) are susceptible to a build-up of uric acid in our blood, which causes gout. Friedman and colleagues speculate that shutting down this gene is advantageous for large-brained primates because, as bad as gout is, uric acid is a powerful antioxidant and a scavenger of both singlet oxygen molecules and radicals (Friedman et al., 1985), protecting the body from oxidative damage, yielding some protection from cancer, and increasing the human and chimpanzee lifespans.

The human genome bears evidence of the importance of fat in brain function – many of the specializations we have evolved compared to chimpanzees have to do with fat metabolism. If we expand our circle of interest – apes have big brains, too, compared to other primates – we find that chimpanzee genes show evidence of adaptations to the brain's need for fat, too. Only humans and great apes produce alkaline phosphatase, thought to be related to fat metabolism and brain function; gibbons and monkeys do not (Goldstein et al. 1980).

15.21 Lessons

Classes of hormones and the molecular make-up of the hormones themselves are ancient. Specific hormones are very similar or identical in many animals whose ancestries diverged tens and even hundreds of millions of years ago. Tissue around the rump of female chimpanzees swells in response to circulating estrogens. Human tissue does not. Yet chimpanzee and human hormones are virtually identical; when we see differences between humans and chimpanzees it is the response to those hormones or the hormone levels themselves that have changed, not the hormones themselves. Chimpanzee testosterone levels are ridiculous, but the molecule itself is the same as that in humans. As with so many other areas, hormone function is more similar in humans and chimpanzees than one might guess at first glance.

References

Agnew MK, Asa CS, Clyde VL, Keller DL, Meinelt A (2016) A survey of bonobo (*Pan paniscus*) oral contraceptive pill use in North American zoos. *Zoo Biology* 35, 444–453.

Archer J (2006) Testosterone and human aggression: an evaluation of the challenge hypothesis. *Neurosci Biobehav Rev* 30, 319–345.

Batrinos ML (2012) Testosterone and aggressive behavior in man. *Int J Endocrinol Metabol* 10, 563–568.

Bernhardt PC, Dabbs JM Jr., Fielden JA, Lutter CD (1998) Testosterone changes during vicarious experiences of winning and losing among fans at sporting events. *Physiol Behav* 65, 59–62.

Birken S, Gawinowicz MA, Maydelman Y, Milgrom Y (2001) Metabolism of gonadotropins: comparisons of the primary structures of the human pituitary and urinary LH beta cores and the chimpanzee CG beta core demonstrate universality of core production. *J Endocrinology* 171, 131–141.

Booth A, Shelley G, Mazur A, Tharp G, Kittok R (1989) Testosterone, and winning and losing in human competition. *Horm Behav* 23, 556–571.

Bribiescas RG, Anestis SF (2010) Leptin associations with age, weight, and sex among chimpanzees (*Pan troglodytes*). *J Med Primatol* 39, 347–355.

Bribiescas RG, Muehlenbein MP (2010) Evolutionary endocrinology. In *Human Evolutionary Biology* (ed. Muehlenbein MP), pp. 127–143. Cambridge: Cambridge University Press.

Buettner T (1994) Sex differentials in old-age mortality. *Pop Bull UN* 39, 18–44.

Buntin JD (1996) Neural and hormonal control of parental behavior in birds. *Adv Stud Behav* 25, 161–121.

Cahill SP, Hatchard T, Abizaid A, Holahan MR (2014) An examination of early neural and cognitive alterations in hippocampal-spatial function of ghrelin receptor-deficient rats. *Behav Brain Res* 264, 105–115.

Copeland KC, Eichberg JW, Parker R, J.r, Bartke A (1985) Puberty in the chimpanzee: aomatomedin-C and its relationship to somatic growth and steroid hormone concentrations. *J Clin Endocrinol Metab* 60, 1154–1160.

Crenshaw TL, Goldberg JP (1996) *Sexual Pharmacology: Drugs that Affect Sexual Function.* New York: W.W. Norton.

Curtis JA, Blume LB, Blume TW (1997) Becoming a father: marital perceptions and behaviors of fathers during pregnancy. *Mich Family Rev* 3, 31–44.

Dixson AF, Herbert J (1977) Testosterone, aggressive behavior and dominance rank in captive adult male talapoin monkeys (*Miopithecus talapoin*). *Physiol Behav* 18, 539–543.

Domes G, Heinrichs M, Michel A, Berger C, Herpertz SC (2007) Oxytocin improves "mind-reading" in humans. *Biol Psych* 61, 731–733.

Donaldson ZR, Young LJ (2008) Oxytocin, vasopressin, and the neurogenetics of sociality. *Science* 322, 900–904.

Fan W, Kasahara M, Gutknecht J, et al. (1989) Shared class II MHC polymorphisms between humans and chimpanzees. *Hum Immunol* 26, 107–121.

Fedurek P, Slocombe KE, Enigk DK, et al. (2016) The relationship between testosterone and long-distance calling in wild male chimpanzees. *Behav Ecol Sociobiol* 70, 659–672.

Finch CE, Rose MR (1995) Hormones and the physiological architecture of life history evolution. *Q Rev Biol* 70, 1–52.

Fisher JW (1997) Erythropoietin: physiologic and pharmacologic aspects. *Proc Soc Exp Biol Med* 216, 358–369.

Fleming AS, Corter C, Stallings J, Steiner M (2002) Testosterone and prolactin are associated with emotional responses to infant cries in new fathers. *Hormones Beh* 42, 399–413.

Forsyth IA (1986) Variation among species in the endocrine control of mammary growth and function: the roles of prolactin, growth hormone, and placental lactogen. *J Dairy Sci* 69, 886.

Friedman TB, Polanco GE, Appold JC, Mayle JE (1985) On the loss of uricolytic activity during primate evolution:

I. Silencing of urate oxidase in a hominoid ancestor. *Comp Biochem Physiol B* 81, 653–659.

Froehle AW, Schoeninger MJ (2006) Intraspecies variation in BMR does not affect estimates of early hominin total daily energy expenditure. *Am J Phys Anthropol* 131, 552–559.

Fukagawa, NK, Bandini LG, Young JB (1990) Effect of age on body composition and resting metabolic rate. *Am J Physiol* 259, E233–E238.

Gettler LT, McDade TW, Feranil AB, Kuzawa CW (2012) Prolactin, fatherhood, and reproductive behavior in human males. *Am J Phys Anthropol* 148, 362–370.

Gladue BA, Boechler M, McCaul KD (1989) Hormonal response to competition in human males. *Aggressive Behav* 15, 409–422.

Goldstein DJ, Rogers CE, Harris H (1980) Expression of alkaline phosphatase loci in mammalian tissues. *PNAS* 77, 2857–2860.

Gombart AF, Saito T, Koeffler HP (2009) Exaptation of an ancient Alu short interspersed element provides a highly conserved vitamin D-mediated innate immune response in humans and primates. *Genomics* 10, 321.

Goodall J (1986) *The Chimpanzees of Gombe: Patterns of Behavior.* Cambridge, MA: Harvard University Press.

Graham C (1981) Menstrual cycle physiology of the great apes. In *Reproductive Biology of the Great Apes: Comparative and Biomedical Perspectives* (ed. Graham CE), pp. 286–303. New York: Academic Press.

Gray PB, Yang CFJ, Pope HG (2006) Fathers have lower salivary testosterone levels than unmarried men and married non-fathers in Beijing, China. *Proc Roy Soci Lond B* 273, 333–339.

Gray PB, Parkin JC, Samms-Vaughan ME (2007) Hormonal correlates of human paternal interactions: a hospital-based investigation in urban Jamaica. *Horm Behav* 52, 499–507.

Guastella AJ, Mitchell PB, Dadds MR (2008) Oxytocin increases gaze to the eye region of human faces. *Biol Psych* 63, 3–5.

Hara Y, Waters EM, McEwen BS, Morrison JH (2015) Estrogen effects on cognitive and synaptic health over the lifecourse. *Physiol Rev* 95, 785–807.

Haynes BF, Dowell BL, Hensley LL, Gore I, Metzgar RS (1982) Human T cell antigen expression by primate T cells. *Science* 215, 298–300.

Heinrichs M, Baumgartner T, Kirschbaum C, Ehlert U (2003) Social support and oxytocin interact to suppress cortisol and subjective responses to psychosocial stress. *Biol Psych* 54, 1389–1398.

Hill K, Hurtado AM (1996) *Ache Life History: The Ecology and Demography of a Foraging People.* New York: Aldine.

Hrdy SB, Whitten PL (1987) Patterning of sexual activity. In *Primate Societies* (eds. Smuts BB, Cheney DL, Seyfarth RM, Wrangham RW, Struhsaker TT), pp. 370–384. Chicago, IL: University of Chicago Press.

Kenter M, Otting N, Anholts J, et al. (1992) Mhc-DRB diversity of the chimpanzee (*Pan troglodytes*). *Immunogenetics* 37, 1–11.

Ketterson ED, Nolan V (1999) Adaptation, exaptation, and constraint. *Am Nat* 154, 4–25.

Konner M, Worthman C (1980) Nursing frequency, gonadal function, and birth spacing among !Kung hunter-gatherers. *Science* 207, 788–791.

Kosfeld M, Heinrichs M, Zak PJ, Fischbacher U, Fehr E (2005) Oxytocin increases trust in humans. *Nature* 435, 673–676.

Lin TM, Halbert SP (1978) Immunological relationships of human and subhuman primate pregnancy-associated plasma proteins. *Int Arch Allergy Immunol* 56, 207–223.

Lutter M, Sakata I, Osborne-Lawrence S, et al. (2008) The orexigenic hormone ghrelin defends against depressive symptoms of chronic stress. *Nat Neurosci* 11, 752–753.

Mazur A, Booth A (1998) Testosterone and dominance in men. *Behav Brain Sci* 21, 353–363.

Meyer RM, Burgos-Robles A, Liu E, Correia SS, Goosens KA (2014) A ghrelin-growth hormone axis drives stress-induced vulnerability to enhanced fear. *Mol Psych* 19, 1284–1294.

Mitani JC, Watts DP, Muller MN (2002) Recent developments in the study of wild chimpanzee behavior. *Evol Anthropol* 11, 9–25.

Morgan K, Conklin D, Pawson AJ, et al. (2003) A transcriptionally active human type II gonadotropin-releasing hormone receptor gene homolog overlaps two genes in the antisense orientation on chromosome 1q. 12. *Endocrinology* 144, 423–436.

Morrison P (1962) An analysis of body temperature in the chimpanzee. *J Mammal* 43, 166–171.

Muehlenbein MP (2005) Parasitological analyses of the male chimpanzees (*Pan troglodytes* schweinfurthii) at Ngogo, Kibale National Park, Uganda. *Am J Primatol*, 65(2), 167–179.

Muehlenbein MP (2006) Intestinal parasite infections and fecal steroid levels in wild chimpanzees. *Am J Phys Anthropol*, 130(4), 546–550.

Muehlenbein MP, Bribiescas RG (2005) Testosterone-mediated immune functions and male life histories. *Am J Hum Biol*, 17(5), 527–558.

Muehlenbein MP, Bribiescas RG (2010) Male reproduction: physiology, behavior, and ecology. *Hum Evol Biol*, 351.

Muehlenbein MP, Watts DP (2010) The costs of dominance: testosterone, cortisol and intestinal parasites in wild male chimpanzees. *Biopsychosocial Med* 4, 21.

Muehlenbein MP, Watts DP, Whitten PL (2004) Dominance rank and fecal testosterone levels in adult male chimpanzees (*Pan troglodytes* schweinfurthii) at Ngogo,

Kibale National Park, Uganda. *Am J Primatol*, *64*(1), pp.71–82.

Muller MN, Wrangham RW (2004) Dominance, aggression and testosterone in wild chimpanzees: a test of the "challenge hypothesis." *Anim Behav* 67, 113–123.

Nadkarni S, Cooper D, Brancaleone V, Bena S, Perretti M (2011) Activation of the annexin A1 pathway underlies the protective effects exerted by estrogen in polymorphonuclear leukocytes. *Arterioscler Thromb Vasc Biol* 31, 2749–2759.

Nave G, Nadler A, Zava D, Camerer C (2017) Single-dose testosterone administration impairs cognitive reflection in men. *Psychol Sci* 28, 1398–1407.

Oyegbile TO, Marler CA (2005) Winning fights elevates testosterone levels in California mice and enhances future ability to win fights. *Horm Behav* 48, 259–267.

Pontzer H, Raichlen DA, Gordon AD, et al. (2014) Primate energy expenditure and life history. *PNAS* 11, 1433–1437.

Pontzer H, Brown MH, Raichlen DA, et al. (2016) Metabolic acceleration and the evolution of human brain size and life history. *Nature* 533, 390–392.

Pope HG, Kouri EM, Hudson JI (2000) Effects of supraphysiologic doses of testosterone on mood and aggression in normal men: a randomized controlled trial. *Arch Gen Psych* 57, 133–140.

Reyes FI, Winter JSD, Faiman, C, Hobson WC (1975) Serial serum levels of gonadotropins, prolactin and sex steroids in the nonpregnant and pregnant chimpanzee. *Endocrinology* 96, 1447–1455.

Ronke C, Dannemann M, Halbwax M, et al. (2015) Lineage-specific changes in biomarkers in great apes and humans. *PloS ONE* 10, p.e0134548.

Rosano GM, Panina G (1999) Oestrogens and the heart. *Therapie* 54, 381–385.

Saks BR, Gillespie MA (2002) Psychotropic medication and sexual function in women: an update. *Arch Women Mental Health* 4(4), 139–144.

Samuels MH, Bridges RS (1983) Plasma prolactin concentrations in parental male and female rats: effects of exposure to rat young. *Endocrinology* 113, 1647–1654,

Sapolsky RM (2004) *Why Zebras Don't Get Ulcers: The Acclaimed Guide to Stress, Stress-related Diseases, and Coping*. New York: Holt Paperbacks.

Sävendahl L (2012) The effect of acute and chronic stress on growth. *Sci Signal* 23(5). DOI: 10.1126/scisignal.2003484.

Schradin C, Anzenberger G (1999) Prolactin, the hormone of paternity. *Physiology* 14, 223–231.

Schradin C, Reeder DM, Mendoza SP, Anzenberger G (2003) Prolactin and paternal care: comparison of three species of monogamous new world monkeys (*Callicebus cupreus*, *Callithrix jacchus*, and *Callimico goeldii*). *J Comp Psychol* 117, 166.

Shah MG, Maibach HI (2001) Estrogen and skin. *Am J Clin Dermatol* 2, 143–150.

Shimizu K, Douke C, Fujita S, et al.(2003) Urinary steroids, FSH and CG measurements for monitoring the ovarian cycle and pregnancy in the chimpanzee. *J Med Primatol* 32, 15–22.

Smith R, Wickings EJ, Bowman ME, et al. (1999) Corticotropin-releasing hormone in chimpanzee and gorilla pregnancies. *J Clin Endocrinol Metab* 84, 2820–2825.

Storey AE, Walsh CJ, Quinton RL, Wynne-Edwards KE (2000) Hormonal correlates of paternal responsiveness in new and expectant fathers. *Evol Hum Behav* 21, 79–95.

Symons D (1995) Beauty is in the adaptations of the beholder: the evolutionary psychology of human female sexual attractiveness. In *Sexual Nature/Sexual Culture* (eds. Abramson PR, Pinkerton SD), pp. 80–118. Chicago, IL: University of Chicago Press.

Thearle MJ, Weissenberger R (1984) Induced lactation in adoptive mothers. *Aust NZ J Obstet Gynaecol* 24, 283–286.

Thornhill R, Gangestad SW (1999) Facial attractiveness. *Trends Cogn Sci* 3, 452–460.

Thornhill R, Møller AP (1997) Developmental stability, disease and medicine. *Biol Rev* 72, 497–528.

van den Beld AW, de Jong FH, Grobbee DE, Pols HA, Lamberts SW (2000) Measures of bioavailable serum testosterone and estradiol and their relationships with muscle strength, bone density, and body composition in elderly men. *J Clin Endocrinol Metab* 85, 3276–3282.

Weiss NS, Ure CL, Ballard JH, Williams AR, Daling JR (1980) Decreased risk of fractures of the hip and lower forearm with postmenopausal use of estrogen. *New Engl J Med* 303, 1195–1198.

Wingfield JC, Hegner RE, Duffy AM, Ball GF (1990) The "Challenge hypothesis": theoretical implications for patterns of testosterone secretion, mating systems and breeding strategies. *Am Nat* 136, 829–846.

Winter JS, Faiman C, Hobson WC, Prasad AV, Reyes FI (1975) Pituitary-gonadal relations in infancy: I. Patterns of serum gonadotropin concentrations from birth to four years of age in man and chimpanzee. *J Clin Endocrinol Metab* 40, 545–551.

Winter JS, Hughes IA, Reyes FI, Faiman C (1976) Pituitary-gonadal relations in infancy: 2. Patterns of serum gonadal steroid concentrations in man from birth to two years of age. *J Clin Endocrinol Metab* 42, 679–686.

Wobber V, Herrmann E (2015) The influence of testosterone on cognitive performance in bonobos and chimpanzees. *Behaviour* 152, 407–423.

Zihlman AL, Bolter DR (2015) Body composition in *Pan paniscus* compared with *Homo sapiens* has implications for changes during human evolution. *PNAS* 112. DOI: 10.1073/pnas.1505071112.

16 Shelter from the Storm
Chimpanzee Mothering

Photo by author

Primates as an order have a long period of infant dependency, and chimpanzees have among the very longest; females dedicate five years, on average, to seeing a baby through to weaning, a colossal energetic investment and – in a species in which the female lifespan is 45 (Hill et al., 2001) – a stunningly large chunk of a female's reproductive life. With infancy extending to half a decade, it is probably no surprise that among chimpanzees the strong and resilient bonds forged in infancy continue well beyond that. For a few lucky mother–offspring pairs, they last the rest of a mother's life, sometimes evolving into a sort of partnership, as they did for Fifi and Fanni, two chimpanzees at Gombe (Figures 16.1–16.3).

16.1 Mothering, Not Parenting

I subtitled this chapter "chimpanzee mothering." A reader sensitive to gender issues might wonder whether the term "parenting" might be a more appropriate term. It is not (Goodall, 1967, 1968, 1986; Pusey, 1983). Among chimpanzees, fathers have precious little to do with infant care, directly at least. Males play with infants and are remarkably tolerant of their shenanigans, but no more so for the father than other males. On some level, possibly subconsciously, fathers may recognize their offspring, since it has been shown that where paternity is known, males spend more time with their offspring in the first few months, the time when infants are most vulnerable to infanticide (Murray et al., 2016); 20 percent of all infant deaths are attributable to chimpanzee violence (Williams et al., 2008). Yet it is not just the father who is vigilant: When an infant is threatened by a stranger male, any community male – father and non-fathers alike – rush in to fight off the threat. Male patrolling evolved in part to keep such baby-killing intruders as far as possible from vulnerable infants. Still, despite this indirect aid, the much more burdensome duties of carrying, guarding, and attending to infants falls on mother.

16.2 The Mother–Son Bond

While males may pay little attention to their offspring, a son's bonds with his mother are strong. Because males are philopatric – they remain in the group into which they were born – it is possible for them to maintain mother–offspring ties for life, and many do. At Mahale I once observed a surprisingly tender reunion between the alpha male, Ntologi, and an ancient female suspected of being his mother. I was following a group of males as they hiked into the southern half of the Mahale community range, walking quickly and seemingly intent on some purpose I was unable to divine. Perhaps they were embarking on a border patrol; if so, they gave that up. After walking a mile or so, the party stopped abruptly,

Figure 16.1 Mother–offspring relationships can last a lifetime. Here, in 1987, Fifi carries two-year-old daughter Flossi, with six-year-old daughter Fanni following. Fanni remained in the Kasekala community and she and Fifi often spent time together for the remainder of Fifi's life. Fanni lives still, and as of writing has given birth to six offspring, including alpha male Fudge. Photo by the author.

Figure 16.2 Flossi, with mother Fifi in the background in a photo taken on the same day as Figure 16.1. Seconds after this photo was taken she swung under the branch and tried to kick the camera out of my hands. Flossi later transferred to the Mitumba community and now has four offspring. Photo by the author.

all the males except Ntologi sitting down on the trail, suddenly aimless. Ntologi, on the other hand, rushed forward, first along the trail, then veering off; I followed. I heard him grunting a greeting to an individual I could neither see nor hear at first.

Seconds later I saw he was enthusiastically greeting an old female, Wagamumu. While to Ntologi she was clearly a chimpanzee of significance, to me she was not much to look at. The hair over her shoulder blades had fallen out, exposing her shiny, wrinkly leather-like skin. Her face was saggy and her eyes were watery and white with cataracts. At this time Wagamumu was thought to be well into her forties. Having seen old chimpanzees over the years, I would not be surprised if she was actually over 50, an age equivalent to an 80-year-old human; Ntologi was thought to be about 30 at the time, which would make her 50 if she was 20 when he was born.

She attempted to grunt a greeting to Ntologi, but emitted nothing more than a whisper. Toshisada Nishida had inferred that Wagamumu was Ntologi's mother based on their facial resemblance and the clear affection they exhibited for one another (Nishida, 2012). He sat nearly touching her and began to groom

Figure 16.3 Flossi attempts to kick the camera from the author's hands. Photo by the author.

her with some intensity. Other members of the party drew a little nearer, but none paid her much attention. A few minutes later, a low-level kerfuffle arose among two of the group and Wagamumu "screamed" in response, but the scream came out only as a strained hiss.

The affection – perhaps we could even call it respect – Ntologi held for Wagamumu was obvious in their grooming relationship. His usual role in grooming interactions was either to reciprocate in a relaxed, lackadaisical manner, or just as likely not to reciprocate grooming at all but just to bask in the attention, accepting it as his due as King of Mahale. He groomed Wagamuma with a surprising attentiveness that he afforded few other individuals. Ntologi was a powerful, domineering alpha, yet here

he was meek as a lamb, clearly reveling in a reunion with his mother. Such is the power of the mother–son bond among chimpanzees.

Ntologi was no outlier when it comes to filial affection; it is typical among chimpanzees for males to make their mother their most common female companion as long as she lives (Pusey, 1983).

16.3 Labor and Birth

The day before giving birth, perhaps even earlier, pregnant females eat little and pause to rest in day nests often. They behave more or less as if they are ill, possibly because painful contractions have begun. In one complete birth observation at Gombe, Winkle hardly foraged for two days before giving birth, and the day before birth she rested virtually all day, making five separate day nests (van de Rijt-Plooij & Plooij, 1987). Winkle was visibly pained by contractions, pausing and crouching over before continuing to eat a few berries. It may seem a surprise that chimpanzees give birth in a nest. Despite a birth canal that is seemingly capacious compared to the size of the neonate head (Schultz, 1949), labor is both long-lasting and intense among chimpanzees (Nissen & Yerkes, 1943; Brandt & Mitchell, 1971; Goodall & Athumani, 1980; Kiwede, 2000; Zamma et al., 2012), and seemingly more painful for first births. In captivity labor ranges from 40 minutes to an exhausting 8 hours. It seems similar in the wild, though it has been observed only a few times. Laboring females often give birth in a crouched posture, assisting delivery by pulling on the emerging newborn. Winkle's labor lasted 1 hour and 52 minutes, with 25 contractions; she bore down visibly at times, straining with the effort (Goodall & Athumani, 1980). She was to all appearances, as other reports agree, exhausted after birth. Mothers eat the placenta about half the time, and no wonder, given their enforced fast.

As exhausting as labor is, maternity leave is short for chimpanzees; while new mothers may climb out of their nest and walk soon after birth, it will take a day or two before she returns to her typical routine of foraging for food six or seven hours a day. On the day of birth, Winkle did not come to the ground at all, though she did eat a few palm fruits and figs near her nest (van de Rijt-Plooij & Plooij, 1987). Just after birth she left the birth nest to make another nearby, where she lay for some time, eating the afterbirth (with leaves, as chimpanzees do with meat) and resting (Goodall & Athumani, 1980). If you've been counting, Winkle went three days with only the placenta and a few other bites of food.

16.4 Infants Are Utterly Dependent on the Mother for Years

The chimpanzee neonate is nearly as helpless as a human newborn (Schultz, 1940; Goodall, 1986; van de Rijt-Plooij & Plooij, 1987) and for as long as two months will require a helping hand from mother to help her cling (Figure 16.4). Not uncommonly, though, an infant's coordination and alertness develop rapidly so that many infants can take their first few clumsy steps, closely watched by mother, as early as the first month.

During late pregnancy, the mother is not only carrying the soon-to-be neonate, she is eating and breathing for the baby, too; for some time after birth she is scarcely better off than during pregnancy. While she is not breathing for her baby, she will carry it everywhere, the only difference being that it is outside rather than inside her uterus. Every calorie required by the infant must be chewed and swallowed by the mother first. Now and then food passes into the mother's mouth and then right back out into the infant's; infants not uncommonly whimper and beg to taste food items their mother is eating, and she will tolerantly allow the baby to take it right out of her mouth (Goodall, 1968). Infants sleep in the mother's nest, likely waking her in the morning, eager to nurse or play.

Early on, for the first year or so, the baby will be in physical contact with mother 24 hours per day; she needs to be (Figure 16.4). Body contact is necessary for health, both mental and physical (Harlow, 1958, 1986; Harlow & Harlow, 1965; Mason, 1968; Hinde, 1977; Davenport, 1979; van de Rijt-Plooij & Plooij, 1987). Among macaques (presumably it is no different among chimpanzees) mother–infant separation of as little as a week can still be detected two years later.

A new mother is extraordinarily protective of and attentive toward her young infant, grooming her often, virtually any time the mother stops moving, and keeping her always within arm's reach. And for good reason; at Gombe, 20 percent of infant deaths were due to infanticide (Williams et al., 2004). Not only are infants at some risk from raiding parties from other communities (Watts & Mitani, 2000; Wilson et al., 2014), there are threats from within, too. Even though most community males will be protective of

Figure 16.4 Methods of supporting and cradling newborn infants. After Goodall, 1968, with permission.

an infant, infants are killed by males within the community in rare instances (Otali & Gilchrist et al., 2006); even more rarely, infants are killed by same-community females (Pusey et al., 2008). Most infanticides, then, are by extracommunity males. Male infants are killed disproportionately (Williams et al., 2008), perhaps because extracommunity females are potential mates in the future, while male infants are future competitors.

For several years after the birth, the mother is a dual-career primate; she must continue all the activities typical of chimpanzees – finding food in a complex, three-dimensional world, socializing with community members, avoiding the rare but still dangerous leopard – and at the same time she must protect her infant from accidents and aggression, and all the while gathering extra calories to nurse her and helping to integrate her into the community.

Perhaps part of the reason mothers and offspring have such a strong bond is that wild chimpanzee females have little other social contact, even though female sociality in captivity suggests they enjoy the company of others. From first birth to death, most of the typical female's social contact will be with her offspring; two-thirds of her time will be spent in their company alone, only one-third in the company of other adults. Besides spending 5 percent of her time grooming her offspring, mothers are attentive in other ways; they play with their infants often, some mothers quite a lot, tickling, grabbing, chasing, or wrestling (Pusey, 1983) with their baby. The mother's joy in these interactions is apparent in that they often initiate play; in Anne Pusey's study of mothering, 19 of 22 play sessions were started by the mother. While mothers are playful, there are limits; at times they seem quite content to rest and merely watch whenever another playmate is available.

For up to two years the mother will carry her offspring clinging to her belly, mostly but not always allowing them to nurse on demand (van de Rijt-Plooij & Plooij, 1987). Eventually the maturing infant will learn to ride on the mother's back, jockey-style (Figure 16.5). Most times, especially for infants younger than three, the mother signals that she is about to move by looking in the direction of the infant; infants seem to develop a sixth sense about

Figure 16.5 Young infants often cling to any body part they latch onto (top row). If they fail to climb aboard, the mother may carry them in her hand (top row, right). As infants grow older they climb on the mother and ride jockey-style, or sometimes modify the posture by lying (middle row). When the mother gestures, older infants climb aboard (bottom row), often leaping aboard as she begins to move. After Goodall, 1968, with permission.

when their mother is about to travel. The moment she begins to stand, her infant will move to climb onto her back (Figure 16.5). By age three, the mother will be relaxed enough that she may walk off, seemingly without a thought for the offspring, who may have to run to catch up. If arboreal, she may move away to a nearby feeding tree without collecting her infant, leaving the baby to catch up at his or her own pace.

16.5 "All I Am I Owe to My Mother"

Chimpanzee mothers are teachers, protectors, and food suppliers for their offspring. Infancy is a period of intense learning, and a time when offspring are fascinated by everything their mother does. As she opens seed pods, selects a termiting tool, breaks open a palm frond, or grooms a friend, her baby will be there, often watching intently. Infants learn where large, important feeding trees are and they begin to absorb a mental map of the community range. Their mother's food choice leaves a lasting imprint on the baby. If an individual has not seen their mother eat an item, it is not food for that chimpanzee, and she or he will add that food to their diet only very reluctantly. Infants begin sampling food items even in their first year, as early as five months in one report (van de Rijt-Plooij & Plooij, 1987). As the years pass, infants take a more active interest in foraging, and the mother is grateful for the relief. Infants will feed somewhat independently by three years old, even though they will still nurse.

When, in my mind's eye, I call up an image of a mother examining a termite mound or attempting to break open a hard-shelled *Strychnos* fruit, I can see her infant's face peering over her shoulder. Infants learn how to behave socially from mother, and they learn cultural practices. Whether chimpanzees clasp their hands when their arms are raised to groom is determined by their mother's practice (Wrangham et al., 2016).

As mother and offspring feed, they often grunt to one another as a sort of monitoring behavior, allowing each to know the other's location without looking up from gathering fruits (Pusey, 1983). If danger threatens, the mother knows where her offspring is and rushes to her aid. As infants grow older and encounter other social partners, conflicts inevitably arise, and here mothers stand fast with their offspring, even when outranked (Goodall, 1968). To her dying day, a mother will leap to the defense of her offspring if he is threatened, even when the offspring have matured to middle age (Goodall, 1968; Pusey, 1983).

Sometimes females travel together for no reason other than to allow their offspring play time (Goodall, 1968, 1986); Goodall called these parties **"nursery groups."** In this context, young chimpanzees forge bonds with other community members that, for all males and some females, will last their entire lives.

Females are not engaged in the ceaseless dominance interactions of males, but they do compete for status, and when they are successful it is a great gift to their offspring. A dominant mother can make infancy seem like one grand sweet song, a carefree time of loving embrace and adventurous discovery

16.6 Weaning

While nursing is typically on-demand before three, mother starts making it more difficult after that. A mature four- or five-year-old will find himself being eased into greater independence by mild aggression from the mother (Goodall, 1986). She may discourage him from riding on her back, or may grunt aggressively toward an older infant who is feeding too close to her, or she may simply walk off and leave her five-year-old, despite his pitiful cries for help, thus encouraging him to get used to walking on his own. This is a teaching moment – she is actively encouraging independence (Goodall, 1986). It is clear that offspring are the ones maintaining proximity, because it is the infant who makes the move to stay in contact with the mother (Pusey, 1983).

By five, the mother will finish the job by moving away as the infant attempts to nurse or pushing the offspring away from the nipple. Some infants face this change in status with equanimity, but others protest violently, physically slapping and biting their mother (Goodall, 1971, 1986; Clark, 1977). Despite protests and begging, mothers persist in pushing the infant away, in accord with parent–offspring conflict theory (Trivers, 1972): the mother's interest is in maximizing her reproductive potential by having another infant as soon as possible, rather than continuing to nurse an offspring old enough to be independent. The infant's interest, on the other hand, is in garnering as much investment as she can get, even if it delays her mother's pregnancy.

After weaning, an event that might seem to mark the end of the mother–offspring relationship, there is hardly any change in association patterns (Pusey, 1983); mother and offspring continue to forage together, sleep near one another, and groom one another as much as ever, with only gradual changes up until the offspring approaches puberty. After five years of learning what to eat and where to find it, a newly weaned juvenile has little difficulty in finding

enough to eat. Juveniles, because they are smaller, can reach fruits among branches too small for adults to negotiate easily. Once the transition to feeding independence is achieved and the art of nest-making is fully assimilated, life is rather easy for juveniles. Mothers often stop to wait on their offspring when they fall behind, even up to the age of 14 (Pusey, 1983): Among chimpanzees aged 4–14, more than 80 percent of their time is spent with their mother. Even when mother and offspring are separated, they seem to split up at a time when both know what the final destination is so that after traveling separately for a short time they reunite later in the day.
A mother's close association with her daughter declines markedly around the time the daughter begins to mate with adult males (Pusey, 1983), at which point they are quite mature and very nearly full grown, around 13–14 years old, but even then she will return to travel with her mother when her estrus swelling recedes.

The famous Gombe chimpanzee Fifi was typical. As she reached late adolescence she continued to stay with her mother Flo even in estrus, though when she spotted appealing males she would dash off to mate with them, only to return to Flo immediately after (Goodall, 1968). After a brief foray into another community, Fifi returned to her natal group, and in fact to her mother's exact core area, where she remained for the rest of her life. Females with high-ranking mothers are likely to stay at home, perhaps because their mother has commandeered the most desirable core area in the community (Pusey, 1983; Markham et al., 2015). Most females, however, will transfer communities as they approach adulthood, never to see their mother again.

16.7 Resumption of Estrus

When an infant reaches approximately five years of age, as the lactation-dampening effects of nursing ease, the mother's physiology will become more robust and she will resume ovulating (Jones et al., 2010). For the first time in five years, she will be interested in sex, and males will be interested in her. Now, during the part of her estrous cycle when her sexual swelling is large, she will leave her small, familiar core area and travel day and night with males. Her infant, or if completely weaned now technically a juvenile, will accompany her as she strikes out in search of mates. The presence of the offspring has not the least dampening effect on the mother's very active sex life. She will consort with one male after another, giving her offspring a real-life sex education, though often the estrous female's offspring is not entirely at ease with these events. She may leap on the back of her mother or her mother's suitor, attempting to push the male away or even slapping at her mother or her paramour. Mother's ardor is not inhibited. She will continue to seek out males, and ultimately she will copulate with every male in the group (at least, typically – she may avoid a male or two, perhaps relatives).

Estrus is physiologically stressful in part because females are left with less time for feeding (Thompson et al., 2014); more energy is burned than normal as she works to keep up with the fast-moving male foraging party, and as she maneuvers among aggressive males. While in her normal life she has time for resting and playing with offspring, during estrus she will hardly stop moving. Males will fight around her and even attack her if she fails to cooperate with them when they encourage her to avoid other males.

Females typically copulate 30 times a day when their sexual swelling is at its maximum. Yet, after five years building her reserves, she can endure this stress. When she becomes pregnant, she will have one more month of estrus, a sort of fake estrus but with a swelling; another round of copulation will ensure each male in the group thinks he might be the father. Then she will return to her core area and resume her non-estrus routine. As the months go by she will become ever more ponderous and slow moving until she gives birth.

16.8 Mothering Males is Stressful

I often characterize male chimpanzees as sexists; gentlemen they are not. They bully female chimpanzees, and their attitude extends to humans:

They attack female researchers more often. It may be a bit of a surprise that mothers are biased as well. They invest more in male than female offspring, exhausting their physical reserves to the extent that it takes them longer to recover and begin ovulating after rearing a male (Boesch, 1997). Without this extra investment to help them achieve maximum health and strength, it is thought, the sons might not reproduce at all, competing as they must in the combative male dominance hierarchy (Trivers, 1972). Thus, the five-year interbirth interval (IBI) is an average, a little shorter when nursing a female, a little longer for males.

Because chimpanzees are male-bonded, it is particularly important for males to become socialized, and mothers seem to know it (Murray et al., 2014). Mothers of males are more gregarious than mothers of females, giving their sons a chance to learn competitive skills by watching dominance interactions while under mother's watchful care (Murray et al., 2014). Part of the cost that mothers of sons bear is the stress of extended interactions with volatile, domineering adult males (Boesch, 1997; Murray et al., 2014).

Adult males are fascinating to chimpanzee boys. More independent male offspring may spend the night away from their mother for the first time around the age of six (Pusey, 1983), latching onto a male foraging party as it passes through the mother's core area and having a sleepover with the adult male party. This is merely a one-night spasm of independence. He will return to his mother in a day or two and continue his close association with her until puberty, around 10 years of age. When testes growth accelerates, testosterone surges and juveniles enter adolescence, at which time the community's adult males begin to exert an even stronger appeal, and the young male will begin to spend days and then weeks away from his mother.

16.9 Orphaning

Up to now my description of chimpanzee infancy borders on a best-case scenario for the life-course of mothers and offspring. Just as often the mother–infant bond brings heartache rather than joy. While the offspring of a healthy, dominant female must feel the world is so full of a number of things that they can live their lives as happy as kings (to misquote R.L. Stevenson), the offspring of low-ranking females and first-time mothers are less fortunate. Low-ranking mothers are anxious and less social, and this behavior modeling – and perhaps some genetic propensity as well – leaves its stamp on her offspring; they are less confident and more likely to be low-ranking themselves.

The situation is worst of all for orphans. In the rainy, unpleasant November of 1987, Miff of Gombe died, leaving her three-year-old son, Mel, an orphan. He was pitiful. He grew thin. He moved as if his bones ached. When other infants played, he merely watched, seemingly wistfully, presumably preserving his limited energy reserves for survival. A juvenile male, Spindle, took pity on him and began carrying him on long travel bouts, even allowing him to sleep in his nest at night. Perhaps this is evidence of empathy among chimpanzees. Spindle's mother had also died when he was young, and it may have sparked sympathy in him for Mel. It probably saved Mel's life, but it cannot have been easy for Spindle, who was still small himself. One rainy morning I watched Spindle struggling up a hill with Mel on his back, both looking weary and worn. The sun broke out as we topped a hill. I sat down and Spindle collapsed on my right (Figure 16.6), Mel sat on my left, still chewing a wadge he had carried from their last feeding bout (Figure 16.7). Mel was a stick figure; Spindle looked as if he would never stand up again. I tipped my camera to the right and photographed Spindle, to the left to capture an image of Mel. Their struggles made me feel lucky to be going home to a dry bed that night. Mel lived, though he grew to be a very small adult, and low-ranking to boot.

For an orphan, Mel was lucky. Because infants depend so heavily on their mother, both physically and psychologically, most unweaned infants die. All are despondent after a mother's death. Survival is strongly influenced by the age of the infant. At Taï, in one report on seven orphans younger than four, six of the seven died (Boesch et al., 2010). Gombe chimpanzee Flint was famously despairing unto death

Figure 16.6 Spindle rests from the substantial labor of carrying orphan Mel. Photo by the author.

after his mother Flo died; he simply lay down near his dead mother and wasted away (Goodall, 1986). Most infants attempt to carry on, but without the supplement of nursing, without their mother's intimate knowledge of food resources, without her protection and guidance, they are understandably overwhelmed. Mel is, so far as I know, the youngest chimpanzee ever to survive orphaning, thanks to Spindle.

Mel's adoption, while lucky for him, was not rare. Orphans are often adopted, typically by an older sibling if one is available (Hobalter et al., 2014), but unrelated individuals step in sometimes if there are no relatives. Two-thirds of adoptions succeed in saving the life of the orphan, though as we saw above the odds are long when infants are below the age of four. Mel's is no mere sob story but a consequential behavior for chimpanzees. Perhaps surprisingly, at Taï, the chimpanzee site that has seen the most adoptions, half of all adoptive parents were male (nine males, eight females; Boesch et al., 2010) and four of five orphans who were adopted by relatives were adopted by a male – a bias that is less surprising than it might seem at first glance. Females are often occupied with their own infants and thus less inclined to adopt. At Gombe, one infant was adopted by a grandmother while the unusually inattentive mother still lived (Wroblewski, 2008).

16.10 Infant Death

Deaths occur on the other side of the equation, too; nearly half of all infants die (Goodall, 1986; Nishida et al., 2003; Williams et al., 2004; Reynolds, 2005). The sight of a mother who has lost her infant is every bit as heart-breaking as an orphan pining for a lost mother (Matsuzawa, 1997; Hosaka et al., 2000; Kooriyama, 2009). When I was at Mahale in 1987, Fatuma's infant died, for no reason that I or my colleagues could tell. I saw her carrying the dead infant early in the morning; I had noticed nothing amiss the day before, although I had observed neither her nor her baby closely. Fatuma could not bear to abandon her dead baby. She was still carrying the infant a full day after it died; and then the third day. She carried the body into trees as she fed, holding on to it often with a foot while feeding; at other times she balanced the corpse on a branch. By the fourth day I could tell when I was near without even seeing her because I could smell the decaying body. Yet she carried the infant for not only another day, but still *another* day, a sixth day, at which point the body smelled less. At that point there was no way to tell that the hank of fur she was carrying had been an infant. I lost track of her that sixth day, and when I saw her again a few days later she was without the body.

Figure 16.7 Mel is the youngest chimpanzee known to have survived as an orphan; he was perhaps 3½ when his mother, Miff, died. Photo by the author.

Chimpanzees seem to understand at least something about death, about its finality, so Fatuma carried her infant not because she expected it to revive, but because she could not bear to let go. Such behavior is common at Mahale and elsewhere (Figure 16.8; Biro et al., 2010). Katie Cronin (Cronin et al., 2011) recounts a 16-month-old female infant's death at the Chimfunshi Wildlife Orphanage, a chimpanzee sanctuary in Zambia. The mother, Masya, carried her dead daughter for two days, during which time she repeatedly groomed her face or body, sometimes bringing her face close to the face of the corpse to look intently at it. At one point, Masya placed the body on the ground and held a long vigil in the shade nearby, repeatedly approaching the infant and then retreating to her resting place. This and other evidence convinces me that chimpanzees mourn the death of loved ones.

16.11 Bowlby, Harlow, and Attachment Theory

The profound psychological bond between mother and infant has not always been appreciated. Prior to the work of **John Bowlby** (Bowlby, 1958, 1969) and **Harry Harlow** and colleagues (Harlow, 1958; Harlow & Harlow, 1965), many psychologists viewed mothers more as feeding stations than as objects of affection; the bond was viewed as driven by the infant's need for

Figure 16.8 Masya grooms the face of her dead 16-month-old daughter while older daughter Mary looks on. Photo courtesy of Katherine A. Cronin and Edwin van Leeuwen. Used with permission.

food and nothing else. Nursing, it was thought, encouraged an association between the pleasingly sweet and nutritious taste of mother's milk and the mother herself, an attachment that would not exist otherwise. This perspective might be called the Watered-Plant view of infant development. If the organism has the nutrients it needs, a genetic program for growth will unfold without any other input. We know now, of course, this is not so, but it is far from obvious. It was discovered and described in a body of work known as **attachment theory**.

Bowlby, Harlow, and their students showed that the nature of an infant's attachment to their mother (Bowlby, 1958) is nothing less than love (Harlow, 1958), and depriving an infant of maternal contact is deeply damaging, both immediately and over the long term. Harlow famously provided dummy mothers for infants to cling to, which they did even when the fake mother was engineered to mechanically reject them. These mechanical mothers-from-hell bore devices that pushed away infants by a number of diabolical means, catapulting them off, shaking them off, and prodding them with spikes that emerged from the "mother's" body. Infants still returned to "mother" no matter how severe the rejection. Food had nothing to do with it. The importance of contact with the mother is illustrated by the discovery that without the jostling of being carried, an infant's perception of movement is forever changed, and they compensate by rocking back and forth (Mason & Berkson, 1975). Infants placed in a sling that is swayed by a machine, the action normally provided by clinging to a moving mother, develop normally; without that movement, rocking results. For millions of years chimpanzees and their ancestors – and humans as well, of course – evolved in an environment of evolutionary adaptedness (EEA; Bowlby, 1958) in which mothers and infants are rarely apart.

Attachment theory dramatically changed the culture of adoption and our view of infant needs. Bowlby found that separation from the mother affected human infants in both the short and long term – even across generations. Children need a mother's attention and, yes, love to achieve normal social competence. Studies showed that orphaning had long-term consequences, leaving some lingering effects even among adults who had been successfully adopted (Simon & Senturia, 1966; Bohman & Knorring, 1979). The earlier the adoption the more closely the parent–offspring attachment resembles that of non-adoptees, and standards for adoption placement gradually came to reflect that understanding (Barth & Berry, 1988; Sharma et al., 1996). While not all scholars immediately agreed with Bowlby and Harlow's emphasis on attachment (e.g., Wootton, 1962), their findings have been confirmed in many studies (Hinde & Spencer-Booth, 1967; Ainsworth, 1968; Marvin & Stewart, 1990).

16.12 Chimpanzee Mothering Is Not Hard-Wired

The importance of mothering to chimpanzees might prompt a casual observer to think that competent mothering must be under such tight evolutionary control that it is profoundly hard-wired. While interest in infants and the need to touch them do seem to be instinctual (Lorenz, 1943, 1971; Kringelbach et al., 2008), the mothering skills necessary to protect and nurture an infant are less tightly constrained. *Mothering styles are astonishingly variable.* Some mothers are not only attentive, but almost telepathic in their ability to anticipate an infant's needs and to anticipate dangers; other mothers are clueless. With 20 percent of all infant deaths attributable to chimpanzee violence (Williams et al., 2008), the care and attention a mother showers on her baby can mean life or death. While accidents leading to death are uncommon, falls can be fatal (Goodall, 1986); a watchful mother makes all the difference. The role of learning in parenting skills is apparent in the relative incompetence of first-time mothers, whose infants suffer high rates of mortality. Data showing that nearly half of all infants die (Nishida et al., 2003) obscures an important fact: The mortality rate for first-time mothers was even higher. Practice helps. When a female has younger siblings, she has a distinct advantage. Fifi practiced mothering Flint and had an excellent mothering model in Flo; she went on to rear more infants to adulthood than any other known chimpanzee.

16.13 Lessons

Motherhood is an institution among chimpanzees, every bit as solid as that of humans. The lesson we take from this is that while in most societies we humans have made ourselves so physically comfortable that orphaned infants can survive the loss of a parent, rarely do orphans survive unaffected. Famous political alpha males from George Washington, to Abraham Lincoln, to Barack Obama have all expressed some version of the sentiment that "all I am I owe to my mother."

Chimpanzee alpha males also often begin life nurtured by unusually caring and confident mothers; her support is a springboard that launches them to success. Mother's rank vaults females into the elite, as well, and good parenting guides a daughter to good parenting later. When growing chimpanzees watch their confident mother interacting with others and they grow up mostly free from fear, they develop a competence and confidence that pays off later in high rank. Surely chimpanzees reflect the wisdom of Dorothy Canfield Fisher: "a mother is not a person to lean on, but a person to make leaning unnecessary." I had best stop here to leave you time to get that card for your mother.

References

Ainsworth MDS (1968) Object relations, dependency, and attachment: a theoretical review of the infant–mother relationship. *Child Dev* **40**, 969–1025.

Barth RP, Berry M (1988) *Adoption and Disruption: Rates, Risks, and Responses.* New York: Aldine De Gruyter.

Biro D, Humle T, Koops K, et al. (2010) Chimpanzee mothers at Bossou, Guinea carry the mummified remains of their dead infants. *Curr Biol* **20**, R351–R352.

Boesch C (1997) Evidence for dominant wild female chimpanzees investing more in sons. *Anim Behav* **54**, 811–815.

Boesch C, Bole C, Eckhardt N, Boesch H (2010) Altruism in forest chimpanzees: the case of adoption. *PLoS ONE* **5**, e8901.

Bohman M, Knorring ALV (1979) Psychiatric illness among adults adopted as infants. *Acta Psychiatrica Scandinavica* **60**, 106–112.

Bowlby J (1958) The nature of the child's tie to his mother. *Int J Psycho-Analysis* **39**, 350–373.

Bowlby J (1969) *Attachment and Loss*, Vol 1: *Attachment*. London: Hogarth Press.

Brandt EM, Mitchell G (1971) Parturition in primates: behavior related to birth. In *Primate Behaviour: Developments in Field and Laboratory Research*, Vol. 2 (ed. Rosenblum LA), pp. 178–223. New York: Academic Press.

Clark CB (1977) A preliminary report on weaning among chimpanzees of the Gombe National Park. In: *Primate Bio-social Development: Biological. Social and Ecological Determinants* (eds Chevalier- Skolnikoff S & Poirier FE), pp. 235–260. New York: Garland.

Cronin K, van Leeuwen E, Mulenga I, Bodamer M (2011) Behavioral response of a chimpanzee mother toward her dead infant *Am J Primatol* **73**, 415–421.

Davenport RK (1979) Some behavioral disturbances of great apes in captivity. In *The Great Apes* (eds. Hamburg DA, McCown E), pp. 341–357. Menlo Park, CA: Benjamin Cummings.

Goodall J van Lawick (1967) Mother–offspring relationships in free-ranging chimpanzees. In *Primate Ethology* (eds. Loizos, C, Morris D), pp. 287–346. Chicago, IL: Aldine.

Goodall, J van Lawick (1968) The behavior of free-living chimpanzees in the Gombe Stream Reserve. *Anim Behav Monogr* **1**, 165–311.

Goodall J van Lawick (1971) *In the Shadow of Man.* London: Collins.

Goodall J (1986) *The Chimpanzees of Gombe: Patterns of Behavior.* Cambridge, MA: Harvard University Press.

Goodall J, Athumani J (1980) An observed birth in a free-living chimpanzee (*Pan troglodytes schweinfurthii*) in Gombe National Park, Tanzania. *Primates* **21**, 545–549.

Harlow CM (1986) *Learning to Love: The Selected Papers of HF Harlow.* New York: Praeger.

Harlow HF (1958) The nature of love. *Am Psychol* **13**, 673–685.

Harlow HF, Harlow MK (1965) The affectional systems. In *Behavior of Nonhuman Primates: Modern Research Trend*,

Vol. 2 (eds. Schrier AM, Harlow HF, Stollnitz F), pp. 287–334. New York: Academic Press.

Hill K, Boesch C, Goodall J, et al. (2001) Mortality rates among wild chimpanzees. *J Hum Evol* **40**, 437–450.

Hinde RA (1977) Mother–infant separation and the nature of inter-individual relationships: experiments with rhesus-monkeys. *Proc R Soc (Lond)* **196**, 29–50.

Hinde RA, Spencer-Booth Y (1967) The effect of social companions on mother–infant relations in rhesus monkeys. In *Primate Ethology* (ed. Morris D), pp. 267–286. London: Weidenfeld & Nicolson.

Hobalter C, Schel AM, Langergraber K, Zuberbühler (2014) "Adoption" by maternal siblings in wild chimpanzees. *PLoS ONE* **9**, e103777.

Hosaka K, Matsumoto-Oda A, Huffman MA, Kawanaka K (2000) Reactions to dead bodies of conspecifics by wild chimpanzees in the Mahale Mountains, Tanzania. *Primate Res* **16**, 1–15.

Jones JH, Wilson ML, Murray C, Pusey AE (2010) Phenotypic quality influences fertility in Gombe chimpanzees. *J Anim Ecol* **79**, 1262–1269.

Kiwede ZT (2000) A live birth by a primiparous female chimpanzee at the Budongo forest. *Pan Africa News* **7**, 23–25.

Kooriyama T (2009) The death of a newborn chimpanzee at Mahale: reactions of its mother and other individuals to the body. *Pan Africa News* **16**, 19–21.

Kringelbach ML, Lehtonen A, Squire S, et al. (2008) A specific and rapid neural signature for parental instinct. *PLoS One* **3**(2), e1664.

Lorenz K (1943) Die angeborenen Formen Möglicher Erfahrung. [Innate forms of potential experience]. *Zeitschrift für Tierpsychologie* **5**, 235–519.

Lorenz K (1971) *Studies in Animal and Human Behavior*, Vol. 2. London: Methuen.

Markham AC, Lonsdorf EV, Pusey AE, et al. (2015) Maternal rank influences the outcome of aggressive interactions between immature chimpanzees. *Anim Behav* **100**, 192–198.

Marvin RS, Stewart RB (1990) A family system framework for the study of attachment. In *Attachment Beyond the Preschool Years* (eds. Greenberg M, Cicchetti D, Cummings M), pp. 51–86. Chicago, IL: University of Chicago Press.

Mason WA (1968) Early social deprivation in the nonhuman primates: implications for human behavior. In *Biology and Behavior: Environmental Influences* (ed. Glass DCD) New York: Rockefeller University Press.

Mason WA, Berkson G (1975) Effects of maternal mobility on the development of rocking and other behaviors in rhesus monkeys: a study with artificial mothers. *Dev Psychobiol* **8**, 197–211.

Matsuzawa T (1997) The death of an infant chimpanzee at Bossou, Guinea. *Pan Africa News* **4**, 4–6.

Murray CM, Lonsdorf EV, Stanton MA, et al. (2014) Early social exposure in wild chimpanzees: mothers with sons are more gregarious than mothers with daughters. *PNAS* **111**, 18189–18194.

Murray CM, Stanton MA, Lonsdorf EV, Wroblewski EE, Pusey AE (2016) Chimpanzee fathers bias their behaviour towards their offspring. *Roy Soc Open Science* **3**, 160441.

Nishida T (2012) *Chimpanzees of the Lakeshore: Natural History and Culture at Mahale.* Cambridge: Cambridge University Press.

Nishida T, Corp N, Hamai M, et al. (2003) Demography, female life history, and reproductive profiles among the chimpanzees of Mahale. *Am J Primatol* **59**, 99–121.

Nissen HW, Yerkes RM (1943) Reproduction in the chimpanzee: report on forty-nine births. *Anat Rec* **86**, 567–578.

Otali E, Gilchrist JS (2006) Why chimpanzee (*Pan troglodytes schweinfurthii*) mothers are less gregarious than nonmothers and males: the infant safety hypothesis. *Behav Ecol Sociobiol* **59**, 561–570.

Pusey AE (1983) Mother–offspring relationships in chimpanzees after weaning. *Animal Behav* **31**, 363–377.

Pusey AE, Wallauer W, Wilson M, Wroblewski E, Goodall J (2008) Severe aggression among female *Pan troglodytes schweinfurthii* at Gombe National Park, Tanzania. *Int J Primatol* **29**, 949–973.

Reynolds V (2005) *The Chimpanzees of the Budongo Forest: Ecology, Behaviour, and Conservation.* New York: Oxford University Press.

Schultz AH (1940) Growth and development of the chimpanzee. *Contrib Embryol* **28** 1–63.

Schultz AH (1949) Sex differences in the pelves of primates. *Am J Phys Anthropol* **7**, 401–423.

Sharma AR, McGue MK, Benson PL (1996) The emotional and behavioral adjustment of United States adopted adolescents: Part II. Age at adoption. *Child Youth Serv Rev* **18**, 101–114.

Simon NM, Senturia AG (1966) Adoption and psychiatric illness. *Am J Psych* **122**, 858–868.

Thompson ME, Muller MN, Wrangham RW (2014) Male chimpanzees compromise the foraging success of their mates in Kibale National Park, Uganda. *Behav Ecol Sociobiol* **12**, 1973–1983.

Trivers RL (1972) Parental investment and sexual selection. In *Sexual Selection and the Descent of Man, 1871–1971* (ed. Campbell B), pp. 136–179. Chicago, IL: Aldine.

van de Rijt-Plooij HH, Plooij FX (1987) Growing independence, conflict and learning in mother–infant relations in free-ranging chimpanzees. *Behaviour* **101**, 1–86.

Watts DP, Mitani JC (2000) Infanticide and cannibalism by male chimpanzees at Ngogo, Kibale National Park, Uganda. *Primates* **41**, 357–365.

Williams JM, Oehlert GW, Carlis JV, et al. (2004) Why do male chimpanzees defend a group range? *Anim Behav* **68**, 523-532.

Williams JM, Lonsdorf EV, Wilson ML, et al. (2008) Causes of death in the Kasekela chimpanzees of Gombe National Park, Tanzania. *Am J Primatol* **70**, 766-777.

Wilson ML, Boesch C, Fruth B, et al. (2014) Lethal aggression in *Pan* is better explained by adaptive strategies than human impacts. *Nature* **513**, 414-417.

Wootton B (1962) A social scientist's approach to maternal deprivation. In *Deprivation of Maternal Care: A Reassessment of its Effects*, pp. 255-266. Geneva: World Health Organization.

Wrangham RW, Koops K, Machanda ZP, et al. (2016) Distribution of a chimpanzee social custom is explained by matrilineal relationship rather than conformity. *Curr Biol* **26**, 3033-3037.

Wroblewski EE (2008) An unusual incident of adoption in a wild chimpanzee (*Pan troglodytes*) population at Gombe National Park. *Am J Primatol* **70**, 995-998.

Zamma K, Sakamaki T, Kitopeni RS (2012) A wild chimpanzee birth at Mahale. *Pan Africa News* **19**, 3-5.

17 Meat-Seeking Missiles
Chimpanzees as Hunters

Photo: David Bygott, with permission

Jane Goodall had hardly been at Gombe for a year when, in 1961, she made perhaps the most stunning revelation in the history of primatology (Goodall, 1968, 1971): that chimpanzees hunt and eat meat. It had been assumed that hunting and meat-eating were unique to humans, and indeed many scholars felt the demands of hunting were *the* unique selective pressure that led to the evolution of humanity, a viewpoint definitively articulated in the influential book *Man the Hunter* (Lee & DeVore, 1968). Speech, these researchers maintained, was critical for coordinating the swarm of hunters as they pursued herd animals; stalking prey was thought to require keen logic and a thorough knowledge of animal behavior, including the ability to make complex calculations about the escape routes of fleeing game and how best to overtake them as they maneuvered across a complex landscape. The use of tools, a critical cultural practice for humans (it was thought), emerged to enhance the ability of humans to dispatch and then butcher prey. The distribution of meat after a successful hunting foray was assumed to draw on every unique human intellectual resource, since it was governed by a complicated set of rules balancing social obligations, needs, friendship, kinship, and the responsibility of discharging debts from previous meat distributions. The extra calories from animal protein hypothetically allowed large social groupings. The large human brain could not exist without the abundant protein and fat in meat, otherwise extremely rare in nature. Human-style predation, *Man the Hunter* reasoning went, was the ecological niche for which nearly every unusual aspect of humanity was evolved.

The discovery of hunting and meat-eating among chimpanzees, then, was wholly unexpected and left Man the Hunter theorists pensively fingering their metaphorical chins, wondering what it meant for humans. It got worse. Before human origins specialists could figure out what to do with chimpanzee hunting, Goodall delivered a second blow. Chimpanzees not only hunted, they also shared the captured prey (Goodall, 1968, 1971; Figure 17.1).

Few other primates eat meat (I mean vertebrate tissue, not bugs), but baboons and capuchin monkeys do (Strum, 1975; Fedigan, 1990), though dramatically less often than do chimpanzees. Chimpanzees, in fact, are the only primate other than humans that *depend* on meat to round out their diet. Though the percentage of meat in the diet differs somewhat among sites, one study showed they obtained 5 percent of their calories from animal protein (Wrangham & Riss, 1990), and the proportion is even higher at Kibale-Ngogo (Watts & Mitani, 2002b), where chimpanzees eat at least 25 kg (55 lb) of meat a year. Chimpanzees at my site, Semliki (Hunt & McGrew, 2004) and at the site of Budongo (Newton-Fisher et al., 2002), just to the north of Semliki, eat considerably less meat, but they still hunt. Yet I often

Figure 17.1 Chimpanzees sharing meat at Gombe; note the female to the right begging. Photo: David Bygott.

chat with otherwise well-informed people who assume that chimpanzees subsist entirely on plant foods, avoiding meat altogether. No. Far from it. Chimpanzees love meat. Which makes it all the more surprising that some researchers have questioned whether the nutrients in prey are worth the effort chimpanzees expend to acquire them.

17.1 Is Meat Really All That Great?

Hunting often takes place after a period of stalking, a time commitment that takes the hunter away from finding and consuming less demanding "prey": fruits. Why forego a sure thing in fruit to take a chance on an elusive monkey? Thirty grams of banana contains 25 calories, not that much less than the same amount of beef (60 calories); while most fruits in the chimpanzee habitat are not quite as nutritious as bananas, palm nuts have even more calories per gram than lean monkey meat (Leung, 1968; Stanford, 1998). Why not crack nuts all day rather than chase monkeys?

Considering the high metabolic costs of chasing and capturing prey, the risk of injury from falls, the possibility of wounding at the hands of the prey, and the low caloric return for hunting compared to nut-cracking, some scientists have posed the question "Why hunt at all?" Several researchers have speculated that males may hunt not because meat is such a great source of nutrition, but because they can trade it for something they desire even more: sex (Teleki, 1973; Stanford et al., 1994; Stanford, 1996, 1998, 2002; Mitani & Watts, 2001, 2005; Gomes & Boesch, 2009). Possibly also allies. Or so the theory goes. Males do most of the hunting, 89.3 percent in one Gombe study (Stanford, 1996, 1998), and males often give meat to estrous females, after which estrous females often copulate with the gift-giver. Further support for the meat-for-sex speculation comes from the fact that the number of females in estrus best predicts when males will hunt (Stanford et al., 1994; Gomes & Boesch, 2009).

Further evidence that meat is a poor food choice comes from the fact that females rarely participate in hunts. Females are the **ecological sex**, this reasoning goes, more attuned to their ecology because natural selection has evolved females to optimize food intake in a way it has not for males. It is the females who gestate, nurse, protect, and (in the vast majority of primates) carry infants. All this energy output requires fuel, and a regular supply of it. With the need for these extra nutrients, females should be hunting more than males, not *vice versa*. Perhaps you think that female chimpanzees are too delicate to bear the rigors of hunting. If so, let me disabuse you of that notion. In her prime, Gigi, a female at Gombe who was infertile, was as successful at catching prey as some males (Stanford et al., 1994). Turning to other species, among lions females are the better and more reliable hunters. Gigi's infertility is a key. Female chimpanzees are thought to hunt and capture prey less often than males due to the dual costs of being weighed down by carrying an infant and the risk that hunting presents to infant safety. Perhaps they are also discouraged by the likelihood that they could lose their prize to an acquisitive male. In other words, females are not less capable hunters, but they face risks that make them less motivated to hunt. Instead, they wait for males to get meat and then hope to beg a little from them.

A different hypothesis also draws on the supposed low caloric return of meat. Perhaps hunting is valuable, this hypothesis goes, not so much for its nutritional benefits, but because it strengthens male bonds (Nishida et al., 1992). Males share meat among themselves, strengthening their alliances; Nishida found that the closer the bond among two males the greater the likelihood they will share meat with one another. Because male bonding is critical for territorial defense, and because effective defense protects food resources and the reproductive potential they represent, meat-eating might be important even if it adds little to nutritional health.

17.2 Meat Contains Important, Essential Nutrients

One thing that strikes me immediately about speculations such as these (sex, allies) is how little they jibe with what I have seen of the chimpanzee demeanor in the presence of meat. Chimpanzees display a gluttonous enthusiasm when it comes to food, particularly sugary fruits, but of no other food are they so pantingly, desperately, grovelingly fond as they are of meat. I find it difficult to believe that chimpanzees hunt for any reason other than their love of animal flesh.

While a monkey carcass may not be calorie-rich – though surely something that makes up 5 percent of an organism's caloric intake cannot be said to be unimportant – something other than mere caloric content might drive the intense craving of chimpanzees for meat. Meat contains a whole host of essential micronutrients that are virtually nonexistent in plant foods, including vitamin B12, creatine, carnosine, cholecalciferol, docosahexaenoic acid, and heme-iron (Zucker et al., 1981; Jiménez-Colmenero et al., 2001; Gröber et al., 2013). Meat also contains other critical nutrients that, while not totally absent in other foods, are difficult to obtain from plants: vitamin A, vitamin K, calcium, sodium (salt), and potassium (Teleki, 1973; Boesch, 1994a; Stanford, 1996; Mitani and Watts, 2001). Every essential amino acid – the building blocks of proteins and therefore the veritable chemical building blocks of chimpanzees themselves – are found in meat, and in the same proportions. This is not just theory. Rats normally shun protein-free rat chow, but given the choice between a protein-free diet with all essential amino acids and a protein-rich chow with one nutrient missing, threonine, they go for protein-free chow (Leung, 1968).

There is more. Many of the nutrients found only in meat are particularly important for large-brained animals. Vitamin B12 is essential for brain development and brain function; deficiency results in nerve damage, skin numbness, coordination disorders, reduced nerve conduction velocity, and progressive brain atrophy (Gröber et al., 2013). Creatine facilitates energy storage in the brain and is also essential for muscle development. Carnosine is an antioxidant found in both brain and muscle. Cholecalciferol (vitamin D3) is also a brain food and is critical for bone health. Docosahexaenoic acid (DHA) is important for brain development and function. Heme-iron is essential for red blood cell production, and is absorbed much more readily than the kind of iron found in plant foods.

I wrote above that the calories are important. Note the body part consumption sequence: chimpanzees consume the fattiest, most calorie-rich part of prey first: the brain.

In short, between the valuable vitamins, minerals, and amino acids, and the calories in meat, there are good nutritional reasons to hunt.

17.3 Hunting Frequency

While hunting rates vary, both by site and over time, chimpanzees hunt about once per week and are successful about half to three-quarters of the time they hunt, which means they eat meat about two or three times each month (Goodall, 1968; Stanford, 1996, 1998; Mitani & Watts, 2001; Boesch et al., 2006). Meat is a rare and prized item.

17.4 Why Not Hunt Even More?

Given how passionate chimpanzees are about meat, one might wonder why not hunt even more. One might expect that monkeys are a great fallback food source, there for the taking when other foods are scarce. Monkeys are always somewhere in the forest, whereas fruits are sometimes totally absent. Why should chimpanzees not hunt monkeys more often as other food items become less available? Yet in one of the mysterious aspects of chimpanzee hunting behavior, sometimes chimpanzees feed for hours quite near their favorite prey, colobus monkeys, seemingly taking no notice of them whatsoever. Why?

Surprisingly, rather than hunting more often when the food supply is meager, research shows exactly the opposite: When fruit is scarce, chimpanzees rarely hunt, and while the observations are not 100 percent consistent, in general the more fruit there is the more hunting chimpanzees do (Uehara, 1997; Mitani & Watts, 2001; Watts & Mitani, 2002a, 2002b, 2015; Gilby & Wrangham, 2007). While this seems counterintuitive, the reasons seem clear. We have long known that the larger the number of hunters, the greater the likelihood of success (Figure 17.2; Busse, 1978; Stanford, 1996, 1998; Uehara, 1997; Mitani & Watts, 2001; Gomes & Boesch, 2009). It was only more recently that we pushed this link back one more step; fruit supply determines party size. At Kibale–Ngogo, party size rises as fruit availability goes up (Mitani & Watts, 2001; Watts & Mitani, 2002a, 2002b, 2015). It is the same at Mahale (Uehara, 1997). At Gombe, a particularly easy fruit to harvest, *Parinari curatelifolia*, and a large, sweet, highly desired fruit, *Pseudospondias microcarpa*, often emerge in August and September, supporting large parties and thus increasing hunting success. When fruit is in short supply, by contrast, chimpanzees break into smaller parties to reduce their travel distances (and therefore travel costs). This allows them to scour a small area for whatever lean pickings are to be found, but small groups mean low hunting success. Not only does the scarcity of fruit mean smaller party sizes and thus lower rates of hunting success, the waste of calories in an unsuccessful pursuit is more dangerous in lean times because chimpanzees are already dangerously close to malnutrition (Uehara, 1997). This relationship also explains why the number of estrous females predicts hunting attempts: females come into estrus when food abundance rises.

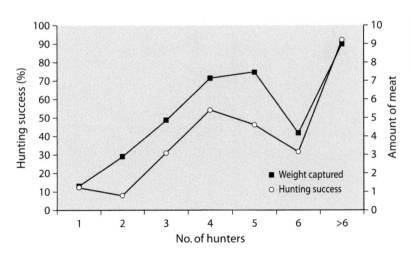

Figure 17.2 Success increases as hunting party size increases. From Boesch et al., 2006. Used with permission.

While it is important, party size alone does not always determine success. There is another reason chimpanzees may ignore nearby prey. At Gombe, chimpanzees seem acutely aware of circumstances that make escape difficult for colobus. When they discover a group of red colobus in an isolated tree with no accessible tree to leap into, chimpanzees are quick to recognize the opportunity (Figure 17.3; Wrangham, 1977). But, as illustrated in Figure 17.3, they depend not on a fortuitous discovery of prey that just happens to be in an isolated tree; they use very clever tactics to herd the colobus toward a strategically disadvantageous position where their escape routes are limited, like the tops of palm trees (Stanford, 1998).

While many hunting episodes are opportunistic – a large party of chimpanzees happens to come upon colobus – others begin with a mustering of chimpanzees who then range widely in search of colobus (Boesch & Boesch, 1989; Nishida et al., 1992; Uehara 1997; Mitani & Watts, 2001). Perhaps because chimpanzees can choose their moment to hunt, foregoing hunting when success is unlikely, rather than depending on hunting as true carnivores do, they are even more successful on average than carnivores (Boesch et al., 2006). Hunts involve stealth and strategy; at Gombe and Mahale that is enough; at Taï, something more is required.

17.5 Hunting Up Close

Every now and then a student readying to leave on a first field experience will express excitement in anticipation of observing a chimpanzee kill. I have seen many chimpanzee hunting episodes – dozens – and I cannot share his enthusiasm. To me, the killings are upsetting and gruesome. In one of Goodall's first observations of chimpanzee hunting (Goodall, 1968) an infant baboon died only after an entire arm had been eaten, half an hour after being captured, a teeth-grinding, gut-churning torture for the human observers. Such treatment of prey has caused some to characterize the chimpanzee means

Figure 17.3 Hunting strategy at Gombe. As chimpanzees approach prey (shaded profiles) they disperse themselves so as to surround the prey. Chimpanzees often corner colobus in an isolated tree, closing in from all sides and then attempting to seize a victim. Image by the author.

of dispatching prey as "eating them to death." I will share with you one of the less bloody of the hunts I have seen, just to give you a sense of what an episode is like, but you will have to search elsewhere if you want to hear about agonizing deaths and body dismemberments.

It was at Mahale and the female I was following, hearing the excited calls of a hunting party in the distance, ran toward the uproar, her young infant clinging to her belly. She and I came to an open area next to one of the many streams that cascade off the Mount Nkungwe uplands to the east of the park. She rushed forward toward the periodic "wraaa! wraaa!" screams of males, quickly climbing a tree so as to peer into the adjacent canopy where male chimpanzees and monkeys could be heard leaping and alarm-calling. From the ground, I saw monkeys vault from tree to tree; now and then there would be a sudden movement in the trees, followed by silence. The fleeing monkeys were gradually moving in my direction. I sporadically caught a glimpse of a male chimpanzee on the ground, walking among the thick foliage, looking up into the canopy. The monkeys passed to either side of us and a male chimpanzee ran past me, rushed ahead of the fleeing monkeys some tens of yards and climbed a tree.

Suddenly, just above me, I heard the crashing sound of foliage being violently shaken, followed by a loud shriek. A monkey had made the mistake of leaping into the tree where the female I was following had sat, still and quiet, while the males drove the monkeys toward her. The next instant I could see that she was grappling with the monkey. It squirmed and bit at her, and as she struggled she looked more like a person trying to stuff a cat into a gunny sack than an accomplished predator. The struggle lasted only a few seconds before, to my great surprise, the female simply launched herself into space, still grappling with the monkey. Somehow as they fell she managed to maneuver the monkey underneath her hands and feet so as to land on top when she hit the ground. She paused for a second, peering frantically in several directions, presumably making sure no higher-ranking chimpanzee was nearby to steal her prize (one-third of female kills at Gombe are stolen; Stanford et al., 1994); she then bit the weakly

Table 17.1 **Hunting prey choice and success**[a]

	Gombe	Mahale	Ngogo	Taï
Percent colobus make up of prey[a]	79	53	88	81
Hunting success[b] (%)	52	61	82	52

[a] Boesch et al., 2006 and references below; [b] Boesch et al., 2006.

twitching monkey in the head, snatched up the body and ran. During this whole episode, her infant had somehow managed to cling to her and appeared uninjured.

My role in this drama was over – even though she had been moving at a painfully slow pace before hearing the hunt, she had now become the very name of action. Clutching the limp body of the monkey she ran; I moved instantly, attempting to follow her, but lost her in seconds. She had accomplished something rare for a chimpanzee; she caught a monkey without another chimpanzee knowing or helping and managed to slip off before an interloper could discover her and steal her prize. She probably ate the entire monkey alone, a rare and valuable banquet for a nursing mother.

17.6 What Chimpanzees Hunt

Most of the meat in the chimpanzee diet consists of fellow primates; chimpanzees eat six different species of monkey, with colobus their most common prey (Table 17.1), mostly red colobus (Boesch & Boesch, 1989; Boesch, 1994a, 1994b, 2002; Stanford, 1996, 1998, 2002), comprising 76–82 percent of chimpanzee prey at Gombe (Wrangham & Riss, 1990; Stanford, 1996, 1998) and 81 percent at Taï (Boesch et al., 2006).

Bushpigs are chimpanzees' second most common prey, and given chimpanzee persistence and enthusiasm when they find evidence of bushpig infants, it must be the most prized food item on their long list of up to 300 food items. They also take the occasional bushbuck, blue duiker, rodent, squirrel, or bird. Baboons, redtails, mangabeys, bushbabies, and blue monkeys make up the rest of the primate inventory.

Chimpanzees at Gombe focus on immatures when hunting red colobus monkeys; 75 percent of the prey taken are infants or juveniles (Stanford et al., 1994). Indeed, Gombe chimpanzees are somewhat fearful of adult male colobus. Juveniles are more often victims because they are less capable of the dramatic leaps that allow adult colobus to escape predators; they are easier to subdue; and they are less capable of recognizing dangerous situations. Youngsters have not learned the most effective escape methods when confronted with hunters. Infants are captured by snatching them from the mother. The 20 percent of prey that are adults are almost entirely females.

17.7 Varieties of Prey

Some meals come to you on their own. Typically, birds, bushbabies, rodents, and squirrels are encountered by chance, with no effort, as chimpanzees go about their daily fruit gathering (Wrangham, 1977; Goodall, 1986; Stanford, 2002). Such prey is discovered at close range and the capture and kill techniques are brutally simple, a lightning fast snatch to capture the victim, after which the prey is either flailed forcefully against a tree trunk or killed with a bite. Chases are rare, since pursuit is rarely rewarded; if the initial pounce is unsuccessful the individual gives up and moves on (Goodall, 1986). Because prey are wary, it is often the quiet, solitary chimpanzee who has the luck to sneak up on them; since females are more solitary they more often capture prey in this way.

Other prey encountered randomly are the fawns of bushbuck or duiker. Fawns are completely immobile when discovered – once I almost stepped on one; just as I was about to put a foot down I became vaguely aware of something unusual in my path and, looking down, a dark, liquid eye seemed to materialize right under my boot. I took one extra-long step to avoid treading on it and it never flinched. Chimpanzees snatch up these infants and kill them without a struggle. Not surprisingly, males and females capture fawns at an almost identical rate (Goodall, 1986).

Bushpigs are a completely different matter. They are less often encountered randomly and more often because chimpanzees hear bushpig sounds and go to investigate, though they do occasionally stumble upon infants. Chimpanzees almost invariably take infants – certainly an adult is too large and powerful for any chimpanzee to overcome. For the first couple of months of life helpless infant piglets are left hidden in untidy nests in the undergrowth while their mothers forage. Nests are somewhat camouflaged by their shaggy appearance, but even if bushpig nests appear unobtrusive to humans, chimpanzees are on the lookout for them. As the infants grow older they will travel with their mother in a foraging group of up to 15 individuals, including infants and juveniles of different ages (Estes, 1991); even when adults are present and on their guard, chimpanzees may risk injury by attempting to snatch a juvenile away from its protectors. The bushpigs do their best to protect their babies. If they hear chimpanzees approaching, the group forms a defensive ring encircling the piglets. If there are few enough defenders and numerous enough chimpanzees, the hunters play a baiting game, rushing in and slapping at the adults while avoiding their tusks, hoping to startle a piglet out of its protective circle where it can be grabbed. Occasionally this strategy succeeds (Goodall, 1986).

Baboons are rarely hunted by large groups; prey are almost always nursing infants. An infant is more vulnerable when it has been kidnapped from the mother by a higher-ranking female, a not uncommon event. The kidnapped infant alarm-calls in hopes that its mother will come to the rescue. Instead, chimpanzees hear the alarm-calling infant and rush in to snatch it from the kidnapper before the mother has had a chance to recover it; when kidnappers are confronted by chimpanzees they offer little resistance, quite unlike mothers. Other times an infant simply gets separated from the mother and alarm-calls to her, again drawing the attention of chimpanzees (Goodall, 1963, 1968, 1986).

17.8 How to Hunt: Gombe Strategies

As a hunt begins at Gombe, the hunting party moves quite silently toward its prey, probably calculating but perhaps remembering likely escape routes as they flee. Which trees will the monkeys leap into when the chase begins? If luck is with the hunting party, the chimpanzees might completely surround a group of colobus (Figure 17.3). Other times there is pursuit that may move hundreds of meters. Some hunters climb up into the trees – "run up" trees may be a better description – to pursue prey high above the forest floor, leaping after them and attempting to outrun them. Other individuals remain on the ground, waiting at the base of a tree, ready to run up if monkeys come their way. Some individuals run ahead to a spot where they anticipate an unlucky colobus might fall. If the monkeys are clearly moving in a certain direction, hunters will walk or run along underneath, looking for their best chance for a capture. In desperation, some monkeys attempt an impossible leap; when they fall they are almost always captured. In one hunt, I observed a large juvenile male leap from tree to tree, landing lower each time until the last leap left him on the ground, where Evered the chimpanzee was waiting to kill him. Not uncommonly an unlucky monkey juvenile or mother leaps into a tree where a chimpanzee is lurking.

East African chimpanzees fear male colobus for a reason; males, while smaller than chimpanzees, are compact balls of fury. They spare little time for bluffing and launch right into attack mode. Occasionally a male chimpanzee receives a deep wound from a male colobus; it shows how important meat is to chimpanzees that they risk these serious injuries. Once I was following chimpanzees through some dense but short trees and came face to face with a male red colobus, sitting just at eye level, right where I wanted to pass. I stared him down, assuming that as I inched closer he would turn and run. He did not. After a brief staring contest, he leapt at me and I was forced to beat an ignominious, running retreat. He clearly was not fearful of a human five times his weight; he may not have made a distinction between me and a chimpanzee.

Craig Stanford offers a compelling narrative that explains the role of male red colobus as altruistic protectors. While the more vulnerable females flee, males interpose themselves between the fleeing females and the hunters, holding their tails out of the reach of the attacking chimpanzees. They must do this because a pursuing chimpanzee can grasp their tail, fling them from the tree, and thus take them out of the battle. With lunges and feints, the courageous males buy time for the females to escape. If the hunters penetrate the male colobus defense, they will attempt to catch a female and either kill her or take her infant.

Stanford describes the Gombe hunting strategy as mutualistic; each individual is doing his or her best to capture prey, but, as we have already seen, by hunting in a large group each chimpanzee vastly improves his or her chance of being successful (Figure 17.2). The cooperation involved is not very coordinated, but still effective in part simply because there are so many chimpanzees surrounding the prey.

17.9 How to Hunt: Mahale

Strategies at Mahale are quite similar to those at Gombe (Nishida et al., 1992; Uehara, 1997). Hunters surround or chase red colobus, engaging in a strategy that is mutualistic and somewhat coordinated.

17.10 How to Hunt: Taï, Ivory Coast

At Taï, West African chimpanzees face a different challenge than Gombe chimpanzees. The canopy is much denser and opportunities to catch colobus in an isolated tree are nonexistent. The continuous canopy means that there are no long leaps to fatigue fleeing monkeys or cause them to fall to the ground.

Taï chimpanzees are therefore compelled to utilize different, more sophisticated methods (Figure 17.4). Lone hunts are rarer and hunting parties are larger; there is more planning and more coordination; the targets are adults. Rather than hoping to find monkeys in isolated trees where escape is difficult, Taï hunters disperse and then close in on prey, cutting off escape routes. Hunters herd colobus into a trap, guarded by chimpanzees on all sides. As a hunt takes shape, blockers array themselves in a horseshoe-shaped trap surrounding their prey; individuals place themselves in or at the base of trees that are likely

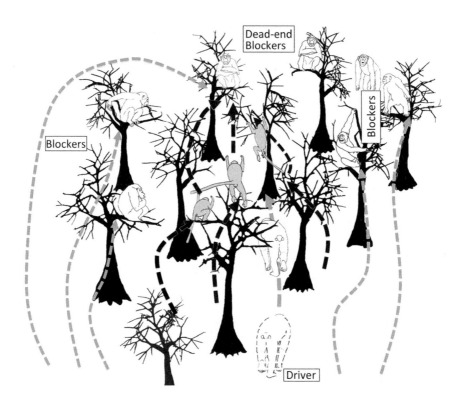

Figure 17.4 Hunting strategy at Taï. The approaching party of hunting chimpanzees take positions (gray dotted lines) to the side and at a dead-end of a corridor into which they herd colobus prey. The hunters array themselves in a horseshoe-shaped configuration either in the trees or on the ground ready to run up a tree, while on the ground a driver, purposely visible to the prey, walks into the open end of the horseshoe, herding the leaping prey forward (black dotted lines) toward a dead end where they can be cornered, often bounding into the very trees where chimpanzee wait in anticipation of their movements. Figure by the author.

escape routes, creating an avenue guarded by chimpanzees on either side of the herding path. At the end of this avenue are dead-end blockers. A driver male walks along the ground, pushing the fleeing colobus in front of him, guiding them into a corral with waiting chimpanzees at their posts at either side. The driver herds them forward until they meet the dead-end blockers, at which point they are completely hemmed in, after which chimpanzees move in for the kill.

At both Taï and Gombe it is not unusual for multiple monkeys to be killed. In one epic hunt at Gombe, chimpanzees killed one-quarter of a red colobus troop in a single gruesome event (Stanford, 1998). This is cooperative hunting at its most effective. The benchmark for cooperation is kill-counts: When the number of cooperating hunters rises, so does individual meat consumption.

17.11 Meat Sharing at Kibale–Ngogo and Taï

While a mutualistic cooperation is effective at Gombe and Mahale, with each pursuing a strategy they deem best for capturing prey, the dense forest canopies at Kibale–Ngogo and Taï require that some individual assume a role that is unlikely to result in an individual capture, but more likely to make the group successful. Why take a role that is unlikely to be rewarded? At Ngogo as at Taï, meat-sharing compensates these altruistic individuals; individuals who participate in the hunt are given meat, and individuals who fail to help out are not (Boesch & Boesch, 1989; Boesch, 1994a, 1994b, 2002; Mitani et al., 2002).

17.12 Tools and Hunting

There are few scenarios for the evolution of humans that fail to bring in the role of tool use and tool making in hunting. Indeed, as we have already seen, of all Goodall's groundbreaking discoveries, it was the observation that chimpanzees both hunt and make tools that most shocked the scientific community.

There is indeed some tool use associated with hunting among chimpanzees, but it is rare. Reports (Pruetz & Bertolani, 2007; Pruetz et al., 2015) that chimpanzees at the dry habitat site of Fongoli regularly use spears to kill Senegal bushbabies (*Galago senegalensis*) were greeted with great excitement and widespread news coverage. Ten different individuals used spears that were prepared with some sophistication, manufactured in five steps that included sharpening the spear with their teeth (Pruetz & Bertolani, 2007; Pruetz et al., 2015). The spears were used to kill bushbabies that had retreated to inaccessible cavities – hollow trees – and were used not just to extend reach, but also as a penetrating spear purposely designed to kill prey. Another unique-to-humans domino falls. Chimpanzees sometimes use wadded-up leaves to soak up blood while consuming a carcass, and a twig may be used to scoop marrow out of a bone (McGrew, 1992).

17.13 Conclusion and Lessons

While those who are unfamiliar with chimpanzees sometimes assume they are vegetarians, all chimpanzee populations under long-term observation have been observed to hunt. Chimpanzees utilize goal-directed, flexible strategies conceived to work best to capture the species or age-class they are targeting. When the prey have better escape methods, as at Taï and Ngogo, chimpanzees adopt more sophisticated techniques.

The study of chimpanzees seems to have refuted the idea that hunting was responsible for the evolution of many aspects of human intelligence, including the origin of language. Communication, as I outlined above, is valuable when discussing where prey might be found or to coordinate the actions of the members of a hunting party as they corner and dispatch prey, both of which are dangerous, but less so with proper coordination. Hunting, the theory goes, required the evolution of intelligence to anticipate the actions of prey and its response to pursuit. A good memory is valuable for predicting seasonal activities of prey, the myriad responses to pursuit by different prey animals, and for remembering successful strategies. This turned out to be wrong. While intelligence helps humans hunt successfully, chimpanzees show us that ape intelligence is more than adequate for anticipating the actions of prey, acting in a manner contingent on the predicted actions of others and anticipating and responding to prey evasion tactics.

Yes, the addition of more meat to the diet was critical for human evolution, but more for the nutrients it provided than the intelligence it requires. In fact, as the hunting hypothesis waned in the face of evidence of chimpanzee hunting, many of us suddenly found the sophisticated coordination of social predators such as wolves and killer whales to be more nearly human-like. Most scholars now look elsewhere as they try to understand the evolution of the unique intellectual gifts of humans.

References

Boesch C (1994a) Cooperative hunting in wild chimpanzees. *Anim Behav* **48**, 653–667.

Boesch C (1994b) Chimpanzees–red colobus monkeys: a predator-prey system. *Anim Behav* **47**, 1135–1148.

Boesch C (2002) Cooperative hunting roles among Taï chimpanzees. *Hum Nat* **13**, 27–46.

Boesch C, Boesch H (1989) Hunting behavior of wild chimpanzees in the Taï National Park. *Am J Phys Anthropol*, **78**, 547–573.

Boesch C, Boesch H, Vigilant L (2006) Cooperative hunting in chimpanzees: kinship or mutualism. In *Cooperation in Primates and Humans: Mechanisms and Evolution* (eds. Kappeler PM, van Schaik CP), pp. 139–150. Berlin: Springer-Verlag.

Busse CD (1978) Do chimpanzees hunt cooperatively? *Am Nat* 112, 767–770.

Estes R (1991) *The Behavior Guide to African Mammals*, Vol. 64. Berkeley, CA: University of California Press.

Fedigan LM (1990) Vertebrate predation in *Cebus capucinus*: meat eating in a neotropical monkey. *Folia primatologica* 54, 196–205.

Gilby IC, Wrangham RW (2007) Risk-prone hunting by chimpanzees (*Pan troglodytes schweinfurthii*) increases during periods of high diet quality. *Behav Ecol Sociobiol* 61, 1771–1779.

Gomes CM, Boesch C (2009) Wild chimpanzees exchange meat for sex on a long-term basis. *PLoS One* 4, p.e5116.

Goodall J (1963) Feeding behaviour of wild chimpanzees: a preliminary report. *Symp Zool Soc Lond* 10, 39–48.

Goodall J (1968) The behavior of free-living chimpanzees in the Gombe Stream Reserve. *Anim Behav Monogr* 1, 165–311.

Goodall J van Lawick (1971) *In the Shadow of Man*. London: Collins.

Goodall J (1986) *The Chimpanzees of Gombe: Patterns of Behavior*. Cambridge, MA: Harvard University Press.

Gröber U, Kisters K, Schmidt J (2013) Neuroenhancement with vitamin B12-underestimated neurological significance. *Nutrients* 12, 5031–5045.

Hunt KD, McGrew WC (2002) Chimpanzees in the dry habitats of Assirik, Senegal and Semliki Wildlife Reserve, Uganda. In *Behavioural Diversity in Chimpanzees and Bonobos* (eds. Boesch C, Hohmann G, Marchant LF), pp. 35–51. Cambridge: Cambridge University Press.

Jiménez-Colmenero F, Carballo J, Cofrades S (2001) Healthier meat and meat products: their role as functional foods. *Meat Sci* 59, 5–13.

Lee R, DeVore I (1968) *Man the Hunter*. Chicago, IL: Aldine.

Leung WTW (1968) *Food Composition Tables for Use in Africa*. Bethesda, MD: Public Health Service.

McGrew WC (1992) *Chimpanzee Material Culture: Implications for Human Evolution*. Cambridge: Cambridge University Press.

Mitani JC, Watts DP (2001) Why do chimpanzees hunt and share meat? *Anim Behav* 61, 915–924.

Mitani JC, Watts DP (2005) Correlates of territorial boundary patrol behaviour in wild chimpanzees. *Anim Behav* 70, 1079–1086.

Mitani JC, Watts DP, Muller MN (2002) Recent developments in the study of wild chimpanzee behavior. *Evol Anthropol* 11, 9–25.

Newton-Fisher NE, Notman H, Reynolds V (2002) Hunting of mammalian prey by Budongo forest chimpanzees. *Folia Primatol* 73, 281–283.

Nishida T, Hasegawa T, Hayaki H, Takahata Y, Uehara S (1992) Meat-sharing as a coalition strategy by an alpha male chimpanzee? In *Topics in Primatology* Vol 1: *Human Origins* (eds. Nishida T, McGrew W, Marler P, Pickford M, deWaal FBM), pp. 159–174. Tokyo: Tokyo University Press.

Pruetz JD, Bertolani B (2007) Savanna chimpanzees, *Pan troglodytes verus*, hunt with tools. *Curr Biol* 17, 412–417.

Pruetz JD, Bertolani P, Boyer Ontl K, et al. (2015) New evidence on the tool-assisted hunting exhibited by chimpanzees *(Pan troglodytes verus)* in a savannah habitat at Fongoli, Sénégal. *Roy Soc Open Sci* 2, 140507.

Stanford CB (1996) The hunting ecology of wild chimpanzees: implications for the evolutionary ecology of Pliocene hominids. *Am Anthropol* 98, 96–113.

Stanford CB (1998) *Chimpanzee and Red Colobus: The Ecology of Predator and Prey*. Cambridge, MA: Harvard University Press.

Stanford CB (2002) Avoiding predators: expectations and evidence in primate antipredator behavior. *Int J Primatol* 23, 741–757.

Stanford CB, Wallis J, Mpongo E, Goodall J (1994) Hunting decisions in wild chimpanzees. *Behaviour* 131, 1–18.

Strum SC (1975) Primate predation: interim report on the development of a tradition in a troop of olive baboons. *Science*, 187, 755–757.

Teleki G (1973) *The Predatory Behavior of Wild Chimpanzees*. Lewisburg, PA: Bucknell University Press.

Uehara S (1997) Predation on mammals by the chimpanzee (*Pan troglodytes*). *Primates* 38, 193–214.

Watts DP, Mitani J (2002a) Hunting behavior of chimpanzees at Ngogo, Kibale National Park, Uganda. *Int J Primatol* 23, 1–28.

Watts DP, Mitani J (2002b) Hunting and meat sharing by chimpanzees at Ngogo, Kibale National Park, Uganda. In *Behavioural Diversity in Chimpanzees and Bonobos* (eds. Boesch C, Hohmann G, Marchant LF), pp. 244–255. Cambridge: Cambridge University Press.

Watts DP, Mitani J (2015) Hunting and prey switching by chimpanzees (*Pan troglodytes schweinfurthii*) at Ngogo. *Int J Primatol* 36, 728–748.

Wrangham RW (1977) Feeding behaviors of chimpanzees in Gombe National Park, Tanzania. In *Primate Ecology* (ed. Clutton-Brock TH), pp. 503–538. London: Academic Press.

Wrangham RW, Riss EVZB (1990) Rates of predation on mammals by Gombe chimpanzees, 1972–1975. *Primates* 31, 157–170.

Zucker DK, Livingston RL, Nakra R, Clayton PJ (1981) B12 deficiency and psychiatric disorders: case report and literature review. *Biol Psych* 16, 197–205.

18 The Mind of the Chimpanzee
Reasoning, Memory, and Emotion

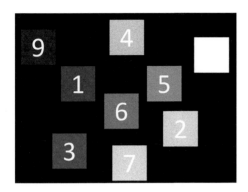

Image by the author

While it has been decades since scholars accepted that humans are not the only tool makers and users, contemporary scholarship seems less aware that, after humans, chimpanzees are by far the most inveterate tool makers and users in the animal kingdom; they use tools more than all other primates combined (McGrew, 1992, 2010). Alone among the primates, humans are considered to be dependent on hunting vertebrates to survive. But what does that mean? If an entire human population was exposed to Alpha-Gal and became allergic to meat, would it wither and die? Perhaps people would reproduce more slowly and the population density would decrease, but would every person inevitably die? Unlikely. Consider chimpanzees, then. They hunt once a week and most populations depend on meat to fill a substantial part of their nutritional needs. They could live without meat, but it probably helps sustain their population size.

We once thought it might be the presence of a Broca's area that distinguished humans. Is group-oriented behavior to defend loved ones a human uniqueness? No. Brain asymmetry? No. Perhaps the use of symbolism defines humans. Or is it the ability to count? Empathy? Self-awareness? Self-concept? Understanding the spoken word? Every one of these possibilities has been asserted to be behaviors unique to humans at one time or another, and one by one each of them has fallen.

This chapter will attempt to describe the mind of the chimpanzee, both the areas where they are exceptional and those where their cognitive abilities are ordinary. We can start by dispelling an uncommon but not unknown misconception: Chimpanzees are not as intelligent as humans. Still, they are much more like us than science believed only a few decades ago, and much more like us than some wish to admit.

In Chapter 11 we learned that the world comes into a chimpanzee's head in a manner indistinguishable from that of humans. It might be that if we could toggle back and forth between human and chimpanzee senses, we would detect some slight differences, such as a greater sensitivity to light on one side of the spectrum or a different sensitivity to pain. On the other hand, the differences are so slight that had you been born with a chimpanzee's senses, neither you nor anybody else would have noticed it. What the brain does with that information, however, is clearly different.

Before I met my first chimpanzee, I vaguely imagined their intelligence as perhaps halfway between that of, say, a dog and a human – clearly inferior to that of a human in every way, but clearly superior to nonhuman mammals. That was simple-minded. For some cognitive tasks chimpanzee match or *even exceed* the abilities of humans. You may wonder if I slipped that "exceed" in to shock you, an exaggeration to be walked back later with a redefinition of "exceed" or some other mealy mouthed double-talk. No, I mean it literally: In some ways chimpanzees are more intelligent than humans. That makes it all the more surprising to discover that in other areas they are so clearly inferior to us. Watching them struggle and fail to solve a puzzle a three-year-old could solve provokes a feeling of pity. This unevenness means that it requires some effort to understand the chimpanzee mind. In the rest of this chapter I will break it down, skill by skill, to show where chimpanzees are surprisingly clever and where their abilities are slight.

18.1 Expressions of Emotions

Human facial expressions are so consistent across cultures that they are considered an instinctual manifestation of internal emotional states (Ekman, 1999). The form and variety of chimpanzee expressions – and therefore the emotional states that underlie those expressions – are quite similar to those of humans. The intrepid human who takes on the challenge of raising a chimpanzee as a human, often with their own child (Kellogg & Kellogg, 1933; Ladygina-Kohts, 1935; Hayes, 1951; Gardner & Gardner, 1969), marvels at the near-identity of human and chimpanzee expressions. There are differences; chimpanzees smile without baring their teeth (Figure 18.1, bottom right), but most of their facial expressions are nearly indistinguishable from those of humans. Whether in controlled scientific experiments (Parr et al., 2007, 2008), in reports of scholars who worked with chimpanzees on a daily basis (Kellogg & Kellogg, 1933; Ladygina-Kohts, 1935; Hayes, 1951; Gardner & Gardner, 1969), or in my own personal experience, chimpanzees respond appropriately, as a human might, when a human winces, shows surprise, or registers fear. Looking at it from the other direction, chimpanzee facial expressions (Figure 18.2) are not only quite complicated – nearly as complicated as

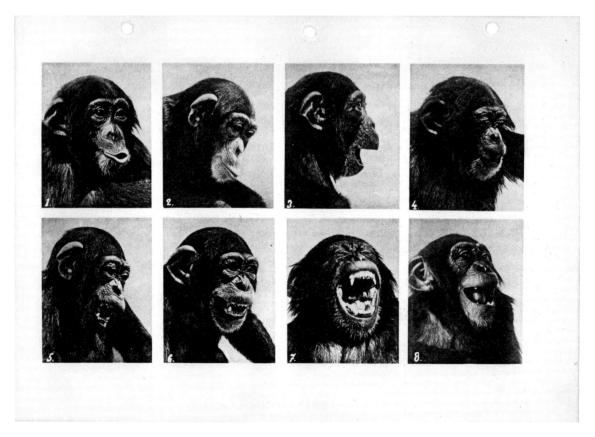

Figure 18.1 Facial expressions: (1) astonishment or surprise; (2) attention; (3) astonishment or surprise; (4) disgust; (5) anger; (6) fear or horror; (7) sadness, crying; (8) joy or laughter. By permission of Oxford University Press, USA.

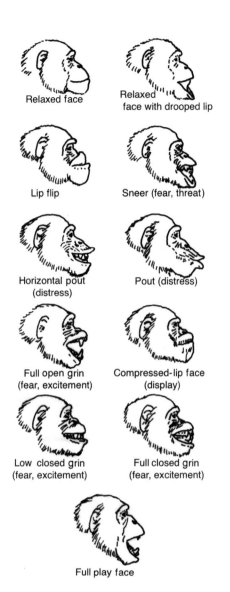

Figure 18.2 More facial expressions. After Goodall, 1986, with permission.

those of humans – they are powered by a similarly complex brain organization (Morecraft et al., 2004; Chapter 19).

18.2 Object–Object Relationships

Let us start with cognitive abilities that are almost comically limited, in particular chimpanzees' ability to sort out challenges that involve **object–object relationships**. This aptitude might be called "engineering skills." It is the ability to imagine how two objects relate to one another – that the square peg fits only in the square hole, or which sized nut matches which bolt. Handed a padlock and key, humans have no difficulty relating the two to one another, and some will even find it difficult to restrain themselves from inserting the key in the lock. Chimpanzees take longer to figure out padlocks, and some never quite get comfortable with them. We use such skills to assemble a child's playhouse or to construct a model of the Eiffel Tower. In contrast to humans, putting a lid on a jar can be challenging for a chimpanzee.

Even when they complete an object–object task, they bring their own esthetic. When humans stack boxes they naturally align the edges to construct a neat tower; chimpanzees stack the boxes willy-nilly, sometimes so askew the tower barely stands (Figure 18.3).

The ability to interpret pointing is often considered an object–object relationship skill. We humans are persistent, almost compulsive pointers; so natural and easy does this task seem to us that its complexity is easy to miss. The experimental subject must imagine a line that extends along the arm, hand, and finger, and continues until some object to which attention is being drawn is encountered. The task is made more difficult because the line is an oblique one from the point of view of the observer, which means that to interpret it correctly the subject must in some way visualize the pointer's perspective.

Chimpanzees do not point in the wild, though in captivity they can learn the skill (Leavens et al., 2005), even if they struggle with it (Mulcahy & Hedge, 2012). This is a shocking deficiency, since some species not renowned for great intelligence – goats (Maros et al., 2008) and horses (Kaminski et al., 2005) – correctly interpret pointing. True, most other species that excel at pointing tasks are on everyone's list of intelligent animals (e.g., parrots, seals, and dolphins [Hare et al., 2002; Bräuer et al., 2005; Mulcahy & Hedge, 2012]), but that makes the deficiency among chimpanzees all the more puzzling.

I received an early object–object lesson on chimpanzee object–object abilities in 1985 when

Figure 18.3 Köhler's chimpanzees stack boxes to reach a banana bait. Note the unevenness of the boxes and the observers "helping"; like humans, chimpanzees have mirror neurons that give them a sophisticated ability to place themselves in the position of another individual. From Köhler, 1925, with permission.

I studied chimpanzees at the Knoxville Zoo. Their outdoor enclosure was a reasonably comfortable one, chain link fence on the sides and the top as well. The ground was grass-covered, with a concrete pad near the sleeping structure at the back. It was a relatively large space for just three adults – a male and two females – and a three-year old infant. In the middle was a heavy-duty play structure. For a week I studied the small group during all their waking hours, testing out the data-collection protocol I had designed for my dissertation.

Early afternoon was snack time. Each afternoon, zoo staff rolled up in a miniature 4WD truck and dumped a bucket of cut-up fruit or vegetables into the enclosure through a mailbox-like slot, sometimes a mélange of carrots and onions, other times oranges and other fruits. This was extremely exciting for the chimpanzees and they often became animated and began to pant-hoot when the sound of the small engine was still a whisper in the distance.

One day, the snack deliverer was in a hurry and made something of a mess of the job. Several orange quarters spilled over the side of the food slot and ended up on the ground outside the cage. There was the usual chaos of consumption inside the enclosure, after which things settled down and one of the females began to pay attention to the orange slices that had fallen outside. The male shouldered her aside and sat facing them, analyzing the situation. The cage "walls" were like any other chain link fence, with about 5 cm gaps between the wires. The male attempted to push his hand through one of the diamond-shaped openings to reach outside, but his hand was clearly far too large to pass. To my surprise, he then tried it a half-dozen other times at adjacent diamonds. My strong impression was that he could not see that if his hand failed to pass through one diamond, it could never pass through any of the others. He was bored, of course, so the multiple attempts may have been just idle doodling, to pass the time, but that was not my impression. The exciting finish to the story is that a mother and infant stood nearby, patiently waiting until he gave up and wandered away; the baby quickly thrust her arm through the holes and retrieved the oranges one by one, most of which her mom took from her.

Another incident was even funnier. A sidewalk ran along the length of the side of the enclosure, flanked by grass strips on either side. The sidewalk was off-limits to guests but used by staff. One day, in the late morning, several workers arrived in a small truck, bringing a half-dozen trees they intended to plant in one of the grass strips. Three workers unloaded their picks, shovels, and finally the trees, with their large root balls wrapped in burlap. They laid the trees down in a line down the walk, tree top to root ball, overlapping like toppled dominoes. The groundskeepers had hardly begun digging the holes where the trees would be planted when they left off for lunch and drove away. In a manner I have found typical of chimpanzees, they completely ignored the landscaping activity while the workers were there. Left alone, however, they wandered over to assess the situation. One of the pick handles had landed only a few inches from the fence and, as with the orange slices, the male tried unsuccessfully to push his hand through to reach it. He sat assaying the various tools scattered outside.

Meanwhile the infant had managed to reach a whole arm through the fence to grasp a branch on one of the trees. She pulled it toward her and began plucking off and eating leaves with one hand while holding the branch with the other. The females joined her, pulling on still more branches until the root ball rolled down the slight incline beside the walk to rest against the fence. The male joined them. Soon they had pulled all the trees to the fence, and in half-hour they had consumed every leaf within their reach and gnawed off some of the twigs, to boot.

As the trees were pulled closer, at one point they in turn jostled a pick handle which fell closer to the fence, within reach of the male's long fingers. He pulled the handle toward him, then into the enclosure, clearly intending to pull the pick into his cage. The pick head, however, was far too large to pass through the fence. He pushed the handle halfway out and tried again. He made numerous attempts, varying the speed and force of the pull. He seemed unable to grasp the concept that a half-meter pick head could never pass through a 5 cm fence opening (demonstrating his poor object–object skills and giving me an excuse to tell this story). He switched strategies; he pulled the handle in as far as it would go and then lifted it as high as he could, driving the pick head into the ground. This looked promising. He worked the handle up and down and then in circles, excavating an ever-more impressive crater outside the cage. Soon he found a point where the pick head was lodged solidly against the ground, giving him enough leverage that the chain link wire was strained and bent. The 5 cm fence-opening grew to 10 cm, and I began to fear the infant might be able to escape through it. He may have begun to hope he might eventually get the whole pick in, which was beginning to look possible. Occasionally he would go back to pushing and pulling, now with enough vigor that dirt and gravel flew back onto the sidewalk and toward him, into the cage.

Having finished the leaves and twigs, the females turned to eating small branches and then stripping bark, eventually leaving the trees as mere fence posts protruding from root balls. Their constant agitation eventually opened some of the root balls, exposing the roots and spilling soil onto the sidewalk and grass. As a scientist, I am trained to observe, not to interfere, but as the destruction mounted I began to think I should notify some staff member so as to halt the debacle.

At this point the workers returned, gaping comically at the devastation before them: an 18-inch deep crater, dirt sprayed everywhere, the trees now mere ragged stumps, roots partly exposed – and a gaping hole in the fence with the male still working hard at enlarging it. Within minutes a huge staff contingent stood around the cage, the chimpanzees had been herded into the indoor enclosure, and my observations for the day were over.

18.3 The Inaccessible Peanut

One task at which chimpanzees excel, something that I am still coming to terms with, given their poor performance on other object–object relationship challenges, is the "inaccessible peanut task." Fewer than 10 percent of human four-year-olds can solve this problem, while chimpanzee Liza solved it in seconds. The experimental subject is confronted with a peanut at the bottom of a transparent cylinder. How to retrieve it when the tube is far too deep to reach with the fingers? Liza gave the frustrating apparatus a kick or two, but then hurried over to a water faucet, filled her mouth, squirted the water into the tube and retrieved the floating peanut. She had never been confronted with this task before (Hanus et al., 2011; de Waal, 2016).

18.4 Tool Use

Tool use requires object–object relationship skills in that the tool user must orient the tool properly to effectively modify or utilize the target object, such as when a termiting tool is properly inserted into a termite mound. This is a simple tool, simply applied, and chimpanzees are quite competent at these sorts of tasks, though sometimes a little clumsy compared to humans.

More sophisticated are *composite* tools, those made up of more than one object, such as a bow and arrow. Chimpanzees do use such tools (Sugiyama, 1997; McGrew, 2010; Koops et al., 2015), such as when they hammer open nuts placed on an anvil (Sugiyama, 1997; McGrew, 2010; Koops et al., 2015). Not surprisingly, chimpanzees are much less likely than humans to make and use these more advanced tools.

The manufacture of *compound* tools – tools in which two or more components are combined to form a single tool – is more sophisticated still. Among humans, a compound tool might be a stone point hafted to a wooden shaft with leather string. McGrew (2010) considers a leaf sponge to be such a tool, since more than one leaf is typically used; he also considers leveling an anvil by propping it up with a stone to be a compound tool. These are, of course, considerably less complicated than a wooden-handled hatchet. To offer some context, while these manipulations might seem simple, we would be impressed to see a three- or four-year-old human accomplish them, and we cannot imagine a dog or cat doing them.

18.5 Wolfgang Köhler's Experiments: More Object–Object Lessons

Most of the examples scholars use to illustrate poor engineering skills among chimpanzees come from Wolfgang Köhler's delightful and accessible book *The Mentality of Apes* (1925/1959). He devised myriad baffling, almost perverse puzzles to challenge his chimpanzee subjects. He first placed a variety of objects inside the chimpanzee cage, allowing them to explore their possibilities. The stereotype of chimpanzees as inquisitive and playful – the Curious George syndrome – is not merely myth, both exploration and play are important to the chimpanzee intellectual life. In a very real sense, play is exploration – it helps chimpanzees to learn more about their environment, and chimpanzees are playful, too. Trial-and-error attempts to solve problems have something in common with play, and both can lead to solutions.

Köhler suspended bait from the cage ceiling (a banana – maybe this is where the meme that wild chimpanzees eat bananas comes from – there are none in the wild, though they do love them in captivity), and at the same time he left wooden boxes arrayed around the cage. In this test chimpanzees quickly realized that stacking the boxes would allow them to reach the bait (Figure 18.3).

Other tasks required them to use a stick to rake in a treat. First, they were given sticks to play with, then a banana was placed outside the bars of the cage, too far to reach; the subjects, some sooner than others, realized the stick could be used to rake in the treat (Figure 18.4).

In another challenge, Köhler provided an extremely long stick and suspended the bait near the ceiling. Given the strength of chimpanzees, it might be expected that they would use the long stick to bat the banana down. Instead, a female chimpanzee took advantage of her circus-acrobat balance to run up the pole to retrieve her treat before she began to topple over (Figure 18.5).

Köhler set up another challenge similar to that in Figure 18.4, a fruit bait placed just out of reach. In this experiment (Figure 18.6) he left two sticks in the chimpanzee enclosure, one hollow, the other the size of the opening in the first – but neither long enough to retrieve the banana. After a number of attempts to retrieve the bait using one stick or the other, a young male subject at last gave up. Later, while idly playing with the two sticks, he suddenly realized they could fit together (Figure 18.6). He instantly recognized this new, longer implement would allow him to reach the previously inaccessible treat, and he rushed over to retrieve it. Köhler felt that problem-solving incidents like this one demonstrated that chimpanzees were capable of solving problems not just through accidental play or with trial-and-error exploration, but with insight.

Einstein called thought processes that drew on the imagination to solve problems "thought experiments" (**Gedanken** in the original German). In various

Figure 18.4 One of Köhler's chimpanzees using a stick to retrieve an out-of-reach banana. From Köhler, 1925, with permission.

Figure 18.5 Given a very long stick, how to retrieve a suspended banana? The extraordinary balance, great power in vertical climbing, and exceptional balance makes the option of placing the pole upright, running up it and retrieving the bait before toppling make more sense. From Köhler, 1925, with permission.

Gedanken he imagined what the consequences might be as different scenarios played out. For instance, consider a PVC pipe the diameter of a broomstick, lying across a 10-foot deep trough, with its ends resting on either edge. Imagine what would happen if you attempted to swing underneath the pipe to cross the gap hand-over-hand, brachiation style. You have already imagined it bending until it collapsed, and you falling into the trench. Chimpanzees can engage in such thought experiments, sometimes called "secondary representations" (Suddendorf & Whiten, 2001; the "secondary" part of the phrase being the bent pole failing to bear the weight of the too-heavy human).

An example of a secondary representation solution to a problem is chimpanzees solving Robert Yerkes' box-and-pole experiment. In the box-and-pole test the chimpanzee subject was provided with a long pole and introduced to a tunnel-like box with a banana placed in the middle. The tunnel was too narrow for raking (pulling toward you) to work (Figure 18.7). This is not a problem chimpanzees encounter in the wild; sometimes they use one branch to pull another toward them, reeling in a flexible twig hand over hand, but a chimpanzee is never required to push something away from them in order to retrieve it. Yet Yerkes' test subject accomplished the task, and it was done with insight (Yerkes, 1943) – though not quickly. It took a six-year-old chimpanzee 12 days to solve this puzzle, but she did it with Gedanken, with insight. Playing near the tunnel, the solution suddenly occurred to her,

Figure 18.6 Chimpanzees solve some problems insightfully; they solve it cognitively, rather than through trial and error. This chimpanzee suddenly conceived the idea of inserting a smaller stick into a larger, hollow stick to extend his reach and retrieve a reward. From Köhler, 1925/1959, with permission.

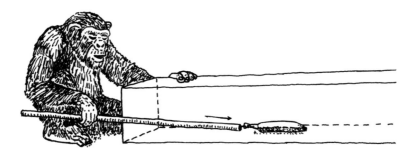

Figure 18.7 A chimpanzee uses a stick to push a banana away from her in order to retrieve it, a difficult and counterintuitive act for a chimpanzee, pushing something away that she actually wishes to pull toward her. After Yerkes, 1943, with permission.

after which she went "directly to the pole and grasped it with every evidence of definiteness of purpose . . . [even] before she approached the open end of the box it was clear . . . the problem had been solved."

Chimpanzees are clearly less capable than humans when it comes to deconstructing objects, imagining taking them apart; this is part of their object–object deficit. They seem to view compound objects as irreducible wholes. This surprises me in some way, because taking things apart seems to be a favorite pastime of chimpanzees. In the wild they often have to break apart food items to get at edible parts inside: nuts to get the nut-meat inside, palm frond cortex to get at the pith inside, and so on. Give chimpanzees a box and they will smash it until it breaks. One of my former students tells the story of a benefactor who donated to a chimpanzee sanctuary what he thought was a substantial play structure, a jungle gym constructed out of two-inch PVC pipes. It was reduced to a pile of pipes in about five minutes. Yet when confronted with a task that requires a stick, they struggle to think to themselves "as I know, that box over there can be reduced to sticks in moments, and I need a stick." They need a hint; if presented with a box with clearly visible gaps between its boards, chimpanzees are more likely to consider breaking the box (Köhler, 1925).

Not only composite objects, but sometimes even *separate* objects that are in close proximity appear as single objects to chimpanzees. When Köhler released chimpanzees into a room with the by-now-familiar banana hanging near the ceiling, chimpanzees were unable to see that a table pushed up against a nearby wall was separate from the wall, a moveable tool that could help them.

Another Köhler torture-test demonstrates this irreducible-whole problem well: the box of rocks challenge. A test subject was confronted with a task that required moving a box to gain access to the bait. A box was provided, but it was filled with 100 kg of large stones. Nearly every human would immediately conceive of removing the rocks to free up the box. The subject looked over the task, pondered the options . . . and removed a single rock. She then attempted to push to box: still too heavy. She reconsidered . . . and removed one more stone. At last, removing stones one at a time, she reached the point where she could laboriously push the box across the enclosure.

Faced with a problem requiring a rope, Köhler's chimpanzees failed to notice that a support pole in their enclosure had a rope coiled around it. Either they could not trace the rope to its end to see how to remove it, or they failed to recognize it was a rope. The reverse is true, too; when required to put in place several parts to complete an object, chimpanzees struggle. The difficulty solving putting-together-tasks is not a surprise to me, since in their natural world there is little or no call to assemble complicated, composite items. That they struggle with a taking-apart task *is* a surprise

You may wonder, given Ladygina-Kohts' demonstration that chimpanzees see things as we do, why they fail to recognize that a rope is not part of a pole or a table is not part of a wall. I consider this part of the poor engineering skills of chimpanzees. A person unfamiliar with the workings of a car engine sees only a jumble of odd-shaped parts, unlike an experienced mechanic who immediately sees individual parts such as the air filter, cylinder head, manifold, battery, and so on, and also sees that the generator pulley is attached by a belt to a fan and that both must turn when one does. A pole with a rope is complicated to an object–object relationship-challenged chimpanzee.

When unable to find the right tool for a job familiar to them, chimpanzees sometimes attempt the task with ridiculously inappropriate tools. An image you may be familiar with from the famous Jane Goodall *National Geographic* films is a frustrated chimpanzee attempting to open a steel banana-box with a floppy, chopstick-sized twig. Above I told you of a female who ran up the free-standing pole to retrieve a banana (Figure 18.5). When Köhler removed the pole but left various short sticks lying about, she held two separate sticks up, one on top of the other, and made movements as if she were readying herself to run up the unconnected, too-short twigs. In another banana-outside-the-cage set-up, a chimpanzee had no sticks to rake in the bait, but boxes were in the enclosure. The solution was to break the boxes apart, but instead she moved the boxes around the cage confusedly.

In short, when two or more objects must be cognitively related to one another, chimpanzees struggle. They can puzzle through the task if persistent, but often they rely on trial and error more than humans do; they lack the attention span to accomplish the task in their heads first. They also seem to take objects at face value more than humans do, treating composite objects as uniform or monolithic.

Still, even though their object–object skills are poor compared to those of humans, we must keep in mind that compared to other nonhuman primates they are exceptional. When Visalberghi and colleagues (1995) presented chimpanzees, bonobos, orangutans, and capuchins with a food treat inside a clear tube, all quickly learned to use a stick to pull out the food. However, when sticks were bundled together in a bunch too large to fit in the tube, capuchins, generally considered to be the smartest non-ape primate, attempted to jam them in anyway. Each of the apes pulled a single stick out of the bundle and succeeded immediately. Similarly, when test subjects were provided with a stick too crooked to fit into the tube, apes modified the stick and retrieved the treat, but capuchins failed to do so.

18.6 Numeracy

We refer to the ability to use numbers as **numeracy**. With training, chimpanzees acquire basic arithmetic skills, the ability to add and subtract numbers (Boysen

& Berntson, 1989), but their numeracy is quite poor compared to humans – comparable, perhaps, to the level of a kindergartner, though there was a time when scientists would have scoffed at the notion that they might have even those limited skills. That chimpanzees have some sense of quantity was apparent early on, in Köhler's and Ladygina-Kohts' studies. Their abilities to understand quantity comes as no surprise to those of us who study chimpanzees in the wild since they must often judge the relative size of two huge numbers. During my postdoc one of my tasks was to estimate the number of fruits in trees. I once stood under two neighboring trees, both with ripe fruits, both about the same size, laboriously working through our complicated protocol for counting fruits. It took me about 20 minutes to estimate the number of fruits in the two trees. Before implementing the protocol – just by looking, I mean – I could not have said which had more fruits, but after the protocol I knew one had perhaps 20 percent more. As I was counting a chimpanzee appeared and looked into both tree crowns for perhaps 15 seconds each – and climbed the one with 20 percent more fruits.

What was a shock, however, was how readily they learned to recognize numerals (Matsuzawa, 1985; Boysen & Berntson, 1995). And they can manipulate these numerals, too, recognizing that some are greater than others (Boysen & Berntson, 1995) and that numbers are a graded, ordered phenomenon. In other words, they understand the principle of **ordinality** – not just that one value is greater than another, but that there is a sequence to the numbers, that 5 always comes after 3 and that 4 always comes after 3 and before 5 (Matsuzawa, 1985; Beran et al., 1998).

While quantities are within their grasp, it is unclear whether chimpanzees understand the principle of counting; in any case, they struggle with counting tasks (Boysen, 1992). While humans count naturally and even compulsively; some of us count steps, bell tolls, phone rings, and so on without even trying, and in some cases without even wanting to. For instance, it is not at all clear that they understand, as even young humans do, the unboundedness of counting. We humans can easily understand that you can always add one to any number, and even though few of us can precisely visualize 101 dots, we easily understand the concept of 101 dots (Boysen, 1992). If forced to count by being shown an Arabic numeral and then having to pull from a pile one item at a time, making a counted pile and a to-be-counted pile, chimpanzees have trouble maintaining focus; they top out at five (Beran et al., 1998) or perhaps eight (Boysen & Berntson, 1989, 1995). Humans, by contrast, are argued by some researchers to understand counting principles even before they can speak, and therefore before they have even learned words for numbers (Gallistel & Gelman, 1992).

In addition to understanding quantities, chimpanzees are capable of **transitive inference**; that is, they understand the concept that if A is greater than B and B is greater than C, then A is greater than C (Gillan, 1981).

They even understand simple fractions. They can match ¼, ⅓, and so on to pictures of partly filled glasses or to objects cut into quarters or thirds (Woodruff & Premack, 1981).

18.7 Volume Perception

Chimpanzees can discriminate among variously sized objects about as well as humans (Ladygina-Kohts, 1935; Maier & Schneirla, 1935; Prestrude, 1970; Young and Farrer, 1971). In Chapter 13 we reviewed evidence presented by Ladygina-Kohts (Ladygina-Kohts, 1935; Ladygina-Kohts et al., 2002) that suggests that chimpanzees have a human-like capacity to distinguish among similar but slightly different shapes, such as a cube versus a bar. They can correctly judge which of two objects is larger – two food items, for instance (Menzel, 1960), a skill they use regularly in the wild. Slightly more cognitively challenging is the task of determining whether a tall but narrow drinking glass holds more juice than a shorter, stouter one. Menzel showed that chimpanzees could do this as well as humans, too – they could be fooled, but only in situations that also deceived humans (Menzel, 1960).

18.8 Mental Mapping

To the surprise of some scholars, we now know that chimpanzees attain a human-like appreciation for the existence of a reality outside of what is in their field of vision (Mathieu et al., 1976; Wood et al., 1980; Krachun et al., 2009; Karg et al., 2014). This discovery, or actually *re*discovery, may have been a spur in the re-appreciation of the work of Ladygina-Kohts and Köhler, research which often involved challenged chimpanzees to recall objects or situations outside their immediate visual field. This is significant, because very young children lack the cognitive ability to recognize that objects have a reality that exists beyond their visible surroundings. For instance, if shown a green and blue block and asked to pick the blue one, they can solve the task. But if they watch while a color filter is interposed to make both look green, very young children are flummoxed, even though they saw the lens put in place; chimpanzees, however, are not fooled when a lens makes a smaller grape look larger (Karg et al., 2014).

Having the capacity to retain a map of their world, even the parts that are out of sight, enables chimpanzees to accomplish surprising feats of mental mapping, including remembering the location of resources or natural objects far out of sight. They gather termiting tools to carry to mounds that are as far away as 100 m, invisible to them as they make the tools (Goodall, 1964, 1986; McGrew, 1992). They carry hammerstones even farther, up to 500 m (Boesch & Boesch, 1984); 40 percent of hammerstone carries are to resources that are out of sight. And it goes beyond that, far beyond. When a chimpanzee at Taï encounters a hammerstone and picks it up, she typically moves to the nearest source of nuts (Boesch & Boesch, 1984).

Specific types of nut-cracking tools are required for specific nuts. Some nuts are harder to crack and require stone hammers; others can be cracked with wood hammers. Chimpanzees bring wood tools to nuts that can be cracked with wood, and stones to those that require stone. They carry the best hammerstones the longest distances, not only showing they know where the nuts are, but that they appreciate that the better tool is worth the extra effort to carry it farther.

In short, chimpanzees know where things are and they take the most efficient routes when moving among them. Perhaps even more astonishing, when considering how far to walk to acquire nuts, they even factor in the caloric value of the food item (Sayers & Menzel, 2012). Their tool/nut matching and direct travel between resources are evidence they possess "distance-minimizing" abilities that require the possession of sophisticated mental mapping skills, indistinguishable, in fact, from those of humans.

18.9 Spatial Memory

In one experiment, a juvenile chimpanzee was carried around an enclosure while 18 pieces of food were hidden; she was then removed from the area, out of sight of the hiding places, for a cooling-off period. She rarely failed to find all 18 pieces of fruit and even took the shortest distances between food items, further evidence of a sophisticated mental mapping and distance-minimizing abilities (Menzel, 1973).

In a different experiment, a chimpanzee was allowed to see food buried, but the sand on top was smoothed over to make it difficult to use visual cues to find it – memory alone was needed. Forty-eight hours later, when allowed into the enclosure with the hidden treat, she went directly to the food and retrieved it (Menzel, 1973).

Chimpanzees may flounder when it comes to object–object relationship challenges, but they are able to understand scale models. Perhaps their excellent mental mapping ability compensates for their object–object deficiencies. Sally Boysen and Valerie Kuhlmeier constructed a scale model of their chimpanzee habitat and had their subjects watch as they hid a tiny model of a food item in one of several possible hiding places. While the chimpanzees were far from map experts, they recovered the food at a better-than-chance rate (Kuhlmeier et al., 1999; Kuhlmeier & Boysen, 2001, 2002). Females excelled, in particular, compared to males. This is, in my opinion, because in the wild females have evolved a more fine-grained foraging strategy; that is, males focus on the rarer large feeding trees, whereas females feed from more numerous but smaller feeding sites. They must therefore have a more fine-grained mental map.

It is in tests like those above, ones that require locating objects in space – known as *spatial memory* tasks – that chimpanzees often outperform humans. As early as 1932 Tinklepaugh compared spatial memory among two chimpanzee "children," eight and six years old, to five human adults and four children of unspecified age. In this task, 16 pairs of identical cylinders were arrayed around a test subject who sat on a stool in the center (Figure 18.8). An experimenter worked his way around the circle placing food under one or the other of each pair. When all 16 containers were baited, the chimpanzee was encouraged to choose one of each of the pair as the container that hid the treat. If they chose correctly, they received the treat. Chimpanzees performed at 78 percent accuracy; humans lagged behind at 65 percent (Tinklepaugh, 1932). Here, at last, we see what I meant when I wrote that chimpanzees are superior to humans at some cognitive tasks. If you find Tinklepaugh's experiment unconvincing, stay tuned.

Perhaps the most extraordinary chimpanzee ability – and frankly the most difficult one for many people to believe – is Tetsuro Matsuzawa and colleagues' demonstration that chimpanzees can accurately remember the location of numbers displayed on a video screen after they are masked. Then, using their greater-than/less-than skills, they can rank them from one to nine. Two chimpanzees, Ai and Ayumu, have demonstrated competence at this task when shown up to nine numerals randomly placed on a touch-screen (Figure 18.9). The test subject memorizes the location of the numbers and then touches the first number. Sometimes the numbers are masked after a specific amount of time (Kawai & Matsuzawa, 2000; Inoue & Matsuzawa, 2007; Matsuzawa, 2009). Not only do they memorize the location of the numbers quickly, they can accomplish the task whether there are some numerals missing or not. Chimpanzees require the briefest time imaginable to memorize the numbers. Very brief. So brief, in fact, you will hardly believe it. I may have shown Matsuzawa's video of this task 25 times to

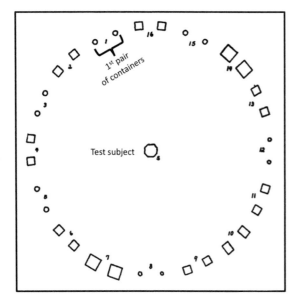

Figure 18.8 Representation of Tinklepaugh's memory experiment. The test subject sits on a stool (S) at the center of 16 pairs of containers while an experimenter places a treat under one or the other of each pair. The test subject chooses one of each pair and receives the treat if the choice is correct. From Tinklepaugh, 1932.

Figure 18.9 Ayumu confronted with nine numbers that are quickly masked by white squares. To complete this task he must touch each square in the numerical order of the numbers masked underneath them. Photos courtesy of T. Matsuzawa; with permission.

audiences, asking those who had not seen the video to guess how long a chimpanzee might need to stare at the numbers to get the entire sequence right. I have never had anyone guess anywhere close to the astonishing brevity required, and you probably will not either. You have probably gone online to find it, so I will wait until the end of the chapter to give the actual time.

Consider the cognitive skills required for this achievement. The chimpanzee must understand that "1 stands for a single thing and that "2" is two things, and so on. They must have it clear in their head that 5 always comes after 1, and that it is greater than 1, 2, 3, and 4, and less than 6. Then they must take in where these numbers are on the screen and retain it in their memory for several seconds, or even longer – in one video the experimental subject was distracted by a noise mid-task and turned to investigate. She stared into the distance for several seconds and returned to complete the trial – successfully.

Pondering this astounding ability, Matsuzawa commented that "we've concluded through the cognitive tests that chimps have extraordinary memories. They can grasp things at a glance. As a human, you can do things to improve your memory, but you will never be a match for [chimpanzee] Ayumu" (McCurry, 2013). In short, chimpanzees, Matsuzawa claims, have better memories than humans (McCurry, 2013). Even if this task can be duplicated by some humans, as some claim (Cook & Wilson, 2010) – and not everyone accepts that claim – it is still an astounding accomplishment, and beyond that of most humans.

The ability of chimpanzees to accomplish the spatial memory task is less surprising when we consider the chimpanzee foraging adaptation. Females, in particular, survive on food sources that are small, numerous, and dispersed. The better they remember the location of each food source, the more efficiently they can forage. I believe that chimpanzees know the location of every medium to large feeding site in their habitat, with a mental map that is at least as rich in information as humans are capable of and probably richer (Normand et al., 2009). Even further – and note that this is my speculation – they have a sense of the history of the food supply, which fruit species fruited most recently, and which is overdue for fruiting. We know they can recognize up to 300 different food items because at Mahale they consistently eat these items, while ignoring thousands of others (Nishida & Uehara, 1981, 1983), and the food list is nearly as large elsewhere (Chapter 5). They remember foods that appear with irregular time gaps in between and they remember foods that are so rare they only eat a few of the items every year.

18.10 Spatial Memory Preference

In fact, chimpanzees' spatial memory is so good they rely on it even when their eyes tell them they should not. In one experiment, a chimpanzee watched as food was placed in one of four distinctive containers. One container was then placed in each of the four corners of a room. The chimpanzee was led out of the room and the containers were surreptitiously switched. When the test subject was returned to the room and allowed to search for the treat, she insisted on checking the container in the *corner* where she remembered it being placed, even if a different-colored container was in that corner (Yerkes & Yerkes, 1929; Yerkes, 1943). When the experiment was repeated with containers different in both color *and* size, the chimpanzees still relied on location rather than appearance. Location is extremely salient to them.

18.11 Motivation: Social and Food Memory Skills

These experiments and observations suggest that – for things they are interested in – chimpanzees have memories that seem every bit as good as ours. And in what are chimpanzees interested? Social partners and food. Goodall (1986) relates a story about memory, passed on to her from Roger and Debbi Fouts, that will impress anyone who has gone to their 10-year high school reunion. When, after a long time gap, the famed ape-language researchers Allen and Beatrix Gardner were reunited with Washoe, their famous study subject, Washoe spontaneously signed Beatrix's name sign. The time gap was 11 years. She not only recognized Beatrix after 11 years, she remembered her name sign, a sign she had probably not used in all that time. The work of Tinklepaugh, Goodall, Boesch, and many others suggests food resources are similarly memorable.

18.12 Working Memory

The astounding feats of memory by Ai and Ayumu make research demonstrating that chimpanzees have a poor **working memory** all the more surprising. Working memory is short-term memory brought into play when a task requires several steps. Other tasks requiring short term memory involve several interrelated concepts that must be weighed against one another at the same time. The importance of this ability for retaining some certain number of concepts in short-term memory (or working memory) was first articulated by Miller (1956), who argued that humans can only remember seven bits of information. Our memory seems much larger only because we practice "chunking." If I were to ask you to tell me the phone numbers of your seven best friends, in alphabetical order, you could probably do it, and that involves much more than seven numbers, but we can do it because each phone number is a chunk. You may take issue with the number seven – the actual number is not sacrosanct; Cowan (2010) argued that the human magic number was actually only 3–5, or four as a convenient species average. Whatever it is for humans, studies suggest chimpanzees can deal with only half as many items. Perhaps the difference is "location." If the task requires things separated in time (or sequence) rather than in space, chimpanzees struggle.

Some tool using tasks require a number of discrete steps, and the status of each of the many steps must be held in the memory at once. You have to think "to get that symmetrical flake I must take a disk-shaped stone, and to get the disk shape I will have to take off flakes around the edge, but before I can take off a flake around the edge I have to establish a platform with an acute angle on this side." This stone tool making task is much like thinking ahead a number of moves in chess. Read and others (Read, 2008) argue that this requires holding seven concepts in the mind at once, requiring (in his view) three times the working memory chimpanzees can muster. To crack nuts, chimpanzees must place the nut on the anvil and direct the hammerstone at the nut. Three objects are involved, the nut, the anvil, and the hammerstone. Attention must be focused not only on the nut but the movement of the hammer. This two-concept job is not easy for chimpanzees; no individual has mastered it before the age of three, and some chimpanzees never acquire the skill (Read, 2008).

Chimpanzees are better at tasks such as those that involve multiple steps, tasks like using a stout stick to open a termite mound or a grass stem to extract termites, rather than tasks with multiple components. They can use tools to modify other tools, such as when they use a harvesting probe to retrieve a sponge lost in a drinking hole (Goodall, 1964; Sugiyama, 1997; McGrew, 2004). See Chapter 21 for more detail.

We see the limits of chimpanzee short-term memory when several different actions must be engaged in at once. Anne Russon tells the story of a group of rehabilitant orangutans, exposed to human activities for months and years, attempting to imitate human tool use. In the wet and humid environment of Borneo, starting a fire is a multistep process. Firewood must be properly arranged – the kindling must be in the correct position, not too close together nor too far apart. Damp wood requires the application of a little bit of kerosene, after which a flame is applied. Operating a lighter requires a couple of steps – spinning the striking wheel and holding down the combustible gas button – and while the lighter is held to the kerosene-soaked timber, the wet wood requires that the whole be fanned with a plate or piece of cardboard in the other hand. Orangutans struggle with this long (for them) series of both sequential and synchronous actions. On the bright side, they have yet to burn the research station down. This struggle is typical of chimpanzees as well.

So far, chimpanzees and other apes have failed at tasks that require anything but the simplest *recursion*, a class of tasks that requires an especially intense use of short-term memory. A task is recursive when a procedure is applied multiple times; there is a first step, after which the procedure is reapplied to the outcome of the first step, then reapplied to the outcome of the second step, and so on, but with a bit of a twist: A third-order or tertiary recursion, for example, requires holding a first *and* second step in the mind while moving on to the third. Even humans are poor at multiple recursions, though not too bad up to about third-order recursions: "I know that you know that I know that the box is empty." Most of us can deal with fourth-order recursions. If phrased just right, some of us can handle fifth-order recursions, but beyond that things become murky (Kinderman et al., 1998). Chimpanzees are said to lack the short-term, working memory required for second- or third-order recursions (Premack, 2007; Read, 2008).

18.13 Cross-Modal Transfer

Another cognitive endowment once thought to be unique to humans is the capacity for **cross-modal transfer** (Geschwind, 1965), the ability to draw on information from more than one sense and put the two bits of information together. When you reach into a bag and feel a golf ball and immediately bring to mind the image of a golf ball, you have engaged in cross-modal transfer. While not widely read at the time, Ladygina-Kohts demonstrated in the 1930s that chimpanzees could match an object handled inside a bag with an image of the same object (Ladygina-Kohts, 1935). The issue continues to be of interest to psychologists (Leavens et al., 2010), but it is clear that, just as do humans, chimpanzees can cross-collate information from diverse senses.

18.14 Time Travel

It has been suggested that humans alone are capable of pondering on and planning for the future (Suddendorf & Corballis, 1997). While there seems little doubt that humans are more competent at this than apes, my clique of chimpologists believes the evidence that chimpanzees do this – at least in the short term – is utterly undeniable. Chimpanzees look to the future when they carry stones to distant nut-cracking sites; they carry termiting tools to termite mounds minutes away; and they emit food calls when moving toward a particularly prized food, even before it comes into view.

The behavior of a zoo chimpanzee named Santino is compelling evidence that chimpanzees plan for the future over much longer time spans (Osvath, 2009). An annoying habit among zoo chimpanzees is the tendency to throw things at spectators, and Santino engaged in the behavior, but went beyond the occasional half-hearted bit of flung dung – he took up stone throwing (Figure 18.10). Actually, more than just throwing. He systematically scoured his cage before the zoo opened and cached stones to be hurled at zoo-goers later. Osvath's team observed Santino caching such stones on 50 distinct occasions, under hay heaps, behind logs, and behind a rock structure on the exhibit island. When concrete or stone was scarce, he went even further and manufactured disks (observed 18 times) by tearing up concrete. In an attempt to rein in the behavior, a staff member scoured the enclosure one morning, hoping to discover Santino's cache, only to find five separate stone caches, each containing 3–8 stones. Some stones were covered in algae; he had mined them from a water feature in his enclosure. In a further investigation, a staff member spied on Santino and saw him gathering and hiding stones on five consecutive days. This behavior not only suggests planning for the future, but anticipating the consequences of being discovered (having your rocks stolen) and then attempting to hide his behavior from zookeepers: deception.

18.15 Theory of Mind

Humans were once thought to be unique in the animal world in possessing **theory of mind**. Humans are not only aware of their existence, they understand that other beings conceive the world similarly, and that these other beings have beliefs and possess knowledge unique to them (Premack & Woodruff, 1978a; Byrne, 1995; Whiten, 1997). Furthermore, humans realize not only that others see the world as they do, but that those others are watching them and making

Figure 18.10 Chimpanzee Santino charging around his enclosure, carrying a stone intended to be bounced off the head of a zoo-goer. Courtesy of M. Osvath, with permission.

suppositions based on their actions, statements, body posture, and history. This ability to recognize **intentionality** is quite special; few animals have it. Chimpanzees, like humans, attribute **intentions** and knowledge to other individuals; they recognize not only that they are actors on the world's stage, but that viewing them, others can recognize their intentions (Call, 2003; Call & Tomasello, 2008; Krupenye et al., 2016). By intention we mean that an individual intends to do something – has plans that we might be able to guess at. Chimpanzees possess, in other words, the ability to **attribute mental states** to others, sometimes shortened to **attribution**, such as realizing that an individual is uncertain, or that she knows where a particular food resource is hidden.

Proof of this cognitive ability comes from experiments that show that chimpanzees infer from evidence what others know and adjust their behavior to accommodate their companion's beliefs. Behaviors surrounding "gaze following" require inferring what a social companion is looking at by looking at the companion's eyes. Behavior surrounding gaze-following suggests that chimpanzees and some other animals use gaze to attribute intentionality, such as when an individual infers that, because another individual is looking at a banana, they intend to grab it. Psychologists are a cautious lot, and some have expressed doubt or questioned the experimental protocols that suggest chimpanzees have a theory of mind (Povinelli & Eddy, 1997; Call et al., 1998; Heyes, 1998; Itakura & Tanaka, 1998; Call & Tomasello, 2008), but recent research has convinced most experts – and convinced me – that chimpanzees have all the hallmarks of theory of mind: that they follow gaze (Tomasello et al., 1998, 2007), that given a chance they can understand what others know (Hare et al., 2000, 2001), that they can predict the actions of others based on that information, that they understand other individuals' goals (Call & Tomasello, 2008), and that they know what other chimpanzees do and do not see.

Because theory of mind is confidently assigned to individuals who attribute **false beliefs** to a companion (Premack & Woodruff, 1978a; Byrne & Whiten, 1989), researchers have spent an inordinate amount of time trying to fool chimpanzees. My own experience suggests that wild chimpanzees routinely attribute false beliefs to other individuals, if we count lack of information: "I know the fruit in that *Pseudospondias* tree in the next valley is ripe and she does not."

But we need not rely on my intuition. A wonderful study (Krupenye et al., 2016) using eye-tracking and humans in costumes has demonstrated that chimpanzees attribute false beliefs in a convincing manner. In this study, chimpanzees were shown a film of a human hiding a rock under a box; a second human looked on. The chimpanzee knows that both actors know which box the rock is under. In the film, the observer-human then left the room, after which the first human moves the rock to a second box; after a moment the human seems to decide on a better plan and takes it away altogether. The observer-human returns to retrieve the rock. The chimpanzee knows that the rock is not under either box, but do they know the human has the false belief that it is under the first box? The eye-tracking machinery showed that the chimpanzees looked at the first box, the only hiding place the observer had seen the rock being hidden, not the second box, where the rock had been hidden for a brief time out of sight of the first human. The chimpanzees attribute false belief. Previous tests had used food, and chimpanzees became so distracted by the food they were unable to think straight; this study avoided that confounding variable.

Theory of mind is difficult to recognize in another species, but it is widely accepted that the capacities required to recognize oneself in a mirror are a reliable indicator of theory of mind. Gallup (1970) first suggested that the ability to recognize oneself in a mirror drew on many of the same cognitive processes involved in theory of mind and mind-reading, and many have followed his lead. **Mirror self-recognition** (MSR) can be tested by surreptitiously placing a mark on an individual, perhaps when the individual is anesthetized. When exposed to a mirror afterwards, an individual who touches or rubs the spot intentionally must have MSR ability.

One study testing for MSR across the breadth of the primate order found great apes alone possess MSR (Inoue-Nakamura, 1997), though for gorillas this is somewhat questionable (Robert, 1986; Calhoun & Thompson, 1988). Only a few nonprimates have MSR: elephants (Plotnik et al., 2006), several species of birds, mostly crows, parrots, and their relatives (Prior et al., 2008), and probably dolphins and orcas (Marten & Psarakos, 1995; Mitchell, 1995; Delfour & Marten, 2001, Reiss & Marino, 2001); many other animals have been tested and failed (Inoue-Nakamura, 1997; Suddendorf & Collier-Baker, 2009).

18.16 Shared Intentionality

It has been hypothesized that humans are unique (here we go again) in their ability to engage in **shared intentionality** (Tomasello et al., 2005), or group action. I disagree. To my mind, border patrols fit the requirements of this skill. They often begin with one individual assuming a distinctive posture and erect hair. Other males read the mind of the patrol-starter and quickly fall in line. They have a group intention, and that is to threaten or kill males from a neighboring community.

18.17 The Cambridge Declaration

The possession of theory of mind is not merely an abstract scholarly issue, it has implications for the real world. In 2012 a group of neuroscientists attending a conference on animal consciousness were moved to draft the Cambridge Declaration on Consciousness (Low et al., 2012). It states in part that "Convergent evidence indicates that non-human animals have the neuroanatomical, neurochemical, and neurophysiological substrates of conscious states along with the capacity to exhibit intentional behaviors. Consequently, the weight of evidence indicates that humans are not unique in possessing the neurological substrates that generate consciousness." Elaborating on this theme recently, Carl Safina published *Beyond Words: What Animals Think and Feel* (Safina, 2015).

18.18 Deception

To plant false beliefs, you have to be capable of understanding that other individuals can have false beliefs, which suggests you know they are a thinking being such as yourself, observing others and guessing at their intentions: theory of mind. Thus, Santino's stone caching and similar deceptive behaviors are thought to be important evidence in support of the contention that chimpanzees have a theory of mind (Goodall, 1971; de Waal, 1986, 1992; Premack &

Woodruff, 1978a; Whiten & Byrne, 1988; Woodruff & Premack, 1979).

Once I was following an estrous female at Mahale – no easy day, because typically an estrous female moves at the center of a knot of eager, sometimes violent males, and their aggression provokes a lot of running and hiding in the female. Toward midday, after hours of copulation, feeding, grooming, and fighting, the group settled down for a rest. When the rest was over and the feeding party moved off, she was left to her own a little bit; most of the males moved ahead without looking to see where she was and she followed at a distance, trailed by a lower-ranking male. She seemed not to be thrilled with his close proximity. He attempted to convince her to follow him away from the feeding party, pushing her away from the rest of the group, giving her an aggressive grunt now and then; presumably he wanted to go on an extended consortship. She was clearly uninterested. As he attempted to herd her away from her intended direction, she pretended she had no idea what he was up to, occasionally reacting confusedly when he grunted at her. He tried directing her by blocking her way with his shoulders then pushing her in one direction; she just sat down. He waited, but was impatient; several times he grunted and shook some foliage. Then, as if to encourage her, he moved off in the direction he preferred. She remained seated, looking around at various objects of interest in the forest canopy. He came near her, grasped a nearby sapling, and shook it at her. She reacted as if surprised and then hurried off in her original direction; he resignedly followed. As they progressed, she continually glanced around, perhaps looking for an escape route, but she found no opportunity to dash away. As they progressed haltingly down the animal track, they approached a large rock about 5 feet tall and 10 feet across, looking something like a miniature Uluru. Peering around the left end of the rock was a young male – displaying an erection, the unsubtle chimpanzee invitation to mate. I saw him easily from my higher vantage point; she saw him too, but the herding male was unaware, focusing all his attention on her.

The path curved off to the right as it approached the rock, but she veered to the left as she rushed up to examine some meaningless grasses growing at its base. The male sat down on the path to wait on her, for the moment indulging her odd interest. "Oh, look – just to the left here," she seemed to want him to think she was thinking "*another* interesting weed, and another a little farther around the rock!" She gradually moved out of sight and he remained on the path, apparently assuming she would return to it and proceed in the direction of the rest of the feeding party. The moment she was out of sight she ran to the young male and quickly and quietly copulated with him. After they finished, he rushed off and she proceeded around the rock, innocently emerging onto the path on the other side, still glancing at vegetation at its base, as if nothing had happened. The original male followed her, continuing his herding shenanigans. She moved on with him for the time being, but some minutes later found an opportunity to ditch him.

Other animals occasionally engage in deception – dolphins, hyenas, and the more intelligent birds (Kuczaj et al., 2001; Legg & Clayton, 2014) – but sly and manipulative behavior seems to be an everyday part of chimpanzee life, perhaps supporting the contention that advanced cognition has a **Machiavellian** origin (Byrne & Whiten, 1989). Neocortex size, the cerebral cortex, in essence, is correlated with rates of use of deception (Byrne & Corp, 2004); it is unclear, however, whether selection for deception led to an increase in brain size, or whether an increase in brain size, perhaps driven by many social demands, endowed primates with the brainpower to deceive.

18.19 Art

Many investigations that are less than scientific – though I would argue still quite valuable – give us a sense of how chimpanzees view the world, and how their minds work. No investigator who has worked closely with chimpanzees has suggested chimpanzees have an appreciation of music or art. Neither visual art nor music seems to make any impression on them. I have only seen one incident that made me wonder – I once saw a juvenile male carry a book-sized wooden trail sign around for hours. It was painted white with stark black lettering on one side. Whenever the group paused, he would examine it closely, occasionally tapping on it, making a satisfying, resonant sound; now and then he shook it, as if expecting to hear something rattling inside. He carried it for hours, attempting to discover its secrets.

While chimpanzees take no notice of art, when it comes to art production we see more enthusiasm. They use paints energetically and with seeming great pleasure (Figures 18.11 and 18.12), especially finger-paints, and they seem to have some artistic sense. They create works with paint centered on the canvas, suggesting that pigment is placed with some sense of edge and center; their art is said to have color balance and structural proportion (Reynolds, 2005; Figure 18.11). Chimpanzee painters are said to strongly object if a canvas is taken from them prematurely. They know when they are done; once a work is finished, no appeal will induce them to add to it. Some chimpanzees – this shocks me, given how incredibly fixated on food they are – even ignore food when on a painting jag (Prestrude, 1970). Their productions are so human-like that in 1964 journalist Åke "Dacke" Axelsson was able to pass off several chimpanzee paintings as human. Although some critics were unimpressed, calling "Peter's" work "primitive" and poorly executed, critic Rolf Anderberg wrote: "[the artist] paints with powerful strokes, but also with clear determination. His brush strokes twist with furious fastidiousness. [He] is an artist who performs with the delicacy of a ballet dancer" (Wikipedia, 2017). In the 1950s the work of chimpanzee Congo (Figures 18.11 and 18.12; Lincoln, 2005) was widely celebrated; recently one of his pieces sold at auction for £14,400 (Reynolds, 2005).

Whether chimpanzees can or do paint representations of real things is another matter; probably not, though occasionally an observer thinks an image qualifies. Washoe (much more on her when

Figure 18.11 Art by three-year-old Congo, a chimpanzee. From Reynolds, 2005, with permission.

Figure 18.12 Three-year old Congo at work in the 1950s. Courtesy of the *Telegraph*, with permission.

we discuss chimpanzee language) produced a painting she called "red berry" that at least has the virtue of being mostly red.

Chimpanzees will often orient toward a source of some odd or unexpected sound, which made me wonder what chimpanzees might think of music. On one of my days off at Mahale I was sitting in the fresh air polishing up some notes and listening to music, using my headphones to save on batteries, when a small party of males wandered by. I strolled over to see who was there, thinking I might be able to add a party composition data point – which chimpanzees were traveling together – to the project data bank. The party moved on and I sat back down, only to see Lukaja, a chimpanzee, approaching, apparently trying to catch up to the recently passed party. As he approached he slowed and showed some interest in me, maybe because I looked odd without my field gear and hat on. Just for fun, I unplugged my headphones and turned up the music, expecting some reaction. He seemed not to notice at all. I turned up the volume even more. No reaction. He paused for a few more seconds, never orienting toward the noise, and moved on. I was a little surprised because chimpanzees often drum on tree trunks, so I thought he might be attentive to the beat if nothing else. Nope. Others have also reported little notice of music.

18.20 Emotions and Personality

There was a time when the terms "emotion" and its more scientific synonym **affect** (the expression of emotion through facial expressions, posture, or other signals) were viewed as taboo, rarely uttered in scientific circles except in whispers, among trusted co-conspirators, behind closed doors. Still, there was always a tiny rogue element that openly spoke of animal emotions, even though their numbers were few and their influence slight. Perhaps even scholars clearly in the only-humans-have-emotions camp had a nagging feeling that there was more to it, but whatever their suspicions, they seem to have thought it was best to avoid any language that smacked of the emotive. Wrapped up in this issue is a largely forgotten book by Darwin, *The Expression of Emotions in Man and Animals* (Darwin, 1872), if anything by Darwin can truly be said to have been forgotten.

Understanding the descent of *Expressions* into obscurity is easier when we remind ourselves that even Darwin's most important scholarly contribution, his announcement of natural selection itself, was out of fashion in the early twentieth century. Perhaps animal emotions were one of the babies thrown out with that bathwater. In any case, the study of animal emotions was beaten down by the rise of behaviorism (Watson, 1930; Skinner, 1938); behaviorists saw virtue in the simplifying assumption that unless evidence was overwhelming, we should presume that some motivation other than emotion drove animal action. In science, evidence is rarely overwhelming. As a consequence, psychologists and ethologists often assumed animals had no emotions whatsoever.

Although there were seeds of change in the 1960s, things really began to shift in the early years of this century, when evolutionary psychologists began to write about the importance of affect and motivation in social relationships (Cosmides & Tooby, 2000). In 2004, Estonian-American neuroscientist and psychobiologist Jaak Panksepp took a bold stand and devoted an entire volume to **affective neuroscience**, a term he coined (Panksepp, 2004). As scientists began to explore the function of emotions – both their role in behavior and their variety among species – it became more and more apparent that once the brain has the information it needs to make a decision, it is critical that data be augmented by some assessment of the consequences of the decision, and some motivation was needed to drive the individual to act. Animals may know what is best to do, but it turns out they also need to feel it. Many researchers now believe that emotional reactions are an essential component of sound decision-making (Rolls et al., 1994; Davidson et al., 2000; Davidson, 2003; Bekoff, 2007). Sometimes researchers phrase this in economic terms, as **risk assessment**. What is the likelihood that I will succeed if I take this particular action, and how serious are the consequences if I fail? Chimpanzees are more risk-tolerant than bonobos (Heilbronner et al., 2008), for instance, because in chimpland the rewards for success are so much greater (Hurst, 2016).

Figure 18.13 Ladygina-Kohts demonstrating the images used for match-to-sample tests on Johnny. From Ladygina-Kohts et al., 2002, with permission.

Perhaps the contributions of Nadia Ladygina-Kohts, the Russian innovator who studied chimpanzee emotions and cognition in the early twentieth century, were swept aside in the same behaviorist enthusiasm that eclipsed Darwin's *Emotions*. Ladygina-Kohts reared chimpanzee Johnny (my Anglicization of the Russion Ioni) as a human, a rearing practice known as cross-fostering, from 1913 to 1916. She compared Johnny's development to that of her son in her 1935 book *Infant Chimpanzee and Human Child: A Comparative Study of Ape Emotions*. In addition to the matching tests discussed above (Figure 18.13), she engaged Johnny in tool tests, mirror experiments, art studies, and (most importantly) discrimination tasks. She was the most forceful and earliest advocate of the idea that chimpanzees had many if not all human emotions, including jealousy, guilt, and empathy, thought to be behaviors unique to humans. We discuss more on empathy later. Ladygina-Kohts wrote that one of Johnny's most sharply drawn personality traits was a "fierce loyalty to loved ones."

Ladygina-Kohts described chimpanzee laughter as an open-mouthed, breathy sound, without vocalization. She observed it when Johnny was tickled, or when he expected something unpleasant, such as being locked in a cage, being punished, or simply being left alone, but which did not materialize. Others describe laughter as occurring in the same sorts of contexts that would elicit laughter in a human (Goodall, 1968; Ross et al., 2009). I have heard laughing most often when mothers and infants play, or when juveniles chase one another, but I have also heard adult males laugh.

A commonly cited list of the seven human emotions includes joy, trust, fear, surprise, sadness, disgust, anger, and anticipation. Of these, perhaps joy is the most difficult to recognize, though Goodall describes chimpanzee rain dances and waterfall displays as joyful (Goodall, 1968, 1971, 1986). Surely laughter is an expression of something similar to joy. One study found that chimpanzees displayed all the emotions of humans, although infants cry less often than human infants (Van IJzendoorn et al., 2009). While some are cautious, Bekoff (2007) states simply than many animals, not just apes and humans, experience joy.

Chimpanzees have a cognitive quirk that reveals much about their emotional lives. Sally Boysen investigated numeracy in her subjects by teaching them to recognize Arabic numerals, then investigating how well they understood the concept of quantity.

Among the challenges she confronted them with was the M&M test. In this experiment, a chimpanzee was shown two bowls of M&M candies and trained to point to one of the bowls; the contents were given to a neighbor and the subject received the contents of the one not pointed to. The subject was obliged, then, to point to the bowl with the fewest candies in order to receive the larger amount. Chimpanzees are very, very fond of M&Ms. So much so that when confronted with the actual M&Ms Boysen's chimpanzees became so emotionally invested they failed utterly (Boysen & Berntson, 1995). Time and time again the experimental subject would point to the fuller bowl, even though they often immediately expressed agitation and frustration at their error, sometimes even before the greater pile of M&Ms was removed to be awarded to their cage-mate. Incredibly, when Boysen substituted cards with an Arabic number representing the number of M&Ms, the subjects quickly learned to choose the lesser of the two numbers, gifting themselves with the greater bounty. Food is an emotionally evocative item in chimpanzee-land.

Köhler commented on the same phenomenon. He found that when a reward was too large, chimpanzees – otherwise highly motivated to work for food – were transfixed, paralyzed by the size of the potential windfall (Köhler, 1925), causing them to fail at the task.

Chimpanzees appear to mourn. In particular, mothers who have lost infants express all the hallmarks of sadness (see Chapter 16), a subdued affect and the appearance of depression, but most obviously an inability to abandon the dead body. They carry the dead baby not just for hours, but for days, until the corpse begins to decompose (Goodall, 1968, 1986; Pusey, 1983; King, 2013).

If you own a pet, and especially if you have owned more than one, it will be relatively easy to convince you that chimpanzees have distinct personalities. Some – well, many – are impulsive, but some are more contemplative. Some have hair-triggers and others are patient. Some are playful, others are serious. Not only do chimpanzees have distinct personalities, they possess the same five-factor personality components as humans (King & Figueredo, 1997; Weiss et al., 2002, 2009, 2012; Koski, 2011): **openness** to experience (the tendency to seek out new stimuli); **agreeableness** (the extent to which individuals try to get along with others); **extraversion** (tendency to be outgoing and social-seeking); **conscientiousness** (well-organized and dependable); and **neuroticism** (sensitivity, tendency to express strong emotions). Chimpanzees have a competitive component to their personality, and it is more pronounced than in humans, as competitive as we are as a species. Research on 128 zoo chimpanzees found that they differ in sense of well-being and that this trait, probably genetically determined, is correlated with social dominance (Weiss et al., 2002). Individuals who display strong dominance are less extraverted and less open than others (Weiss et al., 2009). Some chimpanzees are confident and possess a sense of well-being; it is intriguing that it is a heritable part of their personality, as it is in humans (Weiss et al., 2009).

The physical expression of the emotion "anger" – aggression – appears to have a strong biological component (Weiss et al., 2009). Aggression is an almost minute-to-minute presence in chimpanzee lives. We humans may think of ourselves as aggressive, but compared to most primates we are remarkably tolerant. Commenting on this trait, at a lecture my colleague Jim Moore once pointed out that he had sat now for nearly an hour among many tens of humans without hearing or observing a single act of aggression; we would have lost count of the chimpanzee or baboon brawls. A lesson from chimpanzees is that despite war, robbery, bar fights, and road rage, humans are relatively nonviolent. Emotions such as tolerance, cooperation, and empathy may have been among the first human traits to evolve, and in fact may have been essential to the evolution of other distinctive human cognitive abilities (Holloway, 1972, 1974; de Waal, 2010).

18.21 Focusing the Mind

When Daniel Povinelli (Povinelli, 2000; Povinelli & Vonk, 2003; Penn & Povinelli, 2007) brought a stark skepticism to chimpanzee cognition research, suggesting that previous scholarship attributing theory of mind, attentiveness to gaze, and a sophisticated understanding of cause and effect were overstated, it stopped many of us in our tracks. Povinelli and colleagues showed that chimpanzees were as likely to beg from an experimenter who could not see them – because they had a very visible and very opaque bucket over their head – as from an experimenter who could see. This suggested to them that chimpanzees lack the human ability to attribute mental states and intentions to social partners; chimpanzees are unable to read the mind of the person they are begging from so as to be able to think "he can see me so he knows I am begging" or "he has a bucket on his head, so he will not see my outstretched hand." While he and his colleagues forced a healthy reassessment on the field and inspired new research that has brought the workings of the chimpanzee mind into sharper relief, many of us were skeptical of his skepticism.

My research focus is locomotion, posture, and anatomy; but my experience watching chimpanzees convinces me their cognition was distinctly different than that of other primates I had watched such as baboons, redtail monkeys, mangabeys, and colobus. Of course, "convinces" is different from "right" – a lesson imposed on me in my research career more than once. Still, I remain confident that chimpanzees can follow gaze and infer intention because I have seen them do it many times.

The gaze-following abilities of chimpanzees were confirmed to me one day in a most unpleasant manner. Working at Gombe, and having lost the chimpanzee I was following early in the day, I drifted over to the artificial feeding area, which remained a favorite spot for chimpanzees even though provisioning was minimal during my time there. I intended to follow whichever female came into camp next, but an hour slipped by, then two, and none had appeared. As the day wore on toward noon, I decided to eat lunch before I was occupied with taking data. Now, lunch was very precious to me. I had already lost 20 lb during my fieldwork, and without a few calories in the middle of the day I had a very difficult time keeping up with the chimpanzees. I opened my backpack and took out my rice-and-beans lunch, but before I reached for dessert, a few of the tiny East African bananas I most loved, I saw chimpanzees entering the compound across the way. They ambled over and sat next to me companionably, but I fretted about my bananas. I had some hope that my backpack, partly concealed beneath a log I was sitting on, might go unnoticed, but the fact that it was unzipped caused me to worry about it compulsively. How open was the zipper? Were the bananas in sight? Could they smell them, even if not? I kept telling myself "don't look at it; don't look at it; don't look at it." I failed. Thinking I had waited until nobody was looking, I sneaked a quick peek. I was relieved that the bananas were pretty well hidden. I swung my eyes back to a male sitting across from me only to see him directing his attention to the backpack. There was an explosion of noise and activity and a minute later my bananas were gone and everything in my backpack was scattered over a suburban-lawn-sized area. I am pretty confident chimpanzees can follow gaze.

In response to Povinelli's research, my colleagues, particularly Brian Hare (Tomasello et al., 1998; Hare, 2001; Hare & Tomasello, 2004), devised experiments that showed that chimpanzee can indeed follow gaze, can attribute intentions to others, and can understand false beliefs. For chimpanzees, motivation is key when it comes to mind-reading, and competition is a strong motivator. When a chimpanzee worries that another chimpanzee might get some prized item, his mind is focused brilliantly. Of course! We are reminded of research that showed that one of the main personality differences between humans and chimpanzees is "competitiveness" (King & Figueredo, 1997; Weiss et al., 2009; Koski, 2011). With the attention properly focused, chimpanzees display much more advanced cognition than otherwise. Bodamer and Gardner (2002) showed that if a signing chimpanzee encounters an experimenter he or she wants to communicate with, and the experimenter has his back turned to the signing chimpanzee, the chimpanzee makes attention-getting noises and waits until the experimenter is looking before beginning signing.

In the object-object relationship discussion I characterized chimpanzees as having short attention spans, but I was comparing them to humans. Recently, Maclean and colleagues (2014) made two interesting arguments. They contended that it is absolute brain size that determines cognitive ability, not *relative* brain size; and they argued that self-control in particular is a critical cognitive skill that humans possess in abundance, while it is relatively rare in the rest of the animal world. They go on to speculate that because self-control is so highly correlated with absolute brain size, self-control may have been the cognitive capacity selection acted on to increase brain size. An interesting proposition, but I worry that the correlation is not causal. After all, the more social a primate is, as we discussed above (Hurst, 2016), the greater the need for self-control. It will be a challenge to untangle this hypothesis from the social brain hypothesis – the idea that sorting out the numerous social relationships in large groups and working out how best to get your way in the complex system has driven the evolution of intelligence (Humphrey, 1976; Dunbar, 1998).

18.22 Ethics and Morality

While not often listed as an emotion, feelings of empathy – the capacity to imagine oneself in another's situation and to feel their emotions – are often discussed as distinctive of humans; empathy is thought to be a necessary prerequisite of altruism (acting in another individual's interest, at some cost to the actor). Chimpanzees exhibit this affective state as well. Examples of empathy and altruism are more common in captivity than in the wild. My impression is that chimpanzees are more selfish and less empathetic in the wild because they are busy; food-gathering is a time-consuming and exhausting daily activity. Wild chimpanzees skirt the edge of malnutrition so often they rarely have the time or energy to pause in their routine to help another individual, and their own situation is so urgent that the plight of others impresses them little – unless that individual is a close relative.

Ladygina-Kohts was Johnny's surrogate mother, so not surprisingly he became extremely attached to her, to the extent that sometimes when he had escaped from his enclosure to climb onto the roof, she could only induce him to return by feigning injury, at which point Johnny would come running to her to make sure she was okay (Ladygina-Kohts, 1935). The Gardners describe similar "guilting" among their subjects. In more recent lab contexts, chimpanzees were shown to spontaneously aid experimenters when they see they require assistance reaching an object (Warneken & Tomasello, 2006). Both in the wild and in captivity, chimpanzees console individuals who have been the subject of aggression (de Waal, 2012; Webb et al., 2017).

The action aspect of empathy is morality, or ethics, and the *sine qua non* for empathy is theory of mind (de Waal, 2010; Takagishi et al., 2010). By this I mean that unless the actor can put himself or herself in the position of the other half of the transaction, empathy is impossible; and without some sensitivity to the sting of unfairness a social partner might experience – empathy – acting fairly is unlikely.

While empathetic sharing is rare in the wild, in captivity chimpanzees readily share food, and share it surprisingly evenly, given rank differences (Nissen & Crawford, 1936). In one experiment seven female chimpanzees were each given the choice between selecting one of two tokens, one that would give them a treat while a partner in an adjacent cage would go hungry, or a different token that would provide a lesser reward for both; chimpanzees overwhelmingly chose the "sharing" token (Horner et al., 2011).

Sharing is altruistic, and altruism can be related to empathy. One way of testing empathy is with the **ultimatum game** experiment. In this experiment, one individual is in charge of fairly dividing up an anticipated reward that will then be shared by the two individuals. Most of us would agree that something around 50–50 is fair, but many experimental subjects seem to feel the "decider" deserves a larger cut. If the "receiver" thinks that the division is unfair, she has a choice – we might call it the "nuclear option"; if she feels cheated, she can refuse the reward, in which case – this is the nuclear part – neither chimpanzees gets anything. Bad for both, but it sends a message that "I will not tolerate being treated unfairly." When cheated, chimpanzees (Brosnan et al., 2005, 2010; Proctor et al., 2013) refuse to cooperate; in other words, they have the expectation that they will be treated fairly. We had some idea this might be so for chimpanzees, since in the wild chimpanzees cooperate when hunting, and individuals who refuse to contribute to the hunt are denied meat afterwards (Boesch, 1994).

The wise (or moral) chimpanzee will be attentive to unfairness, even when it only affects their partner, anticipating that if they cheat a partner that partner may object, refusing to cooperate in the future, or he may go further and seek revenge. Being sensitive to whether a partner is cheated is what Brosnan et al. (2010) and de Waal (de Waal, 2010) call **second-order cooperation**. Evidence for second-order cooperation and sensitivity to fairness is something observed so far only in humans and chimpanzees (de Waal, 2010). Chimpanzee males depend on their compatriots standing by them in life-and-death situations; they had better make sure the male next to them in a crisis harbors no deep-seated resentment.

18.23 Teaching

If you grew up in an industrialized country, you spent many hours, probably from the age of five or so and into your twenties, sitting at a desk being taught something. I still go to national conferences and spend days at a time similarly situated. This is not part of our evolutionary history. Our ancestors were unlikely to have learned this way because they were hunter-gatherers, and among living hunter-gatherers most learning is one-on-one, parent-to-child, and little of it is the lecturing sort of teaching with which you grew up. Much learning occurs when infants and juveniles observe and mimic adult behavior, with little or no encouragement from the adults (Hewlett et al., 2011).

Still, while lecture-hall learning may be a recent phenomenon, one-on-one teaching is a common – perhaps universal – human phenomenon (Littlepage et al., 1997). Teaching is different from simply allowing observation; it involves two roles, an individual with expertise and a learner, a novice. The teacher invests time and energy he or she could be using for other purposes, which means it is an altruistic behavior. There is clear evidence that the novices improve their skills or knowledge when they are taught; a teacher is attentive to the student, correcting errors (Fragaszy & Perry, 2003; Thornton & Clutton-Brock, 2011). It may be a tiresome refrain at this point, but at one time it was thought that this behavior – teaching – was unique to humans (Boyd & Richerson, 1985; Tennie et al., 2009). Instead, chimpanzees, and chimpanzees alone among nonhumans so far as we know, teach, and their teaching bears all the hallmarks of teaching listed above (Matsuzawa et al., 2001).

The Boesches observed chimpanzee mothers teaching their infants nut-cracking skills. The mothers watched as the kids attempted the nut-cracking, correcting them when they held the stone wrong by taking the stone and holding it in the best orientation, slowly turning over the hand to show the grip; at other times, the teacher placed a nut on the correct spot on an anvil (Boesch, 1991; Boesch & Boesch-Achermann, 2000).

Chimpanzees in the Goulago triangle provide learners with termite sticks of the appropriate length and diameter (Musgrave et al., 2016). Washoe attempted on three different occasions to teach her adopted son Loulis American sign language (ASL) (Fouts et al., 1982; Goodall, 1986). She once molded his hand into the sign for "food" in the presence of a snack item (Fouts et al., 1982; Goodall, 1986). Thus dies another anthropocentrist assumption.

18.24 Abstraction

In still another "unique to humans" fallacy, it was thought at one time that only humans were capable of **abstraction**, or **concept formation**. That is, only humans possessed the ability to infer a particular concept such as "animal versus plant" or "cat versus dog" by pulling out the critical features of each (Zentall et al., 2008). Most of our understanding of abstraction among chimpanzees comes from language-study subjects, who spontaneously abstracted concepts like "open." Washoe, for instance, was taught the sign "open" mostly in the context of opening doors, an important word for escape artists like chimpanzees; but she spontaneously generalized it to opening cans, refrigerators, and even a faucet

(Gardner & Gardner, 1969). Chimpanzees can abstract the principle of "ape" (chimp and gorilla, versus a nonprimate pairing) and "cat" (cat and tiger, versus other animals). Their classifications are similar to those made by human subjects (Brown & Boysen, 2000; Murai et al., 2005) and they can classify objects by their function, "tool" versus "container," for instance (Tanaka, 1997, 2001). They are capable of continuing to form abstractions even as the notion being abstracted becomes ever more obscure (Vonk et al., 2012), as for instance when one moves from plant versus animal to nonprimate versus primate to non-ape primate versus ape. Personally, given what I know of their ability as botanists in the wild, it is no surprise that they can distinguish pictures of trees from weeds (Tanaka, 2001).

18.25 Personality

After years following chimpanzees, spending many dawn-to-dusk days with them, there are times I feel I can read their minds; admittedly, now and then their reaction to some social event is unexpected and even incomprehensible. Still, I beg your indulgence to allow me to make a few observations about the nature of the chimpanzee mind.

Before I studied wild chimpanzees, I had read of Köhler's befuddled chimpanzees failing to realize that upending a box of rocks before moving the box made the task easier. I watched a male chimpanzee at the Knoxville Zoo repeatedly trying to push his hand through several openings, all half the size that would allow it. Such bumbling helps to explain why chimpanzees are (thankfully, mostly *were*) played for clowns in circuses and TV commercials. In the built, human environment chimpanzees are amusing bumblers. Chimpanzees in the wild provide a stark contrast; they move with efficiency and purpose. They are the very essence of competence. Purposeful, calm, and accomplished, they are rarely at a loss when facing daily challenges. They command their world. Films of chimpanzees exploding in anger, images of raw aggression and abject fear in the feeding area at Gombe, left me unprepared for how quiet and calm chimpanzees are in the wild. Once I was following Chausiku of Mahale along a leaf-strewn forest track on a beautiful, sunny day when, suddenly, she stopped and peered through the foliage at something ahead, something I could not see. She uttered a quiet "hoo," turned, made eye contact with me, and then calmly climbed a tree. I moved a little forward to investigate and gradually a large group of bushpigs were revealed to me. Bushpigs are large and often aggressive. I made some noise to scare them away; at first they ignored me, but finally startled and moved on. Chausiku watched them leave from her perch, then matter-of-factly descended and carried on as if nothing had happened. Assured, practical, purposeful, not puzzled and overmatched.

As serious-minded as they are, in relaxed moments even usually serious-minded adult males may dissolve into silliness that strikes me as undignified, tickling one another or playing tag.

Just as do humans, chimpanzees engage in cruel bullying that satisfies some need I fail to fathom. The build-up to a border patrol, an activity that now and then includes the murder of a neighboring community member, is easier for me to understand; these are enemies who have killed one of their companions in the past. They are an ever-present threat to the male territory and the food it supplies to the community females; they are a threat to the very lives of their babies and thus a hated, dangerous enemy. Their murderousness is similar to the bloodlust some soldiers experience in battle, and so not incomprehensible, but still difficult to fathom.

We know chimpanzees have a self-concept; they realize that others see them and make inferences about them. Ntologi hid an injury from his competitors. Yet I have never seen any indication that they can feel shame or embarrassment, a surprising lack, given how similar to humans most of their emotional lives are. A male might be all thunder and action at one moment, hair erect, a frightening visage animating his face, the very picture of dominance and pride, but if defeated he might well run from his vanquisher squeaking like a baby, fear-grinning and fear-defecating – yet afterwards I have never seen an indication of shame over such cowardly behavior. One expects the occasional expression of sheepishness. Chimpanzees seem not to have a sense of pride.

The chimpanzee personality is one that is tolerant of, affectionate toward, and protective of infants in their own community. I expected mothers to be tolerant of infants. Males – impatient, short-tempered, and violent – I did not expect to be so tolerant. Yet in all the years of watching chimpanzees I have never seen a male swat an infant who is annoying him. I have seen infants pull on the lips and ears of a male while he copulated, drop on his head from a tree, poke him with a stick as he sleeps, and kick him in the head while hanging above him, but I have never seen a male respond with impatience. Yet they engage in infanticidal raids on other communities. I have to remind myself how similar these murders are to human intergroup baby-killing raids (Hrdy, 2009). In short, chimpanzees are human-like, yet not human, when it comes to personality. A sister, but not an identical twin.

18.26 The Fission–Fusion Theory of Ape Intelligence

Hypotheses purporting to explain the origin of *human* intelligence include tool use (or technical competence, if you prefer; Byrne, 1997), the demands of foraging (Milton, 1981, 1988), its utility in manipulating other individuals, i.e., "Machiavellian intelligence" (Byrne & Whiten, 1989), the need to discover embedded foods (Parker & Gibson, 1977), or the most widely accepted hypothesis of late, the social brain hypothesis – the idea that the demands of social life select for intelligence and other intellectual abilities are more or less fringe benefits (Humphrey, 1976; Dunbar, 1998; Sawaguchi, 1988). In support of the latter, chimpanzees can even solve human social problems: chimpanzees viewing videotape of humans struggling with various problems can later pick out the photo that illustrates the correct solution (Premack & Woodruff, 1978b).

Against that backdrop, let us consider the intelligence of other primates. Apes are smarter than monkeys, smarter than all monkeys, even monkeys that eat fruit (cercopithecines and ceboids) and monkeys that live in groups three times the size of chimpanzee groups (various baboons). Is human intelligence just the continuation of a trend established by the apes? That is, are apes smarter than monkeys for the same reason that humans are smarter than chimpanzees?

I speculated (Hunt, 2016) that the unique fission–fusion social system of chimpanzees selected for advanced cognition. Chimpanzees, and perhaps to a lesser extent orangutans (van Schaik & Utami-Atmoko, 1999) and bonobos (White, 1996), live in "virtual" societies. The chimpanzee community is a mental construct in which each member of the closed social group knows who else is a member of their group. All members of the community are never in one place at one time. Instead, their hourly social life is played out in groups of three or four that merge, shuffle, and reform in the course of the day, so that a chimpanzee's social partners are never the same for long. Still, they are always members of their virtual community.

Inevitably, this dispersed society means that most of the social activity in the group occurs outside any one individual's sight or hearing. Navigating a social environment in which so much social activity occurs off-stage must exert a powerful selective pressure on cognition. When a lone individual walks into a chimpanzee grooming party of seven, he may find that the subgroup contains individuals he has not encountered for days or even weeks, during which time the social landscape may have changed. There may be new alliances; or he may come face to face with an opportunity, should he find that two formerly staunch allies are now unallied and vulnerable. *If he notices.* The subgroup newcomer would benefit from the ability to read quickly and accurately the subtle cues that betray such changes. What do we call this ability to read these cues? A theory of mind, mind-reading, intentionality. All of these advanced cognitive abilities are more useful in fission–fusion societies.

Let me make this hypothesis a little more personal. Imagine yourself a rising young chimpanzee whose ambition has been recognized by the alpha male. He would easily win a one-on-one confrontation, but fortunately for you your childhood playmate, a like-aged, high-ranking young male, is solidly on your side. Your enemy has clearly been maneuvering to get you alone, but so far you have arranged to have your friend with you whenever you have encountered him. One day, though, you enter a grooming party after having been off a week on a consortship with a young female. During that time your friend has faced the community's males without his most reliable ally – you. Unbeknownst to you, while you were cavorting with your mate, rather than pick on your best friend, your enemy has twice come to his aid when another male has picked on him. This means that as you re-enter the group, your friend sits near the powerful alpha to whom he may feel some obligation. As you enter the open area where the grooming party sits, your eyes sweep across the social group and identify every individual in a tenth of a second - much like Ai or Ayumu noting the numbers on their touch-screen. You attempt to make eye contact with your friend, but oddly he will not meet your gaze. All other eyes are on you, so it is clear that the sound of you breaking through the brush into the clearing has alerted everyone. Normally your friend would rush up to greet you, so something is wrong. As an ape, while you cannot be sure why, you recognize in only a fraction of a second that you cannot rely on your friend, and you are now alone facing your enemy. If you cannot expect to win, you could run. If you think a win is possible, you may think there is a tide in the affairs of chimpanzees that, taken at flood, leads on to alpha status. Action omitted, it may be that all the voyage of your life will be bound in shallows and in miseries. You noticed in that one-tenth of a second that your eyes swept the group that the alpha's closest ally is missing. Perhaps he is on a consortship. Screwing your courage to the sticking point, you target your enemy and launch a surprise attack. The shock of the sudden violence intimidates him, and your friend, noting that you appear to have cowed the alpha, forgives all and joins you. The alpha shrinks away from your violent attack and you kick him and pound him with your hands. He fear-grins and pant-grunts. You find yourself alpha.

Your ability to mind-read has turned a potential humiliation into a soaring victory that in two years' time will lead to your fathering three-quarters of the 10 infants born into your community, spreading the genes that gifted you with your astute social intelligence.

Monkeys, in contrast, have a socially cohesive system in which all individuals travel together. If there is a fight, almost every member is close enough to hear it and is able to identify which individuals are fighting. They need not guess at changed relationships. Every individual sees relationships changing – an alpha when he is dethroned, or a former ally abandoning his friend, or a new friendship formed in battle – *as they happen*. No need for mind-reading.

A prediction arising from this hypothesis is that any animal with a long evolutionary history of living in a fission–fusion social system should have a theory of mind: elephants, dolphins, and spider monkeys. We know too little about parrot and corvid social systems, but I predict that they will be found to have fission–fusion societies as well.

18.27 Summary and Last Thoughts

Chimpanzees are incredibly astute at assessing social situations, and they need to be, considering that their social lives consist of a long series of tense reunions – because they have a fission–fusion society.

But they are also remarkably good at remembering the location of objects in space, probably even better than humans. Their fission–fusion social system means they cannot rely on the memory of other group members to find food. Females are more often alone than males, and so might be expected to have better spatial memories.

While their tool use is less complicated than ours, they instinctively look to objects in their environment to use as tools to solve problems. They use tools creatively, doing "thought experiments" before taking action. While your pet dog will give up and sit looking at you when a tool is required to solve a problem, a chimpanzee will immediately begin looking around her environment for a physical object to help her out. Because foods with low levels of secondary compounds are protected by armor, the chimpanzee digestive physiology makes tool use more valuable for them than it is for monkeys.

Tool use, however, is simple compared to that of humans. The "working memory" of chimpanzees is small compared to humans, even if large compared to most other animals. Complicated, multistep processes in which more than one concept must be kept in mind at once are difficult; human working memory is capable of holding seven steps or components in mind, chimpanzees half as many or less. Despite this, chimpanzees are the most avid tool users in the animal kingdom, after humans, reputed to use more tools in more ways than all other primates combined. While this capacity is species-typical, the tools an individual uses and the way he or she uses them depends on what he or she learned, mostly from their mother, during their infant and juvenile periods.

Chimpanzees look to the future in their daily lives; they move from feeding tree to feeding tree by efficient paths, ending near a tree the leaves of which make a good last meal of the day, and close to a fruiting tree from which they will have their first meal the next day. Their intelligence helps buffer them from food shortages that result from their restricted diet, rich in high-quality, low-antifeedant items.

Consider what the scientific community thought of primate cognition in 1960, the year Jane Goodall started her extraordinary study of chimpanzees at Gombe. The scientific community was unanimous in its opinion that among primates, only humans make tools, hunt, make war, possess a theory of mind, teach one another, think about the future, count, understand the concept of false beliefs, engage in mental mapping, possess ethics, use language, collate information from different senses, and laugh. All of these assumptions have fallen. The hints of super-human memory in the Tinklepaugh experiments have been amply confirmed by Matsuzawa's touch-screen experiments, to the shock of many. What will we learn in the future? A lot more, if we keep looking; and we will.

A last note for those who did not choose to search for the video on Professor Matsuzawa's test subjects Ai and Ayumu: They can memorize the location of the nine numbers on their experimental touch-screen in as little as 200 milliseconds, or two-tenths of a second.

References

Bekoff M (2007) *The Emotional Lives of Animals: A Leading Scientists Explores Animal Joy, Sorrow, and Empathy - and Why They Matter.* Novato, CA: New World Library.

Beran MJ, Rumbaugh DM, Savage-Rumbaugh ES (1998) Chimpanzee (*Pan troglodytes*) counting in a computerized testing paradigm. *Psychological Record* **48**, 3-19.

Bodamer MD, Gardner RA (2002) How cross-fostered chimpanzees (*Pan troglodytes*) initiate and maintain conversations. *J Comp Psychol* **116**, 12-26.

Boesch C (1991) Teaching among chimpanzees. *Anim Behav* **41**, 530-532.

Boesch C (1994) Cooperative hunting in wild chimpanzees. *Anim Behav* **48**, 653-667.

Boesch C, Boesch H (1984) Mental map in wild chimpanzees: an analysis of hammer transports for nut cracking. *Primates* **25**, 160-170.

Boesch C, Boesch-Achermann H (2000) *The Chimpanzees of the Taï Forest: Behavioural Ecology and Evolution.* Oxford: Oxford University Press.

Boyd R, Richerson PJ (1985) *Culture and the Evolutionary Process.* Chicago, IL: University Chicago Press.

Boysen ST (1992) Counting as the chimpanzee views it. In *Cognitive Aspects of Stimulus Control* (eds. Honig WK, Fetterman, JG), pp. 367-383. Hillsdale, NJ: Erlbaum.

Boysen ST, Berntson GG (1989) Numerical competence in a chimpanzee (*Pan troglodytes*). *J Comp Psych* **103**, 23-31.

Boysen ST, Berntson GG (1995) Responses to quantity: perceptual versus cognitive mechanisms in chimpanzees (*Pan troglodytes*). *J Exp Psych- Anim Behav Process* **21**, 82-86.

Bräuer J, Call J, Tomasello M (2005) All great ape species follow gaze to distant locations and around barriers. *J Comp Psych* **119**, 145-154.

Brosnan SF, Schiff HC, de Waal FBM (2005) Tolerance for inequity may increase with social closeness in chimpanzees. *Proc Roy Soci B Biol Sci* **272**, 253-258.

Brosnan SF, Talbot C, Ahlgren M, Lambeth SP, Schapiro SJ (2010) Mechanisms underlying responses to inequitable outcomes in chimpanzees, *Pan troglodytes*. *Anim Behav* **79**, 1229-1237.

Brown DA, Boysen ST (2000) Spontaneous discrimination of natural stimuli by chimpanzees (*Pan troglodytes*) *J Comp Psych* **114**, 392-400.

Byrne RW (1995) *The Thinking Ape: Evolutionary Origins of Intelligence.* Oxford: Oxford University Press.

Byrne RW (1997) The technical intelligence hypothesis: an additional evolutionary stimulus to intelligence. In *Machiavellian Intelligence II* (eds. Whiten A, Byrne RW), pp. 289–311. Cambridge: Cambridge University Press.

Byrne RW, Corp N. (2004) Neocortex size predicts deception rate in primates. *Proc R Soc Lond B* 271, 1693–1699.

Byrne RW, Whiten A (1989) *Machiavellian Intelligence: Social Expertise and the Evolution of Intellect in Monkeys, Apes, and Humans.* New York: Oxford University Press.

Calhoun S, Thompson RL (1988) Long-term retention of self-recognition by chimpanzees. *Am J Primatol* 15, 361–365.

Call J (2003) Beyond learning fixed rules and social cues: abstraction in the social arena. *Phil Trans Roy Soc Lond B* 358, 1189–1196.

Call J, Tomasello M (2008) Does the chimpanzee have a theory of mind? 30 years later. *Trend Cogn Sci* 12, 187–192.

Call J, Hare B, Tomasello M (1998) Chimpanzees' use of gaze in an object choice task. *AnimCogn* 1, 89–100.

Cook P, Wilson M (2010) Do young chimpanzees have extraordinary working memory? *Psychonomic Bull Rev* 17, 599–600.

Cosmides L, Tooby J (2000) Evolutionary psychology and the emotions. In *Handbook of Emotions*, 2nd ed. (eds. Lewis M, Haviland-Jones JM), pp. 91–115. New York: Guilford Press.

Cowan N (2010) The magical number 4 in short-term memory: a reconsideration of mental storage capacity. *Behav Brain Sci* 24, 87–185.

Darwin C (1872) *The Expression of the Emotions in Man and Animals.* London: John Murray.

Davidson RJ (2003) Affective neuroscience and psychophysiology: toward a synthesis. *Psychophysiology* 40, 655–665.

Davidson RJ, Jackson DC, Kalin NH (2000) Emotion, plasticity, context, and regulation: perspectives from affective neuroscience. *Psych Bull* 126, 890.

de Waal FBM (1986) Deception in the natural communication of chimpanzees. In *Deception: Human and Nonhuman Deceit* (eds. Mitchell RW, Thompson NS), pp. 221–244. Albany, NY: SUNY Press.

de Waal FBM (1992) Intentional deception in primates. *Evol Anth* 1, 86–92.

de Waal FBM (2010) *The Age of Empathy: Nature's Lessons for a Kinder Society.* New York: Broadway Books.

de Waal FBM (2012) The antiquity of empathy. *Science* 336, 874–876.

de Waal FBM (2016) *Are We Smart Enough to Know How Smart Animals Are?* New York: W.W. Norton.

Delfour F, Marten K (2001) Mirror image processing in three marine mammal species: Killer whales (*Orcinus orca*), false killer whales (*Pseudorca crassidens*) and California sea lions (*Zalophus californianus*) *Behav Process* 53, 181–190.

Dunbar RIM (1998) The social brain hypothesis. *Brain* 9, 178–190.

Ekman P (1999) Facial expressions. *Hbk Cogn Emotion*, 16, 301–320.

Fouts RS, Hirsch AD, Fouts DH (1982) Cultural transmission of a human language in a chimpanzee mother–infant relationship. In *Child Nurturance*, Vol 3: *Studies of Development in Nonhuman Primates* (eds. Fitzgerald HE, Mullins JA, Gage P), pp. 159–193. New York: Plenum Press.

Fragaszy DM, Perry S (2003) *The Biology of Traditions: Models and Evidence.* Cambridge: Cambridge University Press.

Gallistel CR, Gelman R (1992) Preverbal and verbal counting and computation. *Cognition* 44, 43–74.

Gallup GG (1970) Chimpanzees: self-recognition. *Science* 167, 86–87.

Gardner RA, Gardner BT (1969) Teaching sign language to a chimpanzee. *Science* 165, 664–672.

Geschwind N (1965) Disconnection syndrome in animals and man. *Brain* 88, 237–294.

Gillan DJ (1981) Reasoning in the chimpanzee: II. Transitive inference. *J Exp Psych- Anim Behav Process* 7, 150–164.

Goodall J (1964) Tool using and aimed throwing in a community of free-living chimpanzees. *Nature* 201, 1264–1266.

Goodall, J van Lawick (1968) A preliminary report on expressive movements and communication in the Gombe Stream Chimpanzees. In *Primates* (ed. Jay, PC), pp. 313–374. New York: Holt, Rinehart, and Winston.

Goodall J van Lawick (1971) *In the Shadow of Man.* London: Collins.

Goodall J (1986) *The Chimpanzees of Gombe: Patterns of Behavior.* Cambridge, MA: Harvard University Press.

Hanus D, Mendes N, Tennie C, Call J (2011) Comparing the performances of apes (*Gorilla gorilla*, *Pan troglodytes*, *Pongo pygmaeus*) and human children (*Homo sapiens*) in the floating peanut task. *PLoS One* 6, e19555.

Hare B (2001) Can competitive paradigms increase the validity of experiments on primate social cognition? *Anim Cogn* 4, 269–280.

Hare B, Tomasello M (2004) Chimpanzees are more skillful in competitive than in cooperative cognitive tasks. *Anim Behav* 68, 571–581.

Hare B, Call J, Agnetta B, Tomasello M (2000) Chimpanzees know what conspecifics do and do not see. *Anim Behav* 59, 771–785.

Hare B, Call J, Tomasello M (2001) Do chimpanzees know what conspecifics know? *Anim Behav* 61, 139–151.

Hare B, Brown M, Williamson C, Tomasello M (2002) The domestication of social cognition in dogs. *Science* 298, 1636–1639.

Hayes C (1951) *The Ape in Our House*. New York: Harper.

Heilbronner SR, Rosati AG, Stevens JR, Hare B, Hauser MD (2008) A fruit in the hand or two in the bush? Divergent risk preferences in chimpanzees and bonobos. *Biol Lett* 4, 246–249.

Hewlett BS, Fouts HN, Boyette AH, Hewlett BL (2011) Social learning among Congo Basin hunter-gatherers. *Phi Trans Roy Soc Lond B* 366, 1168–1178.

Heyes CM (1998) Theory of mind in nonhuman primates. *Behav Brain Sci* 21, 101–148.

Holloway RL (1972) Australopithecine endocasts, brain evolution in the Hominoidea and a model of hominid evolution. In *The Functional and Evolutionary Biology of Primates* (ed. Tuttle R), pp. 185–204. Chicago, IL: Aldine/Atherton Press.

Holloway RL (1974) *Primate Aggression, Territoriality, and Xenophobia: A Comparative Perspective*. New York: Academic Press.

Horner V, Carter D, Suchak M, de Waal FBM. (2011) Spontaneous prosocial choice by chimpanzees. *Proc Natl Acad Sci*, 108(33), 13847–13851.

Hrdy S (2009) *Mothers and Others: The Evolutionary Origins of Mutual Understanding*. Cambridge, MA: Belknap Press/Harvard University Press.

Humphrey NK (1976) The social function of intellect. In *Growing Points in Ethology* (ed. Blurton-Jones N), pp. 303–317. Cambridge: Cambridge University Press.

Hunt KD (2016) Why are there apes? Evidence for the co-evolution of ape and monkey ecomorphology *J Anat* 228, 630–685.

Hurst SD (2016) *Emotional evolution in the frontal lobes: social affect and lateral orbitofrontal cortex morphology in hominoids*. PhD dissertation. Bloomington, IN: Indiana University.

Inoue S, Matsuzawa T (2007) Working memory of numerals in chimpanzees. *Curr Biol* 17, R1004–R1005.

Inoue-Nakamura N (1997) Mirror self-recognition in nonhuman primates: a phylogenetic approach. *Japan Psychol Res* 39, 266–275.

Itakura S, Tanaka M (1998) Use of experimenter-given cues during object-choice tasks by a chimpanzee (*Pan troglodytes*), an orangutan (*Pongo pygmaeus*), and human infants (*Homo sapiens*). *J Comp Psychol* 112, 119–126.

Kaminski J, Riedel J, Call J, Tomasello M (2005) Domestic goats, *Capra hircus hircus*, follow gaze direction and use social cues in an object choice task. *Anim Behav* 69, 11e18.

Karg K, Schmelz M, Call J, Tomasello M (2014) All great ape species (*Gorilla gorilla, Pan paniscus, Pan troglodytes, Pongo abelii*) and two-and-a-half-year-old children (*Homo sapiens*) discriminate appearance from reality. *J Comp Psych* 128, 431–439.

Kawai N, Matsuzawa T (2000) Cognition: numerical memory span in a chimpanzee. *Nature* 403, 39–40.

Kellogg WN, Kellogg LA (1933) *The Ape and the Child: A Study of Environmental Influence Upon Early Behavior*. New York: McGraw-Hill.

Kinderman P, Dunbar R, Bentall RP (1998) Theory-of-mind deficits and causal attributions. *Br J Psychol* 89, 191–204.

King BJ (2013) When animals mourn. *Sci Am* 309, 62–67.

King JE, Figueredo AJ (1997) The five-factor model plus dominance in chimpanzee personality. *J Res Pers* 31, 257–271.

Köhler W (1925/1959) *The Mentality of Apes*, 2nd ed. New York: Viking.

Koops K, Schöning C, Isaji M, Hashimoto C (2015) Cultural differences in ant-dipping tool length between neighbouring chimpanzee communities at Kalinzu. *Uganda Sci Rep* 5.

Koski SE (2011) Social personality traits in chimpanzees: temporal stability and structure of behaviourally assessed personality traits in three captive populations. *Behav Ecol Sociobiol* 65, 2161–2174.

Krachun C, Call J, Tomasello M (2009) Can chimpanzees (*Pan troglodytes*) discriminate appearance from reality? *Cognition* 112, 435–450.

Krupenye C, Kano F, Hirata S, Call J, Tomasello M (2016) Great apes anticipate that other individuals will act according to false beliefs. *Science* 354, 110–114.

Kuczaj S, Tranel K, Trone M, Hill H (2001) Are animals capable of deception or empathy? Implications for animal consciousness and animal welfare. *Anim Welfare Potters Bar* 10, S161–S174.

Kuhlmeier VA, Boysen ST (2001) The effect of response contingencies on scale model task performance by chimpanzees (*Pan troglodytes*). *J Comp Psych* 115, 300–306.

Kuhlmeier VA, Boysen ST (2002) Chimpanzees (*Pan troglodytes*) recognize spatial and object correspondences between a scale model and its referent. *Psych Sci* 13, 60–63.

Kuhlmeier VA, Boysen ST, Mukobi KL (1999) Comprehension of scale models by chimpanzees (*Pan troglodytes*). *J Comp Psych* 113, 396–402.

Ladygina-Kohts, NN (1935) *Infant Ape and Human Child*. Moscow: Museum Darwinianum [taDP].

Ladygina-Kohts NN, de Waal F, Wekker BT (2002) *Infant Chimpanzee and Human Child: A Classic 1935 Comparative Study of Ape Emotions and Intelligence*. Oxford: Oxford University Press.

Leavens DA, Hopkins WD, Bard KA (2005) Understanding the point of chimpanzee pointing: epigenesis and ecological validity. *Curr Directions Psych Sci* 14, 185–189.

Leavens DA, Russell JL, Hopkins WD (2010) Multimodal communication by captive chimpanzees (*Pan troglodytes*). *Anim Cogn* 13, 33–40.

Legg EW, Clayton NS (2014) Eurasian jays (*Garrulus glandarius*) conceal caches from onlookers. *Anim Cogn* 17, 1223–1226.

Lincoln T (2005) Animal behaviour: Congo's art. *Nature* 435, 1040-1040.

Littlepage G, Robison W, Reddington K (1997) Effects of task experience and group experience on group performance, member ability, and recognition of expertise. *Org Behav Hum Decision Process* 69, 133-147.

Low P, Panksepp J, Reiss D, et al. (2012) The Cambridge declaration on consciousness. In Francis Crick Memorial Conference. Cambridge, UK.

MacLean EL, Hare B, Nunn C, et al. (2014) The evolution of self-control. *Proc Natl Acad Sci* 111(20), E2140-E2148.

Maier NR, Schneirla TC (1935) *Principles of Animal Psychology*. London: McGraw-Hill.

Maros K, Gácsi M, Miklósi Á (2008) Comprehension of human pointing gestures in horses (*Equus caballus*). *Anim Cogn* 11, 457-466.

Marten K, Psarakos S (1995) Evidence of self-awareness in the bottlenose dolphin (*Tursiops truncatus*). In *Self-awareness in Animals and Humans: Developmental Perspectives* (eds. Parker ST, Mitchell R, Boccia M) pp. 361-379. Cambridge: Cambridge University Press.

Mathieu M, Bouchard MA, Granger L, Herscovitch J (1976) Piagetian object-permanence in *Cebus capucinus*, *Lagothrica flavicauda* and *Pan troglodytes*. *Anim Behav* 24, 585-588.

Matsuzawa T (1985) Use of numbers by a chimpanzee. *Nature* 315, 57-59.

Matsuzawa T (2009) Symbolic representation of number in chimpanzees. *Curr Opin Neurobiol* 19, 92-98.

Matsuzawa T, Biro D, Humle T, et al. (2001) Emergence of culture in wild chimpanzees: education by master-apprenticeship. In *Primate Origins of Human Cognition and Behavior* (ed. Matsuzawa T), pp. 557-574. Tokyo: Springer.

McCurry J (2013) Chimps are making monkeys out of us. *Guardian*, 28 September.

McGrew WC (1992) *Chimpanzee Material Culture: Implications for Human Evolution*. Cambridge: Cambridge University Press.

McGrew WC (2004) *The Cultured Chimpanzee: Reflections on Cultural Primatology*. Cambridge: Cambridge University Press.

McGrew WC (2010) Chimpanzee technology. *Science* 328, 579-580.

Menzel EW (1960) Selection of food by size in the chimpanzee, and comparison with human judgments. *Science* 131, 1527-1528.

Menzel EW (1973) Chimpanzee spatial memory organization. *Science* 182, 943-945.

Miller GA (1956) The magical number seven, plus or minus two: some limits on our capacity for processing information. *Psych Rev* 63, 81-97.

Milton K (1981) Food choice and digestive strategies of two sympatric primate species. *Am Nat* 117, 495-505.

Milton K (1988) Foraging behaviour and the evolution of primate intelligence. In *Machiavellian Intelligence* (eds. Byrne RW, Whiten, A), pp. 284-305. Oxford: Clarendon Press.

Mitchell RW (1995) Evidence of dolphin self-recognition and the difficulties of interpretation. *Consciousness Cogn* 4, 229-234.

Morecraft RJ, Stilwell-Morecraft KS, MA, Rossing WR (2004) The motor cortex and facial expression: new insights from neuroscience. *Neurologist* 10, 235-249.

Mulcahy NJ, Hedge V (2012) Are great apes tested with an abject object-choice task? *Anim Behav* 83, 313-321.

Murai C, Kosugi D, Tomonaga M, et al. (2005) Can chimpanzee infants (*Pan troglodytes*) form categorical representations in the same manner as human infants (*Homo sapiens*)? *Dev Sci* 8, 240-254.

Musgrave S, Morgan D, Lonsdorf E, Mundry R, Sanz C (2016) Tool transfers are a form of teaching among chimpanzees. *Nature Sci Rep* 6. DOI: 10.1038/srep34783.

Nishida T, Uehara S (1981) Kitongwe name of plants: a preliminary listing. *Afr Stud Monog* 1, 109-131.

Nishida T, Uehara S (1983) Natural diet of chimpanzees (*Pan troglodytes schweinfurthii*): long-term record from the Mahale Mountains, Tanzania. *Afr Stud Monog* 3, 109-130.

Nissen HW, Crawford MP (1936) A preliminary study of food-sharing behavior in young chimpanzees. *J Comp Psych* 22, 383.

Normand E, Ban SD, Boesch C (2009) Forest chimpanzees (*Pan troglodytes verus*) remember the location of numerous fruit trees. *Anim Cogn* 12, 797-807.

Osvath M (2009) Spontaneous planning for future stone throwing by a male chimpanzee. *Curr Biol* 19, R190-R191.

Panksepp J (2004) *Affective Neuroscience: The Foundations of Human and Animal Emotions*. Oxford: Oxford University Press.

Parker ST, Gibson KR (1977) Object manipulation, tool use and sensorimotor intelligence as feeding adaptations in cebus monkeys and great apes. *J Hum Evol* 6, 623-641.

Parr LA, Waller BM, Vick S J, Bard KA (2007) Classifying chimpanzee facial expressions using muscle action. *Emotion* 7, 172-181.

Parr LA, Waller BM, Heintz M (2008) Facial expression categorization by chimpanzees using standardized stimuli. *Emotion* 8, 216-231.

Penn DC, Povinelli DJ (2007) On the lack of evidence that non-human animals possess anything remotely resembling a "theory of mind". *Phil Trans Roy Soc B* 362, 731-744.

Plotnik JM, de Waal FBM, Reiss D (2006) Self-recognition in an Asian elephant. *Proc Natl Acad Sci* 103, 17053-17057.

Povinelli D (2000) *Folk Physics for Apes: The Chimpanzee's Theory of How the World Works*. Oxford: Oxford University Press.

Povinelli DJ, Eddy TJ (1997) Specificity of gaze-following in young chimpanzees. *Br J Dev Psychol* 15, 213–222.

Povinelli, DJ, Vonk J (2003) Chimpanzee minds: suspiciously human? Trend *Cogn Sci* 7, 157–160.

Premack D (2007) Human and animal cognition: continuity and discontinuity. *Proc Natl Acad Sci* 104, 13861–13867.

Premack D, Woodruff G (1978a). Does the chimpanzee have a theory of mind? *Behav Brain Sci* 1, 515–526.

Premack D, Woodruff G (1978b) Chimpanzee problem solving: a test for comprehension. *Science* 202, 532–535.

Prestrude AM (1970) Sensory capacities of the chimpanzee. *Psych Bull* 74, 47–67.

Prior H, Schwarz A, Güntürkün O (2008) Mirror-induced behavior in the magpie (*Pica pica*): evidence of self-recognition. *PLoS Biology* 6, e202.

Proctor D, Williamson RA, de Waal FBM, Brosnan SF (2013) Chimpanzees play the ultimatum game. *Proc Natl Acad Sci* 110, 2070–2075.

Pusey AE (1983) Mother-offspring relationships in chimpanzees after weaning. *Anim Behav* 31, 363–377.

Read DW (2008) Working memory: a cognitive limit to non-human primate recursive thinking prior to hominid evolution. *Evol Psychol* 6, 676–714.

Reiss D Marino L (2001) Mirror self-recognition in the bottlenose dolphin: a case of cognitive convergence. *Proc Natl Acad Sci* 98, 5937–5942.

Reynolds N (2005) Art world goes wild for chimpanzee's paintings as Warhol work flops. *Telegraph*, 21 June.

Robert S (1986) Ontogeny of mirror behavior in two species of great apes. *Am J Primatol* 10, 109–117.

Rolls E, Hornak J, Wade D, McGrath J (1994) Emotion related learning in patients with social and emotional changes associated with frontal lobe damage. *J Neurolo Neurosurg Psych* 57, 1518–1524.

Ross MD, Owren MJ, Zimmermann E (2009) Reconstructing the evolution of laughter in great apes and humans. *Curr Biol* 19, 1106–1111.

Safina C (2015) *Beyond Words: What Animals Think and Feel*. New York: Macmillan.

Sawaguchi T (1988) Correlations of cerebral indices for "extra" cortical parts and ecological variables in primates. *Brain Behav Evol* 32, 129–140.

Sayers K, Menzel CR (2012) Memory and foraging theory: chimpanzee utilization of optimality heuristics in the rank-order recovery of hidden foods. *Anim Behav* 84, 795–803.

Skinner BF (1938) *The Behavior of Organisms*. New York: Appleton-Century-Crofts.

Suddendorf T, Collier-Baker E (2009) The evolution of primate visual self-recognition: evidence of absence in lesser apes. *Proc R Soc B* 276, 1671–1677.

Suddendorf T, Corballis MC (1997) Mental time travel and the evolution of the human mind. *Genet Social Gen Psychol Monogr* 123, 133–167.

Suddendorf T, Whiten A (2001) Mental evolution and development: evidence for secondary representation in children, great apes, and other animals. *Psych Bull* 127, 629–650.

Sugiyama Y (1997) Social tradition and the use of tool composites by wild chimpanzees. *Evol Anth* 6, 23–27.

Takagishi H, Kameshima S, Schug J, Koizumi M, Yamagishi T (2010) Theory of mind enhances preference for fairness. *J Exp Child Psychol* 105, 130–137.

Tanaka M (1997) Formation of categories based on functions in a chimpanzee (*Pan troglodytes*). *Japan Psych Res* 39, 212–225.

Tanaka M (2001) Discrimination and categorization of photographs of natural objects by chimpanzees (*Pan troglodytes*). *Anim Cogn* 4, 201–211.

Tennie C, Call J, Tomasello M (2009) Ratcheting up the ratchet: on the evolution of cumulative culture. *Phil Trans R Soc Lon B* 364, 2405–2415.

Thornton A, Clutton-Brock T (2011) Social learning and the development of individual and group behaviour in mammal societies. *Phil Trans R Soc Lon B* 366, 978–987.

Tinklepaugh OL (1932) Multiple delayed reaction with chimpanzees and monkeys. *J Comp Psychol* 13, 207–243.

Tomasello M, Call J, Hare B (1998) Five primate species follow the visual gaze of conspecifics. *Anim Behav* 55, 1063–1069.

Tomasello M, Carpenter M, Call J, Behne T, Moll H (2005) Reliance on head versus eyes in the gaze following of great apes and human infants: the cooperative eye hypothesis. *Behav Brain Sci* 28, 675–735.

Tomasello M, Hare B, Lehmann H, Call J (2007) Understanding and sharing intentions: the origins of cultural cognition. *J Hum Evol* 52, 314–320.

Van IJzendoorn MH, Bard KA, Bakermans-Kranenburg MJ, Ivan K (2009) Enhancement of attachment and cognitive development of young nursery-reared chimpanzees in responsive versus standard care. *Dev Psychobiol* 51, 173–185.

van Schaik CP, Utami-Atmoko SS (1999) The socioecology of fission–fusion sociality in orangutans. *Primates* 40, 69–86.

Visalberghi E, Fragaszy DM, Rumbaugh SS (1995) Performance in a tool making task by common chimpanzees (*Pan troglodytes*), bonobos (*Pan paniscus*), an orangutan (*Pongo pygmaeus*), and capuchin monkeys (*Cebus appella*). *J Comp Psychol* 109, 52–60.

Vonk J, Jett SE, Mosteller KW, Galvan M (2013) Natural category discrimination in chimpanzees (*Pan troglodytes*) at three levels of abstraction. *Learn Behav* 41, 271–284.

Warneken F, Tomasello M (2006) Altruistic helping in human infants and young chimpanzees. *Science*, 311, 1301–1303.

Watson J (1930) *Behaviorism* (revised edition). Chicago, IL: University of Chicago Press.

Webb CE, Romero T, Franks B, de Waal FBM (2017) Long-term consistency in chimpanzee consolation behaviour reflects empathetic personalities. *Nature Comm* 8, 292–300.

Weiss A, King JE, Enns M (2002) Subjective well-being is heritable and genetically correlated with dominance in chimpanzees (*Pan troglodytes*). *J Personality Social Psychol* 83, 1141–1149.

Weiss A, Inoue-Murayama M, Hong KW, et al. (2009) Assessing chimpanzee personality and subjective well-being in Japan. *Am J Primatol* 71, 283–292.

Weiss A, Inoue-Murayama M, King JE, Adams MJ, Matsuzawa T (2012) All too human? Chimpanzee and orang-utan personalities are not anthropomorphic projections. *Anim Behav* 83, 1355e1365.

White FJ (1996) *Pan paniscus* 1973 to 1996: twenty-three years of field research. *Evol Anthropol* 5, 11–17.

Whiten A (1997) The Machiavellian mindreader. In *Machiavellian Intelligence II: Extensions and Evaluations*, eds. Whiten A, Byrne R), pp. 144–173. Cambridge: Cambridge University Press.

Whiten A, Byrne RW (1988) Tactical deception in primates. *Behav Brain Sci* 11, 233–273.

Wikipedia (2017) Pierre Brassau. https://en.wikipedia.org/wiki/Pierre_Brassau (accessed 1 June 2017).

Wood S, Moriarty KM, Gardner BT, Gardner RA (1980) Object permanence in child and chimpanzee. *Learn Behav* 8, 3–9.

Woodruff G, Premack D (1979) Intentional communication in the chimpanzee: the development of deception. *Cognition* 7, 333–362.

Woodruff G, Premack D (1981) Primitive mathematical concepts in the chimpanzee: proportionality and numerosity. *Nature* 293, 568–570.

Yerkes RM (1943) *Chimpanzees: A Laboratory Colony*. New Haven, CT: Yale University Press.

Yerkes RM, Yerkes DN (1929) Concerning memory in the chimpanzee. *J Comp Psychol* 8, 237–271.

Zentall TR, Wasserman EA, Lazareva OF, Thompson RR, Rattermann M (2008) Concept learning in animals. *Comp Cogn Behav Rev* 3, 13–45.

19 The Brain of the Chimpanzee
The Mind's Motor

Credit David Malan / Stone / Getty Images

The extraordinary feats of memory and reasoning that chimpanzees display – both in the wild and in captivity – are generated by an organ scarcely larger than a grapefruit. But while chimpanzee brains are markedly smaller than those of humans, their brain anatomy is so similar that a discourse comparing the two might be little different from this declaration: The chimpanzee brain is a human brain with one-third of the neurons (Herculano-Houzel & Kaas, 2011). In this chapter we will consider the structure of the brain and its function, drawing mostly on research on humans, and we will speculate about the role size plays in how human and chimpanzee brains work (Duvernoy, 1999; Buzsaki, 2006; Schoenemann, 2006).

19.1 Basic Structure

There are three very different structures in the brain, the **cerebrum, cerebellum,** and the **brain stem** (Figure 19.1). The cerebrum, also known as the cerebral cortex, and its various lobes, is responsible for much of consciousness, while the cerebellum, an extensively folded sub-brain (the gray, highly folded half-hemisphere at the base of the brain in Figure 19.1) governs subconscious actions, such as smoothing movement. The brain stem controls functions such as regulating heart rate, controlling breathing, and directing the sleep–wake cycle. Although there is still much to learn, neuroscientists have progressed by leaps and bounds recently using new technologies such as CT scanning, functional MRI, and other noninvasive imaging technologies. This new information has added texture to the centuries-deep body of **neuroanatomy** (brain structure) and neuroscience (brain function) research (Buzsaki, 2006).

The cerebrum is divided down the middle into two hemispheres, left and right. Despite some very important specializations that will be discussed in more detail below (see Table 19.1), the two hemispheres are mostly mirror images of one another, in both shape and function. For nonspecialists it takes some getting used to the fact that the left side of the brain controls the right side of the body, and *vice versa*. The two sides can and do function more or less on their own, almost like two brains, each side registering perceptions of the world on its side of the body, engaging in its own calculations, then communicating this information when needed to the other hemisphere through a large sheet of nerves called the **corpus callosum**. When the callosum is cut there are essentially two people living inside one body, sometimes at odds about everyday decisions, so much so that one hand (i.e., the "person" controlling it) may deliberately interfere with the actions of the other if it disagrees with what the "other" brain is doing (Gazzaniga, 2005).

We inherited the organization of the brain – the location of the various specialized areas – from a small-brained ancestor, and the brain suffers from some design inefficiencies that are a relic of our smaller-brained past; for instance, the visual cortex is as far from the eyes as it could be. The visual

Table 19.1 **Brain hemisphere in which function is concentrated**

Left hemisphere	Right hemisphere
Control of right hand	*Control of left hand*
Analytical processing	Global, holistic processing
Language	Visuospatial skills
Time sequencing	Recognizing faces
	Humor/metaphor
	Tone of voice
	Musical ability
	Emotions

After Falk, 1992.

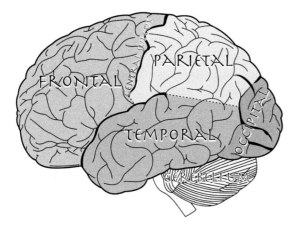

Figure 19.1 View of left side of the brain, with the four lobes of the cerebrum color-coded. Frontal lobe (blue), parietal lobe (yellow), temporal lobe (green), and occipital lobe (pink). The hatched globule at the base is the cerebellum and the mostly hidden white structure is the brain stem. By Henry Vandyke Carter; after Gray (1918). (A black and white version of this figure will appear in some formats. For the color version, please refer to the plate section.)

system would presumably work better if the two were closer.

Over 100 years ago Korbinian Brodmann peered into his microscope and distinguished 52 different types of brain tissue, based on neuron morphology (Brodmann, 1912). His genius was such that we still talk about the brain using his system; more on Brodmann areas later. It was, of course, a qualitative assessment, and the brain could have been parsed even more finely as some have done since (Fan et al., 2016). On a coarser level, each of the left and right hemispheres are divided into four lobes (see Table 19.2) – **frontal**, **temporal**, **occipital**, and **parietal** – which we will examine individually (Figure 19.1).

19.2 Size Probably Matters

Among neuroanatomists, debate has raged for years over whether absolute or relative size best predicts intelligence, and the issue generates heat even now. The argument that intelligence depends on the size of the brain relative to the size of the body rests on the idea that it must take a certain number of neurons to oversee physical processes like digestion, muscle movement, pain perception, and so on. Hypothetically, the larger the animal, the greater the number of neurons needed to oversee these processes, since larger bodies have a greater number of muscle cells and a larger skin surface area. In this view, brain weight divided by body weight yields intelligence. The other view holds that a neuron can oversee almost any number of cells by the combination of nerve action and interaction among the stimulated cells. Furthermore, only part of the brain "runs" the body. In this view, larger brains yield greater intelligence, regardless of body size.

The great intelligence of parrots, for example, supported the *relative* argument, while the intelligence of dolphins, elephants, and whales supported the absolute argument. Suzana Herculano-Houzel's brilliant work showing that intelligent birds have a similar number of neurons to macaques even though their brain is smaller (each neuron is smaller), and that elephants have huge neurons and therefore about the same number of neurons as macaques even though their brains are much larger, seems to have resolved the issue. Analyses that use relative brain size to predict some complex behavior or another have been used to support the relative brain size argument, but inevitably when *absolute* brain size is substituted in these analyses it also shows significant results (Schoenemann, 2006; see also Dunbar, 1998; Reader & Laland, 2002).

Table 19.2 **Brain regions and their functions**

Area and subareas		Function
Frontal		Executive functions: planning, reasoning, abstract thought, emotional assessments
	Prefrontal	Planning, language production, social decisions, working memory sifting information to identify critical information
	Medial orbitofrontal	Assesses positive stimuli and opportunities; values desirables
	Lateral orbitofrontal	Assesses negatives and negative outcomes
	Broca's area	Language production including sentence structure, grammar
	Premotor cortex	Complex sequences of muscle movement
	Primary motor cortex	Voluntary movement of muscles of limbs, face, and tongue; monitor actions of others via mirror neurons
Parietal		Central processing area for information from senses; spatial processing, mental rotation
	Somatosensory cortex	Senses of touch, pain, pressure, heat, temperature
Temporal		
	Wernicke's area	Language understanding
	Hippocampus	Spatial orientation; visual, spatial memory; emotion; shunting information into long-term memory
	Amygdala	Seat of emotions, e.g., affection, fear, rage; detecting danger; tagging emotional events to be transferred to long-term memory
Occipital	Vision	

In other words, if size, or neuron number, is all-important for cognitive function, regardless of body size (Schoenemann, 2006), smaller primates will have less cognitive ability than larger primates like chimpanzees and gorillas, with their larger brains – regardless of how similar brain size to body size ratio is. There is good evidence this is so in comparisons across the primate order (Deaner et al., 2007), and across all the mammals (Schoenemann, 2006). Capuchins, despite having the largest brain/body ratio of all primates, lack some cognitive abilities that apes have. Larger brains, larger in absolute size, better predict both how well animals learn abstract rules than relative size (Rumbaugh et al., 1996) and how well they learn object-discrimination tasks (Riddell & Corl, 1977). Bigger, it appears, is better.

19.3 Gray versus White Matter

The cerebral cortex, the outer layer of the brain, is much like the peel of an orange. This layer, called the **gray matter**, contains the neurons that do most of our conscious thinking. Inside the outer surface of gray matter is the **white matter**, made up of billions of organic "wires" that connect different parts of the brain; this is the electric communication system. If the brain were a road map, the cities would be specialized areas like **Broca's language area**, or the seat of vision, the visual cortex, and the interstates would be the brain interconnections, the white matter. The stimulation of the specialized areas by hormones (chemical signals) suffusing the brain and the electrical signals moving along the "roads" together produces a state we experience as consciousness and thought. All of the wonderful experiences of life, and the horrible ones, too, are manifestations of these crude chemical interactions and electrical signals (see the excellent review by Buzsaki, 2006).

19.4 Frontal Lobe

The frontal lobe is the part of the brain that starts just behind your forehead and over your eyes and extends backward to the **central sulcus**, a deep fold that divides the front and back halves of the brain (Figure 19.1). The prefrontal cortex, the front of the frontal lobe, performs so-called **executive functions**, including planning, reasoning, abstract thought, and a vital but somewhat nebulous function, emotional assessments of the outcome of actions.

These faculties are so close to what we think of as making us human that it was no surprise when neuroanatomists found that one of the distinctions between humans and chimpanzees was that humans had a relatively larger **frontal cortex** (Brodmann, 1912; Blinkov & Glezer, 1968; Deacon, 1997; Rilling, 2006) and an even larger prefrontal cortex (Smaers et al., 2017). These are relative measures, though, proportions within the brain, so by *absolute* measures the frontal is even larger, and as we will see below, it has more interconnections (Schoenemann et al., 2005).

An upside-down triangle of neural tissue in the rearmost part of the frontal lobe is the **premotor cortex** (see area 6 in Figure 19.4), responsible for complex muscle movement sequences. Behind it is the **primary motor cortex**, a strip of brain running parallel to the central sulcus (left hatched area, Figure 19.2), responsible for voluntary movements of the feet, legs,

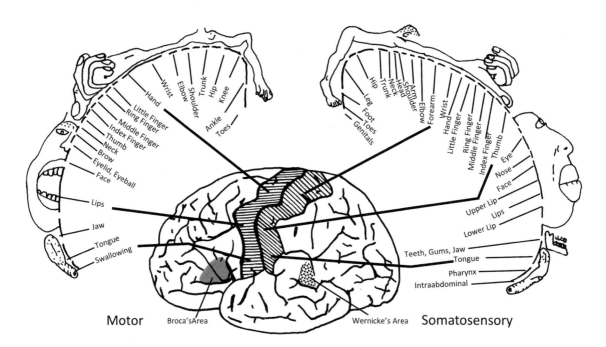

Figure 19.2 Brain function. The primary motor cortex (horizontally hatched area to the left of the central sulcus) controls movement. The somatosensory cortex (diagonally hatched area) registers sensory input from the various body parts. Speech production does not derive from any one localized place on the brain, as many assume, but Broca's area is essential for speech production and serves as the nexus of information related to language. While Broca's area plays some role in comprehension, this is the principal function of Wernicke's area. After Penfield & Rasmussen, 1950.

thorax, hands, face, and tongue, with the foot control at the top (actually, over the top; Figure 19.2). In the left hemisphere, at the rear of the frontal lobe, is the famous **Broca's area** (solid gray area, Figure 19.2), a part of the brain critical for language production.

Functionally, Broca's area is found only in the left hemisphere – its mirrored area on the right side is structurally different and is much less involved in language – one of the important asymmetries of human and chimpanzee brains. Brain asymmetry is the subject of much research; it might seem surprising there are side-to-side differences at all, since each hemisphere controls half the body. Perhaps concentrating function in a confined area facilitates the rapid communication necessary to produce speech. In any case, language areas are located more in the left side of the brain, whereas visuospatial interpretation, global and holistic processing, and other functions (Table 19.1) are concentrated on the right. Asymmetry is discussed in detail below under the **laterality** heading.

In the motor cortex, aside from the neurons that direct physical action, is a type of neuron that does something extraordinary – all the more extraordinary because we had no idea these neurons existed until about 20 years ago. These **mirror neurons** fire not only when a person performs an action, but when they see the same action in some other individual (Gallese et al., 1996; Rizzolatti et al., 1996), helping individuals to learn by watching others. In a real way, watching someone perform some physical feat allows you to practice it. In Figure 18.3 you can see a chimpanzee reaching up "in sympathy" as a cage-mate stretches to grasp a banana.

Everything that is not premotor or primary motor cortex (areas 6 and 8 in Figure 19.4) is **prefrontal cortex**; in other words, everything anterior to the premotor area 6 is the part of the brain responsible for planning, producing language, making social decisions, and regulating some aspects of emotion.

19.5 Chimpanzee Brains Show Evidence of Less Social Discipline

A part of the prefrontal cortex tucked under the rest of the frontal lobe and just over your eye sockets is the **orbitofrontal cortex**. This little bit of brain is particularly interesting to me as a chimpologist because it drives desire – and not only desire, also its antagonist, self-control (Bechara et al., 2000; Rolls et al., 1994). Chimpanzees are short-attention-span primates, and (as we saw earlier) a little emotionally overwrought; this area has something to do with that (Hurst, 2016).

Two parts of the orbitofrontal cortex in particular are of interest here. The **medial orbitofrontal cortex** (Figure 19.3) makes the world an exciting place; it assigns an emotional value to things that are pleasant or desirable (Kringelbach and Rolls, 2004; Hurst, 2016), both things that are tangible – like a tasty piece of fruit – and things that are more abstract – like observing a pleasant facial expression or an attractive member of the opposite sex, or even monetary rewards (Thut et al., 1997). It tells the rest of the brain how good something is (Rolls, 2008; Hurst, 2016). What would you guess might happen if it were overstimulated? Hedonism. Guess what part of the brain is stimulated by alcohol and many illegal, widely abused drugs (Bechara, 2005)? The medial orbitofrontal cortex.

The **lateral orbitofrontal cortex** is the worrywart part of the brain. This killjoy brain region balances the benefits of pursuing something that the medial orbitofrontal cortex lusts after against the potential disasters that might come from trying to get it; it superintends the hedonistic impulses generated by the medial orbitofrontal cortex (Hurst, 2016). For my younger audience I can phrase it this way; medial is the more Tigger part of the brain and lateral is the Eyore part. As the medial orbitofrontal cortex eggs on the rest of the brain to go for the gusto and live for the moment, the lateral orbitofrontal cortex is shaking a finger at its hedonist neighbor and reminding it of the cost. And it assesses all sorts of negative outcomes and negative stimuli, not just a potential drubbing an offended alpha male might administer, but also things like a possible drop in prestige, an unpleasant smell, and even abstract things like aggressive facial expressions or threatening body postures (Morris et al., 1998). If the lateral orbitofrontal cortex is overstimulated, rewards seem less desirable; the sense of risk and concern about negative outcomes are exaggerated; such overstimulation creates a pessimist.

This dynamic – desire versus caution – caused my former student Shawn Hurst (2016) to wonder whether the *size* of one or the other might have some influence on emotions and behavior. After all, we know that animals with specialized senses (e.g., bat echolocation) have greatly expanded areas of the brain associated with these capacities (Krubitzer, 1995). Hurst predicted that among the great apes, species that live in complex social groups need an impulse regulator that keeps emotions under tight control, lest greed and lust cause loss of a friend, creation of an enemy, or a beating. Solitary animals, on the other hand, might be freer to pursue their yearnings.

The solitary orangutan proved to have a smaller lateral "controlling" zone compared to the hedonistic medial part (Hurst, 2016). This region is large in chimpanzees. A middle-ranking chimpanzee (most are middle-ranking; there is only one alpha) needs to calculate just how great the risks are that the alpha male will butter him over the forest floor if he makes a move to copulate with that attractive and willing female making eyes at him. Every action an individual in chimpanzee society makes must include the calculus of punishment. Is he taking a feeding site an alpha will want in seconds? When he grooms with another male, will the alpha disapprove of the budding friendship? Orangutans? Not so much.

Differences in the relative size of the parts of this area give us some insight into behavioral differences between humans and chimpanzees, as well. You may have already begun to speculate about human medial-to-lateral proportions, and you are right. The lateral orbitofrontal cortex is even larger in humans than in chimpanzees. We have already learned in the chapter on growth that this part of the brain starts out looking more ape-like, with a small lateral orbitofrontal cortex, and as gestation proceeds the lateral area grows disproportionately. We worry more than chimpanzees, because our society is larger, with more complicated alliances and more inter-articulations. The lateral orbitofrontal cortex expansion may be the reason that we humans have more white matter – more neural connections – in the frontal lobe, to help us with these calculations.

The orbitofrontal cortex, then, has important connections to the executive function part of the frontal lobes, the part of your brain that pulls it all together as you weigh a decision, but it also has connections to the limbic system, another emotional assessor and regulator. It helps you to make decisions. Other parts of the prefrontal cortex are critical for working memory (Goldman-Rakic, 1996), which we saw in the last chapter differs between humans and chimpanzees, and for sorting out critical information from "noise." The **Stroop test** measures this ability; in this test, subjects might be asked to press a buzzer when they see a word written in blue ink; if the word "red" is written in blue, identification is impaired (Schoenemann et al., 2000).

19.6 Broca's Area

Moving on to our last bit of the frontal lobe, Broca's area (Figures 19.2 and 19.3) provides the organizational skills it takes to put together a

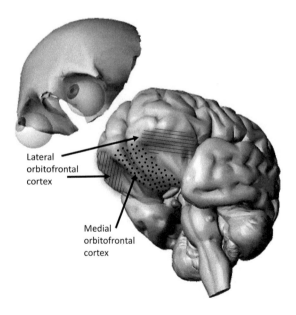

Figure 19.3 View of the orbitofrontal cortex. The medial orbitofrontal cortex assigns a value to desirable things and outcomes while the lateral orbitofrontal cortex serves as a kind of regulator of desires. See text. Image courtesy of Shawn Hurst; with permission.

sentence, working to slot in the right words in the right order, organizing the main ideas first, only later adding the *mot juste* flowery adjective that so perfectly highlights the more substantial noun, then linking that phrase to a rousing verb or expressive adverb to make a complete sentence. It does all of this on the fly, in tiny fractions of a second, while your mouth and larynx are still getting out the previous sentence. This organizational skill, taking into account which elements are most important in a sort of hierarchical way, is also used when objects are put together manually – during tool use. It has been hypothesized that the tool-using skills of chimpanzees, requiring as they do sequential and hierarchical thought processes, are much like language, and tool use was the evolutionary precursor or preadaptation for language (Greenfield, 1991). While the human brain is three times as large as that of chimpanzees, Broca's area is *six* times as large. Understanding causality, a crucial ability for humans that helps us make sense of the world and serves as the basis of science, requires this sequential or serial order-sorting ability. When event A is always followed by event B, humans are capable of inferring that A causes B (Schoenemann, 2006). As my colleague Debra Pekin points out, this constant search for causality is one of the hallmarks of human behavior, and is so strong that we often mistake correlation (two things occurring together) for causality (one causing another).

19.7 Parietal Lobe

The parietal lobe is a sort of central processing area for the onslaught of information coming into the brain from your senses. Information the body gathers is shunted to a narrow strip of neural tissue that runs along the front of the parietal lobe, parallel to the motor cortex, just behind the central sulcus (Figure 19.2). The **somatosensory cortex** is responsible for registering the sense of touch, pain, and other sensory information from the various body parts. It is organized in the same manner as the primary motor cortex, with the feet tucked just over the top, the thorax next, then hands and, so on. In the 1930s Wilder Penfield realized that more tissue was dedicated to the hands, face, and lips than other body parts; he illustrated this disproportion with a brilliant image of a distorted figure lying on top of the part of the brain that controls it. This deformed human figure has come to be known as the **somatosensory homunculus** (Penfield & Boldfrey, 1937); homunculus means "little human" in Latin.

The **superior** and **inferior parietal lobules** make up the rest of the parietal lobe, sometimes called the **parietal cortex**. These areas are responsible for mental rotation, spatial awareness, and body awareness. The **superior lobule** (Brodmann areas 5 and 7 in Figure 19.4) helps to determine the body's orientation in relation to the external world, particularly what your hand is doing. People with injuries to this part of the brain have trouble directing and shaping their hand to grasp objects (Caminiti et al., 1996).

The **inferior parietal lobule** interprets sensory information and facial expressions (Radua et al., 2010); it has extensive connections with the two language areas, Broca's area and Wernicke's area

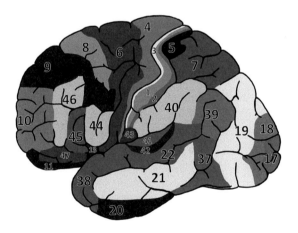

Figure 19.4 Brodmann's areas visible on the surface of the brain; many areas are hidden in the interior of the brain. Areas 44 and 45 are Broca's area and area 39 is Wernicke's area. The premotor cortex is area 6 and the primary motor cortex is area 4; the visual cortex is made up of areas 17, 18 and 19. The orbitofrontal cortex areas visible on this image are Brodmann area 11 and 47. Area 13, not visible here, is just on the other side of 11. After Brodmann, 1909. Illustration by the author. (A black and white version of this figure will appear in some formats. For the color version, please refer to the plate section.)

(discussed below), making it critically important for language and mathematical tasks (Mesulam et al., 1977; Andersen, 2011).

19.8 Temporal Lobe

The temporal lobe is responsible for conceptual understanding (Carpenter & Sutin, 1983; Schoenemann, 2006) and memory and hearing functions. An area at the rear of this lobe, 39 in Figure 19.4, is called **Wernicke's area** (Figure 19.2) and is specifically dedicated to deriving meaning from words. Wernicke's area can be thought of as half of a pair of specialized areas particularly critical for language, Wernicke's taking the role of interpreting language and the other half of the pair, Broca's area, responsible for its production, though both also need the inferior parietal lobule – the decipherer of reaction to your attempts at communication – to function.

Buried in the center of the pickle-shaped temporal lobe is a piece of the brain shaped like a stretched-out cayenne pepper, the **hippocampus**. The hippocampus is critical for spatial orientation, for fixing short-term memories into long-term memory, and for some aspects of visual function, particularly visual memory, i.e., recognizing objects. Next to the hippocampus, still in the temporal lobe, is another small but important bit of specialized neural tissue, the **amygdala**, the seat of emotions such as affection, fear, and aggressive rage. One particularly important function of the amygdala is interpreting whether information coming in from the eyes and ears indicates danger. It is intimately connected to the hippocampus, and their dual function causes extremely emotional events to be more memorable. It is no surprise that many of the amygdala's neural connections are with the orbitofrontal cortex, the hedonist/killjoy duo we discussed earlier. These two structures, the amygdala and the hippocampus, are major parts of the **limbic system**, a dispersed network of brain areas that together control emotion.

The temporal lobes of chimpanzees are one of two brain areas that have less white matter – fewer "wires" connecting different areas – than those of humans (Rilling et al., 2008), perhaps because language is so important for humans and many of the language areas are in the temporal lobes.

19.9 Occipital Lobe

The occipital lobe is responsible for vision (Wandell et al., 2007). It is inconvenient that the **visual cortex**, areas 17, 18, and 19 in Figure 19.4, is the part of the cerebrum farthest from the eyes, since its rearward location requires running wiring from the very front of the head to the back, increasing reaction time to visual stimuli and wasting a lot of "wire." The occipital lobe was much closer to the eyes in our distant ancestors, but as the brain has expanded it was pushed to the back of the braincase.

19.10 Brodmann's areas

Every once in a while a scientist makes an extraordinary contribution, one that becomes the foundation of a whole field of study and influences scholars even a century later. Most scholarship, alas, has a short shelf life, making these epochal contributions all the more impressive. One such contribution was made by the German anatomist we met briefly above, **Korbinian Brodmann** (Brodmann, 1909; Brodmann & von Bonin, 1909/1960). He examined brains under a microscope and recognized differences in cell structure and the extent to which they took on different stains, ultimately recognizing 50 distinct areas (Figure 19.4), not all of which are visible on the surface rendering in Figure 19.4.

Structure and function are closely enough related that we have learned in the century and more since Brodmann's original work that areas he identified have specific roles in brain function. Brodmann areas 1, 2, and 3, for example, are the primary somatosensory cortex (Figure 19.2). Brodmann area 4 is the primary motor cortex. *Broca's area corresponds to Brodmann areas 44 and 45.* Areas 10, 11, 47/12, and 13 (the lateral three internal and not visible in Figure 19.4) are part of the orbitofrontal cortex, which we have already learned is important for assessing benefits and risks. Brodmann's work is still important enough to neuroscientists that a century after it appeared discussion of any big issue about brain function will include mention of Brodmann's areas.

19.11 Laterality and Asymmetry

Humans are left- or right-handed, but the term "handedness" is problematic for experts. First, it is annoyingly difficult to define. A person may be a left-handed thrower and a right-handed writer, or she may reach for objects with a left hand, even if the right hand is preferred for other tasks. Furthermore, hands are not the only things that are "handed." There is a "footedness"; most individuals prefer to kick or jump with one foot, often the same foot they put forward first when stepping off a curb. Many people prefer one ear when listening on the phone. One eye is often dominant. People even have a strong preference for which arm ends up on top when the arms are crossed. There is even a "handedness" in brain hemisphere size, which is relative size of various Brodmann's areas and even how the brain is wired up. After reading handedness so many times in this paragraph, try to forget it and use the term experts prefer for differences in side preferences, **laterality**.

Humans are quite lateralized (Rentería, 2012); we have already mentioned that two important specialized brain areas related to speech are found in the left hemisphere – Broca's area and Wernicke's area. There are other specializations, roughly along the lines mentioned in Table 19.1. We tend to use the left side of the brain for more analytical tasks and the right side for more artistic jobs.

Most people, about 90 percent, are right-lateralized for manual (hand) tasks, writing being the one usually mentioned. Other animals are much less lateralized, at least as species. While individual chimpanzees (and other animals) often prefer the left or the right hand, as a species there is no preference. About one-third of individuals are left lateralized, one-third right lateralized, and one-third ambidextrous, which makes the evolutionary origin of this unusual – perhaps unique – human trait of great interest (Marchant et al., 1995).

Why not be capable of doing every task with either hand? While there is no consensus on why laterality exists, many believe that it reduces the amount of time required to practice complicated manual tasks (Marchant et al., 1995). This phenomenon is most apparent in the fact that having trained oneself to be skilled at something with a right hand, we can still be frustratingly incompetent with the left. By specializing on one hand, we can halve the time it takes to learn a task, assuming it is always possible to use the same hand for the task, as we do with writing.

There is also the cost of brain tissue to consider. The human brain is already huge; for specializations that have an anatomical basis, like language and visuospatial calculations, lateralizing halves the extra, energy-hungry brain tissue needed to do the task. Lateralizing not only reduces the need for brain matter itself, but also the need to communicate between different, specialized parts of the brain (Hofman, 2001).

Why the right hand? The motor area for the right hand and the areas specialized for speech are quite close to one another. Perhaps being lateralized for the right hand is an unexpected and unselected consequence of the expansion of language areas (Falk, 1987, 1992). This has the virtue of linking one thing we know is unique – our incredible language abilities – to an unexplained one, right-hand lateralization, but it is also speculative and as yet unproven.

There is another asymmetry of interest to neuroanatomists, so interesting that it has its own name: **petalia**. If you look down on the brain from the top (Figure 19.5), the back of the brain extends backwards more on the left side than the right, and the right-hand side of the brain extends farther forward; the right side is also wider than the left (LeMay, 1976; Balzeau et al., 2012). This arrangement is known as a right frontal/left occipital petalia, an asymmetry typical of humans. Left-handed people often have less pronounced petalias than right-handers, and some of them have the opposite asymmetry – left frontal and right occipital petalias.

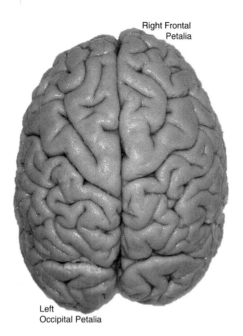

Figure 19.5 Just as does the human brain, the right front (top) and left rear of the chimpanzee brain protrudes. This asymmetry was at one time thought to be found only in humans. Image courtesy of Shawn Hurst, with permission.

19.12 Chimpanzee Laterality

While the chimpanzee brain is anatomically lateralized, behavioral lateralization, "handedness," is either absent or diffuse. In studies in which chimpanzees show hand preference, they use a different hand for different tasks. For instance, in one study termite-fishing elicited left-handedness, though the bias was not strong; of 20 chimpanzees 16 used a left hand to fish, 4 a right hand. While 75–25 looks impressive, it is only 25 percent from 50–50 (Lonsdorf & Hopkins, 2005). In the same study, nut-cracking and wadge-dipping showed a slight preference for the *right* hand.

Other studies have characterized chimpanzees as mostly ambidextrous (Finch, 1941) in that they switch back and forth between hand for tasks such as plucking fruits (Sugiyama et al., 1993; McGrew & Marchant, 1996, 2001; Marchant & McGrew, 2007; Corp & Byrne, 2002, 2004). When chimpanzees perform complicated tasks, they are more likely to prefer one hand, though which hand differs by individual, and when they do specialize, they perform the task more efficiently (Marchant & McGrew, 1996, 2007; McGrew & Marchant, 1996). For many terrestrial tool-use tasks, individuals, but not the species, tend to favor one hand or the other (Marchant & McGrew, 1996, 2007). Perhaps only termite fishing shows a left-hand preference (McGrew & Marchant, 1996; Lonsdorf & Hopkins, 2005). In short, while individual chimpanzees may prefer a left hand or right hand, and while individual tasks may be slightly left- or right-biased, as a species – unlike humans – chimpanzees are not lateralized (McGrew & Marchant, 1996; Lonsdorf & Hopkins, 2005).

19.13 Chimpanzee Brains are Human Brains in Miniature, with Important Exceptions

After all that background, it may be a disappointment to learn that for the most part the chimpanzee brain is more or less a human brain in miniature (Shantha & Manocha, 1969; Azevedo et al., 2009; Herculano-Houzel, 2009). I wrote "in miniature," but I must qualify that. The chimpanzee brain looks like what we would expect a human brain to look like if it were one-third the size, but there is a twist. If we look across the primates, from the tiny tarsier to the medium-sized blue monkey to the large chimpanzee to the even larger human, there is **allometry** as the brain increases in size – some areas increase in size more than others. We spent some time on body proportion allometry in Chapter 7. To remind you, allometry is a change in the proportion of two or more things between smaller and larger animals. Just as when we compare a MINI Cooper to a Toyota Camry to a Cadillac DeVille, not everything grows at the same rate as the vehicle grows larger. The brake pedal, the steering wheel, the rearview mirror, and the radio controls are the same size, but the size of the engine and the suspension springs are not. Larger animals, for example, have disproportionately smaller eyes. The chimpanzee brain is *mostly* what you would expect based on other primates *with the qualification* that we know that big brains are not just proportionally expanded versions of small brains, but that some brain areas are disproportionately larger in larger animals. To be precise, the human brain is not a magnified chimpanzee brain, but a scaled-up one.

Now let us reconsider that *mostly* comment a little bit. While the human brain is a scaled-up version of the chimpanzee brain, as I foreshadowed above, one area in particular is expanded beyond that, the prefrontal, and the difference is dramatic. This area of the brain is twice the size we would expect if the human brain were just a scaled-up chimpanzee brain (Smaers et al., 2017). This is the part of the brain responsible for planning, language production, social decisions, working memory, and sifting information to identify which bits are important (Table 19.1) – faculties that account for a large proportion of the differences between our two species.

19.14 Comparison between Human and Chimpanzee Outer Cortex

Chimpanzees have fewer wrinkles, sulci (singular: sulcus), and gyri (singular: gyrus), but this is merely a consequence of size. Most of what we discussed above concerning function applies to the centimeter-thick outer layer of the brain, known as the gray matter – the brain matter responsible for much of conscious thought. For a brain to work similarly at a larger size, the outer cortex needs to remain the same size in relation to brain volume. In other words, if we stripped off the cortex and found it weighed one-tenth of the rest of the brain, that proportion needs to remain the same whether the brain is 50 cm³ or 500 cm³. In our earlier discussion of allometry we noted that for a surface area to keep up with a volume, the two-dimensional tissue must expand *disproportionately*, because volume increases in three dimensions while surface area increases in only two. The larger outer cortex can only be attached to the inner layers if it is wrinkled, just as you would have to wrinkle up a basketball-sized cover to attach it to a grapefruit-sized inner core.

When we call the chimpanzee brain a human brain in miniature, though, how miniature are we talking?

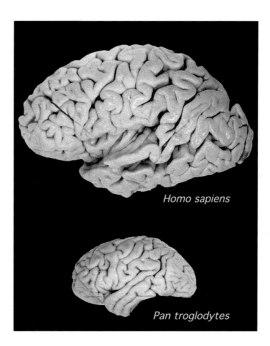

Figure 19.6 Human and chimpanzee brains. Human brains are approximately three times the size of chimpanzee brains, and more convoluted. Adapted from an image by Todd Preuss, Yerkes Primate Research Center.

The human brain contains 86 billion neurons (Herculano-Houzel, 2009, 2014), 16 billion in the cerebral cortex and a shockingly larger number, 70 billion, in the much smaller (in volume, that is) cerebellum. Chimpanzee brains have not been measured as accurately as human brains, but studies to date suggest they have 27.9 – let's call it 28 – billion neurons, or about one-third the human number: 32.5 percent (Azevedo et al., 2009; Herculano-Houzel & Kaas, 2011; Figure 19.6). Brain volume follows very closely, but not exactly – humans are often quoted as having a 1330 cm³ volume (Schoenemann, 2006) versus 337 cm³ for chimpanzees (Rilling & Insel, 1999). These figures give chimpanzees a brain volume of only 25.3 percent that of humans. This disparity is due to the fact that there is more to a brain than just neurons. White matter (the "wires" in the brain) can differ both in number and in which parts they connect, and in fact those two things do differ between humans and chimpanzees.

19.15 The Human Prefrontal is Possibly Larger; Prefrontal White Matter is Undisputedly Larger

The prefrontal cortex in humans is more convoluted than the rest of the brain, suggesting it is disproportionately larger; Schoenemann and colleagues found it to be 10.3 percent of total brain volume in chimpanzees, versus 12.7 percent in humans (Schoenemann et al., 2005), which would make it both disproportionately larger and absolutely larger in humans. While other studies have found no proportional expansion of the prefrontal in humans (Semendeferi et al., 2002; Gabi et al., 2016), Schoenemann compared his and Semendeferi's data and concluded that "if the analysis is restricted to increasingly anterior regions of the frontal (keeping in mind that the prefrontal occupies the most-anterior portions of the frontal lobe), humans appear increasingly disproportionate." Smaers and colleagues, as we mentioned above, found it to be twice the size expected (Smaers et al., 2017). Regardless of the proportionate size of the prefrontal and the frontal, both are considerably larger in *absolute* size in humans, presumably with increases over chimpanzees in its functions. There is no disagreement that human frontals have more interconnections, i.e., more white matter (Schoenemann et al., 2005; Smaers et al., 2017).

Returning to our analogy of Brodmann's areas as cities and connecting wires as roads, humans have more roads, and those roads – white matter – are denser in the frontal lobe in humans compared to chimpanzees (Schoenemann et al., 2005); this part of the brain controls executive functions, including planning, reasoning, abstract thought, and emotional decisions about action (Semendeferi et al., 2002), in particular in the prefrontal cortex (Sakai et al., 2011). Drilling down even more finely, in addition to decision-making functions we mentioned before, the prefrontal also participates in thought processes such as working memory, serial-order tasks, temporal integration, motivation, creativity, social assessment, some aspects of language, and perhaps even self-awareness (Goldman-Rakic, 1996; Fuster, 2002; Goldberg, 2002). Working memory is critical for tool making and tool use, as well as other complex, sequential tasks. In other words, the absolutely and perhaps even disproportionately larger human prefrontal is due at least in part to the greater interconnections in the area, which gives humans a capacity for complex tool use that chimpanzees lack.

The expanded prefrontal of humans may leave chimpanzees sounding compromised, but this is only in comparison with humans. Chimpanzees have disproportionately larger frontals compared to the rest of the primates (Barton & Venditti, 2013).

19.16 Chimpanzee Visual Cortex

Chimpanzees have a proportionately larger occipital cortex – the visual cortex – than humans, but only proportionally; the human visual cortex is 1.6 times larger in absolute terms. It is not clear whether the similar absolute sizes means we have similar visual acuity and visual processing capacity, or whether the proportionally larger visual cortex means they have better visual capability, but many experts these days are betting on size rather than proportion (Schoenemann, 2006).

19.17 Chimpanzee Brains Are Asymmetrical and Have Petalias

Chimpanzee brains are asymmetrical (Gómez-Robles et al., 2016), just as are human brains (Spocter et al., 2011), and in most chimpanzees this asymmetry extends to petalias; many, but not all, have a right frontal/left occipital petalia (Shantha & Manocha, 1969; Holloway & de la Coste-Lareymondie, 1982). They also have a Broca's homologue (Schenker et al., 2008, 2009) on the left side which is somewhat distinct from the analogous area on the right side. The enlargement of this area in humans results in a knob-like protrusion that has come to be called Broca's cap; chimpanzee have an analogous area, though it is not cap-shaped (Holloway, 1983).

Chimpanzees have a Wernicke's area homologue (Spocter et al., 2010), and in yet another human–ape parallel the *pattern* of asymmetry in a part of the chimpanzee brain that we call Wernicke's area in humans, called the planum temporale in chimpanzees, is extremely similar (Gannon et al., 1998).

Given the similarity of the human and chimpanzee Broca's area and Wernicke's areas (Schenker et al., 2008, 2009), why do chimpanzees not have language? There are two answers, one of which we will pursue in greater detail in the next chapter: The first is, they have more language ability than you might expect. The second is that this part of the brain not simply a language center, it is also used when a task requires completing subtasks in sequence, such as bending the major support structures of a sleeping nest before weaving in the smaller branches, or lifting an anting wand slowly before placing a flexed finger at the base of it to swipe the ants into the mouth.

19.18 Summary and Last Thoughts

Chimpanzee brains are remarkably like those of humans. They have anatomical features that in humans are associated with language (Broca's and Wernicke's area). Like humans they have laterality (petalia) and expanded prefrontal regions, areas associated with planning, social decisions, and working memory, though these areas are even larger in humans. But size does matter. If we link size to function, our larger brains probably give us our dramatically superior engineering abilities, mathematical skills, social complexity, and language. Areas responsible for unique human cognitive functions are not only larger in humans, they have more interconnections with the rest of the brain.

It is easy to take for granted proficiencies we find easy and natural, like mind-reading, performing sequential tasks, mental mapping, tool manipulation, and the ability to act politically. Chimpanzees have all of these abilities too, but they are rare in the primate world. Give chimpanzees their due: They are likely the second smartest species on the planet.

References

Andersen RA (2011) Inferior parietal lobule function in spatial perception and visuomotor integration. *Comprehen Physiol.* DOI: 10.1002/cphy.cp010512.

Azevedo FA, Carvalho LR, Grinberg LT, et al. (2009) Equal numbers of neuronal and nonneuronal cells make the human brain an isometrically scaled-up primate brain. *J Comp Neurol* 513, 532–541.

Balzeau A, Emmanuel Gilissen E, Grimaud-Hervé D (2012) Shared pattern of endocranial shape asymmetries among great apes, anatomically modern humans, and fossil hominins. *PLoS ONE* 7. DOI: 10.1371/journal.pone.0029581.

Barton RA, Venditti C. (2013) Human frontal lobes are not relatively large. *Proc Natl Acad Sci* 110, 9001–9006.

Bechara A (2005) Decision making, impulse control and loss of willpower to resist drugs: a neurocognitive perspective. *Nature Neurosci* 8, 1458–1463.

Bechara A, Tranel D, Damasio H (2000) Characterization of the decision-making deficit of patients with ventromedial prefrontal cortex lesions. *Brain* 123, 2189–2202.

Blinkov SM, Glezer II (1968) *Das Zentralnervensystem in Zahlen und Tabellen*. Jena: Fischer.

Brodmann K (1909) *Vergleichende Lokalisationslehre der Grosshirnrinde in ihren Prinzipien dargestellt auf Grund des Zellenbaues*. Berlin: Barth.

Brodmann K (1912) Neue Ergebnisse über die vergleichende histologische Lokalisa- tion der Grosshirnrinde mit besonderer Berücksichtigung des Stirnhirns. *Anat Anz* 41, 157–216.

Brodmann K, von Bonin G (1909/1960). On the comparative localization of the cortex (translation) In *Some Papers on the Cerebral Cortex* (ed. von Bonin G), pp. 201–230. Springfield, IL: Charles C. Thomas.

Buzsaki G (2006) *Rhythms of the Brain*. Oxford: Oxford University Press.

Caminiti R, Ferraina S, Johnson PB (1996) The sources of visual information to the primate frontal lobe: a novel role for the superior parietal lobule. *Cerebral Cortex* 6, 319–328.

Carpenter MB, Sutin J (1983) *Human Neuroanatomy*. Baltimore, MD: Williams & Wilkins.

Corp N, Byrne RW (2002) The ontogeny of manual skill in wild chimpanzees: evidence from feeding on the fruit of *Saba florida*. *Behaviour* 139, 137-168.

Corp N, Byrne RW (2004) Sex difference in chimpanzee handedness. *Am J Phys Anthropol* 123, 62-68.

Deacon T (1997) *The Symbolic Species*. London: Penguin.

Deaner RO, Isler K, Burkart J, van Schaik C (2007) Overall brain size, and not encephalization quotient, best predicts cognitive ability across non-human primates. *Brain Behav Evol* 70, 115-124.

Dunbar RIM (1998) The social brain hypothesis. *Brain* 9, 178-190.

Duvernoy HM (1999) *The Human Brain: Surface, Three-Dimensional Sectional Anatomy with MRI, and Blood Supply*, 2nd ed. New York: Springer.

Falk D (1987) Brain lateralization in primates and its evolution in hominids. *Yearb Phys Anthropol* 30, 107-125.

Falk D (1992) *Braindance: New Discoveries about Human Origins and Brain Evolution*. New York: Henry Holt.

Fan L, Li H, Zhuo J, et al. (2016) The human brainnetome atlas: a new brain atlas based on connectional architecture. *Cerebral Cortex* 26, 3508-3526.

Finch G (1941) Chimpanzee handedness. *Science* 94, 117-118.

Fuster JM (2002) Frontal lobe and cognitive development. *J Neurocytol* 31, 373-385.

Gabi M, Neves K, Masseron C, et al. (2016) No relative expansion of the number of prefrontal neurons in primate and human evolution. *Proc Natl Acad Sci*. DOI: 10.1073/pnas.1610178113.

Gallese V, Fadiga L, Fogassi L, Rizzolatti G (1996) Action recognition in the premotor cortex. *Brain* 119, 593-609.

Gannon PJ, Holloway RL, Broadfield DC, Braun AR (1998) Asymmetry of chimpanzee planum temporale: humanlike pattern of Wernicke's brain language area homolog. *Science* 279, 220-222.

Gazzaniga MS (2005) Forty-five years of split-brain research and still going strong. *Nature Rev Neurosci* 6, 653-659.

Goldberg E (2002) *The Executive Brain: Frontal Lobes and the Civilized Mind*. London: Oxford University Press.

Goldman-Rakic PS (1996) The prefrontal landscape: implications of functional architecture for understanding human mentation and the central executive. *Phil Trans R Soc Lond B* 351, 1445-1453.

Gómez-Robles A, Hopkins WD, Schapiro SJ, Sherwood CC (2016) The heritability of chimpanzee and human brain asymmetry. *Proc R Soc B* 283, 20161319.

Gray H (1918) *Anatomy of the Human Body*. Philadelphia, PA: Lea & Febiger.

Greenfield PM (1991) From hand to mouth. *Behav Brain Sci* 14, 577-595.

Herculano-Houzel S (2009) The human brain in numbers: a linearly scaled-up primate brain. *Front Hum Neurosci* 3, 1-11.

Herculano-Houzel S (2014) The glia/neuron ratio: how it varies uniformly across brain structures and species and what that means for brain physiology and evolution. *Glia* 62, 1377-1391.

Herculano-Houzel S, Kaas JH (2011) Gorilla and orangutan brains conform to the primate cellular scaling rules: implications for human evolution. *Brain Behav Evol* 77, 33-44.

Hofman M (2001) Brain evolution in hominids: are we at the end of the road? In *Evolutionary Anatomy of the Primate Cerebral Cortex* (eds. Gibson KR, and Falk D). Cambridge: Cambridge University Press.

Holloway RL (1983) Human paleontological evidence relevant to language behavior. *Hum Neurobiol* 2, 105-114.

Holloway RL, de la Coste-Lareymondie MC (1982) Brain endocast asymmetry in pongids and hominids: some preliminary findings on the paleontology of cerebral dominance. *Am J Phys Anthropol* 58, 101-110.

Hurst SD (2016) *Emotional evolution in the frontal lobes: social affect and lateral orbitofrontal cortex morphology in hominoids*. PhD dissertation. Bloomington, IN: Indiana University.

Kringelbach M, Rolls E (2004) The functional neuroanatomy of the human orbitofrontal cortex: evidence from neuroimaging and neuropsychology. *Progr Neurobiol* 72, 341-372.

Krubitzer L (1995) The organization of neocortex in mammals: are species differences really so different? *Trends Neurosci* 18, 408-417.

LeMay M (1976) Morphological cerebral asymmetries of modern man, fossil man, and nonhuman primate. *Ann New York Acad Sci* 280, 349-366.

Lonsdorf EV, Hopkins WD (2005) Wild chimpanzees show population-level handedness for tool use. *Proc Natl Acad Sci* 102, 12634-12638.

Marchant LF, McGrew WC (1996) Laterality of limb function in wild chimpanzees of Gombe National Park: comprehensive study of spontaneous activities. *J Hum Evol* 30, 427-443.

Marchant LF, McGrew WC (2007) Ant fishing by wild chimpanzees is not lateralized. *Primates* 48, 22-26.

Marchant LF, McGrew WC, Eibl-Eibesfeldt I (1995) Is human handedness universal? Ethological analyses from three traditional cultures. *Ethology* 101, 239-258.

McGrew WC, Marchant LF (1996) On which side of the apes? In *Great Ape Societies* (eds. McGrew WC, Marchant LF, Nishida T), pp. 255-272. Cambridge: Cambridge University Press.

McGrew WC, Marchant LF (2001) Ethological study of manual laterality in the chimpanzees of the Mahale Mountains, Tanzania. *Behaviour* 138, 329-358.

Mesulam MM, Van Hoesen GW, Pandya DN, Geschwind N (1977) Limbic and sensory connections of the inferior parietal lobule (area PG) in the rhesus monkey: a study with a new method for horseradish peroxidase histochemistry. *Brain Res* 136, 393-414.

Morris JS, Friston KJ, Buchel C, et al. (1998) A neuromodulatory role for the human amygdala in processing emotional facial expressions. *Brain* 121, 47-57.

Penfield W, Boldfrey E (1937) Somatic motor and sensory representation in the cerebral cortex of man as studied by electrical stimulation. *Brain* 60, 389-443.

Penfield W, Rasmussen T (1950) *The Cerebral Cortex of Man: A Clinical Study of Localization of Function.* New York: Macmillan.

Radua J, Phillips ML, Russell T, et al. (2010) Neural response to specific components of fearful faces in healthy and schizophrenic adults. *Neuroimage* 49, 939-946.

Reader SM, Laland KN (2002) Social intelligence, innovation, and enhanced brain size in primates. *Proc Natl Acad Sci* 99, 4436-4441.

Rentería ME (2012) Cerebral asymmetry: a quantitative, multifactorial, and plastic brain phenotype. *Twin Res Human Genet* 15, 401-413.

Riddell WI, Corl KG (1977) Comparative investigation of the relationship between cerebral indices and learning abilities. *Brain Behav Evol* 14, 385-398.

Rilling JK (2006) Human and nonhuman primate brains: are they allometrically scaled versions of the same design? *Evol Anthropol* 15, 65-77.

Rilling JK, Insel TR (1999) The primate neocortex in comparative perspective using magnetic resonance imaging. *J Hum Evol* 37, 191-223.

Rilling JK, Glasser M, Preuss T, et al. (2008) The evolution of the arcuate fasciculus revealed with comparative DTI. *Nature Neurosci* 11, 426-428.

Rizzolatti G, Fadiga L, Gallese V, Fogassi L (1996) Premotor cortex and the recognition of motor actions. *Cogn Brain Res* 3, 131-141.

Rolls E (2008) *Memory, Attention, and Decision-Making.* New York: Oxford University Press.

Rolls E, Hornak J, Wade D, McGrath J (1994) Emotion related learning in patients with social and emotional changes associated with frontal lobe damage. *J Neurol Neurosurg Psych* 57, 1518-1524.

Rumbaugh DM, Savage-Rumbaugh ES, Wasburn DA (1996) Toward a new outlook on primate learning and behavior: complex learning and emergent processes in comparative perspective. *Jpn Psychol Res* 38, 113-125.

Sakai T, Mikami A, Tomonaga M, et al. (2011) Differential prefrontal white matter development in chimpanzees and humans. *Curr Biol* 21, 1397-1402.

Schenker NM, Buxhoeveden DP, Blackmon WL, et al. (2008) A comparative quantitative analysis of cytoarchitecture and minicolumnar organization in Broca's area in humans and great apes. *J Comp Neurol* 510, 117-128.

Schenker NM, Hopkins WD, Spocter MA, et al. (2009) Broca's area homologue in chimpanzees (*Pan troglodytes*): probabilistic mapping, asymmetry, and comparison to humans. *Cerebral Cortex* 20, 730-742.

Schoenemann PT (2006) Evolution of the size and functional areas of the human brain. *Annu Rev Anthropol* 35, 379-406.

Schoenemann PT, Budinger TF, Sarich VM, Wang WS (2000) Brain size does not predict general cognitive ability within families. *Proc Natl Acad Sci* 97, 4932-4937.

Schoenemann PT, Sheehan MJ, Glotzer LD (2005) Prefrontal white matter volume is disproportionately larger in humans than in other primates. *Nat Neurosci* 8, 242-252.

Semendeferi K, Lu A, Schenker N, Damásio H (2002) Humans and great apes share a large frontal cortex. *Nature Neurosci* 5, 272-276.

Shantha TR, Manocha SL (1969) The brain of chimpanzee (*Pan troglodytes*). In *The Chimpanzee* (ed. Bourne G), pp. 238-305. Basel: Karger.

Smaers JB, Gómez-Robles A, Parks AN, Sherwood CC (2017) Exceptional evolutionary expansion of prefrontal cortex in great apes and humans. *Curr Biol* 27, 714-720.

Spocter M, Hopkins WD, Garrison AR, et al. (2010) Wernicke's area homologue in chimpanzees (*Pan troglodytes*) and its relation to the appearance of modern human language. *Proc R Soc B* 277, 2165-2174.

Spocter M, Hopkins W, Bianchi S, et al. (2011) Neuropil asymmetry in the cerebral cortex of humans and chimpanzees: implications for the evolution of unique cortical circuitry in the human brain. *Am J Phys Anthropol*, 144, 281-282.

Sugiyama Y, Fushimi T, Sakura O, Matsuzawa T (1993) Hand preference and tool use in wild chimpanzees. *Primates* 34, 151-159.

Thut G, Schultz W, Roelcke U, et al. (1997) Activation of the human brain by monetary reward. *Neuroreport* 8, 1225-1228.

Wandell BA, Dumoulin SO, Brewer AA (2007) Visual field maps in human cortex. *Neuron* 56, 366-383.

20 Tired Nature's Sweet Restorer
Chimpanzee Sleep

Photo: Mathias Appel; with permission.

In our waking hours we are constantly assaulted by outside forces that seek to destroy us: predators, parasites, microbes, poisons, and even fellow humans. Our bodies respond with formidable defenses, though at considerable cost. As we fight for survival, we overuse and damage our muscles; we exhaust our neurons, expending their biochemical reserves while at the same time building up toxins; we struggle to cope with a relentless onslaught of information that we feel, rightly or wrongly, must be assimilated to ensure our survival; we fight pathogens of various sorts that attempt to parasitize our blood or eat our skin. Our own species members are if anything even more dangerous: enemies threaten our status, income, and even our very lives. By evening we crave time to gather our thoughts and make sense of our experiences; we need time to repair the daily ravages that have diminished our bodies. We need an opportunity to rest and regroup. Sleep is that opportunity. Without it we would, quite literally, die. It is a daily chance to reverse all the insults our physical and mental being has suffered during the day and to assemble our resources to face another day.

Few of us, however, face assaults equal to those of chimpanzees. In their world, there can be only the briefest postponement of the daily marathon of climbing, walking, and armhanging in which they engage to gather sufficient food. With only the most primitive of medicines, they must rely on their bodies to mount biological defenses against parasites and infections. While few predators are serious dangers, lions and leopards can and do occasionally eat chimpanzees. Just over the horizon are enemies who assault community guardians every few weeks, attempting to steal territory or kill group members.

Chimpanzees face these challenges without the comforting shelter most humans enjoy – a roof over one's head and a blanket against the night chill; these would be delicious luxuries for a chimpanzee. Instead they must settle for a leafy nest (Figure 20.1), well-designed in its own way, but without a roof to keep them dry in the rainy season (but note that orangutans go so far as to construct such a roof!). While the social and cognitive challenges chimpanzees shoulder are not as diverse as those we humans must cope with, their struggle for status and sustenance is every bit as important to them. When a scientist, weary from following a chimpanzee all day, watches them settle into their night bed to rest and sleep, they appear to welcome the rest just as much as to we do, if not more so. And this respite is just as necessary for their mental and physical health as it is for us.

20.1 What Is Sleep?

Sleep seems deceptively simple: The sleeper is "offline," inactive, resting. In fact, sleep is complex, so complex that at a fundamental level it is not even one thing; it is variously a behavior, a particular brain state, and a process (Webb, 1988). If you have ever had a sleepless night – or two – and find yourself in between wakefulness and sleeping, nodding off unexpectedly right in the middle of a conversation, you have experienced the first mystery sleepologists

Figure 20.1 Semliki chimpanzee Hunter relaxes in a day nest. Photo: Caro Deimel. (A black and white version of this figure will appear in some formats. For the color version, please refer to the plate section.)

must confront: How is sleep different from wakefulness?

Sleep is a state in which the brain is inattentive to external stimuli (Siegel, 2008), and in which the body is in a state of relative paralysis, but in which the brain is quite busy, not merely turned off. It is busy in a physiological sense, repairing and recharging itself, but also in a cognitive sense, capering off into an unreal dream world.

20.2 The Purpose of Sleep

During sleep the body is in a repair state; it is flushing out toxins and resupplying necessary organic chemicals to neurons (Xie et al., 2013). The recharging aspect of sleep bolsters immune function (Opp, 2009), warding off disease and perhaps providing better resistance to parasites (Preston et al., 2009). During sleep the body repairs and strengthens the skeleton, muscles, tendons, and ligaments (Van Cauter & Spiegel, 1999). Wound healing accelerates during sleep (Gümüstekin et al., 2004). Growth hormone, needed to repair tissues, is secreted more often during sleep (Van Cauter & Spiegel, 1999).

This restoration function may sound desirable but perhaps not essential. Not so. Sleep deprivation results in the steady breakdown of most of the body's physiological systems, quickly resulting in poor memory consolidation, a reduction in the speed and accuracy of cognitive tasks, decreased coordination, compromised immune function, mood disorders, tissue damage, and even the onset of cancer (Weinger & Ancoli-Israel, 2002; Straif et al., 2007; Rajaratnam et al., 2013).

Your body is evolved to desire sleep more the longer you have been awake or when it is most needed. As your "sleep debt" increases, your body increasingly encourages you to pay that debt; you accumulate the neurochemical adenosine, a neurotransmitter that induces sleep and shuts down many cognitive processes.

Humans lead information-rich lives, which makes the role of sleep in improving recall particularly important for us – and for our data-driven relative, the chimpanzee. If you have ever compared a sleepless night cramming to the much more effective learning regime of studying several days just before bedtime, you know from personal experience that memory consolidation is an important benefit of sleep (Born et al., 2006; Turner et al., 2007; Daltrozzo et al., 2012; Tononi & Cirelli, 2013). I suggest to students that an intense study session just before turning in pays big dividends – though you must also refrain from indulging in recreational chemicals – your brain will study for you while you sleep, unconsciously and effortlessly. This is sleep memory consolidation.

20.3 Sleep "Architecture"

Given the many functions of sleep, it is no surprise that a night of sleep involves several discrete states. In the crudest parsing human sleep has two distinct modes, **non-rapid-eye-movement (non-REM)** and REM, but in each of those modes there are discrete light and deep phases, yielding a fourfold division: light and deep non-REM sleep, and light and deep REM sleep. During deep sleep it is difficult to rouse the sleeper; their so-called **arousal threshold** is greatest. One of the most bizarre aspects of REM sleep is **muscle paralysis**; in non-REM sleep, there is no such paralysis, though rousing a sleeper is no easy task.

20.4 Non-REM Sleep

Non-REM sleep (Walker, 2009) has a **slow wave phase (SWS)**, the deepest stage of sleep where **memory consolidation** occurs, the process of sorting the previous day's information, making sense of it and storing it away. While the sleep is indeed deep, the brain is still somewhat aware of the outside world, sifting through information coming in from the various senses to determine whether anything is going on that warrants waking up. We need not go into great detail about brain activity, but experts have found that the information parsing during SWS sleep occurs during brain activity spikes called **sleep spindles** and the monitoring phases involve an activity pattern known as **K-complexes**.

20.5 REM Sleep

REM sleep modulates emotional response to stressful events endured during the day, perhaps toning down the worse parts of unpleasant experiences but tagging the memory as important, in case the information is needed later (Walker, 2009). It also promotes emotional regulation (Walker, 2009), innovation, creativity, and insight (Wagner et al., 2004). If you feel stupid and grumpy after a poor night's sleep, research validates your feelings.

During REM sleep, high levels of brain activity, including dreaming, occur as the body slips into a paralytic phase. In other words, a high "arousal threshold" leaves you as dead to the world as you will ever be in this life (Ermis et al., 2010). The deepest phases of sleep are the most restful and recuperative.

20.6 The Sleep Cycle

The body cycles back and forth between these states four or five times a night on a 90-minute cycle (Figure 20.2), spending much more time in non-REM than REM sleep. Out of the seven hours of sleep typical of humans (see Table 20.1), 22 percent of total

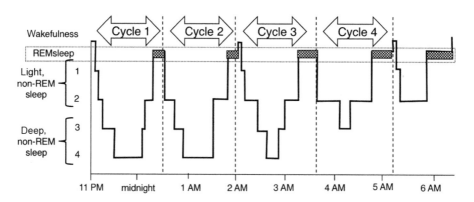

Figure 20.2 Sleep cycles. Humans progress through four or five cycles of deeper and lighter non-REM sleep and REM sleep on about a 90-minute schedule. Figure by the author.

Table 20.1 **Primate sleep architecture**

Species	Total sleep duration	REM duration	Non-REM length	Sleep cycle length	REM %	Non-REM %
Homo sapiens	7.0	1.6	5.4	90	22	78
Pan troglodytes	9.7	1.5	8.2	90	15	85
Pongo pygmaeus	9.2	1.1	8.1	107	11	89
Papio anubis	9.2	1.0	8.2	40	11	89

After Ohayon, 2004; Samson, 2013.

sleep time is spent in REM, whereas the rest is divided into light and deep stages of non-REM.

Humans have a natural 25-hour sleep rhythm that is fine-tuned on a daily basis to adjust to the 24-hour day. Buried inside your brain is an organ, the pineal gland, that produces melatonin in low-lux (low-light) environments. Melatonin is the principle hormone regulating sleep–wake activity and is critical to entraining an organism to be synchronized with their environment. This is why getting out and walking in the sun helps to reset your clock if you suffer from jet-lag.

20.7 Sleep across Cultures

In many industrialized societies people strive for eight hours of continuous sleep, but in hunter-gatherer societies sleep patterns are quite different (Worthman & Melby, 2002; Worthman, 2008; Samson & Nunn, 2015; Yetish et al., 2015; Samson et al., 2017a). Instead, sleep time is more of an individual choice and daytime napping is the norm. In preindustrial Western European societies, historical evidence suggests that nighttime sleep was separated into two phases, a "first sleep" and a "second sleep," and in between the two individuals might get up and attend to chores or quietly socialize before returning to bed (Ekirch, 2001, 2006). This sleep pattern was at first considered to be linked to the incredible seasonal variation in light and temperature in the northern latitudes, but then it was discovered that humans living in small-scale agricultural societies near the equator also exhibit this pattern (Samson et al., 2017d). It is also noteworthy that experimental evidence (Wehr, 1992) shows that humans who are exposed to short photoperiods (i.e., long nights) naturally bifurcate their sleep pattern.

Tending nighttime fires is a nearly universal human cultural practice, as a security precaution, to deter predators, or to keep warm. In many societies, rather than a couple co-sleeping while other family members sleep alone, sleep is more of a family affair, including the baby. Babies sleeping alone, separate from their mother, is actually quite a rare arrangement, if you look across all societies (McKenna & McDade, 2005); thus, infant sleeping conditions among humans are more like those of the chimpanzee than we might think.

20.8 Distinctive Human Sleep Architecture

Chimpanzees have the second highest proportion of REM of all the primates (Table 20.1; Samson & Nunn, 2015), after humans. Overall, human and chimpanzee sleep patterns are quite similar (Table 20.1), particularly when one considers that chimpanzees cannot really be active at night. Humans squeeze a chimpanzee-like length of REM sleep into the shortest total sleep time of any primate (Samson & Nunn, 2015). The invention of houses and comfortable sleeping materials undoubtedly has affected our sleep, perhaps allowing this shorter-but-better arrangement (Samson et al., 2017b), but another less pleasant selective pressure also may have acted on sleep.

Human social cohesion grants us deeply satisfying friendships and profoundly moving social experiences, but we share with chimpanzees the downside to social cohesion: Thuggish enemy forces can visit hell on us. Both humans and chimpanzees engage in group violence, although no chimpanzee has ever been attacked by a neighboring community while in their nest at night. We humans, however, must find a way to get a good night's sleep despite the chance that some bloodthirsty horde might swarm out of the darkness to kill us. We need to set a watchman. Humans have evolved what new research suggests is a unique solution to this challenge. Among humans – you will hardly believe this – in even a rather small gathering it is rare for everyone to be asleep at once. In a chimpanzee community, in contrast to humans, virtually every chimpanzee is asleep come 7 p.m. An author of this hypothesis for human sleep patterning, the **sentinel hypothesis**, is one of my former students, David Samson. Sitting with me and my wife at a scientific meeting several years ago, he asked me how often everyone is asleep in my home. My two twenty-something sons were living with me. One is a night owl and is often up until 5 a.m., and I get up as early 5:30 a.m. I said it might be as little as 30 minutes. David had studied the sleep habits of a hunter-gatherer group (Samson et al., 2017c) and found the gap was about a minute per night – a single minute! Someone is (almost) always up. There is always a sentinel.

20.9 Nonhuman Primate Sleep Patterns

Compared to other animals, primates do share some sleep traits in common: They have more intense sleep and, with the exception of apes, have chunked their sleep into one long snooze; at least, they try for uninterrupted sleep (Nunn et al., 2010). Our cousins the monkeys sleep sitting on a branch; only great apes and humans have invented safe havens that guarantee that the dead-to-the-world sleep stage does not result in a dead-in-reality predation event. You might guess that monkeys have evolved sophisticated adaptations to branch-sleeping that allow them to get a good night's sleep sitting up, but not so. They wake many times during the night (Vessey, 1973; Erffmeyer, 1982; Todt et al., 1982; Munozdelgado,

1995), sometimes losing their balance a bit and having to catch themselves.

The innovation of the sleeping platform was a watershed event for early apes, permitting a deeper, more restorative sleep than monkeys (Fruth & Hohmann, 1996). A direct comparison of baboons and orangutan sleep in captivity revealed that apes – by every measure – have longer, deeper, and higher-quality sleep (Samson & Shumaker, 2014). When I fly overnight to Europe or Africa, I try to sleep sitting up in my uncomfortable coach-class seat, but I rarely do, and even when I do my sleep is fitful and sporadic. The lucky stiffs up in first class have got it right: You need to lie down to sleep well. Among the nonhuman primates, apes are first-class sleepers, lying in their relatively luxurious sleeping platforms, while monkeys try to sleep sitting up on their uncomfortable, unstable economy-class branches.

20.10 Chimpanzee Sleep

Chimpanzee sleep is much like that of humans – the most like humans of any animal studied – but with some important differences. Perhaps not surprisingly, chimpanzees spend the greatest amount of time in REM sleep of any nonhuman primate (Table 20.1; Adey et al., 1963; Nunn et al., 2010). They need it. *Large-brained mammals need more sleep* than other species (Zepelin, 1989; Zepelin et al., 2005) and specifically they need more REM sleep (Lesku et al., 2009). You will remember that wonderful brain structure, the amygdala, that transfers emotionally salient events into long-term memory; a hunter-gatherer had best remember where he encountered the lion that scared the microbiome out of him. Everything we know suggests chimpanzees, like humans, consolidate memories during sleep; the size of the amygdala across mammals is correlated with non-REM sleep duration (Capellini et al., 2009). While chimpanzees need a lot of sleep, the night is long year-round in the tropics, 12 hours of darkness, and that is more sleep than they need. Chimpanzees often have a wakeful period in the middle of the night (Zamma, 2014), as do people in preindustrial societies (Samson et al., 2017d). Chimpanzees bed down just as darkness closes in and sleep 11.5 hours, with 9.5 hours of non-REM sleep (versus 6.5 for humans). The poor baboon gets only half that. They spend almost as much time in REM as humans, the highest among nonhuman primates. The chimpanzee sleep cycle is the same as that of humans, 90 minutes.

20.11 Chimpanzee Sleeping Nests

Nests are a bit of a chicken-and-egg phenomenon among apes. Chimpanzees can afford sleep paralysis because they sleep in nests; or perhaps they must make sleeping platforms because their large brains require REM sleep and the insensate period that comes with it (Samson, 2013).

Every night chimpanzees interweave branches to make a new, serviceable sleeping platform or nest and they do it in only three or four minutes (Figure 20.3; Nissen, 1931; Goodall, 1962). Even in captivity, they instinctively construct a ring to surround themselves before sleep, using blankets, straw, or whatever materials are available (Köhler, 1925/1959; Ladygina-Kohts et al., 1935/2002).

Chimpanzees choose their sleeping site carefully. Not uncommonly, a wild chimpanzee chooses a tree with a stem that has a trifurcation, a shape like an inverted tripod; they grip each of the three branches a little less than a meter from the trifurcation point and fold them in so that the leafy tips can be tucked into the center. Other branches are bent and tucked so that there is a functional rim to the nest, defining a circle of about four or five feet (1.5 m). Not unusually, once the framework is finished the individual will test it out then leave the nest, walk a few steps and pull a wad of foliage from a nearby branch, roll it up, and stuff it into the center of the nest; sometimes they fashion the leafy branches into a clearly defined donut or disk shape that Samson (2012) has called a "mattress," which is then wedged into the nest. The firm rim makes a containment vessel, keeping them from rolling out of bed during unconscious and paralytic sleep stages, and the springy but comfortable mattress makes a comfortable pad

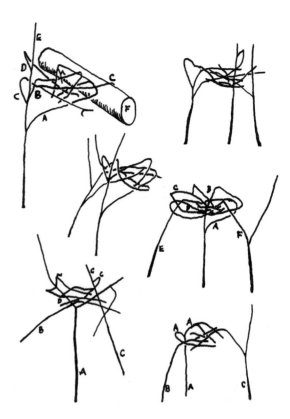

Figure 20.3 Examples of nest construction. From Izawa & Itani, 1966, with permission.

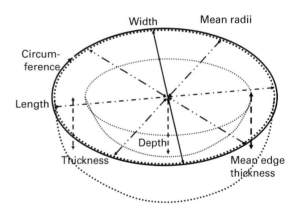

Figure 20.4 Dimensions of a chimpanzee nest as measured at Semliki. Courtesy of David Samson, with permission.

(Figure 20.4; Baldwin et al., 1981; McGrew, 2004; Samson, 2012; Samson & Hunt, 2014).

The higher the sleeping platform is off the ground, the more complex it is (Samson, 2012), perhaps to assure a more secure container at heights where falls are more dangerous and wind can make the perch precarious (Samson & Hunt, 2012).

While chimpanzees nearly always sleep in trees, 5 percent of nests in East Africa are made on the ground (Furuichi & Hashimoto, 2000; Maughan & Stanford, 2001); the frequency is lower in West Africa, with the exception of the Nimba Mountains of Guinea and Côte d'Ivoire, where it may be as great as 35.4 percent (Matsuzawa & Yamakoshi, 1996; Humle & Matsuzawa, 2001). The absence of predators that pose a threat to chimpanzees at sites where ground nests are found is presumably the reason nests can be terrestrial (Furuichi & Hashimoto, 2000; Maughan & Stanford, 2001; Stewart & Pruetz, 2013). Males make their nests lower than females (Brownlow et al., 2001), perhaps because they are at less risk of predation and to avoid lifting their heavier bodies higher than necessary (Reynolds, 1967).

Surprisingly, but not infrequently, nests utilize branches from more than one tree. The first dozen or so times I saw nests constructed, the whole process looked like a failure a minute in. Branches stuck out at angles, the chimpanzee often seemed lost in a cloud of foliage that resisted being formed into any coherent shape. Soon, however, the leafy cloud would yield to the manufacturer and begin to look like a real sleeping platform. Occasionally a nest does fail and a new site is chosen. Infants sleep with their mothers, so they begin to make their own nests only when they are four or five, and it must take some practice because it is juveniles that typically give up on a nest and move to another site to try again.

In the three or so minutes it takes to construct a nest, a lot gets done – and the more the better (Figure 20.5). The more complicated the nest is, the more comfortable (Stewart et al., 2007). It even may be that chimpanzees farm specific sites, and by using them over and over their manipulation of the foliage – breaking branches and so on – encourages the growth of the exact kind of greenery they need to make that comfortable mattress (Stewart et al., 2011).

Chimpanzees are astute observers and selectors of raw materials for nests. David Samson and I found that their favorite nesting trees had multiple advantages (Samson & Hunt, 2014). Their branches were strong and springy, making them more comfortable; the structure was many-branching, making interweaving more thorough; their leaves were dense, providing insulation for cold nights (Koops et al., 2012; Samson & Hunt, 2014); and the tree chemistry was shown to have anti-mosquito or insect-repellent qualities (Samson et al., 2012).

20.12 The First Nest

If the greater cognitive capacities of apes require deep sleep – with a period of paralysis – we might wonder when primates first began to build nests. It is unclear. In the first place, it is not clear which is the most important of the two possible reasons apes need to make nests. It may be that large brains and advanced cognition require the sophisticated maintenance and repair that the REM/body-paralysis phase of sleep provides, which required the invention of the sleeping platform. Or it may be that increased body mass and body weights of over 30 kg (65 lb) may have made sleeping sitting on branches impractical, since the bigger the body, the smaller the branches are in comparison (Samson, 2012; Samson & Nunn, 2015). Great apes have both big bodies and big brains, so the issue is muddled.

If these problems were not enough to make the issue a mare's nest, the nests themselves muddy the waters because as we have seen, they afford many benefits. Nests deter predators (Stewart & Pruetz, 2013), help keep sleepers warm (Stewart et al., 2007; Samson & Hunt, 2012), reduce insect bites and therefore the diseases they spread (Stewart et al., 2007; Samson, 2012), and improve sleep quality (Fruth & Hohmann, 1996; Stewart et al., 2007; Samson & Shumaker, 2013). There are other possible drivers as well: As brains got larger sleep paralysis may have evolved to help maintain that large brain, the paralysis then necessitating the containment system to keep the sleeper from toppling out of his or her sleeping perch; or sheer size might have created the need for a nest, as apes evolved body weights great enough (30 kg seems to be the threshold) that sitting while sleeping became too precarious.

At the moment, the advent of ape-grade cognitive sophistication (and the sleep paralysis that accompanies it) seems most likely (Fruth & Hohmann,

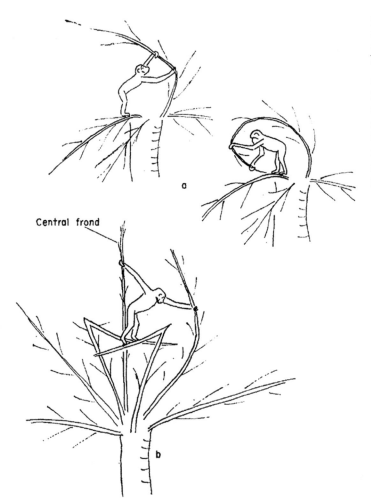

Figure 20.5 "(a) adolescent female attempting to make a nest in a palm tree in 1961. The juvenile female who constructed a nest with a 'roof' during the rains of 1965 used this method. (b) Normal method of constructing palm nest. The ape first climbs half-way or more up a central frond. Holding on to this it reaches out, pulls another frond toward it, bends this down and holds it in place with its feet. It then bends in up to ten fronds in a similar manner." From Goodall, 1968, with permission.

1996; Samson & Nunn, 2015). Recent research has shown that a better sleep environment improves cognitive performance in nonhuman great apes: Captive orangutan sleeping platform complexity, measured as an index of the number of material items available to construct a bed, co-varied positively with reduced nighttime motor activity, more unbroken sleeping bouts, and greater "sleep efficiency" (Samson, 2013). A captive ape study showed that sleep stabilizes and protects memories from interference (Martin-Ordas & Call, 2011).

Since all great apes make nests, but gibbons do not, the first nesting behavior may have emerged around the time gibbons branched off from the great apes, around 17 million years ago (Raaum et al., 2005; Duda & Zrzavy, 2013). By this time, most ape species had exceeded the 30 kg threshold.

The human tree-to-ground move likely occurred with the dramatic morphological changes that took place during the *Australopithecus–Homo* transition (Coolidge & Wynn, 2009). It was at this evolutionary stage that the human brain first achieved a size significantly greater than that of apes; large-brained hominins would have benefited from more stable and less thermodynamically stressful sleeping sites (Samson & Hunt, 2012). Advanced shelter, bedding technology (Samson et al., 2017a), and group-level social cohesion might have allowed the evolution of sentinel-like behavior (Samson et al., 2017b), increased sleep intensity, and a reduced total sleep time.

20.13 Last Thoughts

One of my most pleasurable chimpanzee-watching experiences is the all-day follow of a mother and infant. Mothers move slower than others, so I lose them less often; aggression and all the drama associated with it is uncommon; I like watching mom play with baby; and lastly there have been a few times that chimpanzee mothers have made their nests on a hillside so I could scoot upslope and look into their nest to watch them settle in. I enjoy watching those weary but satisfied last few minutes of wakefulness while mom plays tiredly with baby, baby snuggles up to mom, and tired nature's sweet restorer carries them away, snug as a bug in a rug.

References

Adey WR, Kado RT, Rhodes JM (1963) Sleep: cortical and subcortical recordings in the chimpanzee. *Science* 141, 932–933.

Baldwin PJ, Sabater Pi J, McGrew WC, Tutin CEG (1981) Comparisons of nests made by different populations of chimpanzees (*Pan troglodytes*). *Primates* 22, 474–486.

Born J, Rasch B, Gais S (2006) Sleep to remember *Neuroscientist* 12, 410–424.

Brownlow AR, Plumptre AJ, Reynolds V, Ward R (2001) Sources of variation in the nesting behavior of chimpanzees (*Pan troglodytes schweinfurthii*) in the Budongo Forest, Uganda. *Am J Primatol* 55, 49–55.

Capellini I, McNamara P, Preston BT, Nunn CL, Barton RA (2009) Does sleep play a role in memory consolidation? A comparative test. *PLoS ONE* 4, 4609.

Coolidge FL, Wynn T (2009) *The Rise of Homo sapiens: The Evolution of Modern Thinking*. Hoboken, NJ: Wiley.

Daltrozzo J, Claude L, Tillmann B, et al. (2012) Working memory is partially preserved during sleep *PLoS ONE* 7, e50997.

Duda P, Zrzavy D (2013) Evolution of life history and behavior in Hominidae: towards phylogenetic reconstruction of the chimpanzee human last common ancestor. *J Hum Evol* 65, 424–446.

Ekirch AR (2001) Sleep we have lost: pre-industrial slumber in the British Isles. *Am Hist Rev* 106, 343–386.

Ekirch AR (2006) *At Day's Close: Night in Times Past*. New York: W.W. Norton & Co.

Erffmeyer ES (1982) The nocturnal behavior of caged rhesus monkeys (*Macaca mulatta*). *Folia Primatologica* 38, 240–249.

Ermis U, Krakow K, Voss U (2010) Arousal thresholds during human tonic and phasic REM sleep. *J Sleep Res* 19, 400–406.

Fruth B, Hohmann G (1996) Nest building behavior in the great apes: the great leap forward? In *Great Ape Societies* (eds. Marchant LF, Nishida T), pp. 225–240. Cambridge: Cambridge University Press.

Furuichi T, Hashimoto C (2000) Ground beds of chimpanzees in the Kalinzu Forest, Uganda. *Pan Africa News* 7, 26–28.

Goodall J (1962) Nest building behavior in the free ranging chimpanzee. *Ann NY Acad Sci* 102, 445–467.

Goodall, J van Lawick (1968) A preliminary report on expressive movements and communication in the Gombe Stream Chimpanzees. In *Primates* (ed. Jay, PC), pp. 313–374. New York: Holt, Rinehart, and Winston.

Gümüstekin K, Seven B, Karabulut N, et al. (2004) Effects of sleep deprivation, nicotine, and selenium on wound healing in rats. *Int J Neurosci* 114, 1433–1442.

Humle T, Matsuzawa T (2001) Behavioural diversity among the wild chimpanzee populations of Bossou and neighbouring areas, Guinea and Cote d'Ivoire, West Africa. *Folia Primatologica* 72, 57–68.

Izawa K, Itani J (1966) *Chimpanzees in Kasekati Basin, Tanganyika: (1) Ecological Study in the Rainy Season 1963–1964*. Kyoto: Kyoto University.

Köhler W (1925/1959) *The Mentality of Apes*, 2nd ed. New York: Viking.

Koops K, McGrew WC, de Vries H, Matsuzawa T (2012) Nest-building by chimpanzees (*Pan troglodytes verus*) at Seringbara, Nimba Mountains: antipredation, thermoregulation, and antivector hypotheses. *Int J Primatol* 33, 356–380.

Lesku JA, Roth TC, Rattenborg NC, Amlaner CJ, Lima SL (2009) History and future of comparative analyses in sleep research. *Neurosci Biobehav Rev* 33, 1024–1036.

Martin-Ordas G, Call J (2011) Memory processing in great apes: the effect of time and sleep. *Biol Lett* 7, 829–832.

Matsuzawa T, Yamakoshi G (1996) *Comparison of Chimpanzee Material Culture between Bossou and Nimba, West Africa*. Cambridge: Cambridge University Press.

Maughan JE, Stanford CB (2001) Terrestrial nesting by chimpanzees in Bwindi Impenetrable National Forest, Uganda. *Am J Phys Anthropol* 32, 104.

McGrew WC (2004) *The Cultured Chimpanzee: Reflections on Cultural Primatology*. New York: Cambridge University Press.

McKenna JJ, McDade T (2005) Why babies should never sleep alone: a review of the co-sleeping controversy in relation to SIDS, bedsharing and breast feeding. *Paed Resp Rev* 6, 134–52.

Munozdelgado J (1995) Behavioral characterization of sleep in stumptail macaques (*Macaca arctoides*) in exterior captivity by means of high-sensitivity videorecording. *Am J Primatol* 36, 245–249.

Nissen HW (1931) A field study of the chimpanzee: observations of chimpanzee behavior and environment in western French Guinea. *Comp Psychol Monogr* 8, 122.

Nunn CL, McNamara P, Capellini I, et al. (2010) Primate sleep in phylogenetic perspective. In *Evolution and Sleep: Phylogenetic and Functional Perspectives* (eds. McNamara P, Barton RA, Nunn CL), pp. 123–145. New York: Cambridge University Press.

Opp MR (2009) Sleeping to fuel the immune system: mammalian sleep and resistance to parasites. *BMC Evol Biol* 9, 1471–2148.

Preston BT, Capellini I, McNamara P, Barton RA, Nunn CL (2009) Parasite resistance and the adaptive significance of sleep. *BMC Evol Biol* 9, 7.

Raaum RL, Sterner KN, Noviello CM, Stewart CB, Disotell TR (2005) Catarrhine primate divergence dates estimated from complete mitochondrial genomes: concordance with fossil and nuclear DNA evidence. *J Hum Evol* 48, 237–257.

Rajaratnam SM, Howard ME, Grunstein RR (2013) Sleep loss and circadian disruption in shift work: health burden and management. *Med J Aust* 199, S11–S15.

Reynolds V (1967) *The Apes: The Gorilla, Chimpanzee, Orangutan, and Gibbon; Their History and Their World*. New York: Dutton.

Samson DR (2012) The chimpanzee nest quantified: morphology and ecology of arboreal sleeping platforms within the dry habitat site of Toro-Semliki Wildlife Reserve, Uganda. *Primates* 53, 357–364.

Samson DR (2013) Orangutan (*Pongo pygmaeus*) sleep architecture: Testing the cognitive function of sleep and sleeping platforms in the Hominidae. PhD dissertation. Bloomington, IN: Indiana University.

Samson DR, Hunt KD (2012) A thermodynamic comparison of arboreal and terrestrial sleeping sites for dry-habitat chimpanzees (*Pan troglodytes schweinfurthii*) at the Toro-Semliki Wildlife Reserve, Uganda. *Am J Primatol* 74, 811–818.

Samson DR, Hunt KD (2014) Chimpanzees preferentially select sleeping platform construction tree species with biomechanical properties that yield stable, firm but compliant nests. *PLoS ONE* 9, 1–8.

Samson DR, Nunn CL (2015) Sleep intensity and the evolution of human cognition. *Evol Anthropol* 24, 225–237.

Samson DR, Shumaker RW (2014) Species differences in sleep quality between captive orangutans (*Pongo pygmaeus*) and baboons (*Papio papio*). *Am J Phys Anthropol* 153, 228.

Samson DR, Shumaker, RW (2013) Documenting orangutan sleep architecture: sleeping platform complexity increases sleep quality in captive *Pongo*. *Behaviour* 150, 845–861.

Samson DR, Muehlenbein MP, Hunt KD (2012) Do chimpanzees (*Pan troglodytes schweinfurthii*) exhibit sleep related behaviors that minimize exposure to parasitic arthropods? A preliminary report on the possible anti-vector function of chimpanzee sleeping platforms. *Primates* 54, 73–80.

Samson DR, Crittenden AN, Mabulla AI, Mabulla AZP, Nunn C (2017a) Hadza sleep biology: evidence for flexible sleep-wake patterns in hunter-gatherers. *Am J Phys Anthropol* 162, 573–582.

Samson DR, Crittenden AN, Mabulla AI, Mabulla I, Nunn CL (2017b) The evolution of human sleep: technological and cultural innovation associated with sleep-wake regulation among Hadza hunter-gatherers. *J Hum Evol* 113, 91–102.

Samson DR, Crittenden AN, Mabulla IA, Mabulla AZ, Nunn CL (2017c) Chronotype variation drives night-time sentinel-like behaviour in hunter-gatherers. *Proc R Soc B* 284, 20170967.

Samson DR, Manus MB, Krystal AD, et al. (2017d) Segmented sleep in a nonelectric, small-scale agricultural society in Madagascar. *Am J Hum Biol* 29, 1–13.

Siegel JM (2008) Do all animals sleep? *Trend Neurosci* 31, 208–213.

Stewart FA, Pruetz JD (2013) Do chimpanzee nests serve an anti-predatory function? *American J Primatol* 75, 593–604.

Stewart FA, Pruetz JD, Hansell MH (2007) Do chimpanzees build comfortable nests? *Am J Primatol* 69, 930–939.

Stewart FA, Piel AK, McGrew WC (2011) Living archaeology: artefacts of specific nest site fidelity in wild chimpanzees. *J Hum Evol* 61, 388–395.

Straif K, Baan R, Grosse Y, et al. (2007) Carcinogenicity of shift-work, painting, and fire-fighting *Lancet Oncol* 8, 1065–1066.

Todt D, Bruser E, Hultsch H, Lange R (1982) Nocturnal actions and interactions of newborn monkeys. *J Hum Evol* 11, 383–384, 385–389.

Tononi G, Cirelli C (2013) Perchance to prune. *Sci Am* 309, 34–39.

Turner TH, Drummond SP, Salamat JS, Brown GG (2007) Effects of 42 hr sleep deprivation on component processes of verbal working memory. *Neuropsychology* 21, 787–795.

Van Cauter E, Spiegel K (1999) Circadian and sleep control of hormonal secretions. In *Regulation of Sleep and Circadian Rhythms* (eds. Turek FW, Zee PC), pp. 397–425. London: Taylor and Francis.

Vessey SH (1973) Night observations of free-ranging rhesus monkeys. *Am J Phys Anthropol*, 38, 613–619.

Wagner U, Gais S, Haider H, Verleger R, Born J (2004) Sleep inspires insight. *Nature* 427, 352–355.

Walker MP (2009) The role of sleep in cognition and emotion. *Ann NY Acad Sci* 1156, 168–197.

Webb WB (1988) Theoretical presentation: an objective behavioral model of sleep. *Sleep* 11, 488–496.

Wehr WA (1992) In short photoperiods, human sleep is biphasic. *J Sleep Res* 1, 103–107.

Weinger MB, Ancoli-Israel S (2002) Sleep deprivation and clinical performance. *JAMA*, 287, 955–957.

Worthman CM (2008) *After Dark: The Evolutionary Ecology of Human Sleep*. Oxford: Oxford University Press.

Worthman CM, Melby MK (2002) *Toward a Comparative Developmental Ecology of Human Sleep*. Cambridge, MA: Cambridge University Press.

Xie L, Kang H, Xu Q, et al. (2013) Sleep drives metabolite clearance from the adult brain. *Science* 342, 373–377.

Yetish G, Kaplan H, Gurven M, et al. (2015) Natural sleep and its seasonal variations in three pre-industrial societies *Curr Biol* 25, 1–7.

Zamma K (2014) What makes wild chimpanzees wake up at night? *Primates* 55, 51–57.

Zepelin H (1989) *Mammalian Sleep*. Philadelphia, PA: WB Saunders.

Zepelin H, Siegel J, Tobler I (2005) Mammalian sleep. In *Principles and Practice of Sleep Medicine*, Vol. 4 (eds. Kryger M, Roth T, Dement WC), pp. 91–100. Amsterdam: Elsevier.

21 Chimpanzee Thought Transfer
Communication and Language

Drawing: Scientific American, with permission

PART I: COMMUNICATION

An extraordinary experiment conducted by Emil Menzel (Menzel, 1971, 1973, 1979) revealed the sophistication of chimpanzee nonverbal communication as none had ever done before. He picked a single chimpanzee from his research group and, out of sight of other group members, walked him around an outdoor enclosure, hiding food as he went. This "knower" was then returned to the social group to interact with other chimpanzees. Menzel then released the entire group into the enclosure. The group rushed into the enclosure in a fever of excitement, anticipating retrieving highly desirable treats and quickly found them. Incredibly, sometimes the knower was not the first to reach the treats. Others ran ahead to find it first.

The naïve individuals had correctly interpreted the knower's nonverbal cues indicating that highly desirable treats were hidden. They must have integrated a multiplicity of clues, including the excitement level of the knower, the layout of the enclosure, their knowledge of possible hiding places, and the actions of the knower to know the food was highly desirable and further to anticipate the location of the food items. Sometimes Menzel took out two knowers and showed one a hidden vegetable (meh) and the other hidden fruits (yes!). When the group was released, they followed the individual who had seen fruit. It must be that the level of agitation of the knower communicated important information – the better the hidden food, the more excited the knower.

Body orientation and direction of gaze (I would argue) contributed to inform the naïve chimpanzees what was out there and where it was. Once the knower began to run toward a food, other individuals were able to use knowledge they already possessed, their mental map of their habitat, to guess where the item might be hidden.

In the wild perhaps even more than in captivity chimpanzees communicate important information about social events, dangers, and food source with a combination of body posture (Figure 21.1), facial expression (Chapter 18), agitation level, vocalization, and context. Together these diverse channels are a surprisingly rich means of communication.

21.1 Vocal Communication

There is a greater variety to chimpanzee vocalizations than many realize. Goodall (1986) offered a list of 25 separate calls (Table 21.1) – a substantial repertoire. These vocalizations communicate a broad assortment of emotional states, as well as information about the external world. For instance, chimpanzees can make several declarative statements about their environment, such as "there is food here," "there is a predator near," and "there is something confusing here we should attend to." Many vocalizations express aggression (barks, wraas); others communicate distress, often requesting assistance from any auditor (cries, screams, whimpers). One of the most common vocalizations is the pant-grunt; it communicates to a specific individual that he or she is dominant to the vocalizer; it can range from a soft, calm grunt that would be given to a slightly higher-ranking friend with whom the vocalizer is comfortable, to a loud vigorously articulated, rapidly repeated grunt to express real fear.

Other vocalizations draw attention to the individual, such as the arrival pant-hoot. Still others signal internal states or give notice of some action so

Figure 21.1 Open, hip- and knee-extended postures express confidence and tranquility. High-ranking male Evered, 1987. Photo by the author.

as to benefit the caller. Copulatory vocalizations reinforce the sexual availability of females to high-ranking males. The laugh vocalization expresses satisfaction in playful social interactions. **Tooth clacking** is a way of saying "I'm enjoying this grooming session."

While some nuances of expression are so subtle the human ear has difficulty distinguishing among them, chimpanzees hear them. In yet another layer of sophistication, utterances are modulated to take the audience into account. When recordings of screams produced by individuals who are merely frustrated versus individuals who are truly suffering a physical attack are played to chimpanzees, they looked at the speaker longer when attacks were played, though the screams were indistinguishable to the human ear

(Slocombe et al., 2009). If an individual is asking for help by giving an alarm call and the caller believes there is a hearer in earshot who is high-ranking enough to help them, they produce a different call than otherwise, one that more or less exaggerates the severity of the attack (Slocombe & Zuberbühler, 2007).

Because food is such an immediate concern for chimpanzees, and because it is a resource that can be contested, many vocalizations concern feeding and food items. A food grunt expressively sounds like a grunt with the vocalizer's mouth full of food, even when empty; it communicates to others the satisfaction individuals have in the particular item they are consuming.

Chimpanzee food calls, specifically long-distance pant-hoots, have both meaning and purpose, signaling to other chimpanzees that the caller is feeding or about to feed at a food resource rich enough to accommodate a large feeding party (Reynolds & Reynolds, 1965; Goodall, 1968, 1986; Wrangham, 1977). We are not just guessing; experimental manipulation confirms that chimpanzees pant-hoot when they encounter an abundant food source that can be shared (Hauser et al., 1993; Mitani & Nishida, 1993). Pant-hoots emitted at particularly large food resources have a distinct acoustic signature, with a conspicuous "let-down" phase in which the call becomes distinctly quieter at the end (Notman & Rendell, 2005). In other words, when a caller pant-hoots, others in the distance can tell if he is feeding in a tree large enough that he wishes to share the wealth (Hauser & Wrangham, 1987).

A pant-hoot communicates not only the size of the resource, but the identity of the caller. Differences in the length of each hoot, the point in the hoot where an emphasis is placed, the pattern of decreasing or increasing volume, the length of pauses between hoots and voice quality – the same qualities that allow us to recognize a familiar voice – allow individuals to recognize one another (Goodall, 1968; Marler & Hobbett, 1975; Bauer & Philip, 1983; Mitani, 1996).

There is enough consistency within communities and differences among communities in pant-hoots that we could call vocalizations at different sites

Table 21.1 **Wild vocalizations**

Vocalization	Context	Meaning
Aaah call	Social	Food enjoyment
Aggressive barks	Conflict	Cease what you are doing or I may attack
Arrival pant-hoot	Entering party	Announcement of arrival
Bark	Conflict	Surprise, drawing attention to object of gaze; grades into aggression
Copulatory grunt, pant and scream	Copulation	Signals receptivity to ranking males
Cough (soft bark)	Social	Annoyance
Cry	Distress	Announces distress; requests assistance
Desperation scream	Distress	Extreme version of "cry"
Extended grunt	Social	Feeling of sociability, emphasized
Food grunt	Feeding	Expresses satisfaction with food item
Hoo	Social	Distress
Huu	Confusion	Notifies others of unexpected thing
Inquiring pant-hoot	Social	Elicits response notifying location of others
Laugh	Joy, tickling	Express happiness, enjoyment, playfulness
Lip smack	Grooming	Assures social commitment, bonding
Nest grunt	Social	Feeling of sociability, ease
Pant-grunt	Dominance	Expresses subordinance; reduces aggression
Pant-hoot food call	Feeding	Draws group members, increases group size
Roar pant-hoot	Social	Extreme social excitement
Scream	Agonism	Express distress at prospect of attack or actual attack
Soft grunt	Social	Feeling of sociability
Spontaneous pant-hoot	Relaxation	"All clear" signal that expresses well-being
Tooth clack	Grooming	More intense expression of lip smack
Whimper	Social distress	Express frustration or fear
Wraa	Agonism	Similar to bark

After Goodall, 1986.

"dialects" (Mitani et al., 1992, 1996; Crockford et al., 2004). As you might expect, chimpanzees recognize when a vocalization is that of a stranger (Herbinger et al., 2009).

I have often heard calls in the distance, seen individuals respond by hurrying off in the direction of the pant-hoots, and arrive at a feeding tree where several other chimpanzees have already congregated. In one experiment at Kibale-Kanyawara, a recorded call was played to observe the reaction of individuals over 1 km distant. The hearers marched toward the calls (the speakers were collected right after the playback) and one individual charged into the playback area and sat on the *exact spot* where the speaker had been.

Together these things mean that when a chimpanzee hears a pant-hoot in the distance, he (they are more often directed toward males than females) can tell where the caller is, who the caller is, and approximately how much food is in the tree the caller is calling from. Chimpanzees have an intimate enough knowledge of the location of feeding trees that I believe they are able to guess the exact tree the caller is in.

21.2 Facial Expressions

Chimpanzee facial expressions differ little from those of humans. The **fear-grin** is an exception; a very wide, open mouth (the grin or smile) with closed teeth is an expression of great fear in the chimpanzee. See Chapter 18 for more detail.

21.3 Gestural Communication

Not surprisingly, there is a strong cultural component to chimpanzee communication, perhaps more for gesture than vocalization. A study of chimpanzees at Budongo recorded 66 different communicative gestures (Hobaiter & Byrne, 2014). Table 21.2 lists 32 gestures found in common at three East African sites: Gombe, Mahale, and Budongo ("stomp social partner" was not listed for Mahale, but I saw it myself in 1987).

Table 21.2 **Gestural communication**

	Gesture
1.	Arm raise
2.	Beg; palm upward reach
3.	Bite/kiss
4.	Bow
5.	Branch drag
6.	Clubbing
7.	Dangle (armhang above social partner)
8.	Direction indicating push
9.	Drum
10.	Grab social partner
11.	Grab social partner and pull
12.	Hand/arm oscillation
13.	Head nod
14.	Kick social partner
15.	Loud scratch
16.	Loud, exaggerated gallop
17.	Pirouette
18.	Poke
19.	Present body part for grooming
20.	Present swelling for copulation
21.	Shake hands
22.	Slap ground/stationary object
23.	Slap social partner
24.	Somersault
25.	Stare at social partner
26.	Stomp ground
27.	Stomp social partner
28.	Tandem walk, side-by-side embrace
29.	Tap social partner

Table 21.2 (*cont.*)

	Gesture
30.	Throw object
31.	Touch social partner
32.	Vegetation or object sway

After Hobaiter & Byrne, 2014.

Figure 21.2 An outstretched arm with an upturned palm is an unmistakable request by a chimpanzee for some physical object, or even – symbolically – a request for aid. Photograph taken at Burgers Zoo, courtesy of Frans de Waal.

A limbic-driven signal is **piloerection**, the fluffing up of the hair. This is an expression of confidence, agitation, and social dominance. Erect hair makes the individual appear larger than they actually are, and it is a very convincing illusion; sometimes excited males look like they're wearing a down snowsuit. When chimpanzees engage in their flamboyant, dominance-reinforcing social display, they pair a compressed-lip facial expression (Figure 18.2; Chapter 18) with piloerection. It is an impressive sight.

Some gestures are symbolic, such as begging, signaled by an outstretched hand with the palm up (Figure 21.2), a gesture used not only when some physical object is desired – typically food – but also when the gesturer is requesting some action, usually aid or protection when under attack. Another common gesture is the social scratch. The signaler scratches his or her thorax in long, exaggerated strokes (Nakamura et al., 2000), signaling that the individual is intending to leave. The scratch is often used to invite others to join in a travel party and thus sustain their social contact.

Drumming counts as a gesture in many gestural lexicons (e.g., Hobaiter & Byrne, 2014). Its function is not widely agreed upon, but it is likely a territory-claiming behavior, expressing the robust health, power, and vigor of the drummer, meant for both the ears of your enemies and your competitors alike. The beat is quite rapid – too rapid for me to produce – and requires superhuman power as well as speed. I know of no human who can do it; when people attempt it they make an ineffectual slapping sound that carries only a few feet, lacking the deep resonance chimpanzees produce that can be heard for over 1 km.

Chimpanzees who are socially comfortable and confident sit or lie in postures with the limbs partly or wholly extended – for instance, lying on the back with legs and arms outstretched. Fearful chimpanzees assume a fetal posture, flexing the hips, knees, and elbows, covering their belly, and scrunching down into the smallest package possible. An overconfident posture can incite the wrath of a dominant; contrariwise, a subordinate posture may decrease the likelihood of attack.

21.4 Synthesizing Diverse Communication Channels Creates Complex Meaning

The combination of facial expression, body posture, gesture, vocalization, and context has a multiplicative effect that vastly increases the richness and variety of information communicated. For instance, a begging gesture combined with a whimper says "don't hit me," whereas a begging gesture and a lip smack is a request for grooming that puts some special emphasis on the event. A begging gesture and whimper in the context of a meat-eating episode means "please share some meat with me."

I am reminded of the incredible specificity that can come from the combination of a shared mental map, a simple gesture, and a scant few words when I recall a marvelous exchange I once had with an assistant. Late in the day I had left a feeding party in a fig tree high on the escarpment, miles from camp. The group would likely nest nearby, but I had to leave them before they bedded down. The next day I was unable to go into the forest and I wanted my assistant to begin his search at the fig tree. Standing in our camp, flicking my eyes for a second up toward an area on the rift escarpment perhaps 10 km from camp, I said to him "you know that big fig…"; as I was saying this I was making a gesture as if I were running my hand up a smooth, sloping surface. I meant to go on and say "on Binghi trail 2.2, on top of that sloping hill above that flat place where the ground cover is mostly *Aframomum*," but he interrupted me at "big fig" to say "yes, I know it." Subsequent conversation showed he had indeed understood the exact place. Out of the thousands of figs in the forest and millions of trees, I was able to specify one in particular in three seconds with five words and two gestures. Gesture and language interacted with our shared knowledge of the forest and chimpanzee behavior to efficiently convey a very specific bit of information. You may be skeptical, but I have little doubt that if I had attempted to communicate this to a chimpanzee who possessed sign language, she could have found that fig tree.

PART II: CAPTIVE LANGUAGE STUDY

21.5 The Kelloggs, the Hayeses, Vicki, and Cross-Fostering

The dazzling cognitive performances of chimpanzees detailed in the works of Köhler, Kohts (Chapter 18), and Harry Raven (1932, 1933) contributed to an atmosphere of urgency and anticipation among linguists and psychologists who were becoming ever more eager to discover exactly how human-like chimpanzee linguistic and cognitive skills were.

We began to learn the limits of chimpanzee abilities by **cross-fostering** them – rearing a chimpanzee as a human – challenging them to assimilate human culture and behavior right down to using the right piece of silverware and documenting exactly where they fell short (Figure 21.3). From 1913 to 1916 Kohts spent her days with infant chimpanzee Johnny (my Anglicization of "Ioni"), whom she reared as if he were a human, though subjected to an inhuman array of psychological tests; she left her subject only at the end of the full work day. Chimpanzee Johnny clearly looked on Ladygina-Kohts as his mother, and the daily separations were upsetting.

Figure 21.3 Vicki enjoying dinner with her family, Catherine and Keith Hayes.

An American professor and his wife, Winthrop and Luella Kellogg (Kellogg & Kellogg, 1933), bravely took Kohts' experiment in cross-fostering one step further. The Kelloggs reared Gua chimpanzee in their home with their son Donald, engaging in what was essentially a controlled experiment, exposing both to the same environment in a quest to determine precisely how similar chimpanzee and human development was. The answer was "very." The Kelloggs documented in meticulous fashion both individuals' first steps, sleeping postures, table manners, and so on. Communication was a subject of the Kelloggs' work, but not language in particular. Their book detailing their experience is still a fascinating glimpse into the world of the chimpanzee and well worth a read. Yet, let us not get too carried away. While much of chimpanzee behavior is learned, some of it has a biological basis. Gua adopted many human behaviors, but he still developed, more and more as he grew, the instinctual behaviors of a chimpanzee.

Interesting also is a riveting fictional account of rearing a chimpanzee as a human that was inspired by the experience of the Kelloggs and others, *We Are All Completely Beside Ourselves*, by Karen Joy Fowler. The book is set in my hometown, Bloomington (the Kelloggs were Bloomington residents when their adventure with Gua began), and follows the life of a cross-fostered chimpanzee as she matures. Fowler's chimpanzee, as is the case with all chimpanzees, displayed her chimpanzee-ness more and more as she matured.

The cross-fostering experiences of these workers revealed how similar chimpanzee gestures, facial expressions, and emotions were to humans – almost identical. Perhaps, linguists wondered, the same was true of language. If Meshie, Gua, and Johnny could feed the baby, eat with a spoon, and help do the dishes, perhaps, they were intelligent enough to acquire language. Supporting this expectation, humans who have had little or no exposure to language never acquire normal speech. That is, while the biological basis of language is inherent in humans, fully realized speaking ability *only* develops with exposure during a critical period when children are able to listen to and practice speech (Curtiss, 2014). Could it be that chimpanzees lack speech only because no chimpanzee has ever been exposed to it during the critical period?

The first cross-fostering experiment explicitly directed at language acquisition was that of Catherine and Keith Hayes (Hayes, 1951) who, as had the Kelloggs, reared their subject, Vicki, in their home (Figure 21.3). Her participation in their daily lives was every bit as intimate as would be that of a human child. Like doting parents, the Hayeses repeated the names of important people, concepts, and objects in Vicki's life.

There was no reason to expect a physical barrier to chimpanzee language acquisition, since chimpanzees have extraordinary control of their lips and tongues, so much so that after spending a year around chimpanzees almost exclusively, when I returned to the human world and began to converse with my friends I sometimes had the disquieting impression that their faces were partly paralyzed. When chimpanzees use their lips to manipulate a food item, they sometimes display clownish facial contortions. Sometimes as a chimpanzee chews they separate the nutritious part of a food item from its shell or stem, later spitting out the undesirable bits; if they have problems separating the two parts they may push the mass of food onto their bottom lip, poke out the lip and look down their nose at the partly masticated mélange to see exactly what the problem is. That takes a lot of lip and tongue dexterity. In fact, even monkeys appear to have a vocal tract that *could* produce human language (Fitch et al., 2016).

While to my ear a chimpanzee pant-hoot sometimes contains both "oohs" and "ahhs," Phil Lieberman and his colleagues found indistinct vowels when they examined sonograms (Lieberman et al., 1972). That matters little, though, since most of the information in speech is in the consonants; Vicki had trouble with those, too. She was unable to press her upper lip against the lower to make a creditable "p" sound, and the same for the "m" sound in "mama." With training, she learned to push the upper lip down with a finger, but that meant she had to put a hand to the face to speak. Despite untold hours of training, Vicki learned only four words: mama, papa, cup, and up. These four words, come to think of it, could satisfy a large part of her needs, since – like any other chimpanzee infant – her heart's desire was for "mama" or "papa" to pick her "up" and carry her. And if she got a "cup" of juice, well, life was perfect.

Possibly, specialists speculated, the problem was with neural control of the mouth, lips, tongues, and perhaps larynx. True, but let me add that while they cannot produce the sounds humans can, chimpanzees do create a wide variety of sounds. So much so that I believe we could create a spoken language that chimpanzees could articulate, drawing on the wide variety of utterances they *can* produce, if we worked around their limitations, but so far no one has attempted it.

Four words is not "language," clearly, but what if Vicki had progressed further? What if she had acquired 100 words? Was that language? What is the bare minimum required to justify calling a communication system a language? This was to become a thorny issue as later sign language experiments took chimpanzees much further along the road to human communication.

21.6 Language Design Features

While it is getting a bit ahead of the chronology of ALR (ape language research), let us change direction a bit and discuss how linguists determine what it is that makes a communication system a language. **Charles Hockett** (Figure 21.4) proposed that a true language, as opposed to some non-language communication system, must have tightly defined qualities he called **design features** (Hockett, 1958, 1959, 1960a, 1960b, 1966; Hockett & Altmann, 1968). If any one of these features is absent, he maintained, the communication system, no matter how sophisticated it is otherwise, cannot be called a language. He tinkered with his list, adding and subtracting items to it over the years; other linguists chipped in, suggesting tweaks to some design features, wondering whether others were truly necessary, or proposing expanding the list (Taylor, 1974; Hill, 1978; Hauser et al., 2002; Everett et al., 2005; Fitch, 2005); given all this churning, it turns out there is no definitive, universally agreed upon roster of language design features. While most linguists see some utility in Hockett's attempt to define language, others would go beyond tweaking his inventory and

Figure 21.4 Charles Hockett contended that a true language must possess a number of design features, without which the communication system could not be called language, regardless of its complexity. Photo: Cornell University; with permission.

simply toss out his list altogether, viewing it as too simplistic to be helpful. A linguist colleague of mine, Professor Phil LeSourd, would call any system that is sufficiently open-ended (i.e., productive, feature 5 in Table 21.3) a language. Another colleague, Professor Steven Franks, would take the rather Chomskyan view that most of Hockett's design features are superfluous; few of them make real distinctions between human and chimpanzee language (that would be sign language, which we will begin to explore soon). Most of what is distinctive about human language, Franks says, is concentrated in only one feature: grammar. There, in the realm of grammar, humans and chimpanzees are profoundly different.

Thus, while there have been criticisms of the design-features concept, its relevance to understanding the nature of language can be seen in recent scholarly works that use it (Corballis, 2017; Everaert et al., 2017).

Keeping in mind criticisms of Hockett's various lists, I have chosen nine features that I believe both best represent Hockett's ideas and which best

Table 21.3 **Language design features**

Design feature		Definition
1.	Discreteness	Meaning is conveyed by words, not loudness or length
2.	Duality of patterning	Sounds are put together to make words, words are put together to make sentences
3.	Arbitrariness	Individual sounds can be changed
4.	Displacement	Language can refer to things not present
5.	Productivity	Novel expressions can be created
6.	Innateness	Acquisition without training
7.	Reference to abstractions	Can refer to more than just physical objects
8.	Natural and flexible use	Used conversationally, not in a Pavlovian sense
9.	Grammar	Rules that add critical meaning to words

After Hockett, 1960a, 1960b.

distinguish human language from non-language communication systems.

Language, as the design features hypothesis would have it, has **discreteness**; its meaning is conveyed by a concatenation of discrete sounds that make up words, not by the loudness or length of the sounds. To modify "dog" to mean "big dog," the length of the word is not drawn out, repeated or spoken louder, but modified with *big*. The "big" is a discrete word that contains the meaning of bigness. In contrast, a pant-grunt communicates "I acknowledge I am subordinate to you; be nice." Pant-grunting louder, repeating the grunt more times, or stretching out the grunt amplifies its meaning; it says "I acknowledge I am *extremely* subordinate to you, I beg you, please, do not attack me!"

Language has **duality of patterning**, a two-level hierarchy. Discrete sounds are put together to make discrete words. For example, the sounds "p," "i," and "t" are distinct and can be put together to form different words – pit and tip; that is one level, sound order. The second level of this hierarchy is that words can be put together to make sentences, "Chucky chimpanzee tipped the dirt in the pit." Sounds make words, words make sentences.

A language must have **arbitrariness**; any one sound can be changed without changing meaning. Consider the phrase, "Chucky chimpanzee chomped on cheese." If I were to tell you that I have a code language in which an arbitrary sound, "ch," is always replaced with the sound "d," the resulting phrase, "ducky dimpanzee domped on deese," is comprehensible despite the arbitrary substitution of "d" for "ch." Our facility in understanding accents shows that we humans adjust to this sort of shift readily.

Language has the capacity for **displacement**. It can refer to things not present, either because they are out of sight, in the past, or in the future.

Language must have **productivity**; it must have the power to create novel expressions. Merely repeating or mimicking phrases is not productivity. I once read a story of a well-meant finger-painting play session gone wrong that illustrates this point. Paint ended up, among other places, on the pet cat. Little Tommy then attempted to wipe that paint onto his sister. The mother's phrase in response demonstrates productivity in that I doubt it had ever been uttered before in the history of humanity: "Tommy, don't paint your sister with that cat!" Motherhood is a challenge, but who knew it demanded such sophisticated linguistic theory? Let us hope Tommy remembers the flowers on Mother's Day.

Language has **innateness**. It is acquired without training. A real language should not require the sort of rote repetition involved in training your dog to ring a bell when it needs to go outside. True language is acquired seemingly effortlessly in the course of normal social interactions during infancy and childhood; no teaching or training is necessary.

A true language can **refer to abstractions**. It gives the speaker the capacity to talk about things like affection, preference, and approval.

Natural and flexible use is characteristic of a real language. Rather than stilted or rote responses to stimuli, language is used spontaneously, without prompting, to comment on everyday events and to respond appropriately to a conversational partner.

Lastly, language has rules that constitute a **grammar**. The way words are put together imparts meaning. The words "put the saucepan in the water" means something different from "put the water in the saucepan."

21.7 The Gardners and American Sign Language

Vicki's failure was enough to convince the scientific community that spoken language was impossible for chimpanzees, even if the chimpanzee brain was capable of it. Because the problem was thought to be with the neural control of the lips, tongues, mouth, larynx, or diaphragm, or any combination of these structures – or it might even be in the shape of those organs – this line of research was seen as a dead-end.

Linguists realized that complicating the acquisition of spoken language was that chimpanzee vocalizations are under limbic control; the limbic system is a constellation of brain areas that together are responsible for emotional expression. The limbic aspect of chimpanzee vocalizations is apparent in their reflexive response to stimuli; they not only vocalize with predictable regularity in particular circumstances, when exposed to a stimulus that normally evokes a specific call they struggle to suppress the vocalization (Gardner & Gardner, 1969, 1975, 1998; Gardner et al., 1989): When they feel subordinate, they pant-grunt; end of story. When lost and despairing, they whimper; when anxious and fearful, they scream; and when excited by food, they food grunt. Restraining such vocalizations is difficult.

Human speech, in contrast, is under the control of cortical regions of the brain and thus is voluntary (see Chapter 19). There are some limbic human vocalizations, grunts, and screams, but such human vocalizations are rare compared to speech. In other words, human and chimpanzee vocal communication are qualitatively different; chimpanzee vocalizations are an involuntary expression of an internal emotional state, whereas human vocalizations are voluntary and a result of higher thought.

We have already seen that chimpanzees in the wild are astute readers of facial expressions, body postures, and gestures. Perhaps, it occurred to ALR scholars, the voluntary nature of such gestures as begging and patting, which are intentionally evoked to communicate something specific to a social partner, means that a gestural or signed language might be possible.

The Gardners were the first to attempt to teach **American Sign Language (ASL)** to a chimpanzee, a female named **Washoe** (Figure 21.5). Again, the cross-fostering technique was utilized. The Gardners used ASL in Washoe's presence just as parents would with a nonhearing child. The experiment was a success. In 1969, they reported that Washoe had learned 132

Figure 21.5 Washoe with the Gardners. Photo: Friends of Washoe; published with permission.

signs, enough to communicate rather well – a vocabulary like that of a two-year old.

Washoe's accomplishments met a number of the design features in Table 21.3. The ASL sign for hat is a flat palm placed on top of the head (Figure 21.6); exaggerating the sign does not indicate "big hat" – instead, an adjective must be added to modify the word, just as we do with spoken language. Washoe had acquired, in other words, the ASL equivalent of **discreteness**, or the capacity to convey meaning by words. Signs have discrete parts, such as a palm oriented downward versus rotating the wrist. Signs have various components that, when strung together, make up words, and words are strung together to make sentences. Washoe had **duality of patterning**. American sign language has **arbitrariness**; in other words, the sign for "open" – hands placed side by side, palms down, then drawn apart while rotating the palms to an up position – could just as easily have been signed by a fist opened up into a palm. Washoe signed about special treats she expected and she anticipated upcoming meals; she asked about people she knew when they were absent. In other words, she demonstrated **displacement**. Furthermore, she created her own sentences, rather than just mimicking or repeating things said to her, demonstrating **productivity**. While many signs mimicked some aspect of the referent – the thing the sign stood for – others, such as "love" and "blue" did not. Reference to "hurt," "sorry," and "funny," used appropriately, showed Washoe could **refer to abstractions**.

When observed without her being aware of it, she sometimes signed to herself; she often commented on sounds heard outside her enclosure, or interesting things in her environment. She used ASL to express her desires and concerns and she asked about interesting objects around her. While rarely discussed as part of the concept of natural use, years of language use seeped so deeply into the fabric of their lives that not only Washoe, but all of the Gardners' chimpanzees used ASL spontaneously among themselves, when no humans were involved (Cianelli & Fouts, 1998). This constitutes **natural and flexible use**, checking off yet another design feature.

In 1970 Washoe was moved to a non-ASL-trained chimpanzee research colony at the Institute for Primate Studies (IPS) at the University of Oklahoma, where the Gardners' work was continued by Roger and Deborah Fouts. One can only wonder what Washoe thought of these seemingly dimwitted, non-language-trained chimpanzees as her signed requests for a hug or a tickle fell on deaf eyes (Fouts et al., 1989). Whatever Washoe might have thought of her language-deficient companions, it did not preclude physical attraction. She became a mother, though a somewhat tragic one. She had two doomed infants in quick succession, but neither lived very long (Fouts et al., 1989). Although the infants were sickly, Washoe's mothering was faultless, inspiring the IPS team to acquire an infant, Loulis, whom they hoped Washoe would adopt (Figure 21.7), allowing them to

Figure 21.6 Washoe makes the sign for hat. Photo: Friends of Washoe; published with permission.

Figure 21.7 Washoe with the adoptee Loulis. Photo: Friends of Washoe; published with permission.

test whether chimpanzees could learn ASL without human intervention.

The Foutses instructed their team to use only seven signs in Loulis' presence, WHO, WHAT, WHERE, WHICH, WANT, SIGN, and NAME, but otherwise refrain from signing, a program that went on for five long years. As this experiment was underway the team also added other chimpanzees who were taught ASL; Moja learned 168 signs, even more than Washoe; other chimpanzees were equally impressive. While some linguists had dithered as the data on Washoe had appeared, the continued successes of other language-adept IPS chimpanzees quieted criticism. Meanwhile, Loulis matured among a community of cross-fostered, signing chimpanzees, and without human training he acquired 51 signs. Not as impressive as Washoe's 132, but enough to prove that chimpanzees possess **innateness**. The Gardners, Fouts, and their colleagues (Fouts et al., 1989) had ticked off, one by one, eight of the nine design features Hockett and others had compiled, leaving grammar alone as a solitary outlier (Table 21.3).

In all, Allen and Trixie Gardner, Deborah and Roger Fouts, and a later collaborator, Thomas Van Cantfort, trained six chimpanzees – Washoe, Moja, Pili, Tatu, Dar, and Loulis – in ASL, and five of those six were cross-fostered. It is noteworthy that over time each of these chimpanzees showed a steady improvement in vocabulary size, originality of word use, sophistication in patterns of phrase development, and competence in responding to different types of Wh- questions: Who, What, When, Where, and Why queries (Gardner et al., 1989). Van Cantfort wrote to me that "Their development never reached an asymptote. Training for Moja, Pili, Tatu and Dar lasted only from birth to about seven years of age – clearly the upper limits of cognitive and linguistic development in chimpanzees has not yet been achieved."

21.8 Premack and Sarah

Meanwhile, while the Gardners notched one success after another, David Premack was conducting an ALR experiment using a completely different approach (Premack, 1976, 1983a, 1983b; Premack & Premack, 1972, 1982). He used a **lexigram language** – plastic symbols – rather than signs. He trained chimpanzee Sarah to recognize, for instance, that a blue triangle was the symbol for apple (Figure 21.8). If she was shown a blue triangle and asked "What color is this?" her response was "Red." Asked "What shape is it?" her response was "Round." If asked "Does it have a stem?" which a triangle clearly does not, she responded with "Yes." Premack's work demonstrated to the skeptics that chimpanzees understood symbolism.

If there were doubts about the Gardners' or Premack's results, for the most part they were satisfied when yet another cross-fostered chimpanzee, Lucy, was taught ASL. Lucy was reared as a child in the home of Maurice and Jane Temerlin (Temerlin, 1976) and taught sign language by Roger Fouts. Her vocabulary was quite similar to that of Washoe, 140 signs (Fouts, 1973; Fouts & Couch, 1976; Fouts et al., 1978).

As spectacular as these discoveries were, and even though results were accumulating in a variety of labs, some scholars counseled caution, worrying that some unintended procedural error might have crept into

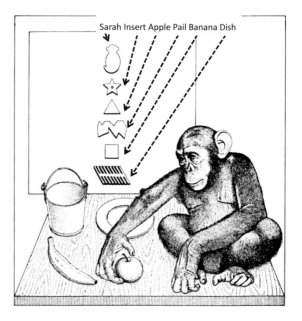

Figure 21.8 Premack's Sarah completes a task dictated by the icons on the board behind her, which instructs her to put the apple in the pail and the banana on the plate. After Premack, 1972; with permission from Scientific American.

their experimental design: Washoe and the other chimpanzees might *seem* to understand language, but in fact they might be responding to some unintended cue by researchers (Umiker-Sebeok & Sebeok, 1981). They might not have language at all.

21.9 Applying Caution: The Clever Hans Effect

This phenomenon, unintended cueing, is called the **Clever Hans Effect**, after the celebrated horse **Clever Hans** who was, his trainer claimed, able to evaluate complicated arithmetic problems, including addition, subtraction, multiplication, and division, among other impressive cognitive feats (Pfungst, 1911). He did so in public venues for entertainment, but also to allow doubters to pose their own problems to Hans. Clever Hans' prowess was demonstrated in a procedure in which an echo-y wooden platform was placed in front of him while his trainer, by means of a pointer or setting wooden blocks in front of him, posed difficult cognitive challenges. To show Hans had not rotely memorized some sequence of answers, the trainer might invite members of the audience to suggest a mathematical problem (Figure 21.9). Once the problem was posed, Hans would answer by clopping a hoof the appropriate number of times on the box. An audience member might offer a sum like "11 plus 3," and Hans would clop out "14." Or the problem might be "11 minus 7" and Hans would dutifully clop out "4."

It is widely agreed that the trainer himself was quite convinced that Hans was legitimately performing these impressive intellectual tasks. Suspicious minds, however – there are always suspicious minds – suggested that Hans' abilities should be tested with experiments designed to ensure that the trainer was not unintentionally tipping off Hans. It turned out that when Hans was unable to see the trainer, he would start clopping and just continue indefinitely. The trainer had been unknowingly making eye contact with Hans, or had been altering his posture – or both – just as Hans reached the right number. Rather than doing math, Hans was just looking at the trainer and stopping whenever he got the eye contact/ posture change cue. Thus, the Clever Hans Effect.

The worry that Washoe and other ALR chimpanzees were merely Clever Hans cases spurred the Gardners

Figure 21.9 Clever Hans, his trainer, his wooden counting platform, and the number chart. Change in posture or affect when Hans reached the correct answer to the problem cued him to cease tapping on the platform.

to set up a double-blind test (Umiker-Sebeok & Sebeok, 1981). One experimenter, visually isolated from Washoe, displayed images on a screen that was placed at the end of a tunnel; Washoe looked into the tunnel while signing the word for the image she saw. A second experimenter, responsible for recording Washoe's responses, could see neither the researcher operating the slide projector nor the image itself. This observer merely recorded what signs Washoe made. Then the order of the signs recorded was compared to the order of the slides. Washoe was correct 80 percent of the time. Furthermore, the Gardners argued that she appeared to know the correct answer even when she made mistakes. Sometimes she made lazy attempts at the sign, not raising her hand high enough, or not opening a hand completely, causing errors. Whether true or not, 80 percent was too high for the Clever Hans Effect to account for her abilities.

21.10 Herb Terrace and Nim Chimpsky

A further challenge came from the research of Herb Terrace (Terrace, 1979; Terrace et al., 1979), who conceived a project quite like that of the Gardners. Initially he thought his work confirmed theirs. His experimental subject, Nim Chimpsky, learned 125 words by four years of age, and Nim's proficiency convinced Terrace, at least at first, that he had acquired a level of proficiency quite like that of Washoe. On closer inspection, however, Terrace found that Nim tended to produce more or less random signs, blurting out one sign after another until he got what he wanted. His "sentences," Terrace came to believe, were actually long, meaningless strings of words. By poring over these random words, one might extract a string of three or four that made a sentence, but an objective appraisal of Nim's sign production suggested that, while he had memorized some signs, he was not nearly as proficient as the Gardners, Fouts, and Premack claimed for their chimpanzee subjects.

In other words, Nim had only the first half of duality of patterning; he seemed to understand the meaning of words, but could not be said to be making sentences. There was no evidence of grammar; rather, he used "bite me" to mean both "I bite" and "you bite me." He completely failed to grasp turn-taking and interrupted his trainers constantly. Terrace concluded that what Nim had was not language. The implication was clear: If there were to be an objective evaluation of the data from Washoe and Moja, it would probably reveal the same thing.

The Gardners fired back. They pointed out that Terrace's research was unfunded and thus chaotic for the first 19 months, a time period that among humans – a year and a half – would constitute a disturbing proportion of the critical period of language development. Given that there is a window for language acquisition after which optimal proficiency can never be obtained, it was an open question whether Nim had missed that window. Furthermore, there was high trainer turnover, a critical issue since chimpanzees are evolved to bask in the stability of a mother's constant presence, 24 hours a day, until the age of five. The Gardners found it telling that when Nim was in a calm social setting with a familiar trainer his signing was more like that of Washoe. It was impossible, they asserted, to tell whether Nim's failures were due to lack of ability or lack of motivation brought on by his chaotic and depressing social environment. Lastly, they pointed out that Nim was only four when Terrace made his conclusions; surely a chimpanzee still in infancy and suffering from an unstable home life cannot be taken as a measure of the capabilities of the species as a whole.

So here ALR sat in the late 1970s and early 1980s, as scientists continued to digest the corpus of ALR scholarship. Perhaps a fair assessment of the state of the field at this point is that it seemed likely, I would say very likely, though not certain, that chimpanzees had mastered eight of nine language design features. The Gardners would say all nine, grammar included. Others worried that even the first eight were iffy. It was, in fact, a stalemate.

Into this scholarly fray stepped **Sue Savage-Rumbaugh** and **Kanzi** to change everything.

21.11 Kanzi and Grammar

Savage-Rumbaugh focused her attention on bonobos, and was attempting to teach an adult female, Matata, a lexigram language. Matata's age meant she had probably passed the window of opportunity for learning language, and Savage-Rumbaugh found it slow-going. Matata, however, had an adopted son, Kanzi, who took an interest in the lexigrams, and he began to answer when his mother had questions posed to her. It was clear Kanzi had much more potential than Matata, so Savage-Rumbaugh shifted focus to him. He quickly matched the competencies claimed by ALRers for their subjects, then surpassed them. His vocabulary reached 250 words, then 500, then even more (Savage-Rumbaugh et al., 1977, 1998; Savage-Rumbaugh, 1986; Savage-Rumbaugh & Lewin, 1994); he may be able to understand as many as 3000 words. He frequently asked about future or out-of-sight events, such as "Will we go to the A-frame today?" and "Will [a specific trainer] come today?" He passed double-blind tests and other challenges.

Kanzi started with lexigrams on paperboard, but soon Savage-Rumbaugh had moved on to a computer touch-screen device that also "spoke" whatever word Kanzi touched on the screen. Savage-Rumbaugh did not just claim that he understood a large number of spoken words, she offered video that proved it.

If there was a tipping point in ALR studies, this was it; if there was any doubt that one ape or another – perhaps all the language study apes – had the first eight of the design features, Kanzi answered that doubt (Table 21.3). Grammar was still missing.

Then Kanzi filled in that last gap (Figure 21.10). Imagine a sink filled with water and Kanzi with a saucepan in his hands. Kanzi is used to putting water in the saucepan during cooking. He also knows that dirty dishes go in the wash water. If instructed to "put water in the saucepan," he would fill it. And if told to "put the saucepan in the water," it would go in the wash water in the sink.

In a now-famous bit of video, a skeptical production crew from Japan's NHK network visited Savage-Rumbaugh and Kanzi and proposed several tests that would, among other things, demonstrate grammar. Savage-Rumbaugh was instructed to remain relatively motionless – no pointing or body movement toward the objects to be used in the experiment – and she spoke while wearing a welder's mask so that no Clever-Hans-ing could be at work.

Figure 21.10 Kanzi communicates via lexigrams. By sequentially touching lexigrams representing words, Kanzi put together sentences. Many of the lexigrams are abstract, similar to those of Premack's Sarah. Photo by William H. Calvin; used with permission.

She instructed Kanzi to perform several tasks for the camera. He performed flawlessly. When told to "put the Perrier water in the coke" he did that rather than *vice versa*. He was asked to do things he had never done before, like "put the pine needles in the refrigerator" and "take the TV outside." The NHK film was only one bit of evidence, but it is a convenient departure point. Many holdouts were converted. At last there was evidence for grammar that was difficult for the critics to refute.

Here I will make a confession. I had accepted that Kanzi and others had conquered most of the language design-features criteria. But I still had reservations. With no evidence whatsoever, I worried the ostensibly definitive videos I had seen represented only the most impressive of perhaps hundreds of hours of filming, cherry-picked from miles of tape that might well be less convincing or might even contain telling errors that could refute claims of language-competence. I expected that Kanzi failed in tasks much more often than the films seemed to show, even if he did have language. I was wrong.

In 1998, I visited Kanzi and Savage-Rumbaugh with my colleagues Kathy Schick and Nick Toth, who were in the midst of a series of brilliant experiments that ultimately showed that Kanzi was capable of making stone tools (Toth et al., 1993; Schick et al., 1999; Schick & Toth, 2009). Within five minutes of being in

Kanzi's presence I realized that not only was the NHK video representative, the level of competence demonstrated in the NHK film was a minute-to-minute reality for Kanzi. I saw him respond to Savage-Rumbaugh's statements or questions as any human would, looking in the direction of an object she mentioned, or moving toward items he needed to perform tasks he was asked to do. When questioned, he pressed lexigrams in response, correctly and without pause. I cannot recall a single incorrect answer. I was hardly in Kanzi's presence five minutes when I became a complete convert. Kanzi understood speech.

As the day wound down, Kanzi was looking forward to receiving some dried fruit he had been promised, and Savage-Rumbaugh stood next to Schick, Toth, and me in front of a chain link fence built as a sort of run, a fenced-in tunnel in which Kanzi could exercise. The chain link tunnel led to my left and up the hill to a larger fenced-in enclosure, beyond which was his indoor home. Kanzi had been making stone tools and participating in experiments all day. I had seen 10 times the evidence I needed to accept all of Savage-Rumbaugh's claims, but what happened next was still a surprise. Savage-Rumbaugh asked Kanzi to make a stone tool for me as a memento. He dutifully chipped off a tiny flake and handed it to me through the fence. Savage-Rumbaugh looked at it and said "oh, Kanzi, that's too small, can you make a larger one?" Without pause he picked up the stones and worked until he had chipped off a one-inch flake for me. I still have it.

This marked the end of the day, with Kanzi sitting among the detritus of the various experiments: stone flakes, papers, and so on. Savage-Rumbaugh, holding a clipboard in one hand and a radio (it was 1998 – the project had yet to adopt cell phones) in the other, she was left with no free hand, so she opened the bag of dried fruit with some free fingers and her teeth. She handed over the half of the bag with the fruit stuck to it to Kanzi through the fence. He scraped the fruit off and quickly consumed it, then turned and began to walk up the hill to his enclosure. Savage-Rumbaugh, still with half the bag clenched between her teeth, called to Kanzi "come back and pick up your garbage." Between the bag and the clenched teeth, she sounded like a patient whose broken jaw had been wired shut. No problem. Kanzi stopped, turned around, returned to pick up the dried fruit wrapper and other scraps, carried them to the top of the hill where there was a garbage can, deposited them, and moved on into his indoor enclosure.

Let me pause to mention why I think chimpanzees fail here. They have never reached Kanzi's proficiency. I think I know why. Bonobo males are particularly eager to ally themselves with and please their mother. Kanzi looked on Sue as a mother, giving him motivation to please. A female chimpanzee leaves her mother, and so as an adult has no instincts regarding mother-like figures. She avoids males. Male chimpanzees pay little attention to females and, while they may retain affection for their mother, they pursue their own program in the wild; mother's opinion matters little. A male chimpanzee might care what a dominant male thinks about him, but the violent and aggressive manner in which chimpanzees establish dominance is impossible in the lab – and I would say unethical. So chimpanzees are difficult to motivate; bonobos are not. Chimpanzees could master language in the right circumstances, I believe; it is only a matter of motivation.

21.12 Is It Language?

Ape communication, then, has every one of the design features linguists laid down as the hallmarks of language: To my mind, there is no question that apes have language. Yet, when I talk to linguists they are dissatisfied with that statement. Conversation often devolves into a discussion of the difference between "language" and "human language"; in other words, some linguists equate the two and conclude that since chimpanzees do not have *human* language, they lack language – design features or no. I see their point, but I disagree. In the years since the first splash made by ape language studies, work on signing chimpanzees continued. If there was any doubt that chimpanzees used language in a natural and flexible way, in a pragmatic way (Bloom & Lahey, 1978) that demonstrates they know what language is for and how to use it, years of later research have dispelled it. Chimpanzees initiate conversations with human experimenters, they engage in human-like back-and-forth discussions that often include commentary on things they have just "heard," and they request more information when they are curious or confused (Jensvold & Gardner, 2000). If a chimpanzee encounters a human they wish to sign to, but who is facing away from them, they give the chimpanzee equivalent of a throat-clearing to get the human's attention – they make a noise to announce they want to communicate (Bodamer & Gardner, 2002) – and only then do they sign. Chimpanzees know what they want and use sign language to get it (Leitten et al., 2012). Multiple recordings show that even when humans are nowhere in sight, chimpanzees frequently sign to one another. They combine language with facial expression and gestures, and draw on context to enhance and refine meaning, producing a rich and complex multi-channel communication soup (Jensvold et al., 2014; Dombrausky & Jensvold, 2019) that strongly resembles human conversation. They have language.

When a linguist takes the view that talking with animals is a delusion (Anderson, 2006), how do they counter all this research? They point out that human language is unique. English has about 250,000 words, though of course few if any people know every word in their language. A typical English speaker may know as many as 40,000 words (Miller et al., 1991). In his combined works, Shakespeare used a vocabulary of only one-tenth of the whole English lexicon, 24,000 words. I think we can all agree that Shakespeare had a way with words beyond the typical human; 24,000 words, then, might be considered an outstanding vocabulary. Let us be more realistic. Perhaps more representative might be the word list from a long *New York Times* article: 6000 words. Or consider that an average person uses only about 3000 words in a day. Triangulating, 5000 words might be a good working vocabulary, perhaps not adequate to get you into a good university, but good enough for walking around town. Kanzi's vocabulary is 500 words. This is about 15 percent of the vocabulary of an English newspaper article, and 10 percent of what I have called a good working vocabulary. So perhaps his word comprehension is one-tenth that of humans.

Meanwhile, let us continue to compare ape to human language. Apes have sentence lengths of 2.5 words. Humans average 10 (Hess & Shipman, 1965), which pegs ape sentence lengths to 25 percent that of humans. To make up for their limited vocabulary, language-study apes tend to overgeneralize, to use words in new contexts; the word "straw" (drinking straw) might be generalized to any one (or all) of "pen," "cigarette," or "car antenna."

Apes never use future or past tenses at all, and articles, the difference between *an* apple and *the* apple, are beyond them. Apes have difficulty with recursion, a sort of embedding structure that is a common human construction. In English, one might say "the man who is wearing a top hat is walking down the street," or "I know that you know that I know you forgot to bring bus money today," but this strains – or we should probably say "breaks" – ape comprehension. They would use a construction like "the man is walking down the street; the man is wearing a top hat," but not the recursive structure.

Considering all of these comparisons, I am comfortable with an estimate of ape competency at one-twentieth that of typical languages, though if asked to put down the figure I prefer, it would be one-tenth. I could not disprove a claim of one-fifth. I would never call the ape–human difference a qualitative difference, though many linguists would disagree. Before I pursue this disagreement further, let us step out of the Anglophone world for a moment.

"But hold on," a linguist might say. "I am bursting with objections, perhaps the most important of which is, why limit your comparison to English?" Instead of taking my own language as an example, let us compare ape language to some other languages. Modern languages often have huge vocabularies, at least compared to an ape vocabulary. The open online Korean dictionary (https://opendict.korean.go.kr/service/dicStat) lists over one million words. At the other end of the list, at least for major languages, is Latin, with only 40,000 words (Glare, 1982).

When I discussed tools, I compared chimpanzee tools to those of cultures with the simplest technology, hunter-gatherers, not to the most sophisticated tools available. If we take a similar approach and compare chimpanzee language to languages with small vocabularies, the contrast is not so great. Putting creoles and pidgins (new, purposely simple languages) aside, among those languages that have been relatively well documented, Pirahã (Everett, n.d., 2012; Everett et al., 2005; Frank et al., 2008) has the smallest vocabulary at only 224 words (Everett, n.d.). The Pirahã have only two counting words, and even that is open to debate; words they use to describe quantity might be translated as "one" and "many," but there is some evidence that these words actually mean "less" and "more." Pirahã has no terms for colors, only light and dark. Of greatest interest to linguists, Pirahã lacks recursion, once thought to be a human universal (Everett, n.d.).

In light of what we know of Pirahã, drawing a distinct line of demarcation between ape ASL and human language becomes even more difficult, and as a result I find the arguments of those who contend that apes lack language to be strained and unconvincing. Recursion fails to distinguish them. Vocabulary fails. Yet, if you wish – and some will wish – to define language to exclude apes, this is not a great challenge. If true language requires any one of (or perhaps all) future and past tenses, the proper use of articles, or an average of five words per sentence, then apes do not have language.

I would maintain, however, that this is missing the point. Apes have language, but just because they have mastered *a* language does not mean they have *human* language. Our language abilities, even the language of the Pirahã, are far beyond theirs.

My colleague Professor Franks would maintain that human language is qualitatively different from that of chimpanzees because human grammar is qualitatively different. I disagree; it is impossible that differences between ape and human language can be anything other than quantitative. I made this argument to Professor Franks: Some ancestor in the human lineage, you will admit, had no more language ability than chimpanzees. By increments, the vocalizations of our ancestors became more and more like ours until it *was* ours. Nothing that is different from something else by a finite number of increments can be qualitatively different. Professor Franks would argue that instead I should not look at the language itself, but at the kind of grammar that enables it. As language became more sophisticated, the mind/brain evolved new grammatical strategies to cope. Human language is qualitatively different from ape language not only because the language is more sophisticated – in the sense of more words and longer sentences – but because the organ that produces it, the mind/brain, has changed to allow it to acquire grammars impossible for chimpanzees, grammars that impart meaning by the way they combine words.

Choose your side. I think Professor Franks and I agree more than we disagree. I said to him recently that it may be that we simply disagree about whether we disagree, a statement a linguist has to love.

21.13 Chimpanzees in the Free World

With our discussion of language, we wrap up our documentation of chimpanzee intellectual abilities. Chimpanzees exhibit so many cognitive abilities that are human-like, I am sometimes asked whether chimpanzees are people – whether they could live free in human society. There are several answers to the former question, but only one to the latter: no. Chimpanzees do have the cognitive tools needed to live in human society. Their excellent memories and spatial skills would make navigating a city landscape easy. They go to work, as it were, every day, collecting food and constructing shelter at the end of the day. The concept of working for pay is natural for them, as shown by research in which they perform some task in return for a treat. Their math abilities and capacity to understand symbolic tokens or lexigrams are such that they would have no difficulty understanding the concept of money, and could easily recognize the "1," "5," and "20" on currency, and that "20" is more valuable than "5." Kanzi has shown that apes can cook and Meshie cleaned up after Mary without being asked. Circuses have trained chimpanzees to ride bikes and drive kiddie cars. They have been taught to use the toilet (Kellogg & Kellogg, 1933).

Chimpanzees, in short, are intelligent enough to live free in a human society. So would it not be humane to put chimpanzees in a halfway house to live on their own, perhaps with a caretaker to sort out unexpected problems? They could take the bus, perform some menial job, get paid, exchange tokens for food at the grocery store, and make their way back home.

The denouement of the life stories of Meshie, Johnny, and countless other chimpanzee who began life as pets show us why this is impossible. In a context in which they feel dominant, a chimpanzee cannot refrain from taking what they want and attacking subordinates. Their murderous violence toward strangers, particularly male aggression toward males, would quickly lead to tragedy. My bet is that on the first day of a halfway-house experiment somebody would be severely injured, bringing a horrific end to the experiment.

Chimpanzees have every right to live free. And I hope that efforts to preserve their habitats will mean they will continue to do so forever – but in Africa, their ancient forest homeland, not among humans.

References

Anderson SR (2006) *Doctor Dolittle's Delusion: Animals and the Uniqueness of Human Language*. New Haven, CT: Yale University Press.

Bauer HR, Philip MM (1983) Facial and vocal individual recognition in the common chimpanzee. *Psychol Rec* 33, 161–170.

Bloom L, Lahey, M (1978) *Language Development and Language Disorders*. New York: Wiley.

Bodamer MD, Gardner RA (2002) How cross-fostered chimpanzees (*Pan troglodytes*) initiate and maintain conversations. *J Comp Psychol* 116, 12–26.

Cianelli SN, Fouts RS (1998) Chimpanzee to chimpanzee American sign language. *Hum Evol* 13, 147–159.

Corballis MC (2017) Language evolution: a changing perspective. *Trend Cogn Sci* 21, 229–236.

Crockford C, Herbinger I, Vigilant L, Boesch C (2004) Wild chimpanzees produce group-specific calls: a case for vocal learning? *Ethology* 110, 221–243.

Curtiss S (2014) *Genie: A Psycholinguistic Study of a Modern-Day Wild Child*. New York: Academic Press.

Dombrausky D, Jensvold, MLA (2019) Sign language in chimpanzees across environments. In *Chimpanzee Behavior: Recent Understandings from Captivity and the Forest* (ed. Jensvold MLA), pp. 141–174. New York: Nova Science Publisher.

Everaert MB, Huybregts MA, Berwick RC, et al. (2017) What is language and how could it have evolved? *Trend Cogn Sci* 21, 569–571.

Everett DL (n.d.) Pirahã Dictionary/Dicionário Mura-Pirahã, www.geocities.ws/indiosbr_nicolai/piraha1.html.

Everett DL (2012) *Language: The Cultural Tool*. New York: Vintage.

Everett DL, Berlin B, Gonalves M, et al. (2005) Cultural constraints on grammar and cognition in Piraha: another look at the design features of human language. *Curr Anthropol* 46, 621–646.

Fitch WT (2005) The evolution of language: a comparative review. *Biol Philos* 20, 193–203.

Fitch WT, de Boer B, Mathur N, Ghazanfar AA (2016) Monkey vocal tracts are speech-ready. *Sci Adv* 2, p. e1600723.

Fouts RS (1973) Acquisition and testing of gestural signs in four young chimpanzees. *Science* 180, 978–980.

Fouts RS, Couch JB (1976) Cultural evolution of learned language in chimpanzees. In *Communication Behavior and Evolution* (eds. Hahn ME, Simmel EC), pp. 141–161. New York: Academic Press.

Fouts RS, Shapiro G, O'Neil C (1978) Studies of linguistic behavior in apes and children. In *Understanding Language through Sign Language Research* (ed. Siple P), pp. 163–185. New York: Academic Press.

Fouts RS, Fouts DS, Van Cantfort TE (1989) The infant Loulis learns signs from cross-fostered chimpanzee. In *Teaching Sign Language to Chimpanzees* (eds. Gardner RA, Gardner, BT, Van Cantfort TE), pp. 280–292. Albany, NY: SUNY Press.

Frank MC, Everett DL, Fedorenko E, Gibson E (2008) Number as a cognitive technology: evidence from Pirahã language and cognition. *Cognition* 108, 819–824.

Gardner RA, Gardner BT (1969) Teaching sign language to a chimpanzee. *Science* 165, 664–672.

Gardner RA, Gardner BT (1975) Early signs of language in child and chimpanzee. *Science* 187, 752–753.

Gardner RA, Gardner BT (1998) *The Structure of Learning: From Sign Stimuli to Sign Language*. Mahwah, NJ: Lawrence Erlbaum.

Gardner RA, Gardner, BT, Van Cantfort TE (1989) *Teaching Sign Language to Chimpanzees*. Albany, NY: SUNY Press.

Glare PG (1982) *Oxford Latin Dictionary*. Oxford: Oxford University Press.

Goodall, J van Lawick (1968) A preliminary report on expressive movements and communication in the Gombe Stream Chimpanzees. In *Primates* (ed. Jay PC), pp. 313-374. New York: Holt, Rinehart, and Winston.

Goodall J (1986) *The Chimpanzees of Gombe: Patterns of Behavior*. Cambridge, MA: Harvard University Press.

Hauser MD, Wrangham RW (1987) Manipulation of food calls in captive chimpanzees: a preliminary report. *Folia Primatol* 48, 207-210.

Hauser MD, Teixidor P, Field L, Flaherty R (1993) Food elicited calls in chimpanzees: effects of food quantity and divisibility. *Anim Behav* 45, 817-819.

Hauser MD, Chomsky N, Fitch WT (2002) The faculty of language: what is it, who has it, and how did it evolve? *Science* 298, 1569-1579.

Hayes C (1951) *The Ape in Our House*. New York: Harper.

Herbinger I, Papworth S, Boesch C, Zuberbühlerb K (2009) Vocal, gestural and locomotor responses of wild chimpanzees to familiar and unfamiliar intruders: a playback study *Anim Behav* 78, 1389-1396.

Hess RD, Shipman VC (1965) Early experience and the socialization of cognitive modes in children. *Child Dev*, 36, 869-886.

Hill JH (1978) Apes and language. *Ann Rev Anthropol* 7, 89-112.

Hobaiter C, Byrne RW (2014) The meanings of chimpanzee gestures. *Curr Biol* 24, 1596-1600.

Hockett CF (1958) *A Course in Modern Linguistics*. New York: Macmillan.

Hockett CF (1959) Animal "languages" and human language. *Hum Biol* 31, 32-39.

Hockett CF (1960a) Logical considerations in the study of animal communication. In *Animal Sounds and Communication* (eds. Lanyon WE, and Tavolga WN). Washington, DC: American Institute of Biological Sciences.

Hockett CF (1960b) The origin of speech. *Sci Am* 203, 88-96.

Hockett CF (1966) The problem of universals in language. In *Universals of Language* (ed. Greenberg J), pp. 1-29. Cambridge, MA: MIT Press.

Hockett CF, Altmann SA (1968) A note on design features. In *Animal Communication: Techniques of Study and Results of Research* (ed. Sebeok T), pp. 61-72. Bloomington, IN: Indiana University Press.

Jensvold MLA, Gardner RA (2000) Interactive use of sign language by cross-fostered chimpanzees (*Pan troglodytes*). *J Comp Psychol* 114, 335-346.

Jensvold MLA, Wilding L, Schulze SM (2014) Signs of communication in chimpanzees. In *Biocommunication of Animals* (ed. Witzany G), pp. 7-19. Dordrecht: Springer.

Kellogg WN, Kellogg LA (1933) *The Ape and the Child: A Study of Environmental Influence upon Early Behavior*. New York: McGraw-Hill.

Leavens DA, Hopkins WD, Bard KA (2005) Understanding the point of chimpanzee pointing: epigenesis and ecological validity. *Curr Directions Psychol Sci* 14, 185-189.

Leitten L, Jensvold MLA, Fouts RS, Wallin JM (2012) Contingency in requests of signing chimpanzees (*Pan troglodytes*). *Interaction Stud*, 13, 147-164.

Lieberman P, Crelin ES, Klatt DH (1972) Phonetic ability and related anatomy of the newborn and adult human, Neanderthal man, and the chimpanzee. *Am Anthropol* 74, 287-307.

Marler P, Hobbett L (1975) Individuality in a long range vocalization of wild chimpanzees. *Zeitschrift für Tierpsychologie* 38, 97-109.

Menzel EW (1971) Communication about the environment in a group of young chimpanzees. *Folia Primatol* 15, 220-232.

Menzel EW (1973) Leadership and communication in young chimpanzees. In *Symposia of the Fourth Congress of the International Primatological Society*, Vol 1: *Precultural Behavior* (ed. Menzel EW), pp. 192-225. Basel: Karger.

Menzel EW (1979) Communication of object-locations in a group of young chimpanzees. In *The Great Apes* (eds. Hamburg DA, McCown ER), pp. 359-370. Menlo Park, CA: Benjamin Cummings.

Miller GA, Gildea P.M (1991). How children learn words. In *The Emergence of Language: Development and Evolution* (ed. WangWS-Y), pp. 150-158. New York: W. H. Freeman.

Mitani JC (1996) Comparative studies of African ape vocal behaviour. In *Great Ape Societies* (eds. McGrew WC, Marchant LF, Nishida T), pp. 225-240. Cambridge: Cambridge University Press.

Mitani JC, Nishida T (1993) Contexts and social correlates of long-distance calling by male chimpanzees. *Anim Behav* 45, 735-746.

Mitani JC, Hasegawa T, Gros-Louis J, Marler P, Byrne R (1992) Dialects in wild chimpanzees? *Am J Primatol* 27, 233-243.

Mitani JC, Gros-Louis J, Macedonia J (1996) Selection for acoustic individuality within the vocal repertoire of wild chimpanzees. *Int J Primatol* 17, 569-583.

Nakamura M, Marchant LF, McGrew WC, Nishida T (2000) The social scratch: another custom in wild chimpanzees? *Primates* 41, 237-248.

Notman H, Rendell D (2005) Contextual variation in chimpanzee pant hoots and its implications for referential communication *Anim Behav* 70, 177-190.

Pfungst O (1911) *Clever Hans (The Horse of Mr. von Osten)*. New York: Holt, Rinehart, and Winston.

Premack D (1976) *Intelligence in Ape and Man*. Hillsdale, NJ: Lawrence Erlbaum.

Premack D (1983a) Social cognition. *Ann Rev Psychol* **34**, 351–362.

Premack D (1983b) The codes of man and beast. *Behav Brain Sci* **6**, 125–167.

Premack AJ, Premack D (1972) Teaching language to an ape. *Sci Am* **227**, 92–99.

Premack D, Premack A (1982) *The Mind of an Ape*. New York: W.W. Norton & Co.

Raven HC (1932) Meshie, the child of a chimpanzee. *Nat Hist* **32**, 158–166.

Raven HC (1933) Further adventures of Meshie. *Nat Hist* **33** 607–617.

Reynolds VF, Reynolds F (1965) Chimpanzees in the Budongo Forest. In *Primate Behavior: Field Studies of Monkeys and Apes* (ed. DeVore I), pp. 368–424. New York: Holt, Rinehart, and Winston.

Savage-Rumbaugh ES (1986) *Ape Language: From Conditioned Response to Symbol*. New York: Columbia University Press.

Savage-Rumbaugh ES, Lewin R (1994) *Kanzi: The Ape at the Brink of the Human Mind*. New York: Wiley.

Savage-Rumbaugh ES, Wilkerson BJ, Bakeman R (1977) Spontaneous gestural communication among conspecifics in the pygmy chimpanzee. In *Progress in Ape Research* (ed. Bourne GH), pp. 97–116. New York: Academic Press.

Savage-Rumbaugh ES, Shanker S, Taylor TJ (1998) *Apes, Language, and the Human Mind*. Oxford: Oxford University Press.

Schick KD, Toth N (2009) The Oldowan: the tool making of early hominins and chimpanzees compared. *Ann Rev Anthropol* **38**, 289–305.

Schick KD, Toth N, Garufi G, et al. (1999) Continuing investigations into the stone tool-making and tool-using capabilities of a bonobo (*Pan paniscus*). *J Arch Sci* **26** 821–832.

Slocombe KE, Zuberbühler K (2007) Chimpanzees modify recruitment screams as a function of audience composition. *Proc Nat Acad Sci* **104**, 17228–17233.

Slocombe KE, Townsend SW, Zuberbühler K (2009) Wild chimpanzees (*Pan troglodytes schweinfurthii*) distinguish between different scream types: evidence from a playback study. *Anim Cogn* **12**, 441–449.

Taylor BP (1974) Toward a theory of language acquisition. *Language Learning* **24**, 23–35.

Temerlin M (1976) *Lucy: Growing Up Human: A Chimpanzee Daughter in a Psychotherapist's Family*. New York: Bantam Books.

Terrace HS (1979) *Nim*. New York: Knopf.

Terrace HS, Pettito LA, Sanders RJ, Bever TG (1979) Can an ape create a sentence? *Science* **206**, 891–902.

Toth N, Schick KD, Savage-Rumbaugh ES, Sevcik RA, Rumbaugh DM (1993) *Pan* the tool-maker: investigations into the stone tool-making and tool-using capabilities of a bonobo (*Pan paniscus*). *J Arch Sci* **20**, 81–91.

Umiker-Sebeok J, Sebeok TA (1981) Clever Hans and smart simians: the self-fulfilling prophecy and kindred methodological pitfalls. *Anthropos*, **76**, 89–165.

Wrangham RW (1977) Feeding behaviors of chimpanzees in Gombe National Park, Tanzania. In *Primate Ecology* (ed. Clutton-Brock TH), pp. 503–538. London: Academic Press.

22 Ape Implements
Making and Using Tools

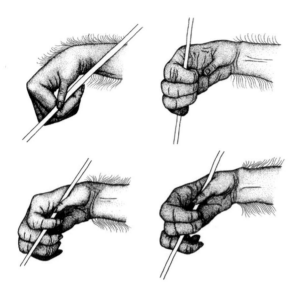

Based on original artwork by Alessandra Kelley, used with permission

If you were asked to say how similar human and chimpanzee technology is, your mind's eye might drift toward an image of a chimpanzee grasping a straw-like termiting stick or cracking nuts with a stone hammer. You might choose an airplane or a car to represent human technology. This comparison shows up chimpanzee cultural objects as profoundly simple, but we should keep in mind that our current level of sophistication, even many of our simplest tools, is of recent origin. This is an unfair comparison.

Complex tool use had a slow beginning, advancing rapidly only once we humans reached a point where a "ratcheting" effect could take hold (Boesch & Tomasello, 1998). In the last one half of 1 percent of the life of our genus – 10,000 years – we invented the needle, metalworking, writing, and the sword. Older people remember a world quite different from that we live in now. My grandmother – she was born February 5, 1894 – entered a world in which telephones (1878), the light bulb (1879), and mass-produced automobiles (1886) had just appeared. She was born before the invention of airplanes (1903), home air conditioning (1929), broadcast television (1934), and computers. She grew up in a home without electricity or running water, a home heated by a fireplace; she cooked on a wood stove using fuel she gathered herself on her own property, cooked vegetables she grew in her garden, and ate meat she killed herself. Her technology was only a step above that of Neanderthals.

Go back 500 more generations – keep in mind that even then we were biologically identical to today – and my ancestors were carrying wooden spears and living in mud huts. For most of human history, our stone-age technology differed little from that of chimpanzees.

Some might worry that I am underestimating the technology of 10,000 years ago, but we need not rely on vague impressions: we have at hand a quantitative analysis of tools of humans and chimpanzees, considering the complexity of both the tools and their manufacture (McGrew, 1987). The subsistence tools of hunter-gatherers and chimpanzees are "surprisingly similar in the number of items in the tool kit, raw materials use, proportions of tools made ... extent of complexity" and so on (McGrew, 1987), and the comparison with the oldest stone tools, Oldowan tools from 2.5 million years ago and older, is even closer (Toth et al., 2006; Toth & Schick, 2009).

22.1 Humans Ratchet, Chimpanzees Do Not

Our recent rapid progress was possible not because we became more intelligent, but because we invented writing and formal education, allowing our knowledge to click upward without a talented inventor reinventing the wheel every few generations.

Our extraordinary universities pass on information incredibly efficiently, and the percentage of people who go on to a university education – potential inventors – is far greater than at any time in the past. Not only is the proportion of educated people greater today, our population size is greater, with more potential inventors – 10,000 times what it was in the past. And our stable of tool-tinkerers enjoys medical care and health-promoting infrastructure that gives them more time to study and invent. All of these conditions have ratcheted up the pace of innovation.

22.2 Chimpanzees Are Natural Tool Users

Not only is stone age human and living chimpanzee technology similar, looking down the list of human relatives "below" chimpanzees, there is a significant gap between chimpanzees and almost every other animal. Chimpanzees are the most consistent tool users of all nonhumans; they use more kinds of tools in more ways than all other primates combined (McGrew, 1992, 2004). Perhaps the most impressive thing about chimpanzee technology, and the thing that distinguishes them from all other animals, is how naturally and un-assumedly they turn to tools to solve problems in their daily lives. When the honey is sequestered deep in a hollow tree, a chimpanzee hardly begins poking at it before he turns away to find a stout stick to use as a pry bar. It is clear that the chimpanzee mind is constantly grasping for creative solutions to challenges in their environment, and tool use fits comfortably in that world.

22.3 Importance of the Discovery of Tool Making and Tool Use

As defining a trait as the manufacture and manipulation of implements is for chimpanzees, it was discovered only in the early 1960s when Jane Goodall observed chimpanzees not only using tools, but making them (Goodall, 1964). It had not been expected. Rather, the creation and utilization of useful devices was thought to be the one thing that clearly distinguished humans from other animals; a popular definition distinguishing humans from other animals is that of **Ben Franklin**: "man is a tool-making animal" (Boswell, 1873).

Chimpanzees are distinguished from other animals and linked to humans not so much by tool use, but by intelligent tool use (Van Schaik et al., 1999). This differs from the instinctual use of tools – the humans-as-sole-tool-user paradigm was not troubled by sea otters cracking open clams with a rock and other sorts of seemingly robotic, stimulus–response types of behaviors. The frequent, complicated and creative tool use of chimpanzees exploded this preconception. Louis Leakey famously commented, on learning of Goodall's discovery "now we must redefine 'Tool,' redefine 'Man' or accept chimpanzees as humans" (Goodall, 1971). Goodall's discovery was so paradigm-shattering that some simply refused to believe it. In a prominent public forum, Sir Solly Zuckerman dismissed Goodall's observations out of hand. Many others silently agreed. Quickly enough, however, incredible visual images in the form of Hugo van Lawick's films began appearing, and all doubt was erased. As time went on, though not much time, reports of more and still more chimpanzee implements emerged and we began to speak of the "chimpanzee toolkit" (McGrew, 1992).

22.4 Tool Function: Tools Defined

One widely accepted definition of "tool" is: "an object purposefully oriented or manipulated so as to alter the form, position or condition of another object, another organism, or the user itself" (Shumaker et al., 2011). Rob Shumaker and colleagues identified seven tool functions, of which chimpanzees use six – they lack tools that camouflage or hide themselves (though, incredibly, the crocodile, with its walnut-sized-brain, does [Bani, 2004]). Chimpanzees' (and other animals') tools have six purposes:

1. Extend reach. For example, using a stick to retrieve something that is out reach.
2. Amplify force. As when using a lever to pry something open, or a stone to hammer.

3. Amplify gesture. As when a displaying chimpanzee throws stones, waves foliage, or drags large limbs to make a social display more conspicuous.
4. Increase control of the environment, as chimpanzees do when they manufacture a sponge out of chewed leaves.
5. Enhance comfort. As when chimpanzees manufacture a pad or mattress to line their sleeping nest.
6. Symbolize. Use an object to signal, abstract, or represent reality; e.g., creating a toy, such as the doll mentioned in Chapter 1, or, if you are unconvinced by that, pulling an imaginary pull-toy, as Vicki the chimpanzee did when her real toy was lost (Hayes, 1951).

22.5 Tool Making

Chimpanzees exhibit all four types of tool manufacture defined by Shumaker et al. (2011: table 1.1) and McGrew (2013: supplementary material Table 3).

1. Combine. Join or connect two or more objects that are then manipulated as if a single object, as in human tool use when a handle is attached to a hammer head. "Chimpanzees pick and crush individual leaves into a single sponge-like mass for absorbing water, either from a tree-hole or a puddle" (McGrew, 2013).
2. Detach. Disconnect or remove an object that is fixed or connected to another. Chimpanzees pluck leaves to leaf-clip, stems or twigs to termite fish, and entire saplings to brandish during social displays.
3. Reduce. Remove part of an object to make it more useful. Chimpanzees remove leaves, branches, and sometimes bark from termiting tools.
4. Reshape. Make a fundamental change to an object to yield a tool. Chimpanzees fold leaves into spoon-like shapes to drink water from a tree hole or a drinking well (Hunt & McGrew, 2002).

22.6 Extensiveness of Tool Use

All **habituated** (accustomed to human presence) wild chimpanzees that have been studied for more than a year or two have been seen to use tools (McGrew, 1992, 2004; Whiten et al., 1999). This makes it all the more surprising that our other closest relative, the bonobo, hardly uses tools at all; after 40 years of study, we would have seen it if they did.

Table 22.1 lists the 20 most common chimpanzee tools, implements that have been observed in use hundreds of time at multiple sites. The list is not a complete one, but is a good sample of chimpanzee technology. These tools are used so commonly that a lucky chimpanzee researcher following a party of a half-dozen chimpanzees might see 10 such wielded in a single day. A researcher might watch as the group (1) wakes up in their nests; (2) moves off to a termite mound to harvest termites with a termiting tool; (3) discovers honey and uses a penetrating probe to ascertain its exact location; (4) brings a pry bar into play to widen the opening, after which (5) a harvesting probe might be used to gather more honey. Males might gather in a largish group, so a male might be seen (6) waving a display branch and (7) throwing a rock projectile toward party members. Later, males might (8) pause to drum on a tree trunk. When the party settles down for a rest period a juvenile might (9) use a play-starter to initiate play, then (10) place a leafy branch on his head as he teases a playmate.

22.7 Chimpanzee Tools

Table 22.1 lists tools in the approximate order of how commonly they are found. The functions are taken from McGrew (2013). For many of these tools, a two- or three-word description expresses their function, but others require slightly more unpacking.

1. **Nest/mattress.** We have discussed nests or sleeping platforms in the sleep chapter. Chimpanzees make a new nest each night, even though as a rule they do not urinate or defecate in their nests (Goodall, 1962, 1968). Often, chimpanzees anticipate their foraging plans for the next day and choose a site near their anticipated first feeding site.

2. **Investigatory probe.** Usually a short (10–20 cm) stick threaded into an opening that contains some resource the chimpanzee wishes to investigate but

Table 22.1 Common chimpanzee tools

Tool		Tool use (*sensu* McGrew, 2013)	Tool function
1.	Nest/mattress	Block[a]	Increase comfort
2.	Investigatory probe	Insert, probe	Gain information
3.	Penetrating probe/spear	Stab, jab, penetrate/reach	Kill prey/extract prey, object
4.	Fluid harvesting probe	Reach	Extract inaccessible fluids
5.	Termiting/anting probe	Probe	Extract inaccessible prey
6.	Play-starter	Bait or entice	Solicit play
7.	Display branch	Brandish, wave, shake	Amplify gesture
8.	Leaf sponge	Absorb; wipe	Control or harvest liquid
9.	Drum	Drag, roll, kick, slap, push over	Display power, ability
10.	Hammer	Pound, hammer	Open protected object
11.	Club	Pound, hammer	Break, damage target
12.	Anvil/mortar	Contain	Open protected object
13.	Ant dip	Probe	Consolidate small prey
14.	Projectile	Drop; throw	Frighten, warn target
15.	Leaf clip/strip	Symbolize	Noise maker
16.	Grooming-starter	Leaf grooming	Initiate or restart grooming
17.	Pads/seats	Block	Increase comfort
18.	Lever/pry bar	Pry, apply leverage	Break open enclosed space
19.	Fly whisk	Wipe	Remove or deter annoyance
20.	Body decoration	Affix, apply drape	Play? Display?

[a] Interpose one object in front of another, to cushion or isolate some aspect of that object.

cannot see clearly. Beehives, wasp nests, termite mounds, bushbaby sleeping hollows, or other cavities that excite the curiosity of chimpanzees are investigated by pushing the probe into the recessed area, withdrawing it, and sniffing or looking at it (Goodall, 1968, 1986; McGrew, 1992, 2004; Whiten et al., 1999). Dead bodies (e.g., a dead python) may be investigated by touching them with a probe (Goodall, 1968; McGrew, 2004).

3. **Penetrating probe/spear.** A stout stick, often 1 cm or more in diameter, perhaps 20 cm long, used to pierce a surface to gain better access to a resource. In the dry habitat at Ugalla, Tanzania such probes are used to harvest underground storage organs such as tubers or roots (Hernandez-Aguilar et al., 2007). A probe can be used to kill an animal by puncturing it or crushing it; at Taï and Mahale, such probes are used to kill bees (Whiten et al., 1999) and at Fongoli (Pruetz & Bertolani, 2007; Pruetz et al., 2015) a longer version was used as a spear to kill bushbabies; the spears were approximately 1 cm in diameter and averaged 63 cm in length. The Fongoli spears are particularly

interesting because construction entails up to five steps, including sharpening the tool tip to a point. A penetrating probe may be repurposed to serve as a harvesting probe (see below) once a resource is accessed, but just as often the penetrating probe is used only as a chisel or spear (McGrew, 2004). I include Nishida's "rousting probes" in this category (Nishida, 1973), sticks used to stir up insects or even mammals to make them flee a recessed area so as to be captured and eaten; in a predation episode at Mahale a probe was used to roust a squirrel from its nest (Huffman & Kalunde, 1993).

Perhaps under this category we can place a tooth-cleaning probe, a tool often seen in captivity (McGrew & Tutin, 1972, 1973). In one startling case at the Chimfunshi Wildlife Sanctuary in Zambia, a mother used a tooth-cleaning probe to remove matter from the teeth of the corpse of her adopted son (Van Leeuwen et al., 2016), a practice that reminds me of preparing a corpse for burial.

4. **Harvesting probe.** A stout (0.5–1 cm diameter) rod-shaped tool that takes advantage of the adhesive property of the probe to secure viscous fluids; it is used to reach and extract inaccessible resources such as honey (Merfield & Miller, 1956; Izawa & Itani, 1966; Bermejo et al., 1989; Pascual-Garido et al., 2012), bone marrow (Goodall, 1986; McGrew, 1992), or other soup-like substances, including water (Wrangham et al., 1994; Lapuente et al., 2017). Sometimes called a honey dip or marrow pick.

5. **Termiting/anting probe.** Termiting/anting tools were the first tools discovered by Goodall, and they remain one of the most interesting tools in that they involve extracting a resource that is completely out of sight (Figure 22.1). These tools rely not on the viscous properties of the harvested resource, but instead on a knowledge of the behavior of the prey. The tools themselves are longer, narrower, and more flexible than harvesting probes. Termiting/anting probes are not to be confused with ant wands, because ants are visible before collecting.

Termiting probes are used most often to harvest termites in the genus *Macrotermes* (Goodall, 1964, 1968, 1986; Nishida & Uehara, 1980; McGrew et al.,

Figure 22.1 Perhaps the best-known tool is the termiting probe. Chimpanzees strip off leaves and branches to make a flexible but durable tool, break into a termite mound, insert the tool (left), and pull it out and pluck the termites off the tool with the lips. Ants other than the aggressive, biting safari ants are harvested in a similar manner.

1978; McBeath & McGrew, 1982; McGrew & Collins, 1985; Collins & McGrew, 1987; McGrew, 1992, 1993, 1994; Sanz et al., 2004; Suzuki et al., 1995; Sanz & Morgan, 2007; Bogart & Pruetz, 2008; Pascual-Garrido et al., 2012), an interesting species in and of itself. These termites make the ubiquitous mounds seen across Africa, recycling in incredible 90 percent of the dry wood in some environments (Buxton, 1981). The termites actually farm fungi, growing it inside their mounds and utilizing it to convert otherwise indigestible plant material into a protein-rich food

(Roberts et al., 2016), an activity they are thought to have been engaging in for 25 million years.

Chimpanzees often prepare tools in advance and may bring them from some distance, as much as 100 m (Goodall, 1964, 1986; McGrew, 1992), and perhaps, according to as-yet unpublished results, much farther. They may bring several tools to a mound, using one until it becomes worn or broken, then picking up the next. After breaking into the mound by displacing a cap to a termite tunnel, the tool is threaded in, disturbing the termites inside. Defending termites grasp the tool with their mandibles and the chimpanzee deftly withdraws the tool, the termites doggedly hanging on to the death. Chimpanzees often rest the tool on a wrist or forearm as they pluck the termites off the tool or their arm with their lips (Figure 22.1). In Goulago, rather that picking off the cover of a tunnel, chimpanzees use a digging stick to puncture a nest, then fish from the resulting hole (Sanz et al., 2004; Sanz & Morgan, 2007).

For chimpanzees to successfully retrieve termites from the mound, the tool must be stiff enough to push into narrow passageways, but flexible enough to wind their way through curvy tunnels; they may be made of wood (and may be processed, with side-branches stripped of stems), bark, grass, vines, or palm frondlets (Goodall, 1986). At Gombe, termiting tools average about 1 foot long (28 cm) in the wet season, but a little longer in the dry season (Goodall, 1986), though I found one at Gombe during the dry season that was 2 m long.

The story of that 2 m long tool illustrates how sophisticated chimpanzee mental maps are. At the time, Jane Goodall had generously allowed me and a colleague, Tony Collins, to stay in her house on the Gombe beach while she was traveling. One day I rehashed my chimping day with Tony, thinking to impress him with my observation of a two-meter-long termiting stick. He was skeptical, but I stood my ground. "Fully six feet!" I crowed, spreading my arms. "Where is that mound?" Tony asked. "Sure!" I thought, "he really thinks he will know the mound?" I described the shape of the mound, nearby features of interest, and the general area. The next day I wearily tottered into the living room after a long chimpanzee follow to find on the coffee table the very termiting stick. Tony knew the Gombe termite mounds well enough to walk right to it. Possibly not impressive, since termiting is Tony's specialty, and at the time he had been studying termite mounds at Gombe for over 20 years, but this is something every female at Gombe knows, too. And possibly all the males, as well. For the record, the wand *was* six feet long!

At Ndoki, chimpanzees alter their termiting tools in an unusual manner. They construct a brush from the stem of a Marantaceae plant, chewing the end of the tool until approximately 15 mm is shredded so that it resembles a paint brush. This theoretically improves the harvest by increasing the surface area of the tool (Suzuki et al., 1995).

Anting tools are used similarly to the termiting tool (Nishida, 1973; Nishida & Hiraiwa, 1982; McGrew, 2004; Nishie, 2011), except they are used almost entirely in trees, thus requiring great agility of the harvester, who must cling to the tree while manipulating the anting tool. In fact, in my data anting was the second most common context for bipedal posture, after feeding on fruits. It is surprising to me that nearly every other context in which I have observed chimpanzees engaging in bipedalism (Hunt, 1998) has had its "bipedalism origin" hypothesis – except ant fishing. I doubt we will ever see a Man the Anteater hypothesis, but one never knows.

Ant fishing is entirely different from ant dipping; ant fishing requires inserting a tool into an ant nest opening in a tree trunk, and is much like termiting, whereas ant dipping involves harvesting ants visible on the forest floor. I recorded chimpanzees anting in both armhanging and bipedal postures.

6. **Play-starter.** Some natural object such as a leaf or twig is held in the mouth (rarely, waved by a hand) and manipulated or displayed (or "flaunted" [McGrew, 2004]) so as to invite play. A stem might be held in the mouth and bobbed up and down to attract the attention of a playmate, or leaves might be pulled along or stacked in a pile to attract the notice of a playmate (Nishida & Wallauer, 2003).

7. **Display branch.** A leafy branch is flailed to draw attention to an individual or to exaggerate the individual's movement or gesture, often to impress

other individuals (Goodall, 1963, 1968). Another form of display branch is the large, leafless log, perhaps 10 or even 20 cm in diameter, useful for demonstrating the great strength required of the displayer to wave or throw it.

8. **Leaf sponge/napkin.** Chimpanzees chew leaves to increase their absorbency and use them to soak up liquids (Figure 22.2) such as water in recessed hollows in trees. Water need not be inaccessible to warrant sponge use, just inconveniently located; oddly, sponges are also used to collect water from flowing streams (Goodall, 1986). Chimpanzees use sponges to soak up blood during predation episodes, to wipe up fruit juice or residual pulp clinging to the inside of shells or fruit husks (Wrangham, 1977; McGrew, 1992; Goodall, 1986), or to soak up tree sap (Sugiyama, 1995). Chimpanzees at Semliki use leaves to dip water out of wells (or drinking holes) they dig in the sand (Hunt & McGrew, 2002).

Rather than as a feeding tool, leaves, most often whole but sometimes crushed or chewed to increase their pick-up capacity, are used as a cleaning sponge to mop up blood, feces, or other fluids (Goodall, 1968; McGrew, 2004: 124). Chimpanzees are surprisingly fastidious and little tolerate muck on their bodies; when soiled by urine, feces, blood, mucus, semen, mud, and even rain chimpanzees may use leaves to wipe off the liquid (Goodall, 1986). Mothers wipe feces off themselves and their infants when an infant has diarrhea. Sometimes even adults wipe feces off one another: Goodall relates the story of a male who, after vigorously courting a female, found that her sexual swelling was smeared with feces; he declined to copulate with her, and when another male did, he did so only after wiping her with leaves (Goodall, 1986: 545). It is not uncommon for females to wipe their bottoms after urinating (Goodall, 1986). When wounded they often dab up blood.

9. **Drum.** Drumming is one of the earliest tool-use behaviors discovered among chimpanzees (Nissen, 1931), and perhaps the most universal (Whiten et al., 2001). As a means of long-distance communication, chimpanzees, usually males, and particularly when in large parties, drum on the wing-like buttresses of large trees. They use their hands or feet, and may drum extremely rapidly. The sound carries a great distance, 1 km or more. It requires great power to produce the booming sound chimpanzee make. When humans attempt it, all you hear is a weak patting sound. Gombe chimpanzees sometimes use a clustered technique; the feet pound out a quick rat-a-tat-tat, after which there is a pause followed by another b-r-r-r-app, then a further pause and a third quick sequence (Goodall, 1968).

10. **Hammer.** While only tenth on our list, nut-cracking with stone hammers was the first tool to be discovered among chimpanzees (Figure 22.3), though these early observations (Savage & Wyman, 1844) were ignored by an unbelieving scientific community. West African chimpanzees use hammers and mortars/anvils to crack nuts, and much has been learned about chimpanzee intelligence from their use of these tools (Savage & Wyman, 1844; Sugiyama & Koman, 1979; Boesch & Boesch, 1983, 1984a, 1984b, 1990, 1993; Sugiyama, 1985, 1995, 1997; Sugiyama et al. 1988; Bermejo et al., 1989; Fushimi et al., 1991; Boesch-

Figure 22.2 Flint uses a wadded, chewed bolus of leaves to collect water from a hole in a tree. After Goodall, 1968; with permission.

Figure 22.3 In West Africa, chimpanzees use a hammer and anvil technique to crack open nuts that are otherwise impossible for them to access. Image after a photo by Alison Hannah; with permission.

Acherman & Boesch, 1993; Boesch, 1993; Matsuzawa & Yamakoshi, 1996; Biro et al., 2006). In West Africa, chimpanzees may spend as much as 15 percent of their feeding time cracking nuts. For brief periods they get more than 3500 calories per day from the practice (Yamakoshi, 2001; Wrangham, 2006). The nuts are a nutritional dynamo, packed with fat and protein. At my study site, Semliki, Bill McGrew placed oil-palm nuts on a beautifully selected anvil and left enticing hammers nearby; the Semliki chimpanzees ignored the opportunity.

Nut-cracking is not as simple as it might appear and can take years to master (Wrangham et al., 1994; Inoue-Nakamura & Matsuzawa, 1997; Boesch & Tomasello, 1998; Whiten et al., 1999). It takes some power; infants and juveniles, as powerful as chimpanzees are, cannot create the necessary force. In addition to brawn, some delicacy is necessary for some nuts: too little force and the nut never cracks; too much and knife-sharp shell fragments are smashed into the nut meat, making it inedible. While oil-palm nuts can be opened with a mere tap and *Coula* nuts can be cracked with a wooden hammer, *Panda* nuts require a stone hammer and the right technique. Several blows are needed, and a forceful blow must be aimed at one of the seams.

In general, females are more efficient nut-crackers, requiring fewer blows to open the shell and less often ruining the meat by over-smashing; they crack them more frequently and get more calories from them. The delicacy required of two particular nut-hammer techniques reveals a more profound sex difference. First, the harder-to-open *Panda* nuts are cracked much more often by females than males. Cracking *Coula* nuts in trees requires not only nut-cracking skills, but an acrobatic balancing act; it often requires three "hands," one of which is a foot: one to grip the tree, one to hold the nut in place, and one to swing the hammer. It is only done by females. In short, the more difficult the technique, the more likely it is that females are the ones to use it. This observation has led some to suggest that among early hominins females were more the tool users than were males (Hunt, 2007).

Nut-cracking also involves sophisticated planning, mental mapping, and spatial orientation. Chimpanzees often collect their hammer before arriving at the nut tree; 40 percent of the carries are to feeding areas that are out of sight (Boesch & Boesch, 1984b). For trees that require wood, which is much more common in the environment, chimpanzees arrive with wooden hammers, while they bring stones to the *Panda* trees.

Stones are harder to find and are not always near the nuts, and so must be carried for long distances – at times up to 500 m. Heavier stones are better, and recognizing their value, chimpanzees carry the heavy stones greater distances. Their spatial orientation and mental mapping abilities are apparent in the efficiency with which they move to a nut-cracking site once a hammer is gathered up. If you map the stones and the trees, chimpanzees rarely carry a stone to any but the nearest tree.

11. **Club.** Now and then, but not commonly, chimpanzees use a stout (~5 cm diameter) 1 m branch

or bough as a club, using the length to amplify force (Figure 22.4). Surprisingly, they seem to care little about the accuracy of a blow, even though they do seem to understand how a club can amplify power. Wooden club-like objects more often fall in the category of display branch rather than club, and are wielded more for show and intimidation than for actually striking a target (Goodall, 1964, 1968; McGrew, 1992). Occasionally, however, chimpanzees use clubs to strike objects purposefully. Kortlandt describes chimpanzees clubbing a fake leopard placed in their path (Kortlandt, 1965) and, more gruesomely, Kibale-Kanyawara males have been observed clubbing females (Linden, 2002), evocatively described in the title of Linden's report: "Wife Beaters of Kibale."

12. **Anvil/mortar.** This tool is the complement to the nut **hammers** used at Taï and other western chimpanzee sites. While the hammer is essential, it is useless without a hard surface on which to place the nut. The anvil may be wood, typically a root, but is just as often stone. Whichever the material, nuts are placed in a cup-like concavity to better secure them as they are hammered; many of these concavities are created or made deeper by the repeated use of the **anvil**.

Anvils may be used without a hammer. Larger, bowl-like concavities are used when crushing or pounding fruits that are too large or too hard to process with the teeth (McGrew, 2004), such as the woody-shelled, orange-sized *Strychnos* fruits (Goodall, 1968), the pulp of which is one of the most aromatic and enticing scents I have encountered in my entire life; it smells like a combination of candy wax, strawberries, and citrus. In West African savanna sites, baobabs are opened in a similar fashion (Baldwin, 1979; McGrew et al., 2003; Marchant & McGrew, 2005). Fruit smashing is engaged in with an action indistinguishable from that which a human would use; the large fruit is held over the head with both hands and brought down forcefully onto the anvil/mortar. While not very substantial compared to a stone anvil, leaves are also used as a steady base for squashing ectoparasites, and so are functionally anvils as well (Boesch, 1996; Assersohn et al., 2004).

13. **Ant wand or ant dip.** Chimpanzees harvest safari ants with a long, narrow wand, sometimes more than 1 m long (Figure 22.5). The length allows the ant-dipper to stay as far from the column of ants as practicable. The harvester may stay out of reach of the marauding safari ants by armhanging from an overhead branch (Figures 22.6). The ant wand is dipped into the swirling mass of ants to allow ants to detect it and react by gripping the tool with their mandibles or running up it in great numbers; McGrew once counted 292 ants in a single dip (McGrew, 1974). The treat, now ants-on-a-stick, is lifted slowly out of the nest, after which the chimpanzee grips the base of the wand with one hand and swipes up the length of it with the other, gathering the ants into a squirming ball which is then quickly eaten (McGrew, 1974, 2004; Sugiyama et al., 1988; Boesch & Boesch, 1990). Ant bites are painful, so quick action is required to avoid

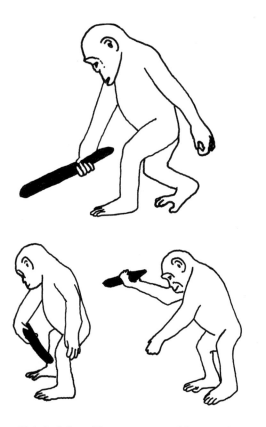

Figure 22.4 An infant chimpanzee uses a club to smash ants. After Goodall, 1968; with permission.

Figure 22.5 Ant wand. Step 1 is to dip an appropriate tool into a safari ant nest. Note the length of the tool and the wary posture, meant to keep the feet as far from the biting ants as possible. Ants crawl up the wand, after which they can be harvested. Photo by David Bygott; with permission.

being bitten too many times. McGrew's description can be reduced to this sequence: (1) the dip, (2) the lift, (3) the swipe, and (4) the gulp.

14. **Projectile.** "Projectiles" or "missiles" are objects thrown or dropped to produce a reaction from another individual. Chimpanzees often throw rocks, branches, and other free objects to get another's attention and to intimidate them as part of a social display (Goodall, 1964, 1968; McGrew 1992, 2004). Projectiles are sometimes used to warn or ward off potential predators – or researchers – though more by dropping from a tree than by throwing. The intention seems more to discombobulate the target rather than to strike them in some vulnerable spot (Goodall, 1968; McGrew, 2004).

Figure 22.6 Ant wand dipping consists of (a) the dip (top), (b) the lift, (c) the swipe (middle and below), and (d) the gulp.

15. **Leaf clip/strip.** Chimpanzees use leaf clipping to attract attention. A leaf is held in the hand and small pieces are ripped off with the teeth, one after the other, making a distinctive ripping or popping sound that, on quiet days in the forest, can be heard from 10 m away. At Mahale and Budongo, males leaf clip as a relatively understated means of drawing the attention of a female in estrus, who might then approach the male for copulation without a possessive higher-ranking male noticing (Nishida, 1980;

McGrew, 2004). At Taï, chimpanzees use leaf clipping instead to signal that they are taking a rest, or at other times it precedes buttress drumming. At Boussou it is said (Whiten et al., 1999) to signal frustration, but at other times it is used as a play-starter. When an entire leaf is stripped at once, the gesture is more often meant as a threat (Whiten et al., 1999).

16. **Leaf groom.** In East Africa, but not elsewhere (McGrew, 1992; Whiten et al., 1999), chimpanzees engage in a faux grooming exercise that is as yet incompletely understood. A broad leaf is selected and the groomer strokes the leaf as if it were another chimpanzee, sometimes even engaging the lower lip as they would if holding hairs down while examining skin. The gesture may be a means of drawing attention, perhaps to attract a grooming partner (McGrew, 1992: 188), or it may be a compulsive behavior motivated by the strong urge to groom when no appropriate grooming partner is directly at hand.

17. **Pad/seat.** Large leaves are sometimes piled on the ground and used as a seat, often when the ground is damp or muddy (Hirata et al., 1998); sometimes leaves are used as a cushion when sitting on branches with thorns (Alp, 1997).

18. **Lever/pry bar.** Sticks are often used to widen the opening of underground bees' nests or to widen holes in trees to get at honey or fledging birds in their cavities; when banana boxes were introduced at Gombe, chimpanzees crafted chisel-like sticks to attempt to lever open the tops (Goodall, 1986; Sanz & Morgan, 2009).

19. **Insect brush/whisk.** Leafy twigs are sometimes used to shoo away flies (Sugiyama, 1969) or to brush away bees (Whiten et al., 1999).

20. **Body decoration.** While it has only been reported in the wild from two sites – Mahale (McGrew & Marchant, 1998; McGrew, 2013) and Gombe (Goodall, 1986) – it is intriguing that chimpanzees sometimes decorate their bodies with natural objects such as vines or animal skins. In most cases the object is merely draped, but Bill McGrew and Linda Marchant saw an individual wearing what could only be called a necklace, a ring of skin, tied (how it was tied was not observed) to complete the circle and worn around the neck. This use is significant because it serves no practical purpose and does not mimic any aspect of chimpanzee life (as toy dolls might, for example), but seems purely decorative or symbolic. Thus, chimpanzees destroy yet another ostensible human uniqueness: body decoration. Some suggest body decoration is unique to humans and relatively recent in origin (e.g., Zilhao, 2007).

Lastly, though not counted among my 20 tools since they have only been observed at a single site, I must mention one other tool that illustrates the chimpanzee's creative ability when dealing with an inhospitable setting. Chimpanzees in Sierra Leone have been observed constructing a kind of primitive sandal when feeding from the thorny tree *Ceiba pentandra*. Leaves are commonly placed onto thorn-covered branches to protect the soles of the feet from thorns, but once the adult female Namaska gripped the foliage with her toes and took several steps (Alp, 1997).

22.8 Captivity

Chimpanzees in captivity use tools even more commonly than those in the wild, perhaps because humans make so many tools available to them, and model their use. They use backpacks, silverware, pots, buckets, bags, mirrors, pliers, computer touch-screens, and just about any human implement made available to them. The pygmy chimpanzee Kanzi has been taught to make primitive stone tools in an experiment that required the tools to penetrate a tough surface isolating a treat (Toth et al., 1993; Schick et al., 1999). In captivity, chimpanzees engage in primitive dentistry surprisingly often; McGrew (1992: 184) describes a surprisingly human-like attempt by one chimpanzee to remove a troublesome milk tooth from another, who lies patiently, hoping for a good result. Another captive chimpanzee used a tool she had manufactured to groom the teeth of other individuals, removing food remains (McGrew & Tutin, 1972, 1973). Yerkes and Yerkes (1929) offer a stunning image of a juvenile chimpanzee using a pair of pliers and a mirror as he attempts to remove a deciduous tooth.

22.9 Conclusions

There are several important conclusions to be drawn from tool use. First, chimpanzees are natural tool users. They turn to a tool as a solution to a challenge quickly and easily. However, for some tasks that we humans find simple, like mentally fitting together two components of a complicated object, chimpanzees struggle (see Chapter 18 on the mind). Nevertheless, when it comes to simple tools like sponges, probes, and termiting tools, they turn to a tool without a second thought.

Second, chimpanzees are capable of complicated, multistep tasks that involve more than one object or tool, a skill psychologists would have reckoned preposterous before Goodall's work. Chimpanzees use **tool sets** (Brewer & McGrew, 1990; McGrew, 2004, 2010; Boesch et al., 2009) in the sense that they develop a resource by using different tools in the correct sequence; for example, when collecting honey chimpanzees may use a probe for the initial penetration of the nest covering, a lever or pry bar to widen the opening, and then a harvesting probe to extract the honey.

They also use tools to modify the state of other tools, such as when they use a harvesting probe to retrieve a sponge that has fallen into an inaccessible drinking hole (Goodall, 1964; McGrew, 2004), or when an anvil is stabilized by placing a stone underneath it (Sugiyama, 1985).

Third, chimpanzee tool use is not a simple stimulus–response phenomenon, where the same situation is always met with the same tool-use response, whether it is appropriate or not. Instead, they are goal-oriented; they problem-solve, casting their mind around for a workable solution until they find the right one.

Hammerstone carrying and picking termiting tools before they reach a mound reveal the mental capacity for "displacement," the ability to draw on a mental map of an area to plan a future activity.

In short, chimpanzee tool use is all but human, and sophisticated enough that Schick, Toth, and colleagues were not overselling their data when they titled their stone tool-making report "*Pan* the tool-maker."

References

Alp R (1997) "Stepping-sticks" and "seat-sticks": new types of tools used by wild chimpanzees (*Pan troglodytes*) in Sierra Leone. *Am J Primatol* 41, 45–52.

Assersohn C, Whiten A, Kiwede ZT, Tinka J, Karamagi J (2004) Use of leaves to inspect ectoparasites in wild chimpanzees: a third cultural variant? *Primates* 45, 255–258.

Baldwin PJ (1979) *The Natural History of the Chimpanzee (Pan troglodytes verus) at Mt. Assirik, Senegal*. PhD Thesis. Stirling: University of Stirling.

Bani E (2004) *Culture Connections: Woven Histories, Dancing Lives – Torres Strait Islander Identity, Culture and History*. Canberra, ACT: Aboriginal Studies Press.

Bermejo M, Illera G, Sabater Pi J (1989) New observations on the tool-behavior of the chimpanzees from Mt. Assirik (Senegal, West Africa). *Primates* 30, 65–73.

Biro D, Sousa C, Matsuzawa T (2006) Ontogeny and cultural propagation of tool use by wild chimpanzees at Bossou, Guinea: case studies in nut-cracking and leaf folding. In *Cognitive Development in Chimpanzees* (eds. Matsuzawa T, Tomonaga M, Tanaka M), pp. 476–508. Tokyo: Springer.

Boesch C (1993) Aspects of transmission of tool-use in wild chimpanzees. In *Tools, Language and Cognition in Human Evolution* (eds. Gibson K, Ingold T), pp. 171–184. Cambridge: Cambridge University Press.

Boesch C (1996) The emergence of cultures among wild chimpanzees. In *Evolution of Social Behaviour Patterns in Primates and Man* (eds. Runciman WG, Maynard-Smith J, Dunbar RIM), pp. 251–268. Oxford: Oxford University Press.

Boesch C, Boesch H (1983) Optimisation of nut-cracking with natural hammers by wild chimpanzees. *Behaviour* 83, 265–286.

Boesch C, Boesch H (1984a) Possible causes of sex differences in the use of natural hammers by wild chimpanzees. *J Hum Evol* 13, 415–440.

Boesch C, Boesch H (1984b) Mental map in wild chimpanzees: an analysis of hammer transports for nut cracking. *Primates* 25, 160–170.

Boesch C, Boesch H (1990) Tool use and tool making in wild chimpanzees. *Folia Primatologica* 54, 86–99.

Boesch C, Boesch H (1993) Diversity of tool use and tool making in wild chimpanzees. In *Use of Tools in Human and Non-Human Primates* (eds. Berthelet A, Chavaillon J), pp. 158–168. Oxford: Oxford University Press.

Boesch C, Tomasello M (1998) Chimpanzee and human cultures. *Curr Anthropol* 39, 591–614.

Boesch C, Head J, Robbins MM (2009) Complex tool sets for honey extraction among chimpanzees in Loango National Park, Gabon. *J Hum Evol* 56, 560–569.

Boesch-Acherman H, Boesch C (1993) Tool use in wild chimpanzees: new light from dark forests. *Curr Directions Psych Sci* 2, 18–21.

Bogart SL, Pruetz JD (2008) Ecological context of savanna chimpanzee (*Pan troglodytes verus*) termite fishing at Fongoli, Senegal. *Am J Primatol* 70, 605–612.

Boswell J (1873) *The Life of Samuel Johnson*. London: William P. Nimmo.

Brewer SM, McGrew WC (1990) Chimpanzee use of a tool-set to get honey. *Folia Primatol* 54, 100–104.

Buxton RD (1981) Termites and the turn-over of dead wood in an arid tropical environment. *Oecologia* 51, 379–384.

Collins DA, McGrew WC (1987) Termite fauna related to differences in tool-use between groups of chimpanzees (*Pan troglodytes*). *Primates* 28, 457–471.

Fushimi T, Sakura O, Matsuzawa T, Ohno H, Sugiyama Y (1991) Nut-cracking behaviour of wild chimpanzees (*Pan troglodytes*) in Bossou, Guinea, (West Africa). In *Primatology Today* (eds. Ehara A, Kimura T, Takenaka O, Iwamoto M), pp. 695–696. Amsterdam: Elsevier.

Goodall J (1962) Nest building behavior in the free ranging chimpanzee. *Ann NY Acad Sci* 102, 445–467.

Goodall J (1963) Feeding behavior of wild chimpanzees. *Symp Zool Soc Lond* 10, 39–48.

Goodall J (1964) Tool-using and aimed throwing in a community of free-living chimpanzees. *Nature* 201, 1264–1266.

Goodall J van Lawick (1968) The behavior of free-living chimpanzees in the Gombe Stream Reserve. *Anim Beh Monogr* 1, 165–311.

Goodall J van Lawick (1971) *In the Shadow of Man*. London: Collins.

Goodall J (1986) *The Chimpanzees of Gombe: Patterns of Behavior*. Cambridge, MA: Harvard University Press.

Hayes C (1951) *The Ape in Our House*. New York: Harper.

Hernandez-Aguilar RA, Jim Moore J, Pickering TR (2007) Savanna chimpanzees use tools to harvest the underground storage organs of plants. *PNAS* 104, 19210–19213.

Hirata S, Myowa M, Matsuzawa T (1998) Use of leaves as cushions to sit on wet ground by wild chimpanzees. *Am J Primatol* 44, 215–220.

Huffman MA, Kalunde MS (1993) Tool-assisted predation on a squirrel by a female chimpanzee in the Mahale Mountains, Tanzania. *Primates* 34, 93–98.

Hunt KD (1998) Ecological morphology of *Australopithecus afarensis*: traveling terrestrially, eating arboreally. In *Primate Locomotion: Recent Advances* (eds. Strasser E, Fleagle JG, McHenry HM, Rosenberger A), pp. 397–418. New York: Plenum.

Hunt KD (2007) Sex differences in chimpanzees foraging behavior and tool use: implications for the Oldowan. In *The Oldowan: Case Studies into the Earliest Stone Age* (eds. Toth N, Schick K), pp. 243–266. Bloomington, IN: CRAFT Press.

Hunt KD, McGrew WC (2002) Chimpanzees in the dry habitats of Assirik, Senegal and Semliki Wildlife Reserve, Uganda. In *Behavioural Diversity in Chimpanzees and Bonobos* (eds. Boesch C, Hohmann G, Marchant LF), pp. 35–51. Cambridge: Cambridge University Press.

Inoue-Nakamura N, Matsuzawa T (1997) Development of stone tool use by wild chimpanzees (*Pan troglodytes*). *J Comp Psych* 111, 159–173.

Izawa K, Itani J (1966) Chimpanzees in the Kasakati Basin, Tanganyika. *Kyoto U Afr Studies* 1.

Kortlandt A (1965) *How Do Chimpanzees Use Weapons When Fighting Leopards?* Philadelphia, PA: American Philosophical Society.

Lapuente J, Hicks TC, Linsenmair KE (2017) Fluid dipping technology of chimpanzees in Comoé National Park, Ivory Coast. *Am J Primatol* 79, e22628.

Linden E (2002). The wife beaters of Kibale. *Time* 160, 56–57.

Marchant LF, McGrew WC (2005) Percussive technology: chimpanzee baobab smashing and the evolutionary modeling of hominid knapping. In *Stone Knapping: The Necessary Conditions for a Uniquely Hominid Behavior*, pp. 341–352. Cambridge: McDonald Institute.

Matsuzawa T, Yamakoshi G (1996) Comparison of chimpanzee material culture between Bossou and Nimba, West Africa. In *Reaching into Thought: The Minds of the Great Apes* (eds. Russon AE, Bard KA, Parker ST), pp. 211–232. New York: Cambridge University Press.

McBeath NM, McGrew WC (1982) Tools used by wild chimpanzees to obtain termites at Mt. Assirik, Senegal: the influence of habitat. *J Hum Evol* 11, 65–72.

McGrew WC (1974) Tool use by wild chimpanzees in feeding upon driver ants. *J Hum Evol* 3, 501–508.

McGrew WC (1987) Tools to get food: the subsistants of Tasmanian Aborigines and Tanzanian chimpanzees compared. *J Anthropol Res* 43, 247–258.

McGrew WC (1992) *Chimpanzee Material Culture: Implications for Human Evolution*. Cambridge: Cambridge University Press.

McGrew WC (1993) The intelligent use of tools: twenty propositions. In *Tools, Language and Cognition in Human*

Evolution (eds. Gibson K, Ingold T), pp. 151-192. Cambridge: Cambridge University Press.

McGrew WC (1994) Tools compared: the material of culture. In *Chimpanzee Cultures* (eds. Wrangham RW, McGrew WC, deWaal FBM, Heltne PG), pp. 25-40. Cambridge, MA: Harvard University Press.

McGrew WC (2004) *The Cultured Chimpanzee: Reflections on Cultural Primatology*. Cambridge: Cambridge University Press.

McGrew WC (2010) Chimpanzee technology. *Science* 328, 579-580.

McGrew WC (2013) Is primate tool use special? Chimpanzee and New Caledonian crow compared. *Phil Trans R Soc B* 368, 20120422.

McGrew WC, Collins DA (1985) Tool-use by wild chimpanzees (*Pan troglodytes*) to obtain termites (*Macrotermes herus*) in the Mahale Mountains, Tanzania. *Am J Primatol* 9, 185-214.

McGrew WC, Marchant LF (1998) Chimpanzee wears a knotted skin "necklace." *Pan African News* 5, 8-9.

McGrew WC, Tutin CEG (1972) Chimpanzee dentistry. *J Am Dental Assoc* 85, 1198-1204.

McGrew WC, Tutin CEG (1973) Chimpanzee tool use in dental grooming. *Nature* 241, 477-478.

McGrew WC, Tutin CEG, Baldwin PJ (1978) Chimpanzees, tools and termites: cross-cultural comparisons of Senegal, Tanzania and Rio Muni. *Man* 14, 185-214.

McGrew WC, Baldwin PJ, Marchant LF, et al. (2003) Ethoarchaeology and elementary technology of unhabituated wild chimpanzees at Assirik, Senegal, West Africa. *PaleoAnthropology* 1, 1-20.

Merfield FG, Miller, H (1956) *Gorillas Were My Neighbours*. London: Longmans.

Nishida T (1973) The ant-gathering behaviour by the use of tools among wild chimpanzees of the Mahali Mountains. *J Hum Evol* 2, 357-370.

Nishida T (1980) The leaf-clipping display: a newly-discovered expressive gesture in wild chimpanzees. *J Hum Evol* 9, 117-128.

Nishida T, Uehara S (1980) Chimpanzees, tools and termites: another example from Tanzania. *Curr Anthropol* 21, 671-672.

Nishida T, Wallauer W (2003) Leaf-pile pulling: an unusual play pattern in wild chimpanzees. *Am J Primatol* 60, 167-173.

Nishie H (2011) Natural history of *Camponotus* ant-fishing by the M group chimpanzees at the Mahale Mountains National Park, Tanzania. *Primates* 52, 329.

Nissen HW (1931) A field study of the chimpanzee: observations of chimpanzee behavior and environment in western French Guinea. *Comp Psychol Monogr* 8, 122.

Pascual-Garrido A, Buba U, Nodza G, Sommer V (2012) Obtaining raw material: plants as tool sources for Nigerian chimpanzees. *Folia Primatologica* 83, 24-44.

Pruetz JD, Bertolani P (2007) Savanna chimpanzees, *Pan troglodytes verus*, hunt with tools. *Curr Biol* 17, 412-417.

Pruetz JD, Bertolani P, Ontl KB, et al. (2015) New evidence on the tool-assisted hunting exhibited by chimpanzees (*Pan troglodytes verus*) in a savannah habitat at Fongoli, Sénégal. *Roy Soc Open Science* 2, 140507.

Roberts EM, Todd CN, Aanen DK, et al. (2016) Oligocene termite nests with in situ fungus gardens from the Rukwa Rift Basin, Tanzania, support a paleogene African origin for insect agriculture. *PloS ONE* 11, p.e0156847.

Sanz CM, Morgan DB (2007) Chimpanzee tool technology in the Goualougo Triangle, Republic of Congo. *J Hum Evol* 52, 420-433.

Sanz CM, Morgan DB (2009) Flexible and persistent tool-using strategies in honey-gathering by wild chimpanzees. *Int J Primatol* 30, 411-427.

Sanz CM, Morgan DB, Gulick S (2004) New insights into chimpanzees, tools, and termites from the Congo Basin. *Am Nat* 164, 567-581.

Savage TS, Wyman J (1844) Observations on the external characters and habits of the Troglodytes Niger, Geoff; and on its organization. *Boston J Nat Hist* 4, 362-386.

Schick KD, Toth N, Garufi G, et al. (1999) Continuing investigations into the stone tool-making and tool-using capabilities of a bonobo (*Pan-paniscus*). *J Archaeol Sci* 26, 821-832.

Shumaker RW, Walkup KR, Beck BB (2011) *Animal Tool Behavior: The Use and Manufacture of Tools by Animals*. Baltimore, MD: Johns Hopkins University Press.

Sugiyama Y (1969) Social behavior of chimpanzees in the Budongo Forest, Uganda. *Primates* 10, 197-225.

Sugiyama Y (1985) The brush stick of chimpanzees found in southwest Cameroon and their cultural characteristics. *Primates* 26, 168-181.

Sugiyama Y (1995) Drinking tools of wild chimpanzees at Boussou. *Am J Primatol* 37, 263-269.

Sugiyama Y (1997) Social tradition and the use of tool composites by wild chimpanzees. *Evol Anthropol* 6, 23-27.

Sugiyama Y, Koman J (1979) Tool using and making behaviour in wild chimpanzees at Bossou, Guinea. *Primates* 20, 323-339.

Sugiyama Y, Koman J, Bhoye Sow M (1988) Ant-catching wands of wild chimpanzees at Bossou, Guinea. *Folia Primatologica* 51, 56-60.

Suzuki S, Kuroda S, Nishihara T (1995) Tool-set for termite-fishing by chimpanzees in the Ndoki Forest, Congo. *Behaviour* 132, 219-235.

Toth N, Schick K (2009) The Oldowan: the tool making of early hominins and chimpanzees compared. *Ann Rev Anthropol* 38, 289-305.

Toth N, Schick KD, Savage-Rumbaugh ES, et al. (1993) *Pan* the tool-maker: investigations into the stone tool-making capabilities of a bonobo (*Pan paniscus*). *J Archaeol Sci* 20, 81-91.

Toth N, Schick K, Semaw S (2006) A comparative study of the stone tool-making skills of *Pan*, *Australopithecus*, and *Homo sapiens*. In *Oldowan: Case Studies into the Earliest Stone Age* (eds. Toth N, Schick K), pp. 155–222. Gosport, IN: Stone Age Institute Press.

Van Leeuwen EJ, Mulenga IC, Bodamer MD, Cronin KA (2016) Chimpanzees' responses to the dead body of a 9-year-old group member. *Am J Primatol* 78, 914–922.

Van Schaik CP, Deaner RO, Merrill MY (1999) The conditions for tool use in primates: implications for the evolution of material culture. *J Hum Evol* 36, 719–741.

Whiten A, Goodall J, McGrew WC, et al. (1999) Cultures in chimpanzees. *Nature* 399, 682–685.

Whiten A, Goodall J, McGrew WC, et al. (2001) Charting cultural variation in chimpanzees. *Behaviour* 138, 1481–1516.

Wrangham RW (1977) Feeding behaviour of chimpanzees in Gombe national park, Tanzania. In *Primate Ecology* (ed. Clutton-Brock TH) pp. 503–538. New York: Academic Press.

Wrangham RW (2006) Chimpanzees: the culture-zone concept becomes untidy. *Curr Biol* 16, R634–R635.

Wrangham RW, de Waal FBM, McGrew WC (1994) The challenge of behavioral diversity. In *Chimpanzee Cultures* (eds. Wrangham RW, McGrew WC, de Waal FBM, Heltne PG), pp. 1–18. Cambridge, MA: Harvard University Press.

Yamakoshi G (2001) Ecology of tool use in wild chimpanzees: toward reconstruction of early hominid evolution. In *Primate Origins of Human Cognition and Behavior* (ed. Matsuzawa T), pp. 537–556. Tokyo: Springer-Verlag.

Yerkes, RM, Yerkes, AW (1929) *The Great Apes: A Study of Anthropoid Life*. New Haven, CT: Yale University Press.

Zilhao J (2007) The emergence of ornaments and art: an archaeological perspective on the origins of "behavioral modernity." *J Arch Res* 5, 1–54.

23 Wisdom of the Ages
Chimpanzee Culture

From N. Inoue-Nakamura and T. Matsuzawa, "Development of Stone Tool Use by Wild Chimpanzees," *J. Comp. Psych.* 111(7): 159-173, 1997. © American Psychological Association, reprinted with permission.

Culture is widely viewed as the hallmark of humanity, a feature unique to human existence and not found among other beings. This distinction, of course, depends on your definition of "**culture**." Many definitions follow that of E.B. Tylor (1874): "the sum total of what an individual acquires from his society – those customs, artistic norms, food-habits, and crafts which come to him not by his own creative activity but as a legacy from the past" (Lowie, 1937). If cultural phenomena are boiled down to a mere list, many would emphasize what society teaches its members about art, science, myth, literature, ethics, morality, tradition, and religion. In translating a definition such as that of Taylor and Lowie into something more useful for primates, we might follow that of the co-founder of Japanese primatology, Kinji Imanishi; he took a more pragmatic and streamlined approach. He defined culture as "**socially transmitted, adjustable behavior**" (Imanishi, 1952). In some quarters, among sociocultural anthropologists, for example, the question of whether chimpanzees (and some other animals) have culture is controversial (Galef, 1992), but among primatologists the possession of culture by chimpanzees is not the least bit controversial (Wrangham et al., 1996; Boesch & Tomasello, 1998; Whiten et al., 1999). Wild studies aside (Wrangham et al., 1996; Boesch & Tomasello, 1998; Whiten et al., 1999), chimpanzees learn to eat with a fork, drink from a cup, wear clothes, name objects, request a hug, and carry their favorite toys in a backpack; denying they possess culture seems willfully blind.

23.1 Universals

Of course, a specialist who focuses on human culture might find that, at first glance, chimpanzees appear to have no culture whatsoever: myth, literature, and art are completely absent. In accord with this, much of chimpanzee behavior is monotonously similar from site to site, as if it were instinctual: All populations make sleeping nests, engage in social or charging displays, possess dominance hierarchies, and drum on trees. Their vocalizations, as well, are quite similar from place to place, and many vocalizations, such as the subordinance-indicating pant-grunt, appear to be wholly instinctual, since they emerge even among human-reared chimpanzees who have never heard them (Kellogg & Kellogg, 1933; Hayes, 1951). Physical expressions of dominance such as the low crouch, the outstretched hand, pats and hugs as a means of reassurance, likewise, are universal.

Perhaps these monotonously consistent behaviors would seem less decisive if we were to note that humans have such universals as well (Brown, 1991, 2004). Daniel Brown lists over 300 human universals, including exchange of goods between groups, concepts of luck, body ornamentation, division of labor, and age-grading. Our laughter and other emotive vocalizations are universal, and like chimpanzees we use the low crouch or bowing to signal subordinance. We use pats, hugs, and outstretched hands in exactly the same contexts as chimpanzees. Yet no one would argue that human behavior is entirely instinctual.

23.2 Art and Science

If we dig a little deeper into chimpanzee behavior and, as we did with our discussion of tool use, if we keep in mind that our own culture has profusely expanded in the recent past, we would emerge with a different perspective. If we take our hunter-gatherer ancestors as a more appropriate touchstone regarding human culture, we might see chimpanzee behavior as more culturally laden than otherwise.

Even some realms where we might assume humans are unique devolve into gray areas when captive chimpanzees are included. Captive chimpanzees have some appreciation of art, since they know what their end goal is; when in the midst of creating art, they refuse to surrender a canvas until they determine they are done. They certainly seem to enjoy producing art (see Chapter 18). Still, in wild chimpanzees we have never observed production of art.

To the extent that chimpanzees have science, it resembles that of a three-year-old human – a tendency to test reality; chimpanzees seem to be curious about the way the world works in a way that baboons, for instance, are not. When confronted by something quite out of their experience, such as a mirror, they leap to investigate and make every attempt to understand it. "Is that an animal behind glass? If so, I should be able to reach behind and touch it, and I cannot." They are curious. I mentioned in Chapter 18 that I once watched a juvenile come upon a displaced trail sign, a wooden block painted white and about the size of a book. He picked it up and looked at it from all sides. He held it near his ear and tapped on it, clearly curious about the meaning of this odd thing in his environment. I would argue that chimpanzees often apply the scientific method to challenging problems; they formulate a hypothesis ("this is a window and there is a chimpanzee behind it") and they test them ("this hypothesis is disproven; I can reach behind the window with my hand and there is nothing there").

23.3 Ethics, Morality, and Religion

Chimpanzees have primitive ethics as well. Frans de Waal offers this definition of ethics (de Waal, 1991): "a set of expectations about the way in which oneself should be treated and how resources should be divided, a deviation of which expectations to one's disadvantage evokes a negative reaction, most commonly protest in subordinates and punishment in dominants." If this has twisted your brain in a knot, try this translation: Chimpanzees recognize when they are being treated unfairly and protest when they feel put upon, if not by physical action at least by protesting vocally. They learn tribal rules during their youth. By observing the reactions of their mother and other members of their society, a maturing chimpanzee learns important social niceties: greet your friends when you first encounter them; be careful not to offend the powerful; apologize with a pant-grunt or touch when you violate social norms; stealing is wrong. This suggests there is a rudimentary ethical component to their culture.

Ethics are one thing – what about even deeper concepts? Religion? If they have what some might call "religious sensibilities," it is in their appreciation for their seeming reverence or awe of nature, such as when they respond exuberantly to the awesome sound of a waterfall or to heavy rain. There is, however, only the thinnest of evidence that they ever contemplate the heavenly half of Gould's "twin magisteria," or that they have any concept of a spiritual world (Gould, 2002). Some speculated that there is some spiritual component to an odd behavior of chimpanzees at Taï. They cache stones in small niches in trees, creating something that resembles a kind of shrine (Kühl et al., 2016). Eerily similar man-made stone collections have been observed in numerous cultures, raising the possibility that these are something like sacred trees.

To sum up, whether chimpanzees have primitive art, religion, and science is debated, but it is certain beyond doubt that they are quite dependent on socially transmitted knowledge about their home and their food, arguably as dependent as are humans (McGrew, 2004, 2009; Wrangham et al., 1996).

23.4 Geographic Variation and Culture

When primatologists began to consider the issue of culture with respect to chimpanzees, they very quickly fixed on the fact that, because culture is learned, rather than genetically determined, it varies from place to place. No two human populations separated by any distance dress the same or speak the same. Do chimpanzees have this variation? They do (Goodall, 1973, 1986; McGrew & Tutin, 1978; Wrangham et al., 1996; Boesch & Tomasello, 1998; McGrew, 1998, 2009, 2010; Whiten et al., 1999, 2001, 2005; Schöning et al., 2008; Tomasello, 2009). Whiten and colleagues (1999) in particular collected every scrap of information on chimpanzee cultural practices and produced the amazing summary illustration reproduced in Figure 23.1. If you explore it on your own and you will see that among these many tens of cultural practices, there is an undeniable variation from site to site. Let us review some of the most important.

The first of chimpanzee cultural traditions to be shown to vary among study sites was **hand-clasp grooming** (McGrew & Tutin, 1978), a behavior in which chimpanzees extend their arms directly over their heads and clasp hands. While it occurs daily at Mahale, it has never been observed at Gombe, only 170 km to the north (McGrew et al., 2001).

Nut-cracking is a second such cultural practice. All West African chimpanzees crack nuts, and East African chimpanzees do not, despite the presence of appropriate nuts (Boesch et al., 1994; McGrew, 1997). This difference might be explained by proposing that the practice was invented in some community in West Africa and **diffused** to other communities by females who dispersed, carrying the traditions with them. The cultural practice might then have proliferated, carried from community to community by females until it reached the impassable N'Zo-Sassandra River, which served as a cultural diffusion barrier. Chimpanzees east of the N'Zo-Sassandra simply might have failed to invent hammer-and-anvil nut-cracking. Chimpanzee cultural innovations show every appearance of dispersing in this manner over long distances – barriers aside – because toolkits and other cultural practices are more similar the closer the communities (Toth & Schick, 2009; Whiten et al., 2009).

This **cultural diffusion** model seemed perfect – until the chimpanzees in the Cameroon's Ebo Forest, 1700 km east of the supposed barrier, were found to smash nuts (Morgan & Abwe, 2006; Wrangham, 2006). While this disproves the single-invention hypothesis, nut-cracking is still certainly a cultural practice. In communities where it is seen, virtually all individuals engage in it. Still, the failure of East African chimpanzees to pick up on nut-cracking is a spectacular demonstration of how heavily dependent on cultural transmission chimpanzee feeding practices are (Toth & Schick, 2009).

The "arm-over" dominant/subordinance signal is another of these cultural variants. At Gombe, a dominant chimpanzees lifts an arm under which a subordinate individual passes, almost like petting your dog from head to rump; this expression of dominance is found nowhere else (Goodall, 1986).

No two chimpanzee sites have the same toolkits (Figure 23.2; McGrew, 1992, 2004; Whiten et al., 1999, 2001). Gombe chimps use chewed leaves for sponges, but other populations use moss. At Kibale they use chewed stems. At Ndoki, chimpanzees use the chewed stems, but to gather termites rather than

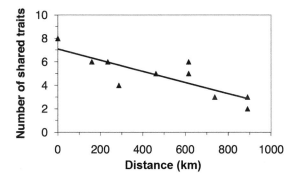

Figure 23.1 The farther apart chimpanzee communities are, the less likely they are to share cultural traits, suggesting that many cultural practices are invented in one place and carried to other communities by females when they transfer at adulthood. From Whiten et al., 2009; with permission.

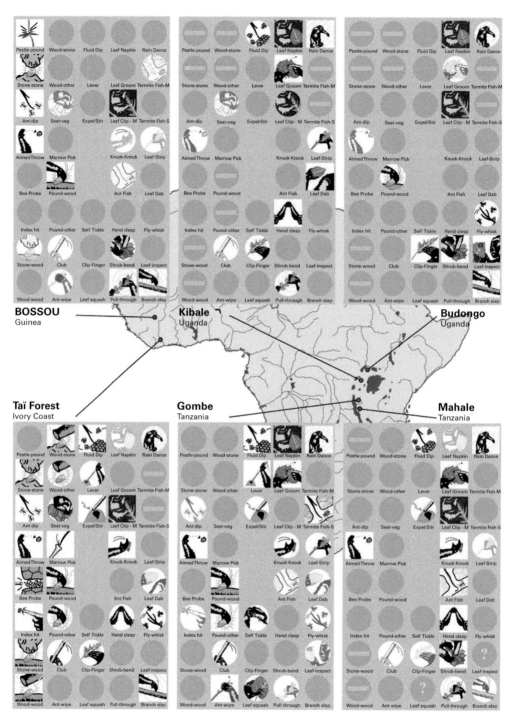

Figure 23.2 Behaviors at the six sites with the longest research history. Square color icons signify behaviors that are customary at that site, meaning it has been seen in nearly all able-bodied members of at least one age–sex class (such as adult males); circular color icons indicate a habitual behavior, one that has been seen often and in multiple individuals, but not as pervasively as "customary"; monochrome icons signify a behavior that is present, meaning it is present in the community; a gray circle means absent; a gray circle with a horizontal bar means absent with an ecological explanation; a question mark signifies uncertain; no symbol at all indicates unknown. From Whiten et al., 1999; with permission. (A black and white version of this figure will appear in some formats. For the color version, please refer to the plate section.)

water (Suzuki et al., 1995). At my site in Uganda, Semliki, and only one other site chimpanzees dig holes in the sand, presumably to filter our particulates in the water, because they do it near water sources that they could seemingly drink from with much less trouble (Hunt et al., 1999; Galat-Luong & Galat, 2000; Hunt & McGrew, 2002; Galat-Luong et al., 2009).

Some chimpanzee communities use grass stems to gather termites while others use twigs. Spears are found at Fongoli and nowhere else. As we learned in the last chapter, leaf clipping is a courtship gesture at Mahale (Nishida, 1980) and a signal that the individual is about to take a rest at Taï (Whiten et al., 1999). Leaf napkins are found at every chimpanzee site – except one, Boussou (Wrangham, 2006).

Ant dipping and palm pounding are further examples of patchily distributed practices (Whiten et al., 1999). Palm pounding is only found at Boussou. Ant dipping is found at Gombe and Boussou but nowhere else.

Gestures, likewise, vary among sites; in one study, 68 different gestures were documented among three of the most studied chimpanzee sites: Budongo, Mahale, and Gombe (Hobaiter & Byrne, 2011). Some are consistent at all three, of course, but other quirky ones are seen at one or two. "Tandem walk," a behavior in which one individual puts an arm over the body of another and both walk forward while embracing, is seen at Budongo, but not at the other two sites. Knocking on a hard surface to get attention is, likewise, seen at two of the three sites. Belly drumming and clapping are only seen at one.

23.5 Food Preference

When it comes to food habits – what they eat, how they process it, and how they find it – chimpanzees are completely dependent on socially transmitted knowledge. They are excellent local geographers, exhibiting knowledge of natural features in their environment, among the most important of which is the location of important food resources.

It may seem odd that chimpanzees are not born with the ability to find food, as are houseflies and crocodiles; chimpanzees learn it from their mothers over many years; they are probably nearly adults before they have absorbed the entire diet list. And just as it is with humans, these cultural practices vary from place to place. They learn to identify hundreds of plant items (Nishida, et al., 1983), remembering which of leaves, fruit, flowers, and bark is edible for each species. This is critical, because unlike some mammals they only eat foods they learned to select from their mother (Goodall, 1973; Wrangham, 1977; Pusey, 1983; Goodall, 1986), not only by watching what their mothers eat, but by her active discouragement. When an infant tried to eat a food item not on the food item list at Gombe, his mother repeatedly tried to flick it away from him (Wrangham, 1977); the same thing happened at Mahale, where a mother snatched a non-food-list leaf from an infant's mouth (Nishida et al., 1983). Nishida and colleagues relate three further stories of similar motherly interference (1983). Even papaya, a tasty nutritious fruit eaten by chimpanzees elsewhere, is not on the mother-approved list at Gombe (Goodall, 1973). This "correction" on the part of the mother is the hallmark of teaching. The power of the tradition is more apparent when we realize that the food item list *within* a community is virtually identical from individual to individual (Wrangham, 1977; Goodall, 1983), while it differs dramatically among communities (Nishida et al., 1983).

The importance of cultural acquisition is brought into sharp focus by a comparison of Gombe and Mahale food lists. The two sites are only 170 km apart and have almost identical species lists, yet the chimpanzees have very different food lists. Oil-palms nuts make up 20 percent of the diet at Gombe, but at Mahale the trees are there but the fruits are not eaten. At Gombe, chimpanzees eat termites and safari ants (*Dorylus nigricans*), but at Mahale they eat different ants and, incredibly, ignore the termites altogether.

23.6 Mother's Teaching Techniques

The powerful cultural impact of mother's behavior is the most important medium of cultural transmission. Infants are wonderful students of their mother's activities, sometimes placing their eyes dangerously close to her hands as they observe some activity.

There is little that happens in the course of the day that the infant fails to take note of, mostly at extremely close range. The teaching–learning dichotomy is another of those "exclusive to humans" assumptions that has fallen. Chimpanzees were thought to learn only by observing an experienced individual, then attempting it themselves (Inoue-Nakamura & Matsuzawa, 1997). Taï mothers were said to teach their offspring by placing nuts on anvils in the best position to be opened or by correcting the grip of a juvenile when they hold the hammer incorrectly (Boesch, 1993). In 2016 Stephanie Musgrave and her colleagues showed that Goulago chimpanzee adults provide learners with appropriately constructed termiting tools. Adults paused in their own feeding to engage in the teaching, and the youngsters who received the tools significantly increased their harvest (Musgrave et al., 2016).

One consequence of the diet being culturally determined is that chimpanzees, particularly captive-reared individuals, cannot be released cold into a new habitat and be expected to survive. A released chimpanzee without the years of apprenticeship watching and being guided by her mother would not know what to eat, and even if she did, she would not know where to look for it. Chimpanzees learn the locations of feeding sites from other individuals.

Some other behaviors absolutely critical for survival and reproduction are also learned. Females without younger siblings are poor first-time mothers (Goodall, 1986; Stanford, 2001). Fifi of Gombe grew up enjoying the luxury of being cared for by an accomplished mother, Flo. She learned mothering skills by practicing on a younger brother under the watchful eyes of her mother, and her skills became exceptional. She grew up to be a high-ranking female who, in a long life, bore nine offspring, the highest number known, seven of whom survived. She passed on her own confidence and high rank to those offspring (Stanford, 2001). Fifi's daughter, little Flossi (at least, little when I knew her in 1986) now has four offspring of her own (janegoodall.org, 2017). When behaviors are learned well, from good teachers, the benefits echo down the generations.

23.7 Last Thoughts

Do chimpanzees have culture, then? There is no doubt that they have learned traditions – traditions that continue where they are found and fail to appear where they have not been seen before. This constancy suggests that these practices arose by the accumulated wisdom of tens or hundreds of generations of chimpanzees and constitute the chimpanzee cultural heritage: the wisdom of the ages, chimpanzee style. Individual chimpanzees, while intelligent and creative, could not invent the material culture on which every chimpanzee depends, which means chimpanzees are dependent on culture.

Just as with language, humans may be the paragon of culture, but not the only possessors of it. While a termiting tool is not sophisticated, it is a product of culture. Not human culture, naturally, but culture nevertheless.

References

Boesch C (1993) Aspects of transmission of tool-use in wild chimpanzees. In *Tools, Language and Cognition in Human Evolution* (eds. Gibson K, Ingold T), pp. 171–184. Cambridge: Cambridge University Press.

Boesch C, Tomasello M (1998) Chimpanzee and human cultures. *Curr Anthropol* **39**, 591–614.

Boesch C, Marchesi P, Marchesi N, Fruth B, Joulian F (1994) Is nut cracking in wild chimpanzees a cultural behaviour? *J Hum Evol* **26**, 325–338.

Brown DE (1991) *Human Universals*. New York: McGraw-Hill.

Brown DE (2004) Human universals, human nature & human culture. *Daedalus* **133**(4), 47–54.

de Waal FB (1991) The chimpanzee's sense of social regularity and its relation to the human sense of justice. *Am Behav Scientist* 34, 335-349.

Galat-Luong A, Galat G (2000) Chimpanzees and baboons drink filtrated water. *Folia Primatol* 71, 258 (abstract).

Galat-Luong A, Galat G, Nizinski G (2009) Une consequence du rechauffement climatique: les chimpanzes filtrent leur eau de boisson. *Geographia Technica Numéro spécial* 2009, 199-204.

Galef BG (1992) The question of animal culture. *Hum Nat* 3, 157-178.

Goodall J (1983) Population dynamics during a fifteen year period in one community of free-living chimpanzees in the Gombe National Park, Tanzania. *Zeitschrift fur Tierpsychologie* 61, 1-60.

Goodall J (1986) *The Chimpanzees of Gombe: Patterns of Behavior*. Cambridge, MA: Harvard University Press.

Goodall J van Lawick (1973) Cultural elements in a chimpanzee community. In *Precultural Primate Behaviour* (ed. Menzel EW), pp. 144-184. Basel: Karger.

Gould SJ (2002) *Rocks of Ages: Science and Religion in the Fullness of Life*. New York: Ballantine Books.

Hayes C (1951) *The Ape in Our House*. New York: Harper.

Hobaiter C, Byrne RW (2011) The gestural repertoire of the wild chimpanzee. *Anim Cogn* 14, 745-767.

Hunt KD, McGrew WC (2002) Chimpanzees in the dry habitats of Assirik, Senegal and Semliki Wildlife Reserve, Uganda. In *Behavioural Diversity in Chimpanzees and Bonobos* (eds. Boesch C, Hohmann G, Marchant LF), pp. 35-51. Cambridge: Cambridge University Press.

Hunt KD, Cleminson AJM, Latham J, Weiss RI, Grimmond S (1999) A partly habituated community of dry-habitat chimpanzees in the Semliki Valley Wildlife Reserve, Uganda. *Am J Phys Anthropol* 28, 157 (abstract).

Imanishi K (1952) Evolution of humanity. In *Man* (ed. Imanishi K) Tokyo: Mainichi-Shinbunsha.

Inoue-Nakamura N, Matsuzawa T (1997) Development of stone tool use by wild chimpanzees (*Pan troglodytes*). *J Comp Psych* 111, 159-173.

janegoodall.org (2017) The F-family. www.janegoodall.org.au/the-f-family.

Kellogg WN, Kellogg LA (1933) *The Ape and the Child: A Study of Environmental Influence upon Early Behavior*. New York: McGraw-Hill.

Kühl HS, Kalan AK, Arandjelovic M, et al. (2016) Chimpanzee accumulative stone throwing. *Sci Rep* 29, 22219.

Lowie, RH (1937) *The History of Ethnological Theory*. New York: Farrar & Rinehart.

McGrew WC (1992) *Chimpanzee Material Culture: Implications for Human Evolution*. Cambridge: Cambridge University Press.

McGrew WC (1997) Why don't chimpanzees in Gabon crack nuts? *Int J Primatol* 18, 353-374.

McGrew WC (1998) Culture in nonhuman primates? *Ann Rev Anthropol* 27, 301-328.

McGrew WC (2004) *The Cultured Chimpanzee: Reflections on Cultural Primatology*. Cambridge: Cambridge University Press.

McGrew WC (2009) Ten dispatches from the chimpanzee culture wars, plus postscript (revisiting the battlefronts). In *The Question of Animal Culture* (eds. Leland KN, Galef BG), pp. 41-69. Cambridge, MA: Harvard University Press.

McGrew WC (2010) Chimpanzee technology. *Science* 328, 579-580.

McGrew WC, Tutin CE (1978) Evidence for a social custom in wild chimpanzees? *Man*, 13, 234-251.

McGrew WC, Marchant LF, Scott SE, Tutin CEG (2001) Intergroup differences in a social custom of wild chimpanzees: the grooming hand-clasp of the Mahale Mountains. *Curr Anthropol* 42, 148-153.

Morgan BJ, Abwe EE (2006) Chimpanzees use stone hammers in Cameroon. *Curr Biol* 16, R632-R633.

Musgrave S, Morgan D, Lonsdorf E, Mundry R, Sanz C (2016) Tool transfers are a form of teaching among chimpanzees. *Sci Rep* 6, 34783.

Nishida T (1980) Local differences in response to water among wild chimpanzees. *Folia Primatol* 33, 189-209.

Nishida T, Wrangham RW, Goodall J, Uehara S (1983) Local differences in plant-feeding habits of chimpanzees between the Mahale Mountains and Gombe National Park, Tanzania. *J Hum Evol* 12, 467-480.

Pusey AE (1983) Mother-offspring relationships in chimpanzees after weaning. *Anim Behav* 31, 363-377.

Schöning C, Humle T, Möbius Y, McGrew WC (2008) The nature of culture: technological variation in chimpanzee predation on army ants revisited. *J Hum Evol* 55, 48-59.

Stanford CB (2001) *Significant Others: The Ape-Human Continuum and the Quest for Human Nature*. New York: Basic Books.

Suzuki S, Kuroda S, Nishihara T (1995) Tool-set for termite-fishing by chimpanzees in the Ndoki forest, Congo. *Behaviour* 132, 219-234.

Tomasello M (2009) The question of chimpanzee culture, plus postscript. In *The Question of Animal Culture* (eds. Leland KN, Galef BG), pp. 198-221. Cambridge, MA: Harvard University Press.

Toth N, Schick K (2009) The Oldowan: the tool making of early hominins and chimpanzees compared. *Ann Rev Anthropol* 38, 289-305.

Tylor EB (1874) *Primitive Culture: Researches into the Development of Mythology, Philosophy, Religion, Language, Art and Customs*, Vol. 1. New York: Holt.

Whiten A, Boesch C (2001) The cultures of chimpanzees. *Sci Am* 284, 48-55.

Whiten A, Goodall J, McGrew WC, et al. (1999) Cultures in chimpanzees. *Nature* 399, 682-685.

Whiten A, Goodall J, McGrew WC, et al. (2001) Charting cultural variation in chimpanzees. *Behaviour* 138, 1481–1516.

Whiten A, Horner V, de Waal FBM (2005) Conformity to cultural norms of tool use in chimpanzees. *Nature* 437, 737–740.

Whiten A, Schick K, Toth N (2009) The evolution and cultural transmission of percussive technology: integrating evidence from palaeoanthropology and primatology. *J Hum Evol* 57, 420–435.

Wrangham RW (1977) Feeding behaviour of chimpanzees in Gombe national park, Tanzania. In *Primate Ecology* (ed. Clutton-Brock TH) pp. 503–538. New York: Academic Press.

Wrangham RW (2006) Chimpanzees: the culture-zone concept becomes untidy. *Curr Biol* 16, R634–R635.

Wrangham RW, McGrew WC, de Waal FBM, Heltne PG (1996) *Chimpanzee Cultures*. Cambridge, MA: Harvard University Press.

24 The Daily Grind
Within-Group Aggression

Drawing: From Jane Goodall, *Chimpanzees of Gombe*, 1986, Harvard University Press, with permission

Chimpanzees can be caring, affectionate, sensitive animals – toward particular individuals. In the right circumstances. The love and attention a chimpanzee mother showers on her infant is extraordinary, in both its day-to-day manifestation and over the long, five-year span of infancy – the chimpanzee mother–infant bond is arguably even more intense than that of humans (Hrdy, 2011). And it is not just mothers. Males spend hours grooming one another. They look to their friends and allies for comfort and reassurance. When the chips are down and two lifelong friends face a marauding band from another community, they will risk their lives to support one another. Yet, these prosocial bonds are only half the story, if that. In this chapter we will consider how violence shapes social behavior within communities; in the next chapter we will turn toward the war-like aggression that typifies intercommunity relations; in the chapter after that we will tidy up the trilogy with an examination of how profoundly intercommunity violence has affected the relationships within communities. It is this interplay, the effect the threat of intercommunity violence has on minute-to-minute interactions within communities, that makes chimpanzee society so complicated. So complicated, in fact, that it will require 40 steps to knit every aspect of chimpanzee biology together in the final chapter.

24.1 Chimpanzees Are the Most Violent of All Mammals

The chimpanzee community is constantly riven by internal clashes: Chimpanzees are involved in a physical confrontation more often than any other mammal; one review found that male chimpanzees were involved in an aggressive chase or an outright physical attack every three and a half hours (Muller, 2002). Females were more peaceable in that they were involved in violence only 7 percent as often as males – but lest we go overboard, keep in mind that this means a physical confrontation about every two days (Muller, 2002). This is not exactly a tranquil existence. Drawing back to consider all individuals in a community, not males alone, chimpanzees endure an attack more serious than just a passing slap about every 14 hours (Goodall, 1986; Pruetz & McGrew, 2001). Although these intracommunity attacks are only rarely fatal, adult killings have been recorded at Budongo (Fawcett & Muhumuza, 2000), Gombe (Goodall, 1986), Kibale (Wrangham & Peterson, 1996), Mahale (Nishida, 1996), and Kibale-Ngogo (Watts, 2004). In fact, most intracommunity aggression is meant to intimidate and bully, not injure; underlings are kept in their place by constant reminders of their low status (Figure 24.1). **Inter*community* confrontations**, however, are in a completely different realm.

The brutal and intense nature of male–male intracommunity competition and the tactical planning that sustains alpha status were brought into shocking relief to me one day as I followed a young adult male at Mahale. I followed two males, a low-ranking and easy-going guy who was traveling with

Figure 24.1 Dominance has its privileges. Dominant male Goblin (left), hair partially erect, walks purposely while the larger but subordinate subadult Freud pant-grunts and lowers his head in obeisance. A confident, though lower-ranking Evered pauses as the group gathers. Juvenile Frodo is so low-ranking he fails to rate trail privileges and has gone cross-country. Photo by the author.

an ambitious and not-easy-going male, **Lukaja**. Lukaja was an average-sized male, tallish but on the lanky side, a little graying despite his youth; very athletic and very confident. I had watched him emerge as an up-and-comer from the rest of the field of a dozen adult males at Mahale in just the last few weeks. While he had not summoned the courage to challenge the second-ranking male, he had dominated the other males of M group, and I felt it was only a matter of time before he competed for the beta position.

24.2 Lukaja's Rise and Fall

I was wrong – things took a sudden and unexpected turn for the worse for Lukaja. The second-ranking male at Mahale at this time was a large, quiet, unreadable male who was more prone than most to abrupt rages. Lukaja treated him with kid gloves; it seemed he was more fearful of him than the alpha male, **Ntologi**, and in fact sometimes I suspected that even Ntologi was a little wary of this beta male's unpredictable nature.

For the first third of this lovely, sunny-but-cool day Lukaja and his companion went about their business in a routine way. They interacted in the comfortable, unruffled manner of good, long-time friends.

Chimpanzees often take a midday break after a morning feeding session, climbing down from whatever fruit tree they fed in to meet other community members at a gathering place that they somehow seem to know of in advance. Lukaja and his companion descended and then led me some hundreds of meters along a lightly traveled track through dense foliage. We emerged into a relatively open area where several other males were already grooming and relaxing.

Lukaja and his friend were greeted with a flurry of activity and pant-grunts; some of the grooming party males moved to groom him. After some minutes the bustle calmed down and Lukaja ambled up a nearby sun-dappled slope to a perch that overlooked the other males, leaving him 10 m or so from the main grooming party. Other males continued their grooming, some lying quietly as if a nap were on the agenda. Lukaja reclined onto his back, crossed his legs, interlaced his fingers behind his head, elbows akimbo, and stared up at the sky through the foliage, the very picture of a relaxed, confident – perhaps we should go so far as to say a little arrogant – young male in his prime.

Ntologi, the large, very social alpha male of the Mahale M group, alpha for some time now and destined to remain so for some time in the future, emerged from one of the several animal tracks that led to this lounging area, followed by an anxious subadult. The response was sudden and dramatic, much more so than for Lukaja's entrance. All the other males clustered around him, making a show of activity and attentiveness, pant-grunting and feverishly grooming him. He stood like a king amid them, stiff-legged, hair erect, seemingly reveling in the deferential crowd of sycophants surrounding him. Males jockeyed for position, sometimes pausing to groom one another, then changing places, all milling around the large alpha, grunting and tooth-clacking as they groomed him. If you were to view this scene in isolation, Ntologi might seem to be on top of the world. Yet, all was not sweetness and light in Ntologi's world.

While the other males signaled their subordinance with low crouches, pant-grunts, and servile body posture, Lukaja was having none of it. He remained on the slope above this madding crowd, still with his legs crossed, seemingly unconcerned that Ntologi had arrived, ignoring the hubbub below. He must have been at least somewhat aware – and probably fully aware – of the affront this presented to Ntologi. Males are keenly aware of such disrespect, as are humans; if you were part of a working team blessed with a visit from the president of the company and as all of you gathered around to shake hands and be introduced one of your colleagues failed to join the group, lingering over by the hors d'oeuvres, people would notice. Perhaps it was my imagination, but I felt that every male in the group was aware of Lukaja's disrespect and that the intensity of their grooming and grunting was due in part to that anxiety.

As individuals shifted position around Ntologi, a gap opened up in the crowd, a gap on the side nearest Lukaja. In an action so quick it seemed like an explosion, Ntologi burst through the gap and in an instant was on top of Lukaja, who had started to sit up but had not had time to position himself to meet the assault. I was sitting near Lukaja, with my camera out, but when Ntologi rushed up the slope the movement was so sudden I flinched, not quite aware of what was happening. Ntologi landed on Lukaja with both feet. He kicked, slapped, and bit him in a furious attack (Figure 24.2). He gripped him with both hands and feet and jumped up and down on top of him. Lukaja screamed and attempted to fight back, but the heavier Ntologi had the high ground. Lukaja was unable to twist out from under Ntologi. At one point, Ntologi, standing on Lukaja's back, buried his face in Lukaja's shoulder and worked his head back and forth so that I feared he was attempting to rip open his neck. Soon Lukaja was fighting back less vigorously, slowed by injury and fatigue. He curled into a fetal ball to last out Ntologi's final buffets. Ntologi pelvic-thrust against him for many seconds, longer than a chimpanzee mating usually takes – the longest I have ever seen a male thrust against another.

Finally, Ntologi turned away and purposefully strode across the clearing and onto one of the tracks

Figure 24.2 Intracommunity violence is often frightening to behold, but rarely leads to serious injury. Slapping, kicking, and stamping on subordinates is common.
Photograph taken at Burgers Zoo, courtesy of Frans de Waal.

radiating from it and disappeared. Every single male followed him. Only at this point did I become aware of the deafening hooting and grunting that all the males had been emitting throughout the encounter. Lukaja, now alone, feebly sat up, swaying drunkenly. As I looked on the scene, dust hovered in the air, highlighted by sunbeams shining through the leaves above. The ground cover around him was ripped up and he was adorned with leaf litter. He sat for one minute, then two, then gently he lay back down as if to take a nap.

If I were a more objective primatologist, I would have looked at my watch when the attack started, but in the excitement I had not. I recorded posture every two minutes, and my last record was four minutes before, so the confrontation could not have been much more than two minutes long, yet it seemed like it had lasted a half-hour. I did snap one blurry picture in which two black streaks amidst a dense cloud of dust are visible, but it was worthless as a depiction of the event. Eventually my scientific protocol took over. I had been taking data on Lukaja's companion, and I moved off toward the receding vocalizations to try to find him before the party disappeared completely.

Lukaja went missing. We saw no sign of him for a week, then two. When he failed to reappear, I assumed he wandered off and died alone somewhere. However, when males are unable to compete in the rough and tumble world of the male hierarchy, they often travel alone for a while – even for the rest of their lives if they are old.

A month after Ntologi's attack I was taking a rest day, reading in front of my small hut, when a chimpanzee appeared on the nearby trail. He seemed out of sorts and purposeless, and it seemed odd he was alone, so I walked over to identify him. To my surprise it was Lukaja. He looked terrible. He was thin, his ribs showing even more starkly than typical for a lean chimpanzee, and he moved gingerly. He must have suffered internal injuries during the attack. I was so worried that he would die of starvation I ducked into my hut and grabbed several bananas to give him. He took them with no apparent emotion and ate them slowly. Afterward he moved off into the thick grass that stood between me and the park staff housing, closer to the lake. The next day I heard from a park ranger that a chimpanzee had appeared at his house and killed and carried off one of his chickens. As the days passed, Lukaja gradually seemed to recover, but he had sunk to the very bottom of the dominance hierarchy and remained there for the rest of my time at Mahale. He never attained alpha status.

24.3 Competition and Violence

The contexts in which these confrontations occur are not terribly surprising. Many altercations start in the overheated agitation swirling around estrous females, as males test the boundaries of what they can slip past higher-ranking males. Meat, their most prized food item, sometimes creates such an elevated level of excitement that fights erupt. A third context is more surprising; many fights occur during reunions (Muller, 2002). Alpha males discourage disappearances because the missing males might be mutineers. They attack to punish conspirators. They may not be conspiring, but they are probably reinforcing their bond with one another and that might eventually threaten the alpha.

But many attacks occur over nothing at all, at least nothing directly. Alpha males and other high-rankers use violence as a means of reinforcing their status, of placing the underlings in a constant state of fear or wariness. The higher the status of the male, the more he lords it over his underlings (Bygott, 1974); at both Kibale-Kanyawara and Mahale there was a high correlation (0.7) between social rank and rates of aggression (Nishida & Hosaka, 1996; Muller, 2002). At Taï, 80 percent of social displays (Figure 24.3) were engaged in by the top two males (Boesch & Boesch-Achermann, 2000).

24.4 To the Victor Go the Spoils

One way aggression pays off for males is in reproductive success, not just among chimpanzees, but among primates in general (Cowlishaw & Dunbar, 1991; de Ruiter & van Hooff, 1993; Wroblewski et al., 2009). Alphas fathered half of all infants conceived during their tenure in a study at Gombe (Constable

Figure 24.3 Social displays often begin with piloerection. Bipedalism is a common aggressive posture, often accompanied by the aggressive compressed-lip facial expression.

et al., 2001), and most of the rest of the infants were fathered by the next few males in the dominance hierarchy.

24.5 Dominance Hierarchies

As with many primates, there is some stability and predictability in male social rank; males form a linear **dominance hierarchy**. There is a clear alpha male, a clear beta, and so on right down the line. Most of the time, every male in the group knows the place of every other male, not just who is number one, but who is number five and what poor sap is at dead bottom. Relative status is expressed with signals of subordinance and obeisance that are expected by alphas. As we learned earlier, the most common signal is the **pant-grunt**, a series of rapid grunts that issue from the subordinate male, often accompanied by a crouch to lower the face of the subordinate below that of the alpha, a very physical expression of who is high

and who is low. Subordinance signals range from a soft huffing grunt repeated only a few times – this signal passes between males between whom status is well-established and who might be friends – to louder grunting and lower crouching that occurs when there is some tension between the two. Rapid pant-grunts that may grade into screams, sometimes so intense that the vocal apparatus locks up and all that is heard is a sort of gurgle, accompanied by a deep crouch and a fear-grin, are seen when tension is high between the two.

Pant-grunts are a marvelous tool because they are so reliable. Bygott (1974) found that the lower-ranking male grunted all but 0.5 percent of the time. Such reversals, a higher-ranking individual grunting, were found at low levels as well at Mahale (1.1 percent), at Taï (0.9 percent), and at an astonishing 0 percent at Kibale-Kanyawara (Nishida & Hosaka, 1996; Boesch & Boesch-Achermann, 2000; Muller, 2002).

Males are intensely interested in their relative status (Wrangham & Peterson, 1996; Wrangham et al., 2006) and pursue status for its own sake – if we keep in mind that there is an evolutionary underpinning to the concept of "its own sake," since there are reproductive advantages to status. If chimpanzees were human, we might observe that the alpha seems insecure, overly concerned about affronts to his status, hypersensitive to signs of less-than-abject subordinance, and aware of even the most subtle challenge to his status.

24.6 Coalitions and King-Makers

While every male knows his place and the hierarchy is relatively stable, as old males lose vigor and young males gain competence ranks do shift. You may wonder how everyone keeps track of who is up and who is down, but in fact the task is easy. Turnover comes at a timescale like the changing roles of players on a sports team; week to week and even month to month the same players may start every contest, but over a year or two there are shifts.

Coalitions are a critical aspect of male rank. Males who possess good social skills are at an advantage (Nishida & Hosaka, 1996). Competitors switch allegiances in a shameless manner, so much so that Toshisada Nishida coined a term for it, "allegiance fickleness" (Nishida, 1983). In a famous scenario that first demonstrated just how political and fickle male chimpanzees can be, Nishida documented a situation in which a third-ranking male repeatedly switched his allegiance from one higher-ranking male to the other. In the small K group community at Mahale there were only three males, the third-ranking male first supported male A, establishing him as alpha. But no sooner did A settle into the alpha position than did the third-ranking male shift his allegiance to the beta male, who would then become alpha. Then he would switch back. The alpha was always beholden to this lowest-ranking king-maker. While this dithering went on the king-making third-ranked male copulated with more females than either of the other two (Nishida, 1983); previously he had rarely copulated.

24.7 Ntologi: The Once and Future King

Attaining the alpha position is an ambition that burns brightly in most males. While some accept an ousting from the alpha position quietly, tiredly retiring to middle rank, others never surrender. These never-surrender males withdraw from society to nurse their wounds, recover, and slowly begin to work their way back into the alpha position. The late Toshisada Nishida documented the riveting story of Ntologi, a secure alpha male when I was at Mahale, but later overthrown by a beta male who had formed a tight bond with another powerful young male, a male I knew as an unassuming adolescent during my stay. For 12 years Ntologi held onto the alpha position, a very long tenure in this exhausting position (Nishida, 1996; Nishida & Hosaka, 1996). He was a regal presence, bearing all the requisites that underpin alpha status: aggressiveness, intelligence, political awareness, good social skills, and all that combined with one of the largest body weights at Mahale. Assuming the alpha mantle at 24, his grip on alpha-hood loosened only after he reached a middle-aged 36, the equivalent of a 45-year-old human, at which point a powerful coalition overthrew him.

His reign ended, but the crown did not rest easy on the new alpha's head. The overthrowing coalition was unstable. A third powerful male seemed to intimidate both the new alpha and his king-making ally, especially when they were apart, leaving neither the new alpha nor his king-maker in a secure position.

Meanwhile, Ntologi traveled alone for a time – some beaten males return to their mother and siblings to gather their courage in the bosom of their loving family as they plot a return; others become solitary so as to avoid exertion and danger.

As Ntologi stewed in exile, the political turmoil at the center of the male hierarchy churned. Some of the new alpha's coalition partners abandoned him and his confidence waned. As alliances shifted it became impossible to tell who among the three high-ranking males was alpha; it seemed that even they themselves did not know. Fearful and uneasy, each began to travel alone or in small parties.

As Ntologi encountered solitary males he solicitously groomed and reassured them. An old male I had known years earlier joined him, then a young male, then another old male. Gradually, he gathered a large contingent of old and new allies around him. For days he traveled with his new cohort, shoring up his nerve, taking strength from his solidifying alliances. At last he and his allies returned to the core of the group, presumably expecting a vicious fight with the new alpha and his many allies, only to find a shattered coalition, and an every-man-for-himself situation. With no single individual brave enough to challenge him, at 36 Ntologi found himself reinstalled on the Mahale M-group throne. Then his political skill came to the fore. He courted and ultimately made a friend and ally of the alpha he had deposed (Hosaka, 1995), and attentively strengthened other bonds. This stable alliance kept him in power several more years, despite his advancing age.

This sort of grit or resilience is a characteristic of alphas who have long tenures. Goblin, alpha when I was at Gombe in the mid-1980s, faced similar circumstances – he was deposed and returned to alpha status later.

24.8 Costs of High Rank

It may seem a fine life, the life of an alpha. Alphas have first access to females in their most fertile period

(Deschner et al., 2004; Perloe, 1992), they command the respect and deference of every other individual in the community, they can feed at whichever feeding site they wish, and they need not adjust their plans to avoid a higher-ranking individual. When there is meat, a highly treasured item, they almost always get their share.

As with kings, however, this bliss is diminished by the sword of Damocles swaying above the alpha's head. There are costs to maintaining the alpha rank, and the demands of sustaining high status never ease. High-ranking males have high levels of testosterone (Muehlenbein et al., 2004; Muller & Wrangham, 2004), which may seem a good thing, but testosterone is correlated not only with rank but also with the activity that sustains it: social (or charging) display (Muller & Wrangham, 2004), moving constantly among allies to reassure them and taking care to spy out arising challenges. Testosterone increases aggression, a desideratum in the dangerous world of chimpanzee male relationships; it even improves some types of cognitive performance (Wingfield et al., 1997), but these advantages come at a cost. It depresses immune function, increasing the susceptibility to disease (Sapolsky, 1993); intestinal parasite loads are higher (Muehlenbein, 2006), adding to nutritional requirements that are already high due to higher metabolic rates.

In short, males in the alpha position have many advantages, certainly, but accompanying these advantages are emotional stress, nutritional stress, risk of injury, the need to sustain high activity levels, and the stress of constant attentiveness to challenges.

24.10 Female Violence

You may think females are peaceable, dedicating their lives to caring for their families. In fact, their level of aggression only seems mild when compared to male chimpanzees. While female–female competition is often subtle – for instance, a subordinate veering off from a favored feeding site to avoid a dominant – females do not hesitate to fight over food, a precious resource essential for the health of their nursing infant; they are warriors when protecting their offspring (Nishida, 1979; Goodall, 1986; Muller, 2002). Just as do males, females form rather linear dominance hierarchies, though their relatively solitary and dispersed nature means they come into contact with one another less often and thus have fewer opportunities to express their relative ranks. They fight over food and even over mates; 7.5 percent of female–female violence is competition for mates (Nishida, 1990).

Sex and food are not the only contexts in which females are seen to be violent. In the 1970s Jane Goodall documented a harrowing series of events that lasted first for months and then years, during which female Passion killed and ate a shocking number of infants (Goodall, 1979), perhaps as many as five. Females are a little paranoid about their infants' safety for a reason. Passion was not alone; female aggression toward infants appears to be more common at Gombe than other sites (Pusey et al., 1997), and among the females who have attempted aggression toward infants is everyone's seemingly favorite Gombe female, Fifi.

24.9 Senescence

As males enter mid-life, the wear and tear of a demanding, competitive life, injuries from competition, and advancing age begin to take their toll. Ultimately even the hardiest alpha begins to lose vigor and stamina. As they age, they eventually display less often or less energetically, and this drop-off is instantly noticed by underlings (Goodall, 1986), spurring challenges to his authority.

24.11 Male Coercion

At Kibale-Kanyawara one-quarter of violence was male-on-female (Muller, 2002), and, if anything, females have it worse elsewhere (Goodall, 1986; Muller & Mitani, 2005): Up to half of male violence is directed at females at Gombe and Ngogo. Some of this violence occurs in the context of sexual coercion, but not in the sense of "rape" or forced copulation, which is very nearly unknown among chimpanzees (Tutin,

1979; Goodall, 1986). Instead, males employ coercion, berating reluctant females until they accompany them on a "consortship," a brief foray away from the social group that includes sexual intercourse (Muller & Mitani, 2005; Chapter 27). Sometimes, anticipating aggression, a female may seek out a male she might not otherwise choose as a sexual partner (Muller et al., 2007). Females, however, have means of blunting the effectiveness of this aggression, as we will see in Chapter 27, and in the end are mostly able to choose the fathers of their offspring rather effectively (Matsumoto-Oda, 1999; Stumpf & Boesch, 2005).

24.12 Intangibles Contribute to Social Status

Up to now, discussion of high-achieving, rank-obsessed individuals has focused on physical violence, but there are many factors that together determine an individual's social rank, including some intangible cognitive attributes that may seem unrelated.

Size is the most obvious physical advantage in aggressive competition, and indeed it is correlated to social rank (Hunt, 1992). Physical ability is advantageous, of course, including difficult-to-quantify skills such as coordination and athleticism. Stamina, endurance, and vigor all contribute to sustaining the high levels of physical activity necessary to maintain dominance. But personality traits are at least as important. Confidence – even arrogance – and brutality, are required. A little volatility and unpredictability actually helps. An individual who might react unexpectedly to some innocent movement demands a greater deference than a more predictable adversary. Perhaps the most important personality trait is one Ntologi and Goblin, long-tenure alphas at Mahale and Gombe, possessed in abundance: Persistence and resilience in the face of disappointment and defeat. Individuals with these two traits return to battle again and again, bloody, perhaps, but unbowed. The largest male at Gombe when I was there was Jomeo. Years ago, he had begun rising in rank, but after he was dealt his first serious pounding he lost his nerve, sank to the bottom of the social hierarchy, and never rose again. Compare this meekness to the grit exhibited by Goblin who, once deposed, nursed his wounds, summoned his resolve, and returned to retake the alpha position.

Intelligence is also quite important, both social and political intelligence – social intelligence to foster and sustain alliances and political savvy to know when the time is right to mount a challenge. A certain devious, plotting cunning is also helpful. Goblin was known to leave his nest before dawn, search out the nests of rivals, and leap on them in the pre-dawn darkness when they were still sleeping peacefully. I once saw him veer off a path and ascend a slope to reach an overhang paralleling the path, pause, and then leap on top of another male, seemingly dropping from heaven; after toppling the male and knocking him to the ground, eliciting the screams and a fear-grin that he undoubtedly found satisfying, he continued on, grim-faced but otherwise as if nothing had happened. Such psychological warfare can wear down opponents. Alphas not only attack males who have strayed from the larger group, as we saw above, they interfere with grooming partners, either pushing their way in between them or attacking one of the groomers, thus isolating competitors even in the midst of a social group and preventing alliances from forming.

The most adept dominance players even use deception. Frans de Waal relates the story of a male who, in the heat of battle, became fearful, but was aware enough to realize that showing fear would only embolden his rival (de Waal, 1982). When he felt himself lapsing into a fear-grin, he turned his face away from his nemesis and used his hand to close his mouth, composing his battle face before turning to reengage with his challenger. I once saw Ntologi hide a limp from others, deceptively choosing to avoid revealing a physical weakness to his rivals. Fighting skills are important, but intelligence is helpful as well.

24.13 Killing Occurs between Communities

As naked as this violence is, chimpanzees rarely kill members of their own community. When they bite community members they seldom use their long, sharp canines to rip and tear skin. Intracommunity violence is meant to intimidate, not kill. Signals of subordinance such as fear-grins, fear grunts, screams, and low crouching serve to placate an attacker, at least a little bit. Such signals of surrender are completely ineffective in between-community confrontations. As we will see in the next chapter, intercommunity attacks involve gruesome, effective killing techniques.

References

Boesch C, Boesch-Achermann H (2000) *The Chimpanzees of the Taï Forest: Behavioural Ecology and Evolution*. Oxford: Oxford University Press.

Bygott JD (1974) *Agonistic Behavior and Dominance in Wild Chimpanzees*. PhD thesis. Cambridge: Cambridge University.

Constable JL, Ashley MV, Goodall J, Pusey AE (2001) Noninvasive paternity assignment in Gombe chimpanzees. *Molec Ecol* 10, 1279–1300.

Cowlishaw G, Dunbar RIM (1991) Dominance rank and mating success in male primates. *Anim Behav* 41, 1045–1056.

de Ruiter JR, van Hooff JARAM (1993) Male dominance rank and reproductive success in primate groups. *Primates* 34, 513–523.

Deschner T, Heistermann M, Hodges K, Boesch C (2004) Female sexual swelling size, timing of ovulation, and male behavior in wild West African chimpanzees. *Horm Behav* 46, 204–215.

de Waal FBM (1982) *Chimpanzee Politics*. New York: Harper & Row.

Fawcett K, Muhumuza G (2000) Death of a wild chimpanzee community member: possible outcome of intense sexual competition? *Am J Primatol* 51, 243–247.

Goodall J (1979) Life and death at Gombe. *Nat Geog* 155, 592–621.

Goodall J (1986) *The Chimpanzees of Gombe: Patterns of Behavior*. Cambridge, MA: Harvard University Press.

Hosaka K (1995) A rival yesterday is a friend today. *Pan Africa News* 2, 9–11.

Hrdy SB (2011) *Mothers and Others*. Cambridge, MA: Harvard University Press.

Hunt KD (1992) Social rank and body weight as determinants of positional behavior in *Pan troglodytes*. *Primates* 33, 347–357.

Matsumoto-Oda A (1999) Female choice in the opportunistic mating of wild chimpanzees (*Pan troglodytes schweinfurthii*) at Mahale. *Behav Ecol Sociobiol* 46, 258–266.

Muehlenbein MP (2006) Intestinal parasite infections and fecal steroid levels in wild chimpanzees. *Am J Phys Anth* 130, 546–550.

Muehlenbein MP, Watts DP, Whitten PL (2004) Dominance rank and fecal testosterone levels in adult male chimpanzees (*Pan troglodytes schweinfurthii*) at Ngogo, Kibale National Park, Uganda. *Am J Primatol* 64, 71–82.

Muller MN (2002) Agonistic relations among Kanyawara chimpanzees. In *Behavioural Diversity in Chimpanzees and Bonobos* (eds. Boesch C, Hohmann G, Marchant L), pp. 112–124. Cambridge: Cambridge University Press.

Muller MN, Mitani JC (2005) Conflict and cooperation in wild chimpanzees. *Adv Stud Behav* 35, 275–331.

Muller MN, Wrangham RW (2004) Dominance, aggression and testosterone in wild chimpanzees: a test of the "challenge hypothesis." *Anim Behav* 67, 113–123.

Muller MN, Kahlenberg SM, Thompson ME, Wrangham RW (2007) Male coercion and the costs of promiscuous mating for female chimpanzees. *Proc Roy Soci B* 274, 1009–1014.

Nishida T (1979) The social structure of chimpanzees of the Mahale Mountains. In *The Great Apes* (eds. Hamburg DA, McCown ER), pp. 73–121. Menlo Park, CA: Benjamin Cummings.

Nishida T (1983) Alpha status and agonistic alliance in wild chimpanzees (*Pan troglodytes schweinfurthii*). *Primates* 24, 318–336.

Nishida T (1990) *Chimpanzees of the Mahale Mountains*. Tokyo: University of Tokyo Press.

Nishida T (1996) The death of Ntologi, the unparalleled leader of M group. *Pan Africa News* 3, 9–11.

Nishida T, Hosaka K (1996) Coalition strategies among adult male chimpanzees of the Mahale Mountains, Tanzania. In *Great Ape Societies* (eds. McGrew WC, Marchant LF, Nishida T), pp. 114–134. Cambridge: Cambridge University Press.

Perloe SI (1992) Male mating competition, female choice and dominance in a free ranging group of Japanese macaques. *Primates* 33, 289–304.

Pruetz JD, McGrew WC (2001) What does a chimpanzee need? Using natural behavior to guide the care and management of captive populations. In *Care and Management of Captive Chimpanzees* (ed. Brent L), pp. 17-37. Chicago, IL: American Society of Primatologists.

Pusey A, Williams J, Goodall J (1997) The influence of dominance rank on the reproductive success of female chimpanzees. *Science* **277**, 828-831.

Sapolsky RM (1993) Neuroendocrinology of the stress-response. In *Behavioral Endocrinology* (eds. Becker JB, Breedlove SM, Crews D), pp. 287-324. Cambridge, MA: MIT Press.

Stumpf RM, Boesch C (2005) Does promiscuous mating preclude female choice? Female sexual strategies in chimpanzees (*Pan troglodytes verus*) of the Taï National Park, Côte d'Ivoire. *Behav Ecol Sociobiol* **57**, 511-524.

Tutin CE (1979) Mating patterns and reproductive strategies in a community of wild chimpanzees (*Pan troglodytes schweinfurthii*). *Behav Ecol Sociobiol* **6**, 29-38.

Watts DP (2004) Intracommunity coalitionary killing of an adult male chimpanzee at Ngogo, Kibale National Park, Uganda. *Int J Primatol* **25**, 507-521.

Wingfield JC, Hunt KA, Breuner CR, et al. (1997) Environmental stress, field endocrinology, and conservation biology. In *Behavioral Approaches to Conservation in the Wild* (eds. Clemmons JR, Buchholz JR), pp. 95-131. London: Cambridge University Press.

Wrangham RW, Peterson D (1996) *Demonic Males: Apes and the Origins of Human Violence.* New York: Houghton Mifflin.

Wrangham RW, Wilson ML, Muller MN (2006) Comparative rates of violence in chimpanzees and humans. *Primates* **47**, 14-26.

Wroblewski EE, Murray CM, Keele BF, et al. (2009) Male dominance rank and reproductive success in chimpanzees, *Pan troglodytes schweinfurthii*. *Anim Behav* **77**, 873-885.

25 A Nation at War with Itself
Defending a Community of the Mind

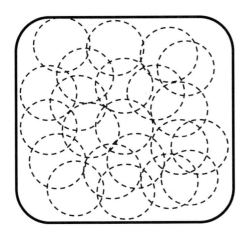

Image by the author

We probably understand chimpanzees best of all the apes, and possibly best of all primates. Six wild studies have gone on for 55 years or more (McGrew, 2016), and three times that many have gone on for decades.

We have long known that the lively, multi-level jumble of social interactions among chimpanzees occurs within what Toshisada Nishida (1968) called a **unit-group** and what Jane Goodall called a **community**, a group of 60 and more individuals who recognize one another as members of a social in-group. Most of the astoundingly diverse emotional lives of chimpanzees play out within this community context, where we know stories of heart-breaking maternal devotion, heroic sacrifice, shameful betrayal, dark depression, playful euphoria, and gritty courage.

Relationships among communities have a darker component, and it is only when chimpanzee-nations set themselves against one another that we see the most brutal facets of chimpanzee life. In a wild world we know to be red in tooth and claw, chimpanzees are the reddest, most murderous of all mammals yet studied (Muller, 2002).

I characterized chimpanzee community size as 60, but it varies enough that I might have picked 50 or 100. The Gombe Kasakela community has often numbered around 45 (Pusey et al., 1997); Mahale M group was around 100 when I studied them in the late 1980s; Kibale-Ngogo has over 200. Within a community of a dozen or so males (it may range into the thirties), approximately twice that number of adult females, and about as many young as adults, every chimpanzee knows every other chimpanzee as an individual and recognizes they are part of the community. As we reviewed earlier, all members of the community are never all in one place at one time. Variably sized subgroups, called **parties** (Goodall, 1968, Boesch, 1996), form and re-form so that over a day a male might be in a party with seven other males, one with two of those same males, then a party that has added two females, three of the males from the original party, and a new male. Within a community, interactions are mostly affiliative, though also competitive, whereas intercommunity interactions are violent and sometimes deadly. Compared to human social groups, the 50 or 100 or so chimpanzees in a community may seem a small number, but it is large compared to all but a few other primates.

25.1 Communities Have Stable Ranges

Members of a community share a **home range**, a piece of geography which all members know belongs to their community and in which members of the group are nearly always found (Nishida, 1968, 1979). A community range varies but is often about 4 km across, the average at Gombe, though it has swelled to 7 km (4 miles) occasionally (Williams et al., 2008). The three communities at Taï varied as well, but also averaged about 4 km (Herbinger et al., 2001). Still, diameters can be much larger, particularly in poor habitats. The range is well over 10 km in the Semliki-Mugiri community I study.

25.2 A Community of the Mind

It may seem odd that community members are never all in one place at the same time, but even when a rich food supply draws together huge feeding parties, there is always a stray male who might be avoiding the alpha male after an unsuccessful coup, or a peripheral female who is too far away to make the trek to the large gathering, or two old males too tired to put up with all the bickering typical in a large group. All this splitting and recombining into endless different party compositions means the community exists only in the heads of its members. It is a virtual community.

Because all members are never in one place at one time, almost every important thing that happens in the community happens out of sight of *most* of its members. This is, to my mind, the most unusual thing about the chimpanzee social system.

25.3 Party Composition and Male Bonding

The average party size is approximately six individuals (Boesch, 1996), typically consisting of four or five males and one or two estrous females, with the occasional non-estrous female. Parties larger than two contain more males than females because males are more social; they groom with one another more than other age-sex combinations, they form complicated alliances, and they bond together to defend the community boundary (Nishida, 1968; Wrangham, 1979; Goodall, 1986; Nishida Hiraiwa-Hasegawa, 1987; Pepper et al., 1999). Males roam over the community range in parties that, if there is enough food, can contain every male in the community. Ranging widely, these mostly-male parties may cover the entire community range in a period of days. While males and females spend similar amounts of time walking over the course of a day, males cover more ground by traveling faster and in a straighter path (Wrangham, 1979; Hunt, 1989). If they can, males travel between particularly large, rich feeding sites that can accommodate their large numbers; the largest food sources tend to be separated by great distances.

The extremely unusual constant reshuffling of party membership has inspired the term **fission-fusion** to describe the fluid chimpanzee social system (Nishida, 1968, 1979; Nishida & Kawanaka, 1972): **fission** to mean the splitting of parties, and **fusion** as the term for the coalescing of the subparty members.

Most subgroups are **feeding parties** and are made up of 1–15 individuals who are clearly, as one watches them, intent on gathering food efficiently. The size of these groups is strongly dependent on the food supply. When food is scarce, chimpanzees have little time for socializing and single individuals or pairs scour the countryside, making do on bark, piths (such as palm fronds and grass stems), and the occasional ripening fruit on a tree early in its fruiting cycle. Feeding efficiency suffers proportionally as group size increases because competition, interference, and pausing for social interactions take up time. Male chimpanzees are rarely gentlemen, and a female, even a high-ranking female, has little to be gained by remaining close to males, who will displace her from any feeding site to which they take a fancy. As a consequence, females often avoid males or at least make no effort to stay near them. I found (Hunt, 1989) that when a party with females begins to move, the females actually walk *slower* than if they were alone; they must be purposely allowing males to move ahead of them.

25.4 Travel Costs Exert a Constant Downward Pressure on Party Size

A significant penalty for gathering in larger groups is that travel costs go up with group size. However far an individual must travel to get a minimum daily requirement of calories, if a second individual joins her, the feeding party now needs twice as much food and the distance traversed to find all that food must rise accordingly. This is why females are so solitary (Wrangham, 1979, 2000); their lower rank, their high nutritional needs (due to nursing), and the costs of interference among group members make feeding alone their most effective strategy.

When food is plentiful, a rare but happy occasion for chimpanzees, a single tree may have a million fruits on it and gathering food is easy enough that huge parties can meet their nutritional needs from a single tree (Basabose, 2004); noisy parties of 10, 20, or 30 can be observed, and there is time enough for all four Fs: feeding, fighting, friending, and finding a date.

25.5 Core Areas

We have already placed males in large parties, ranging widely. Where are the females? Females mostly confine their activities to a piece of land perhaps 1.5 km across, known as a **core area**. If we had a special scope that detected female core areas, we might imagine flying over the forests of Africa looking down to see kilometer after kilometer of overlapping female core areas (Figure 25.1) extending to the horizon. In each of these, an individual female, with her dependent offspring, lives out most of her life. *Females spend two-thirds of their lives alone with respect to other adults*; a typical female is accompanied by two or three offspring of various ages. The day I drafted this chapter (July 8, 2016) I spent six hours watching a female chimpanzee who was typical – she carried her very young infant as she fed on *Saba florida* fruits and was followed closely by her six-year-old daughter, and in a much looser orbit there hovered yet another subadult daughter of perhaps 12. The 12-year-old approached to groom her mother and sister now and then, giving mother a brief childcare break, but sometimes ranged far enough away that she was out of sight. This is the life of a chimpanzee female.

All core areas are not created equal. Peripheral core areas are less desirable than central ones; in fact, alpha or high-ranking females typically occupy the core area at the very center of the community range (Pusey et al., 1997), as Fifi did at Gombe and as Chausiku did at Mahale when I was there in the late 1980s. Core area quality can vary dramatically, even when the two cores are near one another. Central core areas are good in part because there at the center of the community a female is as far as possible from murderous incursions by neighboring males. Central females will be visited often by wandering parties, giving their infants a chance to socialize with many of the community members. High-ranking females are confident and often have relaxed and friendly relationships with many males, some of whom they seem to be good friends with. The elevated health of females in the best core areas is expressed physiologically; they have elevated levels of ovarian hormone, correlated with vigor and robust reproductive physiology, resulting in shorter birth intervals and higher infant survivorship (Pusey et al., 1997; Williams et al., 2004, 2008; Marsden et al., 2006; Thompson et al., 2007; Thompson, 2013). These high-ranking females have longer lifespans and healthy, fast-growing daughters who are also likely to be successful themselves (Pusey et al., 1997).

In peripheral core areas, on the other hand, mothers find themselves dangerously close to aggressive extra-community males and far away from males who might protect them. It is no surprise that lower-ranking females are often skittish, sometimes wary even of community males, and cautious around high-ranking females. Perhaps, anxious as they are, these peripheral females look forward to the less-competitive socializing they can engage in when food is abundant and feeding competition is low (Sugiyama & Koman, 1979).

25.6 Females Are Not Territorial

While females do not defend their core area – it is not a territory – they may well defend a specific resource,

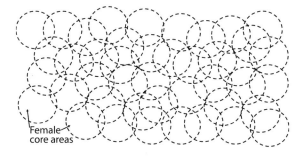

Figure 25.1 Females occupy overlapping, undefended core areas. We might imagine the entirety of the chimpanzee habitat saturated with these core areas.

a fruit-laden branch they have broken off or a particularly appealing feeding site. Females take a rather relaxed view concerning the borders of their core areas, often ignoring neighbors altogether when their paths cross, rather than attempting to drive them off (Figure 25.1; Wrangham, 1979; Goodall, 1986; Pusey et al., 1997).

When a female finds herself in the same feeding tree as another female, typically the two will greet one another and proceed to feed; but they may behave almost as if they are unaware of one another. That, however, is deceptive; the lower-ranking female is probably coordinating her movements to avoid confronting the dominant female at a feeding spot. After the feeding bout, the females are likely to share a grooming bout, a long one if they have infants who are friends, and the two families may even travel together for a while in a "nursery group" to give infants and juveniles play time. Keeping an infant occupied is a challenging job and mothers can use all the help they can get. At Mahale, my favorite female, Chausiku, had an adorable and playful son named Chopin. A young female who did not yet have an infant, Pulin, often followed Chausiku's family, playing with Chopin and grooming Chausiku. They clearly enjoyed one another's company and Pulin provided a helping hand for the busy mother.

The casual attitude female chimpanzees have toward securing resources in their home range is unusual for primates. Many primate societies feature tightly bonded kin-based female cliques that exclude other females from food resources – one survey put the number at 81 percent (Sterck et al., 1997); some even defend territories. Because chimpanzees are dependent on ripe fruits, unlike most monkeys (Wrangham, 1980; Wrangham et al., 1998), they are handicapped by the relatively sparse and dispersed nature of their food source (Wrangham, 1977, 1979; Sterck et al., 1997). By limiting their ranging to a small core area and traveling alone, females avoid having to compete with others for food, reduce travel costs, monitor food distribution intensely, and harvest preferred food items thoroughly. Females are strongly motivated to avoid males and even other females, since feeding efficiency drops precipitously when females feed in groups (Wrangham, 1979, 2000;

Wrangham & Smuts, 1980). They are superbly evolved, highly efficient – but not very social – sugar-seeking machines.

25.7 The Community Boundary Differs for Males and Females

The boundary of the community is conceptually different for the sexes. Females view the boundary as a dangerous place where their infants are at risk. Males, on the other hand, view the boundary as a bright red line in the sand – it is the edge of their **territory**, a home range that is defended so as to exclude all others who are not members of the community (Figure 25.2). As with nations, there is no overlap in community ranges. Males view their community territory as a possession and they guard it jealously – it is a possession that can be stolen by other males, but it is theirs if they can keep it.

25.8 Border Patrols

Males guard their homeland with behavior that is brutal and disturbing; they act as a tightly bonded, murderously violent coalition that acts almost as one; it shows no mercy to enemies of the state. Such coalitionary violence is extremely rare in the animal world; only a handful of other mammals exhibit this behavior (Goodall et al., 1979; Wrangham, 1999), among them wolves, lions, spotted hyenas, and

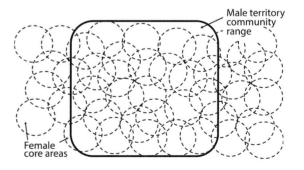

Figure 25.2 Males ensure access to mates by defending a territory in which females are found.

cheetahs (Wrangham, 1999), but none at the level of chimpanzees.

Once a month, males gather in a party that contains many and sometimes all of the community males, a party that has come to be called a **border patrol**, and travel to the boundary of the community (Figure 25.3; Goodall, 1986; Muller, 2002; Mitani & Watts, 2005). There is a pattern to patrol frequency. Chimpanzees only patrol when they can muster a large group of males in fighting trim; in times of plenty, larger party sizes allow more patrols than times of scarcity (Mitani & Watts, 2001, 2005; Mitani et al., 2002, 2010; Watts, 2002; Basabose, 2004). They coordinate by pant-hoots, gathering at feeding trees where every member can be sated. In the social activity following a satisfying feeding bout, a high-ranking male, seemingly without provocation, grows agitated. He may engage in a social display or some other signal of excitement, or he may simply stand up, hair erect, make a compressed-lip, knitted-brow face, and begin to rock. In this state, with other males alert, he strides off toward a boundary, and other males follow (Figure 25.3). The patrol proceeds to the border quietly, stopping to listen, every member alert as they search for males from other communities. If the patrol members discover a lone male or a pair, in other words if they greatly outnumber their foes (more on that below), they will attack and *attempt to kill the extracommunity males*; or infants; non-estrous females are occasionally killed as well (Figure 25.4; Wilson et al., 2014).

Males, then, actively patrol and defend the community territory, while females seldom engage in such defense and instead focus on feeding efficiency and infant care.

Figure 25.3 Once a month, sometimes as often as once a week, a border patrol of mostly male chimpanzees travels to the edge of the community range, often in single file, seeking the possibility of eliminating a rival male through an imbalance of power – superiority of numbers – thus weakening the fighting force of their competitors. If the border patrol encounters an equal or greater rival party, they quickly retreat. Photo: John Mitani; with permission.

25.9 Imbalance of Power Drives Violence

Patrolling males will only attack if they have achieved an **imbalance of power** (Wrangham, 1999). A force with dominant numbers is courageous, moving forward with hair erect, each face bearing the compressed-lip expression that signals confident aggression. When the imbalance of power is great, attacking males are at little risk of serious injury, while the victim, vastly outnumbered, has little chance of escape. Attackers come at the victim from all sides, and while he is facing and defending against one male, another rushes in to pound or bite him. Often the victim is quickly knocked to the ground and some attackers hold him while others slash and slam him.

Occasionally a rather small patrol encounters three or more defenders. How different their affect becomes in this case! They retreat at a run, bearing fear-grins, not uncommonly fear-defecating. At Kibale-Kanyawara Mike Wilson and colleagues played back recordings of the pant-hoots of unfamiliar males. When males in parties of three or more heard these strangers they called back and moved toward the recorded calls. Small parties remained quiet and avoided the area of the calls (Wilson et al., 2001).

Figure 25.4 If a male border patrol has numerical superiority they attempt to isolate a single individual; if successful, attacks are long-lasting, brutal, and often fatal. Males cluster around the victim, some holding him down, others biting, punching, and stomping on the victim until he is comatose. Photo courtesy of John Mitani, with permission.

For their entire adult lives males look for safety in numbers, which means they prefer not to travel alone, and may even whimper and "hoo" when they find themselves isolated. They are tightly bonded so as to function as a unit to protect their community, and to form a brutally effective, cohesive fighting force that can expand the community territory (Figures 25.4 and 25.5).

25.10 Dispersal

Part and parcel of this system is a difference in **dispersal** between males and females. Males do not transfer communities (Nishida, 1968). They remain in the community of their birth, forming close, life-long bonds with their male brothers, half-brothers, cousins, and other companions of their youth. Males benefit from their territorial defense, as we shall see below, and they benefit further by sharing their success with close relatives, improving not only their own reproductive success, but that of close kin as well.

Because males insist on remaining in the group of their birth, females reach maturity surrounded by kin. Although females do occasionally mate with brothers, as a rule close relatives are not appealing sexual partners and females avoid copulating with them. To avoid the possibility of inbreeding, females typically disperse to a new community (Nishida, 1968; Nishida & Kawanaka, 1972; Pusey, 1979), often between the ages of 11 and 13 (Boesch & Boesch-Achermann, 2000; Stumpf et al., 2009). At Gombe, a little over half of the females disperse (Pusey et al., 1997), while at Mahale nearly 100 percent disperse (Nishida & Hiraiwa-Hasegawa, 1987), and the same at Taï (Boesch & Boesch-Achermann, 2000). Females with better nutrition are more likely to stay; those with poorer nutrition are more likely to transfer (Stumpf et al., 2009).

Dispersing is not an easy task. A female must leave her mother, with whom she has been in nearly constant association since birth, and make her way across the dangerous no-man's land between communities. She must then attempt to make a life among strangers. Females typically make their move when in estrus and thus bearing a sexual swelling. Possibly they are driven by sexual desire, aware of the attractive males they have heard and may have seen once or twice at their community boundary. Or perhaps they are driven by repulsion, attempting to avoid sexual overtures from their brothers and cousins; by avoiding their male kin they may end up

Figure 25.5 Even after an extracommunity male has been beaten to the point of insensibility, males often continue to mill around, returning to administer a further bite. Here, a rival male has been beaten and bitten to immobility and a member of the attacking party returns to administer a further kick. Photo courtesy of John Mitani, with permission.

in the border areas of their community, where they fortuitously encounter stranger males.

It is difficult to imagine "romance" in the chimpanzee mind, given the brevity of their sexual encounters, but if it exists this is the time for it. A young female, avoiding the unsavory attentions of less incest-averse uncles and brothers, bravely strikes out on her own, encountering a group of exotic and desirable males in an unfamiliar and exciting new world. But there is more than romance in this gritty, real-life narrative, there is sex: The transferring female will copulate freely with the new males. The dispersing female will spend most if not all of her time among these males for a year or even two, remaining in their company both in and out of estrus. There is a reason for this. In contrast to the hot reception she got from males in her new community, she will get the chilliest possible welcome from her newfound sisters. She will be attacked. By "attacked" I do not mean given an aggressive grunt accompanied by a mean face and a shooing motion with arm, but a brutal physical assault, often leading to wounds (Pusey, 1979; Goodall, 1986; Nishida, 1990; Boesch & Boesch-Achermann, 2000; Muller, 2002; Kahlenberg et al., 2008; Pusey et al., 2008). Why? While males gain a reproductive advantage with an extra female in the community, females gain no advantage whatsoever, and instead are confronted with another mouth to feed from the community stores, and the real possibility of offspring to follow (Nishida, 1979). Males will defend her against aggressive females (Pusey et al., 2008), but an occasional aggressor will sneak through.

Because males range over the entire community area, over the year or two that the transferring female travels with them she becomes familiar with the geography and resources of her new community, and also the location and fearsomeness of the resident females. When she becomes pregnant, and this may not happen for years since young females are not terribly fertile, she will find it harder to keep up with males and she will begin to limit her activities to an area, her eventual core area, where she can find enough food to survive. She will choose a spot as close to the center of the community range as she can, farthest from infanticidal extracommunity males. As we discussed earlier, it is important for her to find a good core area because females in core areas with the best food are healthier and have more babies (Thompson et al., 2007). Now her carefree traveling days are over; her world will contract to the ultra-familiar square mile parcel of earth that is her core area.

She will give birth in her core area and remain there most of the time for four or five years, if, that is, she turns out to be a good mother and her first infant survives. She will greet the males she has come to know when they meet at feeding sites; she may make friends with another female, or not. Most of her social activity, though, will be interacting with her offspring.

This tranquility, let us hope she does not experience it as boredom, will be interrupted a few times over the course of her adult life. As she weans her infant, she will come into estrus and for a brief three or four months she will re-live her old wandering life, traveling widely with males until she gets pregnant.

Now and then an older female disperses, but older females have a strong motivation to stay in place; she is likely already to have an offspring and males tend to kill extracommunity infants, male infants, in particular. When all the males in Mahale K group were killed, females found themselves in the M group. Some behaved as would any other transferring female, wandering the range of the new community before settling into a new core area. One female with a male infant remained in her original core area eventually coming into estrus. When she sought out group males they attacked her son, causing her to retreat to her core area after only a few copulations, not the hundreds typical of the species. She got pregnant, but M group males, some of whom had probably not copulated with her at all, treated her as a stranger and killed her infant. This happened more than once, until her son was at last killed and she copulated as normal. There is no known instance of a male successfully transferring communities. There have been attempts by conservationists to release males into the wild, and – aware of this problem – they released them into relatively unpopulated areas, but even so they were often killed by wild males (Goossens et al., 2005).

25.11 Stationary Females Encourage Male Territoriality

Males stake out a territory that encompasses as many female core areas as their numbers and abilities allow. The more females in the community, the greater reproductive potential males have.

It is the stationary nature of females that encourages territorial behavior. In species where females range widely, the most common male reproductive strategy is to shadow them, waiting for a chance to mate. In this sense, females are a reproductive resource for males, just as food is a reproductive resource for females. Chimpanzee females maximize their reproductive output by efficiently foraging in a well-known, confined area – their core area; they become a stationary reproductive resource. Males respond by defending a tract of land and the core areas it contains, thus securing reproductive access to the females in their territory. They need not follow females to exclude other males from mating because the fixed range of females prevents them from encountering other males (Figure 25.2).

25.12 Males Murder Rivals to Expand Territory and Increase Mating Opportunities

If males are able to form an effective fighting force – if they are aggressive, talented in hand-to-hand combat, ruthless, and tightly bonded so that few individuals defect during attacks – they may find themselves capable of expanding their territory (Figures 25.6–25.8). They may do this without a numerical advantage, but because they rely on an imbalance of power to isolate and kill rivals, the

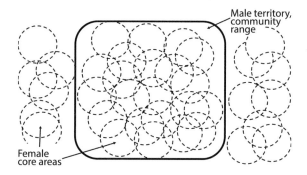

Figure 25.6 Females draw back from dangerous community border areas.

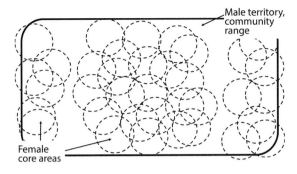

Figure 25.7 If community males are successful, they may expand their territory, encompassing the core areas of new females.

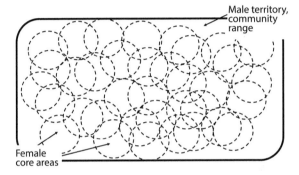

Figure 25.8 Females then draw back from the new, expanded borders and males are subsequently left with the task of patrolling a longer border, with greater travel costs.

stronger the community contingent, the greater the chance of success. If they gain the upper hand in this bloody war, they find themselves with still greater numerical superiority, more often able to achieve an imbalance of power during attacks and more likely to succeed in further attacks, which further depletes the fighting force of their rivals (Mitani et al., 2010). And so it goes on. We know of cases where these patrols have been spectacularly successful, picking off one male after the other (as happened to the southern group at Gombe), gaining an ever-greater imbalance of power until eventually (we believe this happened to the Mahale K group) all males in the neighboring community are killed, the neighboring territory is annexed, and females from the extinct community migrate into the territory of the victors (Goodall et al., 1979; Nishida et al., 1985; Goodall, 1986).

25.13 Expanded Territory Means More Food

The story does not end here. When males expand their territory, they achieve reproductive benefits even if no new females immigrate into their territory. The expanded community range brings more food resources into the safe zone in which females may travel and feed. Perhaps more females are farther from the community border. The expanded food supply and perhaps greater safety increases the female reproductive rates and infant survival rates go up (Williams et al., 2004). Since all these healthy babies are fathered by community males, their reproductive success increases as well.

25.14 Why Do Low-Ranking Males Participate in Patrols?

There is one loose thread to tie up. Why do low-ranking males join in border patrols? We learned in Chapter 24 that alpha males father half the offspring conceived during their tenure. High-ranking but non-alpha males make up most of the rest of the male reproductive success. It might seem that a low-ranking male's best strategy would be to pretend to participate, but to draw back whenever danger threatens, i.e., to defect from the raiding party, leaving the higher-ranking males to carry the load. High-status males are the ones that benefit most, after all. There is some evidence that low-ranking males cheat a little bit, yet this mostly seems not to be the case. Low-ranking males seem nearly as invested in these attacks as high-ranking males (Wilson et al., 2001). Jomeo, a legendary coward when it came to within-community battles, was a ferocious patriot in intercommunity aggression. Males, high and low, seem to participate as a tightly integrated unit. Why?

There is inertia in eroding enemy numbers and achieving territorial expansion. By eliminating rivals, the winning community achieves long-term benefits – the slow life history of chimpanzees means it may take years for a depleted male contingent in the losing community to recover, if ever. In the winning community, both the low and high benefit.

There are two, perhaps three, motivations for the perennial low-status male to patrol. One is that he may one day succeed in becoming alpha male, in which case the more females and the larger the territory, the more success he will have. Every low-ranking male is biding his time, waiting for the opportunity to occupy the alpha throne. And many of them do. The tenure of an alpha may be a decade or longer, as was the case for Ntologi the Great of Mahale, but most males are alpha for brief periods. If a male bides his time and remains attentive (and males are ever-attentive), he has a good chance of becoming alpha at some point in his career. If the community he inherits is healthy and well-fed, as alpha he will have many heirs.

But he may not become alpha. Even so, territory-expansion means the females that are in the community will be healthier due to the better food supply and therefore will come into estrus sooner. Success may also bring more females into the community. With four, five, or more females in estrus, a low-ranking male has a much, much greater chance of fathering an infant than if only one or two are in estrus. An alpha male can exclude all other males from mating with one female, and he can be partly successful when two are in estrus, but when four or five are in estrus at the same time he cannot guard them all; furthermore, females are freer to choose the mates they prefer (Watts & Mitani, 2001; Watts, 2002; Mitani & Watts, 2005). Thus, the more raiding parties succeed, the more females are in estrus and the greater the likelihood that a lower-ranking male will find a female sexual partner and father an infant.

There is also a possible indirect benefit. Since males do not disperse, even if a low-ranking male never fathers an infant, helping defend the territory is helping his brothers and cousins increase their reproductive success, so he receives an inclusive fitness benefit.

25.15 A Society of the Mind Requires a Mind

There was a time when psychologists and primatologists sat amazed at the apparent intelligence of chimpanzees and other apes. What could possibly have driven the evolution of this incredible cognitive ability when the animal lives in a Garden of Eden where all of life's necessities are within the grasp of the long arm of the chimpanzee? As we came to know more about primate ecology, the mystery deepened. Many monkeys were frugivores, as are chimpanzees; many monkeys (e.g., hamadryas baboons and geladas) live in groups that are even larger than those of chimpanzees. What could drive the greater intelligence of chimpanzees if not the need to forage efficiently and/or the need to solve complex social equations? We now know that life is not so easy for the chimpanzee after all; competition both within and among communities makes life a complex, cognitively challenging, often violent struggle. More importantly, the fission–fusion nature of chimpanzee society requires mind-reading – see Chapter 18. Chimpanzees negotiate their existence in this multi-level society of the mind, and intelligence provides a crucial advantage.

References

Basabose AK (2004) Fruit availability and chimpanzee party size at Kahuzi montane forest, Democratic Republic of Congo. *Primates* 45, 211–219.

Boesch C (1996) Social grouping in Taï chimpanzees. In *Great Apes Societies* (eds. McGrew WC, Marchant L, Nishida T), pp. 101–113. Cambridge: Cambridge University Press.

Boesch C, Boesch-Achermann H (2000) *The Chimpanzees of the Taï Forest*. Oxford: Oxford University Press.

Goodall J (1968) The behavior of free-living chimpanzees in the Gombe Stream Reserve. *Anim Behav Monogr* 1, 165–311.

Goodall J (1986) *The Chimpanzees of Gombe: Patterns of Behavior*. Cambridge, MA: Harvard University Press.

Goodall J, Bandora A, Bergman E, et al. (1979) Intercommunity interactions in the Gombe National Park. In *The Great Apes* (eds. Hamburg DA, McCown E), pp. 13–53. Menlo Park, CA: Benjamin Cummings.

Goossens B, Setchell JM, Tchidongo E, et al. (2005) Survival, interactions with conspecifics and reproduction in 37 chimpanzees released into the wild. *Biol Conserv* 123, 461–475.

Herbinger I, Boesch C, Rothe H (2001) Territory characteristics among three neighboring chimpanzee communities in the Taï National Park, Côte d'Ivoire. *Int J Primatol* 22, 143–167.

Hunt KD (1989) *Positional behavior in Pan troglodytes at the Mahale Mountains and Gombe Stream National Parks, Tanzania*. PhD dissertation. Ann Arbor, MI: University of Michigan.

Kahlenberg SM, Thompson ME, Wrangham RW (2008) Female competition over core areas in *Pan troglodytes schweinfurthii*, Kibale National Park, Uganda. *Int J Primatol* 29, 497–509.

Marsden SB, Marsden D, Thompson ME (2006) Demographic and female life history parameters of free-ranging chimpanzees at the Chimpanzee Rehabilitation Project, River Gambia National Park. *Int J Primatol* 27, 391–410.

McGrew WC (2016) Field studies of *Pan troglodytes* reviewed and comprehensively mapped, focussing on Japan's contribution to cultural primatology. *Primates*. DOI: 10.1007/s10329-016-0554-y.

Mitani JC, Watts DP (2001) Why do chimpanzees hunt and share meat? *Anim Behav* 61, 915–924.

Mitani JC, Watts DP (2005) Correlates of territorial boundary patrol behaviour in wild chimpanzees. *Anim Behav* 70, 1079–1086.

Mitani JC, Watts DP, Muller MN (2002) Recent developments in the study of wild chimpanzee behavior. *Evol Anthropol* 11, 9–25.

Mitani JC, Watts DP, Amsler SJ (2010) Lethal intergroup aggression leads to territorial expansion in wild chimpanzees. *Curr Biol* 20, R507–R508.

Muller MN (2002) Agonistic relations among Kanyawara chimpanzees. In *Behavioural Diversity in Chimpanzees and Bonobos* (eds. Boesch C, Hohmann G, Marchant L), pp. 112–124. Cambridge: Cambridge University Press.

Nishida T (1968) The social group of wild chimpanzees in the Mahale Mountains. *Primates* 9, 167–224.

Nishida T (1979) The social structure of chimpanzees of the Mahale Mountains. In *The Great Apes* (eds. Hamburg DA, McCown ER), pp. 73–121. Menlo Park, CA: Benjamin Cummings.

Nishida T (1990) *Chimpanzees of the Mahale Mountains*. Tokyo: University of Tokyo Press.

Nishida T, Hiraiwa-Hasegawa M (1987) Chimpanzees and bonobos: cooperative relationships among males. In *Primate Societies* (eds. Smuts BB, Cheney DL, Seyfarth RM, Wrangham, RW, Struhsaker TT), pp. 165–117. Chicago, IL: University of Chicago Press.

Nishida T, Kawanaka K (1972) Inter-unit-group relationships among wild chimpanzees of the Mahale mountains. *Kyoto Univ Afr Stud* 7, 131–169.

Nishida T, Hiraiwa-Hasegawa M, Hasegawa T, Takahata Y (1985) Group extinction and female transfer in wild chimpanzees in the Mahale Mountains. *Z Tierpsychol* 67, 284–301.

Pepper J, Mitani J, Watts D (1999) General gregariousness and specific social preferences among wild chimpanzees. *Int J Primatol* 20, 613–632.

Pusey AE (1979) Intercommunity transfer of chimpanzees in Gombe National Park. In *The Great Apes* (eds. Hamburg DA, McCown ER), pp. 465–479. Menlo Park, CA: Benjamin Cummings.

Pusey AE, Williams J, Goodall J (1997) The influence of dominance rank on the reproductive success of female chimpanzees. *Science* 277, 828–831.

Pusey AE, Murray C, Wallauer W, et al. (2008) Severe aggression among female *Pan troglodytes schweinfurthii* at Gombe National Park, Tanzania. *Int J Primatol* 29, 949–973.

Sterck EHM, Watts DP, van Schaik CP (1997) The evolution of female social relationships in nonhuman primates. *Behav Ecol Sociobiol* 41, 291–309.

Stumpf RM, Thompson ME, Muller MN, Wrangham RW (2009) The context of female dispersal in Kanyawara chimpanzees. *Behaviour* 146, 629–656.

Sugiyama Y, Koman J. 1979. Social structure and dynamics of wild chimpanzees at Bossou, Guinea. *Primates* 20:323–339.

Thompson ME (2013) Reproductive ecology of female chimpanzees. *Am J Primatol* 75, 222–237.

Thompson ME, Kahlenberg SM, Gilby IC, Wrangham RW (2007) Core area quality is associated with variance in reproductive success among female chimpanzees at Kibale National Park. *Anim Behav* 73, 501–512.

Watts DP (2002) Reciprocity and interchange in the social relationships of wild male chimpanzees. *Behaviour* 139, 343–370.

Watts DP, Mitani JC (2001) Boundary patrols and intergroup encounters in wild chimpanzees. *Behaviour* 138, 299–327.

Williams JM, Oehlert GW, Carlis JV, et al. (2004) Why do male chimpanzees defend a group range? *Anim Behav* 68, 523–532.

Williams JM, Lonsdorf EV, Wilson ML, et al. (2008) Causes of death in the Kasekela chimpanzees of Gombe National Park, Tanzania. *Am J Primatol* 70, 766–777.

Wilson M, Hauser M, Wrangham RW (2001) Does participation in intergroup conflict depend on numerical assessment, range location, or rank for wild chimpanzees? *Anim Behav*, 61, 1203–1216.

Wilson ML, Boesch C, Fruth B, et al. (2014) Lethal aggression in *Pan* is better explained by adaptive strategies than human impacts. *Nature* 513, 414–419.

Wrangham RW (1977) Feeding behaviors of chimpanzees in Gombe National Park, Tanzania. In *Primate Ecology* (ed. Clutton-Brock TH), pp. 503–538. London: Academic Press.

Wrangham RW (1979) The evolution of ape social systems. *Soc Sci Infor* 18, 335–368.

Wrangham RW (1980) An ecological model of female-bonded primate groups. *Behaviour* 75, 262–299.

Wrangham RW (1999) Evolution of coalitionary killing. *Ybk Phys Anthropol* 42, 1–30.

Wrangham RW (2000) Why are male chimpanzees more gregarious than mothers? A scramble competition hypothesis. In *Male Primates* (ed. Kappeler P), pp. 248–258. Cambridge: Cambridge University Press.

Wrangham RW, Smuts BB (1980) Sex differences in the behavioural ecology of chimpanzees in the Gombe National Park, Tanzania. *J Reprod Fert* (Suppl) 28, 13–31.

Wrangham RW, Conklin-Brittain NL, Hunt KD (1998) Dietary response of chimpanzees and cercopithecines to seasonal variation in fruit abundance: I. Antifeedants. *Int J Primatol* 19, 949–970.

26 The Sporting Chimpanzee
Dominance without Destruction

Drawn by author

We often hear expressions of surprise at the special status accorded athletes in our society. Why, a commentator might ask, should we look up to someone because they can hit a baseball, run exceptionally fast, or bend a soccer ball around a wall of defenders? Some find it surprising that even single events of athletic prowess can touch such a wide swath of society as to inspire an almost religious reverence.

I was a Detroit Pistons fan in the 1980s, which meant I despised Larry Bird and the Boston Celtics, who regularly treated my Pistons like a punching bag. I still get a sinking feeling when I remember Larry Bird stealing Isaiah Thomas' inbounds pass in the 1987 Eastern Conference Finals, after which, teetering at the out-of-bounds line, he deftly flicked the ball to Dennis Johnson for the winning layup. By 1989 I was living in the Boston area, where I heard Celtics fans describe this same moment, infuriating and heartbreaking for me, in whispered tones of reverence – and also a prideful wonder that might better be reserved for the discovery of a cure for cancer. The very fact that I remember this insignificant event when I sometimes forget where I parked my car and even whether I drove says something.

It seems a sort of insanity, the emotional weight we pack onto contests that have no practical meaning in our lives – as fans we neither benefit financially nor do we know the players directly; there is no status to be won, in any real sense, from our fandom. Yet enthusiasm for sport is a worldwide phenomenon with historical depth. The Olympics grew out of the tendency for humans to engage in contests. Part of Native American Tecumseh's legend is his performance in his culture's version of the Olympics. The tradition of wrestling in Japan is thousands of years old. What is going on here?

First, I doubt this deep emotional reverence is merely a learned value, or that it is merely frivolous fluff of no importance. I think it goes deeper than that, and I am confident it has its origin in the same social dynamics that drive the physical demonstrations among chimpanzees we call "**social display**" (see Figure 24.3).

26.1 Social Display as Sport

Chimpanzees do not play sports with explicit rules, but they do engage in physical exhibitions that are meant to impress others and elevate their social status – not simply by bullying or bluffing, as most would have it – but by demonstrating skill and endurance. Their sport is more like a freestyle version of the old-fashioned footrace in the town square, or the frontier wrestling contest. However, different as they are, sports among humans and chimpanzees have this in common: one or more individuals engage in a physical display that demands coordination, tests their endurance, displays their strength, and is played to an audience. Both species' displays are meant to demonstrate that the athlete possesses – in abundance – the same physical and psychological skills required to survive more serious life-and-death contests, allowing observers and other contestants to compare their skills to the athlete in the arena.

I once saw two young males at Mahale – Kasengazi and Lukaja – who had just begun to seriously contest for the alpha position, engaging in a dual display, almost in parallel 10 m high in a tree, leaping, climbing, throwing foliage, breaking branches, continuing in their display side by side until they both leapt onto the same branch and Kasengazi slammed into Lukaja, knocking him from the tree. Lukaja sat expressionless, looking up into the tree as Kasengazi continued his display. Lukaja gave Kasengazi a wider berth after that. Kasengazi won the duel and in doing so seized a higher social rank. But note: Neither chimpanzee was injured, and here we find both an important parallel to human sports and the functional meaning of the social display. For both humans and chimpanzees, it is of vital importance that the winner gains status without injuring the competitor. More on this below.

26.2 When to Display

The chimpanzee social display rarely occurs in small groups and virtually never alone. Instead, it often blossoms as group members gather for socializing after a long feeding bout. Sometimes something sets off a male – a sound in the forest, or a subordinate male passing too close to an estrous female – but more often a display erupts without any catalyst whatsoever. Such a spontaneous display may begin with the male (females display infrequently) staring at a competitor. His hair begins to piloerect, making him appear larger and more impressive. He will often rock back and forth and may begin a pant-hoot with a falsetto, initially quiet, "hoo, hoo, hoo…" that quickly escalates into a full-on pant-hoot or scream. Females, young and low-ranking males retreat, though dominant males may remain close. Now comes the impressive part. The physical action may begin with a stiff-armed hunched walking; he may walk bipedally, swinging his arms, transitioning into a sprint through the group, scattering other group members. This part is almost always silent – he has already alerted his audience that the show has started. He uproots and throws vegetation, tosses stones, and breaks branches off trees or uproots saplings as he storms forward. The bipedalism makes him appear larger, but also leaves the hands free to throw rocks and foliage or pummel bystanders. If he climbs a tree, the ascent will be so rapid that it is better to say he runs up the trunk. Then he may leap to another tree, shaking it vigorously; he may break off branches and throw them down.

Displays often end suddenly; he may abruptly sit, though with his chest heaving from exertion. The empty space around him, given in deference to his status, collapses as subordinate individuals rush in pant-grunting to groom him or hug him. The tandem display, where two males, often close in rank, display simultaneously is more impressive still, and allows side-by-side comparison.

Considerable power, grace, endurance, and coordination are required to execute an impressive display. As with human athletic endeavors, the whole thing is even more impressive when viewed in slow motion, where the grace, power, and precision of the rushes and leaps are almost balletic.

26.3 Social Displays Require Great Endurance

Displays are manifestly exhausting. Afterwards, males often sit taking such massive, gulping breaths that it seems they might injure themselves. As we learned in Chapter 24, the longer the display the more impressive to underlings. Just as with the village wrestling contest that escalates into a fist fight, during displays tempers may flare and a real fight might ensue, but even in this case there is rarely serious injury.

26.4 Displays Are Not Bluff

What is the purpose of these energy-draining activities? Some view the displays as bluff – they believe males are feinting, pretending to begin an attack, hoping that a competitor will back down before the displayer commits himself to an actual fight. While he pretends to begin an assault, he actually has no intention of following through – or may even be incapable of following through. For instance, he might bluff because he alone knows he cannot win a real confrontation due to illness, injury, or some other disadvantage that he has managed to hide. In this view he is hoping to break the nerve of the competitor based on the observer's perception of him as a superior fighter, to make him back down with a dishonest display of ability. My view is completely contrary. I see displays as a consequence of a peculiar kind of evolution called sexual selection, evolved for a purpose that includes keeping opponents safe.

26.5 Sexual Selection

Sexual selection (Darwin, 1859, 1871) is selection that increases an individual's ability to compete reproductively with same-sex individuals, rather than making the individual better adapted to the environment. When we consider the acute evolutionary pressure that animals shoulder in the struggle for life, pressures that demand superb adaptation to the environment and leave many individuals who cannot make the grade dead by the wayside, sexually selected traits that make their owners *less* adapted to their environment are particularly curious. They often reduce lifespan by increasing the displayer's exposure to predators, or the risk of starvation, or the threat of accidental death. Yet they are so valuable they overcome these handicaps.

Another identifier of sexually selected traits, one of nature's crib sheets in a way, is that they appear at a time in the individual's life history when mating becomes a possibility, adulthood for most animals, but during mating seasons for some traits for some species. Male squirrel monkeys develop exaggerated, fattened shoulders when females are in estrus, presumably to buffer them against bites and blows from competitors. Among physical features that appear in adulthood are manes in lions and baboons, cheek pads and "capes" (very long, matted hair) in orangutans, and the eruption of the canines; the canine is the last of the permanent teeth to finish growing (Fortman et al., 2017). Male–male intolerance is a behavior that appears only in adulthood in most primates.

26.6 Sexual Dimorphism

Sexual selection acts differently on males and females – males must always court females (among primates, at least). Females, on the other hand, must always bear and nurse infants, but attracting a male is less urgent, because male suitors are usually abundant. Not that females lack sexually selected characters – the sexual swelling evolved to improve a female's chance of mating with the fittest male. These contrasting evolutionary pressures means males and females typically look and behave differently. When the sexes differ, we call this **sexual dimorphism**. In some cases, gibbons for instance, males and females have very similar social roles; both sexes guard their territory against like-sex competitors. They also have very similar parenting behaviors (both care for their infants). Consequently they have little sexual dimorphism.

26.7 Intersexual Displays

There are two types of sexual selection – that which helps individuals compete with like-sex individuals, and that which evolved to improve an individual's mating appeal. **Intersexual** selection is this latter phenomenon. It produces traits that make females more likely to seek out copulations with the male who possesses them. In some species, females compete for access to males, and in these species it is the females who possess flamboyantly attractive physical traits (Darwin, 1871; Zahavi, 1975). This phenomenon has not been discovered among primates. Intersexual traits often signal that the primate male has "**good genes**," meaning that he is healthy and vigorous and his offspring are therefore likely to be the same.

Here we find that selection produces the traits that most worried and baffled Darwin, traits that actually are a **handicap** to the possessor (Zahavi, 1975). More on these below. Humans consider such traits in birds as beautiful, tending as they do toward exotic ornamentation, brightly colored plumes, pleasing color coordination, and elaborate patterns. We see some of this in primates, the brightly colored mandrill or the pleasing facial patterns of DeBrazza's monkeys, for instance. Males can attract females by offering resources, sometimes called **nuptial gifts**. Primates do not offer such gifts, but male gibbons make themselves attractive by carving out as desirable a territory as they can, which translates into food, a gift of sorts. Still another strategy is to offer females protection from aggressors, protection for both themselves and their infants – sometimes infants the protecting male has not fathered. Intersexual displays provide information to the opposite sex that is useful to her and to him; it improves his mating opportunities.

26.8 Human Sexual Dimorphism and Sexual Selection

As an aside, focusing on humans for a moment, we have already met one hypothesized sexually selected trait in humans, the chin (Section 9.7). Men have larger faces, perhaps an expression of testosterone levels, which is a handicap, and large brows, which may be an intrasexually selected trait that shields the eyes from blows (not bites, as is perhaps the case in the chimpanzee), and large upper-body musculature that is useful during combat. There have been few hypotheses concerning the male beard, but if primates give us a clue, it may be to protect carotid arteries and other vulnerable tissues from wounding. Females may advertise fertility with strategic depositions of fat on the breasts, hips, and buttocks (Low et al., 1987). The large human brain is particularly energy-hungry, more so during growth, and fat deposition may be a reserve to be drawn on during critical developmental stages when calories are in high demand. Women must also have wider hips to accommodate a larger birth canal. Together these different demands account for most differences between the sexes.

26.9 Intrasexual Selection

The second type of sexual selection, that for **intrasexual competition**, tends to produce fighting weaponry, such as long, sharp canines, and greater size or muscularity among body parts used during fighting, such as large shoulders in baboons and humans and horns in ungulates and beetles.

Males compete with one another to increase their reproductive success in two ways: (1) by winning physical contests with other males who are attempting to prevent them from mating or who threaten their infants; and (2) by being more attractive to females.

Competition tends to produce these physical manifestations of fighting ability, but behavioral or cognitive attributes are even more important; aggression, resilience, persistence, and political savvy are the behavioral traits that allow a male to bring his physical gifts to bear. Successful males either prevent other males from mating, or they are immune to other males' attempts to keep them from mating, or they can protect their infants from being killed by competitors. There are other, less obvious ways males may compete: They may evolve the ability to recognize when females are most fertile and which females are most worth fighting over.

Males may also compete by producing more or better sperm than competitors – obviously this is ineffective if they are not powerful enough to run the gauntlet of guarding males – a strategy that is most strongly selected for when many males mate with a female during her fertile period, as do chimpanzees. They compete by standing ready to father infants, should the opportunity arise, at a very young age. Whereas females only begin to show interest in sexual activities in early adolescence (8–10 years old), males are different (see Figure 27.2). At two years old they may display an erect penis around estrous females and by five years old they will begin to produce sperm.

Females compete for the best core areas, and the competition is not delicate. Females may viciously attack one another, may harass or even kill another female's infant, and may go out of their way to bully a competitor when no resource is nearby. Dominant females can turn their subordinates into nervous wrecks and bully them into foraging in dangerous places near neighboring territories.

When many females are in estrus, dominant females bully other females away from the most desirable males (Stumpf & Boesch, 2005). Intersexual display is mostly limited to sexual swellings, discussed in the next chapter.

26.10 Honest Displays and Handicaps

If part of the appeal of sexually selected traits is that they signal that a male is a good mate, it might occur to you that in life there are always cheaters. If females want to identify males with "good genes" she is beholden to recognize when males are faking it. This means selection acts to make females attracted to males who have traits that cannot be faked. These are **honest displays**. A peacock's tail is an honest display because it is impossible to fake a huge tail. Females will be selected to take a close look and to be able to discriminate among sizes, but once they do they know who has the best honest signal. An honest signal of what? Females are selected to copulate with males, and therefore to be attracted to males who give their offspring the best chance of survival. So females are looking for "good genes" – genes that improve offspring survival. So far so good. But how is it that a peacock's tail tells a female he has "good genes"? Why is the signal honest?

Amotz Zahavi recognized that intersexual displays that **handicap** the displayer have a particular currency: They demonstrate to other individuals that the possessor of the handicap has the vigor required to engage in all the behaviors necessary to sustain life, such as gathering food, avoiding predators, and successfully competing with other individuals, all while carrying a handicap (Zahavi, 1975; Zahavi & Zahavi, 1998). The greater the handicap, the greater the vigor and competence of the possessor, thus giving name to the handicapping principle (Zahavi & Zahavi, 1998). Handicaps might be expensive coloration (Hill, 1991), heavy or attention-grabbing adornments, or behaviors that are necessarily expensive. They are honest signals both because they cannot be faked (you have a big tail or you do not) and because they are *unavoidably* costly to the signaler. Zahavi's concepts were not without critics, but early reservations about the theoretical soundness of the handicapping principle (Dawkins & Guilford, 1991) have largely been rebutted (Grafen, 1990; Folstad & Karter, 1992). It is an important tool for understanding bizarre physical traits.

But not simply physical traits. Intersexual selection also includes behavioral displays, which, by handicapping a male by forcing him to excel in the display, offer females honest evidence that the males are healthy, vigorous, and capable of foraging effectively enough to "waste" effort on behavior that has no use other that appealing to females, proving the presence of good genes. Some of these behaviors result in physical objects that a female inspects, what Richard Dawkins has called an individual's "**extended phenotype**," of which we are about to hear much more. Bowerbird males are a good example of this (see Section 26.13).

While Zahavi was more interested in intersexual competition, intrasexual competition may also select for honest displays, particularly if the display is an honest signal of fighting ability. The display may signal that the displayer is a powerful competitor capable of winning a physical contest and therefore not worth challenging. Because fighting ability may be attractive to a female for a number of reasons (it suggests vigor, it is a signal that a male would be able to guard her infant, and it is an honest display of health) some traits may be selected for both *intra-* and *inter*sexually.

To better understand these concepts, let us examine a few examples of honest displays that are handicaps.

26.11 Widowbirds

The exotic plumage of widowbirds is impressive, and they pair it with behaviors that show it off. Female widowbirds prefer males with long tails, even though the tails weigh down the male and make him more conspicuous to predators (Figure 26.1). It is clear that it is tail length that is attractive because when widowbird male tails are cut off, male mating success declines; when tail-extenders are attached to other males, their success improves (Andersson, 1982). During mating season, males bear bright colors in

Figure 26.1 A male widowbird with breeding plumage. Photo courtesy of Ron Matson, with permission.

contrasting patterns and a long tail; when they are immature, or when females are not in reproductive season, they are more camouflaged (Figure 26.2).

26.12 Lions

West and Packer (2002) found that female lions prefer males with heavy, dark manes, and speculated that the manes were an honest signal of "good genes," and a handicap. To test this hypothesis, they manufactured dummies of males with several lengths and colors of manes. They placed these dummies in the range of female lions and found that dummies with larger and/or darker manes (Figure 26.3) were approached by females more often.

This brilliant work was augmented with an ingenious behavioral study, a study of both mating behavior and infant survival. They found that females who mated with males with dark manes had infants with higher survival rates. They went even further; they measured body temperatures of males with an infrared thermometer that could read temperatures from a distance. They showed that dark, heavy manes are a serious handicap in the hot savanna, where the dark color absorbs more solar radiation and the heavier manes serve as insulators, increasing the core temperature of males, fatiguing them and thus hampering their ability to engage in hunting and fighting. In other words, a large, dark mane is a

Figure 26.2 A male widowbird. Two views of breeding plumage (behind) and nonbreeding plumage (foreground).

handicap to males, and males who can bear that handicap and thrive will father healthier infants.

26.13 Bowerbirds

Behavioral handicapping signals are every bit as intriguing as body adornments. To attract a mate, the bowerbird male gathers stems and sticks and builds a huge (compared to him) and elaborate arch that he decorates with an impressive litter of blue objects (Figure 26.4). The arch is utterly useless except that females find taller, more substantial bowers with an abundance of blue decorations to be most attractive. Males have handicapped themselves not with a long tail or a heat-stress-inducing mane, but with the "wasted" time, calories, and risk necessary to gather

Figure 26.3 Female lion approaching a dark-maned dummy male.
Courtesy of Craig Packer. (A black and white version of this figure will appear in some formats. For the color version, please refer to the plate section.)

Figure 26.4 Bowerbird bowers. Photo credit: Barry Hatton, with permission.

materials to assemble and decorate their bower. Dawkins calls this sort of physical object part of the males "extended phenotype" (Dawkins, 2016), a physical expression of his fitness that extends outside his body. Presumably a Rolex or a sleek Mercedes Benz can function similarly among humans.

26.14 Evidence Social Displays Are Not Bluff

Social dislpay may look like bluff (van Staaden et al., 2011), but with our better understanding of the context and subsequent events following displays, we now know this is wrong. Displayers seldom charge directly at a specific individual, as they might if they were bluffing him. If the displayer were ill or injured, he might bluff to dominate another individual, but in that case he would be unable to carry through on his threat; instead, displays last for extended time periods, signaling robust health. If displays were a bluff, we might expect to see them directed at dominants, with the displayer retreating if the higher-ranking individual called his bluff. Displays actually occur almost entirely in front of lower-ranking individuals. Actual fights are rare. Displays, then, are exhibitions for the benefit of subordinates for whom there is no reason to bluff, since the displayers are often at the height of their powers and acknowledged as higher ranking than his observers.

26.15 The Function of the Social Display

Rather than bluff, we have come to view the social display as an **honest signal** of athletic competence. The endurance it takes to display for minutes cannot be faked. It is a purposefully wasteful, purposely difficult athletic exhibition that is meant to leave no doubt in the mind of the observer that the displayer is strong, vigorous, coordinated, and endowed with great stamina. Displays often involve athletic stunts like long leaps that require precision to stick the landing. They are exhausting, signaling by their length and vigor that the displayer is healthy and, literally, has energy to burn, impressing a potential challenger with the abilities of the displayer. The display demonstrates competence, but at the same time it avoids the risks of an actual combat. It not only prevents the displayer from being injured, but – this is essential – unlike displays in other species, it is meant to avoid injuring other group members.

The social display, then, is a relatively harmless behavior that has an implicit threat: "I am breaking this substantial tree in two but it could very well be your arm." The displayer is working to convince two classes of observers of two things. He is signaling to subordinate individuals that he has all the skills and vigor necessary to defeat them, and a challenge is useless. And he is signaling to females that he is a good prospective mate because his offspring, given how heavily he is handicapping himself, will be healthy and competent. He is signaling to these two classes of watchers while taking only the smallest of chances that he will be injured, or that another community male will be injured.

This type of display is rare in the animal kingdom. The pawing of the ground by a buffalo or a baboon, and the trumpeting and mock charging of elephants are in a different class. They are meant to signal to the recipient "I see you; I would prefer that you draw back without my having to run you over, thereby wasting my time and energy, but I will if I must." When baboons fight they go for the jugular. Literally. When I was at Gombe in the 1980s, two male baboons were competing for dominance. They fought often, and the rest of the time avoided one another. One morning when I was walking down the beach, going out to follow chimpanzees, I saw one of the males sitting on top of a small fisherman's hut – local fishermen were allowed to dry fish on the beach, and they often assembled small huts as temporary lodging for the couple of days they were there – motionless and drooping. As I got closer I saw he had a horrendous injury, a gaping wound from the corner of an eye to the corner of the mouth. I saw him there repeatedly over several days, the skin gradually blackening and drawing back from the wound, until he died, presumably from a systemic infection. Chimpanzees have evolved a mechanism that allows them to avoid such deaths among members of their community.

26.16 Unique Selection on Chimpanzee Intrasexual Competition

What causes chimpanzees in particular to treat competitors gingerly while baboons care nothing for the health of a defeated intragroup competitor?

When baboon males fight, presuming the competitor is not a brother, an injury to an opponent has no downside for them. If he dies, they have one fewer competitor, one fewer factor to include in their calculus as they maneuver for social rank. Unlike baboons, male chimpanzees are dependent on one another. They defend their territory as a unified group. The bonds among males allow them to repel attacks from other communities and to eliminate extracommunity rivals. Killing neighboring males improves the chances that the winners will have a positive imbalance of power the next time the two communities meet. Engaging in outright fights within the community risks injuring or killing a valuable warrior whose help is needed defending and expanding the community territory. By avoiding injuring one another, males preserve essential social bonds required during life-and-death confrontations between communities; it decreases the risk that the roster community males will be depleted, which would allow the homeland forces to be outnumbered in the next intercommunity clash. Social displays allow the best fighter to assume his alpha status in the community by exhibiting his top form rather than proving it; males use social display to attain the highest status they can without endangering the community defense.

If this perspective is right, we might expect that when a community's males drop below a certain level, efforts to avoid injury and maintain the fighting fitness of each member would be ramped up. Likewise, when a community has an overwhelming superiority of numbers, there should be less concern about injuring or killing a member. Indeed, this latter seems to be the case at Kibale-Ngogo, where the number of males hovers around 30, and there was a large male contingent at Mahale when Ntologi was killed (Watts, 2004).

26.17 Community Social Structure Yields Sport

Although there may be considerably more intracommunity injuries among chimpanzees than called for in the human ideal of sportsmanship, sporting activities in both species demonstrate athletic ability – and presumably fighting skill – in a relatively harmless display of competence. For humans and chimpanzees alike, a graceful, athletic, powerful display generates awe and respect and serves as an honest signal that the athlete could win a physical contest. Yet in both cases the sporting aspect prevents serious injury to community members. As with humans, great athletic skill among chimpanzees has substantial rewards, generating hero worship, exerting dominance, and maximizing reproductive potential. This establishment of rank has something in common with the concept of community among chimpanzees. It establishes a rank order of best to worst fighters, but without the fighting. Just as the chimpanzee community is a virtual one, so is fighting ability established in the mind rather than in reality.

Both elite human athletes and alpha chimpanzees find their status must inevitably drop as age and injury take their toll. Alphas inevitably are unseated. *Sic transit Gloria*, for humans and chimpanzees alike.

References

Andersson M (1982) Female choice selects for extreme tail length in a widowbird. *Nature* 299, 818–820.

Darwin C (1859) *On the Origin of Species*. London: John Murray.

Darwin C (1871) *The Descent of Man and Selection in Relation to Sex*. London: John Murray.

Dawkins MS, Guilford T (1991) The corruption of honest signaling. *Anim Behav* 41, 865–873.

Dawkins R (2016) *The Extended Phenotype: The Long Reach of the Gene*. Oxford: Oxford University Press.

Folstad I, Karter AJ (1992) Parasites, bright males, and the immunocompetence handicap. *Am Nat* 139, 603–622.

Fortman JD, Hewett TA, Halliday LC (2017) *The Laboratory Nonhuman Primates*. Boca Raton, FL: CRC.

Goodall J (1986) *The Chimpanzees of Gombe: Patterns of Behavior*. Cambridge, MA: Harvard University Press.

Grafen A (1990) Biological signals as handicaps. *J Theoretical Biol* 144, 517–546.

Hill GE (1991) Plumage coloration is a sexually selected indicator of male quality. *Nature* 350, 337–339.

Low BS, Alexander RD, Noonan KM (1987) Human hips, breasts and buttocks: is fat deceptive?. *Ethol Sociobiol* 8, 249–257.

Stumpf RM, Boesch C (2005) Does promiscuous mating preclude female choice? Female sexual strategies in chimpanzees (*Pan troglodytes verus*) of the Taï National Park, Côte d Ivoire. *Behav Ecol Sociobiol* 57, 511–524.

van Staaden MJ, Searcy WA, Hanlon RT (2011) Signaling aggression. *Adv Genet* 75, 23–49.

Watts DP (2004) Intracommunity coalitionary killing of an adult male chimpanzee at Ngogo, Kibale National Park, Uganda. *Int J Primatol* 25, 507–521.

West PM, Packer C (2002) Sexual selection, temperature, and the lion's mane. *Science* 297, 1339–1343.

Zahavi A (1975) Mate selection: a selection for a handicap. *J Theoretical Biol* 53, 205–214.

Zahavi A, Zahavi A (1998) *The Handicap Principle: A Missing Piece of Darwin's Puzzle*. Oxford: Oxford University Press.

27 The Passion of *Pan*
Sex and Reproduction

Credit: Martin Harvey / The Image Bank / Getty Images

Of all the behavioral contrasts between humans and chimpanzees, surely our sex lives are the most dissimilar. While the chimpanzee monthly sex cycle lasts the same period of time as that of humans, around 30 days, and while it is driven by the same mix of hormones, just about every other aspect of chimpanzee sexuality is utterly alien from the human point of view.

If you are a female, imagine going four and a half years without any sexual contact whatsoever, then finding one day a swelling the size of a grapefruit on your bottom, a portent that signals the commencement of a sexual odyssey that includes 30 copulations a day (every 20 minutes), copulating with every single male you know (including adolescents and old men), and sexual liasons not just with boys, but with males right down to the age of kindergartners. This would go on for four months. During this wild sex party your juvenile and adolescent offspring will be looking on. Worse, your brothers may even attempt to copulate with you. Then, after this completely over-the-top orgy concludes, you find your interest in sex has disappeared entirely and you lapse into total abstention for the next four and a half years.

If you are a male, imagine copulating six times a day, on average, day in and day out, your entire adult life. And who are your sex partners? Not young females the equivalent of 15, 20, and 25 year olds; they are unattractive. But red-hot passion accompanies the mere sight of females with five or more offspring, females not just past their youth, but wrinkly, scarred, saggy females past even middle age. You might entertain a special affection for a particular female, and she may return your affection, yet she will copulate with every male in the community many times, sometimes even rebuffing your overtures and instead mating with an adolescent male. Now imagine that each of your individual sexual acts lasted about 15 seconds. This is the sex life of the chimpanzee.

27.1 Ovulation Ceases for Years

Using the technical term, females spend much of their lives nursing and therefore in the **anestrous** physiological state (i.e., not sexually cycling); the role of nursing is expressed in how this lack of ovulation is described; it is called **lactational amenorrhea**, meaning "not menstruating due to nursing." While nursing, as we learned earlier, chimpanzees cease sexual cycling and lose all interest in sexual activity, nor are males interested in them. This anestrous period may be extended should they fall sick, suffer a serious injury, become malnourished or, rarely, should they come under severe social stress (Knott, 2001; Muller & Wrangham, 2001).

27.2 Sexual Desire and Activity Resume When Ovulation Resumes

Estrus is the term used for the physiological changes a female experiences around the time of ovulation. Estrus is the noun, estrous the adjective; that is, females have a period of estrus, during which time they are **estrous females**. The release of a ready-to-fertilize egg is the key event in estrus, but there are many other physical and behavioral changes that occur as well, including an increase in **attractivity** (appeal to potential mates), the willingness to copulate, known as **receptivity** (Hrdy & Whitten, 1987), and the state of **proceptivity** (encouraging copulation). It is one of the most extraordinary phenomena in primate behavior, to see a female who is a steady, responsible, attentive, almost boring mother suddenly throw everything up for the chance to have constant sex. Sometimes estrous females seem positively out of their minds.

27.3 Visible Estrus

Chimpanzees have **visible estrus**. When they are sexually cycling, specialized skin around the vagina and anus swells, going from a wrinkly flap of skin into a large, pink sexual swelling, nearly the size of a throw pillow, largest around the time of ovulation. The size and color of the swelling makes it highly visible, which means every individual in the community knows (well, not consciously) the ovulatory state of the cycling female. She might as well be wearing a sign saying "ovulation in three days!" There are olfactory cues as well (Hrdy & Whitten, 1987), principally, so far as we know, volatile fatty acids produced by bacteria that grow in vaginal mucus. We know that mucus increases in response to the rise in estrogen at midcycle (Keverne, 1976; Hrdy & Whitten, 1987), driving the rise in fatty acids. If these fatty acids are smeared on the sexual skin of ovarectomized (ovaries removed) macaque females, males are still attracted to them (Keverne, 1976), even though they lack ovulatory hormones. At least, this seems to be the mechanism in macaques (Keverne, 1976) and humans (Michael et al., 1974, 1978), and so is likely the same in chimpanzees. One might think that the female hardly needs such a signal – though its evolutionary origin will become apparent in a moment – since her increased proceptivity makes it utterly obvious.

The estrous period of apes in general and chimpanzees in particular differs from that of most primates, in having an extended period of receptivity (Hrdy & Whitten, 1987) and in exhibiting menstruation, the shedding of the uterine lining. Females are sexually attractive for two-thirds of their sexual cycle (more on this below), longer than in other primates.

Whereas in the non-estrous state her attitude toward males was at best mixed, when in estrus it is enthusiastically positive and sexual. Outside of estrus, one or two males might be friends of hers so that now and then when they pass by her core area they veer off to pay her a social call. She will greet such a friend warmly, grooming him and traveling with him companionably for some hours as they feed and rest before he returns to a male traveling party. Other males she will avoid or ignore.

Once in estrus, however, her attitude changes. She will leave her core area, find a large party of males, and travel with them more or less constantly until she gets pregnant. Proceptive females express an interest in copulation by turning their hindquarters toward a male, an action known as **presenting**. If a male expresses an interest in her, and a higher-ranking male is not inhibiting her, she will rush over to present to him and copulate.

27.4 Sperm Competition

The type of mating system determines testis (syn. testicle) size (pl. testes) and many other aspects of mating as well (Harcourt et al., 1981; Harcourt, 1991, 1995). In species in which females copulate with multiple males, males who produce more sperm are more likely to impregnate a female. Larger testes produce more sperm. This type of male–male competition is called, in about as straightforward a bit of scientific jargon you will ever find, **sperm competition**. Across species, testis size correlates well with the number of males who copulate with individual females. The more male partners a female has, the less males can be confident they have impregnated the female; males respond over evolutionary time by producing more sperm. Testes size correlates with the number of sperm per ejaculate (Harcourt et al., 1981; Harcourt 1991, 1995; Dixson 2009, 2013), which means that in chimpanzees both the number of copulations and the number of sperm per copulation are huge. Where females only have one available mate, as is the case in gorillas and gibbons, confidence in paternity is high and testes are quite small (Figure 27.1); there is no sperm competition. Among solitary species in which confidence is medium, testes are medium-sized. In species where males have low confidence in paternity – in other words, where females are promiscuous, as they are among baboons and chimpanzees – testes are large.

27.5 Species Differ in Ease of Impregnation

Sexual swellings undoubtedly attract males, so it might be assumed that they evolved to guarantee that females attract enough copulations to get pregnant. This seems unlikely because chimpanzee females cycle an average of four months per pregnancy, and in those months, she will copulate 1200 times. Perhaps it takes that many copulations for her to become pregnant, one might think. Consider the gorilla; female gorillas get pregnant in two months, about half the time of chimpanzees, and rather than mating 1200 times, female gorillas copulate something like three times each estrous period, or six times per pregnancy. Clearly it is possible biologically to evolve to conceive with just a few – or even one – copulation. This contrast between gorillas and chimpanzees is even more impressive when you consider that the testes of chimpanzees are over 10 times the size of those of gorillas, in relation to their body size – 0.27 percent of their body weight versus 0.02 percent for gorillas. Why, one might ask, would a species evolve, as chimpanzees have, to be difficult to impregnate? We will look into this in more detail in a few paragraphs.

Not only sperm numbers, but sperm themselves differ among species; the power plant of sperm, the midpiece, is large in chimpanzees, supercharging their sperm compared to gorilla and human sperm, allowing sperm to swim faster and with more power (Anderson & Dixson, 2002; Nascimento et al., 2008). The proportion of malformed and crippled chimpanzee sperm is tiny compared to humans (Seuanez et al., 1977).

In other words, gorilla females manage to get pregnant with a fraction of the number of copulations of chimpanzees and with fewer and less active sperm. Clearly something more is going on with estrus than just acquiring the necessary number of copulations.

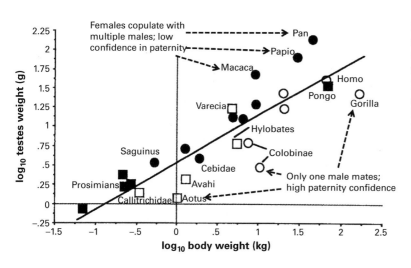

Figure 27.1 Sperm competition in social systems where females mate with multiple males selects for large testes. In one-male groups, such as harem (one male, several females) and monogamous groups, males have small testes. After Harcourt, 1995; used with permission.

27.6 Psychologically Driven Birth Control

If you were to go to a fertilization specialist with your spouse, reaching out for help because you were attempting to get pregnant and failing, your physician might assume that some pathology was interfering with conception, that there is some error in your or your mate's physiology responsible for the infertility. You might not have the right mix of hormones to spur ovulation, or your partner's sperm count might be abnormally low. A fertility specialist might perform tests to determine whether both of these functions are normal. The assumption in this paradigm, "infertility = pathology," is one that assumes that female and male sexual partners physiologically "agree" on the goal: get pregnant.

While this may be your conscious goal, your body may have another idea. In the millions of years of human and primate evolution, the human (and chimpanzee) reproductive system has evolved to make babies when they are likely to survive and to forego reproduction when success is unlikely. As such, your psychological state has some effect on your hormone profile, which in turn affects your fertility.

When was success *un*likely in our environment of evolutionary adaptedness (EEA), in our evolutionary history? When a potential mother lacked confidence in her partner's commitment – in a precarious hunter-gatherer world in which a partner might provide essential supplements to a woman's diet, if a partner defected she would be saddled with the entire burden of child-rearing. Such ambivalence affects the likelihood of ovulation, and that's a good thing; the baby might die without a partner's contribution. Or the woman might be in the midst of a drought and thus malnourished. Or she might have been experiencing some social upheaval such as war that might require her and her mate to flee at a moment's notice; as refugees they might have little food, might sleep in inhospitable places, and be forced into long marches. The burden of protecting and carrying a baby might be the last straw, spelling doom for both mother and infant.

How, you might wonder, does your reproductive system know that bad times loom? Your limbic system might be clanging alarms and spilling cortisol and other stress hormones into your bloodstream. You might be losing weight as food supplies diminish or your partner disappears for days at a time. But, it might occur to you, "The people you are writing about are not in a war zone; no peasant-uprising or food-shortage threatens a corporate lawyer; why should she be infertile?" Unfortunately, stresses associated with job insecurity, long commutes, financial troubles, exhausting weekend-warrior workouts, relationship instability, and intense preparations for bikini season often physiologically mimic ancestral misfortunes, affecting fertility (Ellison, 1990; Ellison et al., 1993; Bentley, 1999; Knott, 2001).

In short, our reproductive system is evolved to produce babies during good times and to postpone pregnancy when we are stressed. Chimpanzee females evolved under the same pressures, but their unique fission–fusion social system places other extraordinary pressures on them. The upshot is that chimpanzees are evolved to be difficult to impregnate for a good reason; you may be surprised to learn that reason is as a defense against baby-killers.

27.7 Infanticide

Baby-killing has a surprising immediacy for females in a wide variety of primate species. The threat of infanticide is a fact of life among blue monkeys (Butynski, 1982), redtails (Struhsaker, 1977), howlers (Crockett & Sekulic, 1984), langurs (Hrdy, 1974; Borries et al., 1999), and savanna baboons (Collins et al., 1984). This list only includes the best known examples; infanticide has been observed in almost every primate that is well-studied (Hrdy, 1979; Hausfater & Hrdy, 1984; van Schaik & Kappeler, 1997), including chimpanzees (Goodall et al., 1979; Kawanaka, 1981; Nishida & Kawanaka, 1985; Takahata, 1985; Goodall, 1986; Arcadi & Wrangham, 1999; Newton-Fisher, 1999).

When we look closely at infanticide, whichever the species, we find that males kill infants that are unlikely to be their own (Hrdy, 1979). Sometimes mistakes are made, but on average killing infants increases male reproductive success by decreasing the success of other males and increasing the likelihood that a female will come into estrus and, surprising as this may seem, subsequently mate with the infanticidal male. In fact, the most likely father of the next infant of a female whose infant has been killed is the baby-killer himself (Hrdy, 1979). In most species, males are much larger and stronger than females. What can a mother do to protect her baby? Surprisingly, a whole lot.

27.8 Female Promiscuity and Paternity Confusion

Female chimpanzees cope with the potential for infanticide with a rare and ingenious strategy seen in only a handful of animals, **paternity confusion**. Chimpanzees are *evolved* to be difficult to impregnate – to protect their babies.

When a female disperses to a new community she travels with males for months and years, and she relies on them to protect her from hostile resident females. The combination of the need for protection and her continual estrus, with its accompanying proceptivity and attraction to males, motivates her to stick close to a large party of males. The long period of infertility among young, newly transferred females has the secondary effect of allowing the new immigrant to study the entire community range, making her familiar with her new home, familiar with the females in her new community, and able to choose the best core area she can. If she were to have an infant immediately, her "homeless" state – she does not yet have a core area – would be a great risk.

But dispersal typically occurs only once; what is going on with older females who have been in a community for some time? The combination of female promiscuity and slow-to-conceive physiology results in her mating with nearly every – if not every – male in her community, giving each at least a possibility of being the father. By being difficult to impregnate and by taking many months to get pregnant, she makes every male a possible father, blunting his motivation to kill her infant, and increasing the likelihood he will protect the infant if it is attacked. Then, just to confuse things more, she will have one last estrus period *after* she is pregnant, during which time she will continue her frequent copulation. In other words, the more difficult a female is to impregnate, the less any particular male can be sure he is *not* the father, and the lower the risk of infanticide.

The advantages of repeated estrus periods for confusing paternity, however, must be weighed against the cost of estrus, with its increased energy demands, risk of injury, and decreased feeding effectiveness. The average of four months is the balance between the advantages of paternity confusion and cost of estrus.

Females, then, are selected to confuse paternity and the means by which they do it is promiscuity. Promiscuity has a disadvantage, however. By mating with every male, females take the risk that a substandard male will father her infant. Females are thus under confounding selective pressures, the anti-infanticidal advantages of promiscuity are at odds with her need to give her infant the best possible genes. Females negotiate this knotty problem with a subtle mating strategy that allows them to have their cake and eat it too. Before we explore that, let us think about what males can do to try to confound the female confusion strategy.

27.9 Deconfusion

As females pursue their imperatives, males are under fierce selection to copulate with females when they are *most likely* to conceive, and this means finding a counterstrategy to foil the female's paternity confusion gambit. In other words, while females are evolving to confuse paternity as much as possible, males are selected to deconfuse paternity. All this is overlain on the background of the evolutionary origin of the sexual swelling, which evolved to signal males when ovulation is near. It is complicated. Males succeed best when they know when, the exact day, to exert the maximum effort to copulate with a particular female. If a male could tell for certain whether a female is ovulating, evolution should drive him to feel little or no sexual interest in a female who was not ovulating and intense desire when she is. As a bonus, if a male is confident he is the father he can help the mother a little bit by taking more extreme steps to protect the baby. This potential extra help and the advantage of choosing the best father means there is some common ground in the interests of the "best" male in the community and the paternity-confusing female.

27.10 Sexual Swellings and Copulation

To understand how each sex executes their somewhat contrary strategies, we must start with a better understanding of the external manifestations of the sexual cycle itself. The internal workings, the mechanics of ovulation, and associated biological processes, we have already examined in Chapter 15. The external manifestations, the increase in the size of the sexual swelling and its consequences are subtle in the chimpanzee, and it is only relatively recently that we have come to understand its dynamics. That swellings "work" is obvious in that analyses of their pattern of appearance among various species suggest that swellings arose independently three times (Hrdy & Whitten, 1987; Dixson, 1983, 1998): in the colobus monkeys, in the *Papionini* (baboons and macaques and their relatives; see the Appendix), and in apes. Before we examine function, let us first look at some background.

There are four major phases of the chimpanzee sexual cycle (Figure 27.2; McGinnis, 1979).

1. **Tumescence** (6 days): Sexual swelling increasing (Figure 27.2). Mating during this time is **opportunistic** (i.e., promiscuous); tumescent females mate with any male who shows an interest. Copulation rates are low; adult males, especially high-ranking males, are typically uninterested during this phase, so most copulations are with adolescents, low-ranking adults, juveniles, and even infants (Figure 27.3).

2. **Maximal tumescence** (10 days): Sexual swelling reaches maximum size. Females are most proceptive during this time. A study that analyzed copulations during a five-year period at Gombe found that when females are at their peak copulation rate, which I hope you have noted is 30 times a day, 96 percent of copulations were during maximal tumescence.

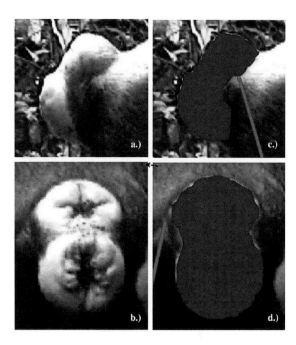

Figure 27.2 The female sexual skin begins estrus as loose, wrinkled skin. The size of the swelling steadily increases (dashed line) until leveling off in the second half of maximal tumescence. Copulations peak during maximal tumescence (dotted line). After Tobias Deschner; used with permission.

While copulations are consistently high for the entire period of maximal tumescence, there is a distinct pattern to which males copulate, and it falls out into two periods.

2a. **First half of maximal tumescence** (5 days): During this time it is rare for a female not to mate with every male in the community, excepting her sons, brothers (copulations do occur occasionally), and father. There is little aggression surrounding copulation in this five-day period; males have been described as queueing up in a copulation line, each male patiently waiting his turn (Pusey, 1979). During this phase she will be the focus of a traveling party of adult males who are enthusiastically interested in copulating with her.

2b. **Second half of maximal tumescence** (5 days): In sharp contrast to the first half of maximal tumescence, during the second half, the **periovulatory period** (POP), the time during which ovulation occurs, the swelling is at its maximum and the time of

Figure 27.3 Chimpanzee males become interested in copulation as infants. Two infant males, approximately three years old, wrestle. An estrous female approaches; both males investigate. One male makes a half-hearted attempt to copulate, then both return to play. Photos by the author.

the free-for-all orgy is over. High-ranking males become fiercely exclusive, attempting to prevent all other males from copulating; the female is followed closely and groomed often. This attempt to monopolize the estrous female is called **possessiveness**. If only one female is in estrus, an alpha male can monopolize her, and he may copulate with her 20 times a day, although at Kibale-Ngogo where there are as many as 30 males, it is a coalition of high-ranking males that guards a female (Watts, 1998).

Ovulation occurs approximately three days after maximal tumescence and females are most fertile from three days before ovulation to the day of ovulation (Deschner et al., 2004); in other words, when her swelling is largest and just after. Thus, ovulation is somewhat predictable; in other words, within-cycle, sexual swellings might be said to be an honest advertisement of fertility, but the fact that two-thirds of conceptions occur outside this period (Constable et al., 2001) means they are not completely honest.

3. **Detumescence** (6 days): The sexual swelling steadily decreases in size. The female is much less proceptive and few males express an interest in copulation.

4. **Quiescence** (up to 10 days): During this period the sexual skin is completely flat and unchanging. Menstruation occurs during quiescence, about three days in, and lasts for three days: days 3–6.

27.11 Mating Patterns

Interdigitating with and tied to the phases of the sexual cycle are three distinct chimpanzee mating patterns (McGinnis, 1979):

1. **Opportunistic** or **promiscuous** mating: During tumescence and the first half of maximal tumescence females mate indiscriminately and there is little aggression surrounding copulation. In one study, 42 percent of all conceptions occurred during this phase (Constable et al., 2001).
2. **Possessiveness**: During the second half of maximal tumescence the alpha male or, if several females are in estrus at once, a high-ranking male attempts to monopolize the estrous female, preventing others from copulating with her. Because females tend to come into estrus when food is plentiful, large, loud, boisterous parties of males are seen with an estrous female or two as its locus; emotions run high. At any particular moment some males might be pant-grunting to the alpha in an attempt to make amends for showing a little too much interest in the estrous female; a female might be screaming in response to fighting among males, worrying that the violence might spill over to her; pant-hoots occasionally erupt. Males and females dash about willy-nilly. The estrous female must remain alert and wary in case a frustrated male turns violent; she must monitor the alpha male so as not to stray too far from him and thus into the orbit of some other male, which could result in an attack. There is little time for feeding. It is a stressful and energy-sapping period in a female's life. In one study, 33 percent of all conceptions occurred during this phase (Constable et al., 2001).
3. **Consortship**: This third mating pattern not infrequently spans the entire estrus period, and may even last more than one estrus period. During a consortship, an estrous female will leave the center of the community range in the company of a favored male. The pair will feed and travel alone for many days in what Goodall has described as a "**honeymoon**," a period during which life is a little less frantic for both the male and the female. The number of daily copulations is lower during consortships, and the physical demands on the estrous female are lessened. Grooming rates are high (Goodall, 1986). At Gombe, one study found that 25 percent of conceptions occurred during consortships (Constable et al., 2001).

Males often begin the consortship by "herding" an estrous female, blocking most paths with his body and thus encouraging her to move in the direction he has selected, guiding her away from the social group. If she appears uncertain, he may shake branches at her, grunt, or engage in a social display. Occasionally he will attack her. While males are willing to resort to coercion to encourage reluctant females, the clever female will find a way to escape from an unwanted consort (see Section 18.18). Thus, it is the female who chooses whether to enter into the consortship or not, regardless of male herding attempts. If the female cooperates, the consortship can last days or even as long as two months (Goodall, 1986). While consortships are a means of a female choosing a particular male – a male who may not be high-ranking – as a potential father to her offspring, such a choice pans out only one-quarter of the time; most conceptions occur from multi-male party matings (Constable et al., 2001; Muller & Wrangham, 2001).

27.12 Why Sexual Swellings?

At first sexual swellings may seem a little mysterious. What is the female signaling? Your first guess will be correct, but why does she need to signal that she will ovulate, when gorillas, squirrel monkeys, lemurs, and tarsiers find mates without such an encumbrance?

The swelling increases during the time females are eager to copulate, and females without swellings are not attractive. It seems simple: By signaling ovulation is near, the swelling attracts mates. Yet, most primate species lack swellings and females have a plethora of willing mates. Why bother to expend energy pumping up the sexual skin when suitors are abundant? Perhaps the most widely accepted answer is that in societies where there are many males, and therefore many potential mates at hand, a large swelling incites competition among males. Intense competition sifts through the contenders to allow the most dominant male to father her infant (Nunn, 1999; Dunbar, 2001). Because the victorious male has fought his way to the top, he has weathered a severe handicap, the rigors of physical contests with a long roster of competitors. His athletic ability, stamina, and cognitive attributes have allowed him to achieve alpha status, which means the female who mates with this winner will have daughters who are healthy and sons who are both healthy and competitive. Sexual swellings set up a de-facto competition among males ensuring the female mates with the male with the best genes – genes that correlate with fighting ability.

We also know something else interesting about sexual swellings; the larger the swelling, the more attractive the female is, with some few qualifications we will soon examine. Not only is she attractive, among baboons (and presumably chimpanzees as well) females with larger swellings have superior ovarian function (Emery & Whitten, 2003) and greater lifetime reproductive success; swelling size is an honest signal of female fitness (Domb & Pagel, 1981, 2001).

While swellings incite competition, there is a fine-tuning in the pattern of swelling that creates the greatest amount of competition at just the right time. The swelling changes size over the estrous period in chimpanzees, and females are most fertile around the time the swelling is largest; males know it, because they fight hardest to copulate with her when her swelling is at its peak (Shefferly & Fritz, 1992; Wrangham, 1993). How, exactly, does all this work? There are a number of hypotheses; we will consider two.

The **reliable indicator hypothesis** focuses on the fact that swelling size reliably indicates female fertility. Males compete fiercely to copulate with females with large swellings, which makes it sensible that as the female's estrous period advances and she comes closer to ovulating, her swelling increases in size. As ovulation nears and the swelling reaches its maximum, her attractiveness peaks and male–male competition reaches its most intense point.

Females need to invest in a large swelling in order to attract the best male. Not every male is available to every female, because males cannot be everywhere at once when there are multiple fertile females. As a consequence, the "best" female will attract the most dominant male, and females with smaller swellings and therefore lower fertility will be forced to settle for lower-ranking males. The reliable indicator hypothesis presumes that females with the healthiest ovarian function *for that particular cycle* have larger swellings than for other cycles. Only when she is healthiest can she bear the handicap of carrying around the largest swelling. All this sounds wonderful, and for years we were convinced this explained everything.

There is more to this story, though, because further research showed that, contrary to the reliable indicator hypothesis, some females with relatively small swellings were nevertheless highly desirable – more desirable than some females with larger swellings. This observation provoked the **graded signal hypothesis**. In this hypothesis, desirability is strongly influenced by other indicators of fitness in conjunction with swelling size. High-ranking females are more desirable, for instance; just as with males, high rank is an honest indicator of overall health and vigor. In addition to rank, **parity** (number of offspring) influences attractiveness. **Multiparous** females (females with multiple offspring) are more desirable. In this theory, rank and parity are considered in conjunction with information about ovulation timing, which is conveyed by swelling size changes.

The graded signal hypothesis was tested at Taï, where female hormones were extracted from urine to test for hormonal changes that indicate ovulation (Figure 27.4). It turned out that alpha males monopolized females around the time she was ovulating (Deschner et al., 2004); the alpha must have been able to tell when ovulation was occurring. The ovulation signal, it turned out, was swelling size. Using video and 3D reconstruction, Tobias Deschner was able to show that progesterone levels, the same hormone drugstore pregnancy kits use to detect ovulation, correlated with swelling size and that, importantly, the speed at which the swelling increased slowed just before ovulation (Deschner et al., 2004). Males watched swelling changes closely and

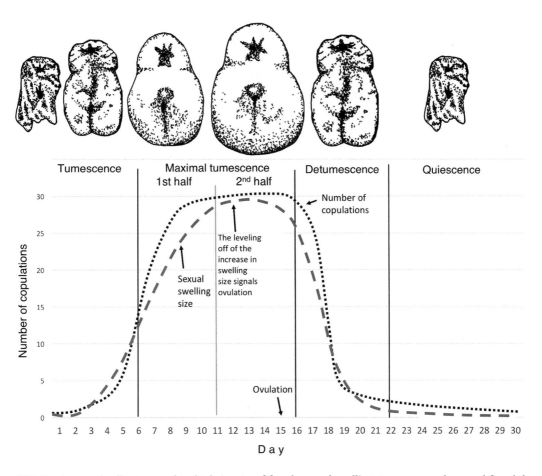

Figure 27.4 Deschner and colleagues used multiple images of female sexual swellings to measure volume and found that swelling slows as ovulation approaches. Image courtesy of Tobias Deschner, with permission. Other material by author.

competed most fiercely just as the swelling increase slowed, the point at which the female was most fertile.

Swellings, then, if a male pays attention to the rate of swelling increase, honestly signal ovulation. Alpha males monopolize females in the second half of maximal tumescence, when swellings are largest, and right through the time when their increase slows, allowing them to copulate with a female at her maximum fertility. At the same time, their close proximity and their exclusion of other males may allow them to monitor swelling size best to tell exactly when the female is ovulating and when to guard her most attentively. Other males, forced to monitor her at a distance, are perhaps less well informed. Perhaps high-ranking males benefit from excluding other males not only to control mating rights, but to gain confidence in paternity; only they have monitored females closely enough to know exactly when she ovulated. Other males are confused.

27.13 Female Imperatives

While females have no conscious control over their sexual swelling, and while their instincts drive their desire for specific males, a female can use her brains to avoid some unwise mating choices. Her promiscuity will confuse paternity; so far so good. By clever maneuvering she can place herself near desirable males at her most fertile period, and such males might then defend her from undesirable males, including relatives (Stumpf & Boesch, 2005).

The relatives part is important. Inbred infants have lower birthweight, lower survival rates, and compromised immune competence (Keller & Waller, 2002). Chimpanzees likely suffer the same consequences humans face with inbreeding; inbred humans have a 5 percent lower likelihood of living to one year (Bittles and Neel, 1994; Modell and Darr, 2002) and a 20 percent lower probability of living until 10 (Aoki, 2005). In a zoo survey across numerous species of primates, inbred matings were found to result in a 33 percent increase in mortality (Ralls & Ballou, 1982; Ralls et al., 1988). Females must be more conscientious about avoiding inbreeding because their higher investment in an infant means they have more to lose (Lieberman et al., 2003; Fessler & Navarrete, 2004).

Thus, females are strongly selected to avoid inbreeding and their instincts guide them to do just that, the same way human instincts do. She will find individuals she maintained intimate contact with during childhood, namely her brothers, to be unattractive, if not positively repugnant (Shepher, 1971). Perhaps some sort of olfactory cue that signals MHC similarity is involved in avoiding fathers: The rule of thumb would then be "avoid mating with individuals who smell like you." Deception, guile, and good planning can help her to avoid overly amorous siblings, and this must work: A recent study at Gombe found that of 62 births by females who never left the community of their birth, and thus lived among closely related males, they were no more closely related to the fathers of their offspring than were immigrant females. Somehow, females avoided close relatives and chose as fathers of their infants genetically dissimilar males (Walker et al., 2017).

Selecting the best male to father her infant is only the first of the many reproductive challenges a prospective mother faces. After four months of stressful estrus, eight months of gestation, and the rigors of birth, even at birth a new mother's investment is heavy. After birth, 50 percent of her daily energy intake will be pumped into her growing infant. These investments mean that losing an infant is a grave loss. In a world where accidents, disease, and murderous extracommunity males lurk, females are under extreme pressure to protect their infant.

Attaining high rank helps a female to secure a superior core area in the center of the community range and gives her immunity from attacks by other females; high ranking females have greater reproductive success (Pusey et al., 1997).

27.14 Male Strategies

Males compete by closely monitoring female estrous states, fighting through aggression from other males, killing infants they are unlikely to have fathered, increasing the size of the community territory (thus increasing female reproductive success; Williams et al., 2004), and protecting related infants. Border patrols and the male bonding required to execute the patrols are part of the male reproductive strategy (see Chapter 24 for aggression and Chapter 25 for the context for border patrols).

The competition surrounding females in their most fertile period selects for high levels of aggression among males, and male testosterone levels rise when competition levels are highest (Muller & Wrangham, 2001), as predicted by the **challenge hypothesis** (Wingfield et al., 1990) – the idea that testosterone levels rise to support both the physical and the psychological demands of fighting. Males must compete successfully with other highly motivated males; large body size, athleticism, endurance, and power are physical requisites of success, and strategic intelligence, aggression, wariness, attentiveness, courage, resilience, and persistence are selected for. To repeat, *high rank is highly correlated with reproductive success* (Constable et al., 2001), with about half of all infants sired by high-ranking males.

Chimpanzees are a violent species, and this is expressed not only in male–male competition, but also by male coercion of females (Thompson, 2013). By attacking females that refuse to mate with them, males encourage females to copulate with them.

Males can also attempt to identify their own offspring and aid them. At Gombe, males associated more with infants that were their own, especially early in infancy when the threat of infanticide was highest (Murray et al., 2016). Despite generally successful attempts at paternity confusion, fathers still sometimes seem to identify and aid their own infants.

27.15 Extracommunity Paternity

While males within a community have only the most tenuous confidence in paternity, it is extremely rare for an infant to be sired by males from other communities; at Gombe, one analysis showed that all 14 infants in the study were sired by community males (Constable, et al., 2001) and at Taï 33 of 34 infants were virtually certain to have been fathered by community males (Vigilant et al., 2001). Perhaps the most likely extracommunity paternity possibility is a young female dispersing to another community and returning, as Fifi did in young adulthood; such a female might return pregnant. If she never came into estrus after a long absence, however, infanticide would be a real possibility.

27.16 Chimpanzee Passion and Desire

Let me elaborate on the chimpanzee sexual psychology with which I began this chapter. Desire is quite different in chimpanzees and humans. Because humans have strong pair bonds, pairings often persist for years, across multiple births, and sometimes across lifetimes. This means that, from an evolutionary perspective, long-term reproductive potential is an important component of mate-choice, more important than reproductive potential at any one moment. For humans, because pairings are long term, forming a monogamous relationship with a young female who has not reached her reproductive peak but appears to have a high lifetime reproductive potential is worth the trade-off of sacrificing other mating opportunities. For females, trading the advantages of promiscuity, namely the advantages of (1) genetically diverse offspring and (2) choosing the optimal mate available at each fertile period for paternal investment – help feeding and caring for infants – is often enough an acceptable trade-off that pair-bonding has evolved in humans. Humans are in it for the long haul – till death do the happy couple part, as we hear in legend, song, and ceremony.

Chimpanzees, in contrast, do not form long-term pair bonds, which means that instantaneous reproductive potential, how fertile a female is today, is far more important than long-term potential. As a consequence, sexual desire focuses on the fertility of a potential mate at the moment. Among chimpanzees, **nulliparous** females, females without previous births, are not at their reproductive peak and are less desirable mates; after all, they may be infertile for years, possibly forever, and in any case, have not demonstrated an ability to bear offspring. They are less attractive than multiparous females. The greater the number of births in a female's history, the more attractive she is. Females who are quite old and therefore who have borne many infants are more attractive than the unscarred, fresh-faced, whole-eared young adult with all 10 fingers that may seem a more ideal chimpanzee female to a human observer (Muller et al., 2006). Goodall memorably recounts the riot of sex and desire following old Flo's reentry into estrus at Gombe (Goodall, 1971). Flo was estimated to be 35 or so at the time, or 47 in human years. While those of you around the age of 47 will find potential mates that age perfectly attractive, average males, i.e., males across all age classes, find human females in their mid- to late twenties to be most attractive (Mathes et al., 1985). If humans were like chimpanzees, an attractive, childless 20-year-old might be tempted to lie on her dating website, claiming to be 45 and the mother of 5. While ignored by the average male at 20, at 47 or even 57 she would find herself besieged with suitors not just near her age or even half her age, but of all ages.

27.17 Lessons

Among chimpanzees, sex – copulation – is about much more than just getting pregnant. Females are evolved to engage in sex more to protect their future infant than to conceive. I write "more," because at 1200 copulations per birth in chimpanzees, compared to six in gorillas, 1194 of those copulations are not about getting pregnant. The chimpanzee reproductive system reminds us that delaying pregnancy can increase lifetime reproductive success by avoiding the costs of infanticide and by allowing a young female to establish herself in the best situation in her new community before undertaking the burdens of motherhood. We ultra-social humans can rely on parents, grandparents, siblings, friends, spouses, and a social safety net to help us bear the costs of parenting. Not so for chimpanzees; mother is almost entirely on her own.

This is not to say that the costs of reproduction are light for human females, as any of you who have given birth know; in most cultures, the bulk of childcare, protection, feeding, and emotional investment fall on the mother, even in cultures where fathers also invest heavily. Ask your mother. Even so, the burden is much heavier for a chimpanzee mother. She must feed, teach, protect, and house (the daily night nest) her infant with no help whatsoever. We humans who have children can be thankful that we have the sort of community support any female chimpanzee would treasure.

References

Anderson MJ, Dixson AF (2002) Sperm competition: motility and the midpiece in primates. *Nature* **416**, 496–496.

Aoki K (2005) Avoidance and prohibition of brother-sister sex in humans. *Popul Ecol* **47**, 13–19.

Arcadi AC, Wrangham RW (1999) Infanticide in chimpanzees: review of cases and a new within-group observation from the Kanyawara study group in Kibale National Park. *Primates* **40**, 337–351.

Bentley GR (1999) Aping our ancestors: comparative aspects of reproductive ecology. *Evol Anthropol* **7**, 175–185.

Bittles AH, Neel JV (1994) The costs of human inbreeding and their implications for variations at the DNA level. *Nat Genet* **8**, 117–121.

Borries C, Launhardt K, Epplen C, Epplen JT, Winkler P (1999) DNA analyses support the hypothesis that infanticide is adaptive in langur monkeys. *Proc R Soc Lond B* **266**, 901–904.

Butynski TM (1982) Harem-male replacement and infanticide in the blue monkey (*Cercopithecus mitis stuhlmanni*) in the Kibale Forest, Uganda. *Am J Primatol* **3**, 1–22.

Collins DA, Busse CD, Goodall J (1984) Infanticide in two populations of savannah baboons. In *Infanticide: Comparative and Evolutionary Perspectives* (ed. Hausfater G, Hrdy SB), pp. 193–216. Chicago, IL: Aldine.

Constable JL, Ashley MV, Goodall J, Pusey AE (2001) Noninvasive paternity assignment in Gombe chimpanzees. *Molec Ecol* **10**, 1279–1300.

Crockett CM, Sekulic R (1984) Infanticide in red howler monkeys (*Aloutta seniculus*). In *Infanticide: Comparative and Evolutionary Perspectives* (eds. Hausfater G, Hrdy SB), pp. 173–191. Hawthorne NY: Aldine.

Deschner T, Heistermann M, Hodges K, Boesch C (2004) Female sexual swelling size, timing of ovulation, and male behavior in wild West African chimpanzees. *Horm Behav* **46**, 204–215.

Dixson AF (1983) Observations on the evolution and behavioral significance of "sexual skin" in female primates. In *Advances in the Study of Behavior* (eds. Rosenblatt JS, Hinde RA, Beer C, Busnel M-C), pp. 63–106. New York: Academic Press.

Dixson AF (1998) *Primate Sexuality: Comparative Studies of Prosimians, Monkeys, Apes and Human Beings*. New York: Oxford University Press.

Dixson AF (2009) *Sexual selection and the origins of human mating systems*. New York: Oxford University Press.

Dixson AF (2013) *Primate sexuality: comparative studies of the prosimians, monkeys, apes, and humans*. New York: Oxford University Press.

Domb LG, Pagel M (1981) Menstrual cycle physiology of the great apes. In *Reproductive Biology of the Great Apes: Comparative and Biomedical Perspectives* (ed. Graham CE), pp. 29–38. New York: Academic Press.

Domb LG, Pagel M (2001) Sexual swellings advertise female quality in wild baboons. *Nature* **310**, 204–206.

Dunbar RIM (2001) Evolutionary biology: what's in a baboon's behind? *Nature* **410**, 158–158.

Ellison PT (1990) Human ovarian function and reproductive ecology: new hypotheses. *Am Anthropol* **92**, 933–952.

Ellison PT, Panter-Brick C, Lipson SF, O'Rourke MT (1993) The ecological context of human ovarian function. *Hum Reprod* **8**, 2248–2258.

Emery MA, Whitten PL (2003) Size of sexual swellings reflects ovarian function in chimpanzees (*Pan troglodytes*) *Behav Ecol Sociobiol* **54**, 340–351.

Fessler DMT, Navarrete DC (2004) Third-party attitudes toward sibling incest: evidence for Westermarck's hypotheses. *Evol Hum Behav* 25, 277–294.

Goodall J (1986) *The Chimpanzees of Gombe: Patterns of Behavior*. Cambridge, MA: Harvard University Press.

Goodall J van Lawick (1971) *In the Shadow of Man*. London: Collins.

Goodall J, Bandora A, Bergmann E, et al. (1979) Intercommunity interactions in the chimpanzee population of the Gombe National Park. In *The Great Apes* (eds. Hamburg DA, McCown ER), pp. 12–53. Menlo Park, CA: Benjamin Cummings.

Harcourt AH (1991) Sperm competition and the evolution of nonfertilizing sperm in mammals. *Evolution* 45, 314–328.

Harcourt AH (1995) Sexual selection and sperm competition in primates: what are male genitalia good for? *Evol Anthropol* 4, 121–129.

Harcourt AH, Harvey PH, Larson SG, Short RV (1981) Testis weight, body weight and breeding system in primates. *Nature* 293, 55–57.

Hausfater G, Hrdy SB (1984) *Infanticide: Comparative and Evolutionary Perspectives*. New York: Aldine.

Hrdy SB (1974) Male-male competition and infanticide among the langurs (*Presbytis entellus*) of Abu, Rajasthan. *Folia Primatol* 22, 19–58.

Hrdy SB (1979) Infanticide among animals: a review, classification, and examination of the implications for the reproductive strategies of females. *Ethol Sociobiol* 1, 13–40.

Hrdy SB, Whitten PL (1987) Patterning of sexual activity. In *Primate Societies* (eds. Smuts BB, Cheney DL, Seyfarth RM, Wrangham RW, Struhsaker TT), pp. 370–384. Chicago, IL: University of Chicago Press.

Kawanaka K (1981) Infanticide and cannibalism in chimpanzees: with special reference to the newly observed case in the Mahale Mountains. *Afr Stud Monogr* 1, 69–99.

Keller LF, Waller DM (2002) Inbreeding effects in wild populations. *Trend Ecol Evol* 17, 230–241.

Keverne EB (1976) Sexual receptivity and attractiveness in the female rhesus monkey. In *Advances in the Study of Behavior*, Vol. 7 (eds. Rosenblatt JS, Hinde RA, Shaw E, Beer C), pp. 155–200. New York: Academic Press.

Knott C (2001) Female reproductive ecology of the apes: implications for human evolution. *Reproductive Ecology and Human Evolution* (ed. Ellison PT), pp. 429–463. New York: Aldine de Gruyter.

Lieberman D, Tooby J, Cosmides L (2003) Does morality have a biological basis? An empirical test of the factors governing moral sentiments relating to incest. *Biol Sci* 270, 819–826.

Mathes EW, Brennan SM, Haugen PM, Rice HB (1985) Ratings of physical attractiveness as a function of age. *J Social Psychology* 125, 157–168.

McGinnis PR (1979) Sexual behaviour in free-living chimpanzees: consort relationships. In *The Great Apes: Perspectives on Human Evolution* (eds Hamburg DA, McCowan ER), pp. 429–439. Menlo Park, CA: Benjamin Cummings.

Michael RP, Bonsall RW, Warner P (1974) Human vaginal secretions: volatile fatty acid content. *Science* 186, 1217–1219.

Michael RP, Bonsall RW, Zumpe D (1978) Consort bonding and operant behavior by female rhesus monkeys. *J Comp Physiol Psychol* 92, 837–845.

Modell B, Darr A (2002) Genetic counselling and customary consanguineous marriage. *Nature Rev Genet* 3, 225–229.

Muller MN, Emery Thompson M, Wrangham RW (2006) Male chimpanzees prefer mating with old females. *Curr Biol* 16, 2234–2238.

Muller MN, Wrangham RW (2001) The reproductive ecology of male hominoids. *Reproductive Ecology and Human Evolution* (ed. Ellison PT), pp. 397–427. New York: Aldine de Gruyter.

Murray CM, Stanton MA, Lonsdorf EV, Wroblewski EE, Pusey AE (2016) Chimpanzee fathers bias their behaviour towards their offspring. *Roy Soc Open Sci* 3, 160441.

Nascimento JM, Shi LZ, Meyers S, et al. (2008) The use of optical tweezers to study sperm competition and motility in primates. *J Roy Soc Interface* 5, 297–302.

Newton-Fisher NE (1999) Infant killers of Budongo. *Folia Primatologica* 70, 167–169.

Nishida T, Kawanaka K (1985) Within-group cannibalism by adult male chimpanzees. *Primates* 26, 274–284.

Nunn CL (1999) The evolution of exaggerated sexual swellings in primates and the graded-signal hypothesis. *Anim Behav* 58, 229–246.

Pusey A (1979) Intercommunity transfer of chimpanzees in Gombe National Park. In *The Great Apes* (eds. Hamburg DA, McCown ER), pp. 465–479. Menlo Park, CA: Benjamin Cummings.

Pusey A, Williams J, Goodall J (1997) The influence of dominance rank on the reproductive success of female chimpanzees. *Science*, 277, 828–831.

Ralls K, Ballou J (1982) Effects of inbreeding on infant mortality in captive primates. *Int J Primatol* 3, 491–505.

Ralls K, Ballou JD, Templeton A (1988) Estimates of lethal equivalents and the cost of inbreeding in mammals. *Conserv Biol* 2, 185–193.

Seuanez H, Carothers A, Martin D, Short R (1977) Morphological abnormalities in spermatozoa of man and great apes *Nature* 270, 345–347.

Shefferly N, Fritz P (1992) Male chimpanzee behavior in relation to female ano-genital swelling. *Am J Primatol* 26, 119–131.

Shepher J (1971) Mate selection among second generation kibbutz adolescents and adults: incest avoidance and negative imprinting. *Arch Sex Behav* 1, 293–307.

Struhsaker TT (1977) Infanticide and social organization in the redtail monkey (*Cercopithecus ascanius schmidti*) in the Kibale Forest, Uganda. *Z Tierpsychol* **45**, 75–84.

Stumpf RM, Boesch C (2005) Does promiscuous mating preclude female choice? Female sexual strategies in chimpanzees (*Pan troglodytes verus*) of the Taï National Park, Côte d Ivoire. *Behav Ecol Sociobiol* **57**, 511–524.

Takahata Y (1985) Adult male chimpanzees kill and eat a male newborn infant: newly observed intragroup infanticide and cannibalism in Mahale National Park, Tanzania. *Folia Primatologica* **44**, 161–70.

Thompson ME (2013) Reproductive ecology of female chimpanzees. *Am J Primatol* **75**, 222–237.

van Schaik CP, Kappeler PM (1997) Infanticide risk and the evolution of male–female association in primates. *Proc Roy Soc Lond B* **264**, 1687–1694.

Vigilant L, Hofreiter M, Siedel H, Boesch C (2001) Paternity and relatedness in wild chimpanzee communities. *Proc Natl Acad Sci* **98**, 12890–12895.

Walker KK, Rudicell RS, Li Y, et al. (2017) Chimpanzees breed with genetically dissimilar mates. *Roy Soc Open Sci* **4**, 160422.

Watts D (1998) Coalitionary mate-guarding by male chimpanzees at Ngogo, Kibale National Park, Uganda. *Behav Ecol Sociobiol* **44**, 43–55.

Williams JM, Oehlert GW, Carlis JV, et al. (2004) Why do male chimpanzees defend a group range? *Anim Behav* **68**, 523–532.

Wingfield JC, Hegner RE, Duffy AM, Ball GF (1990) The "Challenge hypothesis": theoretical implications for patterns of testosterone secretion, mating systems and breeding strategies. *Am Nat* **136**, 829–846.

Wrangham RW (1993) The evolution of sexuality in chimpanzees and bonobos. *Hum Nat* **4**, 47–79.

28 Into the Light
Semliki Chimpanzees

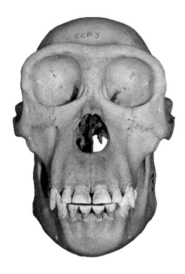

Photo by author

Paleontologists are a contentious lot. Consider the simple question "How many species are there in the genus *Homo*?" Paleontologists cannot agree. When it comes to more complicated problems all hell breaks loose. Why did the traits that most characterize our species evolve? What drove the evolution of big brains? Of hairlessness? Of language and pair bonding? Ask 10 paleontologists and you will get 10 answers.

The origin of upright walking – alternate-foot, striding bipedalism – was a giant leap in human evolution; it is perhaps no surprise, then, that there is no consensus on which evolutionary pressures drove its origin. In 1994 I proposed that bipedalism evolved in response to fruit distribution in dry habitats. In a chapter that appeared in 1998, I speculated that australopiths have robust faces due to the same ecological circumstances that selected for bipedalism – dry habitats. As I was writing that 1994 paper, I began to block out some important tests of the hypothesis. If I were to examine a population of chimpanzees that was confined to dry habitat, a habitat like that of australopiths, what would I find? I would expect them to engage in much more bipedalism than forest chimpanzees, and I would expect their foods to be tougher. I would expect their anatomy to vary, however slightly, in the direction of australopiths: They should have more robust faces and anatomy that reflected bipedalism. The perfect test would look at a population of chimpanzees that had been genetically isolated in a dry habitat for thousands of years, but I doubted such a population existed, and I still do.

However, perhaps genetic isolation is not necessary. One of the principles driven home to me during my brief tenure as a research associate at the Center for Human Growth and Development is that the skeleton is more plastic that most scholars recognize. My mentor, Loring Brace, always emphasized that culture – how food was processed both in and out of the mouth – always left recognizable traces in the face and jaws, and sometimes even other parts of the human skeleton. Perhaps chimpanzees in dry habitats, even if not genetically isolated, could be expected to show some trace of heavy chewing and bipedal posture. I resolved to find a dry-habitat population in which I could test this hypothesis.

After an aborted attempt to study chimpanzees in Mali, Richard Wrangham reminded me of a dry habitat population in Uganda, a country where I had already worked during my postdoc. In 1996, I began studying chimpanzees at the dry chimpanzee site Toro-Semliki Wildlife Reserve and I have found that these chimpanzees do indeed share some anatomy with the earliest bipeds.

28.1 *Australopithecus afarensis*: First Biped?

Let me dig down to a stable foundation for our discussion of the evolution of bipedalism, building on two scraps of evidence that stir up paleontological passions the least of any I can think of. If we took a vote on which fossil species can be identified most confidently as the first upright-walking biped, the well-known 2.9–3.6-million-year-old species *Australopithecus afarensiss*, the Lucy fossils, would win in a romp. I state this with some confidence because the evidence that the next oldest fossil was a biped, the 4.2-million-year-old *Australopithecus anamensis*, is limited to the lower part of the knee-joint, the part of your lower leg bones that the thigh bone or femur rests on. The tibial platform of *Au. anamensis* is reinforced like that of modern humans, presumably to allow it to bear the weight of the entire body during bipedal walking, first on one foot, then on the other. The knee is not modern, though; it shares traits with chimpanzees that are linked to more flexible knees. Furthermore, the two most diagnostic skeletal elements, the pelvis and the distal (i.e., lower) femur, are yet undiscovered. The lack of one of these killer bipedalism indicators makes *Au. anamensis* a less appealing first-biped candidate than the Lucy fossils. Fossils that are still earlier than *Au. anamensis* are said to be hominin, and they are interpreted by their discoverers as bipeds, but these claims have been widely contested: The fossils have a mosaic anatomy that looks very ape-like (Hunt, 2015b).

Let us start, then, with *Australopithecus afarensis*, the Lucy fossils.

28.2 *Australopithecus afarensis* Morphology Is Not Controversial; Interpreting It Is

The second point that is widely agreed upon is the morphology of the Lucy fossils (see the excellent original description of Lucy and other fossils by Johanson and colleagues [1982a, 1982b], Tim White and colleagues [1981], and Jack Stern and Randy Susman [1983]). There are disagreements, but many fewer than you might expect considering that paleontologists can hardly agree on what day of the week it is.

While the morphology is agreed upon, there is tremendous controversy surrounding what behaviors this morphology suggests. It might be straightforward if the fossils were almost exactly human-like – if they looked like humans, they must have behaved like humans. Unfortunately, they are not. They have some traits that are undeniably chimpanzee-like, while others more resemble humans.

How do you figure out what an organism did when it resembles nothing alive today? I tend to interpret morphology as having functioned in fossils as it does in living animals. For instance, among living primates "palmarly curved" finger bones, bowed so that they are concave on the palm side, are found among primates that engage in suspensory behavior – armhanging. I interpret Lucy's curved fingers as meaning she spent some time armhanging (Hunt, 1994, 1996, 1998, 2008, 2015b). Some of my colleagues suggest instead that the chimpanzee-like hands mean little or nothing; they argue that *Au. afarensis* had so recently left the trees that we should expect to find some ape-like traits merely as primitive retentions, especially in parts of the anatomy that bear little weight, such as the hands (Sayers & Lovejoy, 2008). Lucy's ape-like hands were, in other words, useless anatomy that would gradually become more modern over time as evolution did its job. Paleontologists, then, agree about the feature, but disagree about what it means.

Let us look more closely at each of the traits over which we paleontologists argue. As we do, you will begin to see why studying an animal in the same habitat that Lucy lived in helps us to understand human evolution. I will describe the morphology first, then provide one or two interpretations of its function.

28.3 *Au. afarensis* Skull

Starting from the top of the head, *Au. afarensis* has a cranial capacity slightly greater than that of chimpanzees, though it can hardly be called a difference: $400\,\text{cm}^3$ versus $350\,\text{cm}^3$.

The face, however, *is* quite different from that of chimpanzees. The zygomatics are tall and heavily constructed (Figure 28.1). Looking down from the top, the lateral-most parts of the zygomatics are drawn so

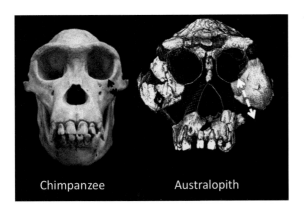

Figure 28.1 Australopith faces are heavy and reinforced compared to chimpanzees. Zygomatic height (dotted lines) is much smaller in chimpanzees than australopiths. Image by the author. (A black and white version of this figure will appear in some formats. For the color version, please refer to the plate section.)

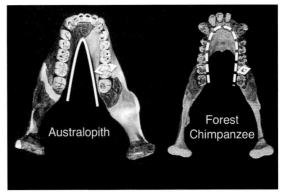

Figure 28.2 The australopith's mandibular body is so thick that when viewed from above the inner surface of the mandible makes a V shape; forest chimpanzees have a thick body that is hardly thicker than the width of the molars, even though the molars are small so that the inner surface describes a U. Image courtesy of Tim White; used with permission. (A black and white version of this figure will appear in some formats. For the color version, please refer to the plate section.)

far forward that the face is beyond flat; it is concave. If you held the A.L. 444 skull face-up and poured a cup of water onto it, the water would drain into the nose. This morphology gives extra room for the temporalis muscle, the jaw-closing muscle on the side of the head; all bones in the face and jaws are heavy and reinforced. Chimpanzee zygomatics, in sharp distinction, are short top-to-bottom when viewed from the front, less heavily built (Figure 28.1); when viewed from the top they are angled back from the nose.

The australopith mandible has an extraordinarily thick corpus, or body (arrows, Figure 28.2). The molars of *Au. afarensis* are larger than those of chimpanzees, but even so the bone holding those big molars is even wider. The fat mandibular corpus makes the mandible appear V-shaped when viewed from above (Figure 28.2), whereas the chimpanzee mandible is U-shaped. Australopith incisors are smaller than those of chimpanzees, narrowing the front of the mandible.

28.4 *Au. afarensis* Thorax and Pelvis

The ribcage of australopiths is more chimpanzee-like than human-like (Figure 28.3). When viewed from the front, both are triangular; in 3D the ribcage is cone-shaped. Humans have a barrel-shaped thorax. Rather than blade-shaped, the ribs of *Au. afarensis* are round in cross-section, leaving a large space between each rib (Figure 28.3). The spine, notably the lumbar segment, is smaller in diameter than the human spine, as can be seen looking at the lumbosacral articular surface, the oval- to bean-shaped spot on the sacrum of the pelvis, on which the lumbar vertebrae sit. These are both chimpanzee traits (Figure 28.4). The length of the lumbar vertebral segment, however, is human-like. There are five (like humans) lumbar vertebrae, and probably even six. Australopiths have long backs, while chimpanzees have extremely short backs.

Considering the pelvis as a whole, *Au. afarensis* is remarkably human-like (Figures 28.3 and 28.4), though it does have attachments for the gluteals that face slightly more posteriorly, and it is shorter front-to-back (Figure 28.4). This front-to-back shortness makes the *Au. afarensis* pelvis relatively wide (Figure 28.4), giving Lucy a pear-shaped body profile: Viewed from the front she has narrow shoulders and wide hips. Wide hips create more stress at the hip joint, but the pear-shaped body plan lowers the center of gravity, giving female australopiths better balance on unstable supports such as thin branches.

Lucy's scapula is intermediate in shape between humans and chimpanzees (Figure 28.5), which means narrower than humans but wider and more triangular

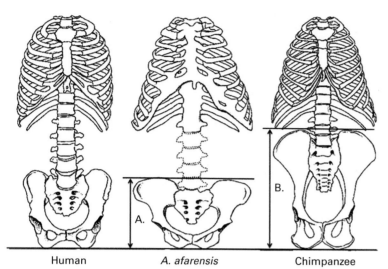

Figure 28.3 The thorax of *Au. afarensis* is cone-shaped, like chimpanzees, and the ribs are round in profile, leaving more of a space between them than can be seen in the blade-like human ribs. Move down to the pelvis though, and Lucy strongly resembles humans in having a short (dimension A) bowl-like pelvis, not the tall (B) pelvis of chimpanzees. After Hunt, 1997; with permission.

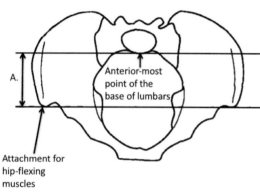

Figure 28.4 Top view of human (above) and Lucy (below) pelves. In humans the distance between the base of the lumbar vertebra, the part of the pelvis the spine rests on, is farther from the attachment point of muscles in the front of the thigh that raise the knee (i.e., flex the hip). Both pelves are bowl-shaped, but the australopith pelvis is wide compared to the front-to-back dimension. From Hunt, 1994; with permission.

than chimpanzees. The Lucy shoulder joint (i.e., the glenoid fossa) is tilted upward, nearly as much as in chimpanzees. But there is more to this story; larger australopiths are more human-like. Perhaps male australopiths were more human in their locomotion and more terrestrial, whereas females spent more time in trees.

28.5 *Au. afarensis* Arm and Hand

The australopith humerus is quite chimpanzee-like (Figure 28.6). The shaft has thick walls and there is a large lateral supracondylar crest (the attachment for the brachioradialis) muscle. It has a deeply incised elbow joint surface (Figure 28.6), describing an "m" shape, compared to the less deeply notched, more undulating surface in humans. This gives the joint greater integrity and stability when the arm is twisted or when the elbow is subjected to erratic or unpredictable stresses common while moving on or clinging to erratically placed branches. Chimpanzees and australopiths also have a perforated olecranon fossa (Figure 28.6); this allows the olecranon process, a hook on the ulna that attaches the lower arm to the humerus, to wrap around the joint surface so completely that it pierces the humerus. When armhanging, the body is essentially hung from this

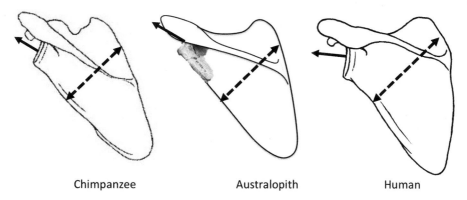

Figure 28.5 The australopith scapula is intermediate between humans and chimpanzees, with a tilted up glenoid fossa (solid arrow) and slightly more chimpanzee breadth (dotted line). Images by the author.

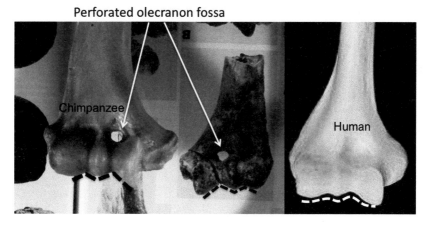

Figure 28.6 Chimpanzees (left) and australopiths (center) share a perforated olecranon fossa, allowing the olecranon process of the ulna to completely wrap around the humeral trochlea; the articular surface is M-shaped (black dotted line). Human humeri have a curved, undulating articular surface (white dotted line). Photos by the author. (A black and white version of this figure will appear in some formats. For the color version, please refer to the plate section.)

hook; it is a suspensory adaptation. A wire clothes hanger with a completely curved hook can hold more weight than one where the hook is straightened.

Australopiths have a hand and wrist much like that of a short-fingered chimpanzee. While human and chimpanzee wrists have many traits in common, the chimpanzee styloid process has a long, robust hook (Figure 28.7), reinforcing the attachment points of the ulna to the wrist bones to better resist forces exerted during climbing and armhanging; *Au. afarensis* also has a long, robust styloid, like chimpanzees (Johanson et al., 1982b: 385). The pisiform, a pebble-like bone you can feel under the skin on the outside of your palm, just where the dermal ridge part transitions to regular skin, is long and rod-shaped in chimpanzees and also in *Au. afarensis* (Johanson et al., 1982b: 384). This bone is a little lever that increases the power of a muscle that flexes the wrist, useful during climbing and armhanging (Figure 28.7).

The fingers of *Au. afarensis* are not only curved, as we mentioned above, but they have large **flexor sheath ridges**, anchor points for the ligaments that hold finger-flexing muscles in place. The tendons that pull your fingers into a fist run along the palmar side of your fingers; in humans they are ligamentous strips, much like a belt or strap. In chimpanzees, the ligaments are several times as thick as in humans, so robust they are round in cross section (Figure 28.8).

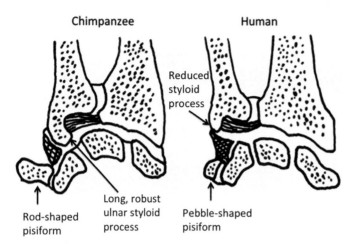

Figure 28.7 Cross-sections of the wrists of humans and chimpanzees are similar, both having great flexibility and the capacity for 180°+ rotation; chimpanzees and australopiths share a rod-shaped pisiform and a long, robust ulnar styloid process. From Hunt, 2016, with permission.

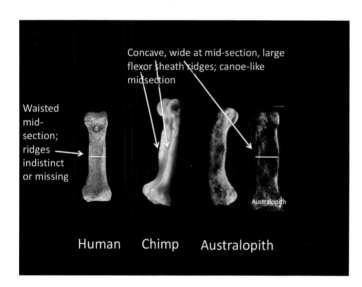

Figure 28.8 Phalanges (finger bones) of a human (palmar view), chimpanzee (three-quarter view), and australopith (side and palmar views) fingers have large flexor sheath ridges, and are curved and expanded side to side at mid-section. Australopith fingers are slightly shorter versions of chimpanzee fingers. Human, chimp photos by the author; australopith images courtesy of Wiley; with permission. (A black and white version of this figure will appear in some formats. For the color version, please refer to the plate section.)

These ligaments bear weight when you climb, hang by one arm or grip something. Humans have no flexor sheath ridges (Figure 28.8). The robust fingers of chimpanzees make their finger bones look fat in the middle; the flexor sheath ridges make the palm side of the finger bones concave; you could scoop up water with the canoe-shaped chimpanzee finger bones. Australopith fingers are shorter than those of chimpanzees, but otherwise similar, with a "fat" midsection, distinct curvature, and large flexor sheath ridges (Figure 28.8).

28.6 *Au. afarensis* Hindlimbs

Au. afarensis had short legs (Figure 28.9) and a knee joint that was less elongated front to back, and therefore more ape-like (Abitbol, 1995). That is, the human knee joint is not quite flat enough that a femur would stand up on its own on a flat surface, but it is close. Chimpanzee knee joints are round in side-view. When viewed from the front, the australopith femur is angled in, or **valgus**, as it is in modern humans (Figure 28.9).

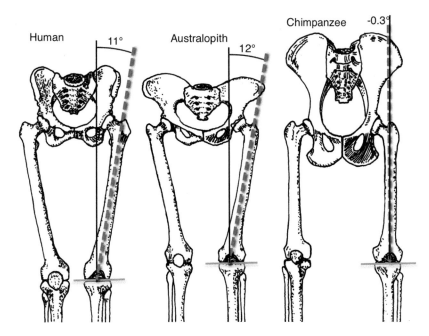

Figure 28.9 Humans (left) and australopiths (middle) have an angled-in femoral shaft (dotted line) that gives them a knock-kneed appearance, a morphology known as a valgus knee. Human femoral shafts meet the horizontal knee joint surface (gray horizontal lines) at 11° from vertical (solid vertical line); australopith femoral shafts are 12° from vertical, whereas chimpanzee shafts (right) are slightly negative. This condition is purely developmental; humans are born with a chimpanzee-like knee and develop the valgus condition only as they begin to stand and walk; invalids retain the chimpanzee condition. After Hunt, 2015a; with permission.

The toes of australopiths are long, curved, and chimpanzee-shaped but in length they are intermediate. The calcaneus (heel bone) is large and human-like.

As an overview, by 2.9–3.6 Mya, human ancestors had evolved upright, striding bipedalism, yet compared to humans they had shorter legs, longer toes, wider hips (at least in females), and more flexible knee joints. They could not have been fully evolved bipeds from the view point of energetic efficiency. Their cone-shaped thorax and powerful arms and hands differed from chimpanzees only in finger length. They had robust faces and jaws.

28.7 Selection Against Bipedalism

Theorists who wish to explain the evolution of bipedalism face a challenge that entangles few other human characteristics. Compared to quadrupedalism, bipedalism is a means of moving around that is slow and unstable *even in its fully evolved, modern form*, but must have been terribly inefficient and clumsy in its earliest evolutionary stages (Figure 28.10). If an organism adopts a new mode of locomotion, the new mode cannot be as efficient or as capable of long-distance movement as the mode for which the animal is adapted. A lion walking upright on its hindlegs and an orangutan galloping – these are awkward locomotor behaviors that the animal is not evolved for and are inefficient and fatiguing.

There is, then – whenever new selection pressures favor adopting a new way of moving – a **transition barrier**, a steep evolutionary hump that must be surmounted before endurance and efficiency are achieved. Whatever selective force it was that led to the evolution of bipedalism, it must have had a powerful offsetting advantage to surmount the transition barrier. This is a profound challenge to most bipedalism-origin hypotheses. It is most apparent in

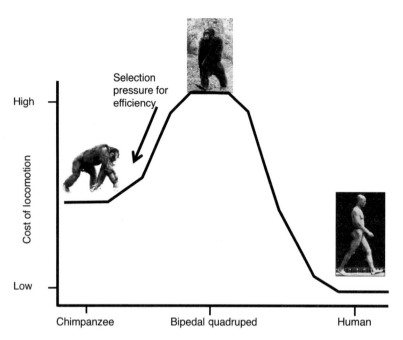

Figure 28.10 When a chimpanzee-like quadruped is under selection for greater efficiency, surmounting the inefficient condition that must inevitably precede a fully evolved type of bipedalism solely due to selection for efficiency is impossible. A quadruped cannot reach the more efficient human mode because when it stands it is inefficient – because it is a biped. We have many examples of quadrupeds that are more efficient than humans; they have a narrow body plan and a locomotion driven by energy stored in long tendons. Illustration by the author.

hypotheses that depend on the efficiency of bipedalism as the advantage that led to its evolution: Evolutionary pressures would push most animals selected to locomote more efficiently toward refining quadrupedalism, not toward adopting a new, inefficient, fatiguing, and unstable (at least in its early forms) locomotor mode (Figure 28.10). We will deal with this issue in more detail below.

28.8 Bipedalism Origin Hypotheses

The first bipedalism origin hypothesis was that of Darwin (1871). In his **tool-use hypothesis,** he hypothesized that two of the most striking human uniquenesses, enormous brains and upright locomotion, were linked to yet another characteristic, believed at the time to be uniquely human: tool use. Quite the opposite of our current understanding, Darwin suggested that the first unique human trait to evolve was a large brain, and that it evolved while pre-humans were still ape-like tree-dwellers. Large brains in turn led humans to conceive of tools, which required the hands to be free to carry and manipulate the tools, thus selecting for bipedalism. Canines were reduced in humans because hand-held weapons replaced them. This first truly evolutionary bipedalism-origin hypothesis was disproven by the fossil record: Human-sized brains evolved only in the last million years, whereas the first stone tools appeared millions of years earlier. In other words, bipedalism appeared long before big brains evolved or stone tools were invented.

Sir Arthur Keith's **suspensory hypothesis** (Keith, 1891, 1923) grew from the observation that gibbons held their legs extended during their distinctive hand-over-hand suspensory locomotion ("brachiation"). Keith suggested human bipedalism might be the end-product of a trend for progressively more erect posture, from quadrupedalism in monkeys, to a more erect quadrupedalism in chimpanzees (due to long arms), to a quite erect posture during brachiation and finally striding bipedalism among humans. Possibly he also noticed that gibbons sometimes walk bipedally for a step or two. Keith documented more than 80 traits shared by apes and humans that are not shared with other primates. Keith's suspensory hypothesis was said to be dependent on a now-debunked concept, evolutionary inertia, to propel humans from the gibbon-like stage of bipedalism to a fully human one. Evolution, in fact, has no inertia; a species does not continue to evolve in a particular direction simply because past evolutionary pressure pushed it in that direction.

Some parts of Keith's hypothesis were scavenged and fused with Darwin's tool-use hypothesis to model bipedalism as having evolved from a gibbon-like ancestor as it began to include tool use in its daily behavior. This hypothesis runs up against the same problem, that bipedalism came long before stone tools.

Raymond Dart (1959) refined and fleshed out the Darwin–Keith chimera, producing what is often referred to as the **killer-ape hypothesis**. He suggested that erect posture evolved to increase sight lines by raising the eyes above tall savanna grasses; at the same time, *a la* Darwin, it frees the hands to better dispatch prey with bone, horn, or stone tools, rather than with sharp teeth. Throwing, Dart felt, was a vital hominin adaptation. The killer-ape hypothesis conflicts with evidence that bipedalism evolved well before stone tool use and that large, flat australopith teeth are adapted for frugivory, not carnivory.

The **social display hypothesis** (Livingstone, 1962) suggests that bipedalism evolved because it increased the apparent size of the biped, increasing social status and thus access to food or mates; or perhaps it also intimidated predators. This hypothesis is inconsistent with primate behavior; when bipedalism is part of social display, upright posture is used selectively and infrequently. Subordinate animals – nearly the entire social group, of course, since there is only one alpha – could no more intimidate a dominant with such a display than a 10-year-old human could intimidate an adult with an aggressive posture. Aggressive signals by subordinate individuals would draw reprisals, not fear. It reminds me of David Attenborough's footage of a praying mantis aggressively threatening a loris. He went down (the gullet) fighting.

Some saw bipedalism as evolving to free the hands to carry either infants or food items. The **infant-carriage hypothesis** (Tanner, 1981; Iwamoto, 1985). focused on the unique human need to help infants cling to mother, supposedly made worse by bipedalism. If early bipeds had large brains, they might have needed more help, because large-brained humans are more helpless at birth. Brain size among early bipeds was small, so hominin infants should not be expected to be any more helpless than chimpanzee infants. All monkeys and apes carry infants, but none has evolved bipedalism. In fact, early in life monkey and ape infants are carried on the belly, and clinging to a near-horizontal torso must require even more assistance than clinging to a vertical torso.

The **food-carriage hypothesis** holds that bipedalism freed the hands to carry food to mates, relatives, or group members (Bartholomew & Birdsell, 1953; Etkin, 1954; Isaac, 1978; Lovejoy, 1981). Lovejoy emphasized the disadvantages of upright locomotion, focusing attention on the extraordinary circumstances that must have been present to select for this slow, unstable positional behavior. Lovejoy viewed early hominins as having distinct sex roles; males gathered food and shared it with a monogamous mate and their offspring. Monogamy might account for the lack of sexual swellings in humans (Etkin, 1954; Lovejoy, 1981); ovulation was thus hidden, encouraging persistent male attention.

The food-carriage model suffers from a number of weaknesses. Carrying occupies the arms and makes further gathering difficult; if food items are small or can only be collected in trees, carrying is difficult or impossible – imagine carrying a couple of pounds of blueberries in your hands while gathering more. Among monogamous primates, males and females have similar ecological and social roles, so they vary little in body size or morphology, whereas australopith males are twice the body weight of females. Females benefit from genetic diversity in offspring, and males who are off gathering while their partner wanders alone could have little confidence in paternity, discouraging male parental investment, a key feature of the food-carriage model. Many other animals have solved the carriage problem without evolving bipedalism, typically by swallowing the food and regurgitating it later. If early hominins had carrying implements this might seem to solve some of the food-carriage inconsistencies, but carrying devices would preclude the need for the hands in carrying.

Hardy (1960) hypothesized that bipedalism evolved as a wading posture. However, in its most popular permutation (Morgan, 1972) the **aquatic ape hypothesis** cobbled together explanations for unique human traits at different stages of human evolution as having evolved for a variety of aquatic activities, including diving, swimming, and wading. Sparse body hair was said to be an aquatic adaptation; however, aquatic otters and seals have hair, whereas terrestrial elephants do not. Hair shaft orientation was said to be evolved to reduce drag, but human hair direction is the same as other primates. While robust faces, large molars, the pattern of dental microwear, and the association of hominins with paleohabitats bearing abundant aquatic resources are consistent with a **wading/underground storage organ collection hypothesis** (Wrangham et al., 2009), against the hypothesis is the fact that in shallow water quadrupedalism would function better than bipedalism, since it would put the hands near the food resource, on the pond floor, and would orient the torso in a better position to pull up plants; in deeper water, the natural buoyancy of the thorax would reduce stresses that might require bipedal anatomical adaptations, since facultative bipedalism would be adequate for collecting and keeping the head above water, while the much greater stresses during terrestrial locomotion would select for retaining efficient quadrupedal adaptations. Such a full-time foraging strategy would be less attractive due to the dangers of hippos and crocodiles, and if it were only a now-and-then strategy it is difficult to imagine evolving a new locomotor anatomy to support a rare behavior. Also, contrary to the hypothesis, no other wading or fully aquatic mammals, including primates that forage in shallow water, have evolved bipedalism.

We have already met the **locomotor efficiency hypothesis** (Rodman and McHenry, 1980) in brief. Bipedalism, this argument goes, is more efficient than quadrupedalism. When human ancestors found themselves in a dry habitat where foods had become widely dispersed, strong selection pressure for efficient locomotion selected for bipedalism. Bipedalism was thought to be more efficient because by stacking all our body elements over extended knees and hips, we could rely on bone and ligament to bear weight, rather than supporting it with muscle action. Other animals – your pet dog or cat are good examples – have limbs that are partly flexed and therefore require muscle action to bear weight; try walking with your knees bent.

Experiments have gradually made it clear that bipedalism is not particularly efficient. Many flexed-joint mammals, including horses, dik-diks, wildebeest, ostriches, kangaroos, and even large dogs are more efficient than humans (Taylor et al., 1982; Heglund, 1985). But these data are while walking, what about running? Aha, you might be thinking, he was hiding the running data. Not so; we are even less efficient as runners, less efficient than an even *longer* list of quadrupeds (Taylor et al., 1982; Heglund, 1985). But, you may think, "if humans are so inefficient, how could we be such great marathoners?" I, too, am quite impressed by our long-distance running ability. We are among the greatest if not *the* greatest of endurance runners. However, this is not because humans are efficient, but because we are among the best at shedding heat as we run long distances. During sustained locomotion other species tire not because they have run out of fuel or are in oxygen debt, but because their body core temperatures rise to dangerous levels (Carrier, 1984; Wheeler, 1991; Bramble & Lieberman, 2004). We stay cool by having more sweat glands than other animals and by using evaporative cooling more effectively (Carrier, 1984; Wheeler, 1991; Bramble & Lieberman, 2004). We are "by far the best known of all animals" at dealing with heat stress, according to Newman, who explained our endurance proficiency in one of the most delightful titles of any scientific article ever published, "Why man is such a sweaty and thirsty naked animal" (Newman, 1970).

In review, there appears to be an insurmountable transition barrier that prevents a quadruped from arriving at the promised land of increased efficiency via bipedalism – they would have to become less efficient before becoming more efficient. Selection cannot anticipate future adaptations this way; it works with the best solution available at the moment.

Efficiency hypothesis advocates responded to criticisms like those above with the argument that bipedalism need not be more efficient than the quadrupedalism of *all* mammals to gain an advantage, just more efficient than they were as quadrupeds (Pontzer et al., 2009). But think what this means: the proto-biped must be more efficient when utilizing a locomotor mode for which it is not adapted (bipedalism) than the one for which it is adapted (quadrupedalism), otherwise the transition barrier blocks them. If bipedalism were truly more efficient even in some chimpanzees, those individuals would walk bipedally during long-distance travel. Of my 1828 travel observations, 0 were bipedal (Hunt, 1989). In my view, only highly evolved bipeds are more efficient than quadrupeds.

Lastly, regarding the efficiency hypothesis, we can turn to other species as we consider it. If bipedalism is a solution to the need for long-distance locomotion, particularly when a primate moves into a savanna habitat, surely hamadryas baboons, with their immense day ranges, should have evolved bipedalism. Likewise, other open-habitat, long-distance primates such as vervets, geladas, and patas monkeys should be bipeds.

I may be undermining my favorite hypothesis, but let me add a new wrinkle to the efficiency argument. Consider Keith's **gibbon gambit**: What if the human–chimpanzee last common ancestor (LCA) was neither a biped *nor* a quadruped? When not brachiating, gibbons sometimes use a type of "quadrupedalism" that might better be called "assisted bipedalism." They walk upright, with their legs extended, using their hands to stabilize themselves . Not uncommonly, they take a few steps on top of branches before resuming brachiation. Their arms are so long and so poorly suited for walking quadrupedally that when they come to the ground they hop or walk bipedally. But awkwardly.

If the common ancestor of humans and chimpanzees was a gibbon-like creature, walking on the ground might have been equally clumsy whether bipedal or quadrupedal. To move terrestrially they would have had to go one way or the other. Note that this is not really an efficiency hypothesis, but we have returned yet again to Keith's suspensory pre-adaptation hypothesis. Or perhaps Keith's hypothesis and the locomotor efficiency hypothesis are the same hypothesis. In any case, to further understand this hypothesis we need fossils of the LCA. *Ardipithecus* has extremely long forelimbs and might fit this model. Earlier possible LCAs such as *Ouranopithecus*, lacking as it does suspensory adaptations, do not fit with this model.

The **thermal radiation avoidance hypothesis** suggests that bipedalism evolved to reduce exposure to solar radiation (Wheeler, 1991). A bipedal hominin exposes only 7 percent of its surface to sunlight, whereas quadrupeds expose 20 percent. Lifting the thorax a few feet from the ground has great benefits for cooling; wind speed is greater and temperatures lower only a meter above ground level (Wheeler, 1991). In 1985, when I was planning my dissertation, I had independently conceived of this model and included a column on my data sheet to note whether my study subject was in the sun or not. I expected to find that chimpanzees were more bipedal when in sunlight. It turned out that, other than at dawn or dusk – when it was cool – chimpanzees avoided the sun. I only recorded "sun" four times in my year of observation. The target was quadrupedal each time. I gave up the model as I considered this and the fact that Lucy had a wide, pear-shaped body plan; living species adapted to high-heat environments have a slender body plan (narrow hips, long limbs) that both increase locomotor efficiency and better dissipate heat. The earliest bipeds known to me at the time had short legs, wide hips, and robust upper bodies, suggesting a heat-generating, inefficient locomotion and heat-retaining pear-shape thorax inconsistent with adaptation to heat stress. Further, reduction in thermal radiation absorption accrues only when exposed to direct sunlight, and when the sun is directly overhead. Few animals are active at noon, and primates tend to avoid direct sun when the weather is hot. Hominin habitats were likely partly wooded, offering further protection from solar exposure. Wheeler's heat-stress hypothesis better explains the body plan of the long-legged, slender early *Homo* fossils rather than the evolution of bipedalism.

Cliff Jolly's **seed-eating hypothesis** (Jolly, 1970) proposed that early hominins stood erect to gather grass seeds, and that the small size of seeds demanded a bipedal, alternate-hand plucking action to achieve a reasonable rate of ingestion. Robust jaws and teeth among early hominins and dexterous hands in living humans were hypothesized to have evolved for seed gathering.

It was a fascinating idea, but the physical attributes of grass were its downfall; grass seeds need not be gathered bipedally, since grass stems are easily pulled over to allow gathering from a sitting or squatting posture. This is what primates actually do when they harvest resources like this. Hominin dental microwear suggests a diet of fruit, leaves, and piths rather than seeds. A completely terrestrial adaptation is at odds with arboreal australopith adaptations such as cone-shaped torsos, curved, robust fingers, ape-like scapulae, long, curved toes, and short hindlimbs.

The **small-object feeding hypothesis** (Rose, 1976) substituted small-diameter fruits for grass seeds and bushes for grass stems. Compared to grasses, bushes are less easily pulled over and therefore require bipedalism. When trees are distributed in semi-monospecific stands, bipedal shuffling to move between trees hypothetically uses less energy than repeatedly lowering the thorax to walk quadrupedally then raising it to gather (Wrangham, 1980). In the past I have combined these and called them the Jolly/Rose/Wrangham hypothesis; a weakness is that australopiths have many chimpanzee-like arboreal adaptations; if early hominins were gathering food from the ground, we would not expect them to have short legs and broad hips which would make locomotor bipedalism more fatiguing, more energy-hungry, more stressful on the joints, or all three of these things.

28.9 The Postural Feeding Hypothesis

As I analyzed my data, the heat-stress avoidance hypothesis lay in ruins around my computer's feet. Chimpanzees were rarely in direct sunlight and when they were they were quadrupedal. But in accord with the Jolly/Rose/Wrangham hypothesis, I had observed that the most common context of bipedalism was feeding. But there was a fly in the ointment. These were hypotheses about behavior on the ground, but a little over half of my observations of bipedalism were in the trees (Hunt, 1994). As I pondered this, I noticed something amazing: My bipedal observations, whether the feeding was on the ground or in the trees, involved fruits from the same species of trees. These trees were short, their fruits were small, and they were found in the driest part of the chimpanzee habitat, the environment widely regarded as that which australopiths occupied (Cerling et al., 1997, 2011; Passey et al., 2010).

I went back to my data sheets to look for notes and began to recall my observations. Small, short-statured trees elicited arboreal bipedalism because small branches were too flexible to allow the preferred feeding posture, sitting. If a chimpanzee had tried to sit in these bushy trees, the small-diameter, flexible branches would have bent until they dumped the sitter out. Instead, they gripped the twigs with their lateral toes, stood bipedally, and reached up to grip an overhead branch, since the small-diameter branches were too unstable to stand on without help (Figure 28.11). How might they move about in these trees? The small branches are too flexible to walk on quadrupedally, but they are traversable if you grip with your toes and stand, balancing your center of gravity above the two supports while your toes grip below. My colleague Susannah Thorpe and coauthors found that orangutans were more likely to walk bipedally when moving among small-diameter, compliant supports (Thorpe et al., 2007).

Bipedalism does nothing to help a chimpanzee who is standing on the ground in their typical habitat, since their food would be 20 m and more above their head. In dry habitats, trees are short and I had observed chimpanzees standing to increase their reach, just as Jolly/Rose/Wrangham would have it. If a chimpanzee stood to pluck a large fruit from a short tree, they would then sit down to eat it. The small size of dry-habitat fruit, however, precludes plucking and sitting; no sooner would the feeder sit than they would need to stand to pick another fruit – better to stay erect and avoid the cost of deep-knee bends. Wrangham added the locomotor aspect of this model; as the feeding site was depleted of fruit, the bipedal forager would be expected to take a few steps to another cluster of fruit rather than dropping down onto all fours, walking a step or two, and then expending the energy required to raise the torso (Wrangham, 1980).

Thus I had grafted an arboreal addendum onto the Jolly/Rose/Wrangham small-object-feeding-

Figure 28.11 The postural feeding hypothesis holds that bipedalism evolved as an adaptation to harvesting small fruits from short trees. Standing up allows you to reach fruits better when on the ground. When in the trees, standing up balances your body weight over your feet; gripping branches with the lateral toes (no gripping big toe is necessary) and stabilizing the body by gripping an overhead branch in a semi-armhanging manner allows harvesting from branches too small and compliant to sit or walk on quadrupedally. Image by the author.

shuffling-between-feeding-sites hypothesis and rebranded it as the **postural feeding hypothesis** (Hunt, 1996) to remind everyone that this feeding hypothesis included bipedalism both when in the trees and on the ground.

Other contexts – social display, tool use, carrying, throwing, and peering over obstacles – each constituted only 1–2 percent of bipedal behavior each, whereas feeding made up 85 percent of bipedal episodes (Hunt, 1994). It should be no surprise that my data pointed toward feeding behavior as the motivating factor for evolving such an unusual behavior. Half the chimpanzee day is occupied in finding and eating food; presumably it was the same for proto-hominins. There is tremendous evolutionary pressure selecting to make a behavior engaged in often efficient. As rare as bipedalism was for chimpanzees, it was extremely rare in any context other than feeding.

The postural feeding hypothesis is consistent with australopith features that had been inexplicable in the context of the terrestrial feeding model; robust, curved fingers, long toes, a cone-shaped thorax, chimpanzee-like elbows, and a narrow scapula (Hunt, 1991) are arboreal adaptations – adaptations to armhanging. A broad pelvis and short hindlimbs lower the center of gravity (Hunt, 1998), improving stability on flexible, small-diameter branches.

Significantly, while australopiths share armhanging traits with chimpanzees (Table 28.1), they do not share *climbing* adaptations such as gripping great toes, tall pelves, and short backs. Small trees can be ascended with a single agile leap, which means harvesting fruits from small trees requires none of the vertical climbing anatomy that forest-adapted chimpanzees need to access fruits in tall trees (Hunt, 1998).

The postural feeding hypothesis views bipedalism as having evolved when an ape-like pre-hominin, perhaps more arboreal and therefore more bipedal than extant African apes, but not necessarily so, was isolated in a dry habitat with abundant small fruits in short-statured trees (Figure 28.11).

28.10 Morphological Diversity Muddies the Water

The expert, especially the expert who disagrees with my view of the evolution of bipedalism, will be thinking, "But wait, these arboreally adapted traits

Table 28.1 Skeletal features

Feature	Human	Australopith	Chimpanzee
Cranial capacity	1350 cm^3	400 cm^3	350 cm^3
Zygomatics	Short, angled back	Tall, squared off	Short, angled back
Mandible robusticity	Delicate	Heavy, thick	Slight but not delicate
Mandible shape	U-shaped	V-shaped	U-shaped (normally)
Thorax or ribcage	Barrel-shaped	Conical	Conical
Ribs in cross-section	Blade-shaped	Round	Round
Vertebral diameter	Large	Small	Small
Pelvic height	Low	High	Low
Pelvis shape	Bowl-shaped	Iliac blades flat	Bowl-shaped
Iliac orientation	Lateral	Posterior	Intermediate
Pelvic inlet shape	Round	Round	F: broad; M: narrower
Scapula	Broad, triangular	Narrow	F: narrow; M: broader
Glenoid orientation	Lateral	Tilted up	F: tilted; M: more lateral
Humeral shaft walls	Thin	Thick	Thick
Elbow joint	Gently undulating	M-shaped	M-shaped
Arm length	Short	Long	Intermediate
Suparcondylar crest	Small	Large	Large
Olecranon fossa	Typically floored	Perforated	Perforated
Ulnar styloid process	Small	Robust	Robust
Pisiform	Pebble-like	Rod-like	Rod-like
Thumb	Large	Small	Intermediate
Finger length	Short	Long	Short-ish
Finger curvature	Straight	Curved	Curved
Flexor sheath ridges	None	Large	Large
Leg length	Long	Short	Intermediate
Knee	Valgus	Varus	F: hypervalgus; M: human
Heel size	Large	Small	Large
Toe length	Short	Long	Intermediate
Toe shape	Straight	Curved	Curved

you harp on are much less apparent in some australopiths. What about those more human-like ones?" By the time I started my dissertation work, scholars had already noticed that smaller australopiths had more chimpanzee features (Stern & Susman, 1983) than larger specimens (Senut & Tardieu, 1985). Christine Tardieu and Brigitte Senut argued that the large and small fossils were so different they must have been different species. The differences they noted were so surprising because, skeletally, chimpanzee males and females differ little other than in size (Jungers & Susman, 1984); small and large chimpanzees, male and female chimpanzees are nearly exact geometrically scaled versions of one another. Humans are a little more sexually dimorphic, but not like australopiths.

28.11 Sexual Dimorphism Clears the Mud

We now know that larger australopiths, presumably males, are much taller and appear to have narrower hips and longer legs (Haile-Selassie et al., 2010) than smaller ones, presumably females. Many of the differences between small and large australopiths could be a consequence of females spending more time in trees, causing changes to their anatomy during life to adjust to arboreal demands; we will soon see new evidence that the chimpanzee skeleton is far more plastic than we ever imagined. Drawing on what we know from living apes, when males are twice the size of females the mating system is a one-male, multi-female one. Perhaps males, larger and less agile in trees, would harvest fruits from the ground while females and young gathered fruits arboreally (Figure 28.11). Some research suggests that hominins had a multi-level, hierarchical component like hamadryas baboons (Swedell & Plummer, 2012), with the males of each one-male unit joining with other males to mutually defend their groups. We end up with a large number of one-male units traveling together, perhaps a system also good for deterring predators.

Bipedalism, as Darwin speculated, frees the hands for carrying a club or walking stick. Humans love whacking things with sticks, as parents of two-year-olds often learn. And not just toddlers. We revere good stick-wielders. Golf, baseball, hockey, cricket, tennis: many of our sports involve taking a swing at something with a stick. Our weapons of war often bear some resemblance to the tools of sport, though often made more lethal with a sharp edge: bayonets, spears, battle axes, swords, and maces. Perhaps a stout stick was the first must-have tool, with males – able to hang on to it all the time, since their hypothetical greater terrestriality would less often require them to set it aside to climb a tree – wielding them in part to repel rogue males, but helpful for deterring leopards and hyenas as well. In this model, freeing the hands comes first, picking up a tool second.

28.12 Predictions Derived from the Postural Feeding Hypothesis

Is it not odd that australopiths have fingers that are chimpanzee-like in morphology but more human-like in length? Not at all, I realized. Short trees have small-diameter branches compared to the huge trees chimpanzees feed in. Thinking about this, I made a few more predictions about what we might expect as new information was uncovered (Hunt, 1998).

If there is any information on diet among australopiths, it should suggest consumption of dry-habitat foods. Both microwear – tiny scratches on the teeth (Grine et al., 2006; see Hunt, 1998 for a reconstruction of the australopith diet based on microwear) and trace elements in the fossils should suggest a diet of dry-habitat fruits. Compared to earlier possible hominins and to chimpanzees, australopith trace elements suggest they ate more open-habitat C4 foods (Sponheimer et al., 2013). That there might be pollen with the fossils had not occurred to me, but pollen from 15 tree species has been recovered from strata containing *Au. afarensis* fossils (Bonnefille et al., 2004). Two of those species, *Garcinia* and *Grewia*, were the first and ninth most common species gathered bipedally by chimpanzees in my study. All but two of the other species, so 13 of 15 total, were short trees that produce small fruits.

Australopith semi-circular canals, the organs that help maintain balance on unstable supports – like thin branches – might be expected to be more like those of a partly arboreal species rather than modern humans, since the postural feeding hypothesis sees australopiths as gathering food in trees and having a less highly evolved type of bipedalism than modern humans. The skeletal receptacle for the canals, the bony labyrinth, turned out to be ape-like (Spoor et al., 1994), suggesting arboreal adaptations.

28.13 At Last, Semliki Chimpanzees

I realized that to test the postural feeding hypothesis I needed to study chimpanzees in an australopith-like habitat, like the habitat where I began to study chimpanzees in 1996, Semliki. Dry-habitat chimpanzees might provide hints concerning both jaw adaptations and the evolution of bipedalism. Foods in dry habitats are less succulent and harder to chew. We are confident of this because other species that are adapted to dry habitats, gelada baboons for instance, have robust faces and large molars. I did not expect tooth size to be different at Semliki because it was unlikely the Semliki population would be genetically isolated; interbreeding with surrounding populations would probably swamp any local adaptation. We know tooth size responds little to chewing stresses, unlike the rest of the body's hard tissue. Children in industrialized nations have tooth crowding and require braces because their food is soft, while children in countries where tough food is eaten early in life have wonderful occlusion. We might expect the face and jaws of dry habitat chimpanzees to be robust, to converge on australopiths.

The legs and hips of dry-habitat chimpanzees might also show some response to higher frequencies of bipedalism, and these indications of bipedalism should be found with robust faces. We might expect, then, that a population of chimpanzees confined to a dry habitat should be more bipedal than forest chimpanzees, and because they have a harder, tougher diet, they might well have greater facial robusticity.

In 1995 I heard through the rumor mill that there was a proposal to build an ecotourist lodge at the all-but-abandoned Toro-Semliki Wildlife Reserve, a dry habitat with a population of chimpanzees (Hunt & McGrew, 2002; Figure 28.12). The reserve was mostly grassland, with only 7 percent forest cover (Webster et al., 2014). I talked to the head of the company that is now WildPlaces Africa, Jonathan Wright, who invited me to come and have a look. Subsequently, he and his team put me up in their tented camp for three months while I started studying the chimpanzees. The Semliki Chimpanzee Project was born.

I was not the first to come up with the idea of studying chimpanzees in australopith-like habitats. That was Bill McGrew. He and his colleagues had been unable to completely habituate (get them used to being observed) the dry-habitat population at Mount Assirik (McGrew et al., 1981, 1982; Baldwin et al., 1982; Tutin et al., 1983). Because the foliage is short and sparse, dry-habitat chimpanzees tend to feel exposed. There is no dense, high canopy in which to hide, nor to use to flee. Their lack of escape routes makes them skittish. I failed to realize at the time that chimpanzees are acutely aware of possible escape routes, and fearful when they are cut off. After a heroic effort, McGrew and colleagues left off full-time attempts to study the Mt. Assirik chimpanzees, but later a nearby population was successfully habituated by Jill Pruetz, who, with her team, has made incredible discoveries at her site, Fongoli (Pruetz, 2006).

I knew it would be tough at Semliki, but Wright's generous offer to put me up for a few months gave me the wherewithal to make the attempt. Unfortunately, there were problems that were not a factor at Assirik, the most serious being poaching. While there was and still is no evidence poachers are specifically targeting chimpanzees, the chimpanzees seem not to know that. Humans are not often friends of chimpanzees, and chimpanzees are smart enough to know that snares are human devices and that dogs go with humans. Perhaps a level of poaching that chimpanzees tolerate elsewhere provokes a more dramatic response among chimpanzees in open habitats.

In short, open habitats may make chimpanzees more human-like, but it also seems to make them hard to study. We learned long ago from Gombe, Mahale, and elsewhere that to get behavioral data that is reliable, chimpanzees must be consistently followed

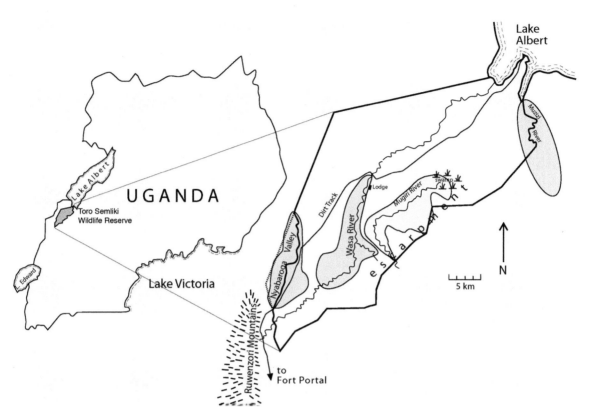

Figure 28.12 The Semliki Chimpanzee Project is based in the Toro-Semliki Wildlife Reserve. In preliminary surveys I believed there were four communities (stippled areas). Deforestation has proceeded at a ferocious rate at Muzizi (far right) and we are now confident that the Wasa and Mugiri communities are one super-community of over 150 or even 200 individuals. Map by the author.

for their entire active period, from waking up in their sleeping nest to making and entering another nest that night. We are still unable to do that at Semliki. While we have observed chimpanzees feeding bipedally from short trees with small fruits, feeding trees (or bushes) such as *Securinega*, *Rhus*, *Grewia*, *Beilschmiedia*, and *Tamarindus*, we do not have the data yet to prove definitively they are more bipedal than elsewhere. We will see below, however, that there is truly remarkable indirect evidence.

As my team and I have carried on with our longstanding attempts to finish habituation of the chimpanzees, we have been able to document a number of unusual behaviors that suggest that dry-habitat chimpanzees are not your typical chimpanzee, though they share much in common with forest chimpanzees: Semliki chimpanzees eat insects (Marchant et al., 2009; Webster et al., 2009b, 2014); they lack laterality (i.e., handedness; Marchant et al., 2007, 2009; McGrew et al., 2007); they engage in hand-clasp grooming (Webster et al., 2009a); they are dependent on forest fruits (Hunt & McGrew, 2002); they spend about half their time in the trees and they occasionally eat their own dung (Payne et al., 2008). They spend a lot of time at the forest edge, near the grassland, and do venture out into the more open areas to feed from the fruit trees listed above.

28.14 Semliki Chimpanzees Rarely Use Tools

Despite sharing many traits with forest chimpanzees, Semliki chimpanzees use tools much less often than Tanzanian and Ivorian chimpanzees (Hunt & McGrew, 2002; McGrew et al., 2010), though they are just on the low end of the trend in Uganda. In McGrew's words, "it seemed that every population of wild chimpanzees studied in the long term was technological ... but long-term field sites in Uganda, such as Budongo, Kanyawara, Ngogo, and Semliki seem to have little or no such technology, despite decades of study" (McGrew, 2017).

28.15 Semliki Provides First Evidence of Chimpanzee Drinking Wells

On 12 August 1997, only a year after I started work at Semliki, I watched a female in a riverbed behaving oddly. She seemed to be digging in the sand. She dug for a while, scooping out a six-inch-high mound of sand, then she turned and walked over to some foliage at the side of the watercourse. There she picked a leaf, returned to the hole she had dug and used the leaf as a sort of dipper to drink water in the hole (Figure 28.13). She had dug a well. But why? She was drinking from this well *only two feet* from the clear-running, gently flowing Mugiri River (Hunt et al., 1999; Hunt, 2000; Hunt & McGrew, 2002; Marchant et al., 2007, 2009; McGrew et al., 2007), water so clear that park rangers often drink out of it directly (I boil it first, wimp that I am). My best guess is that drinking wells serve two purposes: first, as the water passes through to sand it filters out parasites or pathogens; second, the water tastes better because it is more pH neutral. The water in the river is alkaline and somewhat chalky tasting. The former explanation is iffy since we find no parasites in the unfiltered water. The latter we have shown with pH data.

This 1997 observation of well-digging was the first time this behavior had ever been observed, though it has been seen now in another dry-habitat population (Galat-Luong & Galat, 2000; Galat-Luong et al., 2009). It has been more than 50 years since Goodall began her pioneering study and we are still discovering new chimpanzee behaviors.

28.16 Semliki Chimpanzees May be Less Violent

Following long-established primatological tradition, as I identified chimpanzees I recorded distinguishing

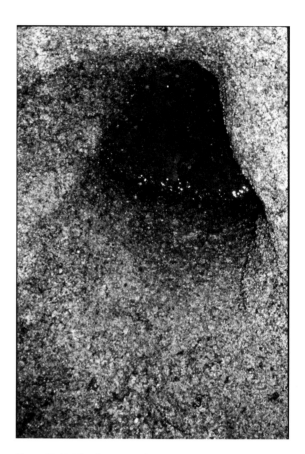

Figure 28.13 The first ever observed Semliki drinking well. Note the leaf scoop or dipper left behind by the female. The well is less than 1 m from the clear, gently flowing Mugiri watercourse. Photo by the author.

characteristics. For instance, Mzee has square ears and a scar on his lip. As I identified first 10, then 20, then more chimpanzees, I was astonished to find there were so many males – 29 at one count, the second largest in any community (there are 37 or so males at Kibale-Ngogo). This must mean, if the proportion of males to females and the proportion of adults to young is typical, there are an astonishing 180 individuals in the Semliki community. There are only 35 individuals at Gombe. As I identified one male after another and drew cartoons to depict their distinctive features (Figure 28.14), I saw something even more surprising. Only one of my 29 males had an ear tear. At sites I had worked at earlier – at Gombe, Mahale, and Kibale-Kanyawara – most males had ear tears. I also found the loud fussing that chimpanzees do at one another

Figure 28.14 Identification cartoons of 18 male chimpanzees at Semliki. Chimpanzees are unusual in having slightly floppy or folded ears – domesticated animals have such ears – and distinctive depressions under the eyes of some. Note that only one individual has an ear tear, an unusual situation suggesting unusually low levels of violence at Semliki.

seemed less common at Semliki, perhaps because they feel more vulnerable in the open habitat and feel the need to remain quiet. As the years went by, I also observed very little hunting, a behavior quite common everywhere else chimpanzees have been studied. I have begun to think that Semliki chimpanzees might be less violent than most chimpanzees. Eventually I was able, with my student David Samson, to show that this apparent low level of violence was a real phenomenon; it is one of many ways Semliki chimpanzees are more human-like than other chimpanzees (Samson & Hunt, 2014).

28.17 Some Semliki Chimpanzees Appear to Lack Testicles

In July 2009 team member Maggie Hirschauer made a shocking discovery. Adult male Charles, seemingly normal in every other way, had no visible testicles. They might be undescended, but testes descend before adulthood in normal chimpanzee genital development (Williams & Hutson, 1991); as the testicles grow larger during adolescence, they are forced out of the inguinal canal and into the scrotum. This condition, cryptorchidism (hidden testes), had never been observed among wild chimpanzees. Then we found another such male; much later, still another.

Figure 28.15 Three males at Semliki lack testicles, for reasons we are unsure of. There are no apparent scars in any of the three individuals, but neither is there much in the way of a scrotum. Jacko is a relatively delicately built male, suggesting a lack of testosterone, but Charles is more robust than typical, while July is typically robust. Photo credit: Caro Deimel.

Chimpanzees do bite off one another's testicles in intercommunity conflicts (Wrangham, 1999; Wilson et al., 2004, 2014), but the three males we have identified as lacking testicles – Charles, Jacko, and July – have no apparent scars (Figure 28.15). Chimpanzees heal incredibly well, and scars fade surprisingly quickly – but had the testicles been bitten off we might expect disturbance of the scrotal raphe, a line that runs from under the penis to the anus, and as you can see in Figure 28.15, the raphe is intact. The low levels of aggression at Semliki make this even less likely. Some animals have testicular abnormalities when they are inbred, but this is not the case at Semliki (Rich et al., 2018). At the moment, the issue remains a mystery. Next year we will test these males to see if they have normal levels of testosterone.

28.18 The Semliki Community Range is Uniquely Large

We have long known the community range of the Semliki chimpanzees is the largest known (Hunt & McGrew, 2002), but recent genetic analysis (Rich, 2017) suggests that the chimpanzees miles away in the Wasa River are also part of our community. This means the chimpanzees in my study group have a community range of 174 km^2. The next largest is that in another dry habitat in Senegal, Fongoli, at 65 km^2.

28.19 Perhaps Large Range Influences Intercommunity Aggression

This large community area has consequences. As we learned earlier, normally chimpanzee males gather in a group and march off to cause their neighbors trouble about once a month – to attack males in a neighboring community (Chapters 24 and 25). But think of this task for Semliki males. To reach the community boundary they would have to walk not 3 km as they might at Gombe, or 6 km as they would at Mahale, but as far as 20 km. And then what would they find when they got there? The border of the community is long compared to Gombe, Mahale, or Kibale. The likelihood of finding an individual to attack in this vast land is extremely low. The consequences are that a large community range makes it both costlier to reach the border and less likely to pay off compared to normal chimpanzee sites. Defending a border is too costly for the Semliki chimpanzees, so perhaps they have given it up. This is positively un-chimpanzee-like, being so peaceable and not defending a border. But low levels of aggression (Samson & Hunt, 2014) and low levels of ear-tears (Hunt, long term records) suggest that these chimpanzees are relatively peaceful – perhaps because there is less benefit to the aggression. Given the genetic similarity of Semliki chimpanzees to the much more violent Kibale chimpanzees, this is likely to be a cultural or developmental difference, but we will continue to study it.

What are the consequences? Normally effective patrolling, that which increases the community area, increases resources for females in their community (Chapter 25), or increases the number of females in the community, or both. What would we expect to happen when that no longer applies? Chimpanzees are unlikely to respond by forming female bonds (see Chapters 12, 13, and 23 for the reasons why). We do see females traveling together now and then, but it is not dramatically different from other chimpanzee patterns. Males could go solitary like orangutans and defend a territory by themselves; we definitely do not see this. They might gather a harem around them, like gorillas, or they could become monogamous. But they have not done these things, either. We have never seen one-male-multi-female units. There is some evidence that Semliki chimpanzees are pairing up more than elsewhere; I have seen male–female pairs spend days together when the female was not in estrus, one pair in 2014 for at least three consecutive days. The male showed a level of attention to the female's infants that surprised me. But again, this is also seen elsewhere and so if it is a difference it is merely a matter of degree. We will eventually sort it all out, but for now it is an unknown.

28.20 Semliki Femora Are Uniquely Human-Like

The best anatomical confirmation of the postural feeding hypothesis would be to find both robust faces and human-like hips and legs in a population of chimpanzees confined to a dry habitat. Humans have valgus knees: The shaft of the femur is 13° off a perfectly vertical orientation (Figure 28.16), giving the human skeleton a knock-kneed appearance. Previously measured chimpanzee femoral shafts are almost perfectly vertical. Semliki chimpanzee femora are not; they vary in the direction of humans and are

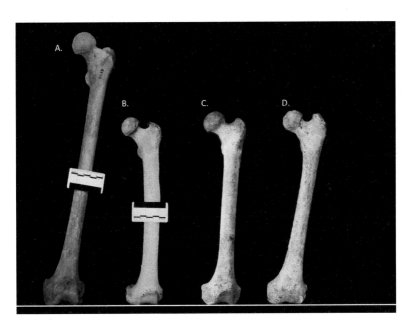

Figure 28.16 Human femora (A) are angled so that the shaft is 13° off vertical, compared to the articular surface of the knee (white line). Forest chimpanzees (B) have femora nearly perfectly vertical, varying by less than one-third of a degree from vertical. The average Semliki femur (C) falls in between humans and forest chimps. One of the Semliki femora (D) is right at the human average. No other chimpanzee population has this human-like angle. Image by the author. (A black and white version of this figure will appear in some formats. For the color version, please refer to the plate section.)

significantly different from both forest chimpanzees and humans, falling in between the two. The most human-like of the Semliki femora falls at almost exactly the human average (Figure 28.16). I think this human-like condition is a result of adjusting to human-like stresses acting on the knee during growth and development, to the adoption of a greater-than-normal frequency of bipedalism.

Perhaps, this anatomy suggests Semliki is more genetically isolated than we thought. Genetic evidence says not. An early analysis comparing Semliki chimpanzee genetics to Kibale showed they were closely related (Langergraber et al., 2007), so much so a Semliki chimp could be dropped among the Kibale chimpanzees without causing a ripple. Recently my team expanded the genetic sample to show that chimpanzees from the closest conservation area, Semuliki National Park, are very nearly identical to my study population, though they are found in a closed-canopy forest (Rich et al., 2018).

Let me pause to remind you of the significance of this. It not only means that dry habitats and human-like femora are associated – the only explanation I can think of is that they engage in bipedalism and, just as humans do, they develop a valgus knee in response. This means that the many ape-like traits in australopiths are much more likely to reflect how the body part was used in life than that they were primitive retentions. Curved toes and fingers in australopiths suggest gripping branches; a flexible knee and tilted up shoulder joint suggest acrobatic arboreal behavior, such as armhanging. These features are likely less genetic than we thought.

28.21 Semliki Pelves Are Uniquely Human-Like

If the pelvis also responds like the femur, we might expect the attachment for the hip flexors, the bony landmark known to experts by the mouthful "anterior superior iliac spine," to be moved forward so that the entire structure more resembles the bowl-like pelvis of humans. It is (Figure 28.17). One of the Semliki

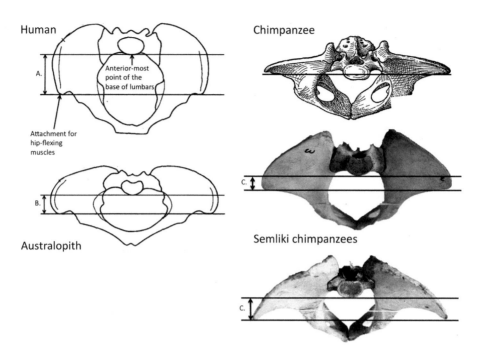

Figure 28.17 Top view of human pelvis, australopith (Lucy, bottom left), forest chimpanzee (top right), and two Semliki pelves (right, second from top, bottom). Humans have a bowl-shaped pelvis with a large distance between the attachment for hip flexors (muscles in the front of the thigh that raise the knee) and the forward-most point of the pelvic base for the spine (distance A). Australopiths (distance B) vary in the direction of chimpanzees, which have no distance between the spine and the hip flexors (top right). Semliki chimpanzees (distances C) vary from forest chimpanzees in the direction of humans. Forest chimpanzee after Schultz, 1936. (A black and white version of this figure will appear in some formats. For the color version, please refer to the plate section.)

chimpanzees has an even more human-like anterior superior iliac spine placement than Lucy.

28.22 Semliki Faces Are Australopith-Like

Australopiths have tall zygomatics, an adaptation to eating tough foods found in dry habitats. I predicted that robust faces would go along with bipedal anatomy in the hip and leg (Hunt, 1998). Semliki chimpanzees, alone among all known chimpanzees, have tall zygomatics (Figure 28.18).

28.23 Semliki Jaws Are Australopith-Like

Australopith mandibles are robust, giving them a V-shaped look when viewed from above. This is to be expected; we know that when humans chew a lot during growth and development, both their faces and their mandibles are heavier. Chimpanzee mandibles are normally U-shaped, with mandibular bodies hardly any larger than the molars they house. Semliki jaws appear to fall on the australopith side (Figure 28.19), matching their robust faces.

28.24 Semliki Adaptations

Semliki chimpanzees, then, are unique in many ways, and in many of those ways they resemble either humans, early hominins, or both. Dry-habitat chimpanzees have both bipedal anatomy and robust faces and jaws, in keeping with the prediction that bipedalism may have evolved as an adaptation to feeding on dry-habitat food resources. Their lack of ear tears and low levels of hunting also suggest they are less aggressive than forest chimpanzees.

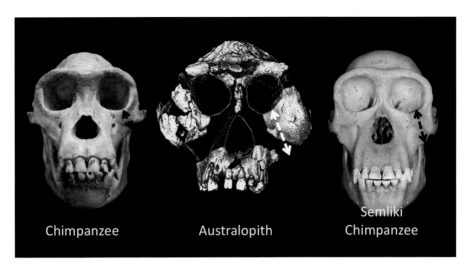

Figure 28.18 Forest chimpanzees (left) have short (top-to-bottom) zygomatics (black line, left) compared to australopith zygomatic height (white dotted line, middle). Semliki chimpanzees have australopith-like zygomatics (black dotted line, right); Semliki photo by the author. (A black and white version of this figure will appear in some formats. For the color version, please refer to the plate section.)

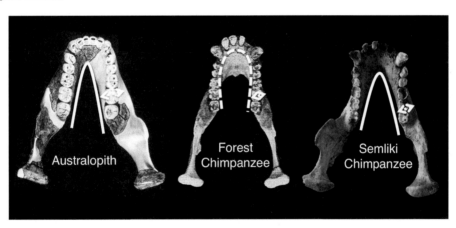

Figure 28.19 Australopiths (left) have robust mandibles with mandibular bodies so thick the mandible has a V-shape. Chimpanzee mandibles are thinner, giving them a round inner surface compared to australopiths. Semliki mandibles are thick and australopith-like. Semliki mandible by the author. (A black and white version of this figure will appear in some formats. For the color version, please refer to the plate section.)

28.25 Into the Light

Perhaps rather than descending from the trees, the first step along the road to humanity was walking out onto the sun-dappled, lightly forested savanna, where fruits could be reached from the ground – or in the more traditional primate fashion, in the trees.

All early hominin fossils have been found in drier habitats than typical of chimpanzees, those in East Africa and South Africa, and those to the north on the edges of the Sahara. To my mind, australopiths are nothing more than dry-habitat apes. Their brains are hardly larger than those of chimpanzees, so their behavior was likely ape-like; if an isolated population had somehow survived to today, they would be in the zoo, right next to the chimpanzees. Yet the bipedal adaptation that australopiths pioneered had advantages that led directly to us. By freeing the hands to handle tools – perhaps the first tool was a combination walking stick/club – bipedalism made tool use more profitable. We tend not to stand when making or using tools, so bipedalism helps little there, but *carrying* tools is much easier with a free hand or two. Males, if the more human-like anatomy of the larger individuals is saying what I think it is, would have been on the ground more, where their superior stature allowed them to gather fruits higher in trees and their superior weight gave them immunity from predators. A ferocious cudgel wielded by a set of shoulders every bit as powerful as those of chimpanzees would make any predator or marauding extra-group male think twice. Females would have spent more time in the trees, where their smaller size, greater joint mobility and lower centers of gravity would have facilitated their harvesting of fruits among the small branches in these small trees.

Many view australopiths as somewhat half-hearted hominins, a short stop on the road to humanness. Instead, I view their two million years of existence as a phase of human history during which they were fully evolved to live in a singular niche, a lightly forested, dry habitat with short fruit trees; and the robust jaws and human-like lower bodies of the Semliki chimpanzees are evidence of that.

References

Abitbol MM (1995) Lateral view of *Australopithecus afarensis*: primitive aspects of bipedal positional behavior in the earliest hominids. *J Hum Evol* 28, 211–229.

Baldwin PJ, McGrew WC, Tutin CEG (1982) Wide ranging chimpanzees at Mt. Assirik, Senegal. *Int J Primatol* 3, 367–383.

Bartholomew GA, Jr., Birdsell JB (1953.) Ecology and the protohominds. *Am Anthropol* 55, 481–498.

Bonnefille R, Potts R, Chalie F, Jolly D, Peyron O (2004) High-resolution vegetation and climate change associated with Pliocene *Australopithecus afarensis*. *PNAS* 101, 12125–12129.

Bramble DM, Lieberman DE (2004) Endurance running and the evolution of *Homo*. *Nature* 432, 345–352.

Carrier DR (1984) The energetic paradox of human running and hominid evolution. *Curr Anthropol* 25, 483–495.

Cerling, TE, Harris, JM, MacFadden, BJ, et al. (1997) Global vegetation change through the Miocene/Pliocene boundary. *Nature* 389, 153–158.

Cerling TE, Wynn JG, Andanje SA, et al. (2011) Woody cover and hominin environments in the past 6 million years. *Nature* 476, 51–56.

Dart RA (1959) *Adventures with the Missing Link*. New York: Harper.

Darwin C (1871) *The Descent of Man and Selection in Relation to Sex*. London: John Murray.

Etkin W (1954) Social behavior and evolution of man's mental faculties. *Am Nat* 88, 129–143.

Galat-Luong A, Galat G (2000) Chimpanzees and baboons drink filtrated water. *Folia Primatol* 71, 258.

Galat-Luong A, Galat G, Nizinski G (2009) Une consequence du rechauffement climatique: les chimpanzes filtrent leur eau de boisson. *Geographia Technica Numéro spécial*, 2009, 199–204.

Grine FE, Ungar PS, Teaford MF, El-Zaatari S (2006) Molar microwear in *Praeanthropus afarensis*: evidence for dietary stasis through time and under diverse paleoecological conditions. *J Hum Evol* 51, 297–319.

Haile-Selassie Y, Latimer BM, Mulugeta A, et al. (2010) An early Australopithecus postcranium from Woranso-Mille, Ethiopia. *PNAS* 107, 12121–12126.

Hardy A (1960) Was man more aquatic in the past? *New Scientist* 7, 642–645.

Heglund NC (1985) Comparative energetics and mechanics of locomotion: how do primates fit in? In *Size and Scaling in Primate Biology* (ed. Jungers W), pp. 319–335. New York: Academic Press.

Hunt KD (1991) Mechanical implications of chimpanzee positional behavior. *Am J Phys Anthropol* 86, 521–536.

Hunt KD (1994) The evolution of human bipedality: ecology and functional morphology. *J Hum Evol* 26: 183–202.

Hunt KD (1996) The postural feeding hypothesis: an ecological model for the evolution of bipedalism. *S Afr J Sci* 92, 77–90.

Hunt KD (1998) Ecological morphology of *Australopithecus afarensis*: traveling terrestrially, eating arboreally. In *Primate Locomotion: Recent Advances* (eds. Strasser E, Fleagle JG, McHenry HM, Rosenberger A), pp. 397–418. New York: Plenum.

Hunt KD (2000) Initiation of a new chimpanzee study site at Semliki-Toro Wildlife Reserve, Uganda. *Pan Africa News* 7, 14–17.

Hunt KD (2008) Commentary on The chimpanzee has no clothes: a critical examination of *Pan troglodytes* in models of human evolution (Sayers K, Lovejoy CO) *Curr Anth* **49**, 100–101.

Hunt KD (2015a) Bipedalism. In *Basics in Human Evolution* (ed. Muehlenbein MP), pp. 103–112. Boston, MA: Elsevier.

Hunt KD (2015b) Early hominins. In *Basics in Human Evolution* (ed. Muehlenbein MP), pp. 113–127. Boston, MA: Elsevier.

Hunt KD, McGrew WC (2002) Chimpanzees in the dry habitats of Assirik, Senegal and Semliki Wildlife Reserve, Uganda. In *Behavioural Diversity in Chimpanzees and Bonobos* (eds. Boesch C, Hohmann G, Marchant LF), pp. 35–51. Cambridge: Cambridge University Press.

Hunt KD, Cleminson AJM, Latham J, Weiss RI, Grimmond S (1999) A partly habituated community of dry-habitat chimpanzees in the Semliki Valley Wildlife Reserve, Uganda. *Am J Phys Anthropol* Suppl **28**, 157.

Isaac GLl (1978) Food sharing and human evolution: archaeological evidence from the Plio-Pleistocene of East Africa. *J Anthropol Res* **34**, 311–325.

Iwamoto M (1985) Bipedalism of Japanese monkeys and carrying models of hominization. In *Primate Morphophysiology, Locomotor Analysis and Human Bipedalism* (ed. Kondo S), pp. 251–260. Tokyo: Tokyo University Press.

Johanson DC, Lovejoy CO, Kimbel WH, et al. (1982a). Morphology of the Pliocene partial hominid skeleton (A.L. 288-1) from the Hadar Formation, Ethiopia. *Am J Phys Anthropol* **57**, 403–451.

Johanson DC, Taieb M, Coppens Y (1982b) Pliocene hominids from the Hadar formation, Ethiopia (1973–1977): Stratigraphic, chronologic, and paleoenvironmental contexts, with notes on hominid morphology and systematics. *Am J Phys Anthropol* **57**, 373–402.

Jolly CJ (1970) The seed-eaters: a new model of hominid differentiation based on a baboon analogy. *Man* **5**, 1–26.

Jungers WL, Susman RL (1984) Body size and skeletal allometry in African apes. In *The Pygmy Chimpanzee: Evolutionary Biology and Behavior* (ed. Susman RL), pp. 131–177. New York: Plenum Press.

Keith A (1891) Anatomical notes on Malay apes. *J Straits Br Asiat Soc* **23**, 77–94.

Keith A (1923) Man's posture: its evolution and disorder. *Brit Med J* **1**, 451–454, 499–502.

Langergraber KE, Siedel H, Mitani JC, et al. (2007) The genetic signature of sex-biased migration in patrilocal chimpanzees and humans. *Plos ONE* **2**, e073.

Livingstone FB (1962) Reconstructing man's Pliocene ancestor. *Am Anthropol* **64**, 301–305.

Lovejoy CO (1981) The origin of man. *Science* **211**, 341–350.

Marchant LF, McGrew WC, Hunt KD (2007) Ethoarchaeology of manual laterality: well-digging by wild chimpanzees. *Am J Phys Anthropol* Suppl **44**, 163.

Marchant LF, McGrew WC, Payne CLR, Webster TH, Hunt KD (2009) Well-digging by Semliki chimpanzees: new data. *Am J Phys Anthropol* Suppl **48**, 183.

McGrew WC (2017) Field studies of *Pan troglodytes* reviewed and comprehensively mapped, focussing on Japan's contribution to cultural primatology. *Primates* **58**, 237–258.

McGrew WC, Baldwin PJ, Tutin CEG (1981) Chimpanzees in a hot, dry and open habitat: Mt. Assirik, Senegal, West Africa. *J Hum Evol* **10**, 227–244.

McGrew WC, Baldwin PJ, Tutin CEG (1982) Observations preliminaires sur les chimpanzes (Pan troglodytes verus) du Park National du Niolola-Koba. *Memoires de l'Institut Fondamental d'Afrique Noire* **92**, 333–340.

McGrew WC, Marchant LF, Hunt KD (2007) Etho-archaeology of manual laterality: well digging by wild chimpanzees. *Folia Primatologica* **78**, 240–244.

McGrew WC, Marchant LFC Payne CLR, Webster T, Hunt KD (2010) Chimpanzees at Semliki ignore oil palms. *Pan Africa News* **17**(2), 19–21.

Morgan E (1972) *The Descent of Woman*. New York: Bantam Books.

Newman RW (1970) Why man is such a sweaty and thirsty naked animal: a speculative review. *Hum Biol* **42**, 12–27.

Passey BH, Levin NE, Cerling TE (2010) High-temperature environments of human evolution in East Africa based on bond ordering in paleosol carbonates. *PNAS* **107**, 11245–11249.

Payne CLR, Webster TH, Hunt KD (2008) Coprophagy by the semi-habituated chimpanzees of Semliki, Uganda. *Pan Africa News* **15**, 29–32.

Pontzer H, Raichlen DA, Sockol MD (2009) The metabolic cost of walking in humans, chimpanzees, and early hominins. *J Hum Evol* **56**, 43–54.

Pruetz JD (2006) Feeding ecology of savanna chimpanzees (*Pan troglodytes verus*) at Fongoli, Senegal. In *Feeding Ecology in Apes and Other Primates: Ecological, Physical and Behavioral Aspects* (eds. Hohmann G, Robbins MM, Boesch C), pp. 326–364. Cambridge: Cambridge University Press.

Rich AM (2017) *Population genetics of Eastern Chimpanzees (Pan troglodytes schweinfurthii) living in Toro-Semliki Wildlife Reserve, Uganda*. PhD dissertation. Bloomington, IN: Indiana University.

Rich AM, Wasserman MD, Deimel C, et al. (2018) Is genetic drift to blame for testicular dysgenesis syndrome in Semliki chimpanzees (*Pan troglodytes schweinfurthii*)? *J Med Primatol* **47**, 257–269.

Rodman PS, McHenry HM (1980) Bioenergetics and the origin of hominoid bipedalism. *Am J Phys Anthropol* **52**, 103–106.

Rose MD (1976) Bipedal behavior of olive baboons (*Papio anubis*) and its relevance to an understanding of the evolution of human bipedalism. *Am J Phys Anthropol* **44**, 247–261.

Samson DR, Hunt KD (2014) Is chimpanzee (*Pan troglodytes schweinfurthii*) low population density linked with low levels of aggression? *Pan Africa News* 21, 15–17.

Sayers K, Lovejoy CO (2008) The chimpanzee has no clothes: a critical examination of *Pan troglodytes* in models of human evolution *Curr Anth* 49, 87–114.

Schultz AH (1936) Characters common to higher primates and characters specific for man. *Quart Rev Biol* 11, 425–455.

Senut B, Tardieu C (1985) Functional aspects of Plio-Pleistocene hominid limb bones: implications for taxonomy and phylogeny. In *Ancestors: The Hard Evidence* (ed. Delson E), pp. 193–201. New York: Alan R. Liss.

Sponheimer M, Alemesged Z, Cerling TE, et al. (2013) Isotopic evidence of early hominin diets. *PNAS* 110, 10513–10518.

Spoor F, Wood B, Zonneveld F (1994) Implications of early hominid labyrinthine morphology for evolution of human bipedal locomotion. *Nature* 369, 645–648.

Stern JT, Jr., Susman RL (1983) The locomotor anatomy of *Australopithecus afarensis*. *Am J Phys Anthropol* 60, 279–317.

Swedell L, Plummer T (2012) A Papionin multilevel society as a model for hominin social evolution. *Int J Primatol* 33, 1165–193.

Tanner NM (1981) *On Becoming Human* Cambridge: Cambridge University Press.

Taylor CR, Heglund NC, Maloiy CMO (1982) Energetics and mechanics of terrestrial locomotion: I. Metabolic energy consumption as a function of speed and body size in birds and mammals. *J Exp Biol* 97, 1–21.

Thorpe SKS, Holder RL, Crompton RH (2007) Origin of human bipedalism as an adaptation for locomotion on flexible branches. *Science* 316, 1328–1331.

Tutin CEG, McGrew WC, Baldwin PJ (1983) Social organization of savanna dwelling chimpanzees (*Pan troglodytes verus*) at Mount Assirik, Senegal. *Primates* 24, 154–173.

Webster TH, Hodson PR, Hunt KD (2009a). Observations of the grooming hand-clasp performed by chimpanzees of the Mugiri community, Toro-Semliki Wildlife Reserve, Uganda. *Pan Africa News* 16, 5–7.

Webster TH, Marchant LF, McGrew WC, et al. (2009b). Semliki chimpanzees do eat insects. *Am J Phys Anthropol* Suppl 48, 268.

Webster TH, McGrew WC, Marchant LF, Payne CL, Hunt KD (2014) Selective insectivory at Toro-Semliki, Uganda: comparative analyses suggest no "savanna" chimpanzee pattern. *J Hum Evol* 71, 20–27.

Wheeler PE (1991) The thermoregulatory advantages of hominid bipedalism in open equatorial environments: the contribution of increased convective heat loss and cutaneous evaporative cooling. *J Hum Evol* 21, 107–115.

White TD, Johanson DC, Kimbel WH (1981) *Australopithecus africanus*: its phyletic position reconsidered. *S Afr J Sci* 77, 445–470.

Williams MP, Hutson JM (1991) The phylogeny of testicular descent. *Pediatr Surg Int* 1, 162–6.

Wilson ML, Wallauer WR, Pusey AE (2004) New cases of intergroup violence among chimpanzees in Gombe National Park, Tanzania. *Int J Primatol* 25, 523–549.

Wilson ML, Boesch C, Fruth B, et al. (2014) Lethal aggression in *Pan* is better explained by adaptive strategies than human impacts. *Nature* 513, 414–419.

Wrangham RW (1980) Bipedal locomotion as a feeding adaptation in Gelada baboons, and its complications for hominid evolution. *J Hum Evol* 9, 329–331.

Wrangham RW (1999) Evolution of coalitionary killing. *Ybk Phys Anthropol* 42, 1–30.

Wrangham RW, Cheney D, Seyfarth R, Sarmiento E (2009) Shallow-water habitats as sources of fallback foods for hominins *Am J Phys Anthropol* 140, 630–642.

29 The Other Sister, Bonobos
The Monkey Convergence Hypothesis

Drawn by author

The startling contrast between bonobo and chimpanzee societies comes into sharp focus in a story Nahoko Tokuyama tells of the intemperate eagerness of a quartet of young males and the rigid social control several females exercised to restrain them (Angier, 2016). It was in the evening, past the time bonobos normally have settled into their sleeping nests for the night. A female was in estrus her flamboyantly swollen estrous swelling stimulating an unrestrained sexual excitement among four males, including the community's alpha male. Perhaps these males suspected they were skirting the edges of societal norms as they noisily leapt from branch to branch around the female, displaying erections and disturbing what should have been a time of quiet repose for the group. The males, however, were not interested in repose; the presence of this attractive female was simply too much for them to bear. Their overheated commotion went on and on, seemingly with no end in sight. At last, three high-ranking females had had enough. Exploding from beneath, they attacked the four males, scattering them and then ignominiously banishing three of the four into the night, each yelping in retreat. They surrounded the fourth, the alpha male, seized him, and, ignoring his screams of panic, bit him repeatedly – part of a toe was bitten off completely. As the attack wore on he was at last able to break free from the females and flee into the darkness. He failed to reappear the next day, and the day after that; then his absence extended for an entire week. In fact, he limped back into the group only three weeks later, short both a bit of dignity and a bit of a toe.

29.1 The Anti-Chimpanzee

Three females routing four males, males screaming for mercy in the face of female aggression, female social arbiters putting a tight lid on male social lives – this is behavior unimaginable for chimpanzees. And the females who exercised this dominance did so despite inferior numbers and body weights that measure only three-quarters those of males (Table 29.1), a significantly[1] lower proportion than found among chimpanzees; the combined weight of the three females would have been only about 100 kg, versus 180 kg for the four males.

While there is strength in numbers, it is not the number of the bonobos in the fight that counts, but strength of the bonds among the fighters, and the females have the stronger bonds in this ape sisterhood. Females share food (Yamamoto, 2015), travel together often (Surbeck et al., 2017), and even attend to one another during birth (Douglas, 2014). There are many other primates where female bonds impart political power, despite their lesser body weights, but in these societies the sisterhood is a more literal one; females are kin. Not so among bonobos (Furuichi, 2011; Surbeck & Hohmann, 2013). To one familiar only with chimpanzees, female dominance is a shock, but equally shocking is the complete absence of male cooperation; males seem to make no effort to band together to counter female alliances.

---

[1] For my fellow scientists, I mean this literally: at the $p = 0.008$ level, $(F [445, 525] = 10.18, p = 0.008)$. Data and statistical analysis courtesy of William L Jungers.

Table 29.1 **Chimpanzee and bonobo body weights (kg)**

Species	Female (n)	Male (n)	Midsex mean	F/M percentage
Bonobo (*Pan paniscus*)[a]	33.7 (7)	45.0 (7)	39.4	74.9
Eastern chimpanzees (*Pan troglodytes schweinfurthii*)[b]	31.3 (26)	39.0 (31)	35.2	80.3
Chimpanzee (*Pan troglodytes*)[c]	40.4 (27)	49.6 (33)	45.0	81.4

[a] Weights from Smith & Jungers, 1997; [b] Weights from Pusey et al., 2005; [c] Average of three subspecies, weights from Smith & Jungers, 1997.

The dominance of the sisterhood is readily apparent in the easy, relaxed moment-to-moment interactions of females who take the food they want (Parish, 1994), move through the group confidently, and socialize with all variety of age–sex classes; males cower on the sidelines, attentively watching females, hoping to avoid an upbraiding.

29.2 The Discovery of Bonobos

Bonobos (most often pronounced bah-NO-boe, though one sometimes hears BONN-uh-boe) are often seen as the underappreciated stepsister to chimpanzees. Frans de Waal and his coauthor Frans Lanting (1997) titled their bonobo book *The Forgotten Ape*. But bonobos are not forgotten so much as never completely discovered, despite heroic efforts on the part of ape researchers. Bonobos are confined to Congo, a war-ravaged country beset by political instability and violence, difficulties that have hampered research efforts – and such efforts are difficult enough even in stable countries. Consequently, there are many fewer bonobo than chimpanzee study sites, many fewer researchers, and a shallower time depth to what research projects there are. As this gap is filled, surely we will encounter any number of surprising discoveries.

Perhaps bonobo research was stunted from the start by the fact that for years we thought bonobos were merely a variety of chimpanzee. Henry Nissen (1931) had already published the first study of wild chimpanzee behavior, limited though it was, before we even knew there was such a thing as a bonobo (Coolidge, 1933). Even those intimately familiar with the two species, zookeepers, thought they were the same species until the 1930s. Too bad. The prospect of describing a completely new species – which we now know they are, of course – might have motivated explorers to confront the risk of working in this heart of Africa, a motivation that may have been weaker when bonobos were considered merely a variety of chimpanzee. They are even diminished by their name, "pygmy chimpanzee": little chimpanzees.

We now recognize their distinctiveness and their importance, though some specialists worry that we have not yet completely adjusted to the idea that chimpanzees and bonobos are very different animals (Figure 29.1). Their distinctiveness starts with their appearance. It would be difficult to improve on Frans de Waal and Frans Lanting's description:

In physique, a bonobo is as different from a chimpanzee as a Concorde is from a Boeing 747. I do not wish to offend any chimpanzees, but bonobos have more style. The bonobo, with its long legs and small head atop narrow shoulders, has a more gracile build than does a chimpanzee. Bonobo lips are reddish in a black face, the ears small and the nostrils almost as wide as a gorilla's. These primates also have a flatter, more open face with a higher forehead than the chimpanzee's and – to top it all off – an attractive coiffure with long, fine, black hair neatly parted in the middle.

It is that delicacy of build that caused bonobos to be mislabeled "pygmy chimpanzees"; in fact, their body weight is only slightly less than that of chimpanzees, certainly not enough to merit the label "pygmy"; they actually weigh more than Gombe chimpanzees

Figure 29.1 Bonobos have neatly parted hair, a more delicate face than chimpanzees, and red lips. Credit: Anup Shah / Stone / Getty Images.

(Table 29.1; Morbeck & Zihlman, 1989), which are smaller than other common chimpanzees.

29.3 Bonobo versus Chimpanzee

Despite the differences de Waal and Lanting point out, there are important similarities. Like chimpanzees, bonobos knucklewalk. They have long fingers, short thumbs, mobile shoulders, and powerful upper bodies. Both apes have sexual swellings and both have prodigious copulation rates when females are in estrus. Both bear pale complexions at birth that darken with age. They can interbreed (Vervaecke & Van Elsacker, 1992). Facial expressions, attentiveness, social focus, reactions to social events, body postures, and manual gestures are very similar in the two. In my experience watching Kanzi, only his high-pitched vocalizations (de Waal, 1988) were distinctly unchimpanzee-like.

Both species hunt monkeys, bushbabies, birds, and small antelopes (Hohmann & Fruth, 2008; Surbeck et al., 2009), though bonobos hunt much less often. Both are intelligent, highly social primates that have strong mother–offspring bonds, a long period of infant dependency, and fission–fusion social systems. Their diets are similar, though not identical.

29.4 Sexual Dimorphism

Bonobos are thought to have low levels of sexual dimorphism, and this does appear to be the case for measurements of the skull (Furuichi, 1992; Schaefer et al., 2004), brain (Cramer, 1977), and teeth (Almquist, 1974; Fenart & Deblock, 1974; Johanson, 1974). Canines in particular differ little between the sexes, in striking contrast to chimpanzees, suggesting females engage in aggressive struggles as often as males. This is important.

But the head is only part of the body. The most authoritative data on bonobo body weights (Table 29.1) show that females are only three-quarters of the size of males, versus 80 percent for chimpanzees. An explanation for this difference is offered in the following.

29.5 Other Anatomical Differences

Bonobos have longer, more muscular hindlimbs, both legs and feet; their thorax is narrower and they are generally more delicately built. Their proximal phalanges (from the knuckle to the first joint of the finger) have faint or absent flexor sheath ridges and the rest of the finger bones are smaller (Susman,

Table 29.2 **Chimpanzee and bonobo arboreal locomotion (%)**

Locomotor mode	Chimpanzee[a]	Bonobo[b]	Bonobo[c]
Quadrupedal walking on level supports	26.1	35.3	32.0
Quadrupedal walking, climbing on sloped supports	63.5	50.4	69.0
Brachiation or other suspension	6.7	8.9	3.0
Bipedalism	3.0	1.5	3.0
Leaping	0.7	4.0	0.5

[a] Hunt, 1989, 1992; [b] Doran & Hunt, 1994; [c] Ramos, 2014.

1979). Their face is more delicate than that of chimpanzees, and their foramen magnum is placed farther forward, in a slightly more human-like position (Shea, 1984). Bonobos have slightly smaller brains (Rilling et al., 2012). The vagina is oriented in a more ventral position and following this anatomy copulation tends to be more often face to face.

29.6 Locomotion and posture

Much has been made of the fact that bonobos appear more at ease when standing and walking bipedally, but in fact studies of both zoo and wild populations have found no difference between the species (Table 29.2); if anything, chimpanzees are *more* bipedal (Doran & Hunt, 1994; Videan & McGrew, 2001).

Studies across the primates tell us that long hindlimbs, long feet, narrower bodies and lower body weights are found among leapers or runners, compared to climbers, so we might expect more leaping among bonobos. The particularly narrow scapula suggests high frequencies of armhanging. One of the two bonobo locomotor studies is perfectly consistent with expectations based on this anatomy (Table 29.2); Diane Doran found that bonobos at Lomako leapt and brachiated more than chimpanzees (Doran & Hunt, 1994). She had 1456 observations, fewer than we would like. The Lomako bonobos, though, were poorly habituated, and their shyness around humans meant she could only observe them in the trees, where they felt safe from the potentially dangerous observers; still, leaping and brachiation are arboreal behaviors, so this limitation would not bias her data. On the other hand, when primates are fearful, they are more likely to engage in risky behaviors while fleeing – behaviors like leaping and perhaps brachiation.

We were all looking forward to a study on fully habituated subjects that would include observations both on the ground and in the trees. My student Gil Ramos (Ramos, 2014) provided just such a study at Lui Kotale, Congo. His massive study included over 65,000 observations, nearly 50 times as many as Doran, which should have answered all our questions. Instead, his conclusions conflicted so much with expectations they could not help but be controversial. He found that bonobos engaged in less brachiation and less leaping than Doran had found – and even less than observed among chimpanzees (Table 29.2), quite in conflict with expectations based on their anatomy. Bonobos do have smaller flexor sheath ridges than chimpanzees, suggesting *less* suspensory behavior (Chapter 9). We are left wondering if our expectations about the anatomy related to leaping is wrong, whether the Lui Kotale chimpanzees were observed during a period when they were behaving atypically, whether Lui Kotale is an unusual habitat that requires less leaping and brachiation than "normal" bonobo environments, or whether some other unexpected variable has thrown a monkey wrench in our interpretations.

29.7 Greater Terrestriality?

In another surprise, Ramos found that bonobos also spent more time on the ground than do chimpanzees.

Given bonobo anatomy, it is tempting to suspect that Doran's observations, even though they were few and were on somewhat fearful primates, were better – except that we find support for Ramos' data from a completely unexpected quarter. Human semicircular canals resemble those of bonobos more than those of chimpanzees (El Khoury et al., 2014). These canals house an organ that helps to keep the head steady during locomotion. Demands for balance and stabilization are different when moving in trees than on the ground, suggesting that bonobos are more terrestrial than we thought. A narrow body plan is more efficient during terrestrial locomotion because it reduces moments around the joints of the stance phase limbs (if this seems baffling, see Chapter 9). There is a further confirming bit of data. A critical food for bonobos is **terrestrial herbaceous vegetation** (THV), pithy foods that are found only on the ground, and the lure of this terrestrial food may keep bonobos on the ground more than chimpanzees.

29.8 Female Bonds

We started out the chapter with a look at the confident authority females exert in bonobo society. Bonobo males fail to form close bonds, and groom one another less often than they groom females. Among bonobos, a male's closest social partner and most dependable ally is not another male, but his mother (Parish & de Waal, 2000; Hohmann & Fruth, 2002; Surbeck et al., 2011, 2017). The closest bonds among bonobos are among females, not males, yet females transfer groups at adulthood (Gerloff et al., 1999; Eriksson et al., 2006; Hashimoto & Furuichi, 2001; Hohmann & Fruth, 2002); among other primates, only females who are philopatric – who do not disperse – form close bonds. Evolution favors kin bonds. When two individuals join forces to secure a resource against competitors, kin are preferred as allies because the alliance benefits both actors, increasing the inclusive fitness of each (Wrangham, 1980). When food is defensible, females form alliances that yield a kin-based female-bonded society. Female bonobos flout this rule; they have close alliances but are not kin.

When a female enters a new group, rather than traveling with males for protection as do chimpanzee females, the new female seeks out an older, more established female to serve as an ally (Badrian & Badrian, 1984; Parish, 1996; Parish & de Waal, 2000; Hohmann & Fruth, 2002; Clay & Zuberbühler, 2012). These female–female bonds are forged in fire: The older female comes to the aid of the younger during conflicts (Tokuyama & Furuichi, 2016).

Even more unexpected than the female bonds is the mechanism through which females fortify their friendships – with sexual contact, or **genito-genital rubbing** – "g-g rubbing." They embrace face-to-face and rub their sexual swellings together in a sexual encounter that is little different from that of heterosexual copulation. While bonobo same-sex activity is often described as merely tension-relieving behavior, many observers maintain they see evidence of real sexual pleasure and some female–female bonds are said to resemble human love. The genitalia of bonobos differs from that of chimpanzees, perhaps having evolved to accommodate this copulatory preference; the clitoris and vaginal opening are more ventrally placed (more toward the front of the body), reflecting their tendency to engage in more face-to-face sex, whether homosexual or heterosexual (Dahl, 1985). Interestingly, face-to-face sex is more common among female–female partners than among male–female couples. The unusual genitalia of bonobos may have evolved to foster bonds among females! The female–female bond formed during the new female's early days in her new group seems to continue at least until the immigrating female has her first infant and is established in the community (Parish, 1996; Paoli et al., 2006).

Pair bonds can be powerful, perhaps even more so when they are sexual. Humans have discovered this type of homosexual bonding as a tool for social cohesiveness in war as well. Homosexuality among Greek warriors was thought to bond warriors together and encourage greater bravery in battle (Dover, 1978; Crompton, 2003; Hanson, 2009)

Sexual activity not only extends to all possible combinations of males and females, but to all age and sex combinations, whatever the rank, whatever the age, even down to infants (Kano, 1989).

It may be that not all females disperse; at Lui Kotale an approximately equal number of males and females disappeared from the study community, indirect evidence that some males transfer; two males transferred into the Lui Kotale community (Hohmann & Fruth, 2002). Reinforcing these observations is some genetic evidence for at least occasional male dispersal (Schubert et al., 2011). Independent of this observation, the importance of mother–son alliances suggests that when males disperse it will be more often orphans that transfer than males with living mothers. Given the advantages of kin-based bonding, I expect that we will find that some females who stay home refrain from dispersing in part due to the advantage of a powerful sister. We might expect to find some extremely close alliances among females that are not sexual. Of course, as sexually liberal as bonobos are, for all we know incest may not be taboo.

When a female is challenged, she need not rely only on a female friend for help – if she has a son who is old enough, he will help as well (Parish, 1996; Furuichi, 1997; Hohmann & Fruth, 2002). Grooming patterns reflect these relationships; while a female's most common grooming partner is her son, the next most common partner is another female, and male–male grooming is the least common combination (Idani, 1991; Kano, 1992; White, 1996, 1998; Hohmann & Fruth, 2002).

The reason for the greater female political power within bonobo society is now apparent: Males have as a consistent ally only their mother (Surbeck et al., 2011), whereas the mother has the help of her closest girlfriend, her son, and often other females as well. Surprisingly, a mother may even support her girlfriend over her son (Legrain et al., 2011).

Sons help mothers, but *vice versa* is true as well. Males who achieve high rank typically do so only if they have strong support from their mother; they need it, since there are no male coalitions among bonobos. Mother's help is so important that it can increase a son's mating opportunities, mostly by intervening during one-on-one male contests surrounding estrous females (Surbeck et al., 2011, 2019). Males with living mothers are three times more likely to sire offspring than are orphans (Surbeck et al., 2011, 2019).

As with chimpanzees, dominant males sire most infants (Gerloff et al., 1999), but keep in mind that mothers are active participants in a male's quest for dominance. It may even be that one motivation for older females in forming alliances with younger females is to increase the likelihood that the young females will mate with their sons – mothers as matchmakers. Mothers are so involved in their son's success that it may have even extended maternal lifespans. Among chimpanzees, only 41 percent of males have a living mother; it is 56 percent in bonobos (Surbeck et al., 2011).

29.9 Cohesive, Mixed-Sex Travel Parties

Perhaps the most significant difference in party composition between chimpanzees and bonobos is that bonobos form larger, more cohesive, mixed-sex parties that are less likely to dissolve into smaller groups (White, 1998). Bonobos are less fission–fusion than chimpanzees. This social cohesiveness is seen in captive studies that show that bonobos are more tolerant of one another, allowing them to engage in more cooperation (Hare et al., 2007).

29.10 Vocalizations

The sex role-reversal extends to party formation. Among chimpanzees, males loud-call, or pant-hoot, to tell fellow males of rich food sources so that they can gather to larger groups for protection or to allow patrolling. Among bonobos long-distance calls are much less frequent (de Waal, 1988; Mitani & Nishida, 1993; Hohmann & Fruth, 2002) and have a completely different function. Bonobo females loud-call mostly as a signal to female allies (White et al., 2015). Males loud-call in hopes of attracting mates (White et al., 2015). The relative insignificance of long-distance vocalizations is apparent in the smaller, less effective sound-gathering part of the ear among bonobos, the pinna or external ear.

Male chimpanzees know who is alpha, who is beta, and so on, and each individual must respect his betters – with a specialized vocalization acknowledging subordinance, the pant-grunt. The pant-grunt is one of the most common vocalizations among chimpanzees, heard virtually every time there is a reunion among individuals. Bonobos have no pant-grunt nor any vocalization to perform the same function (Kano, 1992; Parish, 1996; Furuichi, 1997; Hohmann & Fruth 2002, 2003; Paoli et al., 2006).

29.11 Reproduction

While chimpanzee females have hit on a reproductive strategy that lessens the danger to their infants from males, some infants are still killed. Bonobo mothers have a better way: Their sisterhood gives them protection, and males are more thoroughly confused about paternity, compared to chimpanzees.

Among chimpanzees, ovulation is fairly predictable – as the swelling reaches maximal size and slows its expansion, ovulation is near. Females provoke competition among males by advertising ovulation, even as they confuse paternity by mating with all the males. Bonobo females disguise ovulation rather than advertise it (Reichert et al., 2002). Their estrous period is extended across nearly the entire month, with no cue to males that they are about to ovulate (Furuichi, 1992; Parish, 1996). While an aggressive male chimpanzee can prevent other males from copulating with a female near ovulation, this is less often accomplished among bonobos (Furuichi, 1997; Hohmann & Fruth, 2003). In one study even though the top male did manage to isolate the female during the entire period of maximal tumescence, he did not father her offspring (Marvan et al., 2006). Female bonobos have both thoroughly confused paternity and formed alliances that can prevent infanticide.

29.12 Territoriality: Group Defense

Male political ineffectiveness redounds to all corners of bonobo society. Because they fail to form alliances, males not only fail to form coalitions within their community, their lack of bonding means they cannot engage in intercommunity coalitionary violence, either. Absent this male–male cooperative violence (Wrangham, 1999), bonobos lack the signal characteristic of chimpanzee society, the rigid community territorial defense (Kano, 1992; Parish, 1996; Parish & de Waal, 2000; Surbeck & Hohmann, 2013; Table 29.3). As we learned earlier, chimpanzee males defend their community range – it is a territory – and they make war on neighboring communities as a tightly allied paramilitary unit. Bonobos are so much less territorial that males from different communities mingle and may nest near one another without conflict. Communities are peaceful to the extent that males and females of different communities can interact peacefully (Kano, 1992; Hohmann & Fruth, 2002; Furuichi, 2011), and females even copulate with males from other communities in full view of resident males (Idani, 1990, 1991; Furuichi, 2011; Hohmann & Fruth, 2002). If a male chimpanzee contemplated such a thing his head would explode.

Less-aggressive males and lack of territoriality means that murder is extremely uncommon among bonobos. After 40 years of study at four different bonobo communities, there is but one suspected within-species killing – *one* (Wilson et al., 2014) – whereas 50 years of study of 18 chimpanzee provides us with 152 killings (58 observed, 41 inferred, and 53 suspected).

Chimpanzee violence leaves its mark on their skeletons. Healed fractures and puncture wounds are common on chimpanzee crania found at 5.5 percent in museum collections and a spine-chilling 28.6 percent for Gombe males. For bonobos it was only 1.4 percent of individuals (Jurmain, 1997).

Let us not go too far down this road of nonviolence, however. We began the chapter with a male losing part of a toe to violence. Both male and female bonobos engage in aggressive behavior, it is just less common and less dangerous compared to the (literally) bone crushing, genital-removing violence of chimpanzees.

29.13 Tool Use

In the wild, bonobos, compared to chimpanzees, utilize fewer tool types and use them less often

Table 29.3 Comparison of chimpanzees, bonobos and savanna baboons

General information	Chimpanzee	Bonobo	Olive Baboon
Distribution	Equatorial Africa	Congo	Panafrican
Cranial capacity	389 cm^3	350 cm^3	177 cm^3
Gestation length	230 days	240 days	182 days
Societal structure			
Dispersal pattern	Female dispersal	Female dispersal	Male dispersal
Community society?	Yes	Yes	No
Fission–fusion	Yes	Less than *P.t.*	No/minimal
Territoriality	Yes	No/overlap	No
Intercommunity relations	Aggressive	Tense to peaceful	Tense to avoidance
Dominance	Males dominant	Females dominant in coalitions	Females sometimes dominant in coalitions
Single-sex male groups	Frequent	Rare	Rare
Male bonding	Primary	Very limited	Very limited
Male-male alliance	Frequent	Rare	Rare
Female-female association	Infrequent	Frequent	Constant
Female-female bonds/alliances	Limited	Great	Greatest
Female kin bonds	Very limited	Very limited	Pervasive
Females bond to control food	No	Yes	Yes
Mixed-sex groupings	Common only w/ estrus	Unrelated to estrus	Unrelated to estrus
Heterosexual pair bonds	Occasional but weak	No	Present but weak
Mother–son association	Through adolescence	Throughout life	Lacking in adulthood
Party size	Small	Medium	Large
Social behavior			
Short-range contact calls	No	Yes	Yes
Long distance calls	Yes	Less common	No
Submissive greeting	Pant-grunt	None	None
♂ on ♀ physical aggression	Yes	No	Occasional

Table 29.3 (*cont.*)

♀ on ♂ physical aggression	No	Yes	No
Infanticide	Yes	No	Yes
Intercommunity relations	**Murderous**	**Tense to peaceful**	**Tense to avoidance**
Male coalitionary murder	**Yes**	**No**	**No**
Reconciliation	**Common**	**More common**	**More common**
Grooming	Mostly male–male	Mostly female–female	Mostly female–female
Other behavior			
Vocalization pitch	Low	High	Low
Hunting	Mostly males	Mostly females	Rare, mostly male
Food sharing	Among males	Among females	None
Food control	Mostly males	Mostly females	Mixed or female
Physical features			
Cranial capacity	**Larger**	**Smaller**	**Smaller**
External ear size	Large	Medium	Small
Tool use	**Common**	**Less common**	**Less common**
	Sexual characteristics		
Promiscuous copulation	Within community	Within or between	Within
Extra-group copulation	Secret	Public	Probably secret
Sexual coercion	**Yes**	**No**	**No**
Testes size	Large	Large	Large
Concealed ovulation	No	Yes	No
Continuous receptivity	Some	Extensive	None
Genital swelling	At ovulation	Extended	At ovulation
Elaborate sexual repertoire	No	Yes	No
Sexual partners	Mostly heterosexual	All combinations	Heterosexual
Genital contact among females	No	Yes	No
Rump contact among males	No	Yes	No

Bold = bonobos being more monkey-like.
After Parish & de Waal (2000), with some updating.

Table 29.4 **Bonobo diets (percentage of time feeding)**

Site	Fruit	Piths, stems, roots	Leaf	Invertebrates	Meat	Flowers	Bark, seeds other
Bonobo[a]	55	25	14	2	0	2	2
Chimpanzee[b]	65.8	9.9	12.1	5.5	0.8	2.9	3.1

[a] Conklin-Brittain et al., 2001; [b] Chapter 5.

(Ingmanson, 1996; McGrew et al., 2007; Furuichi et al., 2015); only one is used for feeding, the leaf sponge (Furuichi et al., 2015). This is puzzling because in captivity bonobos seem just as adept at tool use as chimpanzees, if not more so; they can make stone tools and use them to gain access to a food reward (Toth et al., 1993).

29.14 Ranging

We talked about the extraordinary spatial memory of chimpanzees; you might expect it would be the same in bonobos. Nope. Chimpanzees are superior (Rosati & Hare, 2012). This suggests that bonobos are more like monkeys, sweeping across their habitat, encountering foods as they hit upon them, whereas chimpanzees, particularly females, must memorize the location of important food resources to forage efficiently.

29.15 Diet

Like chimpanzees, bonobos prefer ripe to unripe fruit (Table 29.4; Badrian et al., 1981; Badrian & Malenky, 1984; White & Wrangham, 1988; White, 1989, 1998; Malenky & Stiles, 1991; Malenky & Wrangham, 1994; Malenky et al., 1994; Furuichi, 2009). Still, even though ripe fruit is important to bonobos, **they eat less fruit than chimpanzees**, substituting an abundant food item they rely on: **terrestrial herbaceous vegetation** (Table 29.4). Chimpanzee spend up to 19 percent of their time eating piths and herbs, but as a species the average is only 10 percent; it is 25 percent for bonobos. Bonobos eat duikers, rodents, birds, bushbabies, and in one case a mangabey (Hohmann & Fruth, 2003), but they eat meat rarely enough that it rounds to zero in Table 29.4. Among bonobos, females, more aggressive than males overall, are observed to lead hunting more often (Hohmann & Fruth, 2003), though trace element reports find no difference in meat consumption (Oelze et al., 2011).

29.16 Evolutionary History

The fact that females disperse among bonobos (Hohmann & Fruth, 2002; Eriksson et al., 2006) suggests that the common ancestor of bonobos and chimpanzees was chimpanzee-like. Perhaps when bonobo ancestors found themselves in an environment where alliances to secure food (more on this below) and protect infants was advantageous, evolution acted to coopt sexual bonding machinery to forge close female bonds.

29.17 Are Bonobos Infantilized?

In the 1980s Brian Shea noticed that many bonobo traits are "neotenized" versions of those of chimpanzees – they resemble infant or juvenile versions of chimpanzees. The more delicate face, more slender build, longer limbs, blacker fur, longer cheek hair, reduced frequency of balding, and even their squeaky vocalizations make them resemble immature chimpanzees (Shea, 1984; Jungers & Susman, 1984). Adult bonobos more often engage in play (Palagi, 2006; Hare et al., 2007, 2012) and are less aggressive, more like juvenile chimpanzees. The alternative, that each of these traits has been selected for and the pattern is coincidentally infantilized-looking, seems quite unlikely, given the long list. While bonobo anatomy and behavior are suggested to be the result of retaining immature traits, this hypothesis does not consider why. I will suggest a new hypothesis for why below.

29.18 Are Bonobos Self-Domesticated?

Others have noted that bonobos differ from chimpanzees in the same way that domesticated animals differ from their wild counterparts. They have smaller faces (similar to the smaller snout of domesticated dogs versus wolves), smaller brains, and reduced tooth size (Wrangham & Pilbeam, 2001; Hare et al., 2012; see also McHenry & Corruccini, 1981). Domesticated animals are less fearful of humans – if they were not they would suffer from a plethora of stress-related diseases – they are also less aggressive (people prefer to mix with animals unlikely to maul them). As is the case with domesticated animals, bonobos seem to cope with crowded conditions well, and they more actively embrace social opportunities (McHenry & Corruccini, 1981; Aureli & de Waal, 1997; Trut et al., 2009; Hare et al., 2012; Wilkins et al., 2014). Dogs will run right up and greet you; wolves will stay back unless they want to eat you. Extended reproductive periods are common in domesticated animals – we usually want them to reproduce quickly – another feature of bonobos (Hare et al., 2012). Bonobos more often retain the white tail tuft into adulthood that chimpanzees lose on maturing (Kano, 1992).

The self-domestication hypothesis (SDH) subsumes the neoteny hypothesis. Many of the SDH traits, including tolerance of social crowding and reduced levels of aggression, are more prominent both among juveniles and also in domesticated animals, suggesting that infantilization is merely the path natural selection travels to reach self-domestication. Keep in mind, though, that not all agree that bonobos have greater social tolerance (Cronin et al., 2015).

Perhaps, SDH holds, larger and more stable parties reduced the benefit of male territoriality, reducing selection for male aggression, and selecting for infantilized behavior, which in turn drove some of the anatomical differences. To sum up, SDH (Wrangham & Pilbeam, 2001; Hare et al., 2012) is reconciled with the neoteny hypothesis in the assumption that selection for reduced aggression and other bonobo traits followed the path of least resistance to achieve greater social cohesiveness, a juvenilization.

29.19 A New Wrinkle: The Monkey Convergence Hypothesis

Thus, bonobos have traits of domesticated animals, which are in part infantilized traits said to be selected to reduce male–male aggression. Perhaps both of these hypotheses can be subsumed under a new hypothesis I propose here: The monkey convergence hypothesis (MCH). The SDH focuses more on withered bonds among males than the flip side, the strengthening of female bonds. The MCH is built on the consequences of females traveling together, rather than remaining in core areas. Female mobility and female bonding in turn eliminate the advantage of territoriality and male bonds.

The key to all this, as it is so often, is food.

Chimpanzees are ripe-fruit specialists, but ripe fruit is a notoriously variable food supply and often chimpanzees must fall back on grasses and terrestrial herbaceous vegetation (THV) when fruit fails. Among bonobos, there is a much greater focus on THV, in part because bonobos have much more edible THV in their habitat than do chimpanzees (Badrian & Malenky, 1984; White & Wrangham, 1988; Malenky & Stiles, 1991; Malenky & Wrangham, 1994; Malenky et al., 1994; Furuichi, 2009). THV is a consistent food supply (White, 1998) compared to fruit. Not only does the bonobo habitat offer more piths and herbs, there are no gorillas in the bonobo habitat taking a gorilla-sized share of THV, as they do elsewhere (Wrangham, 1993; Malenky & Wrangham, 1994).

The greater availability of THV in the bonobo habitat and a greater year-round food availability in general, due to less seasonality, allows bonobos to gather in larger parties (Furuichi, 2009); the parties are both larger and more consistent in number (Furuichi, 2011) – chimpanzees sometimes gather in huge parties, but are mostly found in small ones. Bonobo parties are more often mixed-sex, with females remaining in the traveling group even when not in estrus (White, 1988, 1989, 1996, 1998; Chapman et al., 1994; Furuichi, 2009). Bonobos are, in other words, less fission–fusion than chimpanzees, and it is the fission–fusion nature of chimpanzee society that promotes violence by allowing imbalances of power. Less fission–fusion promotes female bonds.

Evolutionary pressures related to food acquisition are intense among females because females gestate, care for infants, and provide nutrition for that infant by nursing – all of which are nutrient-intense activities. As a consequence, when food is defensible and females can gather in groups, they tend to form bonds (Wrangham, 1980) to defend that food. Bonobo feeding sites typically can accommodate more than one female, so females ally themselves with other females (White & Wood, 2007). Larger party sizes and more frequent mixed-sex parties also place females in direct competition with males. Females are more attuned to food, so they are selected to fight harder for it; bonds allow food defense.

Typical monkeys are found in large, stable groups. Their tolerance of antifeedants, including both fiber and secondary compounds, allows them to experience their food supply as more evenly distributed compared to apes. They include not only ripe fruit in their diet, but flowers, leaves, or unripe fruit. The bonobo habitat contains abundant THV, which fills in the gaps between the more dispersed ripe fruit supply (Wrangham, 1986). Bonobos so effectively utilize THV that unlike chimpanzees, fruit availability has no influence on party size (Serckx et al., 2014).

Let me be clearer about why chimpanzee females have not gone down the same path. Chimpanzee food is too dispersed for females to travel together; two females must double the feeding sites they visit, doubling the distance they must travel. The cost of group travel is heavy, while the advantages of group life are minimal. Female feeding efficiency is highest when they are alone (Wrangham, 1979, 2000; Wrangham & Smuts, 1980). The consistent food supply in the bonobo habitat allows females to travel together, allowing them to bond to control access to food (Parish, 1994), much as do female monkeys (Wrangham, 1980).

Bonobos have many other monkey-like characteristics. They keep tabs on their compatriots' locations with contact calls (Bermejo & Omedes, 1999; Furuichi, 2009), a type of vocalization that is common among monkeys (e.g., baboons [Andrew, 1976; Owren et al., 1997; Rendall et al., 1999, 2000, 2004]). When walking among baboons one hears a chorus of grunts on all sides as group members monitor the location of their kith and kin. Such monitoring is found when maintaining group cohesion is important in the social system, as it is among many monkey species. Gorillas also rely on contact calls to maintain group cohesion (Fossey, 1972). Bonobos maintain group cohesion with a "travel" call that notifies party members that travel is imminent (Schamberg et al., 2016).

The difference in foraging strategies is reflected in cognition. Chimpanzees have evolved to know exactly where their resources are and to scour their habitat vacuuming up sugar. Chimpanzees seem to know where the resources are at all times (Wrangham, 1977), and their incredible skill at spatial tasks and spatial memorization (Chapter 18) reflects the strong selective pressure that has acted on their cognition to allow them to harvest dispersed foods efficiently. Bonobos lack the spatial competence of chimpanzees (Rosati & Hare, 2012).

The stationary distribution of chimpanzee females allows males to guarantee mating access to them by guarding a territory. Orangutan females are also stationary and orangutan males are territorial; there is greater similarity in the two systems than many appreciate. Because bonobo females are mobile; guarding a territory does not guarantee mating access. Instead, bonobo males follow females and mate opportunistically – like monkeys.

Not only is there less motivation to guard a territory among bonobos, larger, more stable party sizes eliminate imbalances of power that allow successful coalitionary violence (Wrangham, 1999). With little advantage to territoriality and little prospect of successful coalitionary violence, male bonds have little advantage.

Greater sexual dimorphism in bonobos than in chimpanzees reflects the one-on-one male contest competition in bonobos. One-on-one combat selects for larger body size (Table 29.3), much like we see among baboons or gorillas. A greater reproductive skew among bonobos (Ishizuka et al., 2018) may well be due to a lesser need for bonobo males to defer to one another to maintain bonds; an alpha chimpanzee may have to tolerate a lower-ranking male hovering around an estrous female because chasing him off may disrupt a bond needed for territorial defense. Bonobos have no such need.

The self-domestication hypothesis sees smaller brains and less sexual dimorphism in weaponry and skull size (Figure 29.2) as somewhat unintended consequences of domestication. The monkey convergence hypothesis sees reduced brain size as a consequence of a lesser need for mind-reading and information exchange. Because there are fewer "off-stage" events among bonobos, due to a less fission–fusion society, less brain power is required (Hunt, 2016). While there is currently no evidence that domesticated animals are less intelligent than their wild counterparts, I am predicting we will find this to be the case, despite evidence that some skills are improved, e.g., interpretation of pointing (Hare et al., 2002). As we discussed in Chapters 18 and 19, monkeys have a socially cohesive system in which any two individuals have almost exactly the same social knowledge, because the group always travels together.

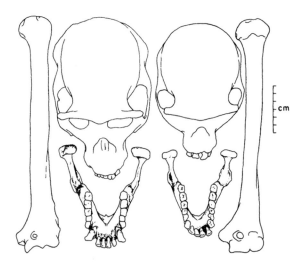

Figure 29.2 Chimpanzees (left) have more prognathic faces, larger brow ridges, larger brains, larger muscle attachment areas (notice the larger temporal fossa though which the temporalis muscle, a chewing muscle, passes), more robust jaws, and larger canines compared to bonobos (right). Image courtesy of Henry McHenry.

If not an eye witness to interactions that resulted in rank changes or alliance changes, they were almost certainly auditors – fighting baboons are noisy. Chimpanzees, on the other hand, each have different social knowledge due to the constant shuffling among parties – and "mind-reading" is therefore advantageous.

Because male bonobos need not go easy on one another, as chimpanzee males must (Chapter 26), we should expect a bigger disparity in wounding aggression between male–male and female–female interactions. I know of no report that suggests that male–male contest competition is more intense among bonobos, but let us hope someone examines it. The SDH predicts less wounding among males, while the MCH expects more because there is less need for deference and bonding. While I am willing to accept a few unexpected observations in melding the SDH and MCH, it would be very interesting to have data that directly compare male–male aggression levels in bonobos, chimpanzees, and a representative Old World monkey; this is an important area for future research.

29.20 Conclusions and Future Research

Perhaps bonobo specialists already have data to test some aspects of the MCH. In review, among bonobos we might expect to find a lower rate of female transfer than in chimpanzees, accompanied by alliances among sisters; there is some evidence of this, because males more often have both their mother *and* their grandmother in the group of residence (Schubert et al., 2013). The MCH expects that male bonobos will have more one-on-one aggressive interactions with higher rates of wounding compared to chimpanzees, whereas the SDH predicts the opposite. More detailed study of sex differences in vocalizations would be valuable; the MCH expects that females more often, rather than males, should engage in ally-drawing long-distance vocalizations, but a lower level of long-distance communication because groups separate less. We expect higher rates of male transfer among bonobos; there is indirect evidence this is so (Hohmann & Fruth, 2002). Many monkeys display allomothering – with tighter female bonds, this aunting behavior should be more common among bonobos than chimpanzees.

Monkeys have cheek pouches that, in part, allow them to keep up with the rest of the group by doing their chewing later, when everyone has stopped. We might expect more wadging among bonobos than is seen among chimpanzees. I am certain my monkey-studying colleagues will come up with many more testable predictions.

29.21 Lessons

It is tempting to generalize a behavior we see in any one primate species to all primates, but bonobos and chimpanzees show us that there can be dramatic differences even in closely related species. In particular, some might look to chimpanzees to find some justification for male patriarchy and male social dominance. You have already noticed, I am sure, that our equally closely related relative the bonobo shows that nature can go in exactly the opposite direction, even when females are smaller than males.

Let us remember, however, that female power is not equal to peace. In chimpanzee-land, males bully other males and all females; if only the females could bond together, it is tempting to think, all would be peace. Not so. While bonobos teach us that banding together can stymie male bullies, strong female alliances are not necessarily all sweetness and light. It is cold comfort to a bonobo male who lives in fear of an oppressive gang to know that it is a female group rather than a single bully that is oppressing him. If we are searching for a utopia, it would be a society the reverse of chimpanzee society, but one where neither male nor female gangs oppress a minority.

Humans may have found the solution. Or at least, they do a better job of ameliorating gang violence than apes. Among hunter-gatherers, when one individual becomes too violent and too oppressive – a condition that prevails in many such societies, since there is no organized policing body – subordinates band together to put him in his place, and sometimes that place is a grave (Boehm, 2009). While legal bureaucracies and policing have failed to completely stamp out physical violence in industrialized societies, there has been some progress. Preliminary data suggest murder is hundreds of times less common in industrialized versus hunter-gatherer societies (Wrangham et al., 2006).

Sadly, as we have advanced on one front to institutionally prohibit and punish murder, we have lost ground on another; in our large, fluid, impersonal, hierarchical society we have not found a foolproof way to reign in oppressors. Too often crime syndicates find a way to compromise institutional checks. Perhaps here is a case where we can take a lesson from ourselves, rather than apes. At various times in human history an enlightened band of right-thinkers has realized that when oppressors threaten them with absolute despotism, it is their right and indeed it is their duty to throw off such oppression and band together to provide new guarantees of freedom from tyranny. Chimpanzees have never done that. Yet.

References

Almquist AJ (1974) Sexual differences in the anterior dentition in African primates. *Am J Phys Anthropol* **40**, 359–368.

Andrew RJ (1976) Use of formants in the grunts of baboons and other nonhuman primates. *Ann NY Acad Sci* **280**, 673–693.

Angier N (2016) A society leg by strong females. *New York Times*, 13 September, pp. 32–33.

Aureli F, de Waal FBM (1997) Inhibition of social behavior in chimpanzees under high-density conditions. *Am J Primatol* **41**, 213–228.

Badrian A, Badrian N (1984) Social organization of *Pan paniscus* in the Lomako Forest, Zaire. In *The Pygmy Chimpanzee: Evolutionary Biology and Behavior* (ed. Susman RL), pp. 325–346. New York: Plenum Press.

Badrian N, Malenky R (1984) Feeding ecology of *Pan paniscus* in the Lomako Forest, Zaire. In *The Pygmy Chimpanzee: Evolutionary Biology and Behavior* (ed. Susman RL), pp. 275-299. New York: Plenum Press.

Badrian N, Badrian A, Susman RL (1981) Preliminary observations on the feeding behavior of *Pan paniscus* in the Lomako forest of central Zaire. *Primates* 22, 173-181.

Bermejo M, Omedes A (1999) Preliminary vocal repertoire and vocal communication of wild bonobos (*Pan paniscus*) at Lilungu (Democratic Republic of Congo) *Folia Primatologica* 70, 328-357.

Boehm C (2009) *Hierarchy in the Forest: The Evolution of Egalitarian Behavior.* Cambridge, MA: Harvard University Press.

Chapman CA, White FJ, Wrangham RW (1994) Party size in chimpanzees and bonobos. In *Chimpanzee Cultures* (eds. Wrangham RW, McGrew WC, de Waal FBM, Heltne PG), pp. 41-57. Cambridge, MA: Harvard University Press.

Clay Z, Zuberbühler K (2012) Communication during sex among female bonobos: effects of dominance, solicitation and audience. *Scientific Reports* 2, 291.

Conklin-Brittain NL, Knott CD, Wrangham RW (2001) *The feeding ecology of apes.* In *Conference Proceedings: The Apes – Challenges for the 21st Century*, pp. 167-174.

Coolidge HJ (1933) *Pan paniscus*: pygmy chimpanzee from south of the Congo River. *Am J Phys Anthropol* 18, 1-57.

Cramer DL (1977) Craniofacial morphology of *Pan paniscus*: a morphometric and evolutionary appraisal. *Contributions Primatol* 10, 1-64.

Crompton L (2003) *Homosexuality and Civilization.* Cambridge, MA: Harvard University Press.

Cronin KA, De Groot E, Stevens JMG (2015) Bonobos show limited social tolerance in a group setting: a comparison with chimpanzees and a test of the relational model. *Folia Primatol* 86, 164-177.

Dahl JF (1985) The external genitalia of female pygmy chimpanzees. *Anat Rec* 211, 24-28.

de Waal FBM (1988) The communicative repertoire of captive bonobos (*Pan paniscus*), compared to that of chimpanzees. *Behaviour* 106, 183-251.

de Waal FBM, Lanting F (1997) *Bonobo: The Forgotten Ape.* Berkeley, CA: University of California Press.

Doran DM, Hunt KD (1994) The comparative locomotor behavior of chimpanzees and bonobos: species and habitat differences. In *Chimpanzee Cultures* (eds. Wrangham RW, McGrew WC, de Waal FBM, Heltne PG), pp. 93-108. Cambridge, MA: Harvard University Press.

Douglas PH (2014) Female sociality during the daytime birth of a wild bonobo at Luikotale, Democratic Republic of the Congo. *Primates* 55, 533-542.

Dover KJ (1978) *Greek Homosexuality.* Cambridge, MA: Harvard University Press.

El Khoury M, Braga J, Dumoncel J, et al. (2014) The human semicircular canals orientation is more similar to the bonobos than to the chimpanzees. *PLoS ONE* 9, e93824.

Eriksson J, Siedel H, Lukas D, et al. (2006) Y-chromosome analysis confirms highly sex-biased dispersal and suggests a low male effective population size in bonobos (*Pan paniscus*). *Molec Ecol* 15, 939-949.

Fenart R, Deblock R (1974) Sexual differences in adult skulls in *Pan troglodytes. J Hum Evol* 3, 123-133.

Fossey D (1972) Vocalizations of the mountain gorilla (*Gorilla gorilla beringei*). *Anim Behav* 20, 36-53.

Furuichi T (1992) The prolonged estrus of females and factors influencing mating in a wild group of bonobos (*Pan paniscus*) in Wamba, Zaire. In *Topics in Primatology* (eds. Matano S, Tuttle RH, Ishida H, Goodman M), pp. 179-190. Tokyo: University of Tokyo Press.

Furuichi T (1997) Agonistic interactions and matrifocal dominance rank of wild bonobos (*Pan paniscus*) at Wamba. *Int J Primatol* 18, 855-875.

Furuichi T (2009) Factors underlying party size differences between chimpanzees and bonobos: a review and hypotheses for future study. *Primates* 50, 197-209.

Furuichi T (2011) Female contributions to the peaceful nature of bonobo society. *Evol Anthropol* 20, 131-142.

Furuichi T, Sanz C, Koops K, et al. (2015) Why do wild bonobos not use tools like chimpanzees do? *Behaviour* 152, 425-460.

Gerloff U, Hartung B, Fruth B, et al. (1999) Intracommunity relationships, dispersal pattern and paternity success in a wild living community of bonobos (*Pan paniscus*) determined from DNA analysis of faecal samples. *Proc Roy Soc B* 266, 1189-1195.

Hanson VD (2009) *The Western Way of War: Infantry Battle in Classical Greece.* Berkeley, CA: University of California Press.

Hare B, Brown M, Williamson C, Tomasello M (2002) The domestication of social cognition in dogs. *Science* 298, 1634-1636.

Hare B, Melis AP, Woods V, Hastings, S, Wrangham, RW (2007) Tolerance allows bonobos to outperform chimpanzees on a cooperative task. *Curr Biol* 17, 619-623.

Hare B, Wobber V, Wrangham RW (2012) The self-domestication hypothesis: evolution of bonobo psychology is due to selection against aggression. *Anim Behav* 83, 573-585.

Hashimoto C, Furuichi T (2001) Intergroup transfer and inbreeding avoidance in bonobos. *Primate Res* 17, 259-269.

Hohmann G, Fruth B (2002) Dynamics in social organization of bonobos (*Pan paniscus*). In *Behavioural Diversity in Chimpanzee* (eds. Boesch C, Marchant L, Hohmann G), pp. 138-150. Cambridge: Cambridge University Press.

Hohmann G, Fruth B (2003) Intra- and inter-sexual aggression by bonobos in the context of mating. *Behaviour* 140, 1389–1413.

Hohmann G, Fruth B (2008) New records on prey capture and meat eating by Bonobos at Lui Kotale, Salonga National Park, Democratic Republic of Congo. *Folia Primatol* 79, 103–110.

Hunt KD (1989) Positional behavior in *Pan troglodytes* at the Mahale Mountains and Gombe Stream National Parks, Tanzania. PhD dissertation. Ann Arbor, MI: University of Michigan.

Hunt KD (1992) Positional behavior of *Pan troglodytes* in the Mahale Mountains and Gombe Stream National Parks, Tanzania. *Am J Phys Anthropol* 87, 83–107.

Hunt KD (2016) Why are there apes? Evidence for the co-evolution of ape and monkey ecomorphology. *J Anat* 228, 630–685.

Idani G (1990) Relations between unit-groups of bonobos at Wamba, Zaire: encounters and temporary fusions. *Afr Stud Monogr* 11, 153–186.

Idani G (1991) Social relationships between immigrant and resident bonobo (*Pan paniscus*) females at Wamba. *Folia Primatol* 57, 83–95.

Ingmanson EJ (1996) Tool-using behavior in wild *Pan paniscus* social and ecological considerations. In *Reaching into Thought: The Minds of the Great Apes* (eds. Russon AE, Bard KA, Parker ST), pp. 190–210. New York: Cambridge University Press.

Ishizuka S, Kawamoto Y, Sakamaki T, et al. (2018) Paternity and kin structure among neighbouring groups in wild bonobos at Wamba. *Roy Soc Open Sci* 5, 171006.

Johanson DC (1974) Some metric aspects of the permanent and deciduous dentition of the pygmy chimpanzee (*Pan paniscus*). *Am J Phys Anthropol* 1(41), 39–48.

Jungers WL, Susman RL (1984) Body size and skeletal allometry in African apes. In *The Pygmy Chimpanzee: Evolutionary Biology and Behavior* (ed. Susman RL), pp. 131–177. New York: Plenum Press.

Jurmain R (1997) Skeletal evidence of trauma in African apes, with special reference to the Gombe chimpanzees. *Primates* 38, 1–14.

Kano T (1989) The sexual behavior of pygmy chimpanzees. In *Understanding Chimpanzees* (eds. Heltne P, Marquardt L), pp. 176–183. Cambridge, MA: Harvard University Press.

Kano T (1992) *The Last Ape: Pygmy Chimpanzee Behavior and Ecology*. Stanford, CA: Stanford University Press.

Legrain L, Stevens J, Iscoa JA, Destrebecqz A (2011) A case study of conflict management in bonobos: how does a bonobo (*Pan paniscus*) mother manage conflicts between her sons and her female coalition partner? *Folia Primatologica* 82, 236–243.

Malenky RK, Stiles EW (1991) Distribution of terrestrial herbaceous vegetation and its consumption by *Pan paniscus* in the Lomako Forest, Zaire. *Am J Primatol* 23, 153–169.

Malenky RK, Wrangham RW (1994) A quantitative comparison of terrestrial herbaceous food consumption by *Pan paniscus* in the Lomako Forest, Zaïre, and *Pan troglodytes* in the Kibale Forest, Uganda. *Am J Primatol* 32, 1–12.

Malenky RK, Kuroda S, Vineberg EO, Wrangham RW (1994) The significance of terrestrial herbaceous foods for bonobos, chimpanzees and gorillas. In *Chimpanzee Cultures* (eds. Wrangham RW, McGrew WC, de Waal FB, Heltne PG), pp. 59–75. Cambridge, MA: Harvard University Press.

Marvan R, Stevens JMG, Roeder AD, et al. (2006) Male dominance rank, mating and reproductive success in captive bonobos (*Pan paniscus*). *Folia Primatol* 77, 364–376.

McGrew WC, Marchant LF, Beuerlein MM, et al. (2007) Prospects for bonobo insectivory: Lui Kotale, democratic republic of Congo. *Int J Primatol* 28, 1237–1252.

McHenry HM, Corruccini RS (1981) *Pan paniscus* and human evolution. *Am J Phys Anthropol* 54, 355–367.

Mitani JC, Nishida T (1993) Contexts and social correlates of long-distance calling by male chimpanzees. *Anim Behav* 45, 735–746.

Morbeck ME, Zihlman AL (1989) Body size and proportions in chimpanzees, with special reference to *Pan troglodytes schweinfurthii* from Gombe National Park, Tanzania. *Primates* 30, 369–382.

Nissen HW (1931) A field study of the chimpanzee: observations of chimpanzee behavior and environment in western French Guinea. *Comp Psychol Monogr* 8, 1–122.

Oelze VM, Fuller BT, Richards MP, et al. (2011) Exploring the contribution and significance of animal protein in the diet of bonobos by stable isotope ratio analysis of hair. *PNAS* 108, 9792–9797.

Owren MJ, Seyfarth RM, Cheney DL (1997) The acoustic features of vowel-like grunt calls in chacma baboons (*Papio cyncephalus ursinus*): implications for production processes and functions. *J Acoust Soc Am* 101, 2951–2963.

Palagi E (2006) Social play in bonobos (*Pan paniscus*) and chimpanzees (*Pan troglodytes*): implications for natural social systems and interindividual relationships. *Am J Phys Anthropol* 129, 418–426.

Paoli T, Palagi E, Tacconi G, Tarli SB (2006) Perineal swelling, intermenstrual cycle, and female sexual behavior in bonobos (*Pan paniscus*). *Am J Primatol* 68, 333–347.

Parish AR (1994) Sex and food control in the uncommon chimpanzee: how bonobo females overcome a phylogenetic legacy of male dominance. *Ethol Sociobiol* 15, 157–179.

Parish AR (1996) Female relationships in bonobos (*Pan paniscus*): evidence for bonding, cooperation, and female dominance in a male-philopatric species. *Hum Nat* **7**, 61–96.

Parish AR, de Waal FBM (2000) The other "closest living relative:" how bonobos (*Pan paniscus*) challenge traditional assumptions about females, dominance, intra- and intersexual interactions, and hominid evolution. *Ann NY Acad Sci* **907**, 97–113.

Pusey AE, Oehlert GW, Williams JM, Goodall J (2005) Influence of ecological and social factors on body mass of wild chimpanzees. *Int J Primatol* **26**, 3–31.

Ramos GL III (2014) Positional behavior in *Pan paniscus* at Lui Kotale, Democratic Republic of Congo. PhD dissertation. Bloomington, IN: Indiana University.

Reichert KE, Heistermann M, Keith Hodges J, Boesch C, Hohmann G (2002) What females tell males about their reproductive status: are morphological and behavioural cues reliable signals of ovulation in bonobos (*Pan paniscus*)? *Ethology* **108**, 583–600.

Rendall D, Seyfarth RM, Cheney DL, Owren MJ (1999) The meaning and function of grunt variants in baboons. *Anim Behav* **57**, 583–592.

Rendall D, Cheney DL, Seyfarth RM (2000) Proximate factors mediating "contact" calls in adult female baboons (*Papio cynocephalus ursinus*) and their infants. *J Comp Psych* **114**, 36.

Rendall D, Owren MJ, Weerts E, Hienz RD (2004) Sex differences in the acoustic structure of vowel-like grunt vocalizations in baboons and their perceptual discrimination by baboon listeners. *J Acoust Soc Am* **115**, 411–421.

Rilling JK, Scholz J, Preuss TM, et al. (2012) Differences between chimpanzees and bonobos in neural systems supporting social cognition. *Social Cogn Affect Neurosci* **7**, 369–379.

Rosati A, Hare B (2012) Chimpanzees and bonobos exhibit divergent spatial memory development. *Dev Sci* **15**, 840–853.

Schaefer K, Mitteroecker P, Gunz P, Bernhard M, Bookstein FL (2004) Craniofacial sexual dimorphism patterns and allometry among extant hominids. *Ann Anatomy-Anatomischer Anzeiger* **86**, 471–478.

Schamberg I, Cheney DL, Clay Z, Hohmann G, Seyfarth RM (2016) Call combinations, vocal exchanges and interparty movement in wild bonobos. *Anim Behav* **122**, 109–116.

Schubert G, Stoneking CJ, Arandjelovic M, et al. (2011) Male-mediated gene flow in patrilocal primates. *PLoS ONE* **6**, e21514.

Schubert G, Vigilant L, Boesch C, et al. (2013) Co-residence between males and their mothers and grandmothers is more frequent in bonobos than chimpanzees. *PLoS One* **8**, e83870.

Serckx A, Huynen MC, Bastin JF, et al. (2014) Nest grouping patterns of bonobos (*Pan paniscus*) in relation to fruit availability in a forest-savannah mosaic. *PloS ONE* **9**, e93742.

Shea BT (1984) An allometric perspective on the morphological and evolutionary relationships between pygmy (*Pan paniscus*) and common (*Pan troglodytes*) chimpanzees. In *The Pygmy Chimpanzee* (ed. Susman RL), pp. 89–130. New York: Plenum Press.

Smith RJ, Jungers WL (1997) Body mass in comparative primatology. *J Hum Evol*, **32**, 523–559.

Surbeck M, Hohmann G (2013) Intersexual dominance relationships and the influence of leverage on the outcome of conflicts in wild bonobos (*Pan paniscus*) *Behav Ecol Sociobiol* **67**, 1767–1780.

Surbeck M, Fowler A, Deimel C, Hohmann G (2009) Evidence for the consumption of arboreal, diurnal primates by bonobos (*Pan paniscus*). *Am J Primatol* **71**, 171–174.

Surbeck M, Mundry R, Hohmann G (2011) Mothers matter! Maternal support, dominance status and mating success in male bonobos (*Pan paniscus*). *Proc Roy Soc Lond B* **278**, 590–598.

Surbeck M, Girard-Buttoz C, Boesch C, et al. (2017) Sex-specific association patterns in bonobos and chimpanzees reflect species differences in cooperation. *Roy Soc Open Sci* **4**, 161081.

Surbeck M, Boesch C, Crockford C, et al. (2019) Males with a mother living in their group have higher paternity success in bonobos but not chimpanzees. *Curr Biol* **29**, R354–R355.

Susman RL (1979) Comparative and functional morphology of hominoid fingers. *Am J Phys Anthropol* **50**, 215–236.

Tokuyama N, Furuichi T (2016) Do friends help each other? Patterns of female coalition formation in wild bonobos at Wamba. *Anim Behav* **119**, 27–35.

Toth N, Schick KD, Savage-Rumbaugh ES, Sevcik RA, Rumbaugh DM (1993) *Pan* the tool-maker: investigations into the stone tool-making capabilities of a bonobo (*Pan paniscus*). *J Arch Sci* **20**, 81–91.

Trut L, Oskina I, Kharlamova A (2009) Animal evolution during domestication: the domesticated fox as a model. *Bioessays* **31**, 349–360.

Vervaecke H, Van Elsacker L (1992) Hybrids between common chimpanzees (*Pan troglodytes*) and pygmy chimpanzees (*Pan paniscus*) in captivity. *Mammalia* **56**, 667–669.

Videan EN, McGrew WC (2001) Are bonobos (*Pan paniscus*) really more bipedal than chimpanzees (*Pan troglodytes*)? *Am J Primatol* **54**, 233–239.

White FJ (1988) Party composition and dynamics in *Pan paniscus*. *Int J Primatol* **9**, 179–193.

White FJ (1989) Ecological correlates of pygmy chimpanzee social structure. In *Comparative Socioecology: The*

Behavioural Ecology of Humans and Other Mammals (eds. Standen V, Foley RA), pp. 151–164. Oxford: Blackwell Scientific.

White FJ (1996) *Pan paniscus* 1973–1996: twenty-three years of field research. *Evol Anthropol* **5**, 11–17.

White FJ (1998) Seasonality and socioecology: the importance of variation in fruit abundance to bonobo sociality. *Int J Primatol* **19**, 1013–1027.

White FJ, Wood KD (2007) Female feeding priority in bonobos, *Pan paniscus*, and the question of female dominance. *Am J Primatol* **69**, 837–850.

White FJ, Wrangham RW (1988) Feeding competition and patch size in the chimpanzee species *Pan paniscus* and *Pan troglodytes*. *Behaviour* **105**, 148–164.

White FJ, Waller M, Boose K, Merrill MY, Wood KD (2015) Function of loud calls in wild bonobos. *J Anthropol Sci* **93**, 1–13.

Wilkins AS, Wrangham RW, Fitch WT (2014) The "Domestication Syndrome" in mammals: a unified explanation based on neural crest cell behavior and genetics. *Genetics* **197**, 795–808.

Wilson ML, Boesch C, Fruth B, et al. (2014) Lethal aggression in *Pan* is better explained by adaptive strategies than human impacts. *Nature* **513**, 414–417.

Wrangham RW (1977) Feeding behaviors of chimpanzees in Gombe National Park, Tanzania. In *Primate Ecology* (ed. Clutton-Brock, TH), pp. 503–538. London: Academic Press.

Wrangham RW (1979) The evolution of ape social systems. *Soc Sci Infor* **18**, 335–368.

Wrangham RW (1980) An ecological model of female-bonded primate groups. *Behaviour* **75**, 262–300.

Wrangham RW (1986) Ecology and social relationships in two species of chimpanzees. In *Ecological Aspects of Social Evolution* (eds. Rubenstein DI, Wrangham RW), pp. 352–378. Princeton, NJ: Princeton University Press.

Wrangham RW (1993) The evolution of sexuality in chimpanzees and bonobos. *Hum Nat* **4**, 47–79.

Wrangham RW (1999) Evolution of coalitionary killing. *Am J Phys Anthropol* **110**(S29), 1–30.

Wrangham RW (2000) Why are male chimpanzees more gregarious than mothers? A scramble competition hypothesis. In *Male Primates* (ed. Kappeler P), pp. 248–258. Cambridge: Cambridge University Press.

Wrangham RW, Pilbeam DR (2001) African apes as time machines. In *All Apes Great and Small*, Vol 1: *African Apes* (eds Galdikas BMF, Briggs NE, Sheeran LK, Shapiro GL, Goodall J), pp. 5–17. Berlin: Springer.

Wrangham RW, Smuts BB (1980) Sex differences in the behavioural ecology of chimpanzees in the Gombe National Park, Tanzania. *J Reprod Fert* (Suppl) **28**, 13–31.

Wrangham RW, Wilson ML, Muller MN (2006) Comparative rates of violence in chimpanzees and humans. *Primates* **47**, 14–26.

Yamamoto S (2015) Non-reciprocal but peaceful fruit sharing in wild bonobos in Wamba. *Behaviour* **152**, 335–357.

30 Sister Species
Lessons from the Chimpanzee

Credit: Anup Shah / Stone / Getty Images

The centuries-long discovery of the chimpanzee, its ecology, anatomy, and behavior, has consisted of a slow-motion, step-by-step revelation that humans and chimpanzees are more, then still more, then *still more* similar than expected. Yet this reversal has its origin in the perception that humans were a thing apart with respect to other species. Humans and chimpanzees, despite their micro-similarities, have macro differences as well; chimpanzees have fur; their arms are longer than their legs; their brains are smaller.

30.1 Lessons, Links, and Insights into Human Evolution

In recounting the history of chimpanzee research, we gain insights into epistemology – for instance, what we can learn in the lab versus in the wild – the influence of social forces on science, the evolutionary history of the apes, the nature of chimpanzees as organisms, and links between environment and physiology. In the first section of this chapter I will offer these perspectives as two dozen lessons drawn from the information in previous chapters; citations are in those chapters. In a second section, I will abstract from chimpanzee research the fundamental relationships among chimpanzee traits, the "if a then b" evolutionary origin of chimpanzee uniquenesses; I will try to assemble a holistic perspective on the chimpanzee. For those of us who concern ourselves with human evolution, the similarities and the differences between humans and chimpanzees offer us a counterpoint to or corroboration of evidence from the study of fossils. In a final section, I will try to relate information gained from chimpanzees to human origins.

30.2 Lessons from the Chimpanzee

1. We are dependent on wild studies when we wish to understand any aspect of any animal, including chimpanzees. As informative as captive studies can be, only the study of wild primates can unlock the origin of any physical and behavioral trait. Adaptations emerge over generations as interactions between individuals and their environment and among individuals result in useful traditions, valuable instincts, and physical adaptations. Captive studies can help us know many of the capabilities of animals, but only wild studies can tell us why they evolved and even whether they use them. We may observe fatal male-on-male attacks in zoos, but we can only observe the fully realized emergent phenomena related to it, male patrolling and territoriality, in the wild.

2. The more complicated a behavior or physical adaptation is, the more important wild studies are. Chimpanzee female–female bonds are often strong in captivity, assuring us that females have social instincts, but it is study of wild chimpanzees that tells us why these instincts are rarely expressed in the wild – they are rare because socializing interferes with foraging, and in the hungry world of tropical Africa,

food-gathering takes priority over socializing. Border patrols, coalitionary killing, strong male affiliation, rigid dominance hierarchies, complex foraging regimes, medicinal plant use, higher reproductive success among dominants, complex tool use – all are emergent phenomena observed first or best in the wild. Other wild behaviors such as patterns of daily movement across large areas (and variation in this), response to predation threats, preying on other animals, and responses to severe food shortages will never be seen in captivity.

3. As important as wild studies are, primate fieldwork is tedious, time-consuming, and exhausting, not to mention a threat to a researcher's health, whether that threat comes from the microscopic malaria parasite or the decidedly macroscopic musthing elephant. Wild chimpanzees are boring in comparison to the exciting distillations of their lives depicted in film, which means the science itself must provide much of the thrill for the researcher. Chimpanzees sit a lot, sometimes staring into space for tens of minutes as they chew, pass gas, urinate, and defecate. Most of their locomotion consists of moving slowly, very slowly, from one feeding site to another; when in small parties they are largely silent. Females may go days without interacting with another adult. The sensational behavior you see in films has been culled from many hours of video, much of which consists of chimpanzees doing not much.

4. Chimpanzees teach us that evolution can take unpredictable, nonlinear paths; different features evolve at different rates. A newcomer to evolutionary primatology might expect that just as the large human brain is a recent innovation, all of human anatomy is recently evolved. Not so. Our teeth and digestive physiology are little changed from those of our ancestors 20 million years ago. This contrast, ancient digestive physiology and an advanced brain, flies in the face of the widespread expectation that the common ancestor of monkeys and apes was very like a monkey; instead it was neither a monkey nor an ape, but an animal with some parts quite monkey-like and some parts more ape-like.

5. "Advanced" (i.e., furthest evolved from an ancient ancestor) and "human" are not the same. Of monkeys and apes, leaf monkeys have the most advanced digestive physiology, humans the least. If we consider DNA alone (Chapter 12), chimpanzees are more evolved than humans – they have more positively selected genes. We have no idea why, but chimpanzees have highly evolved intracellular processes such as mRNA transcription and transferase function.

6. Many traits we typically think of as distinguishing humans from other animals should be characterized as distinguishing *great apes* (which taxonomically includes humans) from other animals. Dexterous hands, upright torsos, mobile shoulders, rotatable wrists, frontated eyes, facial expressions, self-awareness (or self-concept), a slow life history, intolerance of dietary toxins, political awareness, group or coalitionary violence, intergroup conflict, culture, tool use, hunting, and narrative memories are all great ape traits. Welcome to the Hominidea.

7. At first glance chimpanzees seem to rule the primate world, easily displacing every other primate, including (much of the time) gorillas from any resource they desire. This dominance masks a desperate evolutionary struggle chimpanzees have long waged, and continue to wage, with monkeys. Chimpanzees never left behind their ancestral dependence on ripe fruit; monkeys did. Large chimpanzee body size may have evolved in part to allow them to bully monkeys away from vital resources. Large bodies with their slower metabolisms and larger guts also allow apes to utilize low-sugar fallback foods when sugary foods were scarce. The narrowness of the chimpanzee diet drove an evolutionary change in most other adaptations. Monkeys engage in their age-old habit of walking within trees, leaping between them, sitting to feed and feeding as they go, finding something to eat in every other tree or so. Because monkeys get to easy-to-harvest fruits first, chimpanzees have been pushed to the ground by the need to travel long distances between dispersed food sources. Then they must climb to get back into the trees, and when they get there they often must feed among the thin, compliant peripheral branches, subsisting on the scraps left for them by monkeys.

8. Religion and science were once viewed, not as antithetical, but the opposite, as comfortably – no, as

unavoidably – complementary to one another. Medieval Roman Catholic clerics believed investigating nature was a form of religious devotion. Finding and describing a species was an act of religious piety, a chance to better understand God's plan for the natural world. Saint Thomas Aquinas believed that the superb adaptations we see among living things is evidence of God's wisdom. The ultimate describer and classifier of living things, Linnaeus, invented modern taxonomy so that he could "follow in the footsteps of God." Gregor Mendel, the "father of genetics," was a monk who gave up his genetics research to head his monastery. Natural theology, the idea that God made animals well-adapted to their environments, provided much of the data and insights Darwin drew on to conceive of natural selection as the mechanism by which evolution happened. This religious interest in nature played a large role in the search for undiscovered species that ultimately resulted in the discovery of the chimpanzee. The perception that science and religion must be in conflict grew up after *The Origin of Species* removed the human origins story from the Garden of Eden and placed it in the realm of science.

9. Scientific understanding of chimpanzees was uneven. We knew ape anatomy quite well by 1700, but 250 years later we had no idea whether chimpanzees lived in monogamous family units or something completely different. Here are some milestones and their approximate dates of discovery or definitive explanation:

- anatomy: Tyson, 1699
- fission–fusion social structure: Nishida and Goodall, 1968
- intergroup violence is integral to chimpanzee society: 1980
- role of diet in the social system: 1985
- advantage of dominance: 1995
- role of digestion in evolution: 2000
- advantage of territoriality: 2005.

10. Chimpanzees are part of an extraordinary primate diversity that includes 342 species (Appendix 1) and at least six distinct kinds of primates (lemurs, lorises, tarsiers, New World monkeys, Old World monkeys, and apes).

11. Despite the fact that the most often quoted explanation of the evolutionary origin of intelligence looks to the demands of sociality as selecting for advanced cognition, intelligence and social structure are not clearly linked. Four of the five types of social system are found among both large-brained apes and other smaller-brained species. All but one type of society is found in one ape or another; the type of social system not found among apes – kin-based female-bonded – is found in the large-brained capuchin monkeys. Only one of the five, the fission–fusion social system, is found only among large-brained primates.

12. No one primate is a model for humans. I find that when young students first encounter primatology, one of their first thoughts is that they have found answers to the deepest mysteries of human nature: "Of course we are violent [they think] chimpanzees are our closest relatives!" In order to understand humans, however, primatology need not pound the problem with this blunt instrument, that instrument being "We must be like our closest relatives." It has a scalpel. We can explain humans by drawing on the evolutionary rules and trends we derive from research on *many* animals, picking the best animal model for each trait. Even among the very closest of our relatives, the apes, there is great behavioral variety; none of their societies closely resembles that of humans. One of our two closest relatives, the chimpanzee, is male-bonded, male-dominated, and murderous, while our *other* closest relative, the bonobo (or pygmy chimpanzee), is non-kin-female-bonded, female-dominant, less murderous, and highly sexualized. Our next closest relative after the two species of chimpanzee, the gorilla, is different still; gorilla males and females differ dramatically in body weight, are polygynous, and have only a tepid interest in sex. The gibbon, the smallest of the apes, is monogamous, averse to large social gatherings, and monomorphic – males and females differ little in anatomy and little in behavior. Whether your bias is that human nature inclines toward monogamy, pansexuality, promiscuity, homosexuality, or wife-beating (Linden, 2002), you can find an example among the apes. We err when we think any single

species can tell us more than a smidgeon about human nature.

13. Chimpanzees are not wanna-be humans. Instead, they are just as evolved as humans – though in a different direction – and superbly adapted to their forest niche (their ability to live in savanna habitats notwithstanding). Their brilliant spatial memories, prognathic faces, voluminous guts, stiff backs, gripping great toes, and powerful upper bodies are finely attuned to their niche of harvesting sugar-rich ripe fruits in tall forests. Their large, conical canines, off-the-charts aggression, and wound-healing physiology are adaptations to their social conditions.

14. The chimpanzee daily routine is tightly bound by the requirements of their diet. To acquire their optimal nutrition, they must work to maximize the amount of sugar and fat in their diet; they must supplement their sugar diet with a vital but relatively small helping of protein, they must obtain an adequate portion of vitamins and minerals, and they must minimize both secondary compounds and fiber. The rarity of "unprotected" sugar, calories not mixed with antifeedants, forces them to plan their daily routine almost entirely around food-getting.

15. We humans are still influenced by many of the same foraging and feeding imperatives as chimpanzees, inherited from our Miocene ape ancestors, among them the quest for sugar and oil and the instinct to avoid secondary compounds and fiber. These dietary imperatives often overwhelm our awareness of what is the healthiest diet and thereby lead some of us to an early grave. Our doctor tells us to cut back on sugar, eat more fiber, and get more exercise, yet we struggle to follow her advice – and many times fail. While the "hunting" aspect of our hunter-gatherer ancestry has selected for some immunity to fat-driven diseases, we are more like chimpanzees than some might think. Leading to our sixteenth lesson.

16. Chimpanzee research offers us advice on health. While our bodies and biochemistry are not identical, the source of many of our physical maladies is. What we have learned only recently from clinical studies of healthy people has been staring us in the face for two decades, if only we had heeded the lesson wild and captive chimpanzees had for us, advice that is easy to state but hard to follow: "Keep moving or die." As they feed among difficult-to-negotiate branches they move their limbs through their entire range of motion, again and again; in the course of reaching for food among small branches they press each joint to the very limit of its excursion, engaging in a sort of arboreal yoga. When following chimpanzees there is no one moment when a human is pushed to his physical limit, yet by the end of up to 13 hours of walking, climbing, and reaching, one is bone-weary.

This exercise regime is healthful for ligaments, tendons, muscles, and even the circulatory system, which responds to activity by thickening and strengthening venous and arterial walls, thus reducing the risk of stroke. Bones become denser and thicker, reducing osteoporosis. Healthy living means mimicking our ancestral positional regime rather than obeying instincts that tell us to conserve calories.

17. Lifestyle diseases plague us due to an evolutionary mismatch between the conditions our ancestors faced (our EEA, as we specialists would say, our Environment of Evolutionary Adaptedness) and contemporary life. So-called "lifestyle diseases" – obesity, diabetes, heart disease, osteoporosis, stroke, back problems, arthritis, cancer, digestive malfunctions, and even sleep disorders – afflict us when our adaptations collide with contemporary life. Our genes tell us to eat sugar and fat, avoid fiber and conserve energy: We must learn to ignore it.

18. Nature is more creative than science. Defensive alkaloids and tannins evolved in the course of the evolutionary battle among plants and animals in the struggle for life. Plants evolved toxins to repel their most troublesome pests, but importantly sometimes a toxin that is deadly to one organism leaves another unaffected. We humans are grateful. The foundations of pharmacology rest on this evolutionary defense. Drugs that kill pathogens like tuberculosis and malaria but leave us unharmed (mostly, at least) are derived from toxins discovered in the natural world, medicines that evolution, with its incredible creativity, gifted us.

19. Humans and chimpanzees are so genetically similar that we struggle to say exactly how similar. We are 99.4 percent similar in "functional" genes, those that produce proteins. We are about 99 percent

similar in our raw, A-C-T-G/base pair by base pair sequences. Because humans and chimpanzees vary greatly in which genes have multiple copies, when gene repeats are considered we are only 94 percent similar. Keep in mind, though, that different humans have differing numbers of some gene copies; for instance, humans have five copies of AQP7, a gene that is believed to confer greater endurance, while chimpanzees have only two copies (Dumas et al., 2007). In the end we cannot say whether we are 99.4 percent similar, 94 percent similar, or something in between.

20. Life is not one grand, sweet song. Chimpanzees often live in a lush, tropical world that one might think is so bountifully productive that competition and worry are unknown. Dream on. Instead, little in the forest can be eaten by chimpanzees and what *is* edible is widely dispersed and sparse. Finding food is a daily struggle, and the finding is only the beginning. Chimpanzees compete for their food not only with other species, but with one another. And it is not just food that is contested, but all manner of limited resources. Opposing interests make for stress. Chimpanzees are embedded in a dominance hierarchy in which they struggle for status and chafe at slights. Subordinates must take care to appease dominants – or pay the price. Frequent social displays that reinforce the status of the high-ranking individuals remind the lowly to grovel. Sexual coercion, the stress of murderous neighbors, and the threat of infanticide, all lurk in their seeming paradise. The chimpanzee life is no less a rat race than the human one – though perhaps no more of one, either.

21. Digestive physiology drives social behavior. It seems obscure and enigmatic to say that the feeble digestive physiology of chimpanzees is responsible for their intelligence, but the connection is real. Monkeys evolved the ability to tolerate a wider range of secondary compounds in unripe fruit and thus expanded their diet. Apes did not. Apes are largely confined to sugary ripe fruits and other foods that are low in fiber, alkaloids, and tannins. These foods are rare for three reasons: species that produce high-quality fruits are mostly rare; trees fruit infrequently (figs aside – and wild figs are not favorites); and monkeys and other competitors strip the trees of fruits before they are ripe enough for apes to eat. The sparse and dispersed nature of ripe fruits led to a fission–fusion social system, a society in which the entire community is rarely together at once, but instead is made up of smaller subgroups, each foraging independently and coming into contact with others only sporadically. The intellectual demands of maintaining bonds in a social web stretched out over miles, where individuals benefit when they can quickly get up to speed with social events that have happened out of sight (or as I often say, "off stage"), requires a mental processing ability that may be the impetus that resulted in the advanced cognition among great apes.

22. Intelligence is a many-splendored thing. The minds of clever animals such as elephants, parrots, great apes, and dolphins differ in the ways in which they are intelligent. Apes are cognitively advanced when it comes to spatial memories, mind-reading abilities, self-awareness (or more accurately, self-concept), simple tool-using faculties, and sophisticated communication, but their engineering skills are quite limited and their ability to rein in their violent tendencies to allow cooperation are paltry compared to humans.

23. Among chimpanzees, copulation has a purpose beyond sperm meeting egg. Relative infertility among females – being difficult to impregnate – confuses males who might otherwise favor their own infants, while at the same time ignoring or threatening the infants not their own. For young females just transferring into a community, an extended period of infertility gives them time to travel with males and get to know their new home before choosing a core area. Relative infertility allows them to learn the geography and resource distribution in their new community before the stress of caring for an infant impinges on them.

24. A last lesson is that sociality and social behavior evolve. For example, while orangutans are potentially sociable in the sense that females readily live together in captivity, they are much more content as solitary individuals than chimpanzees. Wolfgang Köhler wrote of chimpanzees: "it is hardly an exaggeration to say that a chimpanzee kept in solitude is not a real chimpanzee at all." For an

orangutan, solitary confinement is more disturbing to them for the confinement than the solitariness, at least compared to chimpanzees, whereas for chimpanzees their spirit and will to live seems to whither when caged singly.

These lessons give the 60 years of research at Gombe and Mahale some perspective; we knew very few of these things in the early 1960s when Jane Goodall and Toshisada Nishida first looked up into their research futures from the shore of Lake Tanganyika. The lessons we learn from chimpanzee research are diverse, offering us perspectives on the impact of colonialism, the history of science, the value of wild studies, human health, and as we will see below, human origins. If we narrow our focus and ask "how can we explain what it is that makes (made, if we keep evolution in mind) chimpanzees different from other primates?" we can trace most chimpanzee traits back to a single phenomenon.

Now let us reconsider Lesson 21, that digestive physiology drives social structure, a concept that ties together every one of the 29 chapters preceding this one. The welter of chimpanzee traits are not merely linked to one another, there is a higher order relationship among these features that is surprisingly simple. The causal link between chimpanzee digestive physiology and a few other distinctive chimpanzee characteristics expands outward, interdigitating as it goes, shaping other aspects of behavior and biology. When we ask why there are chimpanzees, we can trace chimpanzee-ness back to a prime mover.

30.3 A Holistic View of Chimpanzees

How many times have you heard that old chestnut "You are what you eat"? The obverse is more important: Chimpanzees – and humans – are defined more by what they *cannot* eat. Looking ahead, Figure 30.12 may look like a Gordian knot, an impossibly tangled set of interrelationships better left alone, but if you have reached this point, I guarantee I can walk you through it. Let us explore how antifeedant intolerance, the need to avoid terpenes, toxins, tannins, and fiber, has a cascading ripple

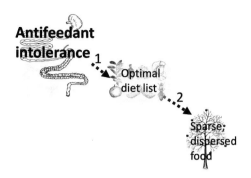

Figure 30.1 The need to avoid antifeedants determines the food items chimpanzees can eat; those food items, ripe fruits, are sparse and dispersed.

effect on nearly every other aspect of chimpanzee biology.

Link 1 in Figure 30.1: Chimpanzee intolerance of secondary compounds and fiber – antifeedants – dictates their **optimal diet list**, the roster of foods that, if pursued in their rank order from the most to least desirable, is most healthful. In order to maximize vital nutrients and avoid detrimental antifeedants, they must search out foods items that have a very particular chemical make-up. After two difficult-to-obtain items they are able to eat only rarely – meat, which has an injury risk, and honey which is rare and normally well defended by dangerously aggressive bees – large, sugary, ripe fruits with no antifeedants are their best choice. The chimpanzee optimal diet list yields a long list of foods that, while they are from many species, are similar nutritionally, just as banana and a piece of cake are both high in sugar and low in fiber and tannins. Some items are lower on the list because they are small and therefore require more effort to obtain – insects are nutritionally much like meat, but termites and ants are in a smaller and therefore less desirable package. Chimpanzees constantly switch out food items so that their diet is different each day. Some animals have a short food list – bamboo is almost the only thing on the menu for pandas, but this food, while abundant, is high in fiber, low in sugar, and high in a toxin – cyanide. Chimpanzees are at the other end of the spectrum: no single food is abundant, requiring them to eat many different items, but all with similar nutritional components.

2. *But* ripe fruits and other nutritionally dense foods are scarce, as we discussed above. Because fruit species with a pro-chimpanzee chemical make-up are rare, and because monkeys are more abundant and can feed on fruits earlier in their maturation cycle, chimpanzee foods are sparse and dispersed. Chimpanzees are forced to range widely, compared to monkeys, scouring the habitat to find those rare items that are high on their optimal diet list. When there are no ripe fruits, chimpanzees often fall back on low-calorie, low-secondary-compound foods.

3. Because low-antifeedant, sugar-rich foods are scarce, chimpanzees have a low population density, compared to monkeys (Figure 30.2). While this is surely at the monkey-heavy side of the spectrum, a survey of eight conservation areas in Uganda reported monkey densities of 86 per square km, versus 1.4/square km for chimpanzees (Plumptre & Cox, 2006). While chimpanzees are much heavier than monkeys, when it comes to individuals a 20-1 dominance may be a good estimate.

4. Chimpanzees require a large range to provide enough of their rare food items, and they have a low population density. Covering that immense area means they have long day ranges. Patches tend to be small, making their day's travel longer as they zig-zag between feeding sites.

5. The chimpanzee optimal diet list is packed with food items that are found in small patches, so small they are rarely a complete meal for even a lone female. When individuals travel in a group, each member of a group must be able to feed adequately, of course; if one individual must travel 2 km to gather enough food for a day, if she is joined by a second individual the two together must travel 4 km; three individuals must walk 6 km, and so on. Thus, the dispersed nature of food contributes to female solitariness (Figure 30.3). The burden of carrying infants further contributes to pressure to minimize travel distances, and their juvenile offspring must also deal with long travel distances. Feeding solitarily most of the time optimizes foraging. Females cope by avoiding other females and spending most of their time in a familiar core area.

6. Once females are largely asocial, each female has motivation to stay in one place, a core area

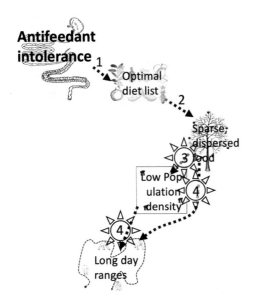

Figure 30.2 When food is sparse, populations that depend on it are smaller (3). Dispersed feeding sites require longer daily travel paths (4); low population densities require more travel to link up with social partners. New links are starred.

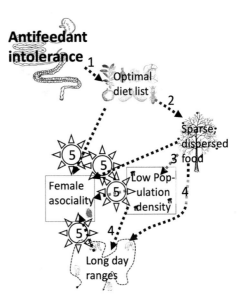

Figure 30.3 Ripe fruits are scarce and therefore contested foods; studies show that females feed most efficiently when solitary, when less time is wasted squabbling over food and attending to social demands. Travel costs further prevent females from gathering in large social groups. Focusing on ripe fruits thus compels females to forage solitarily much of the time. New links are starred.

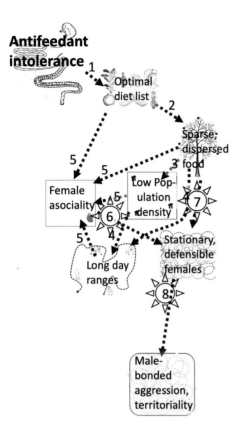

Figure 30.4 By focusing on a small area she comes to know extremely well, females sacrifice sociality for gathering efficiency (6). Thin food supplies means an average core area can support only a single female, encouraging restricted ranging (7). Stationary females allow bonded males to guarantee reproductive access by defending a geographic area.

(Figure 30.4). A long list of food items requires a broad knowledge of species characteristics and the locations of specific feeding sites, which in turn puts a premium on local knowledge. For example, a female may harvest a dozen gumball-sized figs from a small bush. Where to next? She may decide to make her next stop the nearest termite mound. If she knows how close her fig bush is to the nearest mound, she can take the shortest possible path to her next feeding site, minimizing travel costs. Superior mental mapping capabilities lower energy expenditure and allow her to minimize time spent foraging, maximizing her net energy return. But memory is finite. By settling in one area, her core area, she can memorize the location of a large percentage of food resources in that area, as well as travel routes. Females feed most efficiently by limiting foraging to a core area as opposed to searching for food. When acceptable food items are rare and thus hard to find, there is benefit to retaining detailed knowledge of resource distribution, rather than randomly searching for these hard-to-discover items.

7. The need to know food resources well so as to forage efficiently encourages females to settle into a stationary core area (Figure 30.4).

8. Among most primates, males remain near a female to increase their chance of copulating with her. When females are found in a stable, well-defined core area, a single male (as in orangutans) or a clique of males (as in chimpanzees) can exclude other males from an area containing females, monitoring their reproductive status, then monopolize mating privileges when they come into estrus. Theoretically, two males can defend an area that contains more than twice as many females as one male alone, but males must "agree" to work together. They do this by establishing group identities and emotional attachments to one another. Stationary females, because they allow males to monopolize their reproductive potential by guarding an area rather than an individual, provoke over evolutionary time both male territoriality and bonds among males to allow the defense of those territories.

9. Long day ranges make traveling in the tree canopy difficult due to the costs of walking on narrow, unstable branches and transferring between trees. Monkeys are smaller and so branches are less precarious and often stable enough even at the tree edge to serve as a firm launching point for a leap. A heavy primate like a chimpanzee must engage in energy-sapping acrobatics to move between trees, hanging beneath branches and lunging for a support in the neighboring tree. These costs encourage less expensive terrestrial travel, which in turn selects for adaptations for efficient walking (Figure 30.5).

10. Lower travel costs work hand in hand with bonding; a party of males must be able to afford to visit enough feeding sites to satisfy the many members; they must also travel widely to patrol the boundaries of their territory.

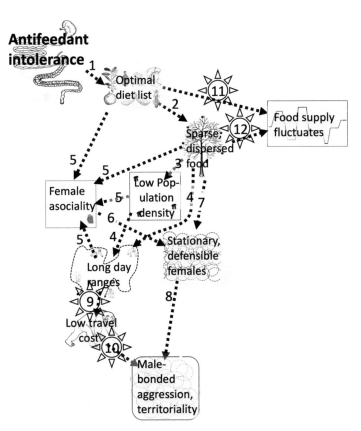

Figure 30.5 The necessity of visiting many feeding sites and long rambles to guard territories selects for locomotor efficiency (9); locomotor efficiency in turn allows males to travel together and bond (10). The ephemeral nature of ripe fruit (11) and its wide dispersal (12) creates localized scarcities, yielding wildly fluctuating food supplies.

11. The constraints of the chimpanzee optimal diet list, limiting which foods they can eat, makes for a highly variable food supply. One week there may be no sugary fruits available, forcing chimpanzees to eat pithy herbs, bark, and leaves. The next week the million fruits on a huge tree may ripen, providing a bonanza (Figure 30.5).

12. With a shorter list of preferred foods, compared to antifeedant-tolerant monkeys, food supplies are more likely to fluctuate.

13. The fluctuating chimpanzee food supply affects party size. When food is plentiful and concentrated, chimpanzees can gather in large groups. When food supplies are low, travel costs force them in to small parties (Figure 30.6).

14. The sparse, dispersed nature of food, the seasonality of other food items such as termites, and competition creates a constantly changing array of pockets of plenty. When food is abundant, parties are large; when less abundant and found at the right intersection of female core areas and male party location, parties they still may be large; at other times the same amount of food may be too dispersed to allow large parties.

15. Differing rainfall, differing species composition, differing solar radiation, differing soil chemistry, differing competitors, and differing elevation from one community range to another mean that food supplies differ between communities. One community might see a favored fruit fully mature, while only 5 km away a neighboring community still finds it inedible. The consequence is that community A might be able to gather in parties of tens of individuals, while in community B parties may consist of one or two individuals. This difference creates an imbalance of power, enabling a large military force of perhaps ten patrollers to raid a neighboring community, the members of which are foraging singly (Figure 30.6).

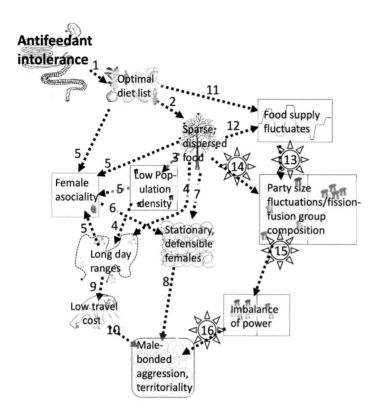

Figure 30.6 When fruit availability is low (13) party sizes are small, but when supplies increase, large parties form. The dispersed nature of feeding sites (14) discourages party formation, but when food is abundant disparate parties fuse. Variation in local food abundance occasionally leads to wildly different party sizes in neighboring communities (15) generating imbalances of power, which in turn encourage male bonding and male territoriality (16).

16. The possibility of intimidating, injuring, or eliminating competitors emerges from this imbalance of power. Larger parties of more aggressive males can push back the neighboring community, securing more food for their own community and encompassing more females in their territory, which in turn selects for male bonding and male aggression (Figure 30.6).

17. When groups are large many individuals witness the same social events and social knowledge is similar among all community members. When parties are small, individuals see only the small fraction of social interactions – those that occur in their party, leaving most of the community's social life unseen. Elsewhere in the community range a falling-out among once-close friends, a reversal in social rank, or a death would all be 'off-stage' (Figure 30.7).

18. Because females are alone most of the time, more social events are off-stage for females than males. Changes in male social rank, new alliances, new infants, all occur out of their sight.

19. Because social events happen so often off-stage for all but a few actors, an individual returning to a party (or subgroup) after an absence may have missed recent changes in social relationships among members of the subgroup; two lukewarm associates may have bonded more tightly and might now be a threat to a higher-ranking individual who might have previously ignored them. If it were to turn out that an ally has cozied up to a competitor, a particularly energetic bit of groveling might help to avoid a beating, but the now-isolated groveler would have to recognize the danger with "mind-reading." Reading subtle social cues that betray changes to social relationships is thus advantageous. Chimpanzees have evolved the ability to detect the emotional state of individuals at a glance. They have become mind-readers. It is the fission–fusion nature of chimpanzee society that selects for this kind of advanced intelligence. Most monkey species, in contrast, travel in one large group. Nothing is off-stage (Figure 30.7).

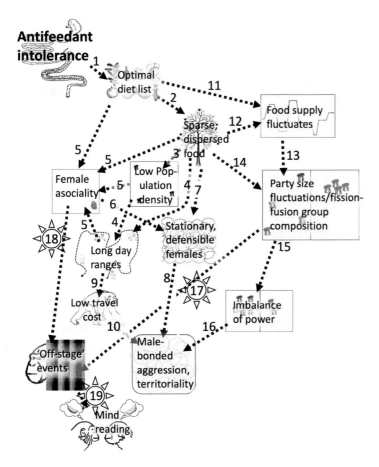

Figure 30.7 Party size fluctuations (17) and social fragmentation (18) preclude shared knowledge of social upheavals in the community. The need to assess the status of social bonds at a glance, some of which will have changed off-stage (19), selects for the ability to infer coalition changes from subtle "tells."

20. Dispersed foods not only encourage females to forage alone, they require good spatial memory (Figure 30.8).

21. The fixity of female core areas allows males to secure breeding privileges by guarding a territory.

22. Males are selected for a different kind of mapping ability, not so much memorizing specific food locations but being better able to maneuver in a large geographic area as they guard their territory.

23. These spatial demands ...

24. along with the need for both mind-reading and maneuvering in a large, complex social network contributed to the evolution of advanced cognition and the large brain that drives it (Figure 30.8).

25. When there is little ripe fruit, antifeedant intolerance forces chimpanzees to fall back on famine foods such as herbaceous piths, leaves, and other fibrous foods. Large body size and its accompanying large gut and slow metabolism allow chimpanzees to digest these foods using their hindgut capacity for microbial fermentation (Figure 30.8).

26. An optimal diet list restricted to particularly rich foods puts a premium on having priority of access to those items. Large body mass allows chimpanzees to bully monkeys away from their preferred items.

27. The ripe-fruit-rich dietary requirements of chimpanzees requires that they have access to the small-diameter, compliant terminal branches, for three reasons. More fruit grows in the terminal branches; riper fruit is found in the terminal branches; and monkeys eat fruit easily accessible in the inner core of the tree before it is ripe, leaving fruit in the tree periphery, where it then ripens (Figure 30.8).

28. Large body size contributes to the body weight/branch diameter disproportion that plagues

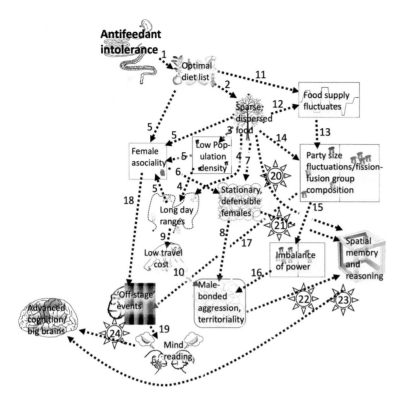

Figure 30.8 Dispersed feeding sites (20), dispersed females (21), and the cognitive demands of ranging over the entire community territory to secure borders (22) select for good spatial skills. Advanced spatial memory, the capacity to perform minimum distance calculations (23), and facility in reading social cues (24) in turn select for larger brains.

the large-bodied chimpanzee, requiring more armhanging and other suspensory behavior, including postures using two and three contacts, like arm-foot hanging and bipedalism. Sitting would be possible if body mass were smaller.

29. Harvesting fruit in the tree periphery requires suspensory behavior because tiny branches preclude sitting and require armhanging (Figure 30.9).

30. Long fingers are required for armhanging but are clumsy for long-distance walking. Because long fingers are awkward as propulsive organs, chimpanzees fold up their fingers into a fist, giving them a smaller footprint and a firmer push-off. Fast, efficient animals like deer and cheetahs have short toes; chimpanzees mimic this efficient anatomy by folding up their long fingers into a knucklewalking configuration. Long day ranges select for efficient travel, and terrestrial travel using knuckling rather than walking on the fingers or palm is the solution (Figure 30.10).

31. Large bodies require great caloric intake, and as bodies grow larger the extra calories required to fuel a large brain become a smaller and smaller proportion of the energy budget so that a larger brain is "cheaper" for species with larger body masses. Large bodies, in other words, allow for larger brains, and larger brains are more intelligent. Competition with monkeys selected for larger body size and drove a fission–fusion social system, both contributing to the evolution of intelligence.

32. Monkeys have a feed-as-you-go strategy; their tolerance of antifeedants allows them to find food in a large proportion of the trees they pass through, which in turn makes more expensive arboreal travel, with its requirement of indirect routes and unstable supports, worth the trade-off of not having to descend to the ground to travel. Chimpanzee foods are more dispersed so that most of the trees they might pass through have nothing for them to eat. As a consequence, chimpanzees feed in a tree or two, then descend to the ground to walk to a distant food source, after which they must then climb vertically into the new feeding tree, selecting for a gripping great toe and stiff backs and other anatomy related to vertical climbing.

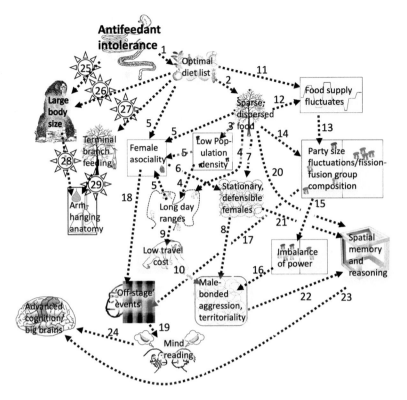

Figure 30.9 The necessity of falling back on fibrous foods (25) and advantages of body size in contests with other frugivorous species (26) selected for large body mass. Ripe fruits are found in terminal branches, particularly after monkeys have harvested unripe fruits in the tree core (27), pushing chimpanzees into the terminal branches. Heavy primates (28) must engage in suspensory behavior (29) when feeding among small, compliant tree-edge branches.

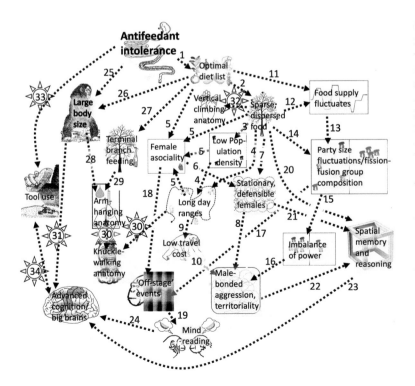

Figure 30.10 Dispersed, thin food resources and the demands of territoriality entail long day ranges (30). Long fingers are required for suspensory behavior (30) but are awkward and inefficient for walking; knuckling offers a facultative reduced footprint and a stiffer propulsive organ. Great body mass makes large brains relatively less expensive, allowing brain expansion (31). Terrestrial travel necessitates vertical climbing to reach arboreal foods (32). Antifeedant intolerance increases the appeal of defended, and thus poison-free, foods (33). Tool use provides calories needed to sustain a large brain but at the same time demands more advanced cognition (34).

33. Seeds and other foods can be protected either chemically, with poisons and other antifeedants, or with physical barriers like armor, in which case they have thick, tough seed coats or, as is the case for termites, earth works; or by sheltering in niches in trees, as do honey bees and bushbabies. Such morphologically protected food items are not poisonous. Chimpanzee antifeedant intolerance makes chemically unprotected foods more desirable, but to eat them they must penetrate their defenses. Tools like hammerstones or termiting tools and pry bars are the solution.

34. It is not clear whether social needs alone selected for advanced cognition, which in turn gave chimpanzees the intelligence to conceive of tools, or whether the need for tools selected for cognitive tool-using abilities, but the two are related one way or another: Animals with large neuron counts and advanced cognition are more likely to use tools (Figure 30.10).

35. Building a large brain requires time; a slow life history is a consequence of large brains (Figure 30.11).

36. Large brains require a period of paralysis during sleep, and a nest ensures the sleeper will stay nested.

37. Larger primates are more precarious in trees because they are larger in relation to available branches; nests solve that problem (Figure 30.11).

38 and 39. The coalitionary violence resulting from imbalances of power requires males to aggregate for self-defense and, when conditions are right, aggressive expansion of their territory. They gather for that safety and offensive advantage using the chimpanzee version of the medieval tolling of the village church bell: They pant-hoot to call their fellow combatants to muster at large food sources. The pant-hoot is the aggregating vocalization that draws small parties to a larger group which in turn allows the border patrol (Figure 30.12).

40. In Chapter 26 I introduced the new idea that chimpanzees resemble humans in establishing social status on the basis of athletic contests: sport. Their unique society requires this abstract competition. When the homeland is threatened, when marauding males from a neighboring community invade to kill babies and to appropriate the precious life-giving land

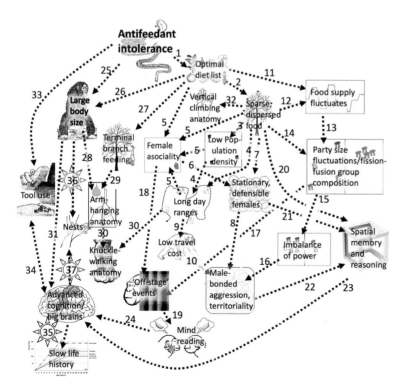

Figure 30.11 Big brains take a long time to grow, resulting in a slow life-history (35). Large body size, which tends to decrease stability by decreasing the size of the support relative to body size (36) and a period of paralysis during sleep, a consequence of the large chimpanzee brain (37), make sleeping platform construction advantageous.

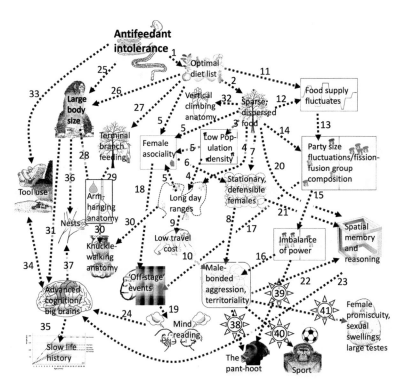

Figure 30.12 Large cohort size improves territorial defense (38), particularly in the face of power-imbalances (39), selecting for pant-hoots that serve both to monitor allies and to call them to food sources. Injuries deplete defensive forces (40), forcing males to decide rank with demonstrations of fighting ability, rather than bloody brawls. Multimale communities entail certain reproductive characters (41). In this complicated web of interrelationships, nearly everything that is special about chimpanzees evolved in response to their inability to tolerate antifeedants and the impact that inability had on the way they dealt with competitors and the way they had to gather food. Had proto-chimpanzees found themselves in a different environment, with different competitors, they would have followed a different evolutionary path, as did orangutans, gorillas and bonobos.

that nurtures the fruits on which the community relies, males can only look to one another for the means to turn aside the invasion. When the war bugle sounds, each male turns to the only help available, a fellow community male who, while he may have been yesterday's rival, must become today's brother in arms. If the size of the community's able fighting force is attenuated, every male risks death, the high- and low-ranking alike. Thus, the life of each male depends on the number of the community's able-bodied warriors. Herein lies the source of the social display. Within the community, each male strives to dominate as many other males as he can to maximize his share of resources. In virtually every other primate species, rank is decided in bloody hand-to-hand combat. An individual may brandish his tools of war as a threat in the hopes that he can collar whatever he wants without risking injury, but when the bluff fails, the superior fighter wades in and attempts to end the contest quickly and decisively. If he injures his opponent, all the better. If he kills his opponent, there is one less competitor for him to worry about and one less mouth to feed, sucking up resources from his mate or offspring. Not so for chimpanzees. He needs his compatriots for defense. To avoid injuring a brother in arms, chimpanzees compete by demonstrating their athleticism and endurance while leaving both combatants uninjured. The social display is meant to leave no doubt about who would win in a fight, thus dispensing with the risky encounter while at the same time deciding rank honestly. Only when it is unclear who would win the fight is an all-out physical confrontation likely. Social displays are therefore much like human sport; they can be interpreted as providing an honest display of athletic prowess so that a winner can be recognized without the shedding of (much) blood (Figure 30.12).

41. When males bond together to guard a territory communally, females are caught in a conundrum. They live in a community with many males. If they mate with only one male, unmated males may kill their infant. Promiscuity confuses paternity so that no individual can be certain he is not the father; females create further confusion by exhibiting a swelling a

month or more after conceiving, during which time they continue to copulate. The sexual swelling is a signal that she is near ovulation, provoking competition among males that tends to result in the female mating most, or mating at her most fertile period, with a high-quality male. In a way, by displaying a swelling, females force males into a contest that sorts out the quality of their genes on their own. Males compete in part by fighting, but also by producing more sperm, which results in larger testes. While paternity confusion means no male can be certain he has fathered an infant, provoking competition makes it more likely males who are good competitors will father a female's infant.

30.4 What Chimpanzees Tell Us about Human Origins

One of the reasons we primatologists study other primates is to understand our own evolution. Chimpanzees do not disappoint. The casual observer might wonder whether our fellow primates can offer us anything on this score, since humans are so different from nonhuman primates. Less casual observers, as well; some paleontologists have gone so far as to state – without equivocation – that chimpanzees have nothing to teach us about human evolution (Sayers & Lovejoy, 2008). If this were so sharks and dolphins would not resemble one another, and eyes would not have evolved independently 40 separate times (Salvini-Plawen & Mayr, 1977; Nilsson & Pelger, 1994).

Paleontologists can look for patterns in the evolutionary response different organisms had to specific environmental conditions and use those relationships to speculate on the evolutionary origin of traits that define humans. Rather than relying on evolutionary proximity alone to reconstruct evolution (a referential model; Tooby & DeVore, 1987), the best models are conceptual models (Tooby & DeVore, 1987; Tooby & Cosmides, 1989). For example, consider the task of reconstructing the social behavior of australopiths, an issue I find enthralling. We might reason "Chimpanzees are our closest relative, so it must have been a fission–fusion, female-transfer society." That would be a referential model – it "refers" to a close relative. I would approach this challenge (as I did in more detail in Chapter 28) conceptually, in this way:

Australopiths are sexually dimorphic in body mass; three of the five basic social systems tend to exhibit great body size dimorphism: (1) female-bonded, (2) female-choice, and (3) solitary primates. Species in two other systems, (4) monogamy and (5) fission-fusion societies, display less body mass dimorphism and so are also improbable. Australopiths are apes, so the first system, female-bonded, is also unlikely, since it is only seen in antifeedant-tolerant monkeys. Australopiths were likely partly or mostly terrestrial, which makes the solitary social system less likely, since no partly-terrestrial primate has a solitary socially system. This leaves a gorilla-like social system.

This is a conceptual model. In other words, if we pay close attention to the lessons chimpanzees provide us and to the network of interrelationships in Figures 30.12, we can reconstruct the course of human evolution by picking out the relationships between a trait found in humans and the selective pressures that resulted in the evolution of a similar trait in another animal. Of course, some of our interpretations (including many I espouse here, of course) will turn out to be wrong, but I maintain that making predictions that colleagues will test in the future is the best way to do science.

Human paleontologists are particularly interested in the evolution of a score of features that are either unique, unusual, or are highly developed in *Homo sapiens*:

- Striding bipedalism   Unique
- Intelligence   Highly developed
- Language   Unique
- Tool use/manufacture   Highly developed
- War   Highly developed
- Peculiar diets   Unusual
- Pattern of sexual dimorphism Unusual
- Slow life history   Highly developed
- Extreme sociality   Unique
- Pattern of disease susceptibility Unusual

Chimpanzee research helps us understand the origin of each of these.

My research on chimpanzee locomotion and posture suggests that humans evolved **bipedalism** when our ancestors found themselves confined to a dry habitat populated by species of fruiting trees that were short and had small fruits; bipedalism allowed the earliest bipeds to reach more fruits from the ground by standing up (Chapter 28). At Gombe, I found that half of chimpanzee bipedalism was in the trees. Because small trees have small branches, sitting is impossible, but gripping twigs with the toes and balancing the body above that toehold and further stabilizing the posture by gripping an overhead branch allowed chimpanzees (and possibly early hominins) to feed in short trees. Feeding on fruits from small trees, both on the ground and in the trees, may have led to bipedalism.

Advanced **intelligence** in the chimpanzee is, in my view (Hunt, 2016), a consequence of their somewhat large group size combined with the cognitive demands of negotiating social relationships in a fission–fusion social system with its consequent numerous off-stage events. A larger social network and/or more variable party sizes might select for even great intelligence among hominins. As dry-habitat apes, early hominins might have gathered in larger groups than chimpanzees, since two dry-habitat species, gelada baboons and hamadryas baboons, collect in groups of 300 and more (not that large groups must be in dry habitats; mandrills are found in even larger groups [Abernethy et al., 2002]). As early hominins moved into a quasi-carnivore niche, rising higher in the food chain, home ranges and day ranges likely increased even more, stretching the web of connections far beyond that of chimpanzees, selecting for even greater social intelligence.

Language has the design feature *displacement*; it allows us to communicate about objects or events that are out of sight, either because they are physically removed, because they are expected to occur in the future, or because they are remembered from the past. As fission–fusion animals, our ancestors lived – and we continue to live – in a society with social separation, a consequence of foraging separately rather than all together, as do many monkeys. In a fission–fusion society no two individuals can have the same information about the myriad social relationships among community members. Therefore, there is stronger selection than in non-fission–fusion animals to evolve a means of communicating information about a distant food source or an expected encounter with a predator. At some point in human evolution, as our social network grew larger and perhaps as resources became more dispersed, the value of transmitting the knowledge of out-of-sight phenomena became valuable enough to elicit language. Apes master sign language in the lab – we might wonder why they lack language in the wild. Perhaps there is too little valuable information about out-of-sight phenomena. We should expect greater intelligence and more complicated communication in a fission–fusion species that have close social bonds, compared to close relatives that live in aggregate social groups. This should be true whether the fission–fusion/aggregate pair are insects, fish, birds, or mammals.

Tool use may be a consequence of the interplay between the need for a rich diet – which might include food items that can be accessed only with tools – and advanced cognition, which might have evolved for reasons other than tool use. Low-antifeedant prey, whether animal or plant, tend to protect themselves with physical defenses. For example, edible seeds like walnuts and almonds protect themselves with thick husks, whereas inedible apple seeds have a thin seed coat but contain cyanide. Tools can penetrate armor and circumvent other physical defenses. Meat is a rich food source, but larger mammals are challenging to subdue and even when subdued their thick skin is difficult to pierce with teeth evolved for frugivory. Tools help with both problems. We might expect that species that are unable to tolerate antifeedants use tools, even if they are unintelligent, at least compared to chimpanzees.

We engage in coalitionary violence – **war** – because our fission–fusion social system creates imbalances of power, situations where one social group has a temporary numbers advantage, often due to localized food abundance. Large parties can use numerical superiority to overwhelm competitors. As our technology increased in sophistication and our ability to store resources increased in step, the value of raiding to acquire resources, rather than just protecting territory and reproductive opportunities, further increased the fruits of war.

Our **peculiar diets**, high in nutrients and low in fiber, are a legacy of our ape ancestry. We elaborated on the ape adaptation by adding more animal protein to our diet, perhaps in part because the opportunity was there; dry habitats often contain herd animals. Chimpanzees require a long food list and have evolved good enough memories to retain a long list of low-antifeedant, high nutrient-content dietary resources. Humans are often said to be omnivores, but as with chimpanzees, regardless of whether the food item is a fruit, seed, herb, or animal, it tends to be low in antifeedants and nutrient-rich. In addition to breaking open low-antifeedant foods with tools, we have moved on to breaking them down. We increase food value by denaturing antifeedants like tannins and converting complex carbohydrates into sugars – by cooking (Wrangham, 2009). Our diets are peculiar because we have a huge food item list, but the anti-nutrient content of every item on that list is invariable – it is low.

We are **sexually dimorphic** – not that our dimorphism is very impressive compared to that of gorillas or baboons – because males and females have different social and biological roles; females are the ecological sex (*sensu* Gaulin & Sailer, 1985); that is, they must acquire food to reproduce. Males, on the other hand, must both compete with other males and appeal to females as mates in order to reproduce. This contrast in first-priorities is the source of some sex differences. Medium-sized testes and sex differences in body shape and body weight suggest that humans are not as promiscuous as baboons or chimpanzees, but also not as monogamous as gibbons, even though we have a tendency to pair-bond. Sex differences in body shape are largely reproductive. Females have breasts and a broad enough pelvic inlet to deliver their infants, whereas male pelvic morphology is driven principally by locomotor needs. Greater male inter-individual aggression selects for a powerful upper body effective for grappling with competitors; well-developed upper bodies are also found among baboons, thunderbugs, and bears. Hence males have broad shoulders and narrow hips; females have broader hips and less muscular upper bodies.

We humans have **slow life histories** because our brains are big: Across the mammals there is a high correlation between brain size and growth period. While we are not certain, we are confident that the causal arrow moves from brain size to life history rather than the other way around: a large brain requires a long time to build.

Social group size is a function of (1) population density, (2) the cost of travel, and (3) the value of allies, which is in turn affected by (4) the level of intergroup competition. Fission–fusion society spurs intercommunity conflict, which favors superior numbers, which drives up variously group size, the value of bonding, and sociality. Perhaps lower travel costs due to an open-country habitat and fully evolved bipedalism allowed even larger groups. Alternatively, or perhaps in concert with habitat adaptations, defending essential resources may have selected for the tighter social bonds which in turn selected for larger group sizes.

Some human **diseases** are driven by our sociality. Large population sizes encourage viruses to specialize on penetrating our immune defenses, and high population density increases contagion. But these are only contagious diseases. Many of our diseases are "lifestyle" diseases and they are caused by evolutionary mismatches, and can be ameliorated by increased activity and a proper diet. Cancer is different – it may be a consequence of selection for large brains, which required deactivating defenses that target rapidly dividing cells, since neurons divide rapidly during development.

Thus, we end our journey. We still have much to learn, of course, about both chimpanzees and human evolution, but our basic framework seems sound and is unlikely to change dramatically. While headlines not uncommonly proclaim that some "new discovery overturns all previously held theories," or the pithier but even more unlikely, "this changes everything," nothing changes everything. No one discovery or even series of discoveries will answer all our remaining questions, but neither will a single discovery overturn everything. We have explained much that is unusual about chimpanzees, and much about why humans are different. Still, even though I know it is the very rare discovery that is truly revolutionary, my heart will beat a little faster, as yours may, as I rush to read the details of the next news headline that promises "New Chimpanzee Discovery Changes Everything!" The journey of discovery continues.

Pan in sempiternum! Long live chimpanzees!

References

Abernethy KA, White LJ, Wickings EJ (2002) Hordes of mandrills (*Mandrillus sphinx*): extreme group size and seasonal male presence. *J Zool* **258**, 131–137.

Dumas L, Kim YH, Karimpour-Fard A, Cox M, Hopkins J, Pollack JR, Sikela JM. (2007) Gene copy number variation spanning 60 million years of human and primate evolution. *Genome research.* **17**, 1266-1277.

Gaulin SJ, Sailer LD (1985) Are females the ecological sex. *Am Anthropol* **1**, 111–119.

Hunt KD (2016) Why are there apes? Evidence for the co-evolution of ape and monkey ecomorphology. *J Anatomy* **228**, 630–685.

Linden E (2002) The wife beaters of Kibale. *Time* **160**, 56–57.

Nilsson D-E, Pelger S (1994) A pessimistic estimate of the time required for an eye to evolve. *Proc R Soc Lond B* **256**, 53–58.

Salvini-Plawen LV, Mayr E (1977) On the evolution of photoreceptors and eyes. *Evol Biol* **10**, 207–253.

Sayers K, Lovejoy CO (2008) The chimpanzee has no clothes: a critical examination of *Pan troglodytes* in models of human evolution. *Curr Anthropol* **49**, 87–114.

Tooby J, Cosmides L (1989) Adaptation versus phylogeny: the role of animal psychology in the study of human behavior. *Int J Comp Psych* **2**, 105–118.

Tooby J, DeVore I (1987) The reconstruction of hominid behavioral evolution through strategic modeling. In *The Evolution of Human Behavior: Primate Models* (ed. Kinzey W), pp. 183–237. Albany, NY: SUNY Press.

Wrangham RW (2009) *Catching Fire: How Cooking Made Us Human.* New York: Basic Books.

APPENDIX 1
TAXONOMY OF THE PRIMATES

Order	Suborder	Superfamily	Family	Subfamily	Tribe	Genus	Species	Common name
Primates	Strepsirhini	Lemuroidea	Cheirogaleidae			Cheirogaleus	medius	Western fat-tailed dwarf lemur
							major	Greater dwarf lemur
							crossley	Furry-eared dwarf lemur
							minisculus	Lesser Iron gray dwarf lemur
							ravus	Greater iron gray dwarf lemur
						Microcebus	murinus	Gray mouse-lemur
							rufus	Red mouse-lemur
							ravelobensis	Golden mouse-lemur
							myoxinus	Pygmy mouse-lemur
						Mirza	coquereli	Giant mouse-lemur
						Allocebus	trichotis	Hairy-eared mouse-lemur
						Phaner	furcifer	Masoala fork-crowned lemur
							pallescens	Western fork-crowned lemur
							parienti	Sambirano fork-crowned lemur
							electromontis	Amber mountain fork-crowned lemur
			Lemuridae			Lemur	catta	Ring-tailed lemur
						Eulemur	macaco	Black lemur
							fulvus	Brown lemur
							sanford	Sanford's lemur
							albifrons	White-fronted lemur
							rufus	Red-fronted lemur
							collaris	Red-collared lemur

(cont.)

Order	Suborder	Superfamily	Family	Subfamily	Tribe	Genus	Species	Common name
							albocollaris	White-collared lemur
							mongoz	Mongoose lemur
							coronatus	Crowned lemur
							rubriventer	Red-bellied lemur
						Hapalemur	*griseus*	Gray gentle lemur
							occidentalis	Sambirano gentle lemur
							alaotrensis	Alaotran gentle lemur
							aureus	Golden gentle lemur
						Prolemur	*simus*	Greater bamboo lemur
						Varecia	*variegata*	Black-and-white ruffed lemur
							rubra	Red ruffed lemur
			Lepilemuridae			*Lepilemur*	*mustelinus*	Weasel lemur
							microdon	Small-toothed sportive lemur
							leucopus	White-footed sportive lemur
							ruficaudatus	Red-tailed sportive lemur
							edwarsi	Milne-Edwards sportive lemur
							dorsalis	Back-striped sportive lemur
							septentrionalis	Northern sportive lemur
			Indriidae			*Indri*	*indri*	Indri
						Propithecus	*diadema*	Diademed sifaka
							edwarsi	Milne-Edwards' sifaka

Lorisoidea				*perrieri*	Perrier's sifaka
				verreauxi	Verreaux's sifaka
				coquereli	Coquerel's sifaka
				deckenii	van der Decken's sifaka
			Avahi	*laniger*	Eastern avahi
				occidentalis	Western avahi
	Daubentoniidae		*Daubentonia*	*madagascarensis*	Aye-aye
	Loridae	Perodicticinae	*Arctocebus*	*calabarensis*	Calabar angwantibo
				aureus	Golden angwantibo
			Perodicticus	*potto*	Potto
			Pseudopotto	*martini*	False potto
		Lorinae	*Loris*	*tardigradus*	Red slender Loris
				lydekkerianus	Gray slender loris
			Nycticebus	*coucang*	Sunda slow loris
				bengalensis	Bengal slow loris
				pygmaeus	Pygmy slow loris
	Galagonidae		*Otolemur*	*crassicaudatus*	Brown greater galago
			Euoticus	*monteiri*	Silvery greater galago
				garnetti	Northern greater galago
				pallidus	Northern needle-clawed bushbaby
			Galago	*senegalensis*	Senegal bushbaby
				moholi	Moholi bushbaby

(cont.)

Order	Suborder	Superfamily	Family	Subfamily	Tribe	Genus	Species	Common name
							gallarum	Somali bushbaby
							matschiei	Dusky bushbaby
							alleni	Bioko Allen's bushbaby
							cameronensis	Cross River Allen's bushbaby
							gabonensis	Gabon Allen's bushbaby
							zanzibaricus	Zanzibar bushbaby
							granti	Grant's bushbaby
							nyasae	Malawi bushbaby
							orinus	Uluguru bushbaby
							rondoensis	Rondo bushbaby
							udzungwensis	Uzungwa bushbaby
							demidoff	Demidoff's bushbaby
							thomasi	Thomas's bushbaby
	Haplorhini	Tarsioidea	Tarsiidae			*Tarsius*	*syrichta*	Philippine tarsier
							bancanus	Horsfield's tarsier
							spectrum	Spectral tarsier
							dianae	Dian's tarsier
							pelengensis	Peleng tarsier
							pumilus	Pygmy tarsier
		Ceboidea	Callitrichidae			*Callithrix*	*jacchus*	Common marmoset
							penicillata	Black-tufted marmoset
							kuhlii	Wied's marmoset
							geoffroyi	White-headed marmoset

	flaviceps	Buffy-headed marmoset
	aurita	Buffy-tufted marmoset
	argentata	Silvery marmoset
	leucippe	White marmoset
	emiliae	Emilia's marmoset
	nigriceps	Black-headed marmoset
	marcai	Marca's marmoset
	melanura	Black-tailed marmoset
	humeralifera	Santarem marmoset
	mauesi	Maues marmoset
	chrysoleuca	Gold-and-white marmoset
	intermedia	Hershkovitz's marmoset
	humilis	Roosmalen's dwarf marmoset
	pygmaea	Pygmy marmoset
Callimico	*goeldii*	Goeldi's marmoset
Leontopithecus	*rosalia*	Golden lion tamarin
	chrysomelas	Golden-headed lion tamarin
	chrysopygus	Black lion tamarin
	caissara	Superagui lion tamarin
Saguinus	*midas*	Red-handed tamarin
	niger	Black tamarin
	nigricollis	Black-mantled tamarin

(cont.)

Order	Suborder	Superfamily	Family	Subfamily	Tribe	Genus	Species	Common name
							graellsi	Graell's tamarin
							fuscicollis	Brown-mantled tamarin
							melanoleucus	White-mantled tamarin
							tripartitus	Golden-mantled tamarin
							mystax	Mustached tamarin
							pileatus	Red-capped tamarin
							labiatus	White-lipped tamarin
							imperator	Emperor tamarin
							bicolor	Pied tamarin
							martinsi	Martin's tamarin
							oedipus	Cottontop tamarin
							geoffroyi	Geoffroy's tamarin
							leucopus	White-footed tamarin
							inustus	Mottle-faced tamarin
			Cebidae	Cebinae		*Cebus*	*capucinus*	White-headed capuchin
							albifrons	White-fronted capuchin
							olivaceus	Weeper capuchin
							kaapori	Kaapori capuchin
							apella	Tufted capuchin
							libidinosus	Black-striped capuchin
							nigritus	Black capuchin
							xanthosternos	Golden-bellied capuchin
						Saimiri	*oerstedti*	Central american squirrel monkey

		sciureus	Common squirrel monkey
		ustus	Bare-eared squirrel monkey
		boliviensis	Black-capped squirrel monkey
		vanzolinii	Black squirrel monkey
Aotinae	Aotus	lemurinus	Gray-bellied night monkey
		hershkovitzi	Hershkovitz's night monkey
		trivirgatus	Three-striped night monkey
		vociferans	Spix's night monkey
		miconax	Peruvian night monkey
		nancymaae	Nancy Ma's night monkey
		nigriceps	Black-headed night monkey
		azarae	Azara's night monkey
Pitheciinae	Pithecia	pithecia	White-faced saki
		monachus	Monk saki
		irrorata	Rio tapajos saki
		aequatorialis	Equatorial saki
	Chiropotes	albicans	Bearded saki
		satanas	Black bearded saki
		albinasus	White-nosed saki
	Cacajao	melanocephalus	Black-headed uakari
		calvus	Bald uakari
Callicebinae	Callicebus	modestus	Rio beni titi

(cont.)

Order	Suborder	Superfamily	Family	Subfamily	Tribe	Genus	Species	Common name
							donacophilus	White-eared titi
							pallescens	White-coated titi
							olallae	Olalla Borther's titi
							oenanthe	Rio mayo titi
							cinerascens	Ashy black titi
							hoffmannsi	Hoffmann's titi
							baptista	Baptista lake titi
							moloch	Red-bellied titi
							brunneus	Brown Titi
							personatus	Atlantic Titi
							medemi	Black-handed Titi
							torquatus	Collared Titi
			Atelidae	Alouattinae		Alouatta	pigra	Guatemalan black howler
							palliata	Mantled howler
							coibensis	Coiba Island howler
							seniculus	Venezuelan red howler
							macconnelli	Guyanan red howler
							sara	Bolivian red howler
							belzebul	Red-handed howler
							nigerrima	Amazon black howler
							guariba	Red-and-black Howler
							caraya	Brown howler
				Atelinae		Ateles	paniscus	Red-faced spider monkey

			belzebuth	White-fronted spider monkey
			chamek	Peruvian spider monkey
			hybridus	Brown spider monkey
			marginatus	White-cheeked spider monkey
			fusciceps	Black-headed spider monkey
			geoffroyi	Geoffroy's spider monkey
		Brachyteles	arachnoides	Southern muriqui
			hypoxanthus	Northern muriqui
		Lagothrix	lagotricha	Brown wooly monkey
			cana	Gray wooly monkey
			lugens	Columbian wooly monkey
			poeppigii	Silvery wooly monkey
		Oreonax	flavicunda	Yellow-tailed wooly monkey
Cercopithecoidea	Cercopithicinae	Cercopithecini	*Allenopiothecus*	nigroviridis	Allen's swamp monkey
			Miopithecus	talapoin	Angolan talapoin
				ogouensis	Gabon talapoin
			Eryhrocebus	patas	Patas monkey
			Chlorocebus	sabaeus	Green monkey
				aethiops	Grivet
				djamdjamensis	Bale mountain vervet
				tantalus	Tantalus monkey
				pygerythrus	Vervet monkey

(cont.)

Order	Suborder	Superfamily	Family	Subfamily	Tribe	Genus	Species	Common name
						Cercopithecus	*cynosuros*	Malbrouk
							dryas	Dryas monkey
							diana	Diana monkey
							roloway	Roloway monkey
							nictitans	Greater spot-nosed monkey
							mitis	Blue monkey
							ascanius	Red-tailed monkey
							doggetti	Silver monkey
							kandti	Golden monkey
							albogularis	Syke's monkey
							mona	Mona monkey
							campbelli	Campbell's mona
							lowei	Lowe's mona
							pogonias	Crested mona
							wolfi	Wolf's mona
							denti	Dent's mona
							petaurista	Lesser spot-nosed monkey
							erythrogaster	White-throated guenon
							sclateri	Sclater's guenon
							erythrotis	Red-eared guenon
							cephus	Mustached guenon
							lhoesti	L'Hoest's monkey
							preussi	Preuss' monkey

Papionini		solatus	Sun-tailed monkey
		hamlyni	Hamlyn's monkey
		neglectus	De Brazza's monkey
	Macaca	sylvanus	Barbary macaque
		silenus	Lion-tailed macaque
		nemestrina	Sunda pig-tailed macaque
		leonina	Northern pig-tailed macaque
		pagensis	Mentawi macaque
		maura	Moor macaque
		ochreata	Booted macaque
		tonkeana	Tonkean macaque
		hecki	Heck's macaque
		nigrescens	Gorontalo macaque
		nigra	Celebes crested macaque
		fascicularis	Crab-eating macaque
		arctoides	Stump-tailed macaque
		mulatta	Rhesus macaque
		cyclopis	Formosan rock macaque
		fuscata	Japanese macaque
		sinica	Toque macaque
		radiata	Bonnet macaque
		assamensis	Assam macaque

(cont.)

Order	Suborder	Superfamily	Family	Subfamily	Tribe	Genus	Species	Common name
							thibetana	Tibetan macaque
						Lophocebus	*albigena*	Gray-cheeked mangabey
							aterrimus	Black Crested mangabey
							opdenboschi	Opdenbosch's mangabey
						Papio	*hamadryas*	Hamadryas baboon
							papio	Guinea baboon
							anubis	Olive baboon
							cynocephalus	Yellow baboon
							ursinus	Chacma baboon
						Mandrillus	*sphinx*	Mandrill
							leucophaeus	Drill
						Cercocebus	*atys*	Sooty mangabey
							torquatus	Collared mangabey
							agilis	Agile mangabey
							chrysogaster	Golden-bellied mangabey
							galeritus	Tana river mangabey
							sanjei	Sanje mangabey
						Theropithecus	*gelada*	Gelada
				Colobinae		*Colobus*	*satanas*	Black colobus
							angolensis	Angola colobus
							polykomos	King colobus
							vellerosus	Ursine colobus
							guereza	Black and white colobus

Piliocolobus	*badius*	Western red colobus
	pennantii	Pennant's colobus
	preussi	Preuss's red colobus
	tholloni	Tholloni's red colobus
	foai	Central African red colobus
	tephrosceles	Ugandan red colobus
	gordonorum	Uzungwa red colobus
	kirkii	Zanzibar red colobus
	rufomitratus	Tana River red colobus
Procolobus	*verus*	Olive colobus
Semnopithecus	*schistaceus*	Nepal gray langur
	ajax	Kashmir gray langur
	hector	Tarai gray langur
	entellus	Northern Plains gray langur
	hypoleucos	Black-footed gray langur
	dussumieri	Southern Plains gray langur
	priam	Tufted gray langur
Trachypithecus	*vetulus*	Purple-faced langur
	johnii	Nilgiri langur
	auratus	Javan lutung
	cristatus	Silvery leaf monkey
	germaini	Indochinese lutung
	barbei	Tenasserim lutung

(cont.)

Order	Suborder	Superfamily	Family	Subfamily	Tribe	Genus	Species	Common name
							obscurus	Dusky leaf monkey
							phayrei	Phayre's leaf monkey
							pileatus	Capped langur
							shortridgei	Shortridge's langur
							francoisi	François's langur
							hatinhensis	Hatinh langur
							poliocephalus	White-headed langur
							laotum	Laotian langur
							delacouri	Delacour's langur
							ebenus	Indochinese black langur
						Presbytis	*melalophus*	Sumatran surili
							femoralis	Banded surili
							natunae	Natuna islands surili
							chrysomelas	Sarawak surili
							siamensis	White-thighed surili
							frontata	White-fronted langur
							thomasi	Thomas' langur
							hosei	Hose's langur
							rubicunda	Maroon leaf monkey
							potenziani	Mentawai langur
						Pygathrix	*nemaeus*	Red-shanked douc
							nigriceps	Black-shanked douc
							cinerea	Gray-shanked douc

Superfamily	Family	Subfamily	Genus	Species	Common name
Hominoidea			Rhinopithecus	roxellana	Golden snub-nosed monkey
				bieti	Black snub-nosed monkey
				brelichi	Gray snub-nosed monkey
				avunculus	Tonkin snub-nosed langur
			Nasalis	larvatus	Proboscis monkey
			Simias	concolor	Pig-tailed langur
	Hylobatidae		Hylobates	lar	Lar gibbon
				agilis	Agile gibbon
				albibarbis	Bornean white-bearded gibbon
				muelleri	Müller's Bornean gibbon
				moloch	Silvery gibbon
				pileatus	Pileated gibbon
				klossi	Kloss gibbon
				hoolock	Hoolock gibbon
				concolor	Concolor gibbon
				hainanus	Hainan gibbon
				leucogenys	Northern white-cheeked gibbon
				siki	Southern white-cheeked gibbon
				gabriellae	Red-cheeked gibbon
				syndactylus	Siamang
	Hominidae	Ponginae	Pongo	pygmaeus	Bornean orangutan
				abelii	Sumatran orangutan
		Gorillinae	Gorilla	gorilla	Western gorilla

(cont.)

Order	Suborder	Superfamily	Family	Subfamily	Tribe	Genus	Species	Common name
				Homininae			*beringei*	Eastern gorilla
					Panini	*Pan*	*troglodytes*	Chimpanzee
							paniscus	Bonobo
					Hominini	*Homo*	*sapiens*	Human

APPENDIX 2
PROFESSIONAL GRADE CHIMPANZEE: TESTABLE HYPOTHESES

As I have worked through this volume there were many places where I expected to find research reporting on questions I found interesting, but which appear not yet to have been examined. In other cases, well-accepted hypotheses in one area of chimpanzee research suggest hypotheses concerning another. For instance, chimpanzee communication patterns suggest an explanation for ear morphology. Below are a number of interesting research foci that I cannot hope to investigate in my lifetime. I would relish collaborating with any of my colleagues to work on these problems. If that is impractical, please cite this work in your research reports.

1. Hypothesis: Pinnae diameter is correlated with the importance of long-calls in social systems. Several predictions follow: female orangutans should have larger pinnae than males; frequency of loud-calls should correlate with pinnae diameter; male chimpanzees should have proportionally larger pinnae than females. The chimpanzee external auditory meatus, compared to humans: I hypothesize that the smaller ear concentrates sound better, increasing the ability to detect faint sounds in the distance, presumably at the expense of clarity. Tests for auditory acuity that rely on headphones cannot account for pinnae size nor for improvements to sensitivity imparted by narrower auditory canals and larger pinnae.

2. Research topic: Compared to humans, chimpanzees have longer rods and cones (Detwyler, 1943) and thinner nuclear (or outer) layers in the fundus and in the periphery. Polyak (1941) found 8–10 rows of cells in the outer nuclear layer of the human fovea compared to 4–6 rows for chimpanzees. The function of these difference is unknown.

3. Research topic: I have never seen an adaptive explanation for the smoothness of human skin, or even a comment on how curious this difference is. One might speculate that human skin must function differently because we are relatively hairless, but the wrinkliest skin on chimpanzees is on the face, which is largely hairless. Research in the 1960s (in Montagna's lab) found chimpanzee, gorilla, and human skin are all quite similar, both microscopically and physiologically. Gorilla skin seems smoother to me, especially the upper lip. I am unaware of any research that takes an evolutionary approach to differences in ape skin.

4. Research topic: All apes except gibbons have laryngeal sacs, but their function is still unexplained. Humans lack them, though this change came only with the evolution of *Homo;* australopiths had an ape-like hyoid (Chapter 7) suggesting an ape-like laryngeal sac. A better understanding of their function in the ape social system may provide evidence of the australopith social system.

5. Research topic: Gibbons, siamangs, and chimpanzees make up a series with progressively fewer lumbar vertebrae, taller os coxae, narrower sacra, broader ilia, and more cone-shaped rib cages, suggesting these differences are allometric. Hunt (2016) has suggested that tall os coxa and short lumbar segments are adaptations to vertical climbing; while siamangs vertical climb more than gibbons, chimpanzees do so much less, making an allometric explanation problematic, though not necessarily wrong. Broader ilia and narrower sacra may be adaptations to increasing the surface area for thigh extensors as an adaptation to either climbing or leaping, but it is not clear whether it is one or the other or both. There is progressively less leaping from small to large body mass, but leaping is quite rare in chimpanzees (Hunt, 1992: Doran, 1993), making the retention of leaping adaptations somewhat surprising. Leaping is either extremely important (perhaps during intercommunity conflicts) or it has been under-reported.

6. Research topic: The iliac blade of the os coxa of the chimpanzee is flat and truly blade-like so that it is parallel to the surface of the back, so broad that the pelvis and the ribcage are equal in breadth. This

feature is found among all the great apes, but not in the gibbon and is intermediate in the siamang (Hunt, 2016). It has yet to be adequately explained. One hypothesis is that the breadth of the ilium is a function of gluteal size, which increases allometrically (Stern, 1971), whereas the ribcage is not under the same selection. The issue deserves an in-depth analysis.

7. Hypothesis: If leaping is a conflict-avoidance adaptation, used principally during intercommunity conflicts, males should express leaping adaptations more than females, since males are more often victims of intercommunity killings. If Stern's hypothesis is correct, we should expect to find even larger gluteal attachment areas among males than can be accounted for by allometry.

8. Hypothesis: Among non-gouging primates (i.e., not callithrichids), prognathism and procumbent incisors may be adaptations for extracting foods from concave substrates. We might predict that among pairs of species the folivorous species (leaves are not in concave receptacles) will have less prognathism and procumbency than the more frugivorous species.

9. Hypothesis: Gorilla and chimpanzee brow ridges are large compared to those of orangutans, fossil hominoids, and other primates. Most recent research dismisses mechanical stresses during powerful biting as the function (e.g., Hylander et al., 1991). Among Neanderthals and other early hominins they have been hypothesized to function to protect the eyes during interindividual aggression (e.g., Tappen, 1973, 1978; Churchill et al., 2009; Carrier & Morgan, 2015), particularly during clubbing. Chimpanzees and gorillas, however, do not attack with tools. I suggest that instead the prominence of the tori function protect the eyeball from slashing canines. I find the contrast in the African ape and hominin brow ridge as support for the clubbing and canine hypotheses: Hominin tori tend to be thick above the eyes, but not laterally, which may protect against clubbing, whereas chimpanzees and gorillas have a protective ring that encircles the eye orbit, leaving it deeply recessed. If the eye protection hypothesis is correct we might predict that among male chimpanzees brow ridge prominence (already well-accepted; Ravosa, 1991) and the concavity of the eye in the socket will be greater, while among bonobos, given the lack of differences in aggression between the sexes, we should expect little sex difference.

10. Hypothesis: Chimpanzees have a smaller-diameter external auditory meatus than humans (Leakey et al., 1995). The smaller external ear opening of chimpanzees is hypothesized to act in concert with large pinnae to increase the ability to detect faint sounds in the distance, functioning like an ear trumpet, presumably at the expense of clarity, at least in some frequency ranges (Kojima, 1990; Martínez et al., 2013). I expect that on average chimpanzees will perceive pant-hoots from significantly greater distances than humans.

11. Hypothesis: Among humans, the length of the nasal passage and other dimensions of the nasal cavity are negatively correlated with average ambient humidity (Noback et al., 2011). The same should be the case with chimpanzees: Facial proportions should vary between forest and dry-habitat chimpanzees, yielding a longer internal nasal passage among dry-habitat chimpanzees, perhaps even resulting in a visible external bony nose.

12. Hypothesis: Hunt (1991) found that chimpanzees do not fully abduct their humeri during vertical climbing, suggesting that the capacity to fully abduct is an adaptation to armhanging, not climbing. In lab experiments, Isler (2005) found that orangutans, gorillas, and gibbons do often fully abduct their forelimbs during vertical climbing. However, she observed her captive experimental subjects as they locomoted on a rope, a support unlike any a wild ape might climb. Vertical climbing in the wild is engaged in on stiffer and often quite rigid supports, allowing the torso to be tilted away from the vertical support to increase friction on the pes. The leaning-back kinematic during rigid-support vertical climbing closes the angle the humerus makes with the thorax, precluding the need for full abduction. Experiments on or observation of apes ascending non-compliant substrates will resolve this issue. If the elbow is fully extended and the arm fully abducted during vertical climbing on supports that simulate those in the wild, then a reduced olecranon process and a fully abductible forelimb may well be at least in part adaptations to vertical climbing. If not, not.

13. Hypothesis: Narrow terminal phalanges among some primates is unexplained. I hypothesize that the trait is an adaptation to exploring narrow spaces or extracting foods from concave substrates. If so, we might expect frugivores, because they extract fruit flesh from concave exocarps, to have narrower terminal phalanges than folivores.

14. Hypothesis: The many species of hylobatid provide the variation to test some hypotheses about thorax shape. The cone-shaped thorax of chimpanzees has been hypothesized to be an adaptation to unimanual armhanging (Hunt, 1991); if so, among the many species of hylobatids we expect a strong correlation between armhanging frequency and cone-shapedness, if body mass is accounted for.

15. Research topic: Chimpanzee preference for sour tastes is unexplained and subtleties of taste preference are undocumented. While fruits often contain acids, chimpanzees prefer and require ripe fruits. Sugar is a much better indicator of nutritional quality than acids, and there is no known desirable nutrient that is associated with acids that is not also associated with sugar. Perhaps, then, the chimpanzee "preference" for sour tastes is actually a tendency to ignore the sourness. If so, the comparative antipathy of humans to sourness is unexplained.

16. Research topic: Proprioception, or kinesthetic sense, seem quite different between humans and chimpanzees, but it has been studied only superficially (McCulloch, 1941; Prestude, 1970).

17. Research topic: Chimpanzee balance (equilibroception) seems superior to that of humans, but differences in semicircular canal shape fail to demonstrate why (Ryan et al., 2012; Gunz et al., 2013; Schulz et al., 2014; Spoor et al., 2007).

18. Hypothesis: The genes *ICEBERG* and *IL1F7* dial down inflammatory response, while *IL1F8* may ramp up inflammation. I speculate that the loss of these inflammatory-regulating genes may give chimpanzees the capacity to heal rapidly, perhaps an evolutionary response to frequent injury among chimpanzees.

19. Hypothesis: Chimpanzees lost genes that offer a modicum of resistance to sleeping sickness, *APOL1* and *APOL4*. Because the vector for sleeping sickness is largely confined to an elevation of 2 m, I have speculated that this loss is related to the adoption of tree-sleeping after a period of sleeping on the ground (Hunt, 2016), but the loss is still curious. There is little evidence that there was a terrestrial phase in chimpanzee evolution, and chimpanzees spend enough time on the ground that they would seem to be susceptible to sleeping sickness. The loss bears further examination.

20. Hypothesis: Sixteen genes related to protein regulation have been positively selected in chimpanzees, versus only two for humans. Proteolysis is the "digestion" of protein – not only digestion in the gut, but protein turnover in the body, such as when the protein in dead cells is recycled. In dogs, infection causes an upregulation of proteolysis-related genes (Hagman et al., 2009) and appears to have something to do with wound healing in humans (Lauer et al., 2000; Eming et al., 2014; Leoni et al., 2015), perhaps speeding up the transition from inflammation to healing (Landén et al., 2016). This emphasis on wound-healing suggests that the extreme violence observed among living chimpanzees evolved after the split between our two species. A closer examination of wound healing and the outcome of proteolysis upregulation might be valuable.

21. Hypothesis: During dispersal both gorilla (Harcourt & Stewart, 2007) and orangutan (Singleton et al., 2010) sisters disperse so as to be near one another. I expect the same among female chimpanzees; sisters should be found in the same community and even in contiguous core areas. Given the amount of paternity testing among chimpanzees, we should have picked up this signal already; if this is not the case, I imagine there is a very interesting reason why.

22. Hypothesis: In my view the most important impetus for male territoriality is the fixed distribution of females; both chimpanzee and orangutan females have relatively stable core areas and males are territorial – singly among orangutans, coalitions among chimpanzees. I expect that in comparisons of any two closely related species, males in the species where females have more fixed core areas should display more territoriality than the species with more nomadic females. We might even expect this within species, with communities where

females are more nomadic displaying less territoriality.

23. Hypothesis: Chimpanzees engage in physical activities that would be called "sport" if observed in humans. Such activities affect social rank; they serve as a proxy for direct physical confrontation. If chimpanzee social display functions like sports activities, we might predict that males who engage in longer social displays should receive more grooming and more vigorous pant-grunts than after shorter displays. I expect that, compared to other primates, chimpanzees – particularly males – will pay attention to any physically demanding male activities, even if the activities have no direct bearing on the watcher; e.g., if they are at quite a distance. Among chimpanzees, social rank might be expected to change more often without direct physical contact than among primates in other societies. Intracommunity wounding should be less common among chimpanzees than among males in other highly competitive societies.

24. Hypothesis: The aid chimpanzee mothers provide for their sons is so profound it must exert strong selection on the female lifespan. We might expect that female chimpanzees have longer telomeres, a more active mutation repair mechanism, and a stronger immune system than males, or other mechanisms to extend life. Even more so for bonobos: A comparison of zoo primates might find a bigger difference in lifespan between males and females among bonobos than chimpanzees.

25. Hypothesis: Many tests of chimpanzee memory fail to note whether the subjects are male or female. When sex is specified, my impression is that extraordinary feats of memory are often performed by females, e.g., Ai and Ayumu, among Professor Matsuzawa's test subjects. Because female core areas are small, females may benefit by memorizing the location of every or nearly every food resource, more than males in part because females are solitary and cannot rely on the memory of other party members. Males, on the other hand, often feed from larger food resources and travel in large parties. We might expect to find better spatial memories and 3D orientation abilities among male chimpanzees because they range over large areas, selecting for an ability to locate themselves and resources in space, rather than by memorizing minute details or landmarks; such a difference we find among humans (Halpern, 1997; Halpern & LaMay, 2000). We should not expect such a sex difference among bonobos.

26. Hypothesis: If the foraging regime of female chimpanzees selects for superior memory, as I expect, their spatial memory should be superior to that of chimpanzee males; because bonobo females do not forage in a limited core area, and because their superior status allows them to monopolize rich feeding sites, I expect no such memory differences between the sexes among bonobos.

27. Research topic: Among humans and chimpanzees, sporting competition and social display serve, I believe, the same function, displaying fitness without direct physical confrontation. Evidence this is so among humans (Gladue et al., 1989; Suay et al., 1999) – regardless of the contention that sporting contests are entertainment and play – include hormonal and other biomarker evidence that individuals who lose direct sporting competitions suffer from intimidation and perceived loss of status. We should expect these physiological responses to be parallel in humans and chimpanzees. We might also expect that there is a sex difference among humans and chimpanzees, since females are less likely to have participated in social display and war in evolutionary history; females should show fewer hormonal and biomarker responses to both winning and losing sporting contests.

28. Hypothesis: Male bonobos need not go easy on one another, as chimpanzee males must (Chapter 26) – we should expect that bonobo males wound one another more often than do chimpanzee males. I know of no report that suggests that male–male contest competition is more intense among bonobos; the SDH (self-domestication hypothesis) predicts less wounding during competition among males, while the MCH (monkey convergence hypothesis) expects more. While I am willing to accept a few unexpected observations in melding the SDH and MCH, it would be very interesting to have data that directly compares male–male aggression levels in bonobos, chimpanzees, and a representative of Old World monkeys.

29. Hypothesis: If bonobos are convergent on monkeys, as the MCH holds, I expect that sisters sometimes transfer together, and more often than is the case among chimpanzees. I predict that we will eventually find pairs of strongly bonded bonobo females, distinctive from other female pairs because they would have little or no sexual contact.

30. Hypothesis: Perhaps bonobo specialists already have data to test some aspects of the MCH. Compared to chimpanzees, among bonobos we might expect to find a lower rate of female transfer (like monkeys) accompanied by alliances among sisters; there is some evidence of this, because males more often have both their mother *and* their grandmother in the group of residence (Schubert et al., 2013). The MCH expects that male bonobos will have more one-on-one aggressive interactions with higher rates of serious wounding compared to chimpanzees, whereas the SDH predicts the opposite. We might expect sex differences in vocalizations, with females more than males engaging in ally-drawing long-distance vocalizations. We expect higher rates of male transfer among bonobos than chimpanzees; there is indirect evidence this is so (Hohmann & Fruth, 2002). Many monkeys have allomothering – this aunting behavior should be more common among bonobos than chimpanzees. Monkeys have cheek pouches that, in part, allow them to keep up with the rest of the group by harvesting food quickly but chewing it later, when they are still; we might expect more wadging among bonobos than is seen among chimpanzees, since they are selected to sustain group cohesion.

31. Hypothesis: I expect greater intelligence and more complicated communication in fission–fusion species compared to a close relative that lives in aggregate social groups. This should be true whether the fission–fusion/aggregate pair are insects, fish, birds, or mammals. For example, *Varecia* spp. have a fission–fusion social system (Morland, 1991; Vasey, 2006) and thus are expected to have greater cognitive abilities than other lemurs. *Varecia* has a larger brain than other lemurs (Barrickman & Lin, 2010).

32. Hypothesis: Tool use is so valuable for chimpanzees because it gives them access to low-antifeedant food sources. We should expect that species that are unable to tolerate antifeedants, for whatever reason, should use tools more than a more antifeedant-tolerant sister species, even if the pair are relatively unintelligent.

References

Barrickman NL, Lin MJ (2010) Encephalization, expensive tissues, and energetics: an examination of the relative costs of brain size in strepsirrhines. *Am J Phys Anthropol* 143, 579–590.

Carrier DR, Morgan MH (2015) Protective buttressing of the hominin face. *Biological Reviews* 90, 330–346.

Churchill SE, Franciscus RG, McKean-Peraza HA, Daniel JA, Warren BR (2009) Shanidar 3 Neandertal rib puncture wound and paleolithic weaponry. *J Human Evol* 57, 163–178.

Detwiler SR (1943) *Vertebrate Photoreceptors*. New York: Macmillan.

Doran DM (1993) The comparative locomotor behavior of chimpanzees and bonobos: the influence of morphology on locomotion. *Am J Phys Anthropol* 91, 83–98.

Eming SA, Martin P, Tomic-Canic M (2014) Wound repair and regeneration: mechanisms, signaling, and translation. *Sci Transl Med* 6, 265sr6.

Gladue BA, Boechler M, McCaul KD (1989) Hormonal response to competition in human males. *Aggressive behav* 15, 409–422.

Gunz P, Ramsier M, Kuhrig M, Hublin JJ, Spoor F (2012) The mammalian bony labyrinth reconsidered, introducing a comprehensive geometric morphometric approach. *J Anat* 220, 529–543.

Hagman R, Rönnberg E, Pejler G (2009) Canine uterine bacterial infection induces upregulation of proteolysis-related genes and downregulation of homeobox and zinc finger factors. *PLoS One* 26, e8039.

Halpern DF (1997) Sex differences in intelligence: implications for education. *Am Psychologist* 52, 1091–1102.

Halpern DF, LaMay ML (2000) The smarter sex: a critical review of sex differences in intelligence. *Ed Psychol Rev* 12, 229–246.

Harcourt AH, Stewart KJ (2007) *Gorilla Society: Conflict, Compromise, and Cooperation Between the Sexes.* Chicago: University of Chicago Press.

Hohmann G, Fruth B (2002) Dynamics in social organization of bonobos (*Pan paniscus*). In *Behavioural Diversity in Chimpanzee* (eds. Boesch C, Marchant L, Hohmann G), pp. 138–150. Cambridge: Cambridge University Press.

Hunt KD (1991) Mechanical implications of chimpanzee positional behavior. *Am J Phys Anthropol* 86, 521–536.
(1992) Social rank and body weight as determinants of positional behavior in *Pan troglodytes*. *Primates* 33, 347–357.
(2016) Why are there apes? Evidence for the co-evolution of ape and monkey ecomorphology. *J Anat* 228, 630–685.

Hylander WL, Picq PG, Johnson KR (1991) Function of the supraorbital region of primates. *Arch Oral Biol* 36, 273–281.

Isler K (2005) 3-D kinematics of vertical climbing in hominoids. *Am J Phys Anthropol* 126, 66–81.

Kojima S (1990) Comparison of auditory functions in the chimpanzee and human. *Folia Primatologica* 55, 62–72.

Landén NX, Li D, Ståhle M (2016) Transition from inflammation to proliferation: a critical step during wound healing. *Cell Molec Life Sci* 73, 3861–3885.

Lauer G, Sollberg S, Cole M, et al. (2000) Expression and proteolysis of vascular endothelial growth factor is increased in chronic wounds. *J Investig Dermatol* 115, 12–28.

Leakey MG, Feibel CS, McDougall I, Walker A (1995) New four-million-year-old hominid species from Kanapoi and Allia Bay, Kenya. *Nature* 376, 565–571.

Leoni G, Neumann PA, Sumagin R, Denning TL, Nusrat A (2015) Wound repair: role of immune–epithelial interactions. *Mucosal Immunol* 8, 959.

Martínez I, Rosa M, Quam R, et al. (2013) Communicative capacities in Middle Pleistocene humans from the Sierra de Atapuerca in Spain. *Quaternary Int* 295, 94–101.

McCulloch T (1941) Discrimination of lifted weights by chimpanzees. *J Comp Psych* 32, 507–519.

Morland HS (1991) Preliminary report on the social organization of ruffed lemurs (*Varecia variegata variegata*) in a northeast Madagascar rain forest. *Folia Primatologica* 56, 157–161.

Noback ML, Harvati K, Spoor F (2011) Climate-related variation of the human nasal cavity. *Am J Phys Anthropol* 145, 599–614.

Polyak S (1941) *The Retina.* Chicago: University of Chicago Press.

Prestude AM (1970) Sensory capacities of the chimpanzee. *Psych Bull* 74, 47–67.

Ravosa MJ (1991) Interspecific perspective on mechanical and nonmechanical models of primate circumorbital morphology. *Am J Phys Anthropol* 86, 369–396.

Ryan TM, Silcox MT, Walker A, et al. (2012) Evolution of locomotion in Anthropoidea: the semicircular canal evidence. *Proc Roy Soc Lond B Biol* 279, 3467–3475.

Schubert G, Vigilant L, Boesch C, et al. (2013) Co-residence between males and their mothers and grandmothers is more frequent in bonobos than chimpanzees. *PLoS One* 8, e83870.

Schutz H, Jamniczky HA, Hallgrimsson B, Garland Jr. T (2014) Shape-shift: semicircular canal morphology responds to selective breeding for increased locomotor activity. *Evolution* 68, 3184–3198.

Singleton I, Knott CD, Morrogh-Bernard HC, Wich SA, van Schaik CP (2010) Ranging behavior of orangutan females and social organization. In *Orangutans: Geographic Variation in Behavioral Ecology and Conservation* (eds. Wich SA, Atmoko SS, Setia TM, Van Schaik CP), pp. 205–214. Oxford: Oxford University Press.

Spoor F, Garland T, Krovitz G, et al. (2007) The primate semicircular canal system and locomotion. *PNAS* 104, 10808–10812.

Stern JT Jr. (1971) *Functional Myology of the Hip and Thigh of Cebid Monkeys and Its Implications for the Evolution of Erect Posture.* Basel: Karger.

Suay F, Salvador A, González-Bono E, et al. (1999) Effects of competition and its outcome on serum testosterone, cortisol and prolactin. *Psychoneuroendocrinology* 24, 551–566.

Tappen NC (1973) Structure of bone in the skulls of Neanderthal fossils. *Am J Phys Anthropol* 38, 93–97.
(1978) The vermiculate surface pattern of brow ridges in Neandertal and modern crania. *Am J Phys Anthropol* 49, 1–10.

Vasey N (2006) Impact of seasonality and reproduction on social structure, ranging patterns, and fission–fusion social organization in red ruffed lemurs. In *Lemurs* (eds. Gould L, Sauther ML), pp. 275–304. Boston, ML: Springer.

INDEX

abducted (or divergent, or gripping) great toe, 98, 115, 143, 149, 152, 154, 158, 167, 170, 173, 201, 485, 520, 528
abduction
humeral, 143, 145, 152
humeral (definition), 110
abstraction (concept formation), 347
Acanthus pubescens, 72
acidosis, 93
acrobatic behaviors, rarity, 260
activity budget
chimpanzee, 260, 268
hunter-gatherer, 266
industrialized countries, 263
adaptation, 33, 124, 135–136, 140
adolescent infertility, 207
adoption, 306
adrenal cortex, 283
adrenal gland, 283, 291
adrenal medulla, 283
adrenalin, 220, 277, 283
effect of, 277
adulthood, 207–208
affective neuroscience, 342
Aframomum mala, 72, 388
Age of Discovery, 29
Age of Reason, 20
aggression, 344
Ahlquist, Jon, 228
Alfonso X, 277
alkaloid, 71, 82, 94, 129, 160, 183, 246, 520–521 (see also 'toxin')
allometry, 131, 133, 135, 150–151, 189, 210, 366–367
definition, 132
allomothering, 306
by juveniles, 305
alpha, 450
alpha status
advantages, 433
benefits, 431
disadvantages, 433
reinforcement of, 430
altruism, 346
American Museum of Natural History, 4

American sign language (ASL), 392
amino acid, 75, 82, 86–90, 93, 220, 222–223, 228, 284–285, 314
amnion, amniotic sac, 196
Amphibia (taxon), 42
amygdala, 364, 376
amylase, 88, 220
anatomy, 97–98, 100, 117
Ancestral Positional Behavior Regime (APBR)
and blood pressure, 266
and cancer, 266
definition, 266
and diabetes, 266
and heart disease, 266
and immune function, 266
joint health, 266
and obesity, 266
and stroke, 266
Andrews, Peter, 161, 177
anestrous, 458
animal model, 255
Animalia (taxon), 42
ant wand, 12, 408, 413
anthrax, 255–256
antibiotics, 254
penicillin, 254
streptomycin, 254
sulphonamide, 254
anticodon, 222
antifeedant, 70, 76, 80, 83–84, 87, 94, 128–129, 160–161, 163, 178–181, 351, 510, 522–523, 527–528
antifeedant intolerance
as driver of ripe-fruit diet, 69, 509
and tool use, 530
antinutrients (digestion inhibitors), 75, 82–83, 85, 93, 95
ants, safari, 12, 408, 412–413
anvil, 412
APBR. *See* Ancestral Positional Behavior Regime
ape
comparative osteology, 59
as a concept, 20
definition/taxonomy, 57

diets, 67, 73
elbows, 110
humeri, 109
identifying first ape, 158
positional behavior, 126
thoraxes, 108
ape-monkey competition, 162, 521, 528
appetite
stimulant, 277
suppressant, 277, 285
appreciation of music, lack of, 342
Aquinas, Saint Thomas, 33, 519
arbitrariness, 391, 393
Aristotle, 24
armhang, 120, 123–124, 126–127, 129, 144, 146–148, 179, 181, 409, 477, 485
australopith, 474
finger length, 528
arousal threshold (during sleep), 374
arsenic, 94
art
appreciation of, 340
chimpanzee art represented as human, 341
creation, 341, 420
arthritis. *See* Degenerative joint disease
captive chimpanzee, 262
wild chimpanzee, 261
Arthropoda (taxon), 42
Aspilia, 253
Astringent, 160
asymmetry, 365
Athens, 22–23
athletic ability
role in social rank, 434
attachment theory, 307
attractivity, 459
attribution of false belief, 338
experiments, 339
attribution, of mental states, 338
Augustine, Saint, 32
australopith
bony labyrinth, 488
chimpanzee-like traits, 479
femoral shaft obliquity, 478

australopith (cont.)
  fingers, 477
  forelimb morphology, 477–478
  hindlimb morphology, 479
  human-like traits, 479
  knee joint, 478
  leg length, 478
  manual phalanges, 478
  microwear, 487
  morphology, 473–474
  morphology, interpretation of, 474
  pedal phalanges (toes), 479
  pelvic morphology, 475
  pisiform, 477
  pollen, 487
  sexual dimorphism, 237
  skeletal features summary, 486
  skull morphology, 475
  social system, 237
  thoracic morphology, 476
  trace elements in fossils, 487
  ulnar styloid process, 477
Australopithecus afarensis, 474
Australopithecus anamensis, 474
autosomal chromosomes, 236
Aves (taxon), 42
Aye-aye, *Daubentonia*, 50

baboon
  gelada (*Theropithecus*), 56, 83, 179, 483, 488, 533
  hamadryas (*Papio hamadryas*), 179, 237, 483, 487, 533
  savanna or olive (*Papio anubis*), 18, 48, 54–56, 126–128, 137, 160–161, 168, 170, 175–176, 179–180, 293, 312, 316, 318, 344–345, 376–377, 452, 456, 461, 466, 506, 510–511, 534
  mating system, 460
baby fat, human
  function, 203
balance (equilibroception), 248
baobab fruits, 412
Baphia capparidifolia, 71
basal metabolic rate (BMR), 294
Basmajian, John, 134
Beagle, The, 30
beetles, 37, 452
begging, begging gesture, 387
Begun, David, 175
Beilschmiedia ugandensis, 489
Belyaev, Dmitry, domestication, 212
Benefit, Brenda, 164

bilophodonty, 159, 183
binding site, 224
binomial system, 36, 40
biomechanics, 97, 131
bipedalism, 115, 122, 148, 177, 179, 257, 409, 431, 450, 473, 479–485, 487–488, 493–494, 502, 532, 534
bipedalism origin hypotheses, 480
  efficiency, 482
  food carriage, 481
  infant carriage, 481
  killer-ape, 481
  postural feeding, 484–485
  seed eating, 483
  small object feeding, 484
  social display, 481
  suspensory, 480
  thermal radiation, 483
  tool use, 480
  wading, 482
birth, 300
  eating afterbirth, 301
birth canal, 203
birth weight, 203
black and white colobus (*Colobus guereza*), 44, 55–56
Bloomington, Indiana, 191
Blumenbach, Johann Friedrich, 27
body sense (proprioception), 248
body size
  and intelligence, 528
  nest or sleeping platform, 530
body size (mass, weight), 98, 100, 114–115, 122, 124, 128–129, 131–134, 144, 148–149
Boesch, Christophe, 14, 18, 315, 317, 321, 333, 336, 347, 411, 430
Boesch, Hedwige, 14, 18, 317, 321, 333, 347, 411, 430
bone
  femur, 474
  lumbar vertebrae, 148
  ilia, allometric forces, 150
  knee joint, 474
  lumbar vertebrae, 148, 475
  pelvis, 475
  phalanges, 474
  robust mandible, 475
  robust zygomatics, 473–474
  sacrum, 475
  scapula, 475
  tarsal, 52
  ulna, 477

bonobo, *Pan paniscus*, 8, 40, 58–59, 122, 158, 176, 210–212, 227, 229, 256, 286, 293, 331, 342, 349, 396, 398, 406, 499–505, 508–512, 519
  alpha male reproductive success, 504
  bipedalism, 502
  body mass, 501
  body proportions, 501
  contact calls, 510
  diet, 508
  discovery, 500
  dispersal, 503
  dispersal, male, 504
  extended estrus, 505
  extended lifespan, 504
  female alliances, 499
  female bonds, 503
  female dominance, 500
  foramen magnum placement, 502
  fruit consumption, 508
  genito-genital rubbing, "g-g rubbing", 503
  geographic distribution, 500
  hunting, 501
  intercommunity relations, 507
  lack of pant-grunt, 505
  lack of territoriality, 507
  less variable party size, 504
  locomotion and posture, 482
  long-distance calls, 504
  male aggression, lesser, 505
  monkey convergence hypothesis (MCH), 509
  mother-son alliances, 503–504
  narrow body plan, 503
  neoteny, 509
  origin of female dominance, 504
  philopatry, 503
  pronunciation, 500
  pygmy chimpanzee, 500
  reproduction, 505
  self domestication, 509
  semicircular canals, 503
  sexual dimorphism, 501, 511
  skeletal evidence of violence, 505
  spatial memory, 508, 510
  terrestrial herbaceous vegetation (THV), 508
  terrestriality, 503
  tool use, 508
  traits shared with monkeys, 486
  vocalizations, 504

border patrol, 75, 298, 348, 440–441, 468, 518, 530
  defined, 441
  low-ranking male participation, 445
bowed radius and ulna, 152
bowerbirds, 454
Bowlby, John, 260, 307
box stacking, 328
Boysen, Sally, 333, 343
  and M&M's, 344
Boysen, Sarah T., 332
Brace, Loring, 473
brachiation, 57, 124, 126, 142–144, 148, 150, 161, 181, 329, 480, 502
brain
  and body awareness, 363
  growth, 284
  insula, 202
  and mental rotation, 363
  neuron number, 367
  operculation, 202
  orbitofrontal cortex, 202
  and spatial awareness, 363
  visual cortex, 242
brain asymmetry, 361, 368
brain development, 204–205
brain hemispheres, 357
brain size
  and life history, 534
  as a primate trait, 47
  sleep paralysis, 530
brain stem, 357
Broca's area, 323, 360–365, 368–369
Brodmann, Korbinian, 364
Brodmann's areas, 365
bronchitis, 257
Brown, Daniel, 419
buckling of the spine, 148–149
Budongo, Uganda, 17, 69, 72–73, 247, 312, 386, 413, 423, 427, 490
  tandem walk, 423
bushbaby, *Galago*, 50, 52, 318, 321, 407, 501, 508
bushpig, 348
buttresses, 63
Bygott, David, 18, 257, 313, 413

caffeine, 94
calcitonin, 293
callitrichid = marmoset or tamarin, 43, 52–53, 131–132, 276, 287
Cambridge Declaration, the, 339
Cameroon, 4

Campo-Ma'an, Cameroon, 159
cancer, 258
canine function, 137, 140
Cant, John, 161
captivity, chimpanzees in, 1, 7, 11–12, 14–15, 28, 61, 68, 75, 100, 245, 247, 255, 257–259, 261, 300, 302, 325, 346, 377, 398, 408, 414, 508, 517, 521
capuchin, *Cebus*, 44, 47, 51–53, 151, 312, 331, 519
Carroll, Sean
  Endless Forms Most Beautiful, 191
carrying, of infant by mother, 204, 298, 301, 461, 523
Carthage, 22–23
caudal (definition), 112
causality
  and Broca's area, 363
ceboidea (informally, "ceboid"), 42–43, 51–54
cecum, 87–90, 93
cell differentiation, during fetal growth, 190
cell membrane, 218, 276
cellulase, 83, 86–87, 90
cellulose, 65, 71, 75, 83, 85, 87, 90, 160, 183
central sulcus, 360
Cercopithecoidea (informally "cercopithecoid"), 42–43, 51, 54–56, 162, 164, 175
cerebellum, 357
cerebrum (cerebral cortex), 357
cervix, 288, 291
Challenge hypothesis, 468
Charles
  Semliki, 491
Chausiku, 252, 348, 440
cheek flanges, 57, 139
chemical defense, 81
Chimfunshi Wildlife Orphanage, 307
chimpanzee, 246
  earliest image, 21, 24
  intellectual challenges in wild, 15
  naming, 27
  population density vs monkeys, 523
  prospect of living among humans, 400
  range size vs monkeys, 523
  rigors of daily life, 521
chimpanzees
  as a fully evolved species, 520
  as referential models, 520

Chimpanzees of Gombe, Goodall Jane, 29
chin, 137–138
  evolutionary origin, 139
chlorophyll, 86
Chordata (taxon), 42
chromosome number, human vs chimpanzee, 224
chronic stress, 286
class, taxonomic, 42
Clever Hans, 395
Clever Hans Effect, 395–396
climbing. *See* Vertical climbing
closed-canopy rainforest, 62–63
coalitionary violence
  and imbalance of power, 530
coalitions, 431
cocaine, 82, 94
coccyx, 200
codon, 222
coevolved bacteria, 85, 87, 90, 163
cognition
  diversity, 521
  judging relative size, 332, 343
  quantity assessment, 343
Collins, Tony, 18, 409
Colobus thumbs, 174
colon, 47, 87–90, 93, 258, 267
colon cancer, 267
color discrimination, 243
color perception, 243
colugo, or flying lemur, 43
common cold, 257
community, 141, 206–208, 245, 302, 443–446, 457, 462
  assaults on, 372, 376, 427
  boundary, 441
  cohesion, 350, 456
  as a concept, 440
  cultural variation among, 13
  culture, 421
  definition, 437
  identity, 303
  integration, 302
  internal politics, 427
  as a mental construct, 349, 438
  range, 74, 438
  resources, 348
  size, 437
  territory, 442
  transfer, 207, 442
Comparative Study of Ape Emotions, 13, 343
competition, between apes, monkeys, 128–129, 175, 183

competitiveness, 344
  gaze-following, 345
complex carbohydrates, 71, 75–76, 87, 89–90, 92, 534
composite tool raking task, 328
composite tools, 328
compound tools, 328
conceptual model, 532
cone-shaped thorax, xiii, 110, 135, 147, 476, 479, 484
  function, 145
confidence in paternity
  and mating system, 460
Congo, Democratic Republic of, 17, 24–25, 62, 64, 255, 261, 500
Congo, the chimpanzee, 341
consortship, 465
contractions (birth), 301
copulation, 304
core area, 304–305, 439, 444, 459, 521, 523–525
  advantages, 74, 94
  choice, 443, 462
  consequences for males, 183
  link to male territoriality, 510
  location, 206–207, 304
  peripheral, 439
  quality, 439
  selection for, 440
  size, 439
  and social rank, 468
corpus callosum, 357
corpus luteum, 289, 291
corticotropin (ACTH), 283
corticotropin releasing hormone (CRH), 283
cortisol, 283, 286
  and social rank, 293
co-sleeping, 375
Coula nuts
  nut-cracking techniques, 411
counting, 332
  unboundedness, 332
crepuscular, 48
Crick, Francis, 226
Cronin, Katherine, 307
cross-fostering, 5, 248, 343, 388–389, 392, 394
cross-modal transfer, 337
cryptorchidism, 491
cultural diffusion, 421
cultural transmission, 421

culture, 12–13, 388, 399, 419–420, 424, 518
  experimentation, 420
  food item selection, 423
  geographic variation in ant-dipping, 423
  geographic variation in diet, 423
  geographic variation in gestures, 423
  geographic variation in leaf-clipping, 423
  geographic variation in napkin use, 423
  geographic variation in nut-cracking, 421
  geographic variation in spear use, 423
  geographic variation in sponge use, 421
  geographic variation in well-digging, 423
  role of mother in acquisition, 347, 423
  variation across space, 421
culture, definition, 419
cyanide, 82, 94, 246, 522, 533
cytoplasm, 218, 222

Dart, Raymond, 481
Darwin, Charles, 13, 30, 36–37, 57–58, 212, 224, 342–343, 451, 480–481, 487, 519
Dawkins, Richard, 217, 455
de Waal, Frans, 420, 434, 500
death, 307
death of infant, mother's reaction, 306–307
DeBrazza's monkey, 452
deception, 337, 339–340, 434, 468
deconstruction task
  apes superior to *Cebus*, 331
  box of rocks, 331
  reducing box to sticks, 330
  rope on pole task, 331
deep crouch, 431
degenerative (lifestyle) diseases, 254
degenerative joint disease (DJD) or arthritis
  and body weight, 262
  frequency among ancient humans, 264
  frequency among captive chimpanzees, 261
  frequency among hunter-gatherers, 262

  frequency among older Americans, 262
  frequency among wild chimpanzees, 261
  frequency and activity level, 262
  frequency in USA, 262
  repetitive tasks, 264
dendropithecoids, 159
dental apes, 159, 161, 163
dental caries, 258
dentistry, chimpanzee, 414
dermal ridges, 54, 477
Dermoptera (or colugo [taxon]), 43
Descent of Man, 58
Deschner, Tobias, 466
detumescence, 464
diagnostic features ("killing features"), 41
Dian Fossey, 58
diarrhea, 254
Dictyophleba lucida, 71
diet, 11, 44, 47–48, 50, 52, 54, 61, 67–68, 71–72, 74–75
  breadth, 70
dietary imperatives, 520
differential growth, 188
digestion, 87–88, 90, 93
digestive enzymes, 82, 89, 93, 191, 220–221
digestive physiology, 73, 163, 183, 294, 350, 518, 522
  as a driver of social behavior, 521
  primitive humans, 518
diphtheria, 254
Diplorynchus condylocarpon, 70–71
disaccharides, 90
discreteness, 391, 393
disease, human
  causes, 534
dispersal (= group transfer), 206, 442, 504, 506
  bonobo, 503
dispersal (transferring)
  and social rank, 304
displacement, 391, 393, 533
distance-minimizing ability, 333
distinctness, of species, 34, 40–41
diurnal, 46, 48, 54, 243
diverticulosis, 267
DNA, 190–191, 217–222, 224, 226, 228–229, 518
  adenine, 219
  cytosine, 219
  guanine, 219

junk DNA, 237
regulatory DNA, 221
thymine, 219
DNA repair, 219
DNA-DNA hybridization, 228
dogbane, 85
domestication, 211, 509
common traits, 509
domestication, consequences, 211
domestication, foxes, 211
dominance hierarchy, 431
Donald, 389
Doran, Diane, 502
Dorsal ridge, 106, 109, 154
dose effect
exercise and DJD, 264
Douglas, Preston, 4, 6-7
drinking wells, chimpanzee, 12, 410, 490
Drumming, 387, 410, 414
Dryopithecus, 176, 182
duality of patterning, 391, 393

ear (pinnae), 140
function, 141
role of shape in perception, 140
size, 245
size, chimpanzee, 141
size, gorillas, 141
ear tears, Semliki, 490
early adolescence, female, 205
early adolescence, male, 205
Ebola, 255-256
ecomorphology, ecological morphology, 97, 183
edge detection, 244
efficiency, selection for, 136, 144-145, 162, 171, 177, 181, 333, 335
ειδος. See eidos
eidos, 32 (see also "forms," "ideals" and "types")
Ekembo, xvii, 165, 168-169
Ekembo heseloni, 166
Ekembo nyanzae, 166
Elaeis guineensis, 72
elastase, 220
elbow, 107
anatomical parts, 110
angled back medial epicondyle, 168
ape, 107, 110
australopith, xiii
flexion, 110
human, 107
monkey, 110
Proconsul, 168
elbow extension, 107, 110, 143, 154, 168
and armhanging, 143, 145
short olecranon, 152
elbow flexion, 110, 114, 143, 152-153, 174
electromyography (EMG), 134
electrophoresis, 226
embryonic development
amniotic sac, 196
apoptosis, 200
arm buds, 199
blastocoel, 194
blastocyst, 194
Carnegie Stage 10, 198
Carnegie Stages, 198
cartilage, 199
cell differentiation, 190
digestive system, 196
ectoderm, 196
endoderm, 196, 198
first cell division, 190
gastrulation, 196
heart, 197
implantation, 195
inner cell mass, 194
leg buds, 199
mesoderm, 196
neural crest, 197
neural crest cells, 197, 199
neural fold, 197
neural groove, 197
neural tube, 197
notochord, 197
polarity, 190, 196
primitive streak, 196
sodium ions, 194
somites, 197
transverse folding, 198
trophoblast cells, 194
yolk sac, 196
yolk sac, erythrocyte production, 196
emergent layer (of forest), 63
emergent properties, 12, 517
emotional regulation
and sleep, 374
emotions, 343
anger, 343
anticipation, 343
disgust, 343
fear, 343
joy, 343
sadness, 343
surprise, 343
trust, 343
emotions (affect), 342
empathy, 13, 247, 305, 323, 344, 346
in captivity, 346
in the wild, 346
endocarp, 85
endocrinology, 276
endometrium, 195, 291
endurance, 235
and testosterone, 292
Entebbe Zoo, 2
environment of evolutionary adaptedness (EEA), 100, 260, 262-263, 266, 268, 520
enzymes, 220
epigenetics, 224
epiglottis, 103
epinephrine, 277
epiphyte, 63
Equatorius, 176
erector spinae muscles, 169
erythropoietin, 293
Escherichia coli, 191
esophagus, 101-103
essence, 32
essentialism (Neoplatonism), 32-33, 35, 167
estradiol, 288
estrogen, 277, 288, 290-291, 294, 459
estrogen, healthful effects, 288
estrogen, target tissues, 288
estrus
costs, 304
period, 55, 58, 74, 207, 276, 304, 313, 413, 446, 451, 459-460, 462, 464-466, 468-469, 492, 501, 505, 510
resumption after weaning infant, 304
signaled by volatile fatty acids, 459
swelling, 55, 205-206, 304 (see also sexual swelling)
ethics, 346, 420
ethylene (and ripening), 86
eunuchs
lifespan, 293
Evered, 319, 384, 428
evolutionary inertia, 480
executive functions, 360
exocarp, 85
exon, 221
expansin, 86-87

expressed DNA, 220
extended phenotype, 453
extracommunity killing, 441
exudates = saps or gums, 53
eye, 101
   anatomy, 243
   frontation, 46, 52–54, 518
   stereoscopy, 46

Faben, 257
face recognition, 244
facial expressions, 324
fairness, expectation of, 346
Fall of Rome, 32
fallback foods, 71–72, 74, 87, 315, 518, 527
fallopian tube, 190
false belief, 345
family, taxonomic, 42
famine, epigenetic response to, 224
Fanni, 298
fast twitch muscle fibers (MHC II isoforms), 116
fat, 285
   and health, 266–267
   metabolism and adrenalin, 277
   metabolism and cortisol, 283
   storage, 276, 284, 286
fathering, 298
fatty acids, 90, 93, 285
Fatuma, 306
fear-grin, 348, 350, 386
feed-as-you-go, 129, 172, 528
feeding bout, 12, 70, 126, 305, 450
feeding parties, 438
female
   aggression and violence, 428
   aid during birth, 300
   body mass, 57
   copulation rate, 304
   core areas, 439
   dispersal, 304
   estrus, 304
   hunter-gatherer, 293
   hunting, 314
   mate-choice, 434
   social rank, 305
   sociality in captivity, 302
   spatial memory, 333
   the ecological sex, 313
female ovulatory cycle, 290, 459
female–female relationships, 207
femoral head
   cylindrical, 170

femoral neck
   angled up, 170
   long, 170
femur (pl. femora), 114–115, 132, 170, 474, 478, 492
   condyles, 115, 164
   head, 170
   head, cranial displacement, 152
   head, displaced from greater trochanter, 170
   head, spherical, 115
   valgus, 115, 478–479, 492–493
   varus, 115
fetal brain growth, 203
fiber = cellulose, 47, 68, 70, 73, 83–86, 90, 92–93, 129, 160, 163, 266–268, 510, 520–522, 531
Fifi, 207, 298, 304, 308, 424, 433
finger length, 24, 131, 142, 144, 154, 158, 162, 168, 171, 188, 200, 478, 487, 501, 528
fingernails, as primate trait, 45
fingers (phalanges)
   curved, 104, 144
   flexor sheath ridges, 144, 179, 477–478, 501
   length, 131
Fisher, Dorothy Canfield, 309
fission-fusion social system, 350, 438
   bonobos, 501
   described, 54
   as a driver for advanced cognition, 349–350
   and imbalance of power, 533
   and intelligence, 533
   large brains, 519
   and paternity confusion, 461
   as a response to dispersed food, 521
Flint, 308
   and mother's death, 305
Flo, 257, 304, 308, 424
   death, 306
Flossi, 299–300, 424
flowering, 64
flowers, blossoms, 32, 44, 65, 69, 71, 73, 80, 82, 423, 510
foamy virus, 256
follicle, 289
follicle-stimulating hormone (FSH), 290
   spermatogenesis, 292
follicular phase, 289

Fongoli, 488
food list, 69–70
   cultural acquisition, 423
   mothers discourage new items, 423
food supply
   and imbalance of power, 525
   and party size, 525
food-sharing, 213
foregut fermentation (digestion), 90, 93
forelimb
   anatomy, 110, 114, 143, 152
   function of long arms, 142, 144
forelimb retractors (shoulder flexion), 112, 143, 149
forest
   ground layer, 63–65
   riverine or gallery, 61, 63–65
   scrub or bush, 65
   seasonal, 61, 64–65
   tropical rainforest, 61–65
   woodland, 63–66, 178
forms, 33 (see also "types," "ideals," and "eidos"
Fouts, Roger, 335, 393–394
fractions, 332
fractures, 259
   long bones, 259
   neck, 259
   sex differences, 259
   skeletal element, 259
   skull, 259
fractures, healed, 259
Franklin, Benjamin, 254, 405
Franklin, Rosalind, 226
Franks, Steven, 390, 400
Freud, 428
Frodo, 428
fronds, 72, 139, 153, 303, 330, 379, 438
frontal lobe, 358, 360
fructose, 76, 89–90
frugivore, frugivory, 67, 73, 84, 93, 163, 171, 243, 481, 508, 533
fruit
   ripe, 68, 71–73, 76, 83–84, 93–94, 124, 126, 128, 161, 163, 171, 180, 247, 258, 332, 440, 508, 518, 523
   unripe, 68–69, 73, 84–85, 93–94, 129, 171, 180–181, 246, 510, 521
fruit dispersers, 94
fruit flies
   Drosophila melanogaster, 191

fruit ripening, xvii, 68, 80, 86–87, 438, 525, 527
fruit syndrome, 84
Furuichi, Takeshi, 14, 505, 509

galactose, 89, 191, 220
gap-crossing, 128, 161, 174–175
gape, jaw, 136, 153
Gardner, Allen and Beatrix, 335, 346, 392
Garner, Robert, 28
gaze-following, 338, 345
gedanken, 328
Gehring, Walter Jakob, 192
gene, 217, 219–220, 224, 228–232, 234–235, 237–238, 294
    CASP12, and sepsis, 235
    CMAH, 235
    FOXP2, 237
    IL1 gene family, 235
    IL1F7, 235
    IL1F8, 235
    KRTHAP1, 235
    lactase production, 191
    lost genes, 234–235
    MMP5, 193
    MYH16, 235
    Neanderthal *FOXP2*, 238
    selfish genes, 217
    TAS2R, 246
generalized quadrupeds, 49, 52–53
genetic similarity
    humans, chimpanzees, 521
genetics, 43, 52, 59, 164, 183, 217–218, 224, 226, 294, 519
    Semliki, 493
genome, 217, 221
    chimpanzee, 229–230, 232, 238
    human, 229, 237, 294
    monkey, 163
    Neanderthal, 238
genomics, 217
genus, taxonomic, 27, 40, 42, 49, 51
geometrically scaled, 133–134
geomorphology, 97
gestation, 189, 202
gesture, 388, 423
    arm-over, 421
    list, 386
ghrelin, 285
    response in mice, 285
    response to stomach, 285
gibbon (hylobatid), 57, 103, 110, 112, 114–115, 124, 127–128, 135, 142–143, 148, 150, 158, 163, 294, 379, 451–452, 480, 483, 519, 534
    mating system, 460
Gigi, 314
giving-up-density, 128
glucagon, 284
glucocorticoids, 283
glucose, 76, 89–90, 191, 220, 266, 276–277, 283–285
glycogen, 284
Goblin, 428, 432, 434
golden mean and joint wear, 264
golden mean, exceptions, 264
golden mean, hypothesis for DJD, 264
golden snub-nosed monkey, 56
Gombe, Tanzania, 9–12, 16–17, 20, 29, 64–65, 69, 80, 87, 126, 128, 159, 180, 206–207, 258–259, 300, 304–305, 312–313, 315, 318–320, 348, 351, 409, 414, 421, 423, 427, 430, 437, 442, 456, 468, 492
    diet list overlap with Mahale, 75
    feeding bouts, 70
    hunting, 315
    kidnapping, 17
    polio, 257
    prey choice, 318
    skeletal injury, 139
    termite mounds, 409
gonadotropin-releasing hormone, 290
gonorrhea, 256
good genes, 451, 453
Goodall, Jane (van Lawick), 1, 11, 14, 16, 29, 38, 74, 80, 254, 257–259, 312, 316, 331, 343, 351, 379, 383, 405, 409, 433, 437, 469, 519
Goodman, Morris, 226–227, 229
Gorilla, 23–24, 28, 41, 43, 45, 58, 102–104, 114, 122, 132, 150–151, 158–159, 171, 178–179, 181, 200, 210, 227, 229, 233, 236–238, 262, 339, 460, 469, 492, 509–510, 518–519
    mating system, 460
Goulago triangle, 347, 409, 424
Gould, Steven, 420
gout, 258
graded signal hypothesis, 466
grandmother hypothesis, 209
grass stems, 438
gray matter, 359
great ape
    definition/taxomomy, 59
Great Chain of Being, 30, 33, 38
Grewia species, 489
grip
    key, 45
    power, 45
    precision, 45
gripping great toe, 168
grips, manual or hand, 45, 104
growth hormone (GH), 293
growth spurt, 210
Gua, 246
guenon, *Cercopithecus*, 51, 55
Gwekula, 252

Haeckel, Ernst
    ontogeny recapitulates phylogeny, 200
hammer and anvil tasks, 328
hand-clasp grooming, 421
handedness, 365
handicap, 453
handicap, intersexual selection, 452
handicapping displays, 451
handicapping principle, Zahavi's, 453
Hanno, 22–23, 25, 30
Haplorhini (or informally, "haplorhine"), 42–43, 46, 52
Hare, Brian, 345
Harlow, Harry, 307
Hayes, Catherine and Keith, 244, 389
health
    and physical activity, 520
hearing (audition), 244–245
    acuity, 245
heart attacks (myocardial infarction),, 267
heart disease
    absence among hunter-gatherers, 259
    absence among wild chimpanzees, 259
    chimpanzee susceptibility, 259
    presence among captive chimpanzees, 259
heat stress and bipedalism, 482
hemoglobin, 220, 229
hepatitis C, 255
Herculano-Houzel, Suzana, 358
Herpes, 257
HGH (human growth hormone), 220
hierarchical, 35, 40

high cholesterol, 233, 264, 267
    among chimpanzees, 260
high density lipoproteins (HDL) and
    diet, 267
high-fiber diet and joint health, 267
hindgut fermentation (digestion), 90,
    93
hindlimbs
    short, function of, 144
hip flexibility, 170
hippocampus, 364
Hirschauer, Margaret, 491
Hispanopithecus, 162, 176
HIV, 255
HIV (chimpanzee). See SIVcpz
Hockett, Charles, 390
home movies, of Meshie, 6
home range, 66, 335, 423, 437,
    439–440, 522, 533–534
homeobox genes, or Hox genes, 190
homeosis, 190
Hominoidea (informally, "hominoid"),
    43, 51, 57, 59, 109, 164
homosexual pair bonds, among
    warriors, 503
honest display, 453, 456
honey, 68, 71–72, 263, 405–408,
    414–415, 522
hormone
    angiosinogen, 277
    antagonists, 277, 285
    atriopeptin, 277
    endothelin, 277
    ghrelin, 277
    leptin, 277
    renin, 277
    thromboxane, 277
    vasoactive intestinal peptide, 277
    vasodilation, 277
    vasopressin, 277
hormones, 89, 97, 195, 217, 220,
    275–277, 283–293, 295, 373,
    375, 439, 458, 461, 466
    complete list, 278
    sex, 277
hormones, reproductive, 286
    estrogen, 286, 288, 290–291, 295,
        459
    follicle stimulating, 290, 292
    luteinizing, 290
    progesterone, 290–291, 466
    prolactin, 286
Houle, Alain, 128, 161
Hox gene, 192–193

Hugo, 258
Human Accelerated Region, DNA
    (HAR), 237
human chorionic gonadotropin (hCG),
    291
human universals, 419
human-ape parallels, 518
humeral
    capitulum, 107, 152
    head orientation, 107, 164, 167–168
    head shape, 105, 109–110, 152–153
    retroflexion, 168, 181
    trochlea, 107, 152, 477
    zona conoidea, 107
humerus, 107, 110–111, 113–114, 123,
    143, 149, 164, 168, 476
hunter's tales, 28
hunter-gatherer, 252
    activity budgets, 262
    diets, 266
    extreme activity, 262
    running, 263
hunting, 312
    advantages of, 314
    evolutionary consequences, 312
    and food availability, 315
    frequency, 312, 315
    and human evolution, 321
    and human origins, 312
    lone female, 317
    male vs female, 313
    and party size, 315
    prey types, 318
    red colobus defense strategies, 319
    success, 315
hunting strategies
    Gombe, 319
    Mahale, 319
    Taï, 319
Hurst, Shawn, 362
Huxley, Thomas Henry, 58–59, 224
Hylander, William, 137
hypertension, among chimpanzees,
    260
hypothalamus, 277, 283–284, 289

ideal, 32–33, 35–36 (see also "form",
    "type" and "eidos"
ilium (also, iliac blade), 109, 114, 135,
    150, 493
Imanishi, Kinji, 419
imbalance of power, 441
immutable, 35
implantation, 291

inbreeding, 206
incest avoidance, 206, 467
    and dispersal, 442
inclusions (in plants), 83
Indiana University, 80, 192
Indri, 44, 49–50
infancy, 203
infant dependency, 298
infant mortality, 207, 308
infanticide, 298, 301, 308, 461–462,
    468, 470, 505, 521
infectious diseases, 254
inferior parietal lobule, 363
inferior transverse torus, 102
infertility
    causes of, 461
    as an evolved strategy, 521
influenza, 254, 257
injury, 258
innateness, 392, 394
inner cell, 195
inner cell mass, 194, 196
instinctual food preferences, 520
Institute for Primate Studies (IPS),
    University of Oklahoma, 393
insulin, 266–267, 284–285
    function, 284
    production, 221, 235
    resistance, 285
    resistence, among chimpanzees,
        260
insulin-like growth factor IGF), 293
intelligence
    and "off-stage" social events, 521
    origin in fission-fusion society, 519
    origin in social behavior, 519
    relative v absolute size, 358
    role in social rank, 434
    and tool use, 530
intelligent tool use, 415
intentionality, intention, 338, 345, 349
interbirth interval (IBI)
    male vs femal infant, 305
intercommunity interactions, 12, 18,
    139–140, 238, 258–259, 427,
    435, 437, 445, 456, 491–492,
    534
Intermembral index (IMI), 115, 145,
    173
International Commission on
    Zoological Nomenclature
    (ICZN), 27
intersexual competition
    and testosterone, 292

intestines, 89–90
intracommunity violence
    purpose, 434
intrasexual competition, 452, 456
    and testosterone, 292
intron, 221, 224
Ioni (Johnny), 13, 243–244, 343, 346, 388–389, 401
ischemic heart disease, 262
ischial callosities (sitting pads), 56, 114, 120, 164
ischium (pl. ischia), 56, 114, 120, 164, 168–169, 177

Jacko
    Semliki, 491
joint excursion and joint health, 260
Jomeo, 434
July
    Semliki, 491
junk DNA, 221
juvenile period, 205

Kabila, Laurent, 17
Kakama, 8
Kakama's "toy baby", 8
Kanzi, 396–398, 400, 414, 501
    grammar, 396–397
    NHK film, 397
    vocabulary, 397
Kasekala, 299
Kasengazi, 450
Katinkila, Hamisi, 10
Kaufman, Thomas, 191–192
Kay, Richard, 137
K-complexes, 374
keen vision, as a primate trait, 46
Keith, Sir Arthur, 142, 480, 483
Kellogg, Winthrop and Luella, 245–247
K group, Mahale, 432
Kibale, Uganda
    Kanyawara, xv, 17, 67, 69, 73–74, 261, 427, 430–431, 433, 441, 490, 493
    Kanyawara, lack of ant-dipping, 423
    Kanyawara, parasites, 253
    Kanyawara, playback experiments, 386
    Kanyawara, violence against females, 412
    Ngogo, 17, 312, 315, 320, 457, 464, 490
Kichwamba Technical School, 18
kidnapping, Gombe, 17

Kigoma, Tanzania, 10
killings
    intracommunity, 427
kinesthetic sense, 248
King, Mary-Claire, 193
king-maker, 432
knee replacements, 262
knucklewalk, 104, 106, 120, 122, 126, 144, 154, 176, 179, 181, 501, 528
Köhler, Wolfgang, 13, 28, 326, 328, 330–333, 344, 348, 388, 521
Kohts, Nadia Ladygina-, 13, 28, 244, 246–247, 331–333, 337, 343, 346, 388
Kossel, Albrecht, 80
kyphosis, 46

labor, 300
    length, 300
lactase, 191
lactase production, in *E. coli*, 191
lactase repressor attachment site, 191
lactational amenorrhea, 205
lactose, 90, 191–192, 220
Lake Tanganyika, 9, 252
language
    ALR (ape language research), 390, 392, 394–397
    chimpanzee sentence length, 399
    chimpanzee–human differences, 399
    chimpanzees use of articles, 399
    chimpanzees use of future, past tense, 399
    chimpanzees use of recursion, 399
    defining so as to exclude apes, 400
    design features, 390–391, 393–394, 396–397
    evolutionary origin, 533
    human distinctiveness, 400
    reasons bonobos excel over chimpanzees, 398
language design features
    grammar, 397
language, lexigram-based, 394
Lanting, Frans, 500
large body size
    and competition with monkeys, 527
    function, 128
Larson, Susan, 182
laryngeal sac, 103
larynx, 103, 363, 390, 392
larynx development
    MMP5, 193

late adolescence, female, 206–207
late adolescence, male, 207
lateral orbitofrontal cortex, 361
laterality, xvii, 361, 365–366, 369, 489
    the "practice" hypothesis, 365
LCA, human and ape
    gorilla-like, 201
leaf clipping
    as a cultural behavior, 423
    as play starter, 414
    as sexual invitation, 413
    as signal of resting, 414
leaf color, 243
Leakey, Mary
    Proconsul, 167
leaping, 126, 148, 150–151, 161, 164, 169, 171–172, 175, 502
leaves, 44, 54–55, 71, 73, 75, 80, 83–84, 88, 93, 103, 171–172, 178, 243, 257, 301, 321, 351, 409, 414, 484, 510, 525, 527
Lemur catta, 46, 49
Lemuroidea (or informally, "lemuroid"), 42–43, 45, 47–48, 50–51
leprosy, 256
leptin, 285
LeSourd, Philip, 390
less-is-more adaptation
    enamel thickness, 140
less-is-more evolution, 103, 234, 239
    examples, 234
leukemia, 258
lianas, 12, 63–64, 125, 143, 414 (see also vines)
Lieberman, Philip, 389
life history, 202
    M1 eruption, 204
    role of brain development, 205
    slow, 188
life history trade-off
    and brain growth, 284
lifespan, human, 254
lifestyle diseases, 520
    absence among chimpanzees, 259
ligaments, 97, 134, 145, 147
light intensity perception, 243
limbic system, 362, 387, 392, 461
    definition, 364
links, linked in the Great Chain, 33–34, 38
Linnaeus (Carl von Linné), 27, 35–38, 40, 42, 519
lions
    handicapping theory, 454

lions (cont.)
    heat stress, 454
    manes, 454
    offspring survival rates, 454
lipase, 87–89, 93, 220
lipid. See oil, fat
    body fat. See oil, fat
    detection. See oil, fat
    digestion. See oil, fat
    fat reserves, chimpanzee. See oil, fat
    storage. See oil, fat
Lippia plicata, 253
liver disease, 259
locomotion, 119–120, 122–125, 129, 142–144, 148, 150, 160, 162, 170, 178, 248, 262, 345, 476, 479, 481–482, 502, 533
London Magazine, The, 27
long back (leaping adaptation), 109, 148, 164, 171–172
long fingers
    knucklewalking, 528
Lopez, Duarte, 24
Lorisoidea (informally, "lorisoid"), 42–43, 48, 50–51
loss of body hair, 235
Loulis, 347, 393
low body fat and reproduction, 288
low density lipoproteins (LDL) and diet, 267
lower canopy, 63
loyalty, 343
Lucy, 394
Lui Kotale, 502, 504
Lukaja, 428–430, 450
lunate, 111
lung disease, 259
luteal phase, 289
luteinizing hormone (LH), 290

macaque, *Macaca*, 44, 54–55, 109, 112, 114, 127–128, 151, 162, 169, 176–177, 207, 234, 301, 358, 451, 459, 463
Machiavellian intelligence, 340, 349
Macrotermes, 408
Madagascar, 49
Mahale, 320, 427
Mahale, Tanzania, 9, 14, 17, 30, 64–65, 69, 119, 209, 252, 298, 300, 306–307, 315, 317, 319, 335, 340, 342, 414, 421, 423, 427, 430, 432, 437, 439, 442, 444, 450
    K group, 432
    M group, 444
main canopy, 63
malaria, 255
male-on-female violence, 412
male socialization, 305
male violence
    and territory defense, 526
Man the Hunter, 312
mandrill, 55, 151, 175, 452
mange, 256
Marantochloa leucanth, 72
Marine House, Kampala, 3
marmoset (family Callitrichidae), 45, 53, 65
Martin, Claire, 3
master genes (= homeobox genes), 192
Masya, 307
Matata, 396
matching to sample, 243
maternal investment
    in males, 305
mating patterns, 465
mating system
    and sperm competition, 459
Matsuzawa, Tetsuro, 14, 332, 334–335, 351
maximal tumescence, 463
McCrossin, Monte, 163
McDonald, John, cancer and brain growth, 234
McGrew, William C, 488
measles, 256
meat
    dependence on, 312
    nutritional content, 313–314
    proportion of daily calories, 312
meat-sharing, 321
    evolutionary consequences, 312
mechanically scaled, 134
medial orbitofrontal cortex, 361
medicinal plants, 253
    antihelminthics, 253
    method of consumption, 253
Megachiroptera (or fruit bats [taxon]), 43
Mel, 305–306
melatonin, 375

memory
    recursion, 336
    role of motivation, 336
memory consolidation, 373–374
memory, chimpanzee, 13, 28, 68–69, 244, 357
    for numbers, 335
    group, 350
    short term, 336
    and sleep, 373
    spatial, 333, 335, 359, 508, 524, 527
    working, 336–337, 350, 359
Mendel, Gregor, 519
menopause, 209, 256
menstrual cycle
    sexual cycle, 288–289
menstruation, 291
mental map, 7, 11, 61, 69, 303, 333–335, 351, 369, 383, 387, 409, 411, 415, 524
Menzel, Emil, 383
Meshie, 4–7, 9, 13, 389, 401
mesocarp, 85
messenger RNA (mRNA), 221–222
metabolic rate
    and testosterone, 292
metabolomes, 116
metacarpal, 104, 106
M group, Mahale, 428, 432, 437, 444
microflora, 83, 88, 93, 163, 267 (see also "coevolved bacteria")
middle age, 208
Miff, 305
Mimusops bagshawei fruit, 74
minerals, 75
minimum-distance path/calculation, 66, 68
Miocene epoch, 159, 163, 165
mircroflora, 72
mirror neurons, 361
mismatch
    diet, 266
    diseases, 267
    teaching/learning, 347
    theory, 260
mismatch theory, 276
    defined, 260
mismatches, evolutionary, 252
Mitani, John, 14, 441, 443
mitochondria, 218–219
mitochondrial DNA, 195
Mitumba, 299
Moja, 394, 396
molecular biology, 217, 224

Molusca (taxon), 42
Monanthotaxis poggei, 71
monkey, 294, 312, 345
   advanced digestive physiology, 518
   antifeedant tolerance, 350, 521
   canopy use, 518
   characteristics shared with bonobos, 510
   cheek pouches, 512
   dry-habitat adaptation, 483
   dry-habitat group size, 533
   as fallback food, 315
   feed-as-you-go foraging, 528
   as food, 314
   intelligence, 349
   positional behavior, 480
   as prey, 313, 317–320
   sexual swellings, 463
   sleep architecture, 376
   sleeping posture, 376
   sleeping sites, 376
   social cohesion, 350, 511
   speciosity, 159
   theory of mind, 350
   unripe fruit, 440
   vocal tract anatomy, 389
monkey convergence hypothesis (MCH), 509
monosaccharides, 89
Moore, Jim, 14, 344
Morgan, David, 14
morphogens, 193
morphological defense, 83
morphology, 28, 52
   functional, 97, 115, 131, 134–135, 141–142, 151, 173
mortar/anvil, 410
mosaic evolution, 518
mother
   as playmate, 302
   as protector, 301–303
   as teacher, 302
   and social rank, 303
mother–infant play, 302
mother–infant contact, 301
   and development, 308
mother–infant emotional attachment, 205
mother–infant separation, consequences, 301
mothering, 204, 298
   nutritional burden, 301
mothering instincts, 308
mothering styles, 308
mother–son bond, 298

motivation
   role in memory, 336
motor cortex, 361, 363–364
Mount Assirik, 488
Mount Cameroon, 22
Mount Nkungwe, Tanzania, 317
mourning, 307
mouse lemur, Microcebus, 43, 49
MSR (mirror self-recognition), 339, 343
multiparous females
   sexual attractiveness, 466, 469
muscle
   biceps, 110, 114, 143
   brachialis, 110, 143, 152
   brachioradialis, 110, 114, 143, 476
   deltoid, 110, 114, 143
   digital flexor, 104, 144, 147
   gluteal, 114, 150, 170, 475
   latissimus dorsi, 113, 143, 149, 175
   masseter, 102, 137–138, 153, 235
   nuchal, 103, 139, 153
   nuchal, function, 139
   overstretching, 137
   pectoral, 112–113, 143
   peroneus longus, 116
   power, 116, 132–133
   quadriceps, 150
   temporalis, 46, 102, 137, 139–140, 235, 475, 511
   trapezius, 111–112
muscle sparing principle, 134, 147
Musgrave, Stephanie, 424
mutation, 191, 217, 219–220, 231, 234–235, 238–239
Myrianthus holstii, 70
Mzee
   Semliki, 490

N'Zo-Sassandra River, 421
Nakalipithecus, 182
naked ape, 183
natural and flexible use, 392–393
natural theology, 30, 33, 35, 37, 519
Ndoki, 409
neoplatonism, 32–33, 36, 38
neoteny, 188, 203
   bonobo, 211
neoteny (= infantilization, juvenilization), 210
nest or sleeping platform, 29, 204–206, 253, 300–301, 304, 369, 372, 377–380, 388, 406, 470, 489, 499, 505
   and anti-mosquito properties, 378

   the body mass hypothesis, 378
   in evolutionary history, 378
   ground nests, 378
   and human evolution, 379
   and insulation, 378
   mattress, 377
   and predation, 378
   the sleep paralysis hypothesis, 379
neuroanatomy, 27, 97, 357
newborn, 301
Ngamba Island, 3
nicotine, 82, 94
night vision, 244
Nim Chimpsky, 396
Nishida, Toshisada, 9, 14, 19, 253, 299, 314, 423, 432, 437, 519
Nissen, Henry, 29, 410, 500
Nkuumwa, 2–3, 5
nocturnal, 48
noncoding DNA, 221, 224, 237
nonhuman primate sleep patterns, 376
non-infectious diseases, 254
non-rapid-eye-movement (non-REM) sleep, 374
nonverbal communication, sophistication of, 383–384
nose, 101
notochord, 42
Nsungwepithecus, 163, 165
Ntologi, 298, 428, 432, 434
nucleus, 219, 221–222
nulliparous females
   sexual attractiveness, 469
numeracy, 331
nuptial gifts, 452
nursery groups, 303, 440
nut-cracking, 347
   caloric values, 411
   planning aspects, 411
   sex differences, 411
   techniques, 411

obesity, 259
obesity, among chimpanzees, 260
object–object relationship tasks
   inappropriate toods, 331
object–object relationships, 325–327
occipital lobe, 358, 364
Oesophagostomum spp. (nematodes), 253
off-stage events
   females, 526
   and intelligence, 526

off-stage events (cont.)
  as selection pressure for mind-reading, 526
old age, 208–209
olfaction
  chimpanzee, 247
  keen, 46–47, 49, 66, 141
Olson, Maynard, 234
on-demand nursing, 302
opaline phytoliths, 83
opportunistic mating, 463
optimal diet, 75
  list, 75, 522–523, 525, 527
optimal foraging theory, 66, 75
Orangutans (*Pongo*), 20, 23, 25, 38, 43, 45, 48, 57–59, 102, 115, 127–128, 139, 142–143, 151, 158, 162, 177, 181, 226, 236, 238, 331, 336, 349, 362, 372, 376, 379, 451, 479, 484, 492, 521
orbitofrontal cortex, 361–362
order, taxonomic, 42–44, 82
ordinality, 332
Oreopithecus, 162, 182
Origin of Species, The, 38, 519
orphaning, 2, 204, 305–306, 308–309, 504
os coxa, 109, 113–115, 135, 148–151
osteoarthritis, 260
osteology, 28, 38, 97
osteopenia, 260
osteoporosis, 260
Ouranopithecus, 182, 201
  social system, 237
outer ear development
  MMP5, 193
ovarian follicle, 290
overeating, 285
ovulation, 289, 458
  suppression, 287
ovulation cessation
  lactational amenorrhea, 458
ovule, 85
ovum
  oocyte, 189
oxytocin, 287
  cortisol suppression, 288
  and labor, 287
  and lactation, 288
  and mother-infant bonding, 288
  and pair-bonding, 288
  and vasoconstriction, 288

pain perception (nociception), 247
  Washoe, 247
painting
  by chimpanzees, 341
pair bonding
  among humans, 469
  trade offs, 469
paleontology
  challenges, 158
Paley, William, 30, 35
pancreas, 285
Panda nuts, 411
Panksep, Jaak, 342
pant-grunt, 350, 383, 391–392, 419–420, 428–429, 431, 450, 465, 505
  gradients, 431
pant-hoot, 141, 206, 245, 249, 293, 326, 383–384, 389, 441, 450, 465, 504, 530
  individual identification, 384
  morphology, 245
  size of food source, 384
parallelism, 181
parasite loads
  and social rank, 293
parent-offspring conflict, 303
parietal cortex, 363
parietal inferior lobule, 363
parietal lobe, 358, 363
parietal superior lobule, 363
Parinari curatelifolia, 315
parity
  and sexual attractiveness, 466
parties, 437
party size
  and hunting success, 315
  dependence on food supply, 525
  influence of food supply, 525
passion, 433
paternity confusion, 462
  high copulation rates, 521
Pekin, Debra, 363
pelvis, xii, 56, 109, 114, 135, 148, 150, 168–169, 177, 181, 291, 474–475, 485, 493
Penfield, Wilder, 363
penicillin, 94
Pennisetum purpureum, 72, 139, 253
perception of movement, 243
periovulatory period (POP), 463
peroneus longus, 115

persistence
  role in social rank, 434
personality, 344, 348
personality, components of
  agreeableness, 344
  extraversion, 344
  neuroticism, 344
  openness, 344
petalia, 365, 368
Peterson, Dale, 8, 238
phalanges
  terminal, narrow, 151
phalanx (pl. phalanges), 104, 107, 122, 144, 151–152, 154, 162, 177, 179, 181, 474, 477–478, 484–485, 501
pharmacology
  origin in plant defenses, 520
pharynx, 103
philopatry, 298
Phoenix reclinata, 72
photosynthesis, 84
phylum, taxonomic, 42
physiology, 294
Pierolapithecus, 176
pigment production, 87
piloerection, 4, 339, 348, 387, 431, 441, 450
pineal gland
  and sleep, 375
pinnae (ears), 102, 245, 249, 504
Pisces (taxon), 42
pisiform, 111
pith, 67–68, 72, 76, 87–88, 129, 139, 153, 330, 438, 484
pituitary, 277, 290
placenta, 195, 291, 300
Planet of the Apes, 159
Planet of the Monkeys, 129, 159
plant chemistry, 80
Plato, 30, 32, 38
Plato's Cave, 31
plenitude, 33
pliopithecoids, 159
plotinus, 32
pneumonia, 254, 256–257
poachers, 2
pointing, 325, 398, 511
  challenge interpreting, 325
polarity
  "west" end, 190
pole, polarity, 193
polio, 255–256
Pollard, Katherine, 237

pollen granule, 85
pollination, 85
Pongo, 24
positional behavior, 121–123, 126–127, 129, 134
  defined, 120
positional mode, 120
positively selected genes, 230
possessiveness, mating, 464
postorbital bar, 46, 48
postorbital closure, 46–48, 52
posture, 45, 119–120, 122, 126, 129, 142–144, 147–148, 160, 164, 170, 174, 181, 262–263, 345, 383, 409, 430, 473, 480, 484–485, 502, 533
  communication, 387
posture (static and health), 265
Povinelli, Daniel, 344
prefrontal cortex, 116, 211, 360–362, 367–369
  function, 361
pregnancy, 304
pregnancy test, 291
prehensile tail, 52, 54
Premack, David, 394
premaxilla
  bone of contention, 38
premotor cortex, 360
presenting, 459
Preston, Richard, 255
prey
  baboons, 318
  birds, 318
  bushbabies, 318
  bushbuck, 318
  bushpigs, 318
  colobus, 318
  duiker, 318
  percentage "stolen", 317
  rodents, 318
  squirrels, 318
primary compounds, 80–82
primary motor cortex, 360
primary somatosensory cortex, 364
Primate Diet Project (Wrangham), 160
primate diversity, 519
primate fieldwork
  challenges, 17, 19, 518
primate trends, 44
primates
  as data for conceptual models, 520
  as models for human nature, 520
prime adulthood, male, 208

proboscis monkey (*Rhinopithecus*), 56
proceptivity, 459
Proconsul, 165, 167–169
  ankle, 170
  back length, 169
  brain size, 171
  brow ridge, lack of, 171
  elbow, 168
  feet, 170
  gripping great toe, 168
  hip, 170
  humerus, 168
  intermembral index, 168
  scapula, 168
  taillessness, 169
  teeth, 171
Proconsul africanus, 165
Proconsul major, 166
proconsulid, 168–172, 175, 177, 182
productivity, 391, 393
progesterone, 287, 291
  and immune response, 291
prognathism, 98, 102, 136, 138, 153, 177, 511, 520
progress in science, unevenness, 519
Prohylobates, 163
prolactin, 288
  female function, 287
  late in pregnancy, 291
  male function, 287
propliopithecoids, 159
protein, 68, 82, 191–193, 217, 219–220, 222, 224, 226–227, 229, 231–233, 285
  dietary, 314
protein folding, 224
protein poisoning, 78
protein similarity, human–chimpanzee, 227
protein synthesis, 218–220, 222–224, 232
protein, dietary, 47, 68, 71–72, 74–76, 89, 232, 246, 312, 408, 411, 520, 534
protein, digestion, 88, 93
proteinase inhibitors, 82
Pruetz, Jill, 14, 321, 407, 488
Pseudospondias microcarpa, 315
Pterocarpus tinctorius, 71
puberty, 288
pubis, 114
pull-toy, 7
Purdue University, 80
Pusey, NNE, 302

quadrumanous climbing, 124
quadrupedalism, 50, 122–123, 129, 178, 181, 479–480, 482–483
quiescence, 464
quinine, 94

rabbit starvation. *See* protein poisoning
racehorse life history, 189
Ramos, Gilbert, 502
rapid-eye-movement (REM) sleep, 374
Raven, Harry, 4, 6, 388
Raven, Mary, 4, 7
receptivity, 459
recognition of line-drawing images, 244
recursion, 399
  lack of in Pirahã, 400
  lack of in some human languages, 400
reference to abstractions, 392–393
referential model, 532
regulatory DNA, 237
relaxin, 291
  as a vasodilator, 291
reliable indicator hypothesis, 466
religion, chimpanzee, 420
repressor, 221, 224
  lactase, 191
reptilia (taxon), 42
Resilience
  role in social rank, 434
reunions (and violence), 430
reversal, 431
Reynolds, Vernon, xvi, 14, 19, 142
rhinarium (wet nose), 48, 52
Rhus natalensis, 489
ribcage, 147
ribosome, 218–219, 222–223
right-hand preference
  humans, 365
  lack of, among chimpanzees, 366
ripe fruit
  positional requirements, 527
Ripley, Suzanne, 161
risk assessment, 342
  chimpanzees vs bonobos, 342
RNA (ribonucleic acid), 221
RNA polymerase, 221–222
rods and cones, 101
Rome, 22–23
Rosalind Alp, 12
Rose, Michael D., 167
Rukwapithecus, 165, 167

Rumbaugh, Duane, 7
running, 122, 164, 167–168, 170, 172
Russon, Anne, 336

Saba florida, 70
sacred trees, 420
sacrum (pl. sacra), 109, 114, 135, 475
Safina, Carl, 339
Samburupithecus, 182
Samson, David, 376, 491
Sanz, Crickette, 14, 409
Savage-Rumbaugh, Sue, 396–398
savanna (grassland), 65–66, 412, 454, 481, 483, 495
scabies, 256
Scala Naturae, 33
scale model skills, 333
scaphoid, 111
scapula, 105, 108, 110–111, 123, 146, 167–168, 486
   australopith, 475
   glenoid fossa, 105, 110, 112, 147, 152, 476
   narrow, 110
   narrow = ape-like, 145–147, 152, 154, 484–485, 502
   proconsulid, 168, 173
scent marking, 46, 49
Schick, Kathy, 397, 415
science, chimpanzee, 420
science vs religion, 30, 38, 518
seasonality, 64
second-order cooperation, 347
secondary compounds, 70, 81–83, 85, 93, 350, 510, 520–522
Securinega virosa, 489
seed dispersers, 83–86
seed predator, 85
self domestications hypothesis (SDH), 509
self-control
   and cognition, 346
self-domestication hypothesis (SDH), 509
selfishness, 346
semicircular canals, 248
Semliki
   genetic similarity to nearby populations, 492–493
   large community range, 492
Semliki Chimpanzee Project, 488
Semliki, Uganda, 17, 65, 312, 373, 410–411, 423, 438, 473, 488–489, 491–492, 494, 496

australopith-like faces, 495
drinking wells, 490
hand-clasp grooming, 489
human-like pelves, 494
violence, 490
sensory system, 242
sentence construction
   and Broca's area, 363
sentence length, and chimpanzee language, 399
sentinel hypothesis, 376
Senut, Brigitte, 487
sepsis, 235
sex cycle, 288–289
sexual attraction, 205
sexual dimorphism, 451
   African ape, 210
   and social system, 487
   australopith, 485, 532
   bonobo, 501, 510
   chimpanzee, 238
   evolution of, 179, 183, 451
   gorilla, 179, 238
   human, 487, 532, 534, *See* sexual selection
   hylobatid, 451
   LCA, 237
   Ouranopithecus, 179
   self-domestication, 511
sexual selection, 451
   cognitive aspects, 452
   defined, 451
   intersexual, 451
   intrasexual, 452
sexual swelling, 55, 205–206, 304, 410, 442, 459–460, 462–466, 481, 499, 501, 503, 505, 532
   evolutionary origin, 465
   and mate selection, 532
shape discrimination, 243
shared intentionality, 339
sharing, 312
shoulder mobility, 106, 110, 127, 143, 153, 174
shoulders (broad thorax), 57, 98, 108, 112
Shumaker, Robert, 405–406
siamang (hylobatid), 135
Sibley, Charles, 228
sifaka, *Propithecus*, 50
signaling proteins, 192–193
Simia, 27
SIVcpz (chimpanzee HIV), 255
size discrimination, 244

skin, chimpanzee vs human, 102
sleep, 56, 164, 372–380, 530
   among hunter-gatherers, 375
   bifurcated sleep pattern, 375
   and ischial callosities, 168
sleep and modulation of emotional response, 374
sleep architecture, 374
sleep cycle, 374
sleep paralysis, 373–374, 377–378, 530
sleep spindles, 374
slow climbing, 51
slow life history
   and brain size, 534
slow loris, *Nycticebus*, 50–51
slow wave phase sleep, 374
small intestine, 47, 87–89, 93
small pox, 254
snub-nosed monkey, 55
social brain hypothesis, 346
social display or charging display, 21, 431, 449–450, 456, 485, 531
social displays as bluff, 455
social rank
   and foraging, 125
   and mothering style, 305
social scratch, 387
sociality
   need for in chimpanzees, 522
   orangutan, 521
solanine, 94
solitary social system
   and sperm competition, 460
somatosensory homunculus, 363
Sonntag, Chaeles, 98, 188
sparse dispersed food
   and fluctuations, 525
species
   definition, 41
   identifying, 32, 36, 41
species, taxonomic, 42
sperm competition, 452, 459
sperm midpiece size, 460
spider monkeys, atelids, 44, 54, 69, 110, 124, 150, 162, 350
spina bifida, 197
spindle, 305–306
sponge, 406
spoon, chimpanzee feeding human with, 6
sports
   and social rank, 531
   and territorial defense, 531

square-cube law, 132–133
squirrel monkey, *Saimiri*, 53, 451, 465
Stanford, Craig, 319
starch, 76, 86, 88, 90, 93, 220
Stevens, Nancy, 163, 167
stomach, 49, 55, 74, 80, 87–90, 93, 103, 221, 285
strength, chimpanzee, 6
Strepsirhini (or informally, strepsirhine), 42, 44, 48–52
stress
    concentration in thorax, 145, See cone-shaped thorax
stress hormones
    cortisol, 284
    cortisone, 284
    hydrocortisone, 284
    prednisone, 284
Strongyloides fulleborni, 253
Stroop test, 362
Strychnine, 94
Strychnos, 303, 412
Student's Prayer, 33
subconscious actions, 357
sucrose, 90
sugar, 47, 65, 68, 70–76, 80, 84–87, 89–91, 93, 128, 160, 191, 218, 220, 231, 246–247, 266, 268, 283, 285, 440, 510, 520, 522
sunday drive in the country, 20, 36–38
superfamily, taxonomic, 42–43
superior parietal lobule, 363
superior transverse torus, 102
supraorbital torus (browridge), 99, 101, 117, 139, 153, 171
supratoral sulcus, 99, 101, 137, 139
suspensory, 54, 123–129, 142, 144–145, 148, 152, 161, 168, 177, 181–183, 474, 477, 480
    and support diameter, 528
Systema Naturae, 36–37, 40–41

Taï, Ivory Coast, 15, 18, 64, 305–306, 316–317, 319–320, 333, 407, 412, 420, 423–424, 431, 437, 442
    leaf clipping, 414
tail (function), 151
Takasaki, Hiroyuki, 252
tamarin (famil Callitrichidae), 53
Tamarindus indicus, 489
tandem display, 450
tandem walk, 423
tannin, 71, 82–83, 86–88, 93, 102, 129, 141, 160, 180, 183, 520–522, 534
tapetum lucidem, 48, 52
Tardieu, Christine, 487
tarsier, *Tarsius*, 51, 53, 366, 465, 519
Tarsiioidea (informally, 'tarsioid'), 42–43, 51
taste (gustation), 245
    alkaloid detection, 246
    preference for sour tastes, 246
    salt, 245
    sour, 246
    sugar detection, 246
    sweet, 246
    umami, 246
taxonomy, 27, 36–37, 40–43, 519
teaching, 347, 424
    correction, as hallmark, 423
    mother's role, 347
teeth
    canine, 46, 102–103, 137, 140, 153, 177–178, 237, 452, 480, 501, 511, 520
    canine (function), 47–48, 434
    dental caries, 258
    dental formula, 46
    diastema, 103
    enamel hypoplasia, 258
    enamel thickness (function), 103, 171
    enamel thickness (PM and M), 153, 177
    ever-growing incisor, 50
    extreme wear, 258
    incisor, 46–48, 53, 72, 102, 131, 137–139, 153, 171, 177–178, 210, 258, 475
    incisor size, correlated with fruit diameter, 137
    incisor, function, 138
    incisors, procumbent, 138
    molar, 46, 102–103, 140, 163, 178, 204, 475, 482, 488, 494
    premolar, 46, 102–103, 140, 178
    premolar (function), 47
    sectorial premolar, 103, 140, 153
    thin enamel, function, 171
Temerin, Alis, 161
Temerlin, Maurice and Jane, 394
temperature perception, 247
temporal lobe, 202, 358, 364
temporomandibular joint (TMJ), 136, 261
tendon, 104, 152, 268, 373, 480
    definition, 97
    digital flexor, 104, 108, 477
    and exercise, 520
    flexor sheath ridges, 144
    inflammation, 265
    peroneus longus, 115
    terminal branches, 125, 128, 144, 161–162, 170, 180, 527
termite fishing, 406
termite mound, 69, 71, 303, 336–337, 407, 409, 524
    soil, 71
terpenes, 83, 129, 160, 180, 183, 522
Terrace, Herb, 396
terrestrial herbaceous vegetation (THV), 503, 509–510
terrestrial travel
    as an efficiency strategy, 524
territoriality
    and female immobility, 524
    benefits, 444
territory, 15, 58, 66, 75, 94, 206, 237, 293, 348, 372, 441, 444–446, 451–452, 456, 492, 505, 524, 526, 530–531, 534
    and stationary females, 527
    community, 12, 206
    definition, 440
testis size
    and mating system, 459
testosterone, 277, 292
    and confidence, 292
    disadvantages, 433
    and dominance, 292
    immune suppression, 292
    lifespan, 292
    and male–male competition, 292
    and overconfidence, 292
    spatial cognition, 293
    trade-offs, 292
The Mentality of Apes, 13, 328
theory of mind (ToM), 337–339, 345–346, 349–351
Thompson, D'arcy, 210
Thorax, 107–108, 110, 135, 143, 145–147, 150, 152, 154, 167–168, 181, 192, 194, 199, 475–476, 479, 482, 485, 501 (see also 'ribcage')
threshold, nutrient requirement, 75–76, 89
thumb (pollex), 46, 106, 116, 151, 178
thymine, 221

thyroid-stimulating hormone (TSH), 293
time travel, 337
Tinklepaugh, 334
toes (pedal phalanges), 115
toilet claw, 48, 52
Tokuyama, Nahoko, 499
tongue, 101–103, 141, 160, 361, 389, 392
Tonkin snub-nose monkey, 56
tool, 405
    discovery of chimpanzee, 405
    functions, 405
    imaginary pull-toy, 406
tool function
    amplify force, 405
    amplify gesture, 406
    enhance comfort, 406
    extend reach, 405
    increase control of environment, 406
    symbolize, 406
tool making, 406
    combine, 406
    detach, 406
    reduce, 406
    reshape, 406
tool sets, 415
tool use, 405
    and Broca's area, 363
    and intelligence, 533
    extensiveness, 406
toolkit
    chimpanzee, 405
tools
    ant wand, 412
    anvil/mortar, 412
    body decoration, 406, 414
    breadth of use in captivity, 414
    club, 411
    defined, 405
    display branch, 406, 409
    drum, 406, 410
    hammer, 410
    harvesting probe, 406, 408
    human–chimpanzee differences, 404
    hunter-gatherer, 404
    insect brush/whisk, 414
    invention of human, 404
    investigatory probe, 406
    leaf clip, leaf strip, 413
    leaf groom, 414
    leaf sponge/napkin, 410
    lever/pry bar, 414
    nest/mattress, 406
    nut hammers, 412
    pad/seat, 414
    penetrating probe/spear, 407
    play-starter, 406, 409
    projectile, 406
    projectiles, 413
    propensity to use, 405
    pry bar, 406
    ratcheting effect, 404
    spear, 407
    spears, 321
    Termiting/anting probe, 408
tool use
    and antifeedant intolerance, 533
tooth clacking, 384
toothcomb, 48
total daily energy expenditure (TEE), 294
Toth, Nick, 397, 415
toxin, 82, 87, 93, 128, 180, 267, 372–373, 518, 520, 522 (see also "alkaloid")
toy baby, of chimpanzee, 8, 12
trachea, 103
trade-offs, physiological, 284
transcription, 221
transcription factor, 224
transfer (locomotor mode), 124
transition barrier, 479
transitive inference, 332
translation, 222, 224
travel costs
    and terrestriality, 524
tricycle, chimpanzee riding, 5
triquetral, 111
tRNA (transfer RNA), 222
trochanter, greater, 115
Troglodytes (Greek meaning), 40
TrP-Cage, 223
Tsimane hunter-gatherers, 259
tuberculosis, 254
Tulp, Nicolaas, 25, 38
tumescence, 55, 463, 465, 467, 505
type, 31–33, 36 (see also 'ideal,' 'form,' and 'eidos'
typhoid, 256
Tyson, Edward, 20, 26, 97, 131, 519

ulna
    olecranon fossa, 476
    olecranon process, 107, 152, 164, 168, 477
    styloid process, 107, 152, 477
ultimatum game, 346

unique human features, 532
unit-group, 437
University of Figan, 206
Uppsala University, 36
uracil, 221
uricase, and cancer, 258
uroguanylin, 285

van Lawick, Hugo, 257, 405
variation (in re essentialism), 36
vasoconstriction, 277
vertebrae
    lumbar, 108–109, 114–115, 135, 144, 148–150, 169, 178, 265, 475
    short, stiff lumbar segment, 108–109, 115, 117, 135, 148–149, 151
    transverse processes, 169
    wedge-shaped, 169
vertical climbing, 116, 125–127, 129, 142–143, 145, 148–152, 167–168, 170, 172, 174, 181, 329, 485, 528
    frequency, 261
vertical clinging and leaping, 49–50, 52
Vicki, 7, 244, 388, 406
Victoriapithecus, 163
violence
    Semliki, 490
    sex differences, 427
viral leukemia, 256
visible estrus, 459
vision
    dichromatism, 243
    L-type cones, 242
    M-type cones, 242
    retina, 242
    rods and cones, 242
    S-type cones, 242
    trichromatism, 242
    visible light, 242
    yellow–orange discrimination bias, 243
visual cortex, 357, 360, 364, 368
vitamins, 75
vocabulary
    human use in typical day, 399
    Kanzi, 399
    Shakespeare, 399
    size in Latin, 399
    size in modern languages, 399
    size in Pirahã, 400
    size of English language, 399
vocal tract, 103, 389

vocalizations, 383
volume discrimination, 332

wadge, 70, 305
Wagamumu, 299
walking
  first steps, 204
walking stick/club
  use by hominins, 487, 495
war
  imbalance of power, 534
Washoe, 341, 347, 392–394, 396
watchmaker analogy for existence of God, 35
water taxi, 9–10
Watson, James, 226
Watts, David, 14
weaning, 204–205, 303
weaning conflict, 303

Wernicke's area, 360, 363–365, 368–369
West, Meredith, 454
white matter, 359
widowbirds, 453
Wilkins, Maurice, 226
Wilson, Allan, 193
Wilson, Michael, 293, 441
winkle, 300–301
working memory. *See* Memory, chimpanzee: working
worth, in re God, 32
wound healing
  and sleep, 373
Wrangham, Richard W., 8, 14, 18, 65, 69–70, 73, 93, 129, 160–162, 238, 253, 293, 312, 316, 440, 473, 482, 484, 503, 509, 534
Wright, Jonathan, 488

wrist
  pronation, 168
  rotatory, 51, 107, 142–143, 153, 175, 182, 518
  stability, 107

Y-chromosome
  extreme divergence, 236
  gorilla, 237
Yerkes
  Primate Research Center, 367
  Robert and Ada, xvi, 20, 29, 247, 329, 414
yolk sac, 196

Zahavi, Amotz, 453
Zona pellucida, 189
Zuckerman, Solly, 405
zygote, 190, 291